APPLIED COMPUTATION

Applied Computational Physics

Joseph F. Boudreau and Eric S. Swanson

with contributions from Riccardo Maria Bianchi

OXFORD
UNIVERSITY PRESS

Great Clarendon Street, Oxford, OX2 6DP,
United Kingdom

Oxford University Press is a department of the University of Oxford.
It furthers the University's objective of excellence in research, scholarship,
and education by publishing worldwide. Oxford is a registered trade mark of
Oxford University Press in the UK and in certain other countries

© Joseph F. Boudreau and Eric S. Swanson 2018

The moral rights of the authors have been asserted

First Edition published in 2018
Impression: 1

All rights reserved. No part of this publication may be reproduced, stored in
a retrieval system, or transmitted, in any form or by any means, without the
prior permission in writing of Oxford University Press, or as expressly permitted
by law, by licence or under terms agreed with the appropriate reprographics
rights organization. Enquiries concerning reproduction outside the scope of the
above should be sent to the Rights Department, Oxford University Press, at the
address above

You must not circulate this work in any other form
and you must impose this same condition on any acquirer

Published in the United States of America by Oxford University Press
198 Madison Avenue, New York, NY 10016, United States of America

British Library Cataloguing in Publication Data
Data available

Library of Congress Control Number: 2017946193

ISBN 978–0–19–870863–6 (hbk.)
ISBN 978–0–19–870864–3 (pbk.)

DOI 10.1093/oso/9780198708636.001.0001

Printed and bound by
CPI Group (UK) Ltd, Croydon, CR0 4YY

Qt® is a registered trademark of The Qt Company Ltd. and its subsidiaries.
Linux® is the registered trademark
of Linus Torvalds in the U.S. and other countries.
All other trademarks are the property of their respective owners.

Links to third party websites are provided by Oxford in good faith and
for information only. Oxford disclaims any responsibility for the materials
contained in any third party website referenced in this work.

For Pascale.
For Lou, Gordy, Suzy, Kris, Maura, Vin.
For Gordon V.

For Erin, Megan, Liam, and Drew.
For Max and Gudrun.

reor ut specular

Preface

New graduate students often experience something like shock when they are asked to solve real-world problems for the first time. These problems can be only rarely solved with pen and paper and the use of computational techniques becomes mandatory. The role of computation in any scientific endeavor is growing, and presents an increasing set of challenges. Numerical algorithms play a central role in theoretical prediction and in the analysis of experimental data. In addition, we see an increasing number of less numerical tasks taking on major importance in the life of young scientists. For example, how do you blend together two computing languages or split a computation between multiple computers? How does one design program libraries of numerical or scientific code for thousands of users? How is functionality added to a million-line reconstruction program? How can complicated datasets be visualized? What goes into a monitoring program that runs in a control room? These tasks are not particularly numerical or even scientific, but they are nonetheless important in the professional lives of scientists. From data acquisition systems to solving quantum field theory or presenting information, students face an intimidating computational environment with many languages and operating systems, multiple users with conflicting goals and methods, and complex code solving subtle and complicated problems.

Unfortunately, the typical student is marginally prepared for the challenges faced in modern computational ecosystems. Most students have had some exposure to a programming language such as C, C++, Java, or Fortran. In their first contact with "real" code, they may well be exposed to a proliferation of legacy software that in some cases is better used as a counterexample of good modern coding practices. Under these circumstances the usual solution is to learn on the fly. In a bygone era when the computing environment was simple this learning process was perfectly satisfactory, but today undirected learning leads to many false starts and some training has become indispensable. The search for help can be difficult because the nearby senior physicist probably grew up in an era preceding the explosive development of languages, paradigms, and computational hardware.

This book aims to fill some of the holes by introducing students to modern computational environments, object-oriented computing, and algorithmic techniques. We will rely on 'canned' code where reasonable. However, canned code is, by definition, incapable of solving research problems. It can at best solve portions of problems. At worst, it can lead the student researcher to false or incomplete conclusions. It is therefore imperative that the student understands what underlies his code. Thus an explanation of the numerical issues involved in common computational tasks will be presented.

Sometimes the numerical methods and applications will be quite technical; for this reason we regard this book as appropriate for newly graduated students. Our examples will be drawn primarily from experimental and theoretical physics. Nevertheless, the book is also useful for students in chemistry, biology, atmospheric science, engineering, or any field in which complex analytical problems must be solved.

This text is meant for advanced (or graduate) students in the sciences and engineering. The physics ranges from advanced undergraduate topics in classical and quantum mechanics, to very advanced subject matter such as quantum field theory, renormalization, and scaling. The concepts of object oriented computing are introduced early in the text and steadily expanded as one progresses through the chapters. The methods of parallel computation are also introduced early and are applied in examples throughout. Since both the physics and the coding techniques can be replete with jargon, we attempt to be practical and provide many examples. We have not made any effort to prune away any discussion of fairly pedestrian material on the pretext that it is not advanced enough for a sophisticated audience. Our criterion is that the topics we discuss be *useful*, not that they be graduate-level, particularly since some topics are interdisciplinary in nature.

The numerical algorithms we consider are those applied in the major domain areas of physics. Classical problems involving a finite number of degrees of freedom are most often reduced to a coupled set of first-order differential equations. Those involving an infinite number of degrees of freedom require techniques used to solve partial differential equations. The study of quantum mechanical systems involves random processes, hence the temporal evolution of the system is handled though simulation of the underlying randomness. The computation of physical processes thus can be generally categorized according to the number of degrees of freedom and the stochastic or deterministic nature of the system. More complicated situations can mix these. For example, to follow a charged particle through a magnetic field in the presence of multiple scattering involves both deterministic and stochastic processes.

The flip side of simulation is data modeling. This is the procedure by which a mathematical description of data, often along with the values of physically interesting parameters, is obtained. Data modeling is an activity that consumes much of the time and creativity of experimental physicists working with datasets, large or small. While many treatises exist on the statistical analysis of data, the goal here is to explore, in somewhat greater detail than is usually found, the computational aspects of this field.

This text is neither a treatise on numerical analysis nor a guide to programming, but rather strives to develop practical skills in numerical and non-numerical methods applied to real world problems. Because of the emphasis on practical skills, students should expect to write programs and to refine and develop their programming techniques along the way. We assume a basic knowledge of C++ (the part of the language taken directly from C, minus the anachronisms), and treat the newer features in dedicated chapters. We do not, however, give a complete lesson on the syntax and symantics of any language, so we advise the reader who has not mastered C++ to learn it in parallel using any one of a number of sources. Our emphasis can then fall on using the language *effectively* for problems arising in physics.

A key ingredient to effective programming nowadays is mastery of object-oriented programming techniques—we strive to develop that mastery within the context of greatest interest to the target audience, namely physics. As noted in *Numerical Recipes*

(Press 2007), object-oriented programming has been "recognized as the almost unique successful paradigm for creating complex software". As a result, object-oriented techniques are widespread in the sciences and their use is growing. The physicist appreciates object oriented programming because his day-to-day life is filled with a rich menagerie of interesting objects, not just integers and real numbers, but also vectors, four-vectors, spinors, matrices, rotation group elements, Euclidean group elements, functions of one variable, functions of more than one variable, differential operators, etc. One usually gets more from a programming paradigm that allows user-defined datatypes to fill in gaps left at the language level. Encapsulation and polymorphism can be effectively used to build up a more functional set of mathematical primitives for a physicist—and also to build up an important set of not-so-mathematical objects such as are found in other nonscientific code.

Many books are devoted to object oriented analysis and design, and while some of these treatises are perfect for their target audience, a typical scientist or engineer most likely gets tired of examples featuring the payroll department and looks for a discussion of object oriented programming that "speaks his language". Accordingly, we include three chapters on object oriented programming in C++: Encapsulation, Polymorphism, and Templates. Other chapters of the book provide excellent examples of object-oriented techniques applied to various practical problems.

A companion web site has been established for this text at:

- http://www.oup.co.uk/companion/acp

The site includes example code (EXAMPLES area), skeletons which students can use as a starting point for certain exercises appearing in the text (SKELETONS area), and raw data for other exercises (DATA area). In addition the site provides user's guides, reference manuals, and source code for software libraries provided free of charge and used within this text. This software is licensed under the GNU Lesser General Public License. In referencing examples, skeletons, data, etc., we generally omit the full URL and refer simply to the directory, e.g. EXAMPLES, SKELETONS, DATA, etc.

Course organization

It is our experience that most of the material in this text can be covered in two terms of teaching. There are three main strands of emphasis: computational, numerical, and physical, and these are woven together, so the reader will find that emphasis alternates, though the book begins with more computational topics, then becomes more numerical, and finally more physical.

Computational topics include: Building programs (Chap 1), Encapsulation (Chap 2), Some useful classes (Chap 3), How to write a class (Chap 6), Parallel computing (Chap 9), Graphics for physicists (Chap 10), Polymorphism (Chap 12), and Templates, the standard library, and modern C++ (Chap 17).

Numerical topics include Interpolation and extrapolation (Chap 4), Numerical quadrature (Chap 5), Monte Carlo methods (Chap 6), Ordinary differential equations (Chap 11).

Topics related to applications in experimental and theoretical physics are Percolation and universality (Chap 8), Nonlinear dynamics and chaos (Chap 14), Rotations and Lorentz transformations (Chap 14), Simulation (Chap 15), Data modeling (Chap 16), Many body dynamics (Chap 18), Continuum dynamics (Chap 19), Classical spin systems (Chap 20), Quantum mechanics (Chap 21 and 23), Quantum spin systems (Chap 22), and Quantum field theory (Chap 24).

Finally, there are nearly 400 exercises of widely varying difficulty in the text. To assist students and instructors in selecting problems, we have labelled those exercises that are meant to be worked out without the aid of a computer as *theoretical* [T]; exercises which are more open-ended and require significant effort are elevated to the status of a *project* and labelled with a [P].

Acknowledgments

We are grateful to many people for encouragement and for lending their expertise. Among these are Jiří Čížek, Rob Coulson, Paul Geiger, Jeff Greensite, Ken Jordan, Colin Morningstar, James Mueller, Kevin Sapp, Søren Toxvaerd, and Andrew Zentner. J. Boudreau wishes to thank Petar Maksimovic and Mark Fishler for inspiration and help with the development of Function Objects (Chapter 3), Lynn Garren for support of the original package which appeared in the CLHEP class library, and Thomas Kittelmann and Vakho Tsulaia for their collaboration over the years. Preliminary versions of this text, and particularly the exercises, have been inflicted on our graduate students over the years–their help has been instrumental in effecting many important revisions. We thank them wholeheartedly.

Contents

1	**Building programs in a Linux environment**	**1**
	1.1 The editor, the compiler, and the make system	3
	1.1.1 Troubleshooting mysterious problems	6
	1.2 A quick tour of input and output	7
	1.3 Where to find information on the C++ standard library	9
	1.4 Command line arguments and return values	9
	1.5 Obtaining numerical constants from input strings	11
	1.6 Resolving shared libraries at run time	11
	1.7 Compiling programs from multiple units	12
	1.8 Libraries and library tools	16
	1.9 More on Makefile	18
	1.10 The subversion source code management system (SVN)	20
	1.10.1 The SVN repository	21
	1.10.2 Importing a project into SVN	22
	1.10.3 The basic idea	22
	1.11 Style guide: advice for beginners	25
	1.12 Exercises	27
	Bibliography	29
2	**Encapsulation and the C++ class**	**30**
	2.1 Introduction	30
	2.2 The representation of numbers	31
	2.2.1 Integer datatypes	32
	2.2.2 Floating point datatypes	33
	2.2.3 Special floating point numbers	34
	2.2.4 Floating point arithmetic on the computer	36
	2.3 Encapsulation: an analogy	37
	2.4 Complex numbers	38
	2.5 Classes as user defined datatypes	43
	2.6 Style guide: defining constants and conversion factors in one place	45
	2.7 Summary	47
	2.8 Exercises	48
	Bibliography	49
3	**Some useful classes with applications**	**51**
	3.1 Introduction	51
	3.2 Coupled oscillations	52

	3.3	Linear algebra with the `Eigen` package	54
	3.4	Complex linear algebra and quantum mechanical scattering from piecewise constant potentials	57
	3.4.1	Transmission and reflection coefficients	59
	3.5	Complex linear algebra with `Eigen`	60
	3.6	Geometry	63
	3.6.1	Example: collisions in three dimensions	64
	3.7	Collection classes and strings	65
	3.8	Function objects	67
	3.8.1	Example: root finding	70
	3.8.2	Parameter objects and parametrized functors	74
	3.9	Plotting	76
	3.10	Further remarks	77
	3.11	Exercises	78
		Bibliography	83

4 Interpolation and extrapolation 84

4.1		Lagrange interpolating polynomial	85
4.2		Evaluation of the interpolating polynomial	87
	4.2.1	Interpolation in higher dimensions	89
4.3		Spline interpolation	90
	4.3.1	The cubic spline	90
	4.3.2	Coding the cubic spline	93
	4.3.3	Other splines	94
4.4		Extrapolation	95
4.5		Taylor series, continued fractions, and Padé approximants	97
4.6		Exercises	102
		Bibliography	106

5 Numerical quadrature 107

5.1		Some example problems	108
	5.1.1	One-dimensional periodic motion	108
	5.1.2	Quantization of energy	110
	5.1.3	Two body central force problems	111
	5.1.4	Quantum mechanical tunneling	112
	5.1.5	Moments of distributions	113
	5.1.6	Integrals of statistical mechanics	115
5.2		Quadrature formulae	118
	5.2.1	Accuracy and convergence rate	124
5.3		Speedups and convergence tests	126
5.4		Arbitrary abscissas	128
5.5		Optimal abscissas	129
5.6		Gaussian quadrature	132
5.7		Obtaining the abscissas	135

		5.7.1 Implementation notes	137
	5.8	Infinite range integrals	139
	5.9	Singular integrands	140
	5.10	Multidimensional integrals	141
	5.11	A note on nondimensionalization	142
		5.11.1 Compton scattering	142
		5.11.2 Particle in a finite one-dimensional well	143
		5.11.3 Schrödinger equation for the hydrogen atom	144
	5.12	Exercises	145
		Bibliography	150
6	**How to write a class**		151
	6.1	Some example problems	152
		6.1.1 A stack of integers	152
		6.1.2 The Jones calculus	152
		6.1.3 Implementing stack	154
	6.2	Constructors	160
	6.3	Assignment operators and copy constructors	162
	6.4	Destructors	165
	6.5	`const` member data and `const` member functions	166
	6.6	Mutable member data	167
	6.7	Operator overloading	167
	6.8	Friends	170
	6.9	Type conversion via constructors and cast operators	171
	6.10	Dynamic memory allocation	174
		6.10.1 The "big four"	176
	6.11	A worked example: implementing the Jones calculus	181
	6.12	Conclusion	190
	6.13	Exercises	191
		Bibliography	195
7	**Monte Carlo methods**		196
	7.1	Multidimensional integrals	197
	7.2	Generation of random variates	198
		7.2.1 Random numbers in C++11	198
		7.2.2 Random engines	199
		7.2.3 Uniform and nonuniform variates	200
		7.2.4 Histograms	202
		7.2.5 Numerical methods for nonuniform variate generation	204
		7.2.6 The rejection method	204
		7.2.7 Direct sampling (or the transformation method)	207
		7.2.8 Sum of two random variables	210
		7.2.9 The Gaussian (or normal) distribution	211
	7.3	The multivariate normal distribution, χ^2, and correlation	212

	7.4	Monte Carlo integration	216
		7.4.1 Importance sampling	219
		7.4.2 Example	220
	7.5	Markov chain Monte Carlo	221
		7.5.1 The Metropolis-Hastings algorithm	223
		7.5.2 Slow mixing	223
		7.5.3 Thermalization	225
		7.5.4 Autocorrelation	227
		7.5.5 Multimodality	228
	7.6	The heat bath algorithm	229
		7.6.1 An application: Ising spin systems	229
		7.6.2 Markov chains for quantum problems	231
	7.7	Where to go from here	232
	7.8	Exercises	233
		Bibliography	239
8	**Percolation and universality**	240
	8.1	Site percolation	241
		8.1.1 The cluster algorithm	241
		8.1.2 Code verification	244
		8.1.3 The percolation probability	244
	8.2	Fractals	249
	8.3	Scaling and critical exponents	252
		8.3.1 The correlation length and the anomalous dimension	253
		8.3.2 Additional scaling laws	256
	8.4	Universality and the renormalization group	258
		8.4.1 Coarse graining	260
		8.4.2 Monte Carlo renormalization group	263
	8.5	Epilogue	264
	8.6	Exercises	265
		Bibliography	272
9	**Parallel computing**	274
	9.1	High performance computing	274
	9.2	Parallel computing architecture	278
	9.3	Parallel computing paradigms	279
		9.3.1 MPI	279
		9.3.2 openMP	285
		9.3.3 C++11 concurrency library	289
	9.4	Parallel coding	299
	9.5	Forking subprocesses	300
	9.6	Interprocess communication and sockets	302
	9.7	Exercises	308
		Bibliography	310

10 Graphics for physicists — 312

- 10.1 Graphics engines — 312
  - 10.1.1 3d libraries and software — 313
  - 10.1.2 Generating graphics — 314
  - 10.1.3 The `Open Inventor/Coin3d` toolkit — 315
- 10.2 First steps in a 3d world–3d visualization — 316
  - 10.2.1 The basic skeleton of a 3d application — 316
  - 10.2.2 A three-dimensional greeting to the world — 318
  - 10.2.3 A colorful spherical world — 321
  - 10.2.4 Deleting nodes in the scene graph — 322
- 10.3 Finding patterns–Testing random number generators — 322
- 10.4 Describing nature's shapes–fractals — 326
  - 10.4.1 Shared nodes — 331
- 10.5 Animations — 335
  - 10.5.1 `Coin` engines: a rotating world — 335
  - 10.5.2 `Coin` sensors: an orbiting planet — 338
- 10.6 The `Inventor` system — 339
- 10.7 Exercises — 340
- *Bibliography* — 342

11 Ordinary differential equations — 343

- 11.1 Introduction — 344
- 11.2 Example applications — 345
  - 11.2.1 Projectile motion with air resistance — 345
  - 11.2.2 Motion of a charged particle in a magnetic field — 346
  - 11.2.3 Simple nonlinear systems: the Lorenz model — 347
  - 11.2.4 The Lagrangian formulation of classical mechanics — 348
  - 11.2.5 The Hamiltonian formulation of classical mechanics — 350
  - 11.2.6 The Schrödinger equation — 351
- 11.3 A high-level look at an ODE solver — 353
  - 11.3.1 A simple integrator class — 354
  - 11.3.2 Example: the harmonic oscillator — 357
- 11.4 Numerical methods for integrating ordinary differential equations — 358
  - 11.4.1 The Euler method — 359
  - 11.4.2 The midpoint method — 361
  - 11.4.3 The trapezoid method — 361
  - 11.4.4 The 4th order Runge-Kutta method — 361
  - 11.4.5 Properties of nth order Runge-Kutta methods — 362
- 11.5 Automated solution of classical problems — 368
  - 11.5.1 Taking partial derivatives with `GENFUNCTION`s — 368
  - 11.5.2 Computing *and* solving the equations of motion — 370
  - 11.5.3 A classical Hamiltonian solver — 371
- 11.6 Adaptive step size control — 373

11.6.1 Step doubling	375
11.6.2 Embedded Runge Kutta methods	379
11.7 Symplectic integration schemes	383
11.7.1 Symplectic transformations	386
11.8 Symplectic integrators of first and higher order	390
11.9 Algorithmic inadequacies	393
11.9.1 Stability	393
11.9.2 Solution mixing	394
11.9.3 Multiscale problems	395
11.10 Exercises	396
Bibliography	401

12 Polymorphism — 402

12.1 Example: output streams	404
12.2 Inheritance	406
12.3 Constructors and destructors	407
12.4 Virtual functions	409
12.5 Virtual destructors	413
12.6 Pure virtual functions and abstract base classes	415
12.7 Real example: extending the GenericFunctions package	416
12.8 Object-oriented analysis and design	421
12.9 Exercises	421
Bibliography	423

13 Nonlinear dynamics and chaos — 424

13.1 Introduction	424
13.2 Nonlinear ordinary differential equations	426
13.3 Iterative maps	429
13.3.1 The logistic map	430
13.3.2 The Hénon map	435
13.3.3 The quadratic map	436
13.4 The nonlinear oscillator	438
13.4.1 The Lyapunov exponent	440
13.5 Hamiltonian systems	443
13.5.1 The KAM theorem	444
13.5.2 The Hénon-Heiles model	445
13.5.3 Billiard models	447
13.6 Epilogue	449
13.7 Exercises	450
Bibliography	453

14 Rotations and Lorentz transformations — 455

14.1 Introduction	455
14.2 Rotations	456

14.2.1 Generators	457
14.2.2 Rotation matrices	458
14.3 Lorentz transformations	460
14.4 Rotations of vectors and other objects	463
14.4.1 Vectors	463
14.4.2 Spinors	464
14.4.3 Higher dimensional representations	465
14.5 Lorentz transformations of four-vectors and other objects	467
14.5.1 Four-vectors	467
14.5.2 Weyl spinors	468
14.5.3 Dirac spinors	470
14.5.4 Tensors of higher order	473
14.6 The helicity formalism	474
14.7 Exercises	479
Bibliography	482

15 Simulation 483

15.1 Stochastic systems	483
15.2 Large scale simulation	484
15.3 A first example	486
15.4 Interactions of photons with matter	488
15.5 Electromagnetic processes	491
15.5.1 Bremsstrahlung	491
15.5.2 Electromagnetic showers	493
15.5.3 The need for simulation toolkits	493
15.6 Fundamental processes: Compton scattering	494
15.7 A simple experiment: double Compton scattering	501
15.8 Heavier charged particles	503
15.9 Conclusion	505
15.10 Exercises	506
Bibliography	510

16 Data modeling 511

16.1 Tabular data	512
16.2 Linear least squares (or χ^2) fit	514
16.3 Function minimization in data modeling	516
16.3.1 The quality of a χ^2 fit	522
16.3.2 A mechanical analogy	523
16.4 Fitting distributions	523
16.4.1 χ^2 fit to a distribution	524
16.4.2 Binned maximum likelihood fit to a distribution	527
16.5 The unbinned maximum likelihood fit	529
16.5.1 Implementation	530
16.5.2 Construction of normalized PDFs	533

16.6 Orthogonal series density estimation	534
16.7 Bayesian inference	538
16.8 Combining data	540
16.9 The Kalman filter	543
16.9.1 Example: fitting a polynomial curve	545
16.9.2 Complete equations	546
16.10 Exercises	550
Bibliography	554

17 Templates, the standard C++ library, and modern C++   556

17.1 Generic type parameters	557
17.2 Function templates	558
17.3 Class templates	560
17.3.1 Class template specialization	561
17.4 Default template arguments	563
17.5 Non-type template parameters	563
17.6 The standard C++ library	565
17.6.1 Containers and iterators	566
17.6.2 Algorithms	575
17.7 Modern C++	577
17.7.1 Variadic templates	578
17.7.2 Auto	580
17.7.3 Smart pointers	581
17.7.4 Range-based for loop	582
17.7.5 Nullptr	583
17.7.6 Iterators: nonmember begin and end	583
17.7.7 Lambda functions and algorithms	584
17.7.8 Initializer lists	586
17.8 Exercises	590
Bibliography	593

18 Many body dynamics   594

18.1 Introduction	594
18.2 Relationship to classical statistical mechanics	595
18.3 Noble gases	597
18.3.1 The Verlet method	598
18.3.2 Temperature selection	603
18.3.3 Observables	605
18.4 Multiscale systems	608
18.4.1 Constrained dynamics	611
18.4.2 Multiple time scales	614
18.4.3 Solvents	616
18.5 Gravitational systems	616
18.5.1 N-Body simulations of galactic structure	618

18.5.2 The Barnes-Hut algorithm	620
18.5.3 Particle-mesh methods	625
18.6 Exercises	627
Bibliography	640

19 Continuum dynamics   642

19.1 Introduction	643
19.2 Initial value problems	643
19.2.1 Differencing	644
19.2.2 Continuity equations	645
19.2.3 Second order temporal methods	648
19.2.4 The Crank-Nicolson method	648
19.2.5 Second order equations	650
19.2.6 Realistic partial differential equations	651
19.2.7 Operator splitting	652
19.3 The Schrödinger equation	655
19.4 Boundary value problems	658
19.4.1 The Jacobi method	659
19.4.2 Successive over-relaxation	662
19.5 Multigrid methods	664
19.6 Fourier techniques	667
19.6.1 The fast Fourier transform	669
19.6.2 The sine transform	670
19.6.3 An application	672
19.7 Finite element methods	673
19.7.1 The variational method in one dimension	674
19.7.2 Two-dimensional finite elements	675
19.7.3 Mesh generation	677
19.8 Conclusions	684
19.9 Exercises	685
Bibliography	699

20 Classical spin systems   701

20.1 Introduction	701
20.2 The Ising model	702
20.2.1 Definitions	703
20.2.2 Critical exponents and finite size scaling	704
20.2.3 The heat bath algorithm and the induced magnetization	706
20.2.4 Reweighting	709
20.2.5 Autocorrelation and critical slowing down	710
20.2.6 Cluster algorithms	713
20.3 The Potts model and first order phase transitions	717
20.4 The planar XY model and infinite order phase transitions	720
20.5 Applications and extensions	724

	20.5.1 Spin glasses	724
	20.5.2 Hopfield model	725
20.6	Exercises	726
	Bibliography	730

21 Quantum mechanics I–few body systems 732

21.1	Introduction	732
21.2	Simple bound states	733
	21.2.1 Shooting methods	734
	21.2.2 Diagonalization	735
	21.2.3 Discretized eigenproblems	735
	21.2.4 Momentum space methods	737
	21.2.5 Relativistic kinematics	739
21.3	Quantum Monte Carlo	740
	21.3.1 Guided random walks	740
	21.3.2 Matrix elements	745
21.4	Scattering and the T-matrix	750
	21.4.1 Scattering via the Schrödinger equation	750
	21.4.2 The T-matrix	752
	21.4.3 Coupled channels	757
21.5	Appendix: Three-dimensional simple harmonic oscillator	761
21.6	Appendix: scattering formulae	763
21.7	Exercises	763
	Bibliography	771

22 Quantum spin systems 773

22.1	Introduction	773
22.2	The anisotropic Heisenberg antiferromagnet	775
22.3	The Lanczos algorithm	778
	22.3.1 The Lanczos miracles	779
	22.3.2 Application of the Lanczos method to the Heisenberg chain	781
22.4	Quantum Monte Carlo	785
22.5	Exercises	791
	Bibliography	800

23 Quantum mechanics II–many body systems 802

23.1	Introduction	802
23.2	Atoms	803
	23.2.1 Atomic scales	803
	23.2.2 The product Ansatz	804
	23.2.3 Matrix elements and atomic configurations	805
	23.2.4 Small atoms	807
	23.2.5 The self-consistent Hartee Fock method	809

23.3	Molecules	814
	23.3.1 Adiabatic separation of scales	815
	23.3.2 The electronic problem	817
23.4	Density functional theory	824
	23.4.1 The Kohn-Sham procedure	827
	23.4.2 DFT in practice	828
	23.4.3 Further developments	830
23.5	Conclusions	831
23.6	Appendix: Beyond Hartree-Fock	831
23.7	Exercises	836
	Bibliography	842

24 Quantum field theory — 844

24.1	Introduction	844
24.2	φ^4 theory	845
	24.2.1 Evaluating the path integral	849
	24.2.2 Particle spectrum	853
	24.2.3 Parity symmetry breaking	856
24.3	Z_2 Gauge theory	857
	24.3.1 Heat bath updates	861
	24.3.2 Average Plaquette and Polyakov loop	863
24.4	Abelian gauge theory: compact photons	865
	24.4.1 Gauge invariance and quenched QED	867
	24.4.2 Computational details	869
	24.4.3 Observables	872
	24.4.4 The continuum limit	875
24.5	$SU(2)$ Gauge theory	877
	24.5.1 Implementation	879
	24.5.2 Observables	881
24.6	Fermions	886
	24.6.1 Fermionic updating	890
24.7	Exercises	893
	Bibliography	903

Index — 905

1
Building programs in a Linux environment

1.1	The editor, the compiler, and the make system	3
	1.1.1 Troubleshooting mysterious problems	6
1.2	A quick tour of input and output	7
1.3	Where to find information on the C++ standard library	9
1.4	Command line arguments and return values	9
1.5	Obtaining numerical constants from input strings	11
1.6	Resolving shared libraries at run time	11
1.7	Compiling programs from multiple units	12
1.8	Libraries and library tools	16
1.9	More on Makefile	18
1.10	The subversion source code management system (SVN)	20
	1.10.1 The SVN repository	21
	1.10.2 Importing a project into SVN	22
	1.10.3 The basic idea	22
1.11	Style guide: advice for beginners	25
1.12	Exercises	27
	Bibliography	29

The goal of this book is to develop computational skills and apply them to problems in physics and the physical sciences. This gives us a certain license to try to teach any (legal) computational skill that we believe will be useful to you sooner or later in your career. The skill set you'll need includes scientific computing and not-specifically-scientific computing. For example, applying statistical techniques in data analysis or solving the Schrödinger equation on the computer are distinctly scientific computing tasks, whereas learning how to work collaboratively with a code management system is a not-specifically-scientific task. But you will use both as you progress in your career and so we will aim to teach you a little of both. We will go back and forth to some extent between scientific computing topics and general computing topics, so that the more generic skill set becomes useful in writing programs of a scientific nature, and the scientific programs provide opportunities to apply and reinforce the full skill set of numerical and not-so-numerical techniques.

Like mathematics, computing often appears to be a collection of tricks, with the well-known tricks elevated to the status of techniques. Deciding which tricks and techniques to teach is a difficult question, and a book like this has no traditional road-map. Our selection criterion is *usefulness*. Many of the topics are concerned with the simulation, classification and modeling of experimental data. Others (like integration or computational linear algebra) provide a basis for some of the later topics in simulation and modeling. Later, application to classical and quantum mechanical problems will be discussed. Like mathematics, computation is an art, and as practitioners we will pass on our own approach. If you learn to play a saxophone from John Coltrane, you will be absorbing John Coltrane's style, but also, hopefully, developing your own style along the way. So it is with the art of computation.

Writing executable programs and toolkits is of course central to this enterprise; since we are not about to describe computational techniques in wholly abstract terms, we have to be specific about which language(s) we are proposing to use. Our coding examples are usually expressed in the modern C++ language, or occasionally in the older, simpler computing language "C". We will sometimes also employ "pseudocode", which is a generic code-like description of any algorithmic process. Like almost any choice in computation, the focus of C++ is not totally obvious or universally acclaimed, but rather involves certain pros and cons–a debate that we will not lead you through here. The motivation for our choice is:

- In physics our lives consist of manipulating objects which are more abstract than scalar data types, including vectors, spinors, matrices, group elements, etc. While calculations involving these objects can be done in many computer languages, our lives will be vastly simpler if our computer languages support the objects of day-to-day life. No language is vast enough to support these at the language level, but languages supporting the object-oriented paradigm do allow you to add user-defined objects to the set of built-in data types. C++ also allows us to define basic operations on these data types, and maintains the speed of a compiled language.
- Most of a typical operating system is written in C and you will find it very easy to integrate specifically scientific software together with a vast body of generic software, particularly lower-level system calls.

Few people learn how to write software by writing programs from the bottom up. The "software stack" of even a very simple program can already involve toolkits that have taken a generation or two of computer scientists and physicists to develop. It is very common to make big scientific contributions by working on a small part of a huge program. Making modifications to an existing program, or filling in a piece of an incomplete program, can be a valuable learning experience. Some infrastructure for building programs is generally required. At a very minimum, a computing platform, operating system, and suite of compilers is needed. More complicated projects may even require a sophisticated set of tools to coordinate the development, distribution, and build of a software system. As more and more software is packaged and distributed for re-use, the build of computer programs becomes more challenging. In this chapter we introduce the basic ideas related to the building of software with some very simple examples.

Our reference operating systems are the popular Ubuntu linux (now at version 17.04) and macOS (version 10.12.6). These are both variants of unix, an operating system written in C and dating back to the 1970s; and we will refer to them generically as such. The commands required for writing, building, and executing programs as well as tailoring the environment will be expressed as required by the bash shell on a Ubuntu linux machine. Because of its low cost and portability, the linux operating system is widely used in scientific computing. Not only can it be used on personal computers (laptops, desktops, and now even smart phones), but it can also be found running in the machine rooms of large computer centers on thousands of high-density rack mount computers. The latest version of Ubuntu linux can always be installed on a PC after downloading from the website www.ubuntu.com. The installation, management and customization of operating systems are not trivial skills, but they are also extremely useful.

> Your first task is to get a laptop or a PC, and equip it with a basic linux operating system. We recommend that you install and maintain the operating system yourself. It is possible to dual-boot desktop and/or laptop computers, preserving the original operating system (e.g. Windows) which then coexists with linux. A Macintosh computer, which runs macOS, will also do for this book, since it runs an operating system similar to linux.

We assume a working knowledge of C++ basics–the part of C++ which is essentially just the C programming language, but minus anachronisms such as `malloc`, `free`, `printf`, `scanf`. In this text we develop in a few chapters that which is necessary to go beyond the ground level and understand classes, inheritance, polymorphism, and templates. For those who need to brush up on the basics, a good, short, but somewhat anachronistic introduction is the famous text of Kernighan and Ritchie (1988). The first few chapters of Capper (1994) or Bronson (2013) also cover the basics and provide a more modern introduction to the same subject. Another good source is the tutorial section of the online reference www.cplusplus.com. While our presentation of the C++ language will be far less formal than other common treatments of this powerful computing language, the physics applications will be more interesting and appropriate for the physical sciences, and you will "learn by doing", though it may be a good idea to refer to the above references occasionally if you prefer a more formal treatment of the language.

1.1 The editor, the compiler, and the make system

You write a program with an editor. There are a large number of these available on linux, and in principle any one will do. The *emacs* text editor (provided by the GNU

Table 1.1 *List of programs (center column) commonly used to compile major computer languages (left column).*

Language	Compiler under linux, OS X	Provided by
Fortran	gfortran	GNU Project
C	cc gcc	GNU Project
C++	c++ g++	GNU Project
Java	javac	Oracle

project; homepage www.gnu.org/s/emacs/) has features such as syntax highlighting, which can be very helpful in writing code, since it can recognize and draw to your attention syntax errors so that you can recognize them before they are reported by the compiler. The *gedit* text editor (provided by the GNOME project, homepage projects.gnome.org/gedit/) also has some of these features, and is perhaps more intuitive though less powerful. A much more basic editor called *vi* is generally pre-installed on even the most basic linux distributions. This is the editor of choice for gnarled veterans. On the other end of the spectrum, interactive development environments such as *Eclipse* (provided by the Eclipse foundation, homepage www.eclipse.org) embed powerful editors in a suite of tools for code development in C++ as well as other languages. There can be a steep learning curve with Interactive Development Environments (IDEs) such as Eclipse, but the effort can be worthwhile.

A single file of instructions is called a ***compilation unit***. A ***compiler*** turns these into ***object code***, and a ***linker*** puts different pieces of object code together into an executable program. Each computing language (Fortran, C, C++, Java) has its own compiler. Since C++ is a superset of C we can and will use the C++ compiler everywhere. Under linux, g++ and c++ are the same program. A table of common compilers is given in Table 1.1.

We look at a simple program which is a single compilation unit. It has a routine called main and like other functions takes arguments and returns an integer value (more on that, later). We call the file containing these lines foo.cpp. ".cpp" is the most common extension for c++ code. The program illustrates the important features of the main program unit, particularly how to write new commands under a unix operating system.

```
int main (int argc, char ** argv) {
   return 0;
}
```

Here are three ways to build an executable program from this source:

1. Compile to object code and then link to make an executable program.

    ```
    $c++ -c foo.cpp -o foo.o
    $c++ foo.o -o foo
    ```

2. Compile/link at the same time to make an executable program in one step.

   `$c++ foo.cpp -o foo`

3. Use make

   `$make foo`

The compilation step transforms human-readable C++ into machine instructions that the CPU can understand, and is called object code. The link step links together object code from various sources into an executable program. Which sources? The example above may give you the impression that there is only one, called `foo.cpp` but that is not true. Your program also contains pieces from the **C standard library** `libc.so` as well as others.

Even when the compilation and link is performed in one single command, there are still two phases to the process, and thus two points of failure. If you get an error message, try to figure out whether the error message is a compile error or a link error. Link errors do not occur at specific instructions, but constitute a failure to assemble the final program from the various pieces of object code, and usually in this case a piece of the program, or the object code containing the piece has been omitted, is missing, or cannot be located.

Once you've built the program you can see which run time libraries have been linked by issuing the command:

`$ldd foo`

which will generate the output

```
linux-vdso.so.1 =>   (0x00007fffc77fe000)
libc.so.6 => /lib/x86_64-linux-gnu/libc.so.6 (0x00007f19f6313000)
/lib64/ld-linux-x86-64.so.2 (0x00007)
```

In addition to these libraries that are included automatically in the link of any C++ program, additional libraries can be linked by mentioning them explicitly in the list of arguments to the compiler as we will soon see, using the `-l` and `-L` flags. In general programs will include program libraries containing useful functions and class libraries containing useful classes (extensions to the basic data types of the language).

Our first example (`foo.cpp`) is extremely simple and not at all a typical project, which these days can consist of many thousands of compilation units. Managing the development and build of large software infrastructure becomes a complicated job. Usually the `make` system (provided by the GNU project, homepage `http://www.gnu.org/software/make/`) is used to build projects containing multiple compilation units. A pedagogical guide can be found in Mecklenburg (2004). The third way of building the program `foo` illustrates the basic principle of the system: `make` knows that a program (`foo`) can be built from its sources (`foo.cpp`) with `g++`, the C++ compiler.

It applies a set of rules to build the target from its prerequisites. One of those rules says that the `g++` command can be used to build an executable with name `foo` from a source code file named `foo.cpp`.

The make system can be extensively customized (and usually is) whenever `make` is used in the context of a large software project. The customization is achieved through *Makefiles*, which are files usually named `makefile`, `Makefile`, or `GNUMakefile` that are placed in directories that contain the source code. In some cases these makefiles are written by hand and in others they may be generated automatically by another tool. We will describe this further as the need arises. Powerful as `make` is, it is often not sufficient on its own to organize the build of large or complicated projects, so additionally, a number of code management systems are available to coordinate and manage the distributed development of software *on top of make*.

1.1.1 Troubleshooting mysterious problems

On a few occasions you are likely to find that properly written code does not compile, link, or execute *because of the configuration of the platform and not the code itself*. Header files (with `.h` or extensions, discussed below) are normally installed in a system directory such as `/usr/include` or `/usr/local/include`; they can also be installed in other locations but then the qualifier

```
-I/path/to/include/area
```

must be added to the command line during the compile step. If the header files are not installed there then obviously the compile step will fail. Libraries, discussed in Section 1.8, are specified during the link step with the `-l` flag, and their search path is specified using the `-L` flag. These libraries must exist, they must be located during the link step, they must actually contain the symbols that they are supposed to provide, and they must be compatible. These symbols are all of the local and global variables known to the compiled code, as well as all the known structures, classes, free subroutines and member functions.

Computing hardware and operating systems exist in both 32 bit and 64 bit architectures, and object code which has been compiled for a 64 bit machine will generally not run on a 32 bit machine. Normally this type of object code would not be built or installed on the wrong architecture, but computers are machines and machines can go wrong. You might find a program or object library on a cross-mounted disk drive shared by machines having different architectures. Even the execution step can fail if bits of object code collected in shared libraries (files with the `.so` extension, discussed below) do not exist, cannot be located, or are incompatible. Incompatibilities can sometimes be caused by linking together object code produced by different compilers, or even the same compiler with different options. If these problems arise, the best approach is to be systematic in investigating and determining the cause.

To that end, it's useful to know about a few utilities in unix to help debug mysterious problems with "perfectly good" code. Table 1.2 summarizes a few of them to help you

Table 1.2 *Table of utilities for examining executable files. This is useful for investigating "mysterious problems" as described in the text.*

Linux command	OS X equivalent	Purpose
ldd	otool -L	check shared libraries required by a program. Prints the location of these libraries and flags any missing libraries.
file	file	classifies files. This command can yield useful information about how object files, and libraries, were compiled.
nm	nm	lists symbols in the object files, archives, shared libraries, and executables (variables and subroutines). Output can be very long!
c++filt	c++filt	Decodes "mangled" C++ symbols (for example, from nm or link error messages) and prints them in human-readable form.

see what you have just built; the unix manual pages for these commands give more information.

1.2 A quick tour of input and output

The first program you will generally write, like the famous Hello, World example in Kernighan and Ritchie (1988), simply echoes a few words to the terminal. In Fortran, input/output (IO) is handled at the language level; in C it is handled at the level of standard C functions, and in C++ it is handled through objects. Three important objects you must learn about are `std::cout` (the standard output), `std::cerr` (the standard error output), and `std::cin` (the standard input). In fact, since C++ is a superset of C, the C standard library routines (`printf` and `scanf` for the cognoscenti) can also be used but they are considered obsolete and should be avoided because they are unable to handle user-defined data types.

Basic usage of the `std::cin`, `std::cout`, and `std::cerr` objects is extremely easy. You will need to include the `iostream` header file at the top of any compilation unit that uses them:

```
#include <iostream>
```

Consider the following line:

```
std::cout << "Hello, World" << std::endl;
```

The "<<" characters in the above line constitute an operator, the left shift operator, which is defined in the C++ standard library for the `std::cout` object. The operator can stream bits of text, `int`s, `float`s, and `double`s, and even user-defined data types (if the designer allows it) to the standard output, i.e. the terminal. You should try this on your own. This is all we will say about `std::cout` for the moment. About `std::cerr`, we will say only that under unix operating systems (i.e. linux, Mac OS, and other variants), it is convenient sometimes to have two streams, both of which will normally end up printing to the terminal, because it is possible to redirect each stream separately[1] to a file or a unix pipe, for example. So, `std::cerr` functions just like `std::cout` except that normally program output is sent to `std::cout` while informational status, warning, and error messages are sent to `std::cerr`.

Your program can read input from the terminal using the `std::cin` class. This class can read in bits of text but also read the values of `int`, `float`, `double` (among others) from text strings from standard input–normally you think of typing these in using your actual fingers, but under unix you can also tell a program to take its "standard input" from a file, like this:

```
$ foo < file.txt
```

We use `std::cin` as follows in our programs:

```
#include <iostream>
...
int i, float f;
std::cin >> i >> f;
```

Input has one unique subtlety which we need to discuss: it can fail for several reasons. One reason is that the input may not be convertible to the right data type. For example, the word "particle" cannot be interpreted as an integer. Another reason is that the input may have reached its end, if the input were a file, or if the typing were terminated by C^D (Control+D). So we normally test that input has worked with the following incantation:

```
int i;
if (std::cin >>i) { // success!
  ...
}
else {              // failure!
  ...
}
```

---

[1] for more information, see the unix manual page on bash, particularly the section REDIRECTION.

1.3 Where to find information on the C++ standard library

The objects `std::cout`, `std::cerr`, and `std::cin` are all part of the C++ standard library, which contains a very large set of functions and objects, many of which will be extremely useful. The most usable documentation that we have encountered so far is an online reference, www.cplusplus.com, particularly the section called "library reference".

The C++ standard library is beyond the scope of our treatment and we will not attempt to explain more than the essentials, just what you'll need to follow this text and complete the exercises. We use it throughout the text, and discuss certain aspects in more detail in Chapters 6, 12, and 17. When you need detailed information, the ultimate authority on the C++ language and the C++ run time library is the ISO C++ standard. An updated standard was published in August 2011 by the International Standards Organization (ISO) and the International Electrotechnical Commission (IEC) as report number ISO/IEC 14882:2011. While this document is authoritative, it has little pedagogical value and is really intended for experts.

The best advice we can give to students confronted with such a large and complex set of standard software tools is to read the documentation as you go along, and try to learn, along the way, some of the tricks that will make you more efficient in your work. In this text we will explain bits and pieces of the C++ standard library as we go along, without any intent or commitment to treat the subject comprehensively.

1.4 Command line arguments and return values

An executable program is, generally speaking, a command like any other command in unix. Like any command, you can pass in *command-line arguments*. These are available through the variable `char **argv`, an array of character strings, whose length is given by `int argc`. Let's modify our program `foo` now so that it simply echoes the command line arguments:

```cpp
#include <iostream>              // Include headers for
                                 // basic i/o
int main (int argc, char ** argv) {  //
  for (int i=0;i<argc; i++) {    // Loop over command
                                 // line args
    std::cout << argv[i] << " "; // Print each argument
                                 // to screen.
  }                              //
  std::cout << std::endl;        // End-of-line.
  return 0;                      // Program successful
}
```

If you build this program and execute it with a few command-line arguments, it will behave as follows:

```
$./foo A B C D
./foo A B C D
```

Notice that the zeroth argument is the command name itself. The $ symbol in the preceding example is the command prompt, echoed by the shell unless the user has configured his system otherwise.

If you are new to unix, you may not know that the operating system looks for commands in a standard set of directories. An ordered list of directories is held in the environment variable PATH. To see what directories are in your path, you can type

```
$echo $PATH
/usr/local/sbin:/usr/local/bin:/usr/sbin:/usr/bin:/sbin:/bin:
/usr/games
```

For that matter, since you have now written a program very much like echo, you can use it to discover your path as well:

```
$./foo $PATH
./foo /usr/local/sbin:/usr/local/bin:/usr/sbin:/usr/bin:
/sbin:/bin
```

You can create your own directories where you can create important programs that you have written:

```
$mkdir ~/bin
$cp ./foo ~/bin
$export PATH=~/bin/:$PATH
```

This latter command can be added to the .bashrc file in your home directory, so that it is executed every time you log in. Then the program foo will appear as a built-in command, just like any other unix command, which, in fact, are built for the most part in exactly the same way. The source code for every command in Linux is free and publicly available–and most are written in the C programming language. For example, the echo command, which echoes the command line in a way very similar to foo, is compiled from source code which can be seen in http://git.savannah.gnu.org/cgit/coreutils.git/tree/src/echo.c; this web page points to a software repository which also contains other linux commands that may be familiar to you. You now have a partial glimpse of how a unix operating system, particularly many of the commands, are constructed.

Now back to our program, foo. The main routine of our program returns an integer value, zero in our example. This is the return status of the program. It's value can be accessed after the program has executed by the "special" parameter "?". Its value is decoded by prepending a "$", as for any environment variable (try it! type echo $?). It is typical to signal successful completion of a command by returning 0, and one or more failure modes with a nonzero value.

1.5 Obtaining numerical constants from input strings

One way of inputting numerical constants to a program is to read them from standard input. This is convenient in some circumstances but not in others, since from the point of view of the user of the program (who is also often the developer) the most useful way to communicate input to the program may be via the command line. However, as we have seen, command line arguments are made available as an array of character strings. How do we extract numbers from these strings?

Part of the C++ standard library is a class called `std::istringstream`. To use this object, you initialize it with a character string and then extract the numerical data through the right shift operator, >>. As with the object `std::cin`, which is closely related to `std::istringstream` objects, you can test for success. Here is an example of how to parse an integer from the second item on the command line:

```cpp
#include <sstream>
...
int main (int argc, char **argv) {
   ...
   int anInteger;
   std::istringstream stream(argv[1])
   if (stream >> anInteger) {   // Success!
   ...
   }
   else {                       // Failure!
   ...
   }
}
```

1.6 Resolving shared libraries at run time

Note that most programs are not complete without a set of shared libraries that are required by the program. Those shared libraries are to be found, normally, in standard locations. By default the area /usr/lib is searched for these libraries first, then /lib. By defining an environment variable called LD_LIBRARY_PATH, which is an ordered list of colon-separated directories, you can specify other libraries to search. If a shared library cannot be found, the system will report an error when you attempt to execute a program:

```
$myprogram
myprogram: error while loading shared libraries:
libMyLibrary.so.1: cannot open shared object file: No such file
or directory
```

Missing libraries will also be reported by the `ldd` utility, described in section 1.1.

1.7 Compiling programs from multiple units

In textbooks and exercises, programs consist of a few lines of code typed into a single source code file. In the real world of physical sciences, computations this simple are rare. In the context of large scientific experiments, often the software infrastructure is a behemoth consisting of multiple millions of lines of code split across tens of thousands of individual files, representing a major financial investment on the part of national governments. This is the ecosystem in which students in the physical sciences, particularly physics and astrophysics, may find themselves trying to operate. It should be quite obvious that organizing and managing software infrastructure involves breaking it down into more manageably sized units. In this and the following sections we will see how to compile a program from multiple compilation units, and then how to archive the units and coordinate the build of the program.

While the necessity of splitting up source code arises from very complicated systems, we will illustrate the basic idea by splitting up a small one. Our program, called `iterate`, is designed to make a certain number of calls to the system routine `sleep`. It takes two command line parameters: the number of iterations, and the duration of each iteration:

```
$ iterate 6 4
0 sleeping for 4 seconds
1 sleeping for 4 seconds
2 sleeping for 4 seconds
3 sleeping for 4 seconds
4 sleeping for 4 seconds
5 sleeping for 4 seconds
$
```

Here is the program, all in one single file, which we call `iterate.cpp`

```cpp
#include <iostream> // for std::cout, & cetera
#include <sstream>  // for istringstream
#include <cstdlib>  // for exit
//
```

```cpp
// Define a data structure
//
struct Control {
  int iterations;
  int seconds;
};
//
// Parse the command line:
//
Control *initialize (int argc, char ** argv) {
  Control *control=NULL;
  if (argc!=3) {
   std::cerr << "Usage: " << argv[0] << " iterations seconds"
       << std::endl; exit(0);
  }
  else {
    control =new Control;
    control->iterations=0;
    control->seconds=0;
    {
      std::istringstream stream(argv[1]);
      if (!(stream >> control->iterations)) return control;
    }
    {
      std::istringstream stream(argv[2]);
      if (!(stream >> control->seconds)) return control;
    }
  }
  return control;
}
//
// finalize:
//
void finalize(Control *control) {
   delete control;
}
//
// execute:
//
void execute(Control *control) {
  if (control) {
    for (int i=0;i<control->iterations;i++) {
      sleep(control->seconds);
```

```cpp
      std::cout << i << " sleeping for " << control->seconds
          << " seconds" << std::endl;
    }
  }
}
//
//
//
int main(int argc, char ** argv) {
  Control *control = initialize(argc, argv);
  execute(control);
  finalize(control);
  return 0;
}
```

The simplest way to build this is to type `make iterate`, which will in turn execute the following command in a child process:

```
$g++ iterate.cpp -o iterate
```

To illustrate the typical way of organizing a project with multiple compilation units, we will separate `iterate.cpp` into several pieces. The main program declares and defines three functions (in addition to `main`): `initialize`, `execute`, and `finalize`. It also declares a single data structure, called `Control`. The functions are defined where they are declared, which must be before they are used. Hence `main` is the last function to be defined, rather than the first.

In C++, as in C, every function and data structure (or class) must be declared before it is used, once, and only once; this amounts to specifying the interface to the function and/or class. The first step to breaking up this program is to separate the declarations from the definitions, putting them in a header file that can be included by each compilation unit. We therefore now create the header file `iterate.h`, and put it in the same directory as the source code:

```cpp
#ifndef _ITERATE_H_
#define _ITERATE_H_

// Data structure controlling the iteration loop:
struct Control {
  int iterations;
  int seconds;
};

// Initialize.  Parse the command line:
Control *initialize(int argc, char ** argv);
```

```
// Execute the iteration loop:
void execute(Control *control);

// Finalize.  Clean up the memory:
void finalize(Control *control);

#endif
```

The preprocessor directives (those including #ifndef, #define, #endif) form an *include guard*, guaranteeing that the declarations occur only once, even if the same header file is included twice, for example, directly, and indirectly.

We can now split our program into four compilation units, which we will call iterate.cpp (containing only main), initialize.cpp, finalize.cpp, and execute.cpp. Each one of these should include the header file iterate.h at the top, along with any other header files that it needs. For example the file execute.cpp looks now like this:

```
#include "iterate.h"
#include <iostream>
void execute(Control *control) {
  if (control) {
    for (int i=0;i<control->iterations;i++) {
      sleep(control->seconds);
      std::cout << i
                << " sleeping for "
                << control->seconds
                << " seconds"
                << std::endl;
    }
  }
}
```

The functions initialize and finalize have been moved into their own files, in a similar way; and the program iterate.cpp now becomes very simple:

```
#include "iterate.h"
int main(int argc, char ** argv) {
  Control *control = initialize(argc, argv);
  execute(control);
  finalize(control);
  return 0;
}
```

At this stage our directory contains four source code files–one of them containing the function main, which is required in any program. We can build the program iterate in a number of different ways from these sources. The first option that we have is to compile and link all of the source code at the same time:

```
$g++ iterate.cpp initialize.cpp execute.cpp finalize.cpp \
   -o iterate
```

This is fine except it has the drawback that if you change one part of this "project" you need to recompile all of its files. This is not too high a price for the program `iterate`, but could be very high in a real-world project. The preferred option is to compile each unit to object code separately, and then link them together at the end. The g++ command can be used for both stages of this procedure (here, the -c flag indicates that g++ should compile to object code and not bother to attempt a link):

```
$g++ -c -o iterate.o iterate.cpp
$g++ -c -o initialize.o initialize.cpp
$g++ -c -o execute.o execute.cpp
$g++ -c -o finalize.o finalize.cpp
$g++ iterate.o initialize.o execute.o finalize.o -o iterate
```

A mixed approach can also be taken:

```
$g++ -c -o iterate.o iterate.cpp
$g++ -c -o initialize.o initialize.cpp
$g++ iterate.o initialize.o execute.cpp finalize.cpp -o iterate
```

Stepping back and looking at what we have done, we notice that the modularity of the program is greatly improved; the protocol for each routine is clear from looking at the header file (where additional documentation can also be collected), and developers can work individually on one piece of the program at a time. On the other hand, while the program development itself is simplified, the price is additional complexity on the side of *building* the program. In the real world the management of very many compilation units is an extremely complicated task often carried out by teams of full-time people. Fortunately, a number of standard tools have been developed to better manage the build of a program. In the following section we will discuss several of them that are almost always at the core of any project: libraries of object code, the make system, and a source code management system.

1.8 Libraries and library tools

Having seen how to break up the source code into more manageable units, we now address the issue of how to keep the compiled code together in order to ease the process of compilation. Files containing object code (with the .o extension) are often collected together into *libraries*, so that the *client* code (`iterate.cpp` in the above example) can link with one library rather than many different object files. In the world of professional programming the library (and associated header files and documentation) is the implementation of what is called the API, or *application programming interface*, which represents a toolkit for the development of applications, vulgarly referred to as "apps" amongst the rabble.

There are two kinds of libraries, called static (files ending with `.a`) and shared (files ending with `.so` on linux systems or with `.dylib` on the mac). The main difference is that at link time, the object code in static libraries becomes part of the executable file; i.e., it is copied in. Programs linked to a shared library contain only pointers to bits of executable code which are not copied into the executable; and therefore the shared library file must be resolved at run time, before the program can be executed (see Section 1.6). Working with static libraries is usually simpler though it does create larger executables.

The linux command `ar` manipulates a static library. To create the static library `libIterate.a`, give the command:

```
$ar rc libIterate.a initialize.o execute.o finalize.o
```

The modifier `rc` stands for "replace" and "create": the library (or archive) is created if it doesn't exist, and if it does exist then any object files within the library are replaced by those given on the command line. The contents of the library can be examined with `ar t`:

```
$ar t libIterate.a
initialize.o
execute.o
finalize.o
```

Also, the contents of the library can be manipulated; for example individual pieces of object code may be extracted (`ar x`) or deleted (`ar d`). More information can be obtained from the `ar` manual page.

Now that you've built the library, you naturally will wonder how to link it to other code. The `g++` command has two important compiler flags that can be used, `-L` and `-l`. The first one, `-L`, specifies a directory to be searched for the library. The working directory is not on that path by default, so if your library lives in your working directory you should add `-L.` or `` -L`pwd` `` to the `g++` command. The second one, -l, gives the name of the library to link, but with an odd rule: to link to `libIterate.a`, you should write `-lIterate`. In other words, transform "lib" into `-l` and drop the extension `.a`. Both flags can be repeated on the command line and specify a search list: if the library is not found in the first directory, the second is searched; likewise if the symbols are not found in the first library on the command line, the second is searched. Therefore the order of `-L` and `-l` arguments is important. In our example, you will use the command:

```
$g++ iterate.o  -L. -lIterate -o iterate
```

To make a shared library, you can use the versatile g++ command again, giving the list of object files on the command line in addition to the `-shared` flag, and specifying a `lib` prefix and an `.so` extension for the output file, like this:

```
$g++ -shared initialize.o execute.o finalize.o -o libIterate.so
```

Some compilers also require the `-fPIC` qualifier during the compilation of object code files like `initialize.o` in order to produce "relocatable" code required in a shared library[2]. The only way to examine the contents of a shared library file is with the nm command, which dumps the symbol table. It is usually best to pipe this through c++filt, so that subroutine names are demangled into a readable form. You can link to the shared library in the same way as you link to a static library, and by default `libIterate.so` (for example) will be take precedence over `libIterate.a` if both files are found. Typically, when you then run the program, the shared object library will need to be resolved as well, as we have discussed in Section 1.6.

The software on a linux system is built from packages and these packages include programs, APIs, and sometimes both programs and APIs. One good (and useful) example is the ***gnu scientific library***, or `gsl`, which can be installed (on a Ubuntu linux system) by typing:

```
$sudo apt-get install libgsl0-dev
```

This installs the libraries `/usr/lib/libgsl.a`, `/usr/lib/libgsl.so`, as well as the headers `/usr/include/gsl/*.h`, and a manual page that can be referenced by typing

```
$man gsl
```

1.9 More on Makefile

The example we have been developing over the last few sections now constitutes a miniproject with a number of steps in the build process, namely (if we are going to use a static library):

- build the object files `initialize.o`, `execute.o`, `finalize.o`
- build the archive `libIterate.a`
- build the executable `iterate`

Extrapolating our experience to a large project, we can see that while building each constituent requires knowledge of a few compiler flags at the very least, building the entire

---

[2] This is particularly the case with the gnu compiler g++.

project interactively quickly becomes prohibitive. One way to automate this is to script the build (in `bash`, `csh`, or a similar scripting language); however with large projects one prefers to skip compiling code that has not changed since the last compilation. The `make` system was invented to build components of a large project from prerequisites. It can detect when a component is stale (because its prerequisites have changed) and rebuild it. It does this from rules, some of them built-in, while others can be specified to the system. The typical way to extend the set of built-in rules is to write a `Makefile` which lives in the same directory as the source code. We demonstrate this by an example. First, notice that the `make` system already knows that a `.cpp` file is prerequisite to an `.o` file with the same base name; typing make `initialize.o` is sufficient for the system to execute

```
g++    -c -o initialize.o initialize.cpp
```

when `initialize.cpp` is found in the working directory and has been modified more recently than `initialize.o`. Ditto for `execute.o` and `finalize.o`. Makefile does not know (yet) that these object files are prerequisite to the archive `libIterate.a`. Thus, we create a file named `Makefile` and add the following lines:

```
libIterate.a:initialize.o execute.o finalize.o
```

This says that `libIterate.a` is a target and that it depends on the three listed object files. Since it is the first target in the `Makefile` (and the only one for the moment), typing make will automatically build the three prerequisite files, from built-in rules. Following this line in the `Makefile`, one can specify which actions to take to build the target. Such lines must begin with a tab. The make system is very sensitive to this and using spaces will cause make to fail. We revise the `Makefile` so that it looks like this:

```
libIterate.a:initialize.o execute.o finalize.o
        ar rc libIterate.a initialize.o execute.o finalize.o
```

This can also be written a little differently, since the syntax of `Makefile` allows you to use the expression $@, which means "the target"; $? which means "the list of prerequisites"; and "$<", which means "the first prerequisite". Thus one can write:

```
libIterate.a:initialize.o execute.o finalize.o
        ar rc $@ $?
```

Macros can be defined in a `Makefile` too, so one can write this as:

```
OFILES=initialize.o execute.o finalize.o

libIterate.a:$(OFILES)
        ar rc $@ $?
```

Additional targets can be added to this list which now only includes `libIterate.a`. A common one to add is `clean` which removes all compiled code (.o files, static and/or shared libraries, the executable) and sometimes the backup files (ending in ~) left by the emacs editor. Such a line would look like

```
clean:
        rm -f *.a *.o iterate *~
```

(obviously you must be extremely careful!) and we can clean our directory now by typing `make clean`, which will rid the directory of everything but source code. Now we still have not told the make system how to build the final executable. We insert (at the top, after macro definition but before any other target) a line that determines how the executable `iterate` will be built; because this line will be the first target in the `Makefile`, `iterate` will be the *default target*. It will be built merely by typing `make` from that directory. The final version of `Makefile` looks like this:

```
OFILES=initialize.o execute.o finalize.o

iterate:iterate.o libIterate.a
        g++ -o $@ $< -L. -lIterate

libIterate.a:$(OFILES)
        ar rc $@ $?

clean:
        rm -f *.a *.o iterate
```

you can build the whole project by typing `make`, or pieces of it:

```
$make execute.o
$make libIterate.a
```

for example.

This probably looks like a nice tool, but even `Makefile` has its limits and many code management systems have been built on top of `Makefile` when things get even more complicated. We will not discuss these in this book, but be prepared for a shock when you enter the world of million-line software projects.

1.10 The subversion source code management system (SVN)

The last important tool that we will discuss in this introduction is the *Subversion system*, or SVN, which is very widely used to manage the process of distributed software development. This is an example of a source code management system. Other examples include the *Concurrent Version System* (CVS), which is now practically obsolete and *git* which is more recent and rapidly gaining popularity. In some ways these systems resemble a dropbox service, which is perhaps more familiar to a general audience. While a source code management system provides a central synchronized repository for code, it contains special features that enable tracking source code development and even restoring an entire project to a specific version and/or date. Some of the exercises in this book will require you to *reposit* the assignment in SVN; and your instructor will provide access to an SVN *repository* on a remote machine. Besides the ability to track changes and restore the code to some previous state, SVN allows for multiple developers to work on the same piece of code. Alternatively, the same individual can work on a project from multiple locations. This, for example, allows you to move potentially valuable pieces of work (including, but not limited to code, perhaps your thesis!) off of your laptop computer and onto a central site. Like many of the tools discussed in this chapter, SVN has a lot of powerful functionality and we will describe just enough of it so that you can get started using it. We assume that for your first experience with the system, somebody else (i.e. your instructor) will set up the SVN repository for you, set up the necessary access mechanisms, and grant you the necessary access rights.

1.10.1 The SVN repository

A repository is an abstract concept that implies a central place where code resides. How central? Repositories can be set up and administered by unprivileged users as files on a single laptop computer, if desired; more privileged administrators can configure servers where contributions from users around the country or around the world are centralized; and now commercial "cloud computing" services operate servers that can be accessed by users at little or no expense.

In the first case the SVN repository is a single file that the user creates:

```
$svnadmin create /path/to/repositoryfile
```

This creates a repository in a file called /path/to/repositoryfile that can now be filled, but *never* by acting directly on the repository file, which is henceforth only to be modified using SVN commands; i.e., those with the format `svn command options`. One useful command, `svn ls`, simply lists the contents of the repository, which, if you have just created it as explained above, is empty. The syntax of the command is

```
$svn ls file:///path/to/repositoryfile
```

If you wish to execute this from a remote machine (assuming that both local and remote machines can communicate via the secure shell, `ssh`), you can issue the following command from the remote machine:

```
$svn ls svn+ssh://user@host.domain/path/to/repositoryfile
```

where `user@host.domain` specified the username, hostname, and domain name of the local machine. Other types of servers, using other protocols, may have other prefixes such as `http://`, `https://`, or `svn://`.

For the following examples we assume that you have access to an SVN repository because either

- You have created a repository, as described above.
- You have requested and received a repository through a web-hosting service such as Sourceforge (www.sourceforge.com) or Cloudforge (www.cloudforge.com).
- Your instructor has set up a repository via one of the above methods, and given you access and instructions.

Web-hosting services have interesting benefits, first because they require no effort from the users to maintain and administer, second because they often add additional useful tools such as web-based code browsing. Whatever solution is adopted will will refer to the repository in the following as `protocol://repository`.

1.10.2 Importing a project into SVN

A project can be imported to SVN as follows. First we assume that the original source code lives in a directory called `/path/to/PROJECT`. To import that code into SVN issue the following command:

```
$svn import /path/to/PROJECT protocol://repository/PROJECT
```

This creates a project within the repository, and copies all of the files in /path/to/PROJECT into the repository. The project then grows only when additional directories are added to it, additional files are added to directories, or files are modified.

Each change to the repository is logged, and all the logged information is accessible via the command `svn log`. If you execute the `svn import` command as it appears above, an editor window will pop up and prompt you for a log message. Alternately, short log messages can be added by adding the option `-m "here is my log message"` to the command line.

1.10.3 The basic idea

We now have both a repository and a project residing in that repository. With this initialization out of the way, we describe the basic SVN operations for source code management.

The main idea behind SVN is that the repository holds a master copy of the project, while developers check out a local copy to their machine and make modifications. When they are satisfied, they check their modifications back into the master copy (repository). This is called a *copy-modify-merge* model. The simplest scenario is illustrated in Figure 1.1. The command

Building programs in a Linux environment 23

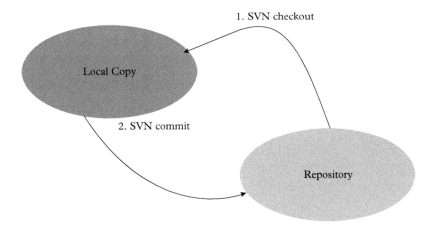

Figure 1.1 *Sketch of the simplest SVN transaction; first a checkout, then a commit following modification to the local copy.*

`$svn checkout protocol://repository/PROJECT.`

checks out the package (step 1 in Figure 1.1), and the command

`$svn commit -m "message"`

puts the modifications back into the repository (and adds the message "message" to the log). This is all that is needed as long as the file content of the project does not change. If files need to be added to the project then this is done with `svn add file1 [file2] [file3] ...`; the command schedules files for addition but does not add them until `svn commit` is issued; directories are added in the same way. `svn remove file1 [file2] [file3] ...` can be issued to remove files from the repository (after the files are removed from the local copy).

With multiple local copies in play (e.g., with multiple developers) the situation is more complicated because the master copy can change while one of the developers is making modifications to his or her local copy. Figure 1.2 shows a diagram of a typical transaction.

1. User 1 checks out the package to his/her local copy.
2. User 2 checks out the package to his/her local copy.
3. User 2 commits a modification to the repository. At this point user 1's local copy is out of date. *User 1 will not be able to commit until he refreshes his own copy with the changes that have gone into the repository.*
4. User 1 refreshes by typing `svn update`.
5. Then user 1 checks his changes into the repository with `svn commit`, after resolving any conflicts.

24 *The subversion source code management system (SVN)*

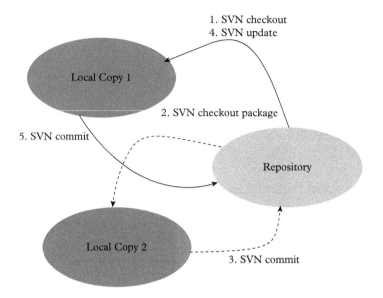

Figure 1.2 *A diagram of a more complicated set of SVN transactions arising from two users simultaneously developing within the same package. See text for details.*

Conflicts can arise when two users simultaneously work on the same package, but usually SVN is smart enough to resolve those without user intervention. SVN will resolve situations in which two users work on two different files in the same project. It will also resolve conflicts when two users have altered different parts of the same file. The only situation that normally requires user intervention is when two users have changed the same part of the same file; this can occur when two people jump in to fix the same bug. Conflicts are clearly indicated in the text and generally do not cause much trouble. We refer the reader to Collins-Sussman *et al.* (2004) for a detailed description of what to do when these cases arise.

The entire set of operations is carried out with very few SVN commands: `checkout`, `commit`, `update`, `add`, `remove`, `status`, and `log`; which suffices for most needs. The svn `checkout` and `update` commands have options that let users checkout, or update to, a tagged version, or a version that was current on a specific date and time. The SVN `log` command allows one to view the whole revision history of a file, including all of the comment messages that were given as arguments to the commit command. Finally, svn `status` displays the status of the project as a whole and/or individual file, depending on the parameters to the command.

It is important to never introduce executable programs, libraries, object code, backup copies, or large binary data files into the repository. These will make the administration of the repository difficult. Besides, these files need not be stored in the repository since they are typically regenerated from source. Putting `Makefiles` in the repository to aid in the build *is* generally a good idea.

After this brief introduction, we leave the reader to explore the full set of `SVN` operations using the references and/or the exercises.

1.11 Style guide: advice for beginners

We close with an offering of advice for beginners that has been gleaned over 80 combined years of programming experience. Presumably these good practices will be set by protocols if you are working in a large group. If you are not, read on.

dive in. Learning a computer language should be like learning French or German. You don't need to have mastered all the nuances of the plus-que-parfait tense to order a burger in Paris. Look at examples, learn some basics, and get coding!

adopt a convention. Adopting a convention for naming your files, libraries, variables, objects, and classes is important. Established conventions do exist, but we will not venture to advocate one over another since this is a matter of personal choice (or a choice that has been imposed by management). Peruse the Wikipedia article on *Hungarian notation* to get an idea of the options, and controversy, in adopting conventions.

use descriptive names. Calling an index `i` is traditional, but it is much better to give it a descriptive name such as `particleNumber`.

document your code. You will not remember the inner workings of your code after a few months. Describe its purpose, method, important variables, compiler directives, and usage somewhere (often at the top of the main file). If you are not using a repository, record revision history.

write clear code. You might think this is obvious, but it is quite easy for code to become unclear and one must be vigilant. Sooner or later you will find some clever compact way to do something and add it to your code. But you will not remember the trick the next time you look at your program and will not feel so clever. If you are going to be clever at least document it.

At a more general level, clear code is the result of clearly understanding the computational issue. Of course, as scientific programmers it is your duty to understand your problem to the best of your ability.

Lastly, if your code is destined to be used by others you should also consider the typical user's conceptualization of the code. In particular, you want the user interface to match the user's expectations. A familiar version of this problem is when a user attempts to change an aspect of text in a word processing program and all sorts of unexpected and frustrating code behavior ensues.

trap errors. A part of structured programming is trapping and dealing with errors when they are generated. This is implemented, for example, in C++ or java with the `try` and `catch` statements. Unfortunately, no such functionality exists in C or Fortran. Do what you can.

avoid magic numbers. Strings of digits should not appear in your code. You may recognize 0.197327 as $\hbar c$ in the appropriate units but the next person to work on your code may not. Better to place a statement such as `static float hbarc = 0.197327` somewhere. While we are on this, why would you ever write `double pi = 3.1415926` when you could write `double pi = 4.0*atan(1.0)`, or even better include the header `<cmath>`, which defines many mathematical constants, including `M_PI`?

allow for parameter input. Students tend to code parameter values directly into their programs. In our experience, programs are never run once and it is therefore important to allow for flexible parameter input. This is typically accomplished in one of three ways: (i) query the user for console input, (ii) via the command line, (iii) via an input file. Option (i) is simple; option (ii) permits easy scripting; and option (iii) is useful when many parameters are required but is more difficult to manage. One approach is to query the user and echo input data into a file. This file can then be modified and redirected as input in subsequent runs: `./foo < input.txt`. A more sophisticated approach might use a graphical user interface or a database.

specify default parameters. It is easy to spend days or weeks determining parameter values that optimize sought behavior. Once these are found they should be documented with the project, preferably as defaults in the code itself. We guarantee that you will not remember the preferred temporal grid spacing in your electromagnetic field computation a few months from now.

document the output. Most scientific code produces output files, sometimes dozens of them. These files should contain header information that specifies the program that created them, parameter values that were used, and the meaning (and units) of the output.

learn a debugger. Simple programs can be debugged with judicious print statements. However more complicated issues will require the use of a debugger, such as `gdb`. The GNU collaboration has created a free graphical front end for a variety of debuggers called *DataDisplayDebugger* that we happily recommend. Integrated Development Environments (such as Eclipse) often have debugging capability as well. It is worth the effort to learn to use these.

squash kludge. As you develop code you will often find yourself adding options to deal with an ever-increasing number of cases. The result is a kludgy mess with multiple layers of `if` or `case` statements. At this stage it is better to redesign the program or, more simply, to clone variant code.

develop systematically. Plan the general code structure before writing anything. Develop code in small logical chunks and debug as you go. Rushing ahead is a guarantee of future headache.

test extensively. If you are writing scientific code, you are doing science, and your code must be reliable. Check your results against known limits and special cases with analytical solutions. If you must strike out on your own, try to make the change from known territory as minimal as possible (for example, substituting a complicated potential for a square well potential in a quantum mechanics problem). Debugging Monte Carlo code is extremely difficult! You need analytic limits and simple cases worked out for comparison.

understand code parameters. Students often confuse *physical* parameters and *algorithmic* parameters. The former define the physics problem that you want to solve. The latter specify the method in which you solve the problem–in principle your answer should be independent of algorithmic parameters. It is up to you to confirm this! In practice this means extrapolating to infinity or zero. You can always double or halve an algorithmic parameter and confirm stability of your answer. Even better is to plot how your answer changes with algorithmic parameters, and better yet is to have a theoretical understanding of the functional form of that dependence so that reliable extrapolations can be made.

Similar advice with additional detail is contained in Wilson (2014).

1.12 Exercises

1. Write and compile an empty program. Name it exp.cpp, and put in in a directory called CH1/EX1. Compile it, and use the four utilities in Table 1.2 to examine the output file. Add to your empty program a call to the math library function exp, which requires you to include the header file:

   ```
   #include <cmath>
   ```

   Run the nm and the ldd utilities on the program and describe carefully what happens before and after the call to exp is added. Try this with both a constant and a variable argument to exp. Pipe the output of nm through c++filt and explain how the output is different. (This is called "name mangling" and "demangling").

2. Now write a variant of this program in a directory called CH1/EX2. Add to your program a single command line argument. Your program should now accept a single command line argument and echo the exponential of that argument, like this:

   ```
   $exp 5
   148.413
   ```

   Examine the program with ldd and nm. Pipe the output of nm through c++filt.

3. In a directory called CH1/EX3, replace the call to your own version of the exp function. Use a Taylor series expansion[3]. Check again with ldd, nm and nm | c++filt. Tabulate x vs. $\exp(x)$ for $x = \{-10, -8, -6....6, 8, 10\}$. By switching between your version of exp, and the one in the math library, determine how nm tells you whether the exp routine is internal or external to the executable program file.
4. In unix the wildcard character is the star or asterisk, "*". In a directory called CH1/EX4, write the program foo described in Section 1.4. Go into your home directory and execute /path/to/foo *. Explain the output.
5. Write a program called plus (in a directory called CH1/EX5) which expects two integers on the command line, adds the two integers and prints out the answer. The expected behavior of this program is:

```
$plus 3 7
3+7=10
```

6. In the previous example, break the code up into three compilation units:

   a) main.cpp – main program
   b) parse.cpp – parses the command line
   c) add.cpp – adds the two input values
   d) print.cpp – prints the result

   Put these four files together a directory called CH1/EX6, and add a Makefile which builds:

   a) a library called libPlus.a
   b) an executable called plus

   Make sure that the Makefile contains a "clean" target to remove all object code and executables generated by the compiler during the make procedure.

7. Clone the directory CH1/EX6 into a directory CH1/EX7; then modify your Makefile so that the shared library libPlus.so is used, rather than a static library. Check and report the size of the final executable in both circumstances.
8. Write Makefiles in the directories (CH1/EX1-CH1/EX7) to build all of the targets in the previous exercises. Make sure each executable has a "clean" target.
9. Your instructor will provide access to a SVN repository. Add the source code and Makefiles for CH1/EX1-CH1/EX7 to the repository.

   a) make sure that you do not reposit any object code.
   b) make sure that you can check out and build each executable from scratch.

[3] In Chapter 4 we will learn a better way to implement an exponential function

BIBLIOGRAPHY

Bronson, G. J. (2013). *C++ for scientists and engineers*. Cengage Learning.
Capper, D. M. (1994). *The C++ programming language for scientists, engineers, and mathematicians*. Springer-Verlag.
Collins-Sussman, B., Brian W. Fitzpatrick, and C. Michael Pilato (2004). *Version control with subversion*. O'Reilly Media; also available freely as http://svnbook.red-bean.com
Kernighan, B. W. and D. M. Ritchie (1988). *The C Programming language*. 2nd ed. Prentice-Hall.
Mecklenburg, R. (2004). *Managing projects with GNU make*. O'Reilly Media; also available as www.oreilly.com/openbook/make3/book
Wilson, G. *et al.* (2014). *Best Practices for Scientific Computing*. PLOS Biol **12(1)**: e1001745.

2
Encapsulation and the C++ class

2.1 Introduction	30
2.2 The representation of numbers	31
2.2.1 Integer datatypes	32
2.2.2 Floating point datatypes	33
2.2.3 Special floating point numbers	34
2.2.4 Floating point arithmetic on the computer	36
2.3 Encapsulation: an analogy	37
2.4 Complex numbers	38
2.5 Classes as user defined datatypes	43
2.6 Style guide: defining constants and conversion factors in one place	45
2.7 Summary	47
2.8 Exercises	48
Bibliography	49

2.1 Introduction

In the previous chapter we concentrated on how computer codes are written, compiled, archived, linked, and executed; how executable systems interact with the linux operating system; how source code is managed; and how build procedures can be automated. In this chapter we will talk more about the code itself, assuming that the reader has familiarity with the basics of C++, namely things like operations, control structures, functions, arrays, and pointers.

While datatypes are likely also familiar to the reader, we discuss integer, floating point, and complex datatypes in some detail for two reasons. The first is that in numerical calculations, one needs to understand the limitations in accuracy arising from the representation of numbers on a computer and the way in which they are acted upon by the CPU during an arithmetic calculation.

The second reason is more subtle. A new feature of C++ (relative to its predecessors C and Fortran), and probably the most important building-block of modern code, is the user-defined datatype or *class*. Some of these datatypes, like those you are already using for input and output, are part of the C++ standard library, while others may be imported from other sources and ultimately written by you.

If the notion of a class is unfamiliar, we will not give a full or accurate definition here, but for our numerically literate target audience it is productive to *initially* think of a class as a new type of numerical datatype, like a floating point number or an integer. For example, in C++ complex numbers are not built into the language; they are represented by their own class in the C++ standard library. Vectors, four-vectors, matrices, quaternions, octonions, and other useful entities can be imported from a variety of sources.

When a skilled programmer implements a class in C++, his or her implementation shares some of the same features that make the built-in datatypes convenient to use. Most notably, there is a clean separation between the representation of the datatype and its interface; i.e., the set of operations to which it responds. This separation is called *encapsulation*. We expect to find this feature in classes as well. Along with inheritance and polymorphism, to be tackled later, encapsulation is one of the key components in the object-oriented style of programming.

2.2 The representation of numbers

A number, floating point or otherwise, is represented by a array of bits held on a magnetic storage device or a silicon flip-flop circuit. Encoding this information is fairly simple in the case of integers but can be quite complicated in the case of floating point numbers. Encapsulation is achieved by hiding all of this internal structure in several datatypes:

- `int`
- `unsigned int`
- `float`
- `double`,

which respond to four basic operations +,-,/,*, as well as a few others. The encapsulation of the internal representation of these numbers is so effective that many programmers seldom consider it. For numerical work it is, however, important to know about the internal structure because the accuracy of numerical computations depends on it. We therefore describe the manner in which numbers are represented and the attendant consequences for computation.

Numbers are stored in arrays of *bits*, which are comprised of 0's and 1's. A collection of eight bits is called a *byte*. All built-in datatypes require at least one byte of storage. The number of bytes used to store a datatype is called the size of the datatype, and it varies from platform to platform. The size can can be determined with the `sizeof` operator in C(++):

```
int sizeofA=sizeof(a);
```

A short program to print out the size of the major built-in datatypes is shown below:

```
#include <iostream>
int main(int argc, char **argv) {
  std::cout << sizeof(bool) << std::endl;
  std::cout << sizeof(char) << std::endl;
  std::cout << sizeof(int) << std::endl;
  std::cout << sizeof(unsigned int) << std::endl;
  std::cout << sizeof(long int) << std::endl;
  std::cout << sizeof(long unsigned int) << std::endl;
  std::cout << sizeof(short int) << std::endl;
  std::cout << sizeof(short unsigned int) << std::endl;
  std::cout << sizeof(float) << std::endl;
  std::cout << sizeof(double) << std::endl;
}
```

You should copy this into an editor and compile and run it on your machine.

2.2.1 Integer datatypes

An unsigned int has 4 bytes or 32 bits. It uses these to express a number in base-2 according to the expression:

$$x = a_{31} 2^{31} + a_{30} 2^{30} + \ldots + a_0 2^0$$

where a_i is either 1 or 0. This is called *fixed-point notation*. Other unsigned integer datatypes are char, with a size of one byte, and long unsigned int, which is 32 bits on some machines and 64 bits on more modern ones. In an int variable the upper bit is reserved for the sign:

$$x = (-1)^{a_{31}} (a_{30} 2^{30} + \ldots + a_0 2^0)$$

A consequence is that there is a maximum and minimum integer that can be represented for every datatype; between those limits integers are stored exactly. Integer operations + (addition), − (subtraction) and * (multiplication) are also exact as long as one does not overflow or underflow the range, but the operation / (division) discards any remainder to obtain an integer result. The range of integer datatypes can be obtained using the std::numeric_limits class:

```
#include <iostream>
#include <iomanip>
#include <limits>
int main( int argc, char ** argv) {
```

```cpp
    std::cout << std::numeric_limits<bool>::min() << std::endl;
    std::cout <<  std::numeric_limits<bool>::max() << std
       ::endl;
    std::cout << (int) std::numeric_limits<unsigned char>
       ::min() << std::endl;
    std::cout << (int) std::numeric_limits<unsigned char>
       ::max() << std::endl;
    std::cout << std::numeric_limits<int>::min() << std::endl;
    std::cout << std::numeric_limits<int>::max() << std::endl;
    std::cout << std::numeric_limits<unsigned int>::min()
       << std::endl;
    std::cout << std::numeric_limits<unsigned int>::max()
       << std::endl;
    std::cout << std::numeric_limits<long int>::min() << std::
       endl;
    std::cout << std::numeric_limits<long int>::max() << std::
       endl;
    std::cout << std::numeric_limits<long unsigned int>::min()
       << std::endl;
    std::cout << std::numeric_limits<long unsigned int>::max()
       << std::endl;
}
```

Good documentation on this class can be obtained from the online reference www.cplusplus.com. Running this code reveals that a 32 bit int has an allowed range of -2147483648 to 2147483647. For an unsigned int the range is $[0, 4294967295]$. A char has a range of $[0, 255]$. These ranges are not big; for larger numbers one needs to resort to floating-point datatypes.

2.2.2 Floating point datatypes

A fixed-length datatype can store a finite number of values, while (even within a finite range of values) the number of rational numbers is infinite, and the number of real numbers is even larger. Thus only certain floating point numbers can be represented exactly, and the result of floating point operations involves some *roundoff error*.

The typical error in a floating point calculation is called the *machine precision*; for float variables this is between 6 and 7 decimal places, and for double variables it is between 15 and 16 decimal places. The class numeric_limits reveals that the range of floats is between $1.17549 \cdot 10^{-38}$ and $3.40282 \cdot 10^{38}$, while that of double variables is from $2.22507 \cdot 10^{-308}$ to $1.79769 \cdot 10^{+308}$.

A modern CPU does not require more time to carry out floating point operations on double variables than on floats, so double will be our preferred floating point datatype (unless memory or disk space forces us to use a more compact representation).

Table 2.1 *Bit allocation for float and double datatypes.*

Type	Sign	Exponent	Fraction
double	1 Bit	11 Bits	52 Bits
float	1 Bit	8 Bits	23 Bits

The IEEE-754 standard format for a floating point number is the following:

$$x = (-1)^s 1.f \times 2^{e-b}$$

where:

- s is the sign bit.
- 1 is always implied. It is called the "ghost bit"
- $1.f$ is the mantissa, with f being the fraction. For doubles, it is stored in 52 bits (23 for float)
- e is the exponent. For doubles, it is stored in 11 bits (8 for float)
- b is the bias. For doubles, it has a fixed value 1023 (127 for float).

Note that zero can have the sign bit set or unset, so the numbers $+0$ and -0 are distinct, though they give the same results in all floating point operations. Table 2.1 shows bit allocation for float and double datatypes.

The exponent of a floating point number is stored exactly, so the precision of the number is the precision of the full mantissa.

In scientific notation the numbers 3.0×10^4 and 30×10^3 are the same. Since a fixed number of bits are allocated for the mantissa it is advantageous to have the mantissa as left-shifted as possible. In this form the binary representation is $1.x \times 2^n$. Since the binary digit to the left of the decimal point is always 1 it is not stored. The resulting number is referred to as *normalized*.

2.2.3 Special floating point numbers

The result of a numerical operation can be undefined if, for example, a division by zero is made. Even if the result of a calculation is defined mathematically, it can still result in an underflow or overflow due to the limited range of floating point variables. For doubles the exponent is stored in eleven bits. The normalized numbers discussed above have an exponent between 0 and 2046 (in binary 11111111110). Underflows, overflows, and other the results of undefined operations are represented by an exponent field with all eleven bits set (i.e., decimal 2047).

Infinity of either sign is represented by a fraction of zero, with the sign determined by the sign bit. When the fraction is nonzero the number represents an undefined quantity, called *NaN* ("not a number"). This could result, for example, from dividing by zero or taking the square root of a negative number. The value of the fraction (called the *payload*) contains extra information that can be decoded by clever programmers.

If the first bit of a fraction is nonzero the invalid number is called a *signalling NaN*. When such variables arise, many floating point routines will raise a signal and exit with an error. If the first bit is not set, the invalid number is called a *quiet NaN*. Computations are expected to fail silently when they encounter such a number. These numbers can be obtained from the `numeric_limits` class:

```
double qNaN=double nan=std::numeric_limits<double>
   ::quiet_NaN();
double sNaN=double nan=std::numeric_limits<double>
   ::signaling_NaN();
double inf =double nan=std::numeric_limits<double>
   ::infinity();
```

The smallest floating point numbers are represented with an exponent $e = 0$. In this case a smaller bias is taken (1022 for `double`, instead of 1023; for `float` 126 instead of 127) and the ghost bit is taken but *subnormal numbers*. Subnormal numbers lie between true underflow and the result of `std::numeric_limits<double>::min()` which in fact returns the smallest *normalized* number. They lack the precision of normal numbers. A routine called `std::fpclassify` is available as part of the runtime library:

```
#include <cmath>
int std::fpclassify(double x);
```

which classifies floating point numbers as zero, NaN, normal, or subnormal. Like all library and system calls, the routine is documented in the linux manual pages and you can read the documentation by typing

```
man fpclassify
```

A related routine called

```
    int std::isfinite(double x);
```

returns 1 if the number is normal or subnormal and 0 if it is infinite or NaN. This is useful when debugging code. For example, the following code illustrates how `isfinite` can be used to track down a divide-by-zero error.

```
#include <sstream>
#include <stdexcept>
...
```

```
double z = y/x;
if (!std::isfinite(z)) {
  std::ostringstream stream;
  stream << "x,y,z=:" << x << "," << y << "," << z << std::
      endl;
  throw std::runtime_error(stream.str());
}
```

This example *throws an exception*, which is a good way for C++ to signal that something has gone wrong; calling routines can *catch* the exception and take appropriate actions. In an interactive debugger, one can not only print out information about the condition that led to an infinite or undefined value of z, but also set a breakpoint at the print statement and examine other variables at that point in the program's execution in order to determine why the abnormal condition has occurred.

Routines called `std::isnormal`, `std::isnan`, and `std::isinf` also exist and are defined in the `<cmath>` header file. Their functionality is pretty obvious; details can be found in the manual pages.

2.2.4 Floating point arithmetic on the computer

When two floating point numbers are added in the CPU, the mantissa of the smaller of the two numbers is right-shifted (divided by two), an operation that is compensated by incrementing the exponent by one unit, until the exponents of the two numbers are equal. As a result the lower-order bits are shifted away and precision is lost. Once this is done, the CPU takes the sum of the two mantissas and uses the common exponent to form the result, which it then normalizes. The summation is carried out in an *arithmetic logic unit* that mostly consists of simple gates capable of carrying out Boolean operations on the bits of the numbers.

A simple example illustrates the loss of precision in base-10 arithmetic.

- We are going to add 1.276×10^6 to 2.852×10^8.
- The smaller number is expressed as 0.012×10^8.
- The sum is 2.864×10^8. Notice that the last two digits in the first operand have been lost.

The same type of rounding error takes place in base-2 arithmetic with floating point operations in the CPU and one has to worry about the cumulative effect of these rounding errors. Notice that the entire mantissa is shifted away when the number of digits in the mantissa is less than the difference in the exponents of two summands. In this case the sum is equal to the larger summand. The largest number for which this occurs is the machine precision mentioned before. The numerical value of the machine precision depends on the data type and can be accessed via the `numeric_limits` class:

```
std::numeric_limits<float>::epsilon()
```

Example: In the Large Hadron Collider (LHC), experiments energy deposits of hundreds of GeV can be placed in the heavy materials of a device called a calorimeter. The deposit consists of very many smaller ionizations called hits, where the hits can consist of deposits down to about one KeV. The full deposit and the hits that contribute to this deposit therefore range over about eight orders of magnitude. When summing all of the smaller hits into one deposit using `floats`, one eventually reaches a point where including additional low energy hits produces no difference, and this energy is lost to the calculation. The simplest resolution is to employ higher precision data types such as `double`. An algorithmic alternative is to sum small numbers in a hierarchical fashion.

A similar loss of precision occurs with subtraction. Subtraction is performed in very much the same way as addition: the two numbers are expressed with a common exponent (shifting the mantissa if needed) and then the mantissas are subtracted using logic gates in the CPU. When two numbers of similar size are subtracted, the result will be expressed in only a few bits. When the result is normalized, those bits will be shifted left as much as possible and the trailing bits will be filled with zeros. The resulting roundoff error can be significant.

Sometimes the problem can be avoided with a little cleverness. For example, consider the stretch factor $\gamma = 1/\sqrt{1-\beta^2}$ for a relativistic particle, where $\beta = v/c$, v is the particle's speed, and c is the speed of light. Since relativistic particles have $\beta \approx 1$, there is a good chance that a loss of precision will make the denominator of this expression zero. One can instead express $1-\beta^2 = (1+\beta)(1-\beta)$. Setting $\epsilon = 1-\beta$ gives $\gamma = 1/\sqrt{(2-\epsilon)\epsilon}$, which will generally be calculated to higher precision.

Machines carry out approximate computations on approximate representations of real numbers. Nevertheless, one must admire the skill with which the developers of the IEEE-754 standard, and our compiler set, have kept these details so effectively hidden from our sight, thereby freeing us to frame the problems of our day-to day life in terms of numbers, rather than bits. We will attempt to achieve the same functionality, via encapsulation, when defining our own datatypes.

2.3 Encapsulation: an analogy

In discussing the notion of encapsulation, we find it germane to draw an analogy to a development in electrical engineering that parallels developments in software engineering. We are referring to the advent of integrated circuitry, which has promoted the proliferation of highly complex circuitry such as sophisticated Central Processing Units (CPUs) with up to 10 billion transistors on a single chip.

Figure 2.1 shows a popular operational amplifier called the LM741. On the right is a circuit diagram, which gives one view of what is inside the neat little package.

Users of the LM741 need to be knowledgeable about the eight pins in the package, which, counterclockwise from the upper left, are the offset null used to cancel voltages that appear as the result of leakage currents, the negative and positive input, the negative supply voltage, an unused pin, the output, the positive supply voltage, and a second offset null. While sometimes a serious designer needs to know about the internal circuitry of this

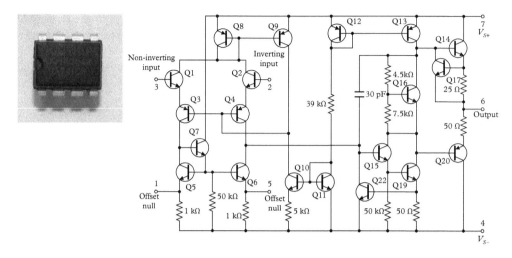

Figure 2.1 *Picture (left) and circuit diagram (right) of the LM741 operational amplifier (adapted from svg file by Daniel Braun from https://en.wikipedia.org/wiki/Operational_amplifier. Creative commons BY SA 3.0).*

device, most of the time he or she can forget those details when building more complex devices, or at least relegate their consideration to the later stages of design.

The LM741 implements a clean separation between what is exposed to the outside world and the messy internal hardware. The choice to hide the details of the circuit is a matter of convenience and practicality. The choice of which parts of the circuit to expose on the chip's interface is a matter for the chip designer, who must develop a clear idea of what is to be achieved with the chip. In this case, the LM741 is an amplifier. Should it contain headphones then? No, because it is not an iPod.

In carrying out computational work we use (and sometimes write) classes for mathematics and physics that help us with these computations. Such classes can represent vectors or matrices, for example. And these classes should, like the integrated circuit, have a trustworthy mechanism on the inside which is not exposed, and a clean, simple, and logical interface on the outside. This is what we call encapsulation. While the LM741 is encapsulated in a plastic case, encapsulation in object-oriented programming is conceptual and is applied to user-defined datatypes, namely the aforementioned classes.

2.4 Complex numbers

Our next example of a numerical datatype is `std::complex`. Unlike `float`, `double`, and the integer datatypes, `std::complex` is a C++ class; actually it is a ***class template***; roughly speaking this is essentially a recipe for generating a class for complex numbers once the internal representation of real and imaginary numbers (`float` or `double`) is fixed. The choice is in your hands. The following incantation will create a complex variable:

```
#include <complex>
std::complex<double> I(0,1);
```

This creates a variable I and assigns it the value $i = \sqrt{-1}$, where i is evaluated to double precision. The statement is intended to closely parallel the declaration of an integer variable i:

```
int i=1;
```

Here, std is a *namespace*, which is a mechanism to avoid clashes between the names we give to constants, variables, and datatypes. If you don't want to type the full namespace-qualified form of the declaration, you can also add the line using namespace std; prior to the declaration, usually at the top of a compilation unit or near the beginning of a particular function.

We introduce some vocabulary which applies to classes. The *declaration* (found in <complex>) of a class does not create any new variables, known as *objects*, but only specifies how those variables are created and what they can do, i.e. their *interface*. New objects are created typically with a statement that looks like the declaration of any other variable. For a class representing a vector the statement might look like

```
Vector m(0,0,0)
```

—think:

```
int i(0);
```

The creation of an object actually results in a function call, to the so-called *constructor*. Vector is the *class*, and m is the *object*. A class is a category of objects which all behave according to a common interface. Each object is said to be an ***instance*** of the class, and the creation of objects is also called ***instantiation***.

Classes typically contain **bound functions**, which are invoked like this:

```
object.function(arguments);
```

rather than the free subroutine calls that you are used to, i.e.,

```
function(arguments);
```

When we invoke a function bound to an object we are said to be ***sending a message*** to the object.

The class template `std::complex` is a built-in part of the C++ standard library; we say that it is parameterized on the data type used to represent the real and imaginary parts of the complex number. To instantiate:

```
std::complex<double> U(1,0);
std::complex<double> I(0,1);
```

where the datatype `double` is called the template parameter, the values used in the constructor call are the real and imaginary parts of the complex number. The first variable declared above (`U`) is the real number 1, while the second variable `I` is the imaginary number $i = \sqrt{-1}$. One way to simplify the declaration of variables such as these is to use a *type definition* to establish a *typedef-name*:

```
typedef std::complex<double> Complex;
```

then, one can use the more convenient typedef-name (`Complex`) instead of the parameterized class (`std::complex<double>`). A number of functions are declared to carry out elementary operations on complex numbers; these are defined in the Table 2.2. The operations in this table can be used to add, subtract, multiply and divide any combination

Table 2.2 *Summary of operations defined on complex numbers.*

Operation	Function	Example
+	operator +()	w=x+y
-	operator -()	w=x-y
*	operator *()	w=x*y
/	operator /()	w=x/y
z^* (complex conjugation)	conj()	w=conj(z)
$Re(z)$	real()	w=z.real()
$Im(z)$	imag()	w=z.imag()
$|z|$	abs()	double x=abs(z)
$|z|^2$	norm()	double x=norm(z)
$arg(z)$	arg()	double phi=arg(z)
addition assignment	operator +=	w+=z
subtraction assignment	operator -=	w-=z
multiplication assignment	operator *=	w*=z
division assignment	operator /=	w/=z

of floating point numbers and complex numbers. A number of important mathematical functions (cos, cosh, exp, log, pow, sin, sinh, sqrt, tan, and tanh) have also been overloaded to work with complex arguments.

Complex numbers find widespread use in physics. We will illustrate how to work with the std::complex class with a simple problem in planar geometry. Complex numbers can be used to represent a point in a plane, if we consider the number $u = a + ib$ to be a representation of the vector $a\hat{x} + b\hat{y}$. Note that the magnitude of the complex number $|u| = \sqrt{a^2 + b^2}$ is the length of the vector. Also, if $v = c + id$, then $uv^* = (ac + bd) + i(bc - ad)$, so that the real part of this number is the dot product of the two vectors and the imaginary part is the magnitude of the cross product, $\vec{v} \times \vec{u}$. To translate a point, we merely have to add a constant (complex) number $t = t_x + it_y$ to the point. Rotations can be carried out by multiplying by $\exp(i\phi)$, where ϕ is the angle of rotation; the sense of the rotation is counterclockwise for $\phi > 0$.

One of the earliest ways in which the value of the constant π was obtained was to compute the perimeter of a polygon inscribed within a circle. Starting with a regular hexagon, one obtains the approximate value of 3.0. One can then consider a regular inscribed polygon of 12 sides, 24 sides, 48 sides, and so forth. In 265 AD the Chinese mathematician Liu Hui found an approximation for π based on a 3,072 sided polygon, and in 480 AD Zu Chongzhi found a more accurate value based on a 12,288 sided polygon. We will compute these approximations to π by starting with a 6-sided polygon centered at the origin, oriented so that one of the vertices is at $\hat{x}$ and the other is at $\frac{1}{2}\hat{x} + \frac{\sqrt{3}}{2}\hat{y}$. The two points will be represented by complex numbers $x_0 = 1$ and $x_1 = \frac{1}{2} + \frac{\sqrt{3}}{2}i$. The length of the segment joining the two vectors is $|x_1 - x_0|$, so the total perimeter of the inscribed polygon is six times this length, which is exactly equal to 3. From this first approximation we proceed to more accurate approximations with the following steps.

- Find the midpoint of the line joining the two vertices. The complex number representing that point is the average of x_0 and x_1.
- This point lies in the middle of a segment of the polygon, but not on the circle. We can put it on the circle by normalizing it to unit length, i.e. transforming

$$\frac{x_0 + x_1}{2} \rightarrow \frac{x_0 + x_1}{2} / |\frac{x_0 + x_1}{2}| = \frac{x_1 + x_0}{|x_1 + x_0|}$$

- At the next step this value becomes the new x_1, and we double the number of sides. The perimeter of the inscribed polygon is again taken.

As the number of sides is increased, the perimeter converges to twice the value of π. The program shown below, called threePi, can be found in CH2/EXAMPLES/THREEPI. It obtains the value of π in three ways. First, it just prints the value of the macro M_PI, defined in the header file cmath. This illustrates the usual way of obtaining the value of π in a program—it's superior, of course, to typing in the value. The second way is the inscribed polygon algorithm, which is essentially an illustration of how to use the class std::complex. The last way uses Euler's famous expression $e^{i\pi} = -1$. Taking the logarithm of both sides gives a number whose real part is zero and whose imaginary part is π. This is included as an illustration of overloaded functions that operate on complex numbers in the C++ standard library. Here is the code example:

```cpp
#include <complex>
#include <iostream>
#include <cmath>
#include <iomanip>
#include <limits>
//
// threePi is a small program to obtain pi in three
// different ways:
//
typedef std::complex<double> Complex;
main () {

  // 1. Just print out the macro:
  std::cout << std::setprecision(16) << M_PI << std::endl;

  // 2. From the area of inner inscribed polygons.  Start
  //    with a hexagon and subdivide the arc into smaller
  //    pieces. Liu Hui(n=3072, 265 AD); Zu Chongzhi
  //    (n=12288, 480 AD)

  Complex x0=1.0;
  Complex x1(1.0/2.,sqrt(3.0)/2.0);
  unsigned int nsides=6;

  while (nsides < std::numeric_limits<int>::max()) {
    double lside = abs(x1-x0);
    double approx=nsides*lside/2.0;
    std::cout << "Sides "
              << nsides
              << "; approx="
              << std::setprecision(16)
              << approx
              << std::endl;

    x1=(x1+x0)/abs(x1+x0);
    nsides *=2;
  }
```

```
    // 3. Take the logarithm of the imaginary number -1:
    std::cout << std::setprecision(16)
              << imag(log(Complex(-1,0)))
              << std::endl;
}
```

The notion of encapsulation applies to classes in C++. In the case of complex numbers, one can see a clean separation between the internal structure of the number and the protocol summarized in Table 2.2 that it obeys. In the next section we describe some of the mechanisms for achieving this; for achieving, in other words, encapsulation.

2.5 Classes as user defined datatypes

C++ classes are major building blocks of modern computer codes, and we shall introduce many examples in the next chapter. What is a class though, exactly? Historically, data structures, which were mere agglomerations of related data, were the precursors to classes. For example, the four components of a quaternion can be agglomerated into a single entity called a ***structure***. A quaternion is like a four-component complex number which is useful for representing rotations in *three* dimensions, among other things. In C and C++ the structure may be declared as:

```
struct Quaternion{
   double a;
   double b;
   double c;
   double d;
};
```

The structure can be passed to functions, returned from functions, and placed into arrays. Since structures are agglomerations of data, they can also agglomerate other structures to build hierarchies of structured data.

Classes add bound functions, a.k.a. ***methods***, to structures. For example, a class called Quaternion may return the magnitude, $\sqrt{a^2 + b^2 + c^2 + d^2}$:

```
class Quaternion{

   public:

   // return the magnitude
   double abs() { return sqrt(a*a+b*b+c*c+d*d); }

   private:
```

```
    double a;
    double b;
    double c;
    double d;

};
```

In furnishing this example, we had to slip in two important keywords in order for our Quaternion class to make sense. These are `public` and `private`, and they enforce encapsulation. These keywords are like switches, switching back and forth between two access modes, public and private. When data or a member functions are public, it means that *client* code can invoke them. In contrast, clients may not access private data or member functions, and the compiler generates an error whenever such access is attempted. This locks away certain parts of the class which the designer does not intend for general purpose use, thereby *enforcing* the separation between interface and implementation.

In the above example, the quaternion's components a, b, c, and d are private so, they can be neither written nor read. At the very least the clients of this class need to initialize them. This is done in a constructor, as shown in this more complete example (which also provides *accessors* to the data members):

```
class Quaternion{

  public:

  // Constructor
  Quaternion (double a, double b, double c, double d):
  a(a),b(b),c(c),d(d){}

  // return the magnitude
  double abs() { return sqrt(a*a+b*b+c*c+d*d);}

  // return the components
  double getA() const {return a;}
  double getB() const {return b;}
  double getC() const {return c;}
  double getD() const {return d;}

  private:

  double a;
  double b;
```

```
    double c;
    double d;

};
```

The syntax of the constructor may strike you as funny. We discuss it in more detail in Chapter 6. Finally, for the sake of accuracy, we mention one fact about structures and classes. While structures were originally simple agglomerations with data without methods, C++ allows developers to add methods to structures, too, making them technically nearly the same as classes. Like classes, the keywords `public` and `private` also function within structure definitions to control access. There is in fact only one difference between structures and classes: the default access, which is `public` for structures and `private` for classes. In our work we generally use structures where simple agglomeration without member functions is required.

2.6 Style guide: defining constants and conversion factors in one place

In the physical sciences we encounter several types of constants that appear frequently in our work: mathematical constants like π and e; fundamental constants like Planck's constant $\hbar$, the speed of light c, the fine structure constant α, the Hubble constant H_0, masses of elementary particles, and so forth. It is not good style to repeat the definitions of these constants in many places in the code. One common solution is to define them in a header file. Namespaces, discussed in Section 2.4, are often used in conjunction with this solution to prevent constants like α from clashing with local definitions. An excerpt from a header file defining physical constants within a namespace is:

```
#ifndef _PHYSICALCONSTANTS_
#define _PHYSICALCONSTANTS_

// This file defines physical constants.  All quantities
// are specified in SI units.

namespace physics {
  static const double hBar  =1.0545718E-34;    // m^2 kg /s
  static const double cLight=299 792 458;      // m / s
  static const double alpha =7.2973525664E-3;  // dimensionless
   ...
};
#endif
```

Client code can then access variables by including the header file and referencing the variables as, for example `physics::cLight`. To input the value of a particle moving at 4/5 the speed of light one could type, for example:

```
#include "physics.h"
...
double velocity = 4.0/5.0*physics::cLight;
```

A related task is converting units. Suppose that one was working cheifly with SI units, but one occasionally needed to input or output the value of some measure of length in mm, cm, feet, or other unit of length. The conversion factors can also be collected in a single header file and scoped within a namespace:

```
#ifndef _UNITS_
#define _UNITS_

// This file contains conversion factors from SI units to
// other common systems

namespace units {
  //
  // Length.
  //
  static const double meter   =1.0;
  static const double cm      =0.01;
  static const double mm      =0.001;
  static const double micron  =1.0E-6;;
  static const double nm      =1.0E-9;
  static const double feet    =0.3048;
  static const        km      =1000.0;
  static const        mile    =1609.34;
  ...

};
#endif
```

To input constants one can then give a number and a unit:

```
#include "units.h"
using namespace units;
double distanceToMoon=238.9E3*mile;
```

while to output a constant one can print something like,

```
#include "units.h"
...
using namespace units;
std::cout << "The distance from earth to moon is "
          << distanceToMoon/mile
          << " miles."
          << std::endl;
```

No single file is likely to be adequate for defining all of the physical constants and conversion factors for all of physics, but centralizing definitions for particular projects can eliminate redundant definitions, confusion, and errors.

2.7 Summary

In this chapter we have surveyed built-in numerical datatypes, and discussed their representations and the limitations to numerical precision that these binary representations unavoidably imply. We also drew attention to the skillful way in which details of the representation are hidden through clever compiler design so that application programmers do not get hung up on these details. This feature, encapsulation, is also evident in the design of the C++ template class `std::complex`, which we introduced in Section 2.4, and will be part of every serious API, some of which will be introduced in the following chapter. The keywords `public` and `private` control access to methods and data within a class and enforce encapsulation. You will need to know about these as you design your own classes and class libraries, but it's also important to distinguish these access patterns while using classes provided within the C++ standard library or by third parties.

Further Reading

Basics on the representation of floating point numbers can be found in Press (2007). As we are now beginning to discuss those features of C++ which are not a part of C, we recommend a good introduction to the C++ language at this point. The online reference

www.cplusplus.com/doc/tutorial

is excellent, and in print works such as Bronson (2013) or Capper (1994) can be consulted. When you get a little more practiced with the basic ideas of C++, you'll start to encounter minor dilemmas about how to write your own classes. We'll scratch the

2.8 Exercises

1. Compute the expected machine precision of float and double datatypes, and then check your calculation with a program.

2. Compute the "stretch-factor" γ for a relativistic particle for speeds approaching the speed of light, i.e. for $\beta = v/c = 0.9, 0.99, 0.999, 0.9999\ldots$. Compute this in two ways, first as $\gamma = 1/\sqrt{1-\beta^2}$ and then as $\gamma = 1/\sqrt{(2-\epsilon)\epsilon}$ where $\epsilon = 0.1, 0.01, 0.001, 0.0001\ldots$. Suppose that the the fractional error in the calculation is required to be one part in one thousand or less. What is the maximum value of β for which this accuracy can be obtained, if one computes it using the former method?

3. Three points labeled $A = 3\hat{x} + 7\hat{y}$, $B = 3\hat{x} + 2\hat{y}$, and $C = 10\hat{x} + 2\hat{y}$ lie in a plane.

   a) Using std::complex, write a program ex1 to compute the area of a triangle whose vertices lie at points A, B, and C. Use the dot product and/or cross product and carry out the computations with complex numbers.

   b) Create a complex number t representing a translation to the point $O = 4\hat{x} + 5\hat{y}$. Apply the translation to A, B, and C, and print out the translated vectors A', B', and C'. Check that the translation leaves your computation of the area of the triangle invariant.

   c) Create a complex number r such that $|r| = 1$ by using the exp function to exponentiate a pure imaginary number. Choose the imaginary number such that r represents a clockwise rotation about the origin through 45°. Apply the rotation to the vectors A', B', and C' from the previous step. Print out the rotated vectors A'', B'', and C'', and check that the rotation leaves your computation of the area invariant.

4. Modify the program in the previous exercise such that instead of computing the area of triangle ABC, it computes and prints out the location of the point of closest approach (in the plane) of segment AC to point B. Use the dot product and/or cross product and carry out the computations with complex numbers. Check that this distance is invariant to rotation and to translation.

5. Using std::complex, compute the position of the points $A = 3\hat{x} + 7\hat{y}$, $B = 3\hat{x} + 2\hat{y}$, and $C = 10\hat{x} + 2\hat{y}$ after they are rotated by +755° about the point $O = 4\hat{x} + 5\hat{y}$.

6. Write a program, based on the example program threePi in this chapter (which is available EXAMPLES/CH2/THREEPI). Your program should compute π from the *areas* of small isoceles triangles making up the polygon inscribed within the radius of a circle rather than the length of their base.

7. Consider the quantum mechanical problem of a particle of energy E incident upon a potential barrier in one dimension of height V, which turns on at $x = 0$. Write

a program that determines the wavefunction for all value of x, and compute the transmission and reflection coefficients, R and T, for different values of the V/E. Do not write a fundamentally different function for the case in which $E < V$ and $E > V$; instead carry out the computation in complex numbers, in a manner which is valid for both cases.

8. Two identities which relate the trigonometric functions to hyperbolic trig functions are (for x real): $\sin(ix) = i\sinh(x)$ and $\cos(ix) = \cosh(x)$. Test these relations for a few values using `std::complex`.

9. Write a function to solve the equation $ax^2 + bx + c = 0$ which returns a pair of complex values, and test this for the following values of a, b, and c:

    - $a = 1, b = 4, c = 0$
    - $a = 1, b = 4, c = 2$
    - $a = 1, b = 4, c = 6$
    - $a = 1, b = 4i, c = -6$
    - $a = 1, b = 4i, c = -4$
    - $a = 1, b = 4i, c = 0$
    - $a = 1, b = 2 + 2i, c = -6$
    - $a = 1, b = 2 + 2i, c = -0$
    - $a = 1, b = 2 + 2i, c = 6$

10. Write a function to diagonalize a 3x3 complex matrix by solving its secular equation. The function should take a 3×3 array of complex numbers as input and should output

    - An array of three complex numbers containing the eigenvalues of the matrix.
    - A 3×3 array of complex numbers, the columns of which contain normalized eigenvectors of the matrix.

    Test your function on the following matrix and provide the result:

    $$\begin{pmatrix} 0.333333 & -0.244017 & 0.910684 \\ 0.910684 & 0.333333 & -0.244017 \\ -0.244017 & 0.910684 & 0.333333 \end{pmatrix}.$$

....

BIBLIOGRAPHY

Bronson, G. J. (2013). *C++ for scientists and engineers*. Cengage Learning.
Capper, D. M. (1994). *The C++ programming language for scientists, engineers, and mathematicians*. Springer-Verlag.

Meyers, S. (2005a). *Effective C++ 55 Specific Ways to Improve your Programs and Designs*. 3rd ed. Addison-Wesley.
Meyers, S. (2005b). *More Effective C++ 35 New Ways to Improve your Programs and Designs*. 3rd ed. Addison-Wesley.
Press, W., S. Teukolsky, W. Vetterling and B. Flannery (2007). *Numerical Recipes, the Art of Scientific Computing*, 3rd ed. Cambridge University Press.

3
Some useful classes with applications

3.1 Introduction	51
3.2 Coupled oscillations	52
3.3 Linear algebra with the `Eigen` package	54
3.4 Complex linear algebra and quantum mechanical scattering from piecewise constant potentials	57
3.4.1 Transmission and reflection coefficients	59
3.5 Complex linear algebra with `Eigen`	60
3.6 Geometry	63
3.6.1 Example: collisions in three dimensions	64
3.7 Collection classes and strings	65
3.8 Function objects	67
3.8.1 Example: root finding	70
3.8.2 Parameter objects and parametrized functors	74
3.9 Plotting	76
3.10 Further remarks	77
3.11 Exercises	78
Bibliography	83

3.1 Introduction

We have seen that one of the reasons classes are useful is because they provide a mechanism by which implementation details can be hidden from the user. In this chapter we continue the exploration of classes by applying a number of high quality classes to typical scientific problems. The practice of writing classes will be taken up again in Chapter 6. Users requiring a more structured introduction to the C++ language are directed to the references at the end of the previous chapter.

Use cases for these classes are taken from classical and quantum physics. In this chapter we carry out numerical solutions to problems which admit, possibly, analytic solutions with the idea that their familiarity makes them good learning examples.

In adopting a class library, we shall always prefer *universally* available software to *widely* available software, and widely available software to *custom* software. Universally available software is that which is found in the C++ standard library. Most of the widely available software can be installed using package management software; on Ubuntu, this would be either the `apt-get` command or the graphical installer `synaptic`.

3.2 Coupled oscillations

Consider the problem of N beads, arranged left to right on a taut frictionless wire. They are free to move in one dimension, but they are connected to their neighbors with ideal springs. The beads are labeled with the index $i = 0, 1, 2...N - 1$; the ith bead has a mass of m_i; it is connected to its left neighbor by a spring constant k_i and to its right neighbor by a spring constant k_{i+1}. The first and last beads are attached on one side only. Neither the masses nor the spring constants need be taken equal in the general case. The position of the ith bead with respect to its equilibrium position is denoted by $x_i(t)$.

The force on the ith bead is related to the stretch of its neighboring springs, and Newton's law yields:

$$m_i \frac{d^2 x_i}{dt^2} = k_i(x_{i-1} - x_i) + k_{i+1}(x_{i+1} - x_i). \tag{3.1}$$

This is a set of coupled differential equations. For the case $N = 4$ (the generalization is clear) it can be written in matrix form as:

$$\begin{pmatrix} m_0 & 0 & 0 & 0 \\ 0 & m_1 & 0 & 0 \\ 0 & 0 & m_2 & 0 \\ 0 & 0 & 0 & m_3 \end{pmatrix} \cdot \begin{pmatrix} \ddot{x}_0 \\ \ddot{x}_1 \\ \ddot{x}_2 \\ \ddot{x}_3 \end{pmatrix} = - \begin{pmatrix} k_1 & -k_1 & 0 & 0 \\ -k_1 & k_1 + k_2 & -k_2 & 0 \\ 0 & -k_2 & k_2 + k_3 & -k_3 \\ 0 & 0 & -k_3 & k_3 \end{pmatrix} \cdot \begin{pmatrix} x_0 \\ x_1 \\ x_2 \\ x_3 \end{pmatrix} \tag{3.2}$$

or

$$\mathbf{M} \cdot \ddot{\mathbf{x}} = -\mathbf{K} \cdot \mathbf{x} \tag{3.3}$$

where $\mathbf{M}$ is the mass matrix, $\mathbf{K}$ is the matrix involving the spring constants, and $\mathbf{x}$ is a column vector containing the displacement of each mass from its equilibrium position.

This matrix equation can be solved in the time-honored tradition of guessing a trial solution:

$$\mathbf{x} = \mathbf{a} e^{-i\omega t} \tag{3.4}$$

where $\mathbf{a}$ is a constant undetermined column vector, and where we imply the real part of the right-hand side. Substituting this into Eq. 3.3, one obtains

$$\mathbf{K} \cdot \mathbf{a} = \omega^2 \mathbf{M} \cdot \mathbf{a} \tag{3.5}$$

The matrix **M** is diagonal and has positive elements only. We can define a matrix $\mathbf{M}^{1/2}$ such that $\mathbf{M}^{1/2}\mathbf{M}^{1/2} = \mathbf{M}$, and $\mathbf{M}^{-1/2}$ to be its inverse, and re-express the previous equation as:

$$\mathbf{M}^{-1/2}\mathbf{K}\mathbf{M}^{-1/2} \cdot \mathbf{M}^{1/2}\mathbf{a} = \omega^2 \mathbf{M}^{1/2} \cdot \mathbf{a} \tag{3.6}$$

We define the matrix $\mathbf{\Omega}^2 \equiv \mathbf{M}^{-1/2}\mathbf{K}\mathbf{M}^{-1/2}$, and $\mathbf{b} \equiv \mathbf{M}^{1/2}\mathbf{a}$, so

$$\mathbf{\Omega}^2 \cdot \mathbf{b} = \omega^2 \mathbf{b} \tag{3.7}$$

and we see that we are faced with an eigenvalue problem. The matrix $\mathbf{\Omega}^2$ is real and symmetric; it is a well-known property of such matrices that they possess N eigenvalues and eigenvectors which can be taken real. These will be denoted as $\lambda_j = \omega_j^2$ and $\mathbf{b}_j$, $j = 0, 1..N - 1$. The λ's are not only real but also *positive* (Exercise 10), and therefore the ω's are real. The eigenvectors are mutually orthogonal in the case of nondegenerate eigenvalues, and they may be chosen to be orthogonal in the case of degenerate eigenvalues by a Gram-Schmidt orthogonalization procedure. They can be normalized[1] by rescaling them such that $\mathbf{b}_j^T \cdot \mathbf{b}_j = 1$. Then, the eigenvectors constitute an orthonormal set, such that:

$$\mathbf{b}_j^T \cdot \mathbf{b}_k = \delta_{jk}. \tag{3.8}$$

Suppose that the initial positions and velocities of all of the beads are known at a give time $t = 0$. The problem is to find the subsequent motion of all of the beads. There are two steps to this: determining the eigenfrequencies $\omega_j = \sqrt{\lambda_j}$ of $\mathbf{\Omega}^2$ and their corresponding eigenvectors $\mathbf{b}_j$ for $j = 0, 1 \ldots N - 1$; and constructing the particular solution from these eigenvalue/eigenvector pairs. This particular solution is the linear combination:

$$\begin{aligned}\mathbf{x}(t) &= \mathrm{Re} \sum_{j=0}^{N-1} \eta_j \mathbf{a}_j e^{-i\omega_j t} \\ &= \mathrm{Re} \sum_{j=0}^{N-1} \eta_j \left(\mathbf{M}^{-1/2} \cdot \mathbf{b}_j\right) e^{-i\omega_j t}\end{aligned} \tag{3.9}$$

where $\mathbf{x}(t)$ is a column vector representing the positions of each of the beads, and η_j ($j = 0, 1..N - 1$) is a set of N complex coefficients which are determined as follows. Using the orthonormality (Eq. 3.8) of the $\mathbf{b}_j$'s, we project out the coefficients:

$$\mathbf{b}_j^T \cdot \mathbf{M}^{1/2} \cdot \mathbf{x}(t) = \mathrm{Re}\left(\eta_j e^{-i\omega_j t}\right) \tag{3.10}$$

---

[1] Numerical methods for diagonalizing symmetric matrices usually yield orthonormal eigenvectors sorted by eigenvalue, freeing the user to think about other issues.

which at $t = 0$ reads

$$\text{Re}(\eta_j) = \mathbf{b}_j^T \cdot \mathbf{M}^{1/2} \cdot \mathbf{x}^0, \tag{3.11}$$

$\mathbf{x}^0$ being the initial position of the beads. Representing the velocity of the beads as a column vector

$$\mathbf{v} = \frac{d\mathbf{x}}{dt} \tag{3.12}$$

and taking the time derivative of Eq. 3.9 we have

$$\mathbf{v}(t) = \text{Re} \sum_{j=0}^{N-1} -i\omega_j \eta_j \left(\mathbf{M}^{-1/2} \cdot \mathbf{b}_j \right) e^{-i\omega_j t} \tag{3.13}$$

which leads, in a similar way, to the equation

$$\text{Im}(\eta_j) = \frac{\mathbf{b}_j^T \cdot \mathbf{M}^{1/2} \cdot \mathbf{v}^0}{\omega_j} \tag{3.14}$$

($\mathbf{v}^0$ indicates the initial velocities) and thus

$$\eta_j = \mathbf{b}_j^T \cdot \mathbf{M}^{1/2} \cdot \mathbf{x}^0 + i \frac{\mathbf{b}_j^T \cdot \mathbf{M}^{1/2} \cdot \mathbf{v}^0}{\omega_j}. \tag{3.15}$$

This relation, when substituted into Eq. 3.9 furnishes a complete solution to the program. To obtain it for a variety of configurations and initial conditions we will use a good class library for linear algebra, which is the topic of our next section.

3.3 Linear algebra with the `Eigen` package

To carry out the computations arising in the previous section, we are going to need a decent class library which provides for matrix algebra, vector algebra, and solutions to eigenvalue problems. It would be nice if powerful linear algebra classes were an integral part of the C++ standard library, because if that were the case one could have a reasonable expectation that any compiler on any platform would compile the code with no extra package installation required. However, no such set of classes has, so far, been integrated with the C++ standard.

Fortunately, a class library known as `Eigen`:

`eigen.tuxfamily.org`

contains a high-quality set of classes for linear algebra and geometry. It can be freely downloaded and installed on your laptop. The package can be installed (on our reference platform, Ubuntu) as follows:

`$sudo apt-get install libeigen3-dev libeigen3-doc`

After the installation is complete, you will find header files in the area

`/usr/include/eigen3`

and there are no libraries to install since all of the classes and functions provided by the `Eigen` library are *inline*; in other words, they are expanded in the body of the calling routine rather than compiled separately and placed into a static or shared library of object code. This area should be added to the include search path (using the `-I/usr/include/eigen3` option) during compilation. Documentation can be found at `eigen.tuxfamily.org`. Eigen is a very complete class library which we will not describe in full detail; our goal for the moment is to familiarize you with a small portion of the library that will prove immediately useful.

Eigen can handle both real and complex vectors and matrices. To begin with, we consider real vectors and matrices, and return to complex vectors and matrices a little further on.

> We assume that you have installed a basic linux operating system on your laptop, desktop, or other computer. Your next task is to install software development kits in order to obtain a minimal environment for compiling and running the examples given in this text, and for conducting the end-of-chapter exercises. The first external package we ask you to install is `Eigen`.

In our first example we use two classes called `Eigen::VectorXd` and `Eigen::MatrixXd` to solve a very simple matrix equation. The `Xd` suffix on these datatypes signifies that the dimensionality of the object (matrix or vector) is determined at runtime rather than compile time, and that the matrix elements are represented by `doubles`. Complex vectors and matrices are denoted (in `Eigen`) by `Eigen::VectorXcd` and `Eigen::MatrixXcd`. The `cd` indicates that `std::complex<double>` is used in the representation of the matrix elements.

Let's start by solving a simple matrix inversion problem. We seek $\vec{x}$ where $A \cdot \vec{x} = \vec{y}$ and:

$$A = \begin{pmatrix} 1.0 & 2.0 \\ 4.0 & 9.0 \end{pmatrix}$$

$$\vec{y} = \begin{pmatrix} 1 \\ 3 \end{pmatrix}$$

Linear algebra with the Eigen package

This can easily be accomplished as follows:

```
#include <Eigen/Dense>
#include <iostream>
int main(int argc, char **argv) {
  Eigen::VectorXd Y(2);
  Y(0)= 1.0;
  Y(1)= 3.0;
  Eigen::MatrixXd A(2,2);
  A(0,0)= 1.0; A(0,1)=2.0;
  A(1,0)= 4.0; A(1,1)=9.0;

  Eigen::MatrixXd AInv= A.inverse();
  std::cout << AInv*Y;
}
```

To obtain a solution to the coupled oscillation problem of the previous section, we need a mechanism for solving eigenvalue problems. A few remarks are necessary before starting. Suppose we have a matrix with real elements, such as the matrix A of the above example. It is still possible that the eigenvalues and eigenvectors are complex. Therefore we would in general be looking for complex return values. However, in the case of coupled oscillations, our matrix is real and symmetric and therefore the eigenvalues and vectors will all be real.

Eigen provides two classes for the solution of eigenvalue problems; the first class is called EigenSolver (namespace Eigen) and can always obtain a solution (but it may be complex), while the second is called SelfAdjointEigenSolver and requires self-adjoint matrices (and returns real eigenvalues and real, orthonormal eigenvectors). Self-adjoint matrices, real or complex, are those for which $A^\dagger = A$, where $A^\dagger \equiv (A^T)^*$. Real symmetric matrices, like the ones we encounter in the coupled oscillator problem, are a subset of self-adjoint matrices.

Like std::complex, the eigenvalue solvers are template classes, you need to furnish the datatype of the matrix which is to be diagonalized as a template parameter. Here is an example, using a simple 2×2 matrix:

```
#include <Eigen/Dense>
#include <iostream>
int main(int argc, char **argv) {

  // Find the eigenvalues and the eigenvectors
  // of the matrix
  // A   = 1.0   2.0
  //       4.0   9.0

  Eigen::MatrixXd A(2,2);
  A(0,0)= 1.0; A(0,1)=2.0;
  A(1,0)= 4.0; A(1,1)=9.0;
```

```
    // initialize an eigensolver with A

    Eigen::EigenSolver<Eigen::MatrixXd> s(A);

    // and solve

    Eigen::VectorXcd  val=s.eigenvalues();
    Eigen::MatrixXcd  vec=s.eigenvectors();
    std::cout << val  << std::endl;
    std::cout << vec  << std::endl;
}
```

The eigenvectors are returned in a complex-valued vector (`VectorXcd val`) and the eigenvectors are returned together as a matrix (`MatrixXcd vec`); each column of the matrix contains a complex eigenvector. Note that if you compute (numerically or analytically) the eigenvalues of the original matrix,

$$\mathbf{A} = \begin{pmatrix} 1 & 2 \\ 4 & 9 \end{pmatrix} \qquad (3.16)$$

you will find that they happen to be real-valued, however `Eigen` cannot determine this in advance so the results are returned in complex datatypes. If on the other hand we have to find the eigenvalues(vectors) of

$$\mathbf{A} = \begin{pmatrix} 1 & 2 \\ 2 & 9 \end{pmatrix} \qquad (3.17)$$

then it is known in advance from the symmetric form of the matrix that the eigenvalues and eigenvectors are real. The `SelfAdjointEigenSolver` can be used in this case; it returns its results in real-valued datatypes, `VectorXd` and `MatrixXd`. In addition, the matrix containing the eigenvectors will be orthogonal, i.e., it will satisfy the condition that $\Lambda^T \cdot \Lambda = \mathbb{I}$, the identity matrix.

You now know everything you need to solve the coupled oscillators problem. Implementing the solution is left to the exercises.

3.4 Complex linear algebra and quantum mechanical scattering from piecewise constant potentials

In our second example, we develop a computational solution to a simple problem in one-dimensional quantum mechanics. This problem is treated in almost every elementary text, for example Gasiorowicz (2003) or Griffiths (2005). Analytical solutions give insights that computational solutions cannot—and vice versa.

A particle with energy $E > 0$ is incident upon a rectangular potential barrier with barrier height of V and a spatial extent from $x = -a$ to $x = a$. The time-independent Schrödinger equation is:

58 *Complex linear algebra and quantum mechanical scattering*

$$\left[-\frac{\hbar^2}{2m}\frac{\partial^2}{\partial x^2} + V(x)\right]\psi(x) = E\psi(x) \tag{3.18}$$

We re-write this as:

$$\left[-\frac{1}{k^2}\frac{\partial^2}{\partial x^2} + v(x)\right]\psi(x) = \psi(x), \tag{3.19}$$

where $v = V/E$ and $k^2 = 2mE/\hbar^2$.

We will assume a particle incident on the barrier from the negative-x direction, and we divide space into region I ($x < -a$), region II ($-a < x < a$), and region III ($x > a$). In regions I and III, where the piecewise constant potential is zero, the Schrödinger equation is:

$$\frac{d^2\psi(x)}{dx^2} = -k^2\psi(x) \tag{3.20}$$

while in region II it is:

$$\frac{d^2\psi(x)}{dx^2} = -n^2k^2\psi(x) \tag{3.21}$$

where

$$n \equiv \sqrt{1-v} \tag{3.22}$$

is a sort of "index of refraction" expressing (for fixed energy) the ratio of wavenumber in a region with constant potential to the corresponding wavenumber in a potential-free region; alternately it is the inverse of the ratio of the wavefunction's phase velocity in the two regions.

The general solution to the Schrödinger equation is:

$$\psi(x) = e^{\pm ikx} \quad \text{regions I and III} \tag{3.23}$$

and

$$\psi(x) = e^{\pm inkx} \quad \text{region II}. \tag{3.24}$$

The particular solution to the Schrödinger equation, describing an incident wave and reflected wave in regions I and II and a transmitted wave in region III can be written as:

$$\psi(x) = Ae^{ikx} + Be^{-ikx} \qquad \text{(region I)}$$
$$\psi(x) = Ce^{inkx} + De^{-inkx} \qquad \text{(region II)}$$
$$\psi(x) = Fe^{ikx} \qquad \text{(region III)}$$

The constants A, B, C, D, and F are constrained by the continuity of the wavefunction and its first derivative at $x = -a$ and $x = a$, whence:

$$Ae^{-ika} + Be^{ika} = Ce^{-inka} + De^{inka}$$
$$Ce^{inka} + De^{-inka} = Fe^{ika}$$
$$ikAe^{-ika} - ikBe^{ika} = inkCe^{-inka} - inkDe^{inka}$$
$$inkCe^{inka} - inkDe^{-inka} = ikFe^{ika}.$$

This can be written in terms of the amplitude of the incoming wave as:

$$\begin{pmatrix} e^{ika} & -e^{-inka} & -e^{inka} & 0 \\ 0 & e^{inka} & e^{-inka} & -e^{ika} \\ -ik\,e^{ika} & -ink\,e^{-inka} & ink\,e^{inka} & 0 \\ 0 & ink\,e^{inka} & -ink\,e^{-inka} & -ik\,e^{ika} \end{pmatrix} \cdot \begin{pmatrix} b \\ c \\ d \\ f \end{pmatrix} = \begin{pmatrix} -e^{-ika} \\ 0 \\ -ike^{-ika} \\ 0 \end{pmatrix} \qquad (3.25)$$

where $b \equiv B/A$, $c \equiv C/A$, $d \equiv D/A$, $f \equiv F/A$. This set of equations is solved analytically in elementary textbooks; here we aim to solve it numerically with the help of complex matrix classes from the package `Eigen`. An example is furnished in the next section.

We remark that one of the advantages is that the technique can be extended easily to deal with more complicated, piecewise-constant potentials. In fact, any potential in one dimension can be treated in this way by dividing it into sufficiently small pieces. Thus the general one-dimensional problem is equivalent to inverting an infinite size matrix equation of the type just developed.

3.4.1 Transmission and reflection coefficients

The wavefunction ψ determines the particle probability density $\rho(x, t) \equiv \psi(x, t)\psi^*(x, t)$ and the probability current:

$$j(x, t) \equiv \frac{i\hbar}{2m}\left(\psi(x, t)\frac{d\psi(x, t)^*}{dx} - \psi(x, t)^*\frac{d\psi(x, t)}{dx}\right). \qquad (3.26)$$

The two are related by a conservation law:

$$\frac{\partial j(x, t)}{\partial x} + \frac{\partial \rho(x, t)}{\partial t} = 0 \qquad (3.27)$$

which follows from the Schrödinger equation. In the barrier potential problem, the ***transmission*** and ***reflection coefficients*** can be computed as the ratio of the magnitude of the probability current of the transmitted and reflected waves to that of the incoming wave. These are:

$$\frac{\hbar k}{m} \begin{cases} |A|^2 & \text{Incident wave} \\ -|B|^2 & \text{Reflected wave} \\ |F|^2 & \text{Transmitted wave} \end{cases}. \tag{3.28}$$

3.5 Complex linear algebra with `Eigen`

In developing a numerical solution to the problem of the previous section, we will need to manipulate complex vectors and matrices. The relevant datatypes are `Eigen::VectorXcd` and `Eigen::MatrixXcd`. The following code can be found in the directory `EXAMPLES/CH3/RECTBARRIER`. It computes the coefficients b, c, d, and f, and also computes the transmission coefficient $|f|^2$ and the reflection coefficient $|b|^2$; these should sum to unity, and this can verified as a sanity check. The inputs are V (specified in units of E) and k (specified in units of a^{-1}):

```
#include <Eigen/Dense>
#include <iostream>
#include <complex>
#include <string>
#include <sstream>
typedef std::complex<double> Complex;
int main(int argc, char **argv) {
  using namespace std;
  string usage = string("Usage: ")
    + argv[0]
    + " [-?] [-v potential] [-k wvnumber]";

  if (argc>1 && argv[1]==string("-?")) {
    cout << usage << endl;
    exit(0);
  }

  // default values
  double v=0.5;
  double k=0.2;

  try {
```

```cpp
    // overwritten by command line:
    for (int i=1; i<argc;i+=2) {
      istringstream stream(argv[i+1]);
      if (string(argv[i])=="-v") stream >> v;
      if (string(argv[i])=="-k") stream >> k;
    }
  }
  catch (exception &) {
    cout << usage << endl;
    exit (0);
  }

  Complex I(0,1.0);
  Complex nk=k*sqrt(Complex(1-v));

  Eigen::VectorXcd Y(4);
  Y(0)= -exp(-I*k);
  Y(1)= 0;
  Y(2)= -I*k*exp(-I*k);
  Y(3)= 0;

  Eigen::MatrixXcd A(4,4);
  // First row:
  A(0,0)= exp(I*k)      ;
  A(0,1)=-exp(-I*nk)    ;
  A(0,2)= -exp(I*nk)    ;
  A(0,3)=0              ;

  // Second row:
  A(1,0)= 0             ;
  A(1,1)= exp(I*nk)     ;
  A(1,2)= exp(-I*nk)    ;
  A(1,3)=-exp(I*k)      ;

  // Third row:
  A(2,0)= -I*k*exp(I*k)    ;
  A(2,1)=-I*nk*exp(-I*nk);
  A(2,2)= I*nk*exp(I*nk)  ;
  A(2,3)=0                 ;

  // Fourth row:
  A(3,0)= 0             ;
  A(3,1)=I*nk*exp(I*nk) ;
```

```
   A(3,2)=-I*nk*exp(-I*nk);
   A(3,3)=-I*k*exp(I*k)     ;

   Eigen::MatrixXcd AInv= A.inverse();
   Eigen::VectorXcd BCDF=AInv*Y;
   Complex B=BCDF(0), F=BCDF(3);
   cout << "Complex coefficients" << endl;
   cout << BCDF << endl;
   cout << endl;
   cout << "Reflection coefficient   : " << norm(B)
        << endl;
   cout << "Transmission coefficient: " << norm(F)
        << endl;
   cout << "Sum:                      " << norm(B)+norm(F)
        << endl;

}
```

In addition to the operations illustrated in the above example, real and complex matrices have a rich set of other operations that can be used. Some of these appear in Table 3.1, and others can be found in the Eigen documentation.

Table 3.2 shows operations that apply to real or complex vectors. Note that transpose and adjoint operations return a $1 \times N$ matrix and not a column vector, which is identical to a $N \times 1$ matrix.

Table 3.1 *Matrix operators.*

Operation	Function	Example
(,)	subscript	A(2,3)
+	operator +()	C=A+B
-	operator -()	C=A-B
*	operator *()	C=A*B
*	operator *()	B=2.0*A
-	unary -	A=-B
tr$A$	trace	x=A.trace()
$|A|$	determinant	x=A.determinant()
A^{-1}	inverse	AINV=A.inverse()
A^T	transpose	AT=A.transpose();
$A^\dagger$	adjoint	ADagger=A.adjoint()

Table 3.2 *Complex and real operations.*

Operation	Function	Example
()	subscript	v(2)
+	operator +()	v=u+w
-	operator -()	v=u-w
*	operator *()	u=2.0*v
-	unary -	u=-v
return norm	norm()	double x=norm(u)
return squared norm	squaredNorm()	x2=squaredNorm(u)
normalize in place	normalize()	v.normalize()
return normalized copy	normalized()	u=v.normalized()
v^T	transpose	vT=v.transpose();
$v^\dagger$	adjoint	vDagger=v.adjoint()

3.6 Geometry

A "vector" has multiple meanings in mathematics, physics, and computer science. On one hand it is a linear algebraic object, an *n*-dimensional tuple in a vector space, acted upon by linear transformations in that space that are represented by matrices. On the other hand (restricting ourselves to Cartesian 3-space) we identify a vector as an object with magnitude and direction, which supports the operations of cross product and dot product, which is affected by a rotation of spatial coordinates, and which are sometimes called *three-vectors*; such three-vectors may be represented on a computer by a triplet of coordinates, but one should not confuse a particular representation with the actual vector, which is a more abstract notion–a thing with magnitude and direction, which we may visualize as an arrow. Thus there are two possible abstractions: an algebraic vector or a geometric vector. In Section 3.7 we will encounter another kind of "vector", representing a computer scientist's notion of the term, and unlikely to be confused with the algebraic or geometric vectors.

Some class libraries provide different classes for each type of object. The Eigen class library stores both types of vector in the same object but provides the extra operation of cross product which is valid when applied to three-dimensional vectors, Vector3f or Vector3d. The former is represented by floats, the latter by doubles; and both classes are typedef'd template classes. The length of these objects (3) is fixed at compile time. The cross product and dot product for Eigen::Vector3d and Eigen::Vector3f are invoked like this:

64   Geometry

```
#include "Eigen/Dense"
using namespace Eigen;
Vector3d u,v;
Vector3d w=u.cross(v);
double   d=u.dot(v);
```

The vectors interoperate with a set of transformations (rotations and affine transformations) that are defined in the header `Eigen/Geometry`. In the following section we provide a few examples of how one can use the vector classes for simple calculations.

3.6.1   Example: collisions in three dimensions

Consider two free particles in three dimensions. Particles 1 and 2 travel from positions $\vec{P}_1$ and $\vec{P}_2$ with velocities $\vec{v}_1$ and $\vec{v}_2$, respectively. Our aim is to compute how close the particles approach each other, and how close their straight trajectories become (these are not the same thing). To compute how close the two particles become, step into the reference frame of particle 1: the velocity of particle 2 in that frame is $\vec{v} = \vec{v}_2 - \vec{v}_1$, and the position of particle 2 is $\vec{r} = \vec{P}_2 - \vec{P}_1$. The distance of closest approach is the quantity $|\vec{r} \times \hat{v}|$. This can be implemented as a short function as follows:

```
#include "Eigen/Dense"
double distance(const Eigen::Vector3d & P1,
                const Eigen::Vector3d & P2,
                const Eigen::Vector3d & v1,
                const Eigen::Vector3d & v2) {
   // Compute the distance of closest approach of particle 2
   // to particle 1:

   return (P2-P1).cross((v2-v1).normalized()).norm();
}
```

To compute the distance between the two trajectories, we designate points $\vec{Q}_1$ as the point on the trajectory of particle 1 which lies closest to the trajectory of particle 2, and $\vec{Q}_2$ as the point on trajectory 2 which lies closest to trajectory 1. We can make a displacement from $\vec{P}_1$ to $\vec{P}_2$ in three steps, first, by moving an unknown distance s from $\vec{P}_1$ to $\vec{Q}_1$ along the direction $\hat{v}_1$; then by moving an unknown distance t from $\vec{Q}_1$ to $\vec{Q}_2$; then by moving an unknown distance u from $\vec{Q}_2$ to $\vec{P}_2$ along the direction $\hat{v}_2$. The distances s, t, and u are signed quantities, and the displacement from $\vec{Q}_1$ to $\vec{Q}_2$ takes place along a direction which is perpendicular to both $\hat{v}_1$ and $\hat{v}_2$, and thus it lies along the direction $\widehat{\vec{v}_1 \times \vec{v}_2}$. We can therefore write:

$$s\hat{v}_1 + t(\widehat{\vec{v}_1 \times \vec{v}_2}) + u\hat{v}_2 = \vec{P}_2 - \vec{P}_1$$

We are interested in knowing the quantity t, which we can obtain by dotting this expression into $\widehat{\vec{v}_1 \times \vec{v}_2}$:

$$t = (\vec{P}_2 - \vec{P}_1) \cdot (\widehat{\vec{v}_2 \times \vec{v}_1})$$

This quantity can be positive or negative, and we only care about the magnitude. In code we could write:

```
#include "Eigen/Dense"
double trajSeparation(const Eigen::Vector3d & P1,
                      const Eigen::Vector3d & P2,
                      const Eigen::Vector3d & v1,
                      const Eigen::Vector3d & v2) {

    // Compute the distance of closest approach of
       trajectory 2
    // to trajectory 1:

    return fabs((P2-P1).dot(v2.cross(v1).normalized()));

}
```

The quantities s and u can also be computed from the simultaneous solution to these two equations:

$$\begin{aligned} s + u(\hat{v}_2 \cdot \hat{v}_1) &= (\vec{P}_2 - \vec{P}_1) \cdot \hat{v}_1 \\ s(\hat{v}_1 \cdot \hat{v}_2) + u &= (\vec{P}_2 - \vec{P}_1) \cdot \hat{v}_2 . \end{aligned}$$

3.7 Collection classes and strings

The classes `std::vector` and `std::string` are similar in some regards to the more basic built-in array-of-object and array-of-characters. However, they do considerably more and are much easier to use. Classes like `std::vector` and `std::string` are sometimes referred to as *first-class* datatypes, while their built-in cousins are called *second-class*. First-class datatypes are so powerful and easy to use that they have practically displaced the use of their second-class cousins in most application programming.

The `std::vector` class holds objects, like an array, but it can increase its size automatically. The class is designed to simplify the storage of objects, and no mathematical operations are defined for it. The most common way to add elements to a vector is by pushing them onto the end of the array. The most common way to access the stored element is by random access using the subscripting operator, [ ].

66 Collection classes and strings

Actually, like `std::complex`, `std::vector` is not a class, but a *class template*, which means it's a recipe for generating a class, and here the class can be generated by specifying as a template argument what kind of object the vector holds. Templates avoid the necessity of writing multiple classes when the only real difference between them is the datatype used in the representation. To declare a vector of `doubles` one writes:

```
#include <vector>
std::vector<double> myVector;
```

To add a few `doubles` to the new vector you could write:

```
myVector.push_back(2.0);
myVector.push_back(3.14159);
myVector.push_back(7.0);
```

The vector can be copied (or passed by reference) between two functions, like typical C++ objects. The size of the vector can be retrieved, as can the elements:

```
for (int i=0;i<myVector.size();i++) {
   std::cout << myVector[i] << std::endl;
}
```

A good exercise is to type the preceding lines of code into a short program, and see what it prints; then change the template parameter from `double` to `int`, recompile, and look again at the output.

Another useful class is `std::string`. This class represents a basic string of characters (words, sentences, paragraphs), and the interface supports searching strings, appending strings (which is done with the "+" operator), and editing strings. A `std::vector` can hold not only `ints`, `floats`, and `doubles`, but any kind of object, such as a string. We modify the code written above, transforming our vector-of-doubles into a vector-of-strings.

```
#include <vector>
#include <string>
main() {
   std::vector<std::string> myVector;
   myVector.push_back("Trout ");
   myVector.push_back("Fishing ");
   myVector.push_back("In ");
   myVector.push_back("America.");
   for (int i=0;i<myVector.size();i++) {
      std::cout << myVector[i] << std::endl;
   }
}
```

which outputs

```
Trout Fishing In America.
```

The class `std::vector` has a lot of other functionality, which you can explore yourself or by consulting a good text such as Weiss (2014). The `std::vector` class is part of the C++ standard library, which also contains more specialized container classes such as `deque`, `list`, `set`, `map`, and `multimap`, which are also very useful, and make up the so-called Standard Template Library(STL). The STL (Stepanov, 1995) was once a separate package but is now fully integrated into the C++ standard library.

3.8 Function objects

Function-objects, or *functors*, are a type of first-class object that overload the function-call operator, `operator()`. The unusual name is not a C++ keyword of any kind, but just a descriptive name given to such classes. Our functors come from the `QatGenericFunctions` library, part of the ***Qat libraries***, which is custom software developed by the authors and available through the companion website:

```
http://www.oup.co.uk/companion/acp
```

We shall discuss and use it throughout the book.

> Go to `http://www.oup.co.uk/companion/acp` and follow instructions to install the Qat packages.

The `QatGenericFunctions` library provides many classes, most of which represent functions of one or more variables. These objects are similar to their second-class cousins, normal C/C++ functions, but they do more. Probably the simplest of these is the class `Variable`, (namespace `Genfun`) which we will use to illustrate the most important use cases. The function-call operator for this class simply returns its argument:

```
#include "QatGenericFunctions/Variable.h"
#include <iostream>
main() {
  Genfun::Variable X;
  std::cout << X(3.14) << std::endl;
}
```

The program prints out 3.14. The object X can be used in expressions, and these expressions can be assigned to a datatype called `Genfun::GENFUNCTION`. The following expression manufactures the function $f(x) = 1 + 2x + x^3$; it prints out a table of the function and its derivative at three points:

```
main () {
  Genfun::Variable X;
  Genfun::GENFUNCTION f=1 + 2*X + X*X*X;
  std::cout << 1 << " " << f(1) << " " << f.prime()(1) << std
      ::endl;
  std::cout << 2 << " " << f(2) << " " << f.prime()(2) << std
      ::endl;
  std::cout << 3 << " " << f(3) << " " << f.prime()(3) << std
      ::endl;
}
```

The class library contains basic functions (Sin, Exp, Gamma) and not-so-basic functions, shown in Table 3.3. These functions can all be used interchangeably and combined in expressions as here:

```
#include    "QatGenericFunctions/Sin.h"
.
.
.
{
  Genfun::Sin sin;
  Genfun::GENFUNCTION f=(1 + sin)/2;
}
```

Table 3.3 *List of the most basic functions in the* QatGenericFunctions *library.*

Name	Description
Variable	Basic building-block of expressions
F1D	General purpose adapter to any function
FixedConstant, Power, Sqrt, Square	Powers
Sigma	Sum of component functions
ArrayFunction	Defined by an input array
Sin, Cos, Tan, ASin, ACos, ATan	Trig and inverse functions
Sinh, Cosh, Tanh, ASinh, ACosh, ATanh	Hyperbolic Trig and inverse functions
Exp, Log, Gamma, LGamma, Erf	exponential, log, gamma, and error functions

Note, the `sin` *functor* is here given the same name as the `sin` *function* from the math library, it **shadows** that function (i.e. makes it invisible). The function `sin` can still be referenced as `std::sin`.

Another more compact way to write the same function is:

```
#include   "QatGenericFunctions/Sin.h"
{
  .
  ..
  Genfun::GENFUNCTION f=(1 + Genfun::Sin())/2;
}
```

How, you may ask, does one express a function like sin 3*x*? Since `QatGenericFunctions` allows function composition, you can express it like this:

```
#include   "QatGenericFunctions/Sin.h"
#include   "QatGenericFunctions/Variable.h"
{
  .
  ..
  Genfun::Variable X;
  Genfun::Sin      sin;
  Genfun::GENFUNCTION f=sin(3*X);
}
```

or more compactly as:

```
#include   "QatGenericFunctions/Sin.h"
#include   "QatGenericFunctions/Variable.h"
{
  .
  ..
  Genfun::GENFUNCTION f=Genfun::Sin()(3*Genfun::Variable());
}
```

To define a function, you can either manufacture it out of more primitive functions as illustrated above; or you can introduce your own extender functors–which we will cover later since this is a little more difficult. Alternately, you can use the adapter class F1D, which endows any function `double f(double x)` with algebraic properties, in particular the operations (+,-,*, /), as well as composition $f(g(x))$ and the derivative. For example, here we turn the hyperbolic tangent, `tanh`, from the standard c++ library, into a functor:

```
#include "QatGenericFunctions/F1D.h"
Genfun::GENFUNCTION tanh=Genfun::F1D(std::tanh);
```

With this, we can now carry out function-arithmetic in the same way that you are used to carrying out floating point arithmetic. For example, the following expression

```
Genfun::GENFUNCTION f=1-tanh*tanh;
```

defines a functor that computes $f(x) = \text{sech}^2(x)$–something that is impossible using raw C/C++ function pointers.

There is one important difference between functors defined with the adapter class F1D, and functors that are predefined in the QatGenericFunctions library: usually the predefined functors return analytical derivatives, while the adapters return a derivative computed numerically. Thus the adapter is inferior to either predefined functors, or properly implemented user-defined extender functors.

3.8.1 Example: root finding

An equation of one variable $F(x) = 0$ can have zero, one, or many solutions, commonly called roots of the equation. Finding these in the most general case of a nonlinear equation can be challenging. One of the commonly used algorithms can be traced back to Isaac Newton and Joseph Raphson, and is now called the *Newton-Raphson method*. It is an iterative method, in which one proceeds from an initial estimate which is then refined until convergence is achieved. One takes the tangent to the curve at the starting point, and intersects it with the *x*-axis. The intersection provides a refined estimate of the root. This continues until the estimate is invariant under further refinement. Note that the algorithm may not converge if a poor starting point is used, so good initial estimates are essential. The Newton-Raphson method is therefore perhaps more accurately described as an algorithm for refining estimates rather then obtaining them from scratch.

The line tangent to $F(x)$ is obtained by taking the first derivative $F'(x)$, and the intersection with the *x*-axis is given by $x - F(x)/F'(x)$. The whole algorithm can be coded like this:

```
double newtonRaphson(double x, Genfun::GENFUNCTION P) {
  double x1=x;
  while (1) {
    double deltaX=-P(x)/P.prime()(x);
    x+=deltaX;
    if (float(x1)==float(x)) break;
    x1=x;
  }
  return x;
}
```

The algorithm iterate until convergence is achieved at float precision.

The case of multiple roots is more complicated, but after each root is found one can simplify the discovery of the remaining roots. The method, called deflation, is to divide the function $f_i(x)$ by $(x - x_i)$ after each root x_i is found: $f_i(x) \to f_{i+1}(x) = f_i(x)/(x - x_i)$. The updated function $f_{i+1}(x)$ is zero wherever $f_i(x)$ is zero, except at $x = x_i$; and one can locate the remaining roots by evaluating $f_{i+1}(x)$. To start the procedure one sets $f_0(x) = F(x)$. The process, applied to a fifth order polynomial, is illustrated in Figure 3.1. This example can be found in EXAMPLES/CH3/NRDEFLATE. The key elements of this code are summarized in this excerpt::

```
using namespace Genfun;
Variable X;
GENFUNCTION F=(X-1)*(X-2)*(X-3)*(X-M_PI)*(X-4);
const AbsFunction * f=&F;
for (int i = 0; i<5; i++ ) {
  double x = newtonRaphson(-1.0, *f);
  GENFUNCTION F1 = (*f)/(X-x);
  if (f!=&F) delete f;
  f=F1.clone();
}
```

Note that at each iteration of the Newton-Raphson step, we have been careful to step away from the root we have just found; that's because evaluation of the deflated function will yield a divide-by-zero error there. Exercise 15 gives some practice with this algorithm, which is frequently required in other exercises throughout the book. A more complete discussion of these methods can be found in Press (2007).

Eigenvalue methods

In Exercise 10 of Chapter 2, you are asked to diagonalize a matrix by solving a secular equation. If you attempted this exercise, you will have noticed that the Eigen classes, particularly EigenSolver<MatrixXd> provide a far simpler mechanism. So, you may ask, rather than finding the roots of a polynomial as a means of diagonalizing a matrix, can we diagonalize a matrix as a means of finding the roots of a polynomial? The answer is yes.

A monic polynomial is a polynomial whose highest order monomial appears with a coefficient of 1, for example $x^4 - 4x^2 - \frac{1}{2}x - 1$. Obviously any polynomial can be expressed as a numerical factor times a monic polynomial. For purposes of root finding we can ignore the factor, while expressing the monic polynomial as:

$$p(x) = x^n + \sum_{i=0}^{n-1} c_i x^i \tag{3.29}$$

Now define the *companion matrix* for the polynomial

$$\mathbf{C} = \begin{pmatrix} 0 & 0 & . & . & . & 0 & -c_0 \\ 1 & 0 & . & . & . & 0 & -c_1 \\ 0 & 1 & . & . & . & 0 & -c_2 \\ . & . & . & & & 0 & . \\ . & . & & . & & 0 & . \\ . & . & & & . & 0 & . \\ 0 & 0 & . & . & . & 1 & -c_{n-1} \end{pmatrix} \tag{3.30}$$

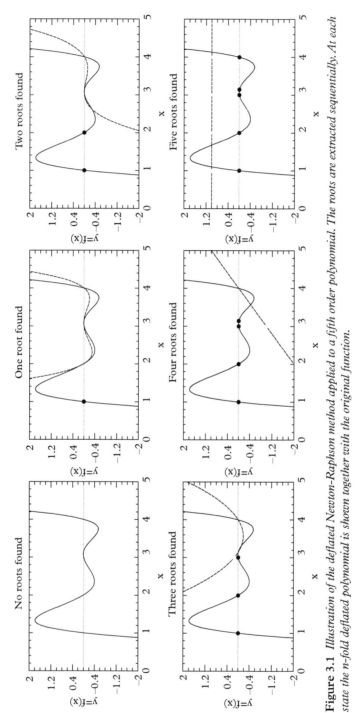

Figure 3.1 *Illustration of the deflated Newton-Raphson method applied to a fifth order polynomial. The roots are extracted sequentially. At each state the n-fold deflated polynomial is shown together with the original function.*

The secular equation which arises in the diagonalization of the companion matrix is:

$$\det(\mathbf{C} - \lambda \mathbf{I}) = 0; \tag{3.31}$$

but it can also be shown that:

$$\det(\mathbf{C} - \lambda \mathbf{I}) = p(\lambda) \tag{3.32}$$

Now,

$$\det(\mathbf{C} - \lambda \mathbf{I}) = \det(\mathbf{A}\mathbf{D}\mathbf{A}^{-1} - \lambda \mathbf{I})$$
$$= \det(\mathbf{D} - \lambda \mathbf{I})$$

where both $\mathbf{A}$ and:

$$\mathbf{D} = \begin{pmatrix} \lambda_0 & 0 & \cdots & 0 & 0 \\ 0 & \lambda_1 & \cdots & 0 & 0 \\ 0 & 0 & \cdot & 0 & 0 \\ 0 & 0 & \cdot & 0 & 0 \\ 0 & 0 & \cdot & 0 & 0 \\ 0 & 0 & \cdot & \lambda_{n-2} & \cdot \\ 0 & 0 & \cdots & 0 & \lambda_{n-1} \end{pmatrix} \tag{3.33}$$

arise from the diagonalization procedure, and the λ_i are the eigenvalues of the companion matrix. So we have:

$$p(\lambda) = \det \begin{pmatrix} \lambda - \lambda_0 & 0 & \cdots & 0 & 0 \\ 0 & \lambda - \lambda_1 & \cdots & 0 & 0 \\ 0 & 0 & \cdot & 0 & 0 \\ 0 & 0 & \cdot & 0 & 0 \\ 0 & 0 & \cdot & 0 & 0 \\ 0 & 0 & \cdot & \lambda - \lambda_{n-2} & \cdot \\ 0 & 0 & \cdots & 0 & \lambda - \lambda_{n-1} \end{pmatrix} = \prod_{i=0}^{n-1}(\lambda - \lambda_i) \tag{3.34}$$

or:

$$p(\lambda) = \prod_{i=0}^{n-1}(\lambda - \lambda_i) \tag{3.35}$$

and we see that the *roots of the polynomial are the eigenvalues of the companion matrix*. The technique is fairly simple to use, handles the case of multiple roots, and even returns complex roots. Example code can be found in EXAMPLES/CH3/COMPANION.

The familiar classical orthogonal polynomials are discussed in the Chapter 5. Closed form expressions for the coefficients of these polynomials can be found, but are usually cumbersome to work with. However, some very slick techniques are also available for these cases. They will be discussed later, in Chapter 5.

3.8.2 Parameter objects and parametrized functors

It will prove to be useful to be able control the shape of our function through one or more adjustable parameters. Indeed, this capability is one of the prime motivations for defining function-objects. Parameter objects from the `QatGenericFunctions` library are complementary to function-objects and are used to parametrize them. The two can appear together in mixed algebraic expressions. A primary use-case is in data modeling, where the shape of a function is adjusted to give the best description to some input data.

Parameters are defined with a constructor that requires a name, value, lower limit, and upper limit:

```
#include "QatGenericFunctions/Parameter.h"
Genfun::Parameter alpha ("alpha", 1/137.0, 1/139.0, 1/135.0);
```

The parameters also obey an algebra that includes +,-,*, and / operations. In our first simple demonstration, we will use `Genfun::Parameters` to parametrize a margarita. A standard margarita is a Mexican cocktail consisting of seven parts tequila, four parts Cointreau, and three parts lemon juice. A more general margarita can be parametrized by the fraction of two of the components (the other fraction will be determined). Using `Genfun::Parameters`, we can write:

```
Genfun::Parameter   pTequila     ("Tequila",   7/14.0, 0, 1.0);
Genfun::Parameter   pCointreau   ("Cointreau",4/7.0 , 0, 1.0);
```

Here, only the tequila is specified in terms of an absolute fraction; the Cointreau content is specified as a fraction of the remainder and the lemon juice is equal to the remainder. The values of `pTequila` and `pCointreau` are allowed to vary over the range [0 : 1]. The absolute fractions can be specified in terms of these parameters:

```
Genfun::GENPARAMETER    fTequila =pTequila;
Genfun::GENPARAMETER    fContreau=(1.0-pTequila)*pCointreau;
Genfun::GENPARAMETER    fLemonJuice=(1.0-pTequila)*
                        (1-pCointreau);
```

Varying the parameters over their allowed range always gives a physical margarita with all absolute fractions positive by construction and summing to unity. Now if we send a message to the parameter `pCointreau`:

```
pCointreau->setValue(0.5)
```

the lemon juice fraction will be automatically updated.

Now suppose that instead of a margarita we mix up a radioactive cocktail of three isotopes, ^{207}Bi, ^{211}Po, and ^{207}Tl, with half-lives of 35 years, 0.52 seconds, and 4.77 minutes, respectively. All of these isotopes decay into ^{207}Pb, which is stable. We would like to build a function representing the fraction of lead in our sample vs. time, but parametrized by the original fractions of Bi, Po, and Tl. We proceed by defining the fractions:

```
Genfun::Parameter     pBi    ("Bi fraction",  7/14.0, 0, 1.0);
Genfun::Parameter     pPo    ("Po fraction",  4/7.0 , 0, 1.0);
```

Next, we make some simple definitions of basic time units and convert half lives to mean lives (dividing by ln(2)):

```
double second = 1.0;            // define second as the
                                   basic unit.
double year   = 365*24*60*60;   // define years in terms of
                                   seconds;
double minute = 60;             // define minutes in terms
                                   of seconds;
double tHalfBi=35*years, tHalfPo=0.52*seconds,
    tHalfTl=4.77*minutes;
double tMeanBi=tHalfBi/log(2);
double tMeanPo=tHalfPo/log(2);
double tMeanTl=tHalfTl/log(2);
```

Then, we make three normalized functions representing the decay probability for each component.

```
#include "QatGenericFunctions/Parameter.h"
#include "QatGenericFunctions/Variable.h"
#include "QatGenericFunctions/Exp.h"
Genfun::Variable X;
Genfun::Exp        exp;
Genfun::GENFUNCTION fBi=(1/tMeanBi)*exp(-X/tMeanBi)
Genfun::GENFUNCTION fPo=(1/tMeanPo)*exp(-X/tMeanPo)
Genfun::GENFUNCTION fTl=(1/tMeanTl)*exp(-X/tMeanTl)
```

Finally we can construct the parametrized function.

```
Genfun::GENFUNCTION fLeadVsTime= 1-(pBi*fBi+(1-pBi)*
    (pPo*fPo+(1-pPo)*fTl));
```

And now, if we send a message to the parameter `fractionBi`:

```
pBi->setValue(0.5)
```

the function `fLeadVsTime` will assume a different shape.

Some functions have "internal" parameters. A normal distribution, or Gaussian, is characterized by a mean and a sigma, or width. We can access these parameters through the interface of `Genfun::NormalDistribution`. The following are legal:

```
#include "QatGenericFunctions/NormalDistribution.h"
#include "QatGenericFunctions/Parameter.h"
main () {
  {
    Genfun::NormalDistribution g;
    g.sigma().setValue(2.0);
  }
  // Also this is legal:
  {
    Genfun::NormalDistribution g;
    Genfun::Parameter sigma ("sigma", 2.0, 0.1, 10.0);
    g.sigma().connectFrom(&sigma);
  }
}
```

We will say more about these `QatGenericFunctions` as the need arises. For now we focus attention on visualizing these functions.

3.9 Plotting

The graphical display of data is critical for all types of science. Unfortunately it drags scientists, at least to some level, into topics, such as graphics, graphical user interfaces, and windowing systems, which are quite removed from their problem domain. For this text we provide a simple plotting system called `Qat`, consisting of the packages `QatPlotting` and `QatPlotWidgets`, that is built on the industry-standard user interface (UI) framework known as `Qt(www.qt.io)`. The system allows you to visualize mathematical functions without mastering the complex `Qt` system, which is beyond the scope of our text. The `Qt` package comes with excellent documentation, typically installed with `Qt` in

`file:///usr/share/qt5/doc/html/index.html`.

Blanchette (2008) is a good reference for those wishing to learn the `Qt` system, but this is not necessary for the exercises in this text. In addition, the `QAT` reference manual is available online through

`http://www.oup.co.uk/companion/acp`

Two working examples of code can be found in

EXAMPLES/CH3/FUNCTIONVIS1

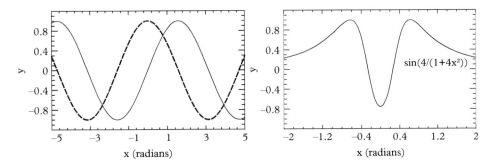

Figure 3.2 *Example plots obtained with the* Qat *system. On the left:* FUNCTIONVIS1 *example; on the right, the* FUNCTIONVIS2 *example referred to in the text.*

and

EXAMPLES/CH3/FUNCTIONVIS2

To build these examples, go to the src subdirectory of FUNCTIONVIS1 or FUNCTIONVIS2, type qmake followed by make. The qmake command is part of the Qt system; it makes makefiles across platforms. The first example plots the functions $\sin x$ and $\cos x$ using function objects; the second plots the user-defined function $\sin(4/(1+4x^2))$ using the previously discussed F1D adaptor. The plots one obtains are shown in Figure 3.2.

3.10 Further remarks

The C++ language appeared in the early 1980s, and was first noticed by the scientific community in the late 80s and early 90s. One of the main benefits of the language was the potential for highly convenient, highly reusable software. This potential was not immediately realized since the scientific community needed time to develop class libraries adapted to its needs. The situation in the early 20th 21st century is very much improved, since library development has flourished over the years and in many cases the development has been standardized, sometimes—as with std::vector and std::string—even to the point of being absorbed into the C++ standard library.

Other packages and programs worth mentioning in this context are listed here:

- The boost library, available at www.boost.org, is a source of portable, peer-reviewed source libraries with a wide range of applications, both numerical and non-numerical. The boost library serves as a *de facto* staging ground for the standard library. Ten packages from boost have now been included in the latest C++11 standard.

- The ROOT system, a data analysis framework by Rene Bruno and Fons Rademackers, available from root.cern.ch. The package contains a wide variety of tools for plotting and statistical analysis and is heavily used in high energy physics. ROOT is now in its sixth major version. Version 7 will comprise a major overhaul of this software package and may be well worth waiting for.
- The GNU Scientific library (GSL), available from http://www.gnu.org/software/gsl/. The package contains a variety of numerical algorithms. Their implementation of special functions is particularly useful. The library consists entirely of C functions which can either be invoked directly or wrapped in C++ objects.

Of course there is significant overlap in many of these libraries. The choice of which library to use often involves issues of functionality, interface quality, modularity, and interoperability. In the early 21st century, scientists and engineers have an enormous choice of numerical algorithms and have only an occasional need to implement their own.

This brief tour of useful classes is designed to familiarize our readers not only with a variety of available software but also with the typical "look and feel" of a high-quality, object-oriented class library. Despite the proliferation of canned solutions to numerical problems, it is inevitable that students will need to carry out their own customization and perhaps even prepare their own specialized class libraries. This requires considerable skill. A little experience in using class libraries is invaluable training for anyone who sets out to build one. The exercises at the end of this chapter help to build that experience. Later we will discuss in more detail some of the principals in writing classes and libraries of classes.

3.11 Exercises

1. A certain crystal has the shape of a parallelepiped and is situated with one vertex at the origin. Determine the volume of this crystal using the `Eigen::Vector3D` class and check the result with the `Eigen::Matrix3D` class. *Hint*: use the triple product and the determinant. The edges of the crystal run in the following ways:

    a) from (0,0,0) to (3,0,0)
    b) from (0,0,0) to (0.5, 2.0,0)
    c) from (0,0,0) to (0.3, 0.2, 1.5).

2. An initially pure sample containing N_0 atoms of $^{213}_{83}$Bi atoms decays according to:

$$^{213}_{83}\text{Bi} \rightarrow {}^{209}_{81}\text{Tl} + \alpha$$

    with a half-life of 45.6 minutes. This is followed by the decay:

$$^{209}_{81}\text{Tl} \rightarrow {}^{209}_{82}\text{Pb} + \beta$$

with a half-life of 132 seconds, then

$$^{209}_{82}\text{Pb} \rightarrow {}^{209}_{83}\text{Bi} + \beta$$

with a half-life of 195 minutes. $^{209}_{83}\text{Bi}$ is so longlived (half-life of 1.9 x 10^{19} years) that it can be considered stable. Letting x_b, x_t, and x_p be the number of atoms of $^{213}_{83}\text{Bi}$, $^{209}_{81}\text{Tl}$, and $^{209}_{82}\text{Pb}$, respectively, and the vector $\vec{x} = \{x_b, x_t, x_p\}$, one can put the equation of growth and decay in the form:

$$\frac{d\vec{x}}{dt} = \Lambda \vec{x}$$

which has a solution is $\vec{x} = \exp(\Lambda t) \cdot \vec{x}_0$.

a) obtain the 3x3 matrix Λ

b) In Eigen, a matrix is exponentiated in the following way:

```
#include "unsupported/Eigen/MatrixFunctions"
Matrix3d M;
Matrix3d expM = M.exp();
```

Use this method to determine the fraction of each species at t=50, 150, 200, 500, 1000 minutes.

3. Make a program to plot the following seven functions.

a) $\sin(x)$
b) $\sin(5x)$
c) $\sin(x^2)$
d) $e^{-x} \sin(x)$
e) The associated Legendre polynomial, $P_3^2(x)$
f) Any function defined by you (using the F1D adapter)
g) The derivative of that function.

The program should take a command line argument from 1-7, which selects the function to be plotted. Print out and save plots from the program.

4. Examine the header file QatGenericFunctions/InterpolatingPolynomial.h. This is the header file for a function that interpolates between points using a function called Lagrange's interpolating polynomial. Construct this function using points from the table below, and make a plot of the function. Also, plot the points themselves using the class declared in QatPlotting/PlotProfile.h.

Point	x	y
0	−2.0	2.0
1	0	4.0
2	1.0	3.3
3	2.0	2.0
4	3.0	6.0

5. Plot the cubic spline function going through all of the points in the table of the previous exercise, also using the class declared in QatGenericFunctions/ CubicSplinePolynomial.h.
6. Plot the probability density, $|\psi(x)|^2$, for the rectangular barrier problem discussed in Section 3.4. Use the function adapter Genfun::F1D.
7. Consider a wavefunction $\psi(x)$ which satisfies the time-independent Schrödinger equation for $E > 0$, impinging upon a potential barrier which has the piecewise-continuous form:

$$V(x) = 0 \quad x < -2a$$
$$V(x) = V_0 \quad -2a < x < -a$$
$$V(x) = 2V_0 \quad -a < x < a$$
$$V(x) = V_0 \quad a < x < 2a$$
$$V(x) = 0 \quad x > 2a \tag{3.36}$$

Solve the Schrödinger equation numerically in all space and determine the reflection and transmission for an interesting range of V_0/E and ka.

8. Plot the solution of the previous problem for $V/E = 1.5$, using PlotView. The value of ka be taken from the command line. Choose a few interesting values and hand in the plots.
9. Imagine a quantum mechanical wavefunction (as in the previous problem) impinging upon a Gaussian potential barrier:

$$V(x) = V_0 e^{-x^2/(2a^2)} \tag{3.37}$$

Break the region from $-4a < x < 4a$ into 2^N different bins of equal size, and approximate the potential as piecewise-constant within those bins. Compute the transmission coefficient as a function of the N and plot it in PlotView, over an interesting range of the parameters V_0 and ka.

10. [T] Show that the eigenvalues of the problem discussed in Section 3.2, $\lambda_i = \omega_i^2$ are not only real, but also positive, so that ω_i is real. This exercise is to be done without a computer.

11. a) Compare the rate of convergence of two iterative methods for the calculation of π: the iterative circumference method of example program `threePi`, in `EXAMPLES/CH2/THREEPI`, and the iterative area method of the Exercise 6 from the previous chapter. Plot the error in the calculation as a function of the number of sides on a log-log plot using `PlotView` and `PlotProfile`.

    b) In the iterative circumference method, the arc length of a small wedge is approximated by the length l of an inscribed triangle. Prove that a better approximation to the arc length is $2z(1 + z^2/6)$, where $z \equiv l/2$. Use this as the basis for an improved iterative circumference method, and plot the error vs. number of sides on the same set of axes as in part (a).

12. Make a function plotting command that reads in a set of points from one or more files and connects them with an interpolating function. The program

    a) has the syntax plot `inputFile1 [inputFile2] -c ...`

    b) uses cubic spline interpolation if the `-c` option is given, otherwise polynomial interpolation.

    c) reads in each input file with points, having the format `x1 y1`
    `x2 y2`
    `x3 y3`

    d) plots the input points and the interpolation function.

13. Five metal balls slide along a frictionless wire which is threaded through holes drilled through the center of each ball. From left to right, the masses of the balls are: m, $3m$, $2m$, m, and $2m$. The five balls are, additionally, connected to each other by four springs whose spring constants are: k, $2k$, $2k$, and k. The five balls are initially at rest, but displaced from their equilibrium positions by -2 cm, 3 cm, 0 cm, -3 cm, and 2 cm. Plot on the same set of axes the position of the five masses as a function of time, where the time is given in units of the characteristic time $\sqrt{m/k}$. Also plot the position of the center of mass of the system. To carry out this calculation, you will find the function

    ```
    Eigen::SelfAdjointEigenSolver<Eigen::MatrixXd>,
    ```

    which diagonalizes a symmetric matrix, to be useful.

14. Like the class `Genfun::F1D`, the class `Genfun::F2D` is an adapter from functions of *two* variables to generic functions of two variables, which function in a similar way[2]. The class `PlotWave1D`, in `QatPlotting`, provides a plotting interface for

---

[2] The `partial(unsigned int i)` method should be used instead of the `prime()` method on functions of two variables.

functions of position and time, and requires a `Genfun::GENFUNCTION` of two variables. Use these two classes to plot the real and imaginary part of $\Psi(x,t) = e^{-i\omega t}\psi(x)$, together with the $|\psi(x)|^2$ from the problem 7–8 or problem 9.

15. The Legendre polynomials $P_l(x)$ for $l = 0, 1, 2, 3, ..$ are polynomials of degree l that are orthogonal under an inner product

$$\langle f|g\rangle = \int_{-1}^{1} f(x)g(x)dx$$

The class `LegendrePolynomial` (in `QatGenericFunctions`, namespace `Genfun`) implements these functions in the Qat library.

Use the Newton-Raphson method to find the zeros of these polynomials, from $l = 1$ through $l = 10$.

- Your program should obtain the order l from the command line.
- It should plot the Legendre polynomial.
- It should mark the zeros on the plot.
- It should print out the zeros to `std::cout`.

If you need a good starting point for the roots, you can use the asymptotic expression

$$x_{l,m} = \left(1 - \frac{1}{8l^2} + \frac{1}{8l^3}\right)\cos\frac{4m-1}{4l+2}\pi + O(l^{-4})$$

where $x_{l,m}$ is the mth zero of $P_l(x)$.

16. The Hermite polynomials $H_n(x)$ for $n = 0, 1, 2, 3 \ldots$ are polynomials of degree n that are orthogonal under an inner product

$$\langle f|g\rangle = \int_{-\infty}^{\infty} f(x)g(x)e^{-x^2}dx,$$

which arise in the study of the quantum harmonic oscillator. In the Qat library, a class `HermitePolynomial` (in `QatGenericFunctions`, namespace `Genfun`) implements these functions. Repeat the previous exercise to find the zeros of the Hermite polynomials. Use the asymptotic expansion

$$H_n(x) \sim \exp(x^2/2)\frac{2^n}{\sqrt{\pi}}\Gamma\left(\frac{n+1}{2}\right)\cos\left(x\sqrt{2n} - n\frac{\pi}{2}\right)$$

to obtain a good starting point.

17. Revisit exercise 10 of Chapter 2; but use the Newton-Raphson method to determine the eigenvalues from the secular equation of the matrix.

BIBLIOGRAPHY

Blanchette, J. and M. Summerfield (2008). *C++ GUI programming with Qt4*. 2nd ed. Prentice-Hall.
Gasiorowicz, S. (2003). *Quantum Physics*. 3rd ed. John Wiley & Sons.
Griffiths, D. J. (2005). *Introduction to Quantum Mechanics*. 2nd ed. Prentice-Hall.
Press, W., S. Teukolsky, W. Vetterling, and B. Flannery (2007). *Numerical Recipes, the Art of Scientific Computing*. 3rd ed. Cambridge University Press.
Stepanov, A. and M. Lee (1995). *The Standard Template Library*. H.P Laboratories Technical Report 95-11(R1).
Weiss, M. A. (2014). *Data Structures and Problem Solving using C++*. 4th ed. Prentice-Hall.

4
Interpolation and extrapolation

4.1	Lagrange interpolating polynomial	85
4.2	Evaluation of the interpolating polynomial	87
	4.2.1 Interpolation in higher dimensions	89
4.3	Spline interpolation	90
	4.3.1 The cubic spline	90
	4.3.2 Coding the cubic spline	93
	4.3.3 Other splines	94
4.4	Extrapolation	95
4.5	Taylor series, continued fractions, and Padé approximants	97
4.6	Exercises	102
	Bibliography	106

While interpolation and extrapolation may seem prosaic, the process of using limited information to make predictions (in this case, the value of a function where it is not known) is rather common in scientific computing. The idea of finding the "true function" from a set of points is of course a fairy tale—there are an infinite number of functions, well-behaved or otherwise, whose values coincide at a fixed number of points. Thus function interpolation requires that we also build in some prior notion (a model) of what the function should look like. In this way, the techniques that we describe arise from the combination of two ingredients:

- a set of points through which the function must pass.
- a model for what the function does between and/or beyond the given points.

For many applications it seems reasonable and will be productive to use some sort of smooth function as a model. In Section 4.1 we discuss an intuitive technique which uses an Nth-order polynomial called *Lagrange's interpolating polynomial* as the model function. This generally works well for $N \lesssim 4$. Above that range, the technique suffers from difficulties that we will illustrate. The cubic spline function, which is a series of polynomials joined smoothly at the interpolation points, is often a better interpolating function if a large number of points is to be used. We discuss it in Section 4.3. More general models are presented in the exercises.

Applied Computational Physics. Joseph F. Boudreau and Eric S. Swanson, Oxford University Press (2018).
© Joseph F. Boudreau and Eric S. Swanson. DOI:10.1093/oso/9780198708636.001.0001

The modeling of functions is key to interpolation, and also plays a role in integration where a certain behavior between mesh points (e.g. trapezoids) is assumed. The subject is also related to fitting techniques that model data points with measurement error. This is the subject of Chapter 16.

4.1 Lagrange interpolating polynomial

The Lagrange interpolating polynomial (Szego, 1939; Davis and Rabinowitz, 1975; Press, 2007) is a key ingredient in a number of numerical techniques that we will develop in this book. The basic idea is that given $N+1$ points in a plane having distinct values of the abscissa $\{x_i\}$ for $i = 0, 1..N$, one can always find a polynomial of order N going through all the points. With two points we obtain a straight line, with three points we obtain a parabola, and so forth. We will develop this idea and apply it immediately to the problem of interpolation and extrapolation.

Suppose we are given the coordinates (x_0, y_0) of a single point in a plane. The zeroth order polynomial going through this point is $y(x) = y_0$. For two points, (x_0, y_0) and (x_1, y_1), one defines a parameter $t(x) = \frac{x-x_0}{x_1-x_0}$; this parameter takes the value $t = 0$ when $x = x_0$ and $t = 1$ when $x = x_1$. Also, let $\bar{t} = 1 - t$. Then the equation of the line can be written as:

$$y(x) = \bar{t} y_0 + t y_1$$
$$= \frac{x - x_1}{x_0 - x_1} y_0 + \frac{x - x_0}{x_1 - x_0} y_1$$

Similarly, for three points one has:

$$y(x) = \frac{x - x_2}{x_0 - x_2} \frac{x - x_1}{x_0 - x_1} y_0 + \frac{x - x_0}{x_1 - x_0} \frac{x - x_2}{x_1 - x_2} y_1 + \frac{x - x_0}{x_2 - x_0} \frac{x - x_1}{x_2 - x_1} y_2.$$

One can easily verify that this expression is

- a second degree polynomial in x, and
- $y(x_0) = y_0$; $y(x_1) = y_1$; $y(x_2) = y(2)$.

Joseph Louis Lagrange in 1795 gave a formula for the general case, which was actually first discovered by Edward Waring in 1779:

$$L(x) = \sum_{j=0}^{N} y_j l_j \qquad (4.1)$$

where

$$l_j = \prod_{0 \le m \le N; m \neq j} \frac{x - x_m}{x_j - x_m}. \qquad (4.2)$$

86 *Lagrange interpolating polynomial*

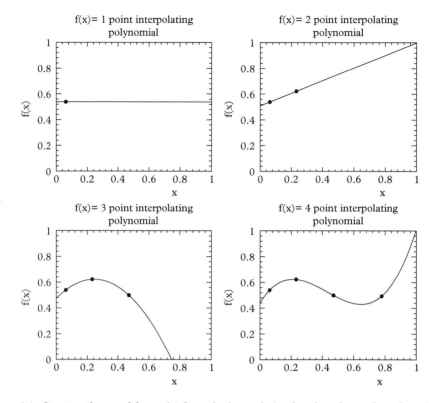

Figure 4.1 *One, two, three, and four point Legendre interpolating functions drawn through various points in the plane.*

$L(x)$ is called the **Lagrange interpolating polynomial**, while the $l_j(x)$s are called the fundamental polynomials for Lagrange interpolation. The plots in Figure 4.1 show a typical 1-point, 2-point, 3-point, and 4-point interpolating polynomial.

A few remarks on the behavior of the polynomial are in order. Imagine that we are using the polynomial to interpolate a function between a few known points. Naively one would expect that the quality of the interpolation is improved as the number of points is increased; i.e., as the order of the interpolating polynomial is increased. This is not necessarily so, as we can see through a little experimentation.

Higher order polynomials can indeed follow the set of interpolation points exactly, but they also often require wild oscillations and wide excursions up and down in order to match these points. Adding new points to the set can cause large changes to the shape of the polynomial. The behavior of the interpolating polynomial becomes almost completely arbitrary as the number of points is increased, unless, of course you happen to be fitting a higher order polynomial. Functions which are sharply peaked, like Gaussian or Breit-Wigner functions, are poorly interpolated by polynomials of high order. However, the behavior for $N \lesssim 4$ is often a reasonable approximation to slowly varying functions.

4.2 Evaluation of the interpolating polynomial

The recommended technique for evaluating the interpolating polynomial is called **Neville's algorithm**, which we describe in this section. We consider a set of $n+1$ points $\{x_0, x_1, \ldots, x_n\}$ interpolated by $L_n(x)$, a Lagrange interpolating polynomial of order n. We remove the jth point from the interpolation, so as to obtain an order $n-1$ polynomial, designated as $L_{n-1,j}(x)$. Now take two such polynomials $L_{n-1,j}(x)$ and $L_{n-1,k}(x)$, where $j \neq k$, and consider the following expression:

$$L_n(x) = \frac{(x-x_j)L_{n-1,j}(x) - (x-x_k)L_{n-1,k}(x)}{(x_k - x_j)}$$

The equality can be established by noting that, on the right hand side, when $x = x_j$, the first term is zero and the second term is y_j; when $x = x_k$, the second term is zero and the first term is y_k, and for any other nodal point $x = x_i$, $i \neq j$ and $i \neq k$, the sum of the two terms gives y_i. The polynomial (of order n) interpolates all n points in the set and is therefore equal to $L_n(x)$, the Lagrange interpolating polynomial of order n. This recurrence relation gives a practical scheme for evaluating the polynomial. To anchor the procedure we evaluate the 0th order polynomials passing through each of the $n+1$ points. These polynomials are used to construct each of the n 1st order polynomials which interpolate between the nearest neighbors. A class to carry this out InterpolatingPolynomial (in the QatGenericFunctions library, namespace Genfun) is excerpted below. The class declaration is:

```
namespace Genfun {

 class InterpolatingPolynomial: public AbsFunction {

 public:

    // Constructor
    InterpolatingPolynomial();

    // Retrieve function value
    virtual double operator () (double argument) const;

    // Add a new point:
    void addPoint(double x, double y);

  private:

    std::vector<std::pair<double,double> > xPoints;
    mutable std::vector<double>            q;
  };
}
```

The implementation of the function call operator is as follows.

```
// Function call operator
double InterpolatingPolynomial::operator() (double x) const {

  if (xPoints.size()==0) throw std::runtime_error
     ("InterpolatingPolynomial: no interpolation points");

  for (unsigned int i=0;i<xPoints.size();i++) q[i]=xPoints
     [i].second;

  for (unsigned int i=1;i<xPoints.size();i++) {
    for (unsigned int j=xPoints.size()-1;j>=i;j--) {
      q[j]=((x-xPoints[j].first)*q[j-1]-(x-xPoints[j-i].
         first)*q[j]) /
      (xPoints[j-i].first-xPoints[j].first);
    }
  }

  return q.back();
}
```

Some classes appear in this implementation that were not previously discussed. The template class `std::pair` is an easy way of agglomerating two variables, here the *x* and *y* coordinate of a point. The keyword ***mutable*** indicates that the variable is used as internal cache or workspace, not representing the state of the object, and so may be changed in a method declared to be `const`. When a designer declares a method to be `const` he/she makes a promise not to change the state, so that the method can be called on an object which is `const`. The compiler then enforces that promise in an admirably rigorous way, but `mutable` provides an often-needed loophole, which is useful when member data represents cache rather than state.

If the difference between *cache* and *state* is confusing, consider a hypothetical matrix object. The values of the matrix coefficients determine the state of the matrix; if they are changed one would say the matrix has changed. The calculation of the determinant is computationally expensive, and one would not want to repeat the calculation if it has already been done. Therefore one could consider caching the result in a piece of member data `mutable double det`. The keyword `mutable` allows the determinant to be (over)written in a constant method e.g. `getDeterminant()`, which is good design because the matrix does not "really" change when the determinant is cached. The cache would be valid as long as the matrix did not change, and a careful designer would have to consider invalidating it if the state (think: coefficients) of the matrix changes. Possibly it could be done by another piece of member data, e.g. `mutable bool stale`.

4.2.1 Interpolation in higher dimensions

We distinguish two cases of interpolation in higher dimensions. In the first case, the function which we interpolate may be a complex-valued, vector-valued, or matrix-valued function of a real variable. This is almost a trivial variant of the previously discussed problem. Instead of a real-valued interpolant constructed from the fundamental polynomials for Lagrange interpolation,

$$y(x) = \sum_{j=0}^{N} y_j l_j \tag{4.3}$$

we have a vector-valued interpolant:

$$\vec{y}(x) = \sum_{j=0}^{N} \vec{y}_j l_j, \tag{4.4}$$

or a matrix-valued interpolant, or something else, according to what we want to interpolate. The fundamental polynomials can be evaluated explicitly if desired, or by generalizing Neville's algorithm. C++ templates (discussed later in this text) are useful for this problem.

A function of two variables $f(x, y)$ can be regarded as a function of y that varies as a function of x. Applying the same logic as above, and designating the interpolating function as z, we can interpolate over x like this:

$$z(x, y) = \sum_{i} l_i(x) f(x_i, y) \tag{4.5}$$

But now suppose that $f(x_i, y)$ happens to be an interpolating function over y, which we can write as

$$f(x_i, y) = \sum_{j} l_j(y) f(x_i, y_j). \tag{4.6}$$

When we substitute this into the previous expression, we obtain:

$$z(x, y) = \sum_{i,j} l_i(x) l_j(y) f(x_i, y_j). \tag{4.7}$$

This expression interpolates over two dimensions, a second kind of higher-dimensional interpolation. When linear (2-point) interpolation is used for both variables, we call it *bilinear interpolation* (Press, 2007), but we can go to higher order if desired. Obvious generalizations of this simple formula handle cases of interpolation over three or more dimensions.

4.3 Spline interpolation

Another approach to interpolation is to break up the full range into smaller intervals and use lower-order polynomials within each interval to interpolate. In principle these can be Lagrange interpolating polynomials, but these would normally have discontinuous derivatives where they are joined. The interpolation can be made smooth by requiring that the first derivative and higher derivatives are continuous over the full range of the interpolation. (Note the similarity between this problem and the problem of piecewise-constant potentials in the Schrödinger equation, e.g. Exercises 7-9, Chap 3.)

The simplest continuous spline connects $N + 1$ points, known as **knots**, with straight lines. Clearly there are no free parameters in this approach and derivatives are discontinuous at the knots. If smoothness of the interpolating function is not an issue, linear splines should be considered: the chief computational effort is on mapping the independent variable to the appropriate interval. If the grid is uniform this can be achieved with a fixed function:

```
i = (int) N*(x-x0)/(xN-x0)  // map [x0..xN] to [0..N-1] (note
    that x=xN is mapped to N)
```

For an irregular grid one could implement a linear search, which is $O(N/2)$, or a bubble search, which is $O(log_2(N))$.

4.3.1 The cubic spline

If a quadratic function is used between knots there are a total of $3N$ spline parameters which are fixed by $2N$ equations specifying function values at the knots and an additional $N - 1$ equations that impose continuity of the first derivatives. Thus quadratic splines form a one-parameter family of interpolating functions. Similarly, cubic functions implement two-parameter interpolating functions called **cubic splines** (Press, 2007). The ambiguity can be resolved by choosing a convention, such as setting second derivatives to zero at the first and last knots. This choice is called a **natural cubic spline**.

To show how this works, first take a single interval and assume that the values of some function are given at its two endpoints x_0 and x_1, so that one is interpolating between the points (x_0, y_0) and (x_1, y_1). Parameterize the position within this interval with parameter $t = (x - x_0)/(x_1 - x_0)$, such that $t = 0 (t = 1)$ when $x = x_0 (x = x_1)$. Also define $\bar{t} \equiv 1 - t = (x - x_1)/(x_0 - x_1)$. Now construct an arbitrary third-order polynomial that passes through the points. Start with the linear interpolant,

$$y(x) = \bar{t} \cdot y_0 + t \cdot y_1$$

which is another way of writing the Lagrange formula. But now we wish to add additional higher-order terms to this which don't disturb the value of the function at the endpoints.

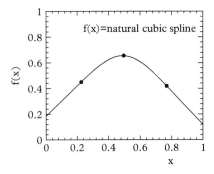

Figure 4.2 *The interpolating cubic spline through three points.*

The new expression is:

$$y(x) = \bar{t} y_0 + t y_1 + t\bar{t}(at + b\bar{t}). \tag{4.8}$$

It is evident that the new terms vanish at $x = x_0$ and $x = x_1$. Notice that $y(x)$ is a cubic function. If we specify additional boundary conditions, such as the first or second derivatives at the endpoints, we will determine the constants a and b.

Figure 4.2 shows three points and their corresponding natural cubic spline. Notice that the spline consists of straight lines outside of the interior region since the second derivatives have been set to zero at the end points

The parameters of the cubic spline can be obtained as follows. Consider a single interval bounded by the points (x_0, y_0) and (x_1, y_1). We adopt a convention in which the curve is expressed in terms of $q_0 \equiv \partial y/\partial x|_{x_0}$ and $q_1 \equiv \partial y/\partial x|_{x_1}$ instead of a and b. In this case

$$y(x) = \bar{t} y_0 + t y_1 + t\bar{t} \Delta x \left((m - q_1)t + (q_0 - m)\bar{t}\right) \tag{4.9}$$

where $\Delta x = x_1 - x_0$, $\Delta y = y_1 - y_0$, and $m \equiv \Delta y/\Delta x$.

The first derivative of y' can be written as

$$y'(x) = m + 2t\bar{t}\left(2m - q_0 - q_1\right) + t^2(q_1 - m) + \bar{t}^2(q_0 - m).$$

It is easily verified that this expression reduces to q_0 and q_1 at x_0 and x_1. The second derivative is:

$$y''(x) = \frac{dy'}{dx} = \frac{\partial y'}{\partial t}\frac{\partial t}{\partial x} + \frac{\partial y'}{\partial \bar{t}}\frac{\partial \bar{t}}{\partial x} = \frac{1}{\Delta x}\left(\frac{\partial y'}{\partial t} - \frac{\partial y'}{\partial \bar{t}}\right)$$

$$= \frac{1}{\Delta x}\left[t\left(4q_1 + 2q_0 - 6m\right) + \bar{t}(-2q_1 - 4q_0 + 6m)\right]$$

so at the endpoints:

$$y''(x_0) = \frac{6m - 2q_1 - 4q_0}{\Delta x} \qquad (4.10)$$

and:

$$y''(x_1) = \frac{-6m + 2q_0 + 4q_1}{\Delta x}. \qquad (4.11)$$

We now generalize to the case of $N+1$ points, $\{(x_0, y_0) \ldots (x_N, y_N)\}$, which divide the range of interpolation into N intervals. A set of unknown slopes $\{q_1, q_2, q_3, \ldots q_{N-1}\}$ are to be determined at points $\{x_1, x_2, x_3 \ldots x_{N-1}\}$. Setting $\Delta x_i = x_i - x_{i-1}$, $\Delta y_i = y_i - y_{i-1}$, and $m_i = \Delta y_i / \Delta x_i$ gives:

$$y''(x_{i-1}) = \frac{6m_i - 2q_i - 4q_{i-1}}{\Delta x_i} \qquad (4.12)$$

and:

$$y''(x_i) = \frac{-6m_i + 2q_{i-1} + 4q_i}{\Delta x_i}. \qquad (4.13)$$

Shifting Eq. 4.12 by substituting $i \to i+1$ and setting the result equal to Eq. 4.13 gives finally:

$$\frac{q_{i-1}}{\Delta x_i} + 2q_i \left(\frac{1}{\Delta x_i} + \frac{1}{\Delta x_{i+1}} \right) + \frac{q_{i+1}}{\Delta x_{i+1}} = 3 \left(\frac{m_i}{\Delta x_i} + \frac{m_{i+1}}{\Delta x_{i+1}} \right) \qquad (4.14)$$

Eq. 4.14 is a linear equation in three unknowns. We get one such equation for each of the intervals from $i = 1$ to $N - 1$ for a total of $N - 1$ equations. But we have $N + 1$ unknown slopes to determine. The two additional conditions are provided by the values of the second derivative at the endpoints. For the natural cubic spline, they can be derived by setting Eq. 4.12 equal to zero for $i = 1$ and by setting Eq. 4.13 equal to zero for $i = N$. We obtain:

$$\frac{2q_0 + q_1}{\Delta x_1} = \frac{3m_1}{\Delta x_1} \qquad (4.15)$$

$$\frac{q_{N-1} + 2q_N}{\Delta x_N} = \frac{3m_N}{\Delta x_N}. \qquad (4.16)$$

These $N + 1$ linear equations can be written in matrix form as:

$$A \cdot V = X \qquad (4.17)$$

where:

$$A = \begin{pmatrix} \frac{2}{\Delta x_1} & \frac{1}{\Delta x_1} & \cdots & & & & & \\ \frac{1}{\Delta x_1} & \frac{2}{\Delta x_1}+\frac{2}{\Delta x_2} & \frac{1}{\Delta x_2} & & & & & \\ & \frac{1}{\Delta x_2} & \frac{2}{\Delta x_2}+\frac{2}{\Delta x_3} & \frac{1}{\Delta x_3} & & & & \\ \cdots & \cdots & \cdots & \cdots & \cdots & \cdots & \cdots & \cdots \\ & & & & \frac{1}{\Delta x_{N-1}} & \frac{2}{\Delta x_{N-1}}+\frac{2}{\Delta x_N} & \frac{1}{\Delta x_N} \\ & & & & & \frac{1}{\Delta x_N} & \frac{2}{\Delta x_N} \end{pmatrix}$$

and:

$$V = \begin{pmatrix} q_0 \\ q_1 \\ .. \\ .. \\ .. \\ .. \\ .. \\ q_N \end{pmatrix} \quad X = \begin{pmatrix} \frac{3m_1}{\Delta x_1} \\ 3(\frac{m_1}{\Delta x_1}+\frac{m_2}{\Delta x_2}) \\ .. \\ .. \\ .. \\ .. \\ 3(\frac{m_{N-1}}{\Delta x_{n-1}}+\frac{m_N}{\Delta x_N}) \\ \frac{3m_N}{\Delta x_N} \end{pmatrix}.$$

The matrix equation can be solved to obtain the $N+1$ values of the slopes q_i at each of the $N+1$ knots. With these in hand, Eq. 4.9 can be used to determine the value of the spline at any point between x_i and x_{i+1}.

4.3.2 Coding the cubic spline

For efficiency one would like to code the interpolator with an initialization step (during which the matrices A and X are constructed and the system of linear equations is solved to determine V), and an evaluation step during which the function is evaluated for a particular value of x. In each step two new tasks arise. The first is how does one solve the matrix equation Eq. 4.17? The second is, given some value x for which we want to evaluate the interpolating spline, how do we quickly determine the index i such that $x_i < x < x_{i+1}$? This is a problem of sorting and searching.

While our numerical programming skills are still under development, we can make an obvious remark here: this is the digital age, and in the real world the lack of specific knowledge about how a solution is obtained is generally no big problem since online information and code libraries are readily available. If you had a real-life problem on your hands, you'd surely find yourself asking questions like: What task do I need to solve? What software product provides solutions to those tasks? *Where can I get it? How can I figure out if it actually works?*

Sorting and searching are very generic computational tasks that are not particularly numerical, although efficient algorithms need to be constructed carefully and logically.

These tasks are so important that many of them have become part of the standard C++ library itself. Ordering a list of data structures and searching the ordered list are among the tasks that C++ now provides. (The standard library is still expanding and it is useful to keep abreast of new developments in the library). We will discuss generic algorithms for tasks like that later in the book. Because the solution of linear equations is important to many scientists and engineers, we recommend using the high quality `Eigen` package, discussed in Chapter 3.

For a tridiagonal system like that of Eq. 4.17 it is faster to find the solution directly rather than to employ matrix inversion. This can be accomplished in the `Eigen` package with the following:

```
Eigen::MatrixXd A(n+1,n+1);
Eigen::VectorXd V(n+1);
Eigen::VectorXd X=A.lu().solve(V);
```

These lines of code invoke the "LU" (lower/upper) matrix decomposition algorithm that is fast and robust. A class (`CubicSplinePolynomial`, namespace `Genfun`) implementing cubic spline interpolation is available in the `QatGenericFunctions` library. The protocol of this class follows closely that of `InterpolatingPolynomial`.

4.3.3 Other splines

Although cubic splines are ubiquitous and powerful, it is a foolish researcher indeed who assumes that they are adequate in all cases. As always, the primary numerical tool is a thorough understanding of the problem at hand. A simple example of the types of problems that can arise is given by attempts to spline a power law function. A ten knot natural cubic spline to the function $1/x$ over the range $[1, 100]$ is presented as the dotted line in Figure 4.3. It is evident that the spline fares very poorly over much of the range, and that it is simply wrong for x larger than 100.

The solution in a case like this is simple: spline the logarithm of the y values and then exponentiate to obtain interpolations. Doing this yields the dashed line of Figure 4.3. The improvement is evident, although the failure for large x remains.

Often a researcher will know the asymptotic behavior of functions she is splining. In this case, for example, one may know that the function is proportional to x^{-b} for some fixed b. It is then expeditious to replace the natural cubic spline with one in which the spline is matched smoothly to the asymptotic form at the last knot:

$$y(x) = y_N \left(\frac{x_N}{x}\right)^b,$$

and:

$$y'(x_N) = -b\frac{y_N}{x_N},$$
$$y''(x_N) = b(b-1)\frac{y_N}{x_N^2}.$$

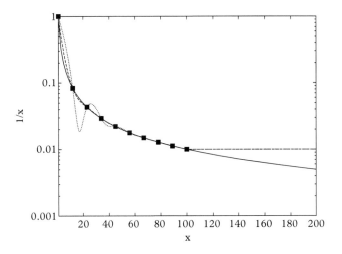

Figure 4.3 *Cubic splines for a power law function.*

If x_N is sufficiently large that the asymptotic form is accurate, the result will be an interpolating function that is reliable over the range (x_0, ∞). This, of course, resolves the issue for large x in Figure 4.3.

4.4 Extrapolation

Numerical work often requires extrapolation to physical limits (volumes to infinity, discretization scales to zero, degrees of freedom to infinity, etc.). Extrapolation of this sort is an art, with some methods performing better in some situations. Several of these are described here. Note that everything said about extrapolation can be applied to series summation, since one can always form a series of partial sums, $S_N = \sum_i^N a_i$.

Sometimes a sequence converges rapidly and it will be possible to estimate the limit directly. One is rarely this lucky, and acceleration methods must be employed. This can be achieved if additional assumptions about the asymptotic form of the sequence are made. Thus, for example, if one knows that the volume dependence of an energy scales as $E + a/V$ then measuring two values of $E(V)$ will allow determination of E and a.

In a similar way, if a sequence behaves as $S_N = S + \alpha q^N$ where $|q| < 1$, then the sequence will converge to S in the limit of large N. The limit can be obtained in an elegant way with ***Shanks transformation*** (Bender, 1999):

$$S = \frac{S_{N+1}S_{N-1} - S_N^2}{S_{N+1} + S_{N-1} - 2S_N} \tag{4.18}$$

as one can easily verify.

Of course, one can consider the transformed sequence to be a new sequence, which we shall denote S'. For example, consider $\log(2)$ evaluated as $\log(1+x)$ at $x = 1$. The

Table 4.1 *Shanks transformations for* $\log(1+1)$. *(Bender, 1999).*

n	S_n	S'_n	S''_n	S'''_n
1	1.000 000 0			
2	0.500 000 0	0.700 000 0		
3	0.833 333 3	0.690 476 2	0.693 277 3	
4	0.583 333 3	0.694 444 4	0.693 105 8	0.693 148 9
5	0.783 333 3	0.692 424 2	0.693 163 3	0.693 146 7
15	0.725 371 9	0.693 113 8	0.693 147 3	0.693 147 2
25	0.712 747 5	0.693 139 7	0.693 147 2	0.693 147 2
35	0.707 228 9	0.693 144 4	0.693 147 2	0.693 147 2
∞				0.693 147 180 6

Taylor series is:

$$\log(1+1) = 1 - \frac{1}{2} + \frac{1}{3} - \frac{1}{4} + \ldots,$$

which converges very slowly. However, a series of Shanks transformations improves convergence dramatically (see Table 4.1).

The form of the denominator in Eq. 4.18 implies that the method is susceptible to roundoff error, especially if the transformation is iterated. In general the method will be affected by roundoff error at the single precision level if double precision arithmetic is used.

Sometimes a Shanks transformation is insufficient to improve convergence. In this case the more general **Richardson extrapolation** can be used. This method is based on the assumption that a sequence behaves as:

$$S_N = q_0 + \frac{q_1}{N} + \frac{q_2}{N^2} + \ldots.$$

Take $M+1$ of these and solve to obtain:

$$q_0 = \sum_{k=0}^{M} \frac{S_{N+k}(N+k)^M (-)^{M+k}}{k!(M-k)!} \qquad (4.19)$$

Table 4.2 *Richardson extrapolants for $\sum 1/n^2$. (Bender, 1999).*

N	M = 0	M = 1	M = 2	M = 4	M = 6
1	1.000	1.500 00	1.625 000 00	1.644 965 277 8	1.644 933 185 185
5	1.464	1.630 28	1.644 166 67	1.644 935 811 1	1.644 934 060 147
10	1.550	1.640 68	1.644 809 05	1.644 934 195 4	1.644 934 066 526
15	1.580	1.642 94	1.644 893 41	1.644 934 089 9	1.644 934 066 812
20	1.596	1.643 78	1.644 916 08	1.644 934 073 2	1.644 934 066 842
25	1.606	1.644 18	1.644 924 58	1.644 934 069 2	1.644 934 066 847

As an illustration, we apply this to the sequence:

$$S_N = \sum_{n=1}^{N} \frac{1}{n^2} \qquad S_\infty = \frac{\pi^2}{6}$$

with the results of Table 4.2. Note that the function $Li_2(z) = \sum z^n/n^2$ has a branch point at $z = 1$, which causes the slow convergence evident in the $M = 0$ column.

Richardson extrapolation is useful in many contexts; we shall see it again in Chapter 5, where it is applied to extrapolate results in the numerical evaluation of integrals.

4.5 Taylor series, continued fractions, and Padé approximants

Representing functions is a central feature of numerical methods. Sometimes one benefits from prior effort, and existing implementations exist. In the absence of such benefit, one can resort to a variety of economical functional representations. For example, it may be possible to write the function in question as an integral and then the methods of Chapter 5 can be employed to implement a representation.

The familiar Taylor series provides another method for computing functions, however the series can be very slow to converge in some cases, and will diverge in others. For example the Taylor series for $\log(1 + z)/z$ converges only for $|z| < 1$. The flaw is due to a branch cut singularity at $z = -1$ that cannot be represented by a polynomial of any degree. This problem, however, can be averted by representing the logarithm as the *ratio* of polynomials–an idea that was first exploited by French mathematician Henri Eugène Padé (1863–1953). We therefore define the ***Padé approximant*** as the ratio

$$P_{NM}(x) = \frac{\sum_{n=0}^{N} a_n x^n}{\sum_{n=0}^{M} b_n x^n}. \tag{4.20}$$

98   *Taylor series, continued fractions, and Padé approximants*

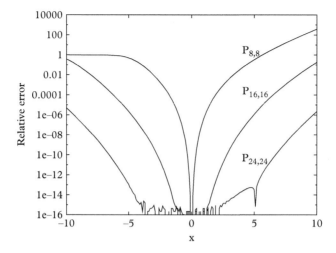

Figure 4.4 P_{NN} *Padé approximants for* $\exp(-x)$.

An algorithm to determine the coefficients will be given below. Notice that b_0 can be set to unity without loss of generality. Also, if P_{NM} is to represent a power series then the first $1 + M + N$ terms of the series representation of the approximant should agree with the corresponding terms in the series.

We illustrate the utility of the Padé approximant by computing the relative error of the [NN] approximant for $\exp(-x)$, shown in Figure 4.4. This double precision calculation reaches an optimal accuracy of order 10^{-16} over an increasingly large range as the Padé order is increased. Importantly, the Nth order approximant is *always* better than the equivalent Taylor series for positive x.

A related method of representing functions is called the ***continued fraction***, which is a fraction of fractions that may or may not terminate. A famous continued fraction is the representation of the golden ratio, $\Phi = (1 + \sqrt{5})/2$:

$$\Phi = 1 + \cfrac{1}{1 + \cfrac{1}{1 + \cdots}}.$$

A standard trick permits one to prove this relationship (which we leave to the reader to discern).

We define a general truncated continued fraction as:

$$\cfrac{a_1}{b_1 + \cfrac{a_2}{b_2 + \cdots + \cfrac{a_n}{b_n}}}. \tag{4.21}$$

This can be re-written without loss of generality as:

$$\frac{c_0}{1 + \frac{c_1 x}{1 + c_2 x + \ldots}} \tag{4.22}$$

where we have also taken the opportunity to introduce dependence on a variable, x.

Any finite continued fraction can be written as a ratio of numbers (or polynomials in the representation of Eq. 4.22) by repeated reduction of the fraction; thus Eq. 4.21 (or Eq. 4.22) can be written as

$$\frac{A_n}{B_n}.$$

A simple recurrence relation permits evaluating the factors:

$$\begin{aligned} A_{n+1} &= b_{n+1} A_n + a_{n+1} A_{n-1}, & A_0 &= 0, A_{-1} = 1 \\ B_{n+1} &= b_{n+1} B_n + a_{n+1} B_{n-1}, & B_0 &= 1, B_{-1} = 0. \end{aligned} \tag{4.23}$$

A similar formula holds in the case of Eq. 4.22. Of course the same formalism can be used to relate a continued fraction function to a Padé approximant.

The conversion from a power series to a continued fraction (and thereafter to a Padé approximant) can be achieved with the **quotient-difference algorithm**.

First we set

$$\sum_{n=0}^{\infty} a_n x^n = \frac{a_0}{1 - \frac{q_1^0 x}{1 - \frac{e_1^0 x}{1 - \frac{q_2^0 x}{1 - \frac{e_2^0 x}{\ldots}}}}}. \tag{4.24}$$

The continued fraction coefficients q_n^0 and e_n^0 are obtained from an iterative process that either differences or divides earlier coefficients (hence the name):

$$\begin{aligned} e_0^n &= 0 \\ q_1^n &= \frac{a_{n+1}}{a_n} \\ e_m^n &= q_m^{n+1} - q_m^n + e_{m-1}^{n+1} \\ q_{m+1}^n &= \frac{e_m^{n+1}}{e_m^n} q_m^{n+1}. \end{aligned} \tag{4.25}$$

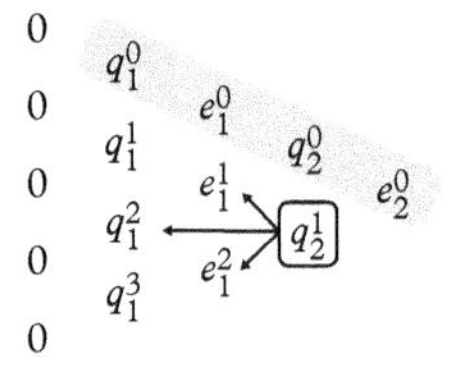

100   Taylor series, continued fractions, and Padé approximants

$$
\begin{array}{cccccc}
0 & & & & & \\
& q_1^0 & & & & \\
0 & & e_1^0 & & & \\
& q_1^1 & & q_2^0 & & \\
0 & & e_1^1 & & e_2^0 & \\
& q_1^2 & & \boxed{q_2^1} & & \\
0 & & e_1^2 & & & \\
& q_1^3 & & & & \\
0 & & & & & \\
\end{array}
$$

Figure 4.5 *Relationship of quotient-difference coefficients.*

This represents a rather intricate relationship wherein starting values are iterated in a triangular shape with the goal of obtaining the top row (the shaded region of Figure 4.5). Code to achieve this is given below to spare the student from wading through too many loop structures.

```
double f(const std::vector<double> & a, double x) {

  // convert sum_{n=0}^2N a_n x^n to continued fraction
  // coefficients cf_i, i=0 ... 2N-1

  unsigned int N=(a.size()-1)/2;

  // Allocate and zero the arrays:
  std::vector<double> e(2*N+1,0.0);
  std::vector<double> q(2*N,0.0);
  std::vector<double> eold(2*N+1,0.0);
  std::vector<double> qold(2*N+1,0.0);
  std::vector<double> cf(2*N,0.0);

  for (int j=0;j<=2*N-1;j++) {qold[j] = a[j+1]/a[j];}
  cf[0] = a[0]; cf[1] = -qold[0];

  // Evaluate the coefficients:
  for (int ell=1;ell<N;ell++) {
    for (int j=0;j<2*N-2*ell;j++) {e[j] = qold[j+1] -
        qold[j] + eold[j+1];}
    for (int j=0;j<2*N-2*ell-1;j++) {q[j] = e[j+1]/e[j]*
        qold[j+1];}
    for (int j=0;j<2*N-2*ell;j++) {eold[j] = e[j];}
    for (int j=0;j<2*N-2*ell-1;j++) {qold[j] = q[j];}
    cf[2*ell] = -e[0];
    cf[2*ell+1] = -q[0];
  }
```

```
    // Evaluate the continued fraction:
    double di= 0;
    for (int i=cf.size()-1;i>0;i--) {
      double ai=cf[i]*x;
      double bi=1;
      di = ai/(bi+di);
    }
    return cf[0]/(1+di);
}
```

A useful illustration of the application of these techniques is provided by the problem of diagonalizing a tridiagonal matrix. We shall make this explicit by using the notation of Eq. 4.26:

$$\begin{pmatrix} \alpha_1 & \beta_1 & & & \\ \beta_1 & \alpha_2 & \beta_2 & & \\ & \beta_2 & \alpha_3 & \beta_3 & \\ & & & \beta_{N-1} & \alpha_N \end{pmatrix} \vec{x} = \lambda \vec{x}. \tag{4.26}$$

The first of these equations is

$$\alpha_1 x_1 + \beta_1 x_2 = \lambda x_1, \tag{4.27}$$

with a similar equation for the last row. The remaining equations read:

$$\beta_{n-1} x_{n-1} + \alpha_n x_n + \beta_n x_{n+1} = \lambda x_n. \tag{4.28}$$

Eq. 4.28 is a linear three-term recurrence relation. Our strategy will be to convert this to a continued fraction, convert that to a Padé approximant, and then solve the resulting secular equation with bisection. The conversion to continued fraction can be achieved with a trick: divide Eq. 4.28 by x_n, solve for x_{n-1}/x_n, and invert. The result is:

$$\frac{x_n}{x_{n-1}} = \frac{\beta_{n-1}}{\lambda - \alpha_n - \beta_n \frac{x_{n+1}}{x_n}}$$

Set $n = 2$ and use the previous equation repeatedly:

$$\frac{x_2}{x_1} = \frac{\beta_1}{\lambda - \alpha_2 - \beta_2 \frac{x_3}{x_2}} = \frac{\beta_1}{\lambda - \alpha_2 - \cfrac{\beta_2^2}{\lambda - \alpha_3 - \cfrac{\beta_3}{\cdots}}}.$$

Finally, equate this to Eq. 4.27 to obtain the eigenvalue equation:

$$\lambda = \alpha_1 + \cfrac{\beta_1^2}{\lambda - \alpha_2 - \cfrac{\beta_2^2}{\lambda - \alpha_3 - \cfrac{\beta_3^2}{\dots}}}. \qquad (4.29)$$

This can be reduced to a polynomial equation in λ by employing Eqs. 4.23. It is wise to cross multiply by $B(\lambda)$ at this stage. Finally, eigenvalues can be obtained by determining the roots of the resulting secular equation.

4.6 Exercises

1. [T] Show that the natural cubic spline connecting two points (x_0, y_0) and (x_1, y_1) is a straight line.
2. The function $y = \sqrt{1 - x^2}$ for $-1 < x < 1$ is a semicircle in the upper half-plane. The points $x = \cos\theta$, $y = \sin\theta$ lie on this semicircle. Using x,y for $\theta = j\pi/6, j = \{0, 1, 2, 3, 4, 5, 6\}$ as interpolation points, plot the Lagrange interpolating polynomial and the cubic spline polynomial which interpolate these points, together with the original function.
3. The P1 bus in the city of Pittsburgh leaves the port authority in the city center every half hour, and more frequently during rush hour. It serves Penn, Heron, Negley, East Liberty, Homewood station, Wilkinsburg, Hamnet, Roslyn and Swissvale stations along the Martin Luther King Jr. east busway.

   a) The directory SKELETONS/INTERPOLATION/P1 contains the skeleton of a program for you to fill out. The skeleton creates a PlotView and co-displays it with a map of the P1 bus route. Check this package out, build it, and run it.
   b) By trial and error, mark (using a PlotPoint object, in the QatPlotting package) the position of each station in the above list in pixel units on the plot.
   c) Interpolate the bus route between these stations. You must omit the last two from the interpolation if you use Cartesian coordinates (why?). Plot the interpolated trajectory in the PlotView of the skeleton program.

4. In the previous problem, you can extend the interpolation to include the last two points by choosing a convenient point on the concave side of the bus route as an origin and reformulating the problem in polar coordinates. After interpolating the function in polar coordinates, the class PlotOrbit (in the QatPlotting library) can be used. Carry out the analysis in this way and plot the trajectory in the PlotView as before.
5. Chateau Lafitte Rothschild, one of the most renowned wines in the world, is a first growth premier grand cru classé from the original classification in 1855. During the early years of the 21st century its price increased spectacularly. In 2007, the average

price of a bottle was $346; in 2008 it had risen to $654; in 2009, it was $724; in 2010, $873; and in 2011, $1410. Plot the price as a function of year. What is your estimate of one bottle of Chateau Lafitte Rothshild today? How much money would you make if you bought one bottle now, and resold it one year later?

6. Interpolation in Higher Dimensions.
   In this exercise you will interpolate a color value over a surface. Color values have four components (red, green, blue, and the transparency, alpha) which vary on the range [0,1]. The surface is parameterized by x and y, also taking values on the range [0,1]. The directory SKELETONS/CH4/INTENSITY contains a program to display a colored blob and to interactively set the interpolation points on along a two-dimensional grid. This program uses a class Interpolator, whose code (.h and .cpp files) in the src directory. This class has one major defect:

   RBGAlphaData Interpolator::interpolate(double x,double y);

   contains only a nonworking implementation! Your task is to replace the bogus implementation with a method which works. To do this: modify the code that you will find in Interpolator.cpp and recompile. You need not touch any other compilation unit in the directory.

7. Data on the dispersion of light in water is given in Table 4.3. Interpolate a smooth function for these data. Plot the function together with the input data.

8. Figure 4.6 is a plot from an measurements of the response of the human retina carried out in the last century (Bowmaker, 1980). There are four types of photoreceptors in the human retina, called *rods*, and three types of *cones* called red, green, and blue cones. The rods have a response peaking at 498 nm, the red, green, and blue cones have responses peaking at 564, 534, and 420 nm, respectively. Color perception is carried out by the rods. Suppose that it is desired to import these curves into a calculation.

   a) Using any available computer program, measure the pixels coordinate of the center of each dot and convert that to a wavelength and an absorbance.
   b) Write functions which interpolate these data between 400–700 nm, for each type of photoreceptor, and plot these functions together as in the figure.

9. Use the result of the previous exercise to compute the red/green/blue response to monochromatic light of a given wavelength, and from there generate and plot a continuous band of color from $\lambda = 400$ nm to $\lambda = 700$ nm. To do this you could use the classes QImage and QLabel from the Qt library. The first of these classes allows you to manipulate individual pixels, the second allows you to display the result.

10. Consider the Taylor series of:

    $$f(x) = \frac{1}{(1+x)(2+x)}.$$

    How many terms are needed in the sum to obtain $f(0.99)$ to six decimal places?

Table 4.3 *Index of refraction of water vs. wavelength. Source: Hale (1973).*

Wavelength (nm)	Index of refraction
200	1.396
250	1.362
300	1.349
350	1.343
400	1.339
450	1.337
500	1.335
550	1.333
600	1.332
650	1.331
700	1.331
750	1.330
800	1.329
850	1.329
900	1.328
950	1.327
1000	1.327

11. Verify that the Shanks transformation works exactly when summing
$$\sum_i (-x)^i.$$

12. Evaluate
$$\sum_{n=0}^{\infty} \frac{(-1)^n}{\sqrt{n}}$$
to seven digits.

13. Evaluate the following continued fraction representation of π using P_{NN} Padé approximants. Plot the error vs. N up to $N = 30$. Comment on your plot. You may use

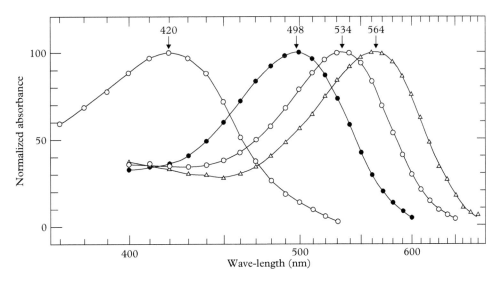

Figure 4.6 *Response curves of photoreceptors in the human retina. From Bowmaker (1980) ©1980 The physiological society. Reprinted with permission.*

$$\frac{\pi}{4} = \cfrac{1}{1+\cfrac{1^2}{3+\cfrac{2^2}{5+\cfrac{3^2}{7+\ldots}}}}$$

14. The following eigenvalue problem arises in solving SU(2) lattice gauge theory (see Chap. 24):

$$(x^2 - 1)\psi'' + 3x\psi' + \beta^2(2 - 2x - \frac{aE}{\beta})\psi = 0$$

where a and β are constants and E is to be solved for. Obtain the lowest 8 eigenvalues by

(i) expanding ψ in Chebyshev polynomials of the second kind;
(ii) establishing a tri-diagonal form for the eigenvalue problem;
(iii) converting this to a continued fraction;
(iv) solving this using Padé approximants and bisection root finding.

Set $a = \beta = 1$.

You may find the following properties of Chebyshev polynomials useful:

$$U_{n+1}(x) = 2xU_n(x) - U_{n-1}(x),$$

$$\int_{-1}^{1} \sqrt{1 - x^2}\, U_n U_m\, dx = \frac{\pi}{2}\delta_{nm},$$

and

$$(1 - x^2)U_n'' - 3xU_n' + n(n+2)U_n = 0.$$

15. Use the quotient-difference algorithm and Padé approximants to approximate $\exp(-x)$. Plot the relative error vs. x for the 5th and 10th approximants.
16. Plot $\exp(-10)$ obtained from $P_{NN}(x)$ vs. N.
17. [T] Determine the coefficients of the continued fraction for

    (i) $\exp(z)$.
    (ii) $\int_0^\infty dt \, \frac{\exp(-t)}{1+zt}$
    (iii) $\log(1+z)/z$.

BIBLIOGRAPHY

Bender, C. M. and S. A Orszag (1999). *Advanced Mathematical Methods for Scientists and Engineers.* Springer-Verlag.

Bowmaker, J. K. and H. J. A. Dartnall (1980). *Visual Pigments of Rods and Cones in the Human Retina.* J. Physiol. **298**, 501–511.

Davis, P. J. and P. Rabinowitz (1975). *Methods of Numerical Quadrature.* Academic Press.

Hale, G. M. and M. R. Querry (1973). *Optical constants of water in the 200-nm to 200-μm wavelength region.* Appl. Opt. 12, 555–563.

Press, W., S. Teukolsky, W. Vetterling, and B. Flannery (2007). *Numerical Recipes, the Art of Scientific Computing.* 3rd ed. Cambridge University Press.

Szego, Gabor (1939). *Orthogonal Polynomials.* Colloquium Publications–The American Mathematical Society V23.

5
Numerical quadrature

5.1	Some example problems	108
	5.1.1 One-dimensional periodic motion	108
	5.1.2 Quantization of energy	110
	5.1.3 Two body central force problems	111
	5.1.4 Quantum mechanical tunneling	112
	5.1.5 Moments of distributions	113
	5.1.6 Integrals of statistical mechanics	115
5.2	Quadrature formulae	118
	5.2.1 Accuracy and convergence rate	124
5.3	Speedups and convergence tests	126
5.4	Arbitrary abscissas	128
5.5	Optimal abscissas	129
5.6	Gaussian quadrature	132
5.7	Obtaining the abscissas	135
	5.7.1 Implementation notes	137
5.8	Infinite range integrals	139
5.9	Singular integrands	140
5.10	Multidimensional integrals	141
5.11	A note on nondimensionalization	142
	5.11.1 Compton scattering	142
	5.11.2 Particle in a finite one-dimensional well	143
	5.11.3 Schrödinger equation for the hydrogen atom	144
5.12	Exercises	145
	Bibliography	150

Integrals fall into two categories: indefinite integrals and definite integrals. Both arise frequently in physics, and quite often cannot be computed analytically, either because the integrand is resistant to the host of specialized techniques that can be applied (change of variable, integration by parts, etc); or even because the integrand itself is defined by data and has no analytic form.

Numerical techniques exist to compute both indefinite and definite integrals. In either case we are given some function of a single variable, expressed mathematically as $f(x)$ and in code as `double f(double x)`, though functors (i.e. classes that behave like a function) may also be used. The goal of indefinite integration is to compute another function $F(x)$, called the antiderivative, such that

$$f(x) = \frac{dF(x)}{dx}. \tag{5.1}$$

This is the more common problem and arises particularly in the integration of classical equations of motion. It is discussed later, in Chapter 11, which deals with the solution of ordinary differential equations.

The goal of definite integration is to produce a single real number from $f(x)$ given an upper and lower bound a and b, namely the definite integral

$$I = \int_a^b f(x)\,dx, \tag{5.2}$$

equal to the "area under the curve".

Mathematically, the two concepts are related by the fundamental theorem of calculus, which states that

$$I = \int_a^b f(x)\,dx = F(b) - F(a). \tag{5.3}$$

Conceptually, given the definite integral I, one could consider it as a function of the upper limit b, such that $I(b)$ defines a function, the antiderivative; and likewise given an antiderivative $F(x)$, one could evaluate Eq. 5.3 to determine I for given values of a and b. Numerically, however, it is preferable (for performance reasons) to use dedicated techniques in either case. In this chapter we will survey some problems in physics requiring the numerical computation of definite integrals, known, alternatively, as "quadrature"–an ancient term derived from the effort to express the area of a circle in units of length squared. We will then develop the numerical techniques to address this type of problem.

5.1 Some example problems

Applications of numerical quadrature are frequent in the practice of experimental and theoretical physics. We choose a few examples from classical, quantum, and thermal physics for their simplicity and because they are broadly familiar to students of physics. Numerical solutions to some of these are the subject of the exercises at the end of the chapter.

5.1.1 One-dimensional periodic motion

Consider a classical particle free to move in one dimension; its position at time t is given by $x(t)$. It is subject to a conservative force described by a potential $V(x)$, and has a conserved energy E, a constant. The particle moves between two classical turning points $x = x_{min}$ and $x = x_{max}$, where $V(x) = E$. Let us start the particle at position x_{min}, at rest, and ask ourselves, what is the period T of the motion? This is the time it takes for the particle to move from $x_{min} \to x_{max}$ and then from $x_{max} \to x_{min}$, or twice the time it takes to move from $x_{min} \to x_{max}$. Conservation of energy tells us that:

$$E = \frac{1}{2}mv^2 + V(x) = \text{constant} \tag{5.4}$$

so:

$$v = \frac{dx}{dt} = \sqrt{\frac{2(E - V(x))}{m}} \tag{5.5}$$

The period is obtained by rearranging the equation,

$$dt = \sqrt{\frac{m}{2(E - V(x))}} dx \tag{5.6}$$

and integrating,

$$T = 2 \int_{x_{min}}^{x_{max}} \sqrt{\frac{m}{2(E - V(x))}} dx, \tag{5.7}$$

or alternatively,

$$T = \oint \sqrt{\frac{m}{2(E - V(x))}} dx. \tag{5.8}$$

For some simple cases, depending on the form of the potential $V(x)$, the integral can be performed analytically, but for the general case one must resort to numerical methods.

This simple treatment can be generalized to other one-dimensional periodic, conservative systems governed by a classical Hamiltonian $H = H(p, q)$ where p is a generalized momentum and q is a generalized coordinate (Goldstein, 2001). In that case, the time evolution of the generalized coordinate is given by:

$$\dot{q} = \frac{dq}{dt} = \frac{\partial H}{\partial p} \tag{5.9}$$

The trajectory through phase space is determined by this equation, and by two constants of the motion, which can be taken as the initial time, t_0, and the energy E. Integrating Eq. 5.9 gives

$$T = \oint \frac{dq}{\left(\frac{\partial H}{\partial p}\right)}$$

The denominator will be a function of both p and q, but if the system is conservative, then $H = E$, p can be eliminated,

$$\left(\frac{\partial H}{\partial p}\right)^{-1} = \frac{\partial p}{\partial E} \tag{5.10}$$

and

$$T = \oint \left(\frac{\partial p}{\partial E}\right) dq \tag{5.11}$$

where $\left(\frac{\partial p}{\partial E}\right)$ is to be expressed as a function of q and the constant energy E. The reader should check that this gives Eq. 5.8 when $H = p^2/2m + V(x)$.

5.1.2 Quantization of energy

Eq. 5.11 can be written as:

$$T = \frac{\partial}{\partial E} \oint p\, dq$$

The integral appearing in the equation,

$$\mathcal{J} \equiv \oint p\, dq \tag{5.12}$$

is called the *action*; it is a function that does not depend upon the time but only upon the energy E. We can rewrite the equation as:

$$T = \frac{\partial \mathcal{J}}{\partial E}. \tag{5.13}$$

In the days of the "old quantum theory", the action also played a prominent role in the determining the quantized energy levels of a system. The original Bohr-Sommerfeld quantization rule,

$$\mathcal{J} = nh \tag{5.14}$$

where $n = 1, 2, 3 \ldots$ applies to rotational motion. Planck's constant h was originally known as the *quantum of action*. A similar quantization rule can be obtained from the Schrödinger equation, using a method known as the WKB approximation (Gasiorowicz, 2003; Griffiths, 2005), a quasi-classical approximation in which a particle, moving in a slowly-varying potential, is viewed as slowing down in regions of space where the potential $V(x)$ is high; because the kinetic energy depends upon position, so does the momentum $p = \sqrt{2m(E - V(x))}$ and the de Broglie wavelength $\lambda = h/p$. For vibrational motion in one dimension, one can derive from the WKB approximation the quantization condition:

$$\mathcal{J} = (n + 1/2)h \tag{5.15}$$

The action variable $\mathcal{J}$ is a definite integral taken over one complete cycle of motion between two classical turning points–even though the wavefunction penetrates to some extent into the classically forbidden region. This action $\mathcal{J} = \mathcal{J}(E)$ therefore implicitly determines energy quantization. For a particle in one dimension confined to a region between two turning points x_{min} and x_{max}, the rule can be rewritten as

$$2 \int_{x_{min}}^{x_{max}} \sqrt{2m(E - V(x))} dx = (n + 1/2)h; \tag{5.16}$$

(the factor of two on the left hand side arises because the integral is to be taken over one complete cycle of the classical motion).

To find the quantized energy level for a particle of mass m in a potential $V(x)$, one first needs to determine the turning points from the condition that $V(x) = E$, which is an exercise in root-finding. The quantity on the left hand side can then be determined and solved for the variable n. By varying the energy until n takes on integer values 0, 1, 2, one can determine the "WKB eigenenergies", approximate solutions for the quantized energy levels.

A plot of action $\mathcal{J}(E)$ vs. energy E is extremely powerful and informative. The slope of this curve at any value of E determines the period of oscillation given E, for the classical problem; and the intersection of the curve $\mathcal{J}(E)$ with fixed values $(n + 1/2)h$ determines the quantized energies in the quantum mechanical problem. In the simple case of the harmonic oscillator, the relationship happens to be linear. The harmonic oscillator therefore has an amplitude independent of the energy (in the classical problem) and equally spaced energy levels (in the quantum mechanical problem).

5.1.3 Two body central force problems

In the previous two sections we've discussed the classical and quantum mechanical motion of a particle in a one-dimensional potential. The methods are also applicable to two-body central force problems. On one end of the distance scale, two mutually gravitating massive stars furnish an example of a classical system; on the other end of the scale, a bottom quark and anti-bottom quark held together by the strong interaction in a "bottomonium" system furnishes a quantum mechanical example.

The theoretical treatment of quantum and classical two-body central force problems is mathematically different, but the "bottom line" is the same: the center-of-mass of the system evolves independently as a free particle with mass $M = m_1 + m_2$, while the coordinates describing the relative motion, $r \equiv \vec{r}_1 - \vec{r}_2$, evolves as in an *equivalent one-body problem*. The equivalent one-body problem has the following features:

- The equivalent particle has a mass of $\mu \equiv \frac{m_1 m_2}{m_1 + m_2}$, called the *reduced mass*.
- The equivalent particle moves in a modified potential $V'(r) = V(r) + \frac{L^2}{2mr^2}$, the original potential $V(r)$ modified by an additional term $\frac{L^2}{2mr^2}$ called the centrifugal barrier, where L is the angular momentum. The angular momentum is conserved in the two-body central force problem. Classically, therefore, it is a constant of the motion, while, in quantum mechanics it is quantized according to $L^2 = l(l+1)\hbar^2$, for $l = 0, 1, 2, \ldots$.

These observations allow us to include three-dimensional two-body central force problems along with one-dimensional classical and quantum mechanics problems in the set of applications of numerical quadrature. In the context of the quantization rules previously discussed, a detailed treatment of the WKB approximation (Quigg, 1979) leads to a modification of the quantization rule, Eq. 5.16. For the two-body central force problem, the rules are:

$$\int_0^{x_{max}} \sqrt{2\mu(E - V'(x))}\,dx = (n + 3/4)\pi\hbar \qquad x_{min} = 0,$$

$$\int_{x_{min}}^{x_{max}} \sqrt{2\mu(E - V'(x))}\,dx = (n + 1/2)\pi\hbar \qquad x_{min} > 0$$

where x_{max} and x_{min} are the classical turning points, and $n = 0, 1, 2, \ldots$

5.1.4 Quantum mechanical tunneling

We have mentioned already the WKB approximation; for bound states it produces a quantization rule that approximates (and sometimes, fortuitously, even gives exact solutions) to the quantized energy levels; and in addition to that it can be used to compute quantum mechanical transmission through potential barriers, giving us yet another use case for numerical quadrature.

The quantity that appears in WKB approximations to solutions of transmission problems is:

$$\tau \equiv \exp \int_{x_{min}}^{x_{max}} \kappa\, dx \qquad (5.17)$$

where $\kappa = \sqrt{2m(V(x) - E)/\hbar^2}$, and $x_{min} < x < x_{max}$ defines the classically forbidden region, or barrier; over this interval, $V(x) > E$ and κ is therefore real. It can be shown that the transmission coefficient through the barrier is:

$$T = \frac{1}{r^2} \tag{5.18}$$

and the reflection coefficient is $R = 1 - T$.

5.1.5 Moments of distributions

Distributions of mass, electric charge, electric current, and probability arise frequently in physics. From these distributions one can extract *moments*, which are real numbers characterizing important features of the shape of the distribution. Common examples are the moment of inertia which abstracts important features of a mass distribution; the electric or magnetic monopole[1], dipole, and quadrupole moments describe charge distributions and the important features of the field they create; quantities such as the mean, variance, skewness and kurtosis are commonly used to describe features of probability distributions.

Moments of probability distributions

A probability density function is a positive definite function $\rho(x)$ normalized such that:

$$\int \rho(x)dx = 1 \tag{5.19}$$

where the integration is carried out over the entire domain of the random variable x, a continuous random variable. The probability that the variable x takes a value between a and b is given by:

$$P[x_{min} \le x \le x_{max}] = \int_{x_{min}}^{x_{max}} \rho(x)dx \tag{5.20}$$

Moments of the distribution characterize its shape. The *mean* of the distribution is:

$$\bar{x} = \int \rho(x)x\,dx; \tag{5.21}$$

the *variance* is:

$$\sigma^2(x) = \int \rho(x)(x - \bar{x})^2 dx; \tag{5.22}$$

---

[1] Magnetic monopoles have been sought but not found.

the *skewness* is:

$$\gamma_1(x) = \int \rho(x) \left(\frac{(x-\bar{x})}{\sigma}\right)^3 dx; \qquad (5.23)$$

and the *kurtosis* is

$$\beta_2(x) = \int \rho(x) \left(\frac{(x-\bar{x})}{\sigma}\right)^4 dx. \qquad (5.24)$$

Mass Distributions and The Inertia Tensor

The *inertia tensor* $\mathbf{I}$ is a 3×3 real, symmetric tensor characterizing the properties of a rigid body under rotation about a point fixed with respect to the body. It relates the rotation of the body to its angular momentum and kinetic energy. The rotation is described by a vector $\vec{\omega}$, whose magnitude is the rate of rotation, and whose direction is chosen to lie along the axis of rotation. The angular momentum $\vec{L}$ is given by:

$$\vec{L} = \mathbf{I} \cdot \vec{\omega} \qquad (5.25)$$

while the kinetic energy K is:

$$K = \frac{1}{2} \vec{\omega}^T \cdot \mathbf{I} \cdot \vec{\omega}. \qquad (5.26)$$

A coordinate system is chosen so that the fixed point of rotation lies at the origin. The nine components of the inertia tensor in this coordinate system are the integrals:

$$(\mathbf{I})_{ij} = \int \rho(\vec{r})(\delta_{ij} r^2 - r_i r_j) dV. \qquad (5.27)$$

where $\vec{r}$ is the position of a mass point in the body, and $\rho(\vec{r})$ is the mass density at that point.

If the axis of rotation is along the unit vector $\hat{n}$ so that $\vec{\omega} = \omega \hat{n}$, then we have:

$$K = \frac{\omega^2}{2} (\hat{n}^T \cdot \mathbf{I} \cdot \hat{n}), \qquad (5.28)$$

and the quantity in parenthesis

$$I = \hat{n}^T \cdot \mathbf{I} \cdot \hat{n} \qquad (5.29)$$

is called the *moment of inertia* about $\hat{n}$. Three definite volume integrals need to be done to compute the moment of inertia about a specific axis of rotation. In many cases, an axis of symmetry can be found and happens to be the axis of rotation; in such circumstances only one of the three quantities is nonzero.

Multipole moments of charge distributions

The *multipole expansion* (Jackson, 1998) is a systematic approach to describing the electric field produced by a localized charge distribution $\rho(\vec{r})$. The scalar potential Φ produced by such a potential is given by the integral:

$$\Phi(\vec{r}) = \int \frac{\rho(\vec{r}')dV'}{|\vec{r}' - \vec{r}|}. \tag{5.30}$$

The numerator can be expanded as follows:

$$\frac{1}{|\vec{r}' - \vec{r}|} = 4\pi \sum_{l=0}^{\infty} \sum_{m=-l}^{l} \frac{1}{2l+1} \frac{r'^l}{r^{l+1}} Y_l^{m*}(\theta', \phi') Y_l^m(\theta, \phi) \tag{5.31}$$

where θ, ϕ and θ', ϕ' are the polar and azimuthal angles of $\vec{r}$ and $\vec{r}'$, respectively. Substituting into Eq. 5.30 and interchanging the order of summation and integration, we have

$$\Phi(\vec{r}) = 4\pi \sum_{l=0}^{\infty} \sum_{m=-l}^{l} \frac{1}{2l+1} \left[\int \rho(\vec{r}') r'^l Y_l^{m*}(\theta', \phi') dV' \right] \frac{Y_l^m(\theta, \phi)}{r^{l+1}}$$

$$= \sum_{l=0}^{\infty} \sum_{m=-l}^{l} q_{lm} \frac{4\pi}{2l+1} \frac{Y_l^m(\theta, \phi)}{r^{l+1}}$$

where the coefficients

$$q_{lm} \equiv \int \rho(\vec{r}') r'^l Y_l^{m*}(\theta', \phi') dV' \tag{5.32}$$

are called the multipole moment of the charge distribution. These coefficients form an infinite set, but usually one is only interested in the first few, since the potential generated by a multipole moment of order l falls off as r^{l+1}. The first few multipole moments are referred to as the monopole ($l = 0$), dipole ($l = 1$), quadrupole ($l = 2$), and sextupole($l = 3$). These and higher order multipole moments can be determined by integrating the charge distribution, numerically if necessary.

5.1.6 Integrals of statistical mechanics

A quantum gas is a gas of identical particles at low temperature, high density, or both. A volume containing the particles has a set of orbitals, which in the simplest case, in which interactions between the particles are absent or approximated away, are those of a particle-in-a-box, i.e.

$$\epsilon_n = \frac{n^2 \pi^2 \hbar^2}{2mL^2}. \tag{5.33}$$

In quantum statistical mechanics, a sharp difference in macroscopic behavior results from the spin quantum number of the particles within the gas, according to the spin-statistics theorem. A system comprised of *fermions*, which are particles with half-integer spin including protons, neutrons, and electrons, has a wave function that is antisymmetric under the interchange of any two particles. A system comprised of *bosons*, such as He^4 atoms, is symmetric. Fermions obey the Pauli exclusion principle.

Real systems are often approximated quite crudely as a noninteracting quantum gas. Electrons in a metal, electrons in a white dwarf star, or neutrons in a neutron star can be treated as a Fermi gas. Bose gases are rarer in nature; liquid He^4 is a superfluid below a critical temperature of 2.18 K, but is neither noninteracting nor a gas. Pairs of conduction electrons in certain metals form a bosonic condensate of "Cooper pairs" at low temperature, which gives rise to superconductivity. Atomic systems consisting of up to a few hundred thousand atoms behave like a Bose gas, and in particular exhibit the phenomenon of *Bose-Einstein condensation*, first observed in 1995 (Jin, 1996), and which earned a Nobel prize for its discoverers, Carl Wieman, Eric Cornell, and Wolfgang Ketterle.

The treatment of Fermi and Bose gases, which arises in elementary statistical mechanics (Kittel, 1980), is a problem that leads to an integral which resists an analytic solution. We model a quantum gas as a set of noninteracting fermions, or bosons in a box of length L and volume $V = L^3$. The mean number of particles in a given orbital, or *occupancy*, is given by a distribution function,

$$\bar{n}(\epsilon_i) = \frac{1}{\exp\left((\epsilon_i - \mu)/\tau\right) \pm 1} \tag{5.34}$$

where the upper sign applies to a Bose gas and the lower sign applies to a Fermi gas.

This distribution is governed by two parameters: $\tau = k_B T$ is the temperature in energy units, where T is in Kelvin and $k_B = 1.381 \times 10^{-23}$ J/K is the Boltzmann constant, and μ, which is the chemical potential. The chemical potential (unlike the temperature) is not an external parameter that one can control: it is determined implicitly by the density of the gas and the temperature. Our task is to find its value, which, once determined, defines the distribution function.

We note that the total number of particles in the box is given by:

$$N = \sum_i \bar{n}(\epsilon_i) = \sum_i \frac{1}{\exp\left((\epsilon_i - \mu)/\tau\right) \pm 1} \tag{5.35}$$

The sum can be approximated by an integral, whose integrand consists of two factors,

- The number of orbitals with energy between ϵ and $\epsilon+d\epsilon$. This factor is called $D(\epsilon)$, the *density of states*.
- The mean number of particles in an orbital with energy ϵ.

The former is worked out in many elementary texts:

$$D(\epsilon) = sV \frac{(2m/\hbar^2)^{3/2}}{4\pi^2} \epsilon^{1/2} \tag{5.36}$$

(where s is a constant spin-degeneracy factor) while the latter is given by Eq. 5.34. Combining these two factors gives us

$$N = \int_0^\infty D(\epsilon)\bar{n}(\epsilon)d\epsilon$$
$$= sV \frac{(2m/\hbar^2)^{3/2}}{4\pi^2} \int_0^\infty \frac{\epsilon^{1/2}}{\exp\left((\epsilon-\mu)/\tau\right) \pm 1} d\epsilon.$$

This approximation is adequate for a Fermi gas, but for a Bose gas it fails because the density-of-states rolls off to zero while the distribution function $\bar{n}(\epsilon)$ is growing large. This cannot happen in a Fermi gas because of the exclusion principle. In fact there is one state with $\epsilon = 0$ (s states if the spin states are s-fold degenerate)–the ground state, and for Bose gases at low enough temperatures, where Bose-Einstein condensation occurs, it can contain most of the particles.

We remedy this by counting the particles in the ground state $N_0 = 1/(\exp(-\mu/\tau)-1)$ explicitly and adding them back into the approximation:

$$N = \frac{s}{\exp-(\mu/\tau) - 1}$$
$$+ sV \frac{(2m/\hbar^2)^{3/2}}{4\pi^2} \int_0^\infty \frac{\epsilon^{1/2}}{\exp\left((\epsilon-\mu)/\tau\right) - 1} d\epsilon \quad \text{(Bose gas)}$$
$$N = sV \frac{(2m/\hbar^2)^{3/2}}{4\pi^2} \int_0^\infty \frac{\epsilon^{1/2}}{\exp\left((\epsilon-\mu)/\tau\right) + 1} d\epsilon \quad \text{(Fermi gas)}. \tag{5.37}$$

These expressions can be written in term of rescaled dimensionless variables $x = \epsilon/\tau$ and $\lambda = \exp(\mu/\tau)$:

$$N = \frac{s}{\lambda^{-1} - 1} + \frac{2s}{\pi^{1/2}} sV \left[\frac{m\tau}{2\pi\hbar^2}\right]^{3/2} \int_0^\infty \frac{x^{1/2}}{\lambda^{-1}e^x - 1} dx \quad \text{(Bose gas)}$$
$$N = \frac{2s}{\pi^{1/2}} sV \left[\frac{m\tau}{2\pi\hbar^2}\right]^{3/2} \int_0^\infty \frac{x^{1/2}}{\lambda^{-1}e^x + 1} dx \quad \text{(Fermi gas)}. \tag{5.38}$$

The quantity $V_q \equiv \left[\frac{2\pi\hbar^2}{m\tau}\right]^{3/2}$, which clearly depends upon temperature τ, is called the quantum volume. In terms of the rescaled dimensionless variable $v \equiv V/V_q$,

$$N = \frac{s}{\lambda^{-1} - 1} + \frac{2sv}{\pi^{1/2}} \int_0^\infty \frac{x^{1/2}}{\lambda^{-1}e^x - 1} dx \quad \text{(Bose Gas)}$$

$$N = \frac{2sv}{\pi^{1/2}} \int_0^\infty \frac{x^{1/2}}{\lambda^{-1}e^x + 1} dx \quad \text{(Fermi Gas)}. \tag{5.39}$$

At fixed temperature, particle number, and volume, and known particle mass, Eq. 5.39 determines λ and therefore μ. However, since the integral cannot be performed analytically, no closed form expression can be obtained, and one has recourse to further approximation, or to a numerical solution.

5.2 Quadrature formulae

Given a function $f(x)$ and two points a and b along the x-axis, we want to compute, numerically, the value of integral $\int_a^b f(x)dx$. This is equivalent to finding the area under the curve. A simple approach is to divide the interval $[a, b]$ up into small subintervals and then to approximate each subinterval with a shape of known area: a rectangle, a trapezoid, or some other shape. Some of these are illustrated in Figure 5.1. Using n rectangles with a base of equal size, and setting the height of the rectangle to the value of the function at the midpoint of the interval, will give us an approximation called the ***midpoint rule***, $M_n(f)$. It is given by the Riemann sum:

$$M_n(f) = \sum_{i=1}^n hf(x_i)$$

where $h = (b-a)/n$, and $x_i = a + (i - 1/2)h$ is the abscissa of the midpoint of the ith rectangle, as illustrated in Figure 5.1 (lower left). The true value of the integral is the limit:

$$\int_a^b f(x)dx = \lim_{n \to \infty} M_n(f)$$

The computer cannot take the limit, of course, but $M_n(f)$ can be taken as an approximation, the quality of which depends upon n as well as the function being integrated. The convergence of this approximation is quite slow for most functions, as we shall see. Two other quadrature rules, the left-rectangular and right-rectangular rules, which come from the diagrams in the top row of Figure 5.1, are even worse.

We can view $M_n(f)$ in another way by writing it as:

$$M_n(f) = (b-a)\left[\left(\sum_{i=0}^{n} f(x_i)\right)/n\right]$$
$$\equiv (b-a)\bar{f} \qquad (5.40)$$

where $\bar{f}$ denotes the mean value. This emphasizes that the sum is the product of the length of the interval $[a, b]$ times an average value of the function over that interval.

To obtain another important approximation, we can use trapezoids within each subinterval rather than rectangles, as in Figure 5.1 (lower right). Doing so defines the

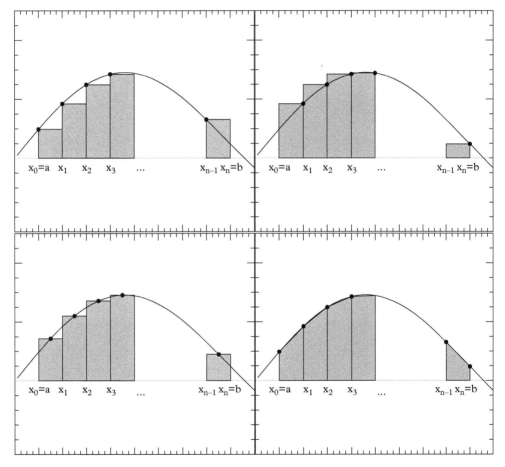

Figure 5.1 *Four plots illustrating quadrature rules. Left and right rectangle rules (top row) and the midpoint rule (bottom left) are obtained by using a set of rectangles to approximate the area of the curve. The trapezoid rule (bottom right) uses trapezoids.*

trapezoid rule, $T_n(f)$. The area of the first trapezoid is $\frac{1}{2}[f(x_0) + f(x_1)](x_1 - x_0)$. Assuming equally spaced points such that $x_{i+1} - x_i = h$, the area under the curve is approximated by

$$T_n(f) = h\left(\frac{f_0}{2} + f_1 + f_2 \ldots + \frac{f_n}{2}\right)$$

$$= \sum_{i=0}^{n} w_i \cdot f(x_i)$$

where $h = (b-a)/n$, $x_i = a + ih$, $w_0 = w_n = \frac{(b-a)}{2n}$, and $w_i = \frac{(b-a)}{n}$ for $i \neq 0, n$. Since

$$\sum_{i=0}^{n} w_i = (b-a),$$

one can also express this as:

$$T_n(f) = \sum_{i=0}^{n} w_i \cdot f(x_i) \frac{\sum_{i=0}^{n} w_i}{\sum_{i=0}^{n} w_i}$$

$$= (b-a) \frac{\sum_{i=0}^{n} w_i f_i}{\sum_{i=0}^{n} w_i}$$

$$= (b-a)\bar{f}.$$

With the trapezoid rule the integral is the the product of the length of the interval $[a, b]$ times a *weighted* average value of the function.

Both approximations $M_n(f)$ and $T_n(f)$ are Riemann sums which can be put into the form of:

$$I_n(f) = \sum_{i=0}^{n} w_i \cdot f(x_i) \qquad (5.41)$$

with:

$$\sum_{i=0}^{n} w_i = (b-a).$$

There are many more classical quadrature formulae like this. Only the abscissas (i.e., the points x_i at which $f_i = f(x_i)$ is evaluated) and the weights, differ. The pertinent question is how the *performance* differs. One can easily imagine that the accuracy of the approximation improves with the number of intervals n in the approximation, converging to (actually, defining) the true value of the integral in the limit $n \to \infty$. The rate of convergence is an important distinguishing feature.

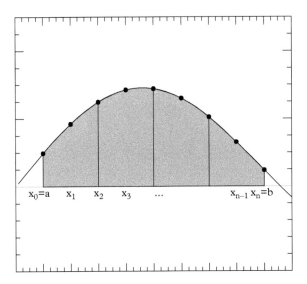

Figure 5.2 *Simpson's rule uses shapes delimited by a three point Lagrange interpolating polynomial to estimate the area under the curve.*

Before comparing convergence rates, we extend the set of approximate integration formulae, starting with an important formula called **Simpson's rule**. In Simpson's rule the subinterval $[x_0, x_2]$ is evaluated at three points, namely, the lower bound x_0, the midpoint $x_1 = (x_2 - x_0)/2$, and the upper bound x_2. The area of the subinterval is approximated by a shape bounded at the top end by Lagrange's three-point interpolating polynomial, which is of order two, as illustrated in Figure 5.2. Simpson's rule is a logical next step if one regards a trapezoid as shape delimited by a first-order polynomial, and a rectangle as a shape delimited by a zeroth-order polynomial.

We compute the area of the subinterval by situating it symmetrically about $x = 0$, such that $x_0 = -h$, $x_1 = 0$, and $x_2 = h$. The interpolating polynomial is:

$$y(x) = \frac{x(x-h)}{2h^2} f_0 - \frac{x^2 - h^2}{h^2} f_1 + \frac{x(x+h)}{2h^2} f_2$$

and its integral is:

$$\int_{-h}^{h} y(x)dx = h \left(\frac{f_0}{3} + \frac{4f_1}{3} + \frac{f_2}{3} \right).$$

From this we can build the extended formula by adding up a series of subintervals:

$$S_n(f) = h \cdot \left(\frac{f(x_0)}{3} + \frac{4f(x_1)}{3} + \frac{2f(x_2)}{3} + \frac{4f(x_3)}{3} + \ldots + \frac{4f(x_{n-1})}{3} + \frac{f(x_n)}{3} \right)$$

122 *Quadrature formulae*

with $h = (b-a)/n$ and $x_i = a + ih$. One identifies the weights as:

$$\{w_i\} = \left\{ \frac{(b-a)}{3n}, \frac{4(b-a)}{3n}, \frac{2(b-a)}{3n} \cdots \frac{4(b-a)}{3n}, \frac{(b-a)}{3n} \right\},$$

which again sum to $(b-a)$.

To test an integration algorithm it is useful and instructive to apply it to a function that can be integrated analytically; in such circumstances the error can actually be measured as a function of the step size (alternately, the number of intervals into which the integral is divided). Typically the algorithms converge as some power of the step size. The behavior can be seen by plotting the error versus step size on a log-log plot, as in Figure 5.3. Exercise 1 explores a few examples, and in a moment we attempt to explain the behavior that can be seen in Figure 5.3 and in those examples.

There are a few classes from the `QatGenericFunctions` library that can be used to integrate using a fixed-step-size classical integration formula. The class `SimpleIntegrator` (in `QatGenericFunctions`, namespace Genfun) provides a means of computing a definite integral using one of the classical quadrature formulae. `MidpointRule`, `TrapezoidRule`, and `SimpsonsRule`, all inheriting from `QuadratureRule`, can be used interchangeably with the `SimpleIntegrator`; these classes all specify abscissas and weights. Figure 5.4 shows a class tree diagram. The constructor to `SimpleIntegrator`,

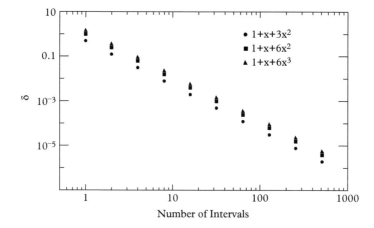

Figure 5.3 *The trapezoid rule is applied to three functions with nonvanishing second derivative. Plotted is the error vs. the number of intervals used in the calculation. The graph describes a power law with exponent -2 because the trapezoid rule is a second order quadrature rule.*

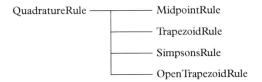

Figure 5.4 *Class tree diagram for* `QuadratureRule` *and subclasses. The diagram indicates that any of the subclasses may be used where a* `QuadratureRule` *is required.*

```
    SimpleIntegrator(double a,
                     double b,
                     const QuadratureRule    &rule =
                        TrapezoidRule(),
                     unsigned int            nIntervals=64);
```

requires a lower limit (a), an upper limit (b), a quadrature rule, and a number of intervals. The function call operator,

```
    virtual double operator () (GENFUNCTION function) const;
```

computes the estimate of the integral. An example of the usage is as follows:

```
#include "QatGenericFunctions/Sqrt.h"
#include "QatGenericFunctions/QuadratureRule.h"
#include "QatGenericFunctions/SimpleIntegrator.h"
.
.
.
{
  using namespace Genfun;
  GENFUNCTION f=Sqrt()
  MidpointRule rule;
  //
  // Integrate from 0 to 1 with 128 intervals.
  //
  SimpleIntegrator integrator(0,1,rule, 128);
  double integral = integrator(f);
}
```

The classes `MidpointRule`, `TrapezoidRule`, and `SimpsonsRule` are defined in the `QuadratureRule.h` header file. Other classical quadrature rules can be implemented as extensions to the base class `QuadratureRule`.

5.2.1 Accuracy and convergence rate

Now we return to the issue of the accuracy of these formulae and the rate of convergence. The first comment which is in order is that these properties depend as much on the integrand as on the approximation method; in particular the degree of smoothness of the integrand. A function with a continuous n^{th} derivative on the interval $[a, b]$ is said to belong to $C^n[a, b]$. An analytic function on $[a, b]$ has a convergent Taylor series expansion on that interval, and is therefore infinitely differentiable. A polynomial of degree n belongs to this class.

Any rule based on rectangles will integrate a zeroth order polynomial exactly. A rule based on trapezoids will integrate a first-order polynomial exactly, and a rule based upon a three point Lagrange interpolating function integrates a second order polynomial exactly. These occur because of the way in which the elementary inscribed shapes are chosen. The midpoint rule, however, is exact even for first-order polynomials, while Simpson's rule is exact even for third order polynomials. In both cases the extra order comes from a fortuitous choice of the abscissa. We shall return later to this point.

For functions of higher order, or having a higher-order expansion, the accuracy depends on the number of points and on the higher derivatives of the integrand. Formulae for the error in the approximation can be obtained by first expanding the function in a Taylor series and integrating term-by-term over a single subinterval. Consider a function $f(x)$ which is integrated between two points at $x = -h$ and $x = +h$. Expanding $f(x)$ in a Taylor series and integrating term-by-term gives us:

$$f(x) = f(0) + xf'(0) + \frac{x^2}{2}f''(x) + \frac{x^3}{6}f'''(0) + O(x^4)$$

$$\int_{-h/2}^{h/2} f(x)dx = hf(0) + \frac{h^3}{24}f''(0) + O(h^5)$$

$$= h\left(f(0) + \frac{h^2}{24}f''(0) + O(h^4)\right). \qquad (5.42)$$

The latter expression gives the actual integral of the function within a single subinterval. The integral $\int_a^b f(x)dx$ is the sum the contributions from each subinterval. There are three pieces:

1. $\sum_{i=0}^{n} hf(x_i) = M_n(f)$, which is the midpoint rule approximation.
2. $\sum_{i=0}^{n} \frac{h^3}{24}f''(x_i)$ which is the leading error in the midpoint rule approximation.
3. $O(h^4)$ which are the subleading errors.

The leading error is:

$$E_n(f) \equiv \sum_{i=1}^{n} \frac{h^3}{24}f''(x_i) = \sum_{i=1}^{n} \frac{1}{24}\left(\frac{a-b}{n}\right)^3 f''(x_i)$$

$$= \sum_{i=1}^{n} \frac{1}{24} \frac{(a-b)^3}{n^2} \frac{f''(x_i)}{n}$$

$$= \frac{1}{24} \frac{(a-b)^3}{n^2} \bar{f}''$$

where $\bar{f}''$ is an average value of the second derivative. It can be expressed as $f''(\xi)$ where $a < \xi < b$, so that the leading error estimate is

$$E_n(f) = \frac{1}{24} \frac{(a-b)^3}{n^2} f''(\xi) \qquad a < \xi < b. \tag{5.43}$$

This error approaches zero as $1/n^2$, so the approximation is said to have an order of convergence of two; more informally it is said to be a second-order quadrature formula.

A formula for the leading error in the trapezoid rule can be derived with the help of two Taylor series expansions for the value of the function at $x = \pm h/2$:

$$f\left(-\frac{h}{2}\right) = f(0) - \frac{h}{2}f'(0) + \frac{h^2}{8}f''(0) - \frac{h^3}{48}f'''(0) + O(h^4)$$

$$f\left(+\frac{h}{2}\right) = f(0) + \frac{h}{2}f'(0) + \frac{h^2}{8}f''(0) + \frac{h^3}{48}f'''(0) + O(h^4)$$

and their average:

$$\frac{1}{2}\left(f\left(+\frac{h}{2}\right) + f\left(-\frac{h}{2}\right)\right) = f(0) + \frac{h^2}{8}f''(0) + O(h^4).$$

Using this in Eq. 5.42 we obtain:

$$\int_{-h/2}^{h/2} f(x)dx = h\left[\frac{1}{2}\left(f\left(+\frac{h}{2}\right) + f\left(-\frac{h}{2}\right)\right) - \frac{h^2}{12}f''(0) + O(h^4)\right]$$

and a set of steps paralleling those from Eq. 5.43 to Eq. 5.43 shows that:

$$E_n(f) = -\frac{1}{12} \frac{(a-b)^3}{n^2} f''(\xi) \qquad a < \xi < b. \tag{5.44}$$

One can show in a similar way that Simpson's rule is fourth order: the fractional error converges to zero as $1/n^4$. Some of these results, which are taken from Davis (1975), are tabulated in Table 5.1, for formulae with equally spaced abscissas.

Note, finally, an important loophole: the derivatives that appear in Table 5.1 and the expressions above are required to exist over the full interval $[a, b]$. This means, for example that $M_n(f)$ is second order if $f \in C^2[a, b]$. The integral $\int_0^1 x^{-1/2} dx$ (which

Table 5.1 *Leading error terms in several quadrature formulae with equally space abscissas and n subintervals. In this table, $h = (b-a)/n$.*

Name	Abscissas	$i \in$	Weights	Error
Right Rect. $R_n(f)$	$x_i = a + ih$	$1..n$	$w_i = h$	$-\frac{(b-a)}{2n}f'(\xi)$
Left Rect. $\bar{R}_n(f)$	$x_i = a + ih$	$0..n-1$	$w_i = h$	$+\frac{(b-a)}{2n}f'(\xi)$
Midpoint $M_n(f)$	$x_i = a + (i-1/2)h$	$1..n$	$w_i = h$	$\frac{(b-a)^3}{24n^2}f''(\xi)$
Trapezoid $T_n(f)$	$x_i = a + ih$	$0..n$	$h/2, h, h...$	
			$...h, h, h/2$	$-\frac{(b-a)^3}{12n^2}f''(\xi)$
Simpson $S_n(f)$	$x_i = a + ih$	$0..n$	$h/3, 4h/3, 2h/3$	
			$..4h/3, 2h/3, 4/3..$	
			$..2h/3, 4h/3, h/3$	$-\frac{(b-a)^5}{2880n^4}f''''(\xi)$

cannot be integrated by the trapezoid rule because of the singular point at $x = 0$) has an integrand that does not belong to $C^2[0, 1]$, so convergence of the midpoint rule is worse than second order. This type of situation is more frequent than one might expect.

5.3 Speedups and convergence tests

For the trapezoid rule, the error goes as the *square* of the mesh spacing h. By doubling the number of intervals (one can add additional points midway between x_i and x_{i+1} in order to do this), the error decreases by a factor of 4, at leading order. The trapezoid rule sum computed with $2n$ intervals S_{2n} is thus better than the sum S_n computed with n intervals, but in addition to that it establishes a trend. When plotting the sum versus the square of the mesh spacing, as in Figure 5.5, one sees that successive refinements lie close to a straight line, converging finally towards S_∞.

The extrapolation to zero mesh spacing is simple:

$$3(S_{2n} - S_\infty) = (S_n - S_{2n})$$

or

$$S_\infty = \frac{4}{3}S_{2n} - \frac{1}{3}S_n$$

or

$$S_\infty = \frac{4h}{3}\left(\frac{f_0}{2} + f_1 + f_2 \ldots + f_{n-1} + \frac{f_n}{2}\right) - \frac{2h}{3}\left(\frac{f_0}{2} + f_2 + f_4 \ldots + f_{n-2} + \frac{f_n}{2}\right)$$

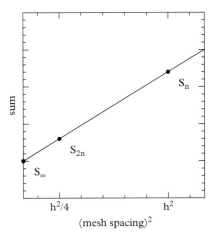

Figure 5.5 *A trapezoid rule sum is evaluated with n intervals, giving a result S_n, then the number of intervals is double by bisecting again each interval, giving a new result S_{2n}. The result is plotted vs. the (mesh spacing) squared, because the error at leading order is proportional to this quantity, so that continued reduction of the interval will give results lying along a straight line joining the two points.*

which is:

$$S_n = h\left(\frac{f_0}{3} + \frac{4f_1}{3} + \frac{2f_2}{3} + \frac{4f_3}{3} \ldots \frac{4f_{n-1}}{3} + \frac{f_n}{3}\right).$$

Which happens to be Simpson's rule. We can continue this indefinitely until convergence is reached. This is what is called **Romberg's method**.

In Romberg's method (Davis, 1975; Press, 2007), one starts with a single trapezoid covering the entire interval. One subdivides, keeping track of S_1 and S_2, calling the function to be integrated just for the new points. Polynomial extrapolation is called in order to extrapolate the result to an infinitely small grid spacing–as we have seen this is equivalent to Simpson's rule at the first step in the algorithm. One further subdivides, recalculates S_4, and re-extrapolates, this time using a three-point extrapolating polynomial. Each extrapolation is compared with the previous extrapolation. When the difference is within a preset tolerance, the method can be considered to have converged. In Romberg's method one obtains higher and higher order quadrature formulae while simultaneously reducing the mesh spacing, and the method converges surprisingly fast. When integrating a polynomial of order two, it converges to the exact result after just three function evaluations–an additional two are needed to detect convergence.

An implementation of Romberg's integration is provided in the `QatGeneric Functions` library by the class `RombergIntegrator` (in `QatGenericFunctions`, namespace Genfun). One instantiates a `Genfun::RombergIntegrator` using a constructor taking two arguments, the lower limit a and the upper limit b. The integral can be taken by calling `operator ()` giving the function to be integrated as an argument.

In the following example the routine is used to compute $\frac{1}{\pi}\int_0^\pi \sin^2(x)dx$, which evaluates to 0.5. A default tolerance of 10^{-6} on the fractional error is taken in this calculation. It can be overridden if desired.

```
#include "QatGenericFunctions/Sin.h"
#include "QatGenericFunctions/RombergIntegrator.h"
#include <iostream>
#include <cmath>   // Defines M_PI
int main(int argc, char **argv) {
  Genfun::Sin sin;
  Genfun::RombergIntegrator  Integrate(0, M_PI);
  double integral=Integrate(sin*sin/M_PI);
  std::cout << "Integral=" << integral << std::endl;
  std::cout << "Fraction deviation from true value " <<
      (integral-0.5)/0.5 << std::endl;
}
```

The program produces the following output:

```
Integral=0.5
Fraction deviation from true value -1.11022e-16
```

Indicating that the procedure has converged to machine precision.

5.4 Arbitrary abscissas

In the previous section we considered several approximations based upon Eq. 5.41, with equally spaced abscissas. In this section we extend our treatment to include the general case in which the x_i are arbitrary points lying in the interval $[a, b]$. We shall see that the additional freedom permits powerful extensions of the quadrature methods considered above. Our guiding principle will be the exact integration of polynomials of order n or lower, where n is the number of points. The resulting formulae can be applied either to subintervals or to entire intervals.

Let us demand that an n-point quadrature rule exactly integrates all polynomials of order n or less. This can be expressed as a condition on the weights:

$$\sum_{i=0}^{n-1} w_i f(x_i) = \int_a^b f(x)dx$$

for $f(x) = x^k$; $k = 0, 1, 2\ldots$. There are n equations in all, and also n weights w_i to determine. The equations are linear, so we will generally be able to solve them using matrix methods, e.g., using Eigen. For the relatively simple case of $n = 3$ they can be obtained by hand–we expect to recover Simpson's rule. We place our three points at $x_0 = a = -h$, $x_1 = 0$, and $x_2 = b = h$. There are three equations to solve, one for $f(x) = 1$, one for $f(x) = x$, and one for $f(x) = x^2$. The first is:

$$\sum_{i=0}^{2} w_i = (b - a)$$

or $w_0 + w_1 + w_2 = (b - a) = 2h$. This is the normalization equation, and it is always part of any set of equations that are derived with this method. The full set of equations is:

$$w_0 + w_1 + w_1 = 2h$$
$$-w_0 h + w_2 h = 0$$
$$w_0 h^2 + w_2 h^2 = \frac{2}{3} h^3$$

which can be solved to obtain $w_0 = \frac{1}{3}$, $w_1 = \frac{4}{3}$, and $w_2 = \frac{1}{3}$, which gives us Simpson's rule again (Gibbs, 2006).

A four-point quadrature formula for the interval $[a, b]$ with any four abscissas can be found from the equation:

$$\begin{pmatrix} 1 & 1 & 1 & 1 \\ x_0 & x_1 & x_2 & x_3 \\ x_0^2 & x_1^2 & x_2^2 & x_3^2 \\ x_0^3 & x_1^3 & x_2^3 & x_3^3 \end{pmatrix} \cdot \begin{pmatrix} w_0 \\ w_1 \\ w_2 \\ w_3 \end{pmatrix} = \begin{pmatrix} (b-a) \\ \frac{b^2-a^2}{2} \\ \frac{b^3-a^3}{3} \\ \frac{b^4-a^4}{4} \end{pmatrix} \quad (5.45)$$

The quantities a, b, x_0, x_1, x_2, and x_3 will generally be known. Once they are furnished the solution to this equation is quite trivial. The matrix on the left hand side is known as a ***Vandermonde matrix***. The procedure is useful for computing weights when the abscissas are specified in advance or fixed by some practical constraint. For example, the values of $f(x)$ could consist of measurements taken by some experimental apparatus at n fixed values of x.

5.5 Optimal abscissas

On the other hand, if the abscissas are not fixed we are at liberty to choose them and it is worth asking if the choice can be made in an optimal way. To simplify matters, we can take $b = -1$ and $a = 1$. Any proper integral can be transformed to this interval by a simple change of variable, so there is no loss of generality here.

The Legendre polynomials $P_l(x)$ play an important role in this discussion. They are defined on the interval $[a, b] = [-1, 1]$. The first four Legendre polynomials are shown in Figure 5.6. Important properties of the Legendre polynomials are that $P_n(x)$ has n

Optimal abscissas

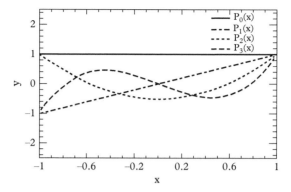

Figure 5.6 *The four lowest-order Legendre polynomials $P_0(x)$, $P_1(x)$, $P_2(x)$, and $P_3(x)$.*

zeros in the range $[-1, 1]$ and the set of Legendre polynomials of degree less than or equal to n forms a basis for polynomials of degree less than or equal to n. Lastly, the polynomials obey an orthogonality condition:

$$\int_{-1}^{1} P_l(x) P_{l'}(x) dx = \frac{2}{2l+1} \delta_{ll'} \tag{5.46}$$

and a recurrence relation:

$$P_{l+1}(x) = \frac{2l+1}{l+1} x P_l(x) - \frac{l}{l+1} P_{l-1}(x). \tag{5.47}$$

We continue to use the basic equation of approximate integration, Eq. 5.41, but we now require that it exactly integrate all Legendre polynomials up to order $n - 1$; this is completely equivalent to requiring that all monomials or polynomials up to order $n - 1$ be exactly integrated.

The integral of any function $f(x)$ can be written as:

$$\int_{-1}^{1} f(x) dx = \int_{-1}^{1} P_0(x) f(x) dx \tag{5.48}$$

(since $P_0(x) = 1$) and so the error is:

$$E_n(f) = \int_{-1}^{1} P_0(x) f(x) dx - I_n(f)$$

$$= \int_{-1}^{1} P_0(x) f(x) dx - \sum_{i=0}^{n-1} w_i f(x_i).$$

Substituting the functions $P_l(x)$ for $f(x)$ in this expression gives:

$$E_n(P_l(x)) = \int_{-1}^{1} P_0(x)P_l(x)dx - \sum_{i=0}^{n-1} w_i P_l(x_i)$$

$$= 2\delta_{l0} - \sum_{i=0}^{n-1} w_i P_l(x_i).$$

Requiring the error to be zero gives:

$$\sum_{i=0}^{n-1} w_i P_l(x_i) = 2\delta_{l0}. \tag{5.49}$$

This can be regarded as a matrix equation:

$$\begin{pmatrix} P_0(x_0) & P_0(x_1) & P_0(x_2) & P_0(x_3) & \cdots \\ P_1(x_0) & P_1(x_1) & P_1(x_2) & P_1(x_3) & \cdots \\ P_2(x_0) & P_2(x_1) & P_2(x_2) & P_2(x_3) & \cdots \\ P_3(x_0) & P_3(x_1) & P_3(x_2) & P_3(x_3) & \cdots \\ \cdot\cdot & & & & \\ \cdot\cdot & & & & \end{pmatrix} \cdot \begin{pmatrix} w_0 \\ w_1 \\ w_2 \\ w_3 \\ \cdot \\ \cdot \end{pmatrix} = \begin{pmatrix} 2 \\ 0 \\ 0 \\ 0 \\ \cdot \\ 0 \end{pmatrix}, \tag{5.50}$$

which can be solved to obtain the weights w_i once the abscissas x_i are known. The equation

$$\sum_{i=0}^{n-1} w_i P_l(x_i) = 0 \tag{5.51}$$

holds for $0 < l < n$ and implies exact integration for all polynomials of order greater than zero and less than n. Now we try choose the n abscissas x_i so that the equation is also satisfied for higher-order monomials as well. For $l = n$, it can be easily satisfied by choosing the x_i to be the zeros of $P_n(x)$; the weights w_i then follow from Eq. 5.50. The condition 5.51 is then true for $l = n$. We now check Eq. 5.51 with $l = n+1$ by evaluating:

$$\sum_{i=0}^{n-1} w_i P_{n+1}(x_i) = \sum_{i=0}^{n-1} w_i \left(\frac{2n+1}{n+1} x_i P_n(x_i) - \frac{n}{n+1} P_{n-1}(x_i) \right).$$

The first term vanishes because $P_n(x_i) = 0$, and the second term vanishes (if $n > 0$) because of condition 5.51, which was just shown to be valid for $l < 0 < n+1$. So the previous expression is zero and condition 5.51 is also valid for $l = n+1$. Repeating for $l = n+2$, we have

$$\sum_{i=0}^{n-1} w_i P_{n+2}(x_i) = \sum_{i=0}^{n-1} w_i \left(\frac{2n+3}{n+2} x_i P_{n+1}(x_i) - \frac{n+2}{n+3} P_n(x_i) \right)$$

$$= \sum_{i=0}^{n-1} w_i \frac{2n+3}{n+2} x_i P_{n+1}(x_i)$$

$$= \sum_{i=0}^{n-1} w_i \frac{2n+3}{n+2} x_i \left(\frac{2n+1}{n+1} x_i P_n(x_i) - \frac{n}{n+1} P_{n-1}(x_i) \right)$$

$$= \sum_{i=0}^{n-1} w_i \left(\frac{2n+3}{n+2} \frac{2n+1}{n+1} x_i^2 P_n(x_i) - \frac{2n+3}{n+2} \frac{n}{n+1} x_i P_{n-1}(x_i) \right)$$

$$= \sum_{i=0}^{n-1} w_i \left(\frac{2n+3}{n+2} \frac{2n+1}{n+1} x_i^2 P_n(x_i) \right.$$

$$\left. - \frac{2n+3}{n+2} \frac{n}{n+1} \left\{ \frac{n}{2n-1} P_n(x_i) + \frac{n-1}{2n-1} P_{n-2}(x_i) \right\} \right)$$

This will be zero, too, as long as the very last term in the expression has degree greater than or equal to zero. One can see by iterating this formula further that polynomials of decreasing degree appear in the final expression. When finally the minimum degree is zero, the argument fails and the expression is nonzero. This occurs for $l = 2n$. We conclude that condition 5.51 is valid for $0 < l < 2n$, and that therefore all polynomials of degree up to $2n - 1$ are exactly integrated. This is quite an exploit! The technique is called **Gauss-Legendre integration**. It makes the integration of functions *even more* efficient.

One can ask, is it really sensible to expect, for example, fourth-order accuracy in an integral by evaluating the integrand at only two points? The answer is "maybe". The foundation of this algorithm is that one can determine all that one needs to know about the function from probing it at a few places. This assumption depends on *how well the actual function is described by a fourth-order polynomial*. If the function is wildly different from the model, the integral can be far from the calculated value.

5.6 Gaussian quadrature

Gauss-Legendre integration is only one example of a set of techniques called **Gaussian quadrature** that provide very efficient approximations of definite integrals. To motivate the developments which follow, consider an integral of the form:

$$\int_{-\infty}^{\infty} e^{-x^2} f(x) \, dx$$

where $f(x)$ is a smooth function (e.g., a polynomial). This type of integral occurs frequently in problems involving the harmonic oscillator, for example. In the study of the quantum Coulomb problem, integrals having the form:

$$\int_0^\infty x^a e^{-x} f(x) dx$$

are common. These examples both have the form $\int W(x)f(x)dx$, where $W(x)$ is a weight function; the main restriction on the function is that it be positive-definite over the range of integration.

We seek an n-point integration formula:

$$\int W(x)f(x)dx \approx \sum_{i=0}^{n-1} w_i f(x_i)$$

which we require to be exact when $f(x)$ is a polynomial of degree $n-1$ or less. We will build such a formula using polynomials $p_n(x)$ of degree n, which are orthogonal under the inner product:

$$\langle f(x)|g(x)\rangle \equiv \int W(x)f(x)g(x)dx$$

where the integration is carried out over some specified range of x, which may be bounded or unbounded. These functions, like the Legendre polynomials, form a basis for their function space. Table 5.2 lists a few of them.

All of the polynomials in this table possess a recurence relation similar to that of Eq. 5.47, which can be put in the following form:

$$p_{n+1}(x) = (a_n x + b_n)p_n(x) - c_n p_{n-1}(x). \tag{5.52}$$

Table 5.2 *Common weight functions $W(x)$ and their corresponding orthogonal polynomials.*

Weight function, W	Interval	Orthogonal polynomial
1	$-1 < x < 1$	Legendre $P_l(x)$
$(1-x^2)^{-1/2}$	$-1 < x < 1$	Chebyshev $T_j(x)$
$(1-x^2)^{1/2}$	$-1 < x < 1$	Chebyshev $U_j(x)$
$x^\alpha e^{-x}$	$0 < x < \infty$	Associated Laguerre $L_j^\alpha(x)$
e^{-x^2}	$-\infty < x < \infty$	Hermite $H_j(x)$

The recurrence can be anchored by the definitions $p_0(x) = 1$ and $p_1(x)$, which in fact is the basis for most practical schemes for computing these functions. The functions so defined are orthogonal but not normalized, instead the normalization is given by a relation:

$$\int W(x)(p_n(x))^2 dx = N_n^2, \tag{5.53}$$

where N_n is a factor depending upon the type and order of the polynomial. The values of constants in the recurrence formula and the normalizing factors are given in Table 5.3. The polynomials $\tilde{p}_n(x) = p_n(x)/N_n$ are therefore, by construction, orthonormal.

We proceed as in Section 5.5 by noting that we can write:

$$\int_{-1}^{1} f(x)W(x)dx = \int_{-1}^{1} p_0(x)f(x)W(x)dx; \tag{5.54}$$

the error in the approximate integral, evaluated for one of the polynomials in the applicable set is:

$$E_n(p_l(x)) = \int p_0(x)p_l(x)W(x)dx - \sum_{i=0}^{n-1} w_i p_l(x)$$

$$= \delta_{l0} \cdot \int W(x)dx - \sum_{i=0}^{n-1} w_i p_l(x).$$

Requiring the error to be zero gives:

$$\sum_{i=0}^{n-1} w_i p_l(x) = \delta_{l0} \cdot \int W(x)dx \tag{5.55}$$

Table 5.3 *Recurrence coefficients and normalizing factors for common orthogonal polynomials. See Eq. 5.52.*

a_n	b_n	c_n	N_n^2	Orthogonal polynomial
$\frac{2n+1}{n+1}$	0	$\frac{n}{n+1}$	$\frac{2}{2n+1}$	Legendre $P_n(x)$
2	0	1	$\pi/2$ ($n \neq 0$)	Chebyshev $T_n(x)$
			π ($n = 0$)	
1 ($n=0$)	0	1	$\pi/2$	Chebyshev $U_n(x)$
2 ($n \neq 0$)				
$\frac{-1}{n+1}$	$\frac{2n+\alpha+1}{n+1}$	$\frac{n+\alpha}{n+1}$	$\frac{\Gamma(n+\alpha+1)}{n!}$	Associated Laguerre $L_n^\alpha(x)$
2	0	$2n$	$\sqrt{\pi}2^n n!$	Hermite $H_n(x)$

which can be used to solve for the weights, given the abscissas. The abscissas, for n-point Gaussian quadrature, are taken to be the zeros of $p_n(x)$, which must be determined through some means. The recurrence formula, Eq. 5.52 guarantees that the approximation integrates, exactly, all polynomials of degree less than $2n$, as we saw in the previous section.

The weights can then be determined, by solving the matrix equation:

$$\begin{pmatrix} p_0(x_0) & p_0(x_1) & p_0(x_2) & p_0(x_3) & \cdots \\ p_1(x_0) & p_1(x_1) & p_1(x_2) & p_1(x_3) & \cdots \\ p_2(x_0) & p_2(x_1) & p_2(x_2) & p_2(x_3) & \cdots \\ p_3(x_0) & p_3(x_1) & p_3(x_2) & p_3(x_3) & \cdots \\ \cdots \\ \cdots \end{pmatrix} \cdot \begin{pmatrix} w_0 \\ w_1 \\ w_2 \\ w_3 \\ \cdot \\ \cdot \end{pmatrix} = \begin{pmatrix} \int W(x) dx \\ 0 \\ 0 \\ 0 \\ \cdot \\ 0 \end{pmatrix} \quad (5.56)$$

Alternately, it can be shown (Szego, 1939) that

$$w_i = -\frac{\lambda_{n+1}}{\lambda_n} \frac{1}{\tilde{p}_{n+1}(x_i) \tilde{p}'_n(x_i)} \quad (5.57)$$

where λ_n is the coefficient of the leading term in $\tilde{p}_n$. This can be written in several ways, including

$$w_i = -a_n \frac{N_n}{N_{n+1}} \frac{1}{\tilde{p}_{n+1}(x_i) \tilde{p}'_n(x_i)} \quad (5.58)$$

with a_n and N_n as shown in Table 5.3 The roots and weights can be computed up-front, once and for all, or obtained from a table, i.e. most of the actual computation is in the initialization of the abscissas and weights; once these are known the actual integration involves only a simple weighted average, as with the classical formulae.

For the scheme to be effective, the function f should be smooth; which means generally that weighting function $W(x)$ should absorb most of the variation in the integrand. The technique has an added advantage when considering integrals with infinite upper or lower limits, which can be computed without introducing any arbitrary cutoffs.

5.7 Obtaining the abscissas

Gaussian integration resolves many problems in integration in an elegant way, but it leaves us with an unsolved problem: how to determine the abscissas (roots of the appropriate orthogonal polynomial), and how to determine the weights. A little experience with the Newton-Raphson method will discourage you from using this, as the dependence upon the initial guess makes the method too unreliable. The following simple method due to Golub and Welsh (1969) is based upon the three-term recurrence formula, Eq. 5.52.

Obtaining the abscissas

Divide both sides of Eq. 5.52 by the coefficient a_n:

$$\frac{1}{a_n}p_{n+1}(x) = \left(x + \frac{b_n}{a_n}\right)p_n(x) - \frac{c_n}{a_n}p_{n-1}(x) \tag{5.59}$$

Szego (1939, p. 42) shows that:

$$\frac{c_n}{a_n} = \frac{1}{a_{n-1}} \tag{5.60}$$

so we define:

$$\alpha_n \equiv \frac{1}{a_n} \qquad \beta_n \equiv -\frac{b_n}{a_n}. \tag{5.61}$$

Rewrite Eq. 5.59 as:

$$\alpha_n p_{n+1}(x) = (x - \beta_n)p_n(x) - \alpha_{n-1}p_{n-1}(x) \tag{5.62}$$

and rearrange:

$$\alpha_{n-1}p_{n-1}(x) + \beta_n p_n(x) + \alpha_n p_{n+1}(x) = x p_n(x). \tag{5.63}$$

To find the zeros of the orthogonal polynomial $P_N(x)$, we have to consider this set of equations for the polynomials $p_n(x)$ for $n = 0, 1, 2..N - 1$. This is a set of equations that can be expressed as a single matrix equation,

$$\mathbf{J} \cdot \mathbf{p} = x\mathbf{p} + \alpha_{N-1}\mathbf{q} \tag{5.64}$$

where:

$$\mathbf{J} = \begin{pmatrix} \beta_0 & \alpha_0 & 0 & 0 & . & 0 & 0 \\ \alpha_0 & \beta_1 & \alpha_1 & 0 & . & 0 & 0 \\ 0 & \alpha_1 & \beta_2 & \alpha_2 & . & 0 & 0 \\ 0 & 0 & \alpha_2 & \beta_2 & . & 0 & 0 \\ . & . & . & . & . & . & 0 \\ 0 & 0 & 0 & 0 & \alpha_{N-3} & \beta_{N-2} & \alpha_{N-2} \\ 0 & 0 & 0 & 0 & 0 & \alpha_{N-2} & \beta_{N-1} \end{pmatrix} \tag{5.65}$$

$$\mathbf{p} = \begin{pmatrix} p_0 \\ p_1 \\ .. \\ .. \\ .. \\ .. \\ p_{N-1} \end{pmatrix} \quad \mathbf{q} = \begin{pmatrix} 0 \\ 0 \\ .. \\ .. \\ .. \\ .. \\ p_N \end{pmatrix}. \tag{5.66}$$

When x is a zero of $P_N(x)$, Eq. 5.64 simplifies to the *eigenvalue* equation

$$\mathbf{J} \cdot \mathbf{p} = x\mathbf{p} \tag{5.67}$$

which may be solved to determine values of x. Having done that, we can use Eq. 5.56, 5.57, or 5.58 to determine the weights.

5.7.1 Implementation notes

Two classes called `GaussIntegrator` and `GaussQuadratureRule` (namespace Genfun) implement Gaussian quadrature. There are several kinds of GaussQuadratureRule. We introduce here the concept of a ***class tree diagram***. Figure 5.7 is a class tree diagram for classes related to `GaussQuadratureRule`.

Four classes,

- `GaussLegendreRule`
- `GaussHermiteRule`
- `GaussLaguerreRule`
- `GaussTchebyshevRule`

are called ***subclasses*** of the class `GaussQuadratureRule`, which is called the ***superclass***. The subclasses can be substituted for the superclass when needed. This pattern is known as ***polymorphism*** and will be covered in detail in a later chapter. To apply Gaussian integration you need to create a `GaussIntegrator`; the `GaussIntegrator` then requires a `GaussQuadratureRule`. You will find, if you try, that you cannot create a `GaussQuadratureRule` *per se*, but you can create one of the four specific types.

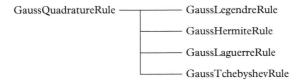

Figure 5.7 *Class tree diagram illustrating the relationship between the various kinds of* `GaussQuadratureRules`.

138 *Obtaining the abscissas*

(A more precise vocabulary for defining this state of affairs will be introduced later). Here is an example of a program that applies four point Gauss-Legendre integration to compute $\frac{1}{(2\pi)^{1/2}} \int_{-1}^{1} \exp(-x^2/2) dx$ (this function is called a normal distribution, or a Gaussian distribution, after the prolific Mr. Gauss):

```
#include "QatGenericFunctions/NormalDistribution.h"
#include "QatGenericFunctions/GaussQuadratureRule.h"
#include "QatGenericFunctions/GaussIntegrator.h"
//
// Use 4pt Gauss Legendre Quadrature to compute the integral
// of a normal distribution from [-1,1].
//
int main() {

  using namespace Genfun;
  NormalDistribution f;

  GaussLegendreRule rule(4);
  GaussIntegrator integrator(rule);

  std::cout << integrator(f) << std::endl;

  return 0;
}
```

The computation gives a result (0.68268) which differs from the accepted value (0.68269) only in the fifth decimal place, after only four evaluations of the function.

We mention now two issues. First, imagine that we are to compute the integral $\int f(x)w(x)dx$. In some cases one may find it convenient to provide the entire integrand $f(x)w(x)$ to the integral, in others it may be desired to provide only $f(x)$. A switch is provided so that the user can choose:

```
  GaussLegendreRule rule(4);

  // Default: the function to be integrated is the entire
     integrand f(x)w(x):
  GaussIntegrator integrator1(rule,GaussIntegrator::
     INTEGRATE_DX);

  // Optional: w(x) is implied, and only f(x) need be given
     to the integrator.
  GaussIntegrator integrator2(rule,GaussIntegrator::
     INTEGRATE_WX_DX);
```

A second point concerns the limits of integration. These are fixed in our implementation, and users therefore requiring integration over another range are first required to change variable in order to apply these methods. While it is common to carry out the change of variable automatically, in our implementation we have not taken these steps.

5.8 Infinite range integrals

While the `GaussIntegrator` class largely automates the evaluation of finite range integrals, some care is required when considering integrals over infinite ranges. As discussed, the preferred method for integrals of the form:

$$\int_0^\infty x^\alpha e^{-x} f(x)\, dx$$

is Gauss-Laguerre quadrature and for integrals of the form:

$$\int_{-\infty}^\infty e^{-x^2} f(x)\, dx$$

is Gauss-Hermite quadrature. It is important to remember that "of the form" means that the function f is smooth with respect to the weight functions.

Another simple-minded approach is to cut off the integral at some large value, Λ, and consider the double limit $h \to 0$ and $\Lambda \to \infty$ using the extrapolation methods of Chapter 4. Implementing this method without care is *not* recommended. The issue is the shape and location of the support of the integrand. For example, the integrand may be smooth and small except between $x = 10^{23}$ and $x = 10^{24}$; numerical exploration of the integral would completely miss this feature if inappropriate cut offs were employed.

One can also consider transforming the interval $[0, \infty)$ to a finite range; for example $x \to x' = x/(1+x)$ brings the half plane to $[0, 1)$. This can either be done in the integral itself, or the abscissas and weights can be adjusted in the quadrature formula:

```
// generate Gauss-Legendre weights in [0,1)
...
// and remap to [0,infinity)
for (int i=0;i<nGL;i++) {
  xprime[i] = x[i]/(1.0-x[i]);
  wprime[i] = w[i]/((1.0-x[i])*(1.0-x[i]));
}
```

This method has the benefit of only requiring extrapolation in one variable (the number of Gaussian grid points); however, care must still be exercised since the mapped Gauss-Legendre abscissas will cluster near $x = 1$ and will only weakly sample the integrand for large or small x. This is shown in the left panel of Figure 5.8. The squares are mapped

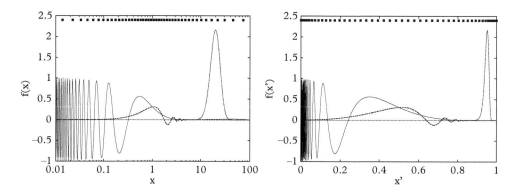

Figure 5.8 *Infinite range functions (left). Infinite range functions mapped to [0:1] (right).*

Gauss-Legendre grid points ($N = 50$). It is evident that the dashed function (with support near unity) will be sampled, while the solid and dotted functions will need an alternative method. The right panel displays the same functions mapped to the interval $[0, 1)$ with the commensurate Gauss-Legendre abscissas. In this case the dashed and dotted functions may be accurately sampled, while the solid function requires further attention.

5.9 Singular integrands

It is not unusual for integrands to feature singularities somewhere in the region of integration. Integrable power law singularities at end points can be handled by a judicious change of variables. For example, if a function has a square root singularity at a:

$$f(x) \to \frac{1}{\sqrt{x-a}} \quad \text{as } x \to a$$

then one can use the identity

$$\int_a^b f(x)\,dx = \int_0^{\sqrt{b-a}} 2u f(a + u^2)\,du$$

to numerically evaluate the integral. For more information on this situation, see the discussion in Chapter 4 of Press (2007).

An integrable singularity such as:

$$\int_0^2 \frac{1}{1-x}\,dx$$

presents no problems numerically as long as quadrature abscissas and weights that are symmetric in the region [0, 2] are used. Sometimes this is difficult to arrange; in this case we recommend subtracting zero as follows:

$$\int \frac{f(x)}{x - x_0} dx = \int \frac{f(x) - f(x_0)}{x - x_0} dx \qquad (5.68)$$

which follows for any region of integration symmetric about x_0 (typically $(-\infty, \infty)$). The result is a *regulated* integral that can be evaluated with impunity by a number of methods.

In a similar fashion:

$$\int_0^\infty \frac{f(x)}{x^2 - x_0^2} dx = \int_0^\infty \frac{f(x) - f(x_0)}{x^2 - x_0^2} dx \qquad (5.69)$$

and:

$$\int_0^\infty \frac{f(x)}{\sqrt{x^2 + A^2} - \sqrt{x_0^2 + A^2}} dx = \int_0^\infty \left(\frac{f(x)}{\sqrt{x^2 + A^2} - \sqrt{x_0^2 + A^2}} - \frac{2\sqrt{x_0^2 + A^2} f(x_0)}{x^2 - x_0^2} \right) dx.$$

$$(5.70)$$

(Consult the exercises for proofs of these relationships.) A working example of this sort of tricks is presented in Section 21.4.2 where the quantum mechanical scattering amplitude is discussed.

Finally, we remark that *not* being clever sometimes pays off. For example:

$$\lim_{\epsilon \to 0^+} \int_0^\infty \frac{1}{x^2 - 1 - i\epsilon} dx = i\frac{\pi}{2}.$$

Directly evaluating the integral with mapped naive quadrature gives $\pi/2$ to five digits accuracy when ϵ is set to 0.01 and 3,000 grid points are used. The real part of the integral goes to zero linearly in ϵ while the error in the imaginary part goes as ϵ^2. Thus there is no need for tricks if precise answers are not required.

5.10 Multidimensional integrals

Multidimensional integrals are a formidable and common foe faced by the typical numericist. Simply specifying the boundary is often a daunting task. For low dimensional integrals we recommend iterating one-dimensional techniques.

Dimensions between three and eight are perhaps best handled with a Monte Carlo program called *vegas* (Lepage, 1978). This program uses Monte Carlo methods (the subject of Chapter 7), which stochastically sample the integrand. The chief problem with this approach is that finite regions can get lost in spaces of large dimension. This will be familiar to students of statistical mechanics, who learn that the surface of a large-dimensional space completely dominates its volume. The vegas algorithm deals with this problem by adaptively exploring regions of large weight and concentrating function evaluations in these regions.

Problems with a large number of dimensions typically require Monte Carlo methods that are specifically tailored for those problems. Several chapters of this text deal with precisely this situation: the central task when computing with classical spin systems is evaluating high-dimensional sums in partition functions (Chapter 20); the same issue applies to quantum spin systems (Chapter 22); finally evaluating the path integrals that arise in quantum field theory require extrapolating to the limit of infinite dimensions (Chapter 24).

5.11 A note on nondimensionalization

Equations arising in physics depend on quantities like mass and charge as well as fundamental constants like $\hbar$ and c. Often, however, a problem can be reformulated in terms of reduced quantities that are dimensionless variables representing ratios of length, energy, frequency, etc. to characteristic quantities specific to the problem. **Nondimensionalization** is the process by which the appropriate scales are identified and used to reformulate the problem in simpler terms. If, for example, a characteristic distance x_0 happens to arise in the problem, then one eliminates x in favor of $\xi = x/x_0$.

The first step is the identification of scales arising from the system under study. This is often important in its own right—it allows, for example in the case of the hydrogen atom, to identify the angstrom (10^{-10}m) as a relevant distance scale and the electron volt (eV) as the relevant energy scale. The subsequent analysis is then simpler and less error prone since some or all of the physical constants have been removed. In addition, the interpretation of the solution and the identification of its limiting behavior is also simpler, since words like "large" and "small" can only be given meaning when the scales have been identified. A benefit of nondimensionalization is that entire classes of problems that correspond to a given set of parameter ratios can be solved at a stroke. This seemingly trivial observation can lead to deep insight, as we shall demonstrate in Section 8.3.

In fairness, we are compelled to mention that there is a downside to nondimensionalization; namely, the powerful method of error checking via dimensional analysis is no longer available.

We illustrate the methods of nondimensionalization with three examples.

5.11.1 Compton scattering

A photon interacts with an electron at rest in a process called Compton scattering. The photon has an wavelength λ, the scattered photon has a wavelength λ'. The relation between the wavelengths of the incident and scattered photon is $\lambda' = \frac{h}{m_e c}(1 - \cos\theta) + \lambda$. One can see that a natural distance scale arises from the problem

$\lambda_c = \frac{h}{m_e c} = 2.43 \times 10^{-12}$m, called the Compton wavelength; this distance scale is present even though the electron it is a pointlike particle, having no finite spatial extent. The problem can be reformulated, without reference to the three quantities h, m_e, and c, as $\lambda' = \lambda_c(1 - \cos\theta) + \lambda$. Letting $x = \lambda/\lambda_c$ and $x' = \lambda'/\lambda_c$, this is equivalent to measuring all distances in terms of the Compton wavelength, and we can write:

$$x' = (1 - \cos\theta) + x$$

This equation is now completely devoid of physical constants.

The Compton wavelength is a natural length scale which can be used to nondimensionalize the Schrödinger equation. In one dimension:

$$-\frac{\hbar^2}{2m}\frac{\partial^2}{\partial x^2}\psi(x) + V(x)\psi(x) = E\psi(x)$$

We denote the reduced Compton wavelength $\bar{\lambda} = \frac{\hbar}{mc}$ of the particle in the potential. Let $\xi = x/\bar{\lambda}$, and substitute $x = \bar{\lambda}\xi$ into the Schrödinger equation:

$$-\frac{1}{2}mc^2\frac{\partial^2}{\partial \xi^2}\psi(\xi) + V(\xi)\psi(\xi) = E\psi(\xi)$$

The term mc^2 will be recognized as the rest energy of the particle which moves in the potential. Now rescale the energies such that $v(\xi) = V(\xi)/(mc^2)$ and $\varepsilon = E/(mc^2)$ and we have:

$$-\frac{1}{2}\frac{\partial^2}{\partial \xi^2}\psi(\xi) + v(\xi)\psi(\xi) = \varepsilon\psi(\xi)$$

In this form, the Compton wavelength of the particle is used as the length scale, and the rest mass of the particle is the energy scale. Other length scales may also arise, depending upon $v(\xi)$, so it may be advantageous, as in the following examples, to revisit the choice of scale.

5.11.2 Particle in a finite one-dimensional well

A particle is confined a one dimensional box of length a and depth V_0. We take a to be the characteristic length and write the Schrödinger equation as:

$$-\frac{\hbar^2}{2ma^2}\frac{\partial^2}{\partial \xi^2}\psi(\xi) + V(\xi)\psi(\xi) = E\psi(\xi)$$

where $\xi = x/a$ and $V(\xi) = -V_0$ for $0 < \xi < 1$ and $V(\xi) = 0$ otherwise. The equation can be further reduced by noticing that $\frac{\hbar^2}{ma^2} = E_0$ has the dimensions of energy, and write the equation as:

$$-\frac{\partial^2}{2\partial\xi^2}\psi(\xi) + v(\xi)\psi(x) = \varepsilon\psi(x) \tag{5.71}$$

where $v(\xi) = V(\xi)/E_0$ and $\varepsilon = E/E_0$. The equation has been simplified and the relevant energy and length scales have been identified. The equation can now be solved analytically or numerically. One point of interest is how many bound state solutions exist for this equation, and one can see that it is determined solely by the quantity $v_0 = V_0/E_0$. One can anticipate that the distance scale a typifies wavelengths of the solutions, even without finding them, and E_0 typifies the bound state energies.

5.11.3 Schrödinger equation for the hydrogen atom

For an attractive Coulomb force, the potential is $V(r) = -\frac{e^2}{4\pi\epsilon_0 r}$ and the time-dependent Schrödinger equation is:

$$-\frac{1}{2m}\nabla^2\psi(x) - \frac{e^2}{4\pi\epsilon_0 r}\psi(x) = E\psi(x). \tag{5.72}$$

If we begin as in Eq. 5.71 by taking the Compton wavelength as the characteristic length, we have:

$$-\nabla_\xi^2\psi(\xi) + v(\xi)\psi(\xi) = \varepsilon\psi(\xi)$$

where $\vec{\xi} = \frac{\vec{r}}{\lambda}$ and

$$v(\xi) = \frac{1}{\xi}\left(\frac{e^2}{4\pi\epsilon_0\hbar c}\right).$$

Since $v(\xi)$ and ξ have been made dimensionless, the quantity in parenthesis is dimensionless and is denoted as α, the fine structure constant:

$$\alpha = \left(\frac{e^2}{4\pi\epsilon_0\hbar c}\right) \approx \frac{1}{137}.$$

The Schrödinger equation is:

$$-\nabla_\xi^2\psi(\xi) - \frac{\alpha}{\xi}\psi(\xi) = \varepsilon\psi(\xi).$$

Another rescaling is useful in eliminating α from the formulation. If we let $\vec{\eta} = \alpha\vec{\xi}$ and substitute we obtain:

$$-\nabla_\eta^2\psi(\eta) - \frac{2}{\eta}\psi(\eta) = \epsilon\psi(\eta) \tag{5.73}$$

where $\epsilon = \frac{\varepsilon}{\alpha^2} = \frac{E}{\frac{1}{2}\alpha^2 mc^2}$ is the energy measured in units of the characteristic energy $E_0 = \frac{\alpha^2}{2}mc^2 = 13.6$ eV, which happens to be the well-known ionization energy for hydrogen. As for the dimensionless quantity

$$\eta = \alpha\xi = \frac{e^2}{4\pi\epsilon_0\hbar c} \cdot \frac{mcr}{\hbar} = \left(\frac{me^2}{4\pi\epsilon_0\hbar^2}\right)r,$$

it can be regarded as the position r divided by the characteristic distance:

$$a_0 = \frac{4\pi\epsilon_0\hbar^2}{me^2} = 0.529\text{Å},$$

which is known as the Bohr radius. The important distance scales and energy have been extracted already, without solving the Schrödinger equation, merely by nondimensionalizing it. The details can be sought in a solution to Eq. 5.73, which is a purely mathematical problem since physical constants are no longer present in the equation.

5.12 Exercises

1. Compute the integral $\int_0^1 \tanh(x)dx$ analytically and also using the rectangle rule, the trapezoid rule, and Simpson's rule. For each quadrature rule, perform the calculation in two, four, six, eight, intervals and so forth and plot the difference between the approximate result and the true result on a logarithmic scale vs. the mesh spacing $1/n$. Then, do the same for the integral $\int_0^1 \sqrt{\coth(x)}dx$. Label the plot axes.

2. A simple pendulum consisting of a pointlike mass at the end of a massless string of length l has a period which can be expressed as $T(\theta_{max})$ where θ_{max} is the amplitude of the motion. For small amplitudes $T_0 \approx 2\pi\sqrt{\frac{l}{g}}$. For larger amplitudes one can write $T = T_0 f(\theta_{max})$, where $f(\theta_{max})$ is a slowly varying function. Integrate the equations of motion of the pendulum, in order to obtain $f(\theta_{max})$ at the points $\theta_{max} = \frac{\pi}{12}, \frac{\pi}{6}, \frac{\pi}{3}, \frac{\pi}{2}, \frac{2\pi}{3}, \frac{5\pi}{6}, \frac{11\pi}{12}$. Plot those functions on a graph, making sure to label both axes.

3. Compute the WKB eigenenergies for a particle of mass m in a potential $\frac{1}{2}kx^2$. This is the quantum harmonic oscillator problem, whose exact solution is well known. The energies obtained from approximation 5.15 can also be computed analytically and are known to coincide with the exact solution. In this exercise you are going to compute them numerically.

    a) Write the Schrödinger equation for the system, and nondimensionalize it. Identify your characteristic length scale and energy scale.
    b) Using Eq. 5.15, compute the energy levels of the system for $n = 0 \ldots 10$
    c) Plot E/E_0 vs. n for your solutions.

Exercises

4. A particle of mass m moves in a one-dimensional potential given by $V(x) = -V_0 e^{-x^2/(2a^2)}$ where V_0 is the depth of the potential well and a is its width. The system has a finite number of bound states, having negative energies. This number is determined by some parameter which is a combination of the constants V_0, m, and a.

   a) Find an expression for the aforementioned parameter in terms of the aforementioned constants, possibly involving other fundamental constants of nature. Also, determine an energy scale E_0 for the problem.

   b) Then, calculate numerically the number of bound states vs. this parameter and plot it over an interesting range.

5. A particle of energy E is incident upon a potential barrier $V(x) = V_0 e^{-x^2/(2a^2)}$. Nondimensionalize the problem as in the previous exercise; compute and graph the transmission coefficient T as a function of V_0/E_0. Use the quasiclassical techniques of Section 5.1.4.

6. The functions:

$$\phi_n(x) \equiv \frac{\pi^{-1/4}}{\sqrt{2^n n!}} \exp(-x^2/2) H_n(x) \tag{5.74}$$

arise as solutions to the harmonic oscillator problem with nondimensionalized Hamiltonian:

$$H = -\frac{1}{2}\frac{d^2}{dx^2} + \frac{1}{2}x^2 \tag{5.75}$$

These are not polynomials, but they are orthonormal functions under an inner product:

$$\langle f|g \rangle = \int_{-\infty}^{\infty} f(x)g(x)dx \tag{5.76}$$

Check this orthonormality numerically by computing and printing out the matrix $\mathbf{A}$ where $(\mathbf{A})_{ij} \equiv \langle i|j \rangle = \langle \phi_i(x)|\phi_j(x) \rangle$, using Gauss-Hermite integration. Do this for $i,j = 0,1\ldots 5$. What is the minimum number of points needed for this computation?

7. The functions ϕ_i identified in the previous problem constitute a basis for a vector space of functions in the interval $[-\infty, \infty]$. An *operator* is an object that takes a function, "grinds on it for a while (Richard Feynman)", and spits out another function. In quantum mechanics all observables are represented by operators (though not vice-versa). Examples of operators are $\hat{x}$ (the position operator), $\hat{x}^2$, and

$\hat{D} \equiv \frac{d}{dx}$. Finally, the Hamiltonian itself is an operator, $H_{SHO} \equiv -\frac{1}{2}\hat{D}^2 + \frac{1}{2}\hat{x}^2$. It can be extremely useful in quantum mechanics, as you may already know, to represent these operators as matrices. The actual matrices are infinite dimensional but, with a basis, we can compute finite dimensional submatrices whose elements are:

$$(\mathbf{A})_{ij} \equiv \langle i|A|j\rangle \equiv \langle i|Aj\rangle \tag{5.77}$$

a) Compute and print out the lowest 5x5 submatrices starting at row 0 and column 0 for operator $\hat{x}$, in the basis of the previous problem.

b) Repeat for the operator $\hat{x^2}$; also compare the matrix you obtain with the square of the matrix found in part (a) and explain any discrepancies you observe.

c) Repeat parts (a) and (b) for the operator $\hat{D}$.

d) Compute and print out the 5×5 submatrix, starting at row 0 and column 0, of the Hamiltonian operator $\hat{H}$. Do the elements of this matrix make sense to you?

Use Gauss-Hermite integration to compute the matrix elements and use care when determining the number of points required for accurate values.

8. A deuteron is a bound system of neutron (mass 939.6 MeV/c^2) and proton (mass 938.3 MeV/c^2). The orbital angular momentum ℓ is known empirically to be zero. The potential between proton and neutron has been parametrized as:

$$V(r) = -(635\,\text{MeV} \cdot \text{fm})\frac{e^{-(1.55\,\text{fm}^{-1})r}}{r} + (1458\,\text{MeV} \cdot \text{fm})\frac{e^{-(3.11\,\text{fm}^{-1})r}}{r}.$$

- Make a plot of this potential.
- Look for solutions to Eq. 5.15 in the range of 0-5 MeV. How many solutions can you find? Give their WKB eigenenergies (put it directly into the plot).
- Draw (as a constant line on your plot) the energies and indicate the corresponding classical turning points with two points.

Consider the proton and neutron to be spinless particles for the purposes of this exercise.

9. The charmonium system is a bound state of charm quark (c) and its antiquark ($\bar{c}$); as in the previous exercise, consider the two quarks to be spinless. Take the mass of each constituent to be $m_c = 1370$ MeV/c^2 and let the potential be

$$V(r) = -\alpha\frac{4\hbar c}{3r} + \frac{r}{\hbar c a^2} \tag{5.78}$$

where $\alpha = 0.38$ is a dimensionless constant and $a = 2.43 \times 10^{-3}$ MeV^{-1}.

a) Use the appropriate semiclassical quantization rule to determine the energies of the $c\bar{c}$ system for the S states ($\ell = 0$). Do this for the ground state and the first two excited states.

b) Now do the same for the P-wave states ($\ell = 1$).

c) You should find that these energies are significant in comparison to the mass of the c and $\bar{c}$. Each state (ground state and excited state) has a mass equal to the rest energy of the two quarks plus the excitation energy. Calculate this for the six states you determined in the previous two parts to this question and plot an energy level diagram.

10. A sphere of radius a has a total charge of Q spread on its surface in a nonuniform way. The charge distribution is azimuthally (or ϕ-) symmetric and varies with polar angle according to

$$\sigma(\cos\theta) \propto (\cos\theta)^{1/2}(1 - \cos\theta)^{3/2} \tag{5.79}$$

a) Determine a normalizing factor such that the charge distribution integrates to Q over the full surface of the sphere. Note, this does not need to be done numerically.

b) Determine the monopole, dipole, quadrupole, sextupole, and octupole moments of the charge distribution.

11. [P] Diatomic Molecule.
In diatomic molecules the *Morse potential*, given by $V(x) = D_e(1 - e^{-\beta(r-r_e)})^2$ is used to describe the interaction between the two atoms in the molecule. For the ground state diatomic nitrogen, for example the dissociation energy is $D_e = 9.904$ eV and the width parameter is $\beta = 2.85\,(\text{Å}^{-1})$. The equilibrium bond distance r_e is 1.080 Å.

- Expand the Morse potential as a Taylor series in powers of $\rho = r - r_e$, and retain the terms up to $O(\rho^2)$.
- Choose a length scale and an energy scale to nondimensionalize Schrödinger equation for the approximation obtained in the previous part, such that the rescaled problem has the same form as the Hamiltonian of Problem 7. What are the characteristic length ρ_0 (in Å) and the characteristic energy E_0 (in eV)? Give both algebraic and numerical answers.
- Re-express the original Morse potential in units of the characteristic energy, and in terms of the characteristic length, and plot it on a graph.
- Using the orthogonal functions of Problem 6 as a basis, compute the matrix elements $\langle i|\hat{H}|j\rangle$, where $\hat{H}$ includes the full Morse potential. Print out the 5×5 submatrix starting from row 0, column 0.

- Determine the eigenvalues of this Hamiltonian, and express them in units of the characteristic energy.

12. [P] Radioactive Decay.
    A radioactive source contains N atoms of an unstable isotope with a mean life of τ. The radioactive substance decays isotropically. It is embedded a distance z into the flat surface of an infinitely thick, dense substance with a photon attenuation length of λ; i.e., the probability density for a gamma ray to be absorbed at distance x from emission in such a substance is:

    $$P(x) = \frac{1}{\lambda} \exp(-x/\lambda)$$

    a) Show that the number of gamma rays escaping to the surface of the substance is a function $F(z/\lambda)$. Compute and graph this function.
    b) Instead, assume that the substance itself is uniformly contaminated by the radioactive species, and compute the flux of gamma rays from the surface, as a function of the mean lifetime τ and number density n of the radioactive contaminant and the attenuation length λ of the substance.
    c) Granite, with a density of 2.7 g/cm^3 contains about 4% potassium oxide (K_2O) by weight. Some of the potassium is the radioactive isotope K^{40}, with a natural abundance of 0.012% and a half life of 1.2 billion years, but only about 11% of the decays give a 1.46 MeV gamma ray. The attenuation length of the gamma ray is 7.5 cm. How many gamma rays are emitted per cm^2 from the surface of an infinitely thick slab of granite?

13. [P] van der Waals Gas.
    A *van der Waals gas* obeys the equation of state

    $$(P + 3/V^2)(3V - 1) = 8T \qquad (5.80)$$

    where the variables P, T, and V are pressure, temperature, and volume of the gas expressed in terms of critical values for these quantities. In this exercise you will carry out the *Maxwell construction* which is used to determine the vapor pressure of the gas at different values of the temperature, in the context of the van der Waals model. A good discussion of this model and of the physics of phase transformations in fluids can be found in Kittel and Kroemer (1980).

    a) Solve the equation of state for P, and plot P vs. V on the same set of axes for $T = 0.8, 0.85, 0.90, 0.95, 1.00, 1.05, 1.10$.
    b) For $T < 1$, the function $P(V)$ has two extrema at $V = V_0$ and $V = V_1$. For $T = 0.8, 0.85, 0.90, 0.95$, determine these values numerically together with the corresponding pressures P_0 and P_1. (The average of these two pressures,

$P_{ave} = (P_0 + P_1)/2$, constitutes a decent starting point for the determination of the vapor pressure, P_V, of the fluid at temperature T). You may wish to check your computation by plotting the calculated extrema on the same set of axes as in part (a).

c) For $T < 1$, and $P_0 < P_V < P_1$ the equation $P(V) = P_V$ has three solutions ordered by volume: $V = V_{min}$, $V = V_{mid}$, and $V = V_{max}$. Compute these points and also compute

$$I(P_V) = \int_{V_{min}}^{V_{max}} (P(V) - P_V)\, dV \tag{5.81}$$

setting $P_V = P_{ave}$. Note that the integral of the previous step depends on P_V both through the integrand and through the limits of integration.

d) Now vary P_V until the value of the integral $I(P_V)$ is exactly zero. This condition determines the vapor pressure.

e) Plot the vapor pressure thus determined vs. the temperature for $T = 0.8, 0.85, 0.90, 0.95$.

14. [T] Regulated Integrals.
Establish Eqs 5.69 and 5.70 by considering the behavior of the denominators near their zeros.

....

BIBLIOGRAPHY

Davis, P. J. and P. Rabinowitz (1975). *Methods of Numerical Quadrature*. Academic Press.
Gasiorowicz, S. (2003). *Quantum Physics*. 3rd ed. John Wiley & Sons.
Gibbs, W. (2006). *Computation in Modern Physics*. 3rd ed. World Scientific.
Goldstein, H., C. P. Poole, Jr, and J. L. Safko (2001). *Classical Mechanics*. 3rd ed. Pearson Education Limited.
Golub, G. H. and J. H. Welsh (1969). *Calculation of Gauss quadrature rules*. Math. Comp. **23**, 221–230.
Griffiths, D. J. (2005). *Introduction to Quantum Mechanics*. 2nd ed. Prentice-Hall.
Jackson, J. D. (1998). *Classical Electrodynamics*. 3rd ed. John Wiley and Sons.
Jin, D. S., J. R. Ensher, M. R. Matthews, C. E. Wieman, and E. A. Cornell (1996). *Collective Excitations of a Bose-Einstein Condensate in a Dilute Gas*. Phys. Rev. Lett. 77 3 420.
Kittel, C. and H. Kroemer (1980). *Thermal Physics*. 2nd ed. W. H. Freeman.
Lepage, G. P. (1978). *A New Algorithm for Adaptive Multidimensional Integration*. J. Comput. Phys. **27**, 192.
Press, W., S. Teukolsky, W. Vetterling, and B. Flannery (2007). *Numerical Recipes, the Art of Scientific Computing*. 3rd ed. Cambridge University Press.
Quigg, C. and J. L. Rosner (1979). *Quantum mechanics with applications to quarkonium*. Physics Reports **56** 167.
Szego, G. (1939). *Orthogonal Polynomials*. Colloquium Publications-The American Mathematical Society V23.

6
How to write a class

6.1	Some example problems	152
	6.1.1 A stack of integers	152
	6.1.2 The Jones calculus	152
	6.1.3 Implementing stack	154
6.2	Constructors	160
6.3	Assignment operators and copy constructors	162
6.4	Destructors	165
6.5	`const` member data and `const` member functions	166
6.6	Mutable member data	167
6.7	Operator overloading	167
6.8	Friends	170
6.9	Type conversion via constructors and cast operators	171
6.10	Dynamic memory allocation	174
	6.10.1 The "big four"	176
6.11	A worked example: implementing the Jones calculus	181
6.12	Conclusion	190
6.13	Exercises	191
	Bibliography	195

You may consider this chapter, *How to Write a Class*, as long-delayed since we have been dealing with classes from the very beginning of the book and because classes are central to C++. Why did we wait so long?

In the first instance, the main motivation is to develop some level of intuition about what a class should be and how it should behave from the perspective of clients of the class. Now that this is accomplished we are ready to go further and describe what goes into making a class, in particular the syntax of class declaration and definition. Once you begin to get the basic idea, we encourage you to consult the references, particularly Meyers (2005a and 2005b), which go much further into the topic of "how to write a class" than we do here, and can contribute significantly to the development of a powerful programming style.

In some cases (like `std::complex`) it may be possible to think of a class in isolation, without needing to think about other classes that have to exist to make the entire system work. In other cases, it may seem meaningless to define a single class and no others. A rotation, for example does not have much meaning without something to rotate, like a vector. In practical work it is not uncommon to design entire categories of interoperating classes. This requires considerable thought and practice, and leads inevitably to the concept of object oriented analysis and design (Booch, 2007). While OOAD is important for software engineering (particularly for professionals) we largely ignore the topic in this book. Instead, this chapter is devoted to the first steps arising in the implementation of "typical" classes.

6.1 Some example problems

Two example problems will be discussed. The first involves one single class, `Stack`, which implements a stack of integers. A `Stack` is a collection class (like `std::vector`) but with different storage and access mechanisms. The second example involves two interoperating classes, `JonesVector` and `JonesMatrix`, which together implement a calculational tool for simulating the effect of optical elements on the polarization of light (Hecht, 2001).

6.1.1 A stack of integers

A stack is a collection class that is designed to collect integers. To add an item to the stack you "push" it onto the stack, and to access it you "pop" it off the stack. A "stack" of books (actual books, not "book objects") is an appropriate mental model, assuming that books are added only to the top of the stack and can only be removed from the top. "Popping" has two functions: it fetches the item which is at the top of the stack and it removes the object from the stack at the same time.

We want to implement an abstract datatype (call it `Stack`) representing a stack of integers. In addition to the two functions (`push` and `pop`) mentioned above, the `Stack` should support copy and assignment. For convenience we expect to print out the `Stack`. This specifies our interface, which must, however, be formalized in a class declaration. In this chapter we will describe a possible implementation of the class `Stack`. The companion website:

http://www.oup.co.uk/companion/acp

contains a directory `SKELETONS/CH6/STACK` under which you will find the files `STACK/Stack.h` and `src/Stack.cpp`; these are mere skeletons and it will be for you to fill them with "implementation".

6.1.2 The Jones calculus

The second problem we take up is the polarization of light, a problem that is mathematically equivalent to many aspects of a quantum mechanical two-state system. The basis states will be $|H\rangle$, which designates horizontal polarization, and $|V\rangle$, which designates vertical polarization. In a vectorial representation these can be written as:

$$|H\rangle = \begin{pmatrix} 1 \\ 0 \end{pmatrix} \quad |V\rangle = \begin{pmatrix} 0 \\ 1 \end{pmatrix}.$$

This is not the only type of polarization that is possible; for example "diagonal" light is linearly polarized at +45° from the x-axis:

$$|D\rangle = \frac{1}{\sqrt{2}} \begin{pmatrix} 1 \\ 1 \end{pmatrix},$$

and "antidiagonal" light is polarized at −45° from the x-axis:

$$|A\rangle = \frac{1}{\sqrt{2}} \begin{pmatrix} 1 \\ -1 \end{pmatrix}.$$

Right handed circularly polarized light is like diagonal light except the vertical component lags the horizontal component by 90°:

$$|R\rangle = \frac{1}{\sqrt{2}} \begin{pmatrix} 1 \\ i \end{pmatrix},$$

while left-handed circularly polarized light has the vertical component leading the horizontal component by 90°:

$$|L\rangle = \frac{1}{\sqrt{2}} \begin{pmatrix} 1 \\ -i \end{pmatrix}.$$

Optical filters of various sorts can be used to transform one type of polarized light into another. Because of the superposition principle, filters act linearly on the polarization vectors and their action can be represented by matrices. These matrices transform the polarization vector, changing both the square magnitude of the vector (which represents the relative intensity of light, or alternatively, the survival probability of a photon) and the direction. The most common are polarizing filters with a polarization axis along the horizontal axis $\begin{pmatrix} 1 & 0 \\ 0 & 0 \end{pmatrix}$; along the vertical axis $\begin{pmatrix} 0 & 0 \\ 0 & 1 \end{pmatrix}$, and along an arbitrary axis making an angle θ with respect to the x-axis:

$$\begin{pmatrix} \cos^2 \theta & \sin \theta \cos \theta \\ \sin \theta \cos \theta & \sin^2 \theta \end{pmatrix}.$$

These filters are physically realized in plastics that contain long, resistive polymer chains, all oriented in the same direction. Another useful optical element is a birefringent material, which has a "slow" transmission axis and a "fast" transmission axis due to the material's crystal structure. By selecting the right material thickness one can produce a phase shift in the *y*-component of the electric field (relative to the *x*-component). A slab of material having the right thickness to shift the phase by 90° (one-quarter wavelength) is called a quarter-wave plate. A quarter-wave plate with the fast axis along the vertical axis is represented by the matrix: $\begin{pmatrix} 1 & 0 \\ 0 & -i \end{pmatrix}$, while a quarter-wave plate with the fast axis along the horizontal axis is represented by the matrix $\begin{pmatrix} 1 & 0 \\ 0 & i \end{pmatrix}$. Circular polarizers are represented by the matrices $\frac{1}{2}\begin{pmatrix} 1 & i \\ -i & 1 \end{pmatrix}$, which produces right-handed circularly polarized light, and $\frac{1}{2}\begin{pmatrix} 1 & -i \\ i & 1 \end{pmatrix}$, which produces left-handed circularly polarized light. One can verify that the former matrix annihilates left-handed circularly polarized light while transmitting right-hand circularly polarized light–and vice versa for the latter matrix.

Light passing through an optical system consisting of multiple elements undergoes successive transformations by the matrices representing the appropriate element, thus the cumulative effect of a series of elements is the matrix product

$$T = T_n \ldots T_3 T_2 T_1,$$

where T is the matrix of the full optical system while T_i is the matrix for the ith optical element. For example, we can build a right circular polarizer by a combination of three elements, the first being a quarter-wave plate with the fast axis along the vertical direction, the second being a linear polarizer with a polarizing axis at +45°, and the third being a quarter-wave plate with the fast axis along the horizontal direction. This follows from

$$\begin{pmatrix} 1 & 0 \\ 0 & i \end{pmatrix} \cdot \frac{1}{2}\begin{pmatrix} 1 & 1 \\ 1 & 1 \end{pmatrix} \cdot \begin{pmatrix} 1 & 0 \\ 0 & -i \end{pmatrix} = \frac{1}{2}\begin{pmatrix} 1 & -i \\ i & 1 \end{pmatrix}.$$

We want to build a small routine to keep track of the state of polarized light as it passes through a series of filters. To keep track of the intensity of the light, we shall need to compute the magnitude of a Jones vector; to compute the polarization we will need an inner product. The composition of Jones matrices needs to be defined, as well as their action on Jones vectors. However addition, subtraction, trace, and determinant are unimportant for the task and should not be part of the interface.

6.1.3 Implementing stack

A Stack class, and an example program that uses the Stack class, has been placed in the directory

SKELETONS/CH6/STACK.

The first thing you should do is to load that directory and build the skeleton and example. This is *not* working code! The example will not work until *you* have implemented the skeleton. The set of commands below builds the `Stack` class, creates a library called `libStack.a` containing the partially implemented `Stack` class, compiles the example client and links it against the `Stack` class, and installs it within the directory `SKELETONS/CH6/STACK/local/bin`.

```
cd SKELETONS/CH6/STACK/src
make
```

First, have a look at the client code, in `client.cpp`, a listing of which is shown below.

```cpp
// This is a program to test the stack. It will work as soon
//     as you have
// finished implementing the stack class, which means filling
//     in all of the
// function definitions in Stack.cpp
#include <iostream>
#include "STACK/Stack.h"
int main (int argc, char **argv) {

  // Create a new Stack called a;
  Stack a;

  // Fill it with the powers of 2, through 128
  a.push(1);   a.push(2);   a.push(4);   a.push(8);
  a.push(16);  a.push(32);  a.push(64);  a.push(128);

  // OK now how big is it?
  std::cout << "Size of stack " << a.size() << std::endl;

  // Print the stack out:
  a.print();
  // Now pop two element from the stack:
  std::cout << "Popping " << a.pop() << " from the stack"
     << std::endl;
  std::cout << "Popping " << a.pop() << " from the stack"
     << std::endl;

  // See what it looks like now:
  a.print();

  // Create another stack:
  Stack b;
```

```
    while (!a.isEmpty()) b.push(a.pop());

    // Now a is empty...
    a.print();

    // And b contains a, in reverse order, right?
    b.print();

    // But now let's assign the stack b to a:
    a=b;

    // and finally we end up with identical stacks:
    a.print();

    return 0;
}
```

From a quick read of this code, you get the idea of what the Stack class is supposed to do. In general, you should have a pretty good idea of what the class ought to do before you set about writing it![1]

The next step is to make a list of all of the functions that your class should fulfill and write the class declaration. The declaration will live in a header file–we have put ours in STACK/STACK/Stack.h, which is shown here:

```
#ifndef _Stack_h_
#define _Stack_h_
#define STACK_MAX 1024
class Stack {
 public:

  // Constructor:
  Stack();

  // Push:
  void push(int i);

  // Pop:
  int pop();

  // How big is the stack?
  unsigned int size() const;
```

---

[1] You will learn as you go along how much effort can actually be involved with such preliminaries as this.

```cpp
    // Print (to std::cout)
    void print() const;

    // Test if the stack is empty
    bool isEmpty() const;

private:

    unsigned int    _count;              // Actual length of
                                         //    stack.
    int             _data[STACK_MAX];    // An array to store
                                         //    data.

};

#endif
```

There are two parts of the class, data and functions, which are called **member data** and **member functions** to distinguish them from ordinary functions, which are called *free subroutines*. These members are bound to a particular object (call it a), and they are accessed or invoked with a statement like `a.data` or `a.function()`. For member data, this is equivalent to accessing the members in a `struct`. Apart from their object binding, member functions are to a large extent like any other functions in C++; they have arguments (with default values if desired); they have return types, and they can be **overloaded** (i.e., they can exist in different versions that take different arguments). When we invoke a bound function, e.g. `a.print()`, we say that we are "sending a message to an object." Besides the constructor, other member functions (push, pop, print, size, and isEmpty) are defined in the `Stack` class.

Member data and member functions have two different types of access, *public* and *private*[2], with keywords `public` and `private` that act like switches in the class definition. They switch the access to the following data or functions on (`public`) or off (`private`). Public access means that any client can access the data or call the function. Private access means that access to the data or function is only granted to the code that implements the class itself. The default access is private. The two access mechanisms should be used judiciously to achieve a reasonable level of encapsulation.

The implementation of the class is the definition of the six member functions, and this implementation is in the file `src/Stack.cpp`. The version of this file you will check out of the repository contains empty hooks for each of these functions. It is now your

---

[2] A third kind of access, protected, is also possible but should be used in special circumstances and sparingly, so we do not describe that here.

158 *Some example problems*

job to go in and fill in each of these hooks in order to obtain a working `Stack` class (Exercise 1).

The subject of the following sections is the syntax of classes in C++ using the `Stack` class as an example. Note that the `Stack` class is fairly limited, because it cannot grow past a fixed size. A proper `Stack` class will dynamically allocate the memory that it needs in order to solve that problem. The `std::vector` class from the C++ standard library already does this kind of thing, so the only size limitation you have to worry about is the physical memory of your machine. One conceptually difficult part of designing C++ classes has to do with avoiding the pitfalls arising from dynamic memory allocation, so we will discuss a fixed-size `Stack` class first, then introduce the operators `new` and `delete`, used for dynamic memory allocation and deallocation, before finally discussing the implications of memory allocation for C++ class design.

Here is the empty implementation of the class `Stack`, in the file `Stack.cpp`, which we shall now discuss:

```cpp
#include "STACK/Stack.h"

// The constructor is called every time an object is created.
// This ensures that the object starts from a well defined
      state
// ... without requiring further initialization from the
      client.

// unimplemented!
Stack::Stack() {
  // ==> Insert code here:
}

// The push function increases the length of the stack by one
      element
// add the value to the end of the stack.

// unimplemented!
void Stack::push(int i) {
  // ==> insert code here
}

// The pop function decreases the length of the stack by one
      element,
// and returns the value that is popped off

// unimplemented!
int Stack::pop(){
```

```
  //   ==> Change the following line to something sensible...
  return 0;
}

// The size function returns the size of the stack. The stack does
// not change so this function is "const".

// unimplemented!
unsigned int Stack::size() const {
  //   ==> Change the following line to something sensible...
  return 0;
}

// The print function dumps the entire stack to std::cout, in a
// human-readable format.  The stack does not change, so this function
// is "const"

// unimplemented!
void Stack::print() const {
  // ==> insert code here
}

// The isEmpty() query function returns true or false, depending on
// whether the stack is empty (length 0).  The stack does not change,
// so this function is "const"

// unimplemented!
bool Stack::isEmpty() const {
  // Change the following line to something sensible
  return 0;
}
```

These definitions start with the scope operator, `Stack::`, which means that the function being implemented is a member function of the class `Stack`. Within the definition, the member functions and member data of the bound object can be accessed without putting an additional scope operator, i.e. simply as _count, or _data. The scope of all variables within member functions is the same as for ordinary functions: arguments to the function have local scope, as well as any variables declared within the function, and global variables—generally frowned upon—can be accessed from anywhere. There is also a third kind of scope besides local and global, called *class scope*. Member data has class scope, which means that it is visible within the functions implementing the class.

6.2 Constructors

Looking first at the client code (`client.cpp`), one sees the following declaration:

```
Stack a;
```

This is a declaration that creates a single `Stack`. `Stack` is said to be the class; `a` is the object. The object `a` is said to be of class `Stack`; one can also say that `a` is an instance of the class `Stack`, and the creation of an instance of a class is also called instantiation. Unlike the creation of an instance of a built-in datatype like `int`, a function call occurs during the instantiation of a `Stack`. The function is called the constructor. The declaration of the constructor can be seen in the file `Stack.h` and the definition in `Stack.cpp`.

There are different ways to write a constructor, and indeed a single class can have several forms of the constructor. Constructors are functions which:

- have the same name as the class (eg. `Stack`).
- do not have any return type (not even void).
- have no parameters, one parameter, or multiple parameters.

The purpose of a constructor is object initialization, such that all objects are born in a well-defined state. One might imagine having a function called `init()` which the client would call to initialize an object, but the constructor is a better solution because it's automatic.

What should we do to make sure our `Stack` comes up in a proper state? Well, one should certainly set the stack counter equal to zero, indicating that nothing is initially in the `Stack`. Note that if no constructor is defined in the class, then the compiler generates one by default. This may not be the constructor you want. For example, a compiler-generated default constructor would not properly initialize the stack counter. Constructors, like other functions, can be overloaded and they can have default values for their arguments.

Consider a class that represents rational numbers, which we call `Rational`. This class, let us say, has three constructors, and excerpts for its header file might look like:

```
class Rational {
public:
```

```
  // Default constructor, constructs 0/1
  Rational();

  // Construct a rational number from a real number:
  Rational (double x);

  // Construct with numerator and denominator
  Rational (int num, int den);

private:

  double _num, _den;
};
```

There are therefore three ways to construct a `Rational`:

```
Rational c;
Rational c=4.0;   // or Rational c(4.0)
Rational c(4.0,2.0);
```

Alternatively, the designer of the class `Rational` could have enabled all three of the preceeding declarations by defining a single constructor having default arguments, like this:

```
class Rational {
  .
  .
  .
  // Constructor:
  Rational(double num=0.0, double den=1.0);
  .
  .
  .
}
```

If more than one constructor is defined, the compiler tries to locate the correct one at compile time. It will complain if it cannot resolve a constructor from the arguments, or if an ambiguity is detected. A line like:

```
Rational c=4;
```

may appear to have the syntax of assignment, wherein first a default constructor is called and then an assignment is made; however this is actually initialization and is treated by the compiler in the same way as

```
Rational c(4);
```

or, for that matter

```
Rational c=Rational(4);
```

The implementation of a constructor should try to initialize its member data where possible rather than assign them. In case the member data is `const`, there is no choice, since `const` data can never be altered! The initialization of member data has a special syntax in C++. Here is an example from class `Rational`:

```
Rational::Rational(double num, double den):
_num(num),_den(den) {}
```

This constructor initializes the two data members _num and _den before entering the function body. In the function, nothing more is done. The initialization is carried out *in the order that the data is declared in the header file, not in the order that it appears to be initialized within the function definition*. So it is best to make these two correspond. If they do not, a compiler warning will be generated.

Note that the compiler-generated default constructor is *not* generated if *any other* constructor is defined in the class. This includes the copy constructor which is discussed next.

6.3 Assignment operators and copy constructors

It is possible to initialize a `Stack` object with the state of another `Stack`; the code for doing this would look like this:

```
Stack a;
.
.
.
Stack b=a;
```

On the other hand, it is also possible to assign an existing `Stack` so that it gets the value of another `Stack`:

```
    Stack a;
.
.
.
    Stack b;
```

```
// Now let's assign the stack b to a:

a=b;
```

For user-defined datatypes the distinction between initialization and assignment is important because functions are invoked in each case, and they are *different functions*. For initialization the function is the so-called *copy constructor*, for assignment it's the *assignment operator*. These two functions are automatically generated by the compiler, but in certain cases, particularly when using dynamic memory allocation, the compiler-generated function simply will not do and you have to write your own. You need to know the syntax of these functions so that you can write your own (and thereby turn off their automatic generation). Here are the signatures for these two functions for the Stack class:

```
class Stack {
.
.
.

  // Copy constructor
  Stack (const Stack & stack);

  // Assignment operator:
  Stack & operator= (const Stack & stack);


  .
  .
  .
};
```

The assignment operator has a syntax that might strike one as unusual because it returns a value. The purpose for this is to permit expressions like:

```
Stack a,b,c,d;
a=b=c=d;
```

which are allowed for built-in datatypes and which (usually) also make syntactical sense for user-defined datatypes.

The copy constructor is invoked automatically in certain circumstances and it is important to know about that in order to understand the programs you write, particularly with regard to performance. The first circumstance is when an object is passed-by-value to a function; and the second circumstance is when an object is returned from a function. Consider the following code, which generates a sawtooth pattern within a stack of integers:

```
Stack sawtooth (Stack s)
{
  int nTeeth=5;
  nStep=10;
  for (int iTooth=0;iTooth<nTeeth;iTooth++) {
    for (int iStep=0;iStep<nStep;iStep++) {
      s.push(iStep);
    }
  }
  return s;
}
```

In this example, since the argument s to the function sawtooth is passed by value, a copy constructor is invoked. Furthermore, since the function sawtooth returns a Stack, a copy constructor is also invoked at the end of the function. In C++ we generally prefer to copy pointers rather than data, so a better way to implement sawtooth is like this:

```
void sawtooth (Stack & s)
{
  int nTeeth=5;
  nStep=10;
  for (int iTooth=0;iTooth<nTeeth;iTooth++) {
    for (int iStep=0;iStep<nStep;iStep++) {
      s.push(iStep);
    }
  }
}
```

Here we have written a void function that takes a *reference* to a Stack. The function can then modify the Stack variable in the calling routine, rather than return a copy.

In a little while we'll implement a function

```
Stack operator+ (const Stack & a, const Stack & b);
```

that will return the sum of two Stacks. A copy constructor is called when the function returns. The compiler then allows statements such as:

```
Stack a,b,c,d;
Stack e=a+b+c+d;
```

You should be aware of what is happening when adding four Stacks. First, the compiler adds c+d, creates a *temporary* to hold the result, and calls a copy constructor to initialize the temporary. Then this temporary is added to b, creating a second temporary and initializing it with a call to the copy constructor. The `Stack a` is then added to the second temporary, and again a copy constructor is called to initialize the temporary. Depending then on the optimization level and the compiler, a fourth call to the copy constructor can occur while copying this temporary to the `Stack e`. Sometimes, the so-called return value optimization does away with this call. Exercise 2 explores some of the behavior of constructors, assignment operators, and destructors, which is our next topic.

The creation and destruction of temporaries can be a serious source of inefficiency; it particularly affects vector and matrix operations. Some program libraries, and in particular the highly recommended `Eigen`, eliminate this by using an advanced technique called *expression templates*, described in Vandevoorde (2015).

6.4 Destructors

When objects go out of scope (at the end of a control loop or function definition, for example) the object is destroyed and its memory is reclaimed. At this point; however, it is sometimes necessary to take certain additional actions to clean up. For example, the class `std::ofstream` represents an output file stream. It can be constructed from a file name and used very much like `std::cout`, except that the output is directed to the file rather than the terminal. Here is how you can use it:

```
#include <fstream>
{
  std::ofstream stream("outfile.txt");
  stream << "HELLO from Trout Fishing in America"
      << std::endl;
}
```

The client initializes the stream and writes something to it. The stream goes out of scope at the second curly bracket at which point the memory for the class is released. But something else happens there too: *the output file is closed*.

A special hook is provided in C++ for taking actions like this when a class is destroyed; that hook is called a ***destructor***. The most common use of a destructor is to clean up memory that has been previously allocated at some point during the lifetime of the object. But whatever the reason, one can override the default compiler-generated destructor by declaring and defining his or her own destructor. The syntax for that is:

```
class Stack {
.
.
.
  //Destructor
  ~Stack();
};
```

and in the implementation,

```
Stack::~Stack() {
 . . .
}
```

One does not generally "call" a destructor, at least not explicitly, rather the destructor is called automatically when a variable goes out of scope.

6.5 `const` member data and `const` member functions

Consider the declaration of a `const int` in C++, which looks like:

```
const int i=5;
```

This integer constant can be read, but not written. Since it cannot be written, obviously it must be initialized where declared. One can also define an analogous user-defined datatype (e.g. `Stack`) to be `const`:

```
Stack a;
 .
 .
 .
Stack b=a;
const Stack c=a;
```

Having declared the object c to be `const`, certain member functions (those that are declared `const`) can be invoked on the object. In the `Stack` class, one such example is `size()`, declared as follows:

```
unsigned int size() const;
```

Such functions are not supposed to change the state of the object. In the following we present some rules governing `const` member functions.

If a member function is declared `const`, it can be invoked on a `const` object; otherwise the compiler will disallow the operation. For the `Stack` class, for example, the operations `push()` and `pop()` are not `const`, they cannot be invoked upon the object c above, which is `const`. That is reasonable, since `push()` and `pop()` change the state of the object c, either growing it or shrinking it by one element. The `size()` method, on the other hand, does not change the state of the `Stack` but merely retrieves its size; therefore it is declared to be `const`, and can be invoked upon the `const` object c. The functions `isEmpty()` and `print()` also do not change the state of the `Stack`, and are declared `const`.

Functions that change the state of an object are called *mutating*; those that do not are called *non-mutating*. By declaring the latter const, you are 1) promising not to change the state of the bound object within the function and as a result 2) allowing the use of the function when the bound object is const. Note that the compiler will not allow you to violate your promise! A member function that changes any of the member data cannot be declared as const and neither can a function that calls non-const member functions.

Only if a function does not change the member data directly or through function calls can it be declared const; and only if it is declared const can it be called for const objects. Remember this rule: *use* const *whenever you can* (Meyers, 2005a). A class lacking const-correctness has a negative knock-on effect throughout the code; and unfortunately many novices do not take the time to understand this important aspect of C++. The reader is referred to (Meyers, 2005a and 2005b) or (Press, 2007) for more information.

6.6 Mutable member data

There is, in certain circumstances, a good motivation to relax the restriction that const methods do not modify member data. That is when the member data is conceptually divided into variables which represent the *state* of the object, and those which represent a *cache* of quantities that can be returned upon demand. For example, a member function could compute the determinant of a matrix–likely an expensive operation. If the determinant were previously computed, it could simply be stored in an object's cache, avoiding a duplicate calculation at the expense of a little extra memory.

The keyword mutable can be combined with the declaration of any member data, i.e. : mutable int, mutable double, mutable std::complex<double>. Suppose the cache of some matrix object contained the member data, mutable double _determinant, and that a method called determinant() computed the determinant and stored it in the cache together, possibly, with another variable indicating whether the cache was stale or not. The mutable qualifier indicates that the _determinant is cache and is allowed to be altered in a const operation. The compiler then allows the declaration

```
double determinant() const;
```

6.7 Operator overloading

Practically any operator that is available in C or C++ can be *overloaded* to work on classes. These include the binary operators of Table 6.1, the shortcuts of Table 6.2, the relational operators of Table 6.3, the increment and decrement operators of Table 6.4, the unary operators of Table 6.5, and the logical operators of Table 6.6.

Table 6.1 *Binary operators.*

Symbol	Name
+	addition
−	subtraction
*	multiplication
/	division
%	modulus
^	bitwise exclusive OR
&	bitwise AND
|	bitwise OR
<<	left shift
>>	right shift

Table 6.2 *Shortcuts.*

Symbol	Name
+=	increase
−=	decrease
*=	multiply
/=	divide
%=	remainder
^=	exclusive OR into
&=	AND into
|=	OR into

All of these functions can be implemented as member functions. For the binary operators, the first operand is the bound object, and the second operand is the one and only argument to the function, like this:

```
class Stack {
    .
    .
    .
```

How to write a class **169**

Table 6.3 *Relational operators.*

Symbol	Name
==	Equality
!=	Inequality
<	Less than
>	Greater than
<=	Less than or equal to
>=	Greater than or equal to

Table 6.4 *Increment and decrement.*

Symbol	Name
++i	preincrement
i++	postincrement
--i	predecrement
i--	postdecrement

Table 6.5 *Unary operators.*

Symbol	Name
+	positive (no-op)
-	negative
*	dereference
&	reference (address-of)
~	ones complement

```
// Addition of two stacks:
Stack operator+ (const Stack & op2) const;
    .
    .
    .
};
```

Table 6.6 *Logical operators.*

Symbol	Name
||	OR
&&	AND
!	NOT
[]	Index
()	Function Call

There are two ways of invoking this operator, either

```
Stack a, b;
Stack c = a.operator+(b);
```

which is unnatural looking, or

```
Stack a,b;
Stack c=a+b;
```

which is fully equivalent- the compiler, in fact, makes no distinction between the two. The operator can also, however, be implemented as a free subroutine, which looks like this:

```
Stack operator+(const Stack & opA, const Stack & opB);
```

and now the second form looks identical and the first form is written as

```
Stack a,b;
Stack c=operator+(a,b);
```

The specific implementation depends largely on how much access is required to implement the operation. With our `Stack` class, the only access to the stack contents that we have provided through the public interface is through the `pop()` method, which makes it very difficult to splice two `Stacks` together in the right order in a free subroutine. However member functions have access to public and private member data, so in the case of our `Stack` class we are most likely to write the overloaded operator as a member function.

6.8 Friends

It is possible to relax the access restriction imposed by the keyword `private` if desired. For each class you are allowed to make a list of other classes, other functions, and other member functions that can have full access to all of the private member functions and member data of the class. These are called *friend classes* and *friend functions*. One simply provides a list of the authorized classes and functions in the class definition:

```
class Stack {
  .
  .
  .
  friend class classA;
  friend void classA::print();
  friend Stack operator+(const Stack &, const Stack &);
  .
  .
  .
};
```

Friend declarations should be used sparingly or not at all, since too much friendliness breaks encapsulation. If you are adding more than an occasional friend to the list, it might be a sign that your classes are poorly designed. Experienced programmers do not use the `friend` keyword frequently, if at all.

6.9 Type conversion via constructors and cast operators

Built-in datatypes have built-in conversion rules. For example it is legitimate to assign an integer to a floating point number–the integer is then promoted to a higher datatype. It is also legitimate to assign a floating point number to an integer–the floating point number is truncated in the operation.

User-defined datatypes can also have user-defined conversion rules. The conversion rule can be defined using two mechanisms:

- through a constructor taking a single argument
- through an explicit cast operator.

To illustrate this, we define a class called `Rational`, meant to store exact rational numbers. Internally the class is represented by an integer numerator and an integer denominator. The directory `EXAMPLES/CH6/RATIONAL` contains a minimal interface for the class, some example client code, and a `Makefile`. The header file is:

```
#ifndef _RATIONAL_H_
#define _RATIONAL_H_
#include <cmath>
#include <limits>
class Rational {
//
```

```cpp
// Here is an incomplete class representing rational
// numbers.  It can be constructed from numerator and
// denominator or with a double precision number.
//

public:

  // Construct from numerator and denominator:
  Rational(int num, int den):_num(num),_den(den){}

  // Construct from a double:
  Rational(double d):_num(1),_den(1){
    static int digits=std::numeric_limits<double>::digits;
    _num=(d*=(1<<(digits-1)));
    _den*=(1<<(digits-1));
  }

  // accessors:
  long int num() const { return _num;}
  long int den() const { return _den;}

  // compute the value as a double precision number:
  double value() const { return double(_num)/double(_den);}

private:

  long int _num;   // store the numerator
  long int _den;   // store the denominator

};
#endif
```

The client code is in fact a small test program called `client.cpp` located in the same directory. This code constructs three rational numbers and prints them out, using a routine called `print`:

```cpp
#include "Rational.h"
#include <iostream>
double print(Rational r) {
  std::cout            <<
     "Numerator :"     <<
     r.num()           <<
     " denominator "   <<
     r.den()           <<
```

```
                " value "              <<
                r.value()              <<
                std::endl;
}

int main() {
    Rational r=3.14159;
    Rational a(1,2);
    Rational b(1,6);
    print(r);
    print(a);
    print(b);
    return 1;
}
```

This code produces the following output:

```
Numerator :3294195 denominator 1048576 value 3.14159
Numerator :1 denominator 2 value 0.5
Numerator :1 denominator 6 value 0.166667
```

You can check this code out, compile it, and experiment with it a bit. The first experiment to perform is to replace `print(r)` with `print(3.14159)`. You may be surprised to see that the program works the same way. Why?

The explanation is the second form of the constructor `Rational(double d)`. Any constructor taking a single argument can and will be used to effect a type conversion. The `print` routine requires a `Rational` argument. The client code provides a double precision number. The compiler looks for a conversion, and finds one: the constructor. The constructor is called, the conversion is applied, and the resulting `Rational` is passed to the `print` routine. This illustrates one type of conversion.

Incidentally, if you don't want automatic type conversion you can turn it off with the `explicit` keyword, like this:

```
explicit Rational (double d);
```

Here is the next experiment: add the following lines to the client code (after the instantiation of a and b):

```
print(a+b);
```

This will not compile as is; the error will look something like:

```
error: no match for operator+ (operand types are Rational and
    Rational)   print(a+b);
```

However, you can add the following lines to the interface of `Rational`; they define the so-called *cast operator* (to double):

```
operator double() const {
  return _value();
}
```

And your compilation problems are solved. Moreover, the program also functions as expected, i.e., the output now has the extra line:

```
Numerator :699050 denominator 1048576 value 0.666666
```

the correct value for the variable c. How has this happened?

The cast operator is commonly used in a C-style cast, allowing you to explicitly convert one type of variable to another. For example:

```
Rational a(3,2);
double v=(double) a;
```

However, the cast operator is also employed at compile time to automatically convert one type of variable to another. A class can therefore convert in both directions, from a specified source and to a specified target. If we look at our example, we see that addition is defined on two `doubles` but not on `Rationals`, however the cast operator is invoked automatically to convert the `Rationals` to `doubles` for addition, then the result of the operation (a double) is converted again to a `Rational` and passed to the `print` routine. This type of behavior can be programmed into the interface of the class, or omitted–the decision is up to the designer.

6.10 Dynamic memory allocation

Now we come to a critical tool that we will use (Exercise 3) to turn our `Stack` class from an amateur job using a fixed array size to a more professional stack class that can grow to an arbitrary size, up to the available memory of the machine, taking up only as much memory at any time as it actually needs. The mechanism for allocating memory in C++ is `operator new`. Memory can be freed later with `operator delete`. The requisite syntax is shown here:

```
int     *i = new int;       // makes a single int and returns
                            //             a pointer
double  *d = new double;    // makes a new double and returns
                            //             a pointer
Stack   *s = new Stack;     // makes a new stack and returns
                            //             a pointer;
```

```
Rational *c = new Rational; // makes a new Rational and
                                        returns a pointer;
Rational *x = new Rational(1.0,2.0); // makes a new Rational
                                        1/2
```

operator new allocates space for a particular data type, allocating the needed space and, for user-defined datatypes, calling the specified constructor. The operator returns a pointer to the newly created object. You can play with the new object for as long as you like, and when you no longer need it you call delete on it:

```
delete i;
delete d;
delete s;
delete c;
delete x;
```

Deletion calls the destructor for the doomed object.

A pointer is often set to NULL (=0) to indicate that no object is allocated. When the time comes to allocate the object, the object is new'd and the pointer is set to the address of the new object. The pointer is then tested to see whether or not any object exists. Upon deletion, the object is reset to NULL. Since it does no harm to delete zero, delete can be called as a matter of course. The pattern looks something like this:

```
Stack *stack=NULL;
if (someCondition) {
  stack=new Stack;
}

if (stack) {
  stack->push(0);
  stack->pop();
  stack->print();

  // and whatever else you'd like to do...
}

delete stack;
stack=NULL;
```

The keywords new and delete can also be used to allocate entire arrays of either built-in data types like int or double, or user-defined data types like Stack or Rational. For the latter to work, the class should have a default constructor, which is called for each element of the new array. Here are a few examples:

```
int N=100;

int     *i = new int[N];        // array of int and returns a
                                //   pointer
double  *d = new double[N];     // array of double and
                                //   returns a pointer
Stack   *s = new Stack[N];      // array of stack and returns
                                //   a pointer;
Rational *c = new Rational[N];  // array of Rational and
                                //   returns a pointer;
```

If you have allocated an array, you must delete it using a form of delete that is special for arrays: `delete [] pointer`. Failure to do this will delete only one element of the array. Here are a few examples of the proper syntax for deletion of an array:

```
delete [] i;
delete [] d;
delete [] s;
delete [] c;
```

One must be aware of two dangers associated with dynamic memory allocation. The first is called **memory leak** and the second is called **memory overwrite**. A memory leak (Figure 6.1) occurs when a piece of memory that is allocated is not released when its job is done. If the allocation is performed within a loop, this can occur repeatedly. The usual punishment is that the size of the program increases in memory, exhausting the pool of available free memory, and causing the program to crash. A memory overwrite (Figure 6.2) occurs when a piece of memory that is still in use is accidentally deleted. In this case this piece of memory becomes available for re-assignment when another request for memory occurs somewhere in the program. The re-assigned memory is then overwritten. These two problems can be extremely difficult to track down. Sophisticated tools called *purify* and *valgrind* are sometimes used to identify and locate memory problems.

6.10.1 The "big four"

Obviously the task of class design becomes more complicated when memory is dynamically allocated. A rule of thumb that is easy to remember is that whenever you have dynamic memory allocation you can no longer get by with the compiler generated versions of the "big four" member functions, namely the constructor, the copy constructor, the destructor, and the assignment operator.

We will now adapt the `Stack` class to dynamically allocate its memory rather than using a fixed-size array (Exercise 3). The first thing we do is to comment out the fixed size array and implement a pointer instead:

MEMORY LEAK

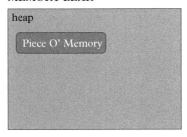
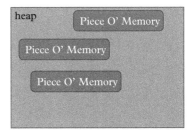
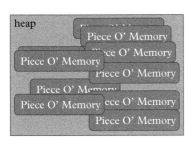
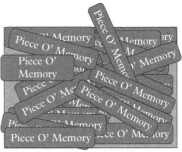

Figure 6.1 *In a memory leak memory that is allocated is not released.*

MEMORY OVERWRITE

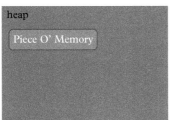

You allocate a piece of memory. Before you are really done using the object(s) you put there, you inadvertently release it.

That memory is no longer reserved for you. Elsewhere (in the program) another memory allocation takes place and uses your memory for its own purposes.

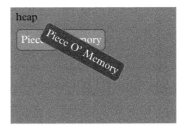

Figure 6.2 *Memory overwrites can occur when a piece of memory that is still in use is accidentally deleted.*

```
class Stack {
  .
  .
  .
  // Old definition:
  // int                _data[STACK_MAX];

  // New definition:
  int                *_data;
  .
  .
  .
}
```

When the `Stack` is born, the pointer `_data` should be initialized to zero, and whenever a new `int` is pushed onto the `Stack` we check the allocation, and if it is insufficient, we allocate more memory, then we copy the data from the old array to the new array, delete the old array, and set the `_data` pointer to point to the new array. This is an expensive operation, so we should generally allocate enough memory for a few more `push()` operations. One good strategy is to double the existing allocation each time more memory is needed. The initialization of the `_data` pointer requires us to write our own default constructor.

When the `Stack` goes out of scope we have to clean up any memory that has been allocated, by calling `delete []` on the pointer `_data`; we use the array form of `delete` because we will be allocating arrays of `int`s. If the `Stack` has never been used the `_data` pointer is zero (most people use the macro NULL from the header file `cstddef` for pointers, which is defined to be zero), but it does no harm to delete the NULL pointer anyway. Thus we are required to write a constructor and a destructor. For reasons we are about to explain, we also need to write a copy constructor and an assignment operator.

When a statement like

```
   Stack A=B;
```

is executed the compiler-generated copy constructor leaves the two objects in a state illustrated by Figure 6.3. The member-to-member copy of data has copied the pointer to the allocated memory but it has not cloned the memory. Thus both objects point to the same dynamically allocated memory. Now, if an `int` is added to A, B's memory is modified and may even be relocated. Worse, when A goes out of scope, B's memory is destroyed. To avoid these problems, a proper user-defined copy constructor clones any dynamically allocated objects owned by the source object, setting its own pointers to the address of cloned objects. The state of the two objects following this sequence of operations is shown in Figure 6.4.

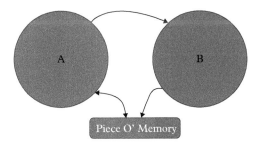

Figure 6.3 *An illustration of what the compiler-generated copy constructor does to a class with dynamically allocated memory. After the copy, both objects share the same memory.*

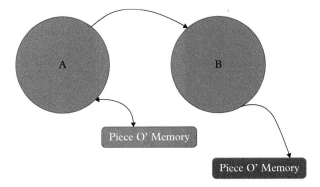

Figure 6.4 *A user-defined copy constructor should clone the memory of the source, by allocating its own space and then copying the contents of the dynamically allocated memory within the source object to its space.*

A compiler-generated assignment operation also wreaks havoc in a class with dynamically allocated memory. Consider the statement

```
Stack A, B;

B=A;
```

Prior to this set of operations, the two objects are in a state illustrated by Figure 6.5. The default assignment operator sets B's pointer to point to A's memory and loses the pointer to memory previously owned by B. This sets us up for, first, a memory leak, and second, a memory overwrite. You have to write your own assignment operator to avoid this disaster. The assignment operator has the signature:

```
Stack & operator= (const Stack & stack);
```

180 *Dynamic memory allocation*

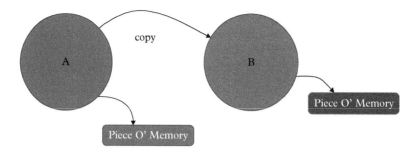

Figure 6.5 *Two objects before the assignment of A to B. The default assignment operator sets B's pointer to point to A's memory and loses the pointer to memory previously owned by B.*

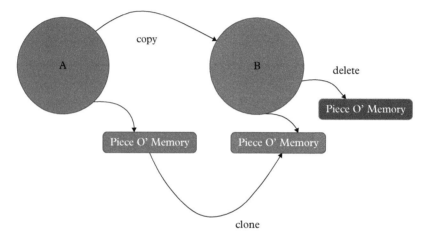

Figure 6.6 *The correct sequence of events in an assignment operator is, first, check for assignment to self; second, delete any existing dynamically allocated objects; third, clone the memory of the source; fourth, switch the pointer to point to this memory. Here, A is assigned to B; A is the source, and B is the target.*

which indicates that the operator has two functions; it must, first, carry out the assignment and, second, return the object to which it is bound. In the implementation:

```
Stack & Stack::operator= (const Stack & stack) {
}
```

you might well be puzzled by what to return. The answer is provided by a special pointer to bound objects called this in C++. You should dereference this pointer and return the result:

```
return *this;
```

As to the actual assignment the correct sequence of operations is illustrated in Figure 6.6.

There is one subtlety with assignment operators: what if an object is assigned to itself, i.e., A=A? This may seem crazy, but it is easily possible whenever objects are manipulated through pointers or references since these may obscure the true identify of the object. A check for assignment-to-self should be carried out and the actual assignment should be skipped if the condition is detected. It can be done using the `this` pointer, as follows:

```
if (&stack != this) {
   // then do assignment
}
return *this
```

In Exercise 3 you will get a chance to try implementing dynamic memory allocation in the `Stack` class as well as the "big four" member functions.

6.11 A worked example: implementing the Jones calculus

In this section we are going to write two classes that implement the Jones calculus, described earlier. These will use `std::complex` internally. For pedagogical purposes we will implement all the functions in-line. This involves defining a complex two-dimensional vector called a `JonesVector` and a complex 2x2 matrix called a `JonesMatrix`. The code we show below is also available in the EXAMPLES/CH6/JONESCALC area of the companion site, http://www.oup.co.uk/companion/acp.

The project contains the following files:

```
JONESCALC/JonesMatrix.h
JONESCALC/JonesMatrix.icc
JONESCALC/JonesVector.h
JONESCALC/JonesVector.icc
src/polarization.cpp
```

The two classes (`JonesVector`, `JonesMatrix`) are declared in the *.h files and defined in the *.icc files. This is a typical structure for classes which include only inline functions; inline functions are not actually "called" but rather expanded in the body of the calling routine, which can increase performance; but note that this behavior is switched on or off by compiler flags, particularly the optimization level (-On). The file `JonesVector.h` looks like this:

```
#ifndef _JONESVECTOR_H_
#define _JONESVECTOR_H_
#include <complex>
#include <iostream>
typedef std::complex<double> Complex;
class JonesVector {
```

```cpp
public:

  enum Type {Horizontal,Vertical,Diagonal,Antidiagonal,Right,
     Left};

  // Construct a zero vector:
  inline JonesVector();

  // Construct a type from a predefined type
  inline JonesVector(Type type);

  // Construct on two complex numbers:
  inline JonesVector (const Complex &x0, const Complex & x1);

  // Compute the magnitude square (=intensity) :
  inline double  magsq() const;

  // Access to individual elements:
  inline Complex & operator () (unsigned int i);
  inline const Complex & operator () (unsigned int i) const;
private:
  Complex x0,x1;

};
inline std::ostream & operator << (std::ostream & o, const
   JonesVector & v);
inline JonesVector operator* (const JonesVector & v, const
   Complex &c);
inline JonesVector operator* (const Complex & c,    const
   JonesVector & v);
#include "JONESCALC/JonesVector.icc"
#endif
```

In C++, a class should be declared once and only once, or else a compiler error will occur. It's possible–almost inevitable, particularly in a large project–that the include file containing the class definitions are included both directly by some compilation unit, or indirectly via another header file. To prevent the compiler from seeing two definitions, the entire header file is enclosed in what is called a ***guard macro***:

```cpp
#ifndef _JONESVECTOR_H_
#define _JONESVECTOR_H_
...
#endif
```

Directives starting with the pound sign (#) are interpreted by a preprocessor, which can define macros (here, _JONESVECTOR_H_) and, using #ifndef and #define, switch in or out various sections of the code. The preprocessor is run as part of the compiler step, using a command called cpp–you can, if you are interested, find more information about that by typing "man cpp" at the command line.

Include statements like

```
#include <iostream>
#include <complex>
...
```

are also processed by the C preprocessor cpp. The class above defines a nested enumerated type JonesVector::Type, which is like an integer but it can only take a limited set of values, namely JonesVector::Horizontal, JonesVector::Vertical etc. It also defines three overloaded constructors, which means that it permits three ways for the JonesVector to be initialized, namely, one of the following syntaxes:

```
JonesVector v0;                              // constructs a
                                             //    NULL vector
JonesVector v1(JonesVector::Horizontal);     // constructs a
                                             //    vector with
                                             // horizontal
                                             //    polarization

JonesVector v2(1,Complex(0,1));              // constructs a
                                             //    vector (1,i)
```

In addition, another way is implied (the compiler generates what is called a copy constructor to do this):

```
JonesVector v3(v2); // Is completely equivalent to:
JonesVector v3=v2;
```

We are completely satisfied with the compiler-generated version of the copy constructor, as with the compiler-generated assignment operator. The method magsq() returns the square magnitude of the vector. The method has the const qualifier, which as we have previously discussed, means that when this method is called, the object does not change state (i.e., the member data do not change). The declaration guarantees that the implementation cannot change the state of the class, and the guarantee allows the method to be called even on objects that are const JonesVectors.

Here we are using the function call operator, operator()(int i) for accessing an element of the vector. Generally speaking, any method that gives access to the data members is called an *accessor function*. The two forms of this accessor we have declared above have two differences: the second form is a const method while the first form isn't; and the second form gives only const (readonly) access while the second form gives read/write access. So when you type something like:

```
Complex u=myVector(0);
```

which method is called? Well, this depends on how the class `myVector` is defined, and there are two possibilities:

```
JonesVector myVector;
```

or

```
const JonesVector myVector;
```

With the second form, you cannot ever change the object `myVector`. The method which is declared to be `const` is the one which is invoked on a `const JonesVector`. Now, the first form of the accessor returns a non-const reference, and that allows us to carry out the following operation:

```
myVector(0)=Complex(0,1);
```

Should this be also allowed for a `const` object? No, because it changes the value of the object. But the second form of the accessor returns a const reference, and explicitly forbids this kind of access. We have here an example of *overloading on const*. Several discussions in the literature go more deeply into this subject: see for example (Meyers, 2005a) and (Press, 2007, Section 1.5.2).

While we are discussing the accessor we just wrote, note that we are also allowed to write:

```
myVector(0)=1.0;
```

even though the type of `myVector(0)` is `Complex`. This line of code is permitted and properly interpreted by virtue of automatic type conversion, discussed in Section 6.9.

Two `Complex` member data are part of the class, and they are hidden, encapsulated by placing them in the private part of the code. Several operators are defined on the `JonesVector` class, but they are not part of the class; rather they are *free subroutines*. The functions can print the `JonesVector` to the terminal (`operator <<`), or multiply the `JonesVector` by a scalar. We wrote two forms of this so that either `u=v*3.0` or `u=3.0*v` will work (for another example see Section 18.3.1). The header file includes the file `JonesVector.icc`, as is common practice when inline functions are declared. One can also declare the functions in the `.h` file at the same place they are declared, but we generally prefer to avoid this. The reason is that a header file can serve as pretty good documentation for the class, and the less cluttered the file is with implementation code, the better it is for documentation purposes.

We now examine the file JonesVector.icc.

```
#include <stdexcept>

inline JonesVector::JonesVector():x0(0.0),x1(0.0) {}

inline JonesVector::JonesVector(type) {
  const static Complex I(0,1);
  const static double s2=sqrt(2.0);
  if (type==Horizontal)         {x0=1.0;   x1=0.0;}
  else if (type==Vertical)      {x0=0.0;   x1=1.0;}
  else if (type==Diagonal)      {x0=1/s2;  x1=1/s2;}
  else if (type==Antidiagonal)  {x0=1/s2;  x1=-1/s2;}
  else if (type==Right)         {x0=1/s2;  x1=I/s2;}
  else if (type==Left)          {x0=1/s2;  x1=-I/s2;}
}
inline JonesVector::JonesVector(const Complex & x0, const
    Complex & x1):x0(x0),x1(x1) {}

inline double JonesVector::magsq() const {
  return norm(x0)+norm(x1);
}

inline Complex & JonesVector::operator () (unsigned int i) {
      if (!((i==0) || (i==1))) throw std::range_error
          ("Index out of range");
      return i==0 ? x0 : x1;
}

inline const Complex & JonesVector::operator () (unsigned
    int i) const{
  if (!((i==0) || (i==1))) throw std::range_error("Index out
      of range");
  return i==0 ? x0 : x1;
}

inline std::ostream & operator << (std::ostream & o, const
    JonesVector & v) {
  return o << "[" << v(0) << "," << v(1) << "]";
}

inline JonesVector operator * (const JonesVector & v,  const
    Complex & c) {
```

```
  return JonesVector(v(0)*c, v(1)*c);
}

inline JonesVector operator * (const Complex & c, const
   JonesVector & v) {
  return JonesVector(v(0)*c, v(1)*c);
}
```

The header file `<stdexcept>` is included because an exception is thrown on lines 22 and 27 when an index is out of range. Exceptions are an important part of a built-in mechanism for error handling in C++, called *exception handling*. For now we will remark that the header file `stdexcept`, which is part of the C++ runtime library, contains a set of predefined exceptions. These fall into two categories, `logic_error` and `runtime_error`. Logic errors are used to signal conditions that *should never happen*. Example: a programmer has taken great pains to ensure that two arrays, `double x[]` and `double y[]`, have the same length, but checks anyway, and if it is found that the two lengths do differ then a logic error is thrown. Logic errors include `domain_error`, `invalid_argument`, `length_error`, and `out_of_range`. The other category is called `runtime_error`, and includes `range_error` (used above), `underflow_error` and `overflow_error`. A class tree diagram is shown in Figure 6.7. All of the member functions in the class `JonesVector` use the scope operator `::` in their definitions to indicate, by prepending `JonesVector::` to each function that it is a member function of the `JonesVector` class. The default constructor uses a peculiar syntax for initialization. To initialize member functions in a constructor, a colon separates a list of initialization statements. Our default constructor makes a null vector. The third

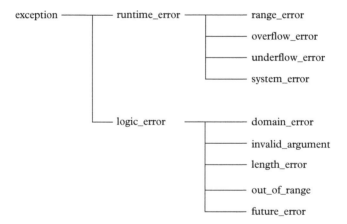

Figure 6.7 *Class tree diagram illustrating the relationship between the various kinds of* `exceptions`. *These classes require the inclusion of the header file* `stdexcept`. *They are declared within the namespace* `std`.

constructor is even more peculiar because the arguments to the constructor have the same name as the member data. The member data are said to **shadow** (take precedence over) the constructor arguments, but during the initialization the member data do not yet exist! So the syntax is permitted.

Finally, note that the three free subroutines that we define are not scoped within the `JonesVector` class since they do not belong to that class.

For completeness we also include the header files `JonesMatrix.h`:

```
#ifndef _JONESMATRIX_H_
#define _JONESMATRIX_H_
#include <complex>

typedef std::complex<double> Complex;

class JonesMatrix {
 public:
   enum Type { Horizontal,Vertical,Diagonal,
               Antidiagonal,Right,Left,
               Identity, FastHorizontal, FastVertical};

   // Construct the identity matrix:
   inline JonesMatrix();

   // Construct a type from a predefined type
   inline JonesMatrix(Type type);

   // Construct on two complex numbers:
   inline JonesMatrix (const Complex &a00, const Complex &
                       a01,
                       const Complex &a10, const Complex &
                       a11);

   // Access to individual elements:
   inline Complex & operator      () (unsigned int i,
                                      unsigned int j);
   inline const Complex & operator () (unsigned int i,
                                       unsigned int j) const;

 private:

   Complex a00,a01,a10,a11;
};
inline std::ostream & operator << (std::ostream & o, const
    JonesMatrix & v);
```

```cpp
inline JonesMatrix operator* (const JonesMatrix & m1, const
   JonesMatrix & m2);
inline JonesVector operator* (const JonesMatrix & m, const
   JonesVector & v);
#include "JONESCALC/JonesMatrix.icc"
#endif
```

and JonesMatrix.icc:

```cpp
#include <stdexcept>
inline JonesMatrix::JonesMatrix():a00(1.0),a01(0.0),a10(0.0),
    a11(1.1) {}

inline JonesMatrix::JonesMatrix(JonesMatrix::Type type) {
  const static Complex I(0,1);
  if (type==Horizontal)          {a00=1.0 ;a01=0.0;  a10=0.0;
                                  a11=0.0;}
    else if (type==Vertical)     {a00=0.0 ;a01=0.0;  a10=0.0;
                                  a11=1.0;}
    else if (type==Diagonal)     {a00=0.5 ;a01=0.5;  a10=0.5;
                                  a11=0.5;}
    else if (type==Antidiagonal) {a00=0.5 ;a01=-0.5;
                                  a10=-0.5; a11=0.5;}
    else if (type==Right)        {a00=0.5 ;a01=0.5*I;
                                  a10=-0.5*I; a11=0.5;}
    else if (type==Left)         {a00=0.5 ;a01=-0.5*I;
                                  a10=0.5*I; a11=0.5;}
    else if (type==Identity)     {JonesMatrix();}
    else if (type==FastHorizontal) {a00=1.0 ;a01=0.0; a10=0.0;
                                  a11=I;}
    else if (type==FastVertical) {a00=1.0 ;a01=0.0; a10=0.0;
                                  a11=-I;}
}

inline JonesMatrix::JonesMatrix(const Complex & a00,
                                const Complex & a01,
                                const Complex & a10,
                                const Complex & a11):
       a00(a00),a01(a01),a10(a10),a11(a11 ) {}

inline Complex & JonesMatrix::operator () (unsigned int i,
    unsigned int j) {

  if (!((i==0) || (i==1))) throw std::range_error("Index out
      of range");
```

```
    if (!((j==0) || (j==1))) throw std::range_error("Index out
       of range");

    return i==0 ?      (j==0 ? a00: a01) : ( j==0 ? a10: a11);
}

inline const Complex & JonesMatrix::operator () (unsigned
   int i, unsigned int j)   const {
   if (!((i==0) || (i==1))) throw std::range_error("Index out
      of range");
   if (!((j==0) || (j==1))) throw std::range_error("Index out
      of range");

   return i==0 ?      (j==0 ? a00: a01) : ( j==0 ? a10: a11);
}

inline std::ostream & operator << (std::ostream & o,
   const JonesMatrix & m) {
   return o << "[" << m(0,0) << " " << m(0,1)     <<  std::
      endl
          << " " << m(1,0) << " " << m(1,1)     << "]";
}

inline JonesMatrix operator* (const JonesMatrix & m1,
   const JonesMatrix & m2) {
   Complex a00= m1(0,0)*m2(0,0) + m1(0,1)*m2(1,0);
   Complex a01= m1(0,0)*m2(0,1) + m1(0,1)*m2(1,1);
   Complex a10= m1(1,0)*m2(0,0) + m1(1,1)*m2(1,0);
   Complex a11= m1(1,0)*m2(0,1) + m1(1,1)*m2(1,1);
   return JonesMatrix(a00,a01,a10,a11);
}

inline JonesVector operator* (const JonesMatrix & m,
   const JonesVector & v) {
   Complex v0= m(0,0)*v(0) + m(0,1)*v(1);
   Complex v1= m(1,0)*v(0) + m(1,1)*v(1);
   return JonesVector(v0,v1);
}
```

Operations that take JonesVectors and JonesMatrixs pass those objects by reference rather than by value. Operations which return JonesVectors or JonesMatrixs will be slowed down by the fact that many constructor calls—including the constructor of the Complex class and constructors of the JonesVector (or JonesMatrix class); moreover those routines pass-by-value, which means memory-to-memory copies are carried out. In practice one may need to avoid this type of routine if it becomes a bottleneck in a CPU-bound program. Alternatives can sometimes be found: the addition increment operator for example, `operator +=`, avoids the memory-to-memory copy. Inlining function calls can avoid the overhead of a subroutine call.

Finally we come to a small test program. The program below creates a `JonesVector` representing horizontally polarized light, and passes it through a set of three filters that together make a right circular polarizer. The intensity of the light going through the filters (relative to the input intensity) is computed, as is the polarization vector of the emerging light.

```
#include "JONESCALC/JonesVector.h"
#include "JONESCALC/JonesMatrix.h"
#include <iostream>
#include <string> int main (int argc, char ** argv) {

  JonesVector v0(JonesVector::Horizontal);
  JonesMatrix PH(JonesMatrix::FastHorizontal);
  JonesMatrix DG(JonesMatrix::Diagonal);
  JonesMatrix PV(JonesMatrix::FastVertical);

  // Make a circular polarizer:
  JonesMatrix CP=PH*DG*PV;
  JonesVector v1=CP*v0;

  std::cout << "Initial= " << v0 << std::endl;
  std::cout << "Filter= " << std::endl;
  std::cout << CP << std::endl;
  std::cout << "Final intensity= " << v1.magsq()
            << std::endl;
  std::cout << "Final polarization= "   << v1*(1/sqrt(v1.
     magsq())) << std::endl;
  return 1;
}
```

This is a lot of work to compute something as simple as the effect of some polarizers on a beam of polarized light. The virtue, however, clearly lies in reusability of the classes in the context of related problems.

6.12 Conclusion

In this chapter we have given a very quick introduction to the basic syntax of class declaration and definition. We have also discussed dynamic memory allocation and how it can be hidden behind the interface of a class like `Stack`. Now you are ready to write simple classes as your programs call for them. While it may seem complicated, with a little practice you will master issues like those discussed here, but only to find yourself thinking about bigger problems like how to design whole categories of classes to operate together. There is still one aspect of C++ that needs to be discussed before you can do that, which is the feature that allows whole categories of objects of different class to be handled identically. This is called *polymorphism*; a cornerstone of object-oriented programming. We shall return to this topic later; in the next chapter we will instead turn to a few applications in the physical sciences.

6.13 Exercises

1. Finish the class Stack by implementing the six functions. Check, whenever you push a new integer onto the Stack, that you do not exceed the fixed size of the Stack, and if you do, throw an exception. When you are done, create a subdirectory of your own area called STACK. If you are using a source code management system such as SVN, commit your work.

2. Edit the program client.cpp, adding a statement which initializes a third Stack and prints it out:

   ```
   Stack c=b;
   c.print();
   ```

   In the class Stack, make your own copy constructor by replacing the compiler-generated constructor. Make sure that it works! Check by printing the Stack! Do the same for the assignment operator. Also, add an empty destructor. Then, add short messages "HELLO from the copy constructor," "HELLO from the default constructor," "ASSIGN from the assignment operator," and "GOODBYE from the destructor" to the copy constructor, default constructor, assignment operator, and destructor in order to track the invocation of these functions. Recompile client.cpp and watch what happens. Create another directory called STACKTRACK and place the modified code there. If you are using a source code management system such as SVN, commit your code.

3. Rewrite the Stack class so that it uses dynamic memory allocation rather than a fixed array size. Do not forget to implement a constructor, copy constructor, destructor, and assignment operator. Create another subdirectory called STACK-DYNAMIC, place the modified code there, and if you are using a source code management system such as SVN, commit your work.

4. The Jones calculus is applicable to problems involving polarized light, but cannot handle situations involving partially polarized light. For these situations the polarization state is described by a four dimensional *Stoke's vector* and the action of optical elements is described by 4×4 *Mueller matrices*. Hecht (2001) gives a complete

description. Design and implement two classes implementing the Mueller calculus, together with a small test program, as we have done for the Jones calculus.

5. Tensor Product Spaces.
   The tensor product of two vectors, denoted as $\vec{v}_1 \otimes \vec{v}_2$, defines a vector space called the tensor product space; if $\dim \vec{v}_1 = M$ and $\dim \vec{v}_1 = N$, then $\dim \vec{v}_1 \otimes \vec{v}_2 = MN$. Define a class (call it `StateVector`) that supports the operation of the tensor product, as well as the more common operations: addition, subtraction and dot product. These operations should throw `std::runtime_error` when applied to incompatible operands. The class should be defined over a complex number field.

6. Tensor Product Operators.
   Write a class called `Operator` which acts linearly upon a `StateVector` to return a `StateVector` having the dimension of the original S. `Operators` should also support the tensor product operation, defined to act in the following way upon `StateVectors`:

$$(\mathbf{M}_1 \otimes \mathbf{M}_2) \cdot (\vec{v}_1 \otimes \vec{v}_2) \equiv (\mathbf{M}_1 \cdot \vec{v}_1) \otimes (\mathbf{M}_2 \cdot \vec{v}_2)$$

and on themselves according to:

$$(\mathbf{M}_1 \otimes \mathbf{M}_2) \cdot (\mathbf{A}_1 \otimes \mathbf{A}_2) \equiv (\mathbf{M}_1 \cdot \mathbf{A}_1) \otimes (\mathbf{M}_2 \cdot \mathbf{A}_2)$$

Implement this class defining the operations of addition, subtraction, scalar multiplication, composition, and action on `Operators`. These operations should throw `std::runtime_error` when applied to incompatible operands. It should be implemented over a complex number field.

7. Simultaneous Diagonalization of Commuting Operators.
   The commutator theorem states that two Hermitian operators $\mathbf{A}$ and $\mathbf{B}$ can be simultaneously diagonalized if their commutator $[\mathbf{A}, \mathbf{B}] \equiv \mathbf{AB} - \mathbf{BA} = 0$. Enhance the fledgling class library of the previous two problems to include a commutator operation, and a method for the simultaneous diagonalization of two Hermitian operators. The latter should be robust enough to diagonalize the operators when one or both operators are singular.

8. [P] Addition of Angular Momentum.
   In this problem we use the notation $|\chi\rangle$ to denote a vector in a complex two-dimensional vector space (also called a spinor). Choose, as basis for this space, the vectors denoted as

$$|\uparrow\rangle \equiv \begin{pmatrix} 1 \\ 0 \end{pmatrix}, \qquad |\downarrow\rangle \equiv \begin{pmatrix} 0 \\ 1 \end{pmatrix},$$

and as a basis for the tensor product space the vectors:

$$|\uparrow\uparrow\rangle \equiv |\uparrow\rangle \otimes |\uparrow\rangle, \quad |\downarrow\downarrow\rangle \equiv |\downarrow\rangle \otimes |\downarrow\rangle, \quad |\uparrow\downarrow\rangle \equiv |\uparrow\rangle \otimes |\downarrow\rangle, \quad |\downarrow\uparrow\rangle \equiv |\downarrow\rangle \otimes |\uparrow\rangle$$

We use the Pauli spin matrices,

$$\sigma_x = \begin{pmatrix} 0 & 1 \\ 1 & 0 \end{pmatrix} \quad \sigma_y = \begin{pmatrix} 0 & -i \\ i & 0 \end{pmatrix} \quad \sigma_z = \begin{pmatrix} 1 & 0 \\ 0 & -1 \end{pmatrix}.$$

Use the classes you wrote in the previous exercises to build the spin operators $S_i \equiv \frac{1}{2}\sigma_i$ ($i = \{x, y, z\}$) in a complex two-dimensional space as well as the following tensor product operators:

a) $\Sigma_x \equiv S_x \otimes I + I \otimes S_x$
b) $\Sigma_y \equiv S_y \otimes I + I \otimes S_y$
c) $\Sigma_z \equiv S_z \otimes I + I \otimes S_z$
d) $\Sigma_z^2 = \Sigma_x \cdot \Sigma_x + \Sigma_y \cdot \Sigma_y + \Sigma_z \cdot \Sigma_z$.

Determine the commutation relations between the operators Σ_i and Σ_j, ($i, j = \{x, y, z\}$) and between Σ_i and Σ^2. For each pair of commuting operators, determine the common eigenvectors in the basis stated above. (Note, the components of these vectors are the famous Clebsch-Gordan coefficients).

9. [P] Three Spin-1/2 Particles.
   Instead of adding two spin-1/2 particles as in the previous exercise, add three spin-1/2 particles. To do this generalize the Σ operators of the previous problem in the following way:

   a) $\Sigma_x \equiv S_x \otimes I \otimes I + I \otimes S_x \otimes I + I \otimes I \otimes S_x$
   b) $\Sigma_y \equiv S_y \otimes I \otimes I + I \otimes S_y \otimes I + I \otimes I \otimes S_y$
   c) $\Sigma_z \equiv S_z \otimes I \otimes I + I \otimes S_z \otimes I + I \otimes I \otimes S_z$

   with, as before, $\Sigma_z^2 = \Sigma_x \cdot \Sigma_x + \Sigma_y \cdot \Sigma_y + \Sigma_z \cdot \Sigma_z$. Determine the eigenvalues and common eigenvectors of the operators Σ_z and Σ^2; and express the eigenvalues in terms of the basis given by $|\uparrow\uparrow\uparrow\rangle, |\uparrow\uparrow\downarrow\rangle, |\uparrow\downarrow\uparrow\rangle$, etc.

10. Systems with one unit of angular momentum can be represented by a set of three basis vectors:

    $$\begin{pmatrix} 1 \\ 0 \\ 0 \end{pmatrix}, \begin{pmatrix} 0 \\ 1 \\ 0 \end{pmatrix}, \begin{pmatrix} 0 \\ 0 \\ 1 \end{pmatrix},$$

    while the angular momentum operators are defined as:

    $$\mathcal{J}_x \equiv \frac{1}{\sqrt{2}} \begin{pmatrix} 0 & 1 & 0 \\ 1 & 0 & 1 \\ 0 & 1 & 0 \end{pmatrix}, \quad \mathcal{J}_y \equiv \frac{1}{\sqrt{2}} \begin{pmatrix} 0 & -i & 0 \\ i & 0 & -i \\ 0 & i & 0 \end{pmatrix}, \quad \mathcal{J}_z \equiv \frac{1}{\sqrt{2}} \begin{pmatrix} 1 & 0 & 0 \\ 0 & 0 & 0 \\ 0 & 0 & -1 \end{pmatrix}.$$

Repeat the previous problem, but this time taking the tensor product of a) two angular-momentum 1 matrices, and b) an angular-momentum 1/2 matrix and an angular-momentum 1/2 operator. Determine, as in the previous problem, the simultaneous eigenvectors of Σ_z and Σ^2, defined as in the previous problem.

11. [P] Clebsch-Gordan Coefficients: the general case.
    In quantum mechanics, for a particle with spin quantum number j, the operator S_z has eigenvalues $m = j, j-1, j-2 \ldots -j+2, -j+1, -j$. Integer or half-integer values of the j are allowed in nature. For a given value of j, the angular momentum operators $\mathcal{J}_x$ and $\mathcal{J}_y$ can be expressed as:

    $$\mathcal{J}_x = \frac{1}{2}(\mathcal{J}_+ + \mathcal{J}_-)$$
    $$\mathcal{J}_y = -\frac{i}{2}(\mathcal{J}_+ - \mathcal{J}_-)$$

    where $\mathcal{J}_\pm$ has matrix elements $\langle m|\mathcal{J}_\pm|m'\rangle = \sqrt{(j \mp m')(j \pm m' + 1)}\delta_{m,m'\pm 1}$ When the basis is ordered such that $(1, 0, 0, \ldots)^T$ represents $m = j$, $(0, 1, 0 \ldots .)$ represents $m = j - 1$, and so forth, the matrix representing $\mathcal{J}_\pm$ is upper (lower) triangular with nonzero entries only immediately above (below) the main diagonal.

    a) From this prescription rederive the matrices $\mathcal{J}_x, \mathcal{J}_y$ and $\mathcal{J}_z$ for spin 1/2 and spin 1.
    b) Write a generic routine based on the classes `StateVector` and `Operator` which constructs the operators $\mathcal{J}_x$, $\mathcal{J}_y$ and $\mathcal{J}_z$ for arbitrary total angular momentum.
    c) By taking the tensor products as before, and performing a simultaneous diagonalization, perform the Clebsch-Gordan decomposition of the sum of any two angular momenta.

12. [P] SU(2) Group Elements.
    SU(2) is the group represented by 2 × 2 unitary matrices of unit determinant. An element of SU(2) can be represented by a unit four-vector. Thus if $a \in SU(2)$, then:

    $$a = \mathbb{1}\, a_0 + i\vec{a}\cdot\vec{\sigma},$$

    where σ is a Pauli matrix and:

    $$a^2 \equiv a_0^2 + \vec{a}\cdot\vec{a} = 1.$$

    Write a class to realize this group. Your code should implement:

    a) overloaded constructors
    b) a function returning the inverse of an element

c) a function returning the trace of a group element
d) group multiplication
e) multiplication of a group a element by a scalar s. Pay attention to $s*a$ and $a*s$.

See Chapter 24 if you need help.

....

BIBLIOGRAPHY

Booch, G. (2007). *Object Oriented Analysis and Design*. 3rd ed. Benjamin Cummings.
Bronson, G. J. (2013). *C++ for scientists and engineers*. Cengage Learning.
Capper, D. M. (1994). *The C++ programming language for scientists, engineers, and mathematicians*. Springer-Verlag.
Hecht, E. (2001). *Optics*, 4th ed. Pearson.
Meyers, S. (2005a). *Effective C++ 55 Specific Ways to Improve your Programs and Designs*. 3rd ed. Addison-Wesley.
Meyers, S. (2005b). *More Effective C++ 35 New Ways to Improve your Programs and Designs*. 3rd ed. Addison-Wesley.
Press, W., S. Teukolsky, W. Vetterling, and B. Flannery (2007). *Numerical Recipes, the Art of Scientific Computing*. 3rd ed. Cambridge University Press.
Vandevoorde, D. and N. M. Josuttis (2015). *C++ Templates, the Complete Guide*. Pearson.

7
Monte Carlo methods

7.1	Multidimensional integrals	197
7.2	Generation of random variates	198
	7.2.1 Random numbers in C++11	198
	7.2.2 Random engines	199
	7.2.3 Uniform and nonuniform variates	200
	7.2.4 Histograms	202
	7.2.5 Numerical methods for nonuniform variate generation	204
	7.2.6 The rejection method	204
	7.2.7 Direct sampling (or the transformation method)	207
	7.2.8 Sum of two random variables	210
	7.2.9 The Gaussian (or normal) distribution	211
7.3	The multivariate normal distribution, χ^2, and correlation	212
7.4	Monte Carlo integration	216
	7.4.1 Importance sampling	219
	7.4.2 Example	220
7.5	Markov chain Monte Carlo	221
	7.5.1 The Metropolis-Hastings algorithm	223
	7.5.2 Slow mixing	223
	7.5.3 Thermalization	225
	7.5.4 Autocorrelation	227
	7.5.5 Multimodality	228
7.6	The heat bath algorithm	229
	7.6.1 An application: Ising spin systems	229
	7.6.2 Markov chains for quantum problems	231
7.7	Where to go from here	232
7.8	Exercises	233
	Bibliography	239

Monte Carlo, the capital city of the Principality of Monaco, is Europe's premiere gambling destination. Monte Carlo methods were given their name because they are implemented by throwing computer-generated random numbers, similar to "virtual dice". These methods are useful in two areas. First, they are used in the simulation of stochastic processes, including quantum mechanical processes, which reflect the inherent randomness of nature, and thermodynamic processes which result from the randomness with which nature organizes staggeringly large systems of particles. Second, they are used to compute the numerical value of integrals. Both of these topics are quite generic, for example, 'simulation' may mean something as prosaic as following the path of an electron through a gas or as abstract as the path of a quantum system through Hilbert space. Similarly, 'integration' can range from evaluating the volume of a complex region to computing a field-theoretic Feynman path integral. In this chapter we discuss the building blocks of the Monte Carlo method and applications to integration. A more wide-reaching discussion of simulation is given in Chapter 15.

7.1 Multidimensional integrals

In Chapter 5 we discussed a variety of techniques for integrating one-dimensional functions. We did not have an extensive discussion of multidimensional integrals. Nesting integrals provides an obvious technique for the computation of approximate multidimensional integrals, but one should be aware that the rate of convergence decreases with dimensionality. To see this, suppose we seek to estimate the D-dimensional integral

$$I = \int dx_1 \int dx_2 \int dx_3 \ldots \int dx_D f(\vec{x})$$

by repeatedly using a classical quadrature formula for the each integral. The integral I is estimated as:

$$I \sim \sum_{i_1=1}^{n} \sum_{i_2=1}^{n} \sum_{i_3=1}^{n} .. \sum_{i_D=1}^{n} w_{i_1} w_{i_2} w_{i_3} .. w_{i_D} f(x_{i_1}, x_{i_2}, x_{i_3} \ldots x_{i_D})$$

where x_{i_j} are the weights of the ith point in the jth dimension and the w_{i_j} are the corresponding weights. These are determined by the classical quadrature formula which we wish to apply, as discussed in Chapter 5. The total number of points at which the function is evaluated is $N = n^D$. If the order of convergence of the classical quadrature formula is p, then the error decreases as $(1/n)^p = N^{-p/D}$. Thus a fourth-order method, such as Simpson's method, becomes a first-order method when applied to a four-dimensional integral.

In Monte Carlo integration, fixed abscissas are replaced with abscissas chosen at random; for multidimensional integrals the abscissas are points chosen at random throughout the multidimensional domain of integration. An estimate of the integral is derived from the value of the function evaluated at the randomly chosen points. Independent estimates evaluated with different random sequences of points will yield different estimates. The error in the estimate is statistical in nature, and like all statistical errors, it falls as the square root of the number of points used in the evaluation. While this may strike you as a slow convergence rate, it is independent of the number of dimensions, and superior to other methods in large dimensional spaces. In addition, Monte Carlo integration is usually very simple to code.

Monte Carlo integration will be the first application we discuss, and we will see how to set up the integrals, estimate their statistical errors, and understand how to optimize the calculation for CPU efficiency. The first task is to establish a method to draw random numbers according to a chosen probability distribution.

7.2 Generation of random variates

Random variables can be discrete or continuous. Discrete variables are characterized by a *probability distribution*, $P(x_i)$, where x_i ranges over all of possible values of x. Continuous variables are characterized by a ***probability density function*** (PDF), $\rho(x)$, defined such that $\rho(x)dx$ represents the probability of obtaining a value between x and $x + dx$. We take this to be normalized such that:

$$\int \rho(x)dx = 1,$$

with the integral taken over the range of possible outcomes.

A *variate* is a random number drawn from a probability distribution or a probability density function. Both simulation and integration require a source of random variates, which is the first computational challenge to be addressed in this chapter. This task is also referred to as *sampling* the distribution, or generating a realization of the distribution. In some cases we may know the analytic form of the PDF, and in other cases we may lack knowledge about the analytic form, but instead we have a clear idea of the steps which produce the PDF. This is particularly true when the PDF arises by following a stochastic or deterministic/stochastic system–like a gamma ray penetrating a slab of lead, where it can suffer one or more Compton scattering events, photoelectric absorption, or electron-positron pair production.

7.2.1 Random numbers in C++11

Since the advent of the new C++ standard, C++11, adopted by the International Organization for Standards (ISO) in August 2011, the situation with random number generation has changed substantially. The new standard library includes random number engines plus uniform and nonuniform variate generators of many types. The original source of this code was the Boost project (www.boost.org). Because these developments are very recent, the new standard may not be fully implemented by your compiler, or the implementation may be "experimental". In recent versions of the g++ compiler (4.8 and greater) the additional compiler flag -std=c++11 is required. All the class declarations and definitions for random number generation are defined in the header file <random>. The best documentation we have found for this is online at:

www.cplusplus.com/reference

In most implementations, including that of the C++ standard library, the generation of random variables is separated into two steps. In the first step, a random sequence of bits is generated by an *engine*. In the second step, these bits are turned into boolean, integer, or floating point variates following a specified probability distribution or PDF.

7.2.2 Random engines

The output of an engine is a completely deterministic series of bits that depends on the internal state of the engine; the state may consist of a single integer or an array of integers. The initial values of the integer(s) are called the *seed(s)*. Once the seed is given, the random engine will produce the same "random" pattern of bits each time the program is executed. Because the behavior is entirely deterministic, these generators are referred to as *pseudorandom* number generators. In some cases, this reproducibility is desired. In other cases, like when statistically independent random numbers are required from the same program, it is not what you want. To get a statistically independent sample, you need to change the seed(s) used to initialize the random engine at the beginning of each independent run. Alternately, you can use a genuine source of random numbers.

We will not discuss the many different algorithms used to generate random bits. The subject is vast, often touching on number theory, and the research is quite active since cryptography relies critically on good generators. For a basic discussion we refer the interested reader to (Press, 2007, Chapter 7). The most widely used random number generator in the world is the *Mersenne twister* engine (Matsumoto, 1998). Its standard library implementation is the class `std::mt19937`, which we will use in our examples. The following code instantiates a Mersenne twister engine and prints out the random bits from that engine:

```
#include <iostream>
#include <random>
typedef std::mt19937 EngineType;
int main() {

  // A way of generating random bits:
  EngineType e;
  for (int i=0;i<10;i++) std::cout << e() << std::endl;
}
```

Note that we have first defined the `EngineType` to be `std::mt19937` which can be easily changed to any other engine, e.g. `std::ranlux24` or `std::knuth_b`. All of the these generators are pseudorandom number generators: run the program again and you will see that the same set of random bits is produced.

After a very long time, the random numbers start to repeat the same sequence. The number of numbers that can be drawn from the generator before repetition occurs is called the *cycle length*, and for a good random number generator this is very long. For `std::mt19937` the cycle length is $2^{19937} - 1$.

The seed is a `long unsigned int` and it can be specified in the constructor to the engine, like this:

```
EngineType e(seed)
```

As an exercise, try this, and verify that the sequence of random numbers varies with the value of the seed.

If you need a source of truly random and always independent numbers, you can use the `std::random_device` engine. It functions like any other random engine in the library but it provides genuine random numbers instead of pseudorandom numbers, taking these numbers from occurrences that are detected on the host computer. You can switch to this engine with the declaration:

```
typedef std::random_device EngineType;
```

Unlike seeded pseudorandom number generators, this generator will provide a different stream of random bits every time the program is run. It can also be used to generate a single truly random number which one then uses to seed a pseudorandom number generator.

7.2.3 Uniform and nonuniform variates

Once the random bits are generated, they are easily transformed to a uniform variate by letting N bits represent an unsigned integer, then dividing this integer by $2^N - 1$ (which is the largest integer representable by N bits). The standard library provides a (template) class for this,

```
std::uniform_real_distribution
```

which has an interface similar to that of other distributions available in the library.

The following example produces and prints ten uniform variates between 0 and 1:

```
#include <cstdlib>
#include <iostream>
#include <string>
#include <random>
typedef std::mt19937 EngineType;
int main (int argc, char **argv) {

  EngineType e(100);
  std::uniform_real_distribution<double> u;

  for (int i=0;i<10;i++) {
    std::cout << u(e) << std::endl;
```

```
    }
    return 1;
}
```

You can change the range of the uniform variate in the constructor call, for example the following line

```
std::uniform_real_distribution<double> u(-1,1);
```

will instantiate a uniform distribution on the interval [-1,1]. The distribution u produces a new result each time the function call operator u(e) is invoked. The default interval is [0,1]. Generating nonuniform variates is a common task. The techniques we discuss later use the uniform variate as a key component in the generation of nonuniform variates.

A variety of nonuniform variable generators are available in the C++ standard library. A complete description can be found at www.cplusplus.com/reference. These are called "distributions", which are classes that generate a random variate using a random engine. They include:

- `uniform_real_distribution` and `uniform_int_distribution`, generating floating point or integer variates spread uniformly over an interval [a,b]. (Parameters like a and b are generally specified in the constructor call, here and throughout this list.)
- The `bernoulli_distribution`, generating the boolean result of a single "yes" or "no" experiment that is parametrized by the probability p of succeeding.
- The `binomial_distribution`, generating the integer number N of successes in an ensemble of M experiments parametrized by the probability p of succeeding. The corresponding probability distribution is:

$$P(N|M,p) = \frac{M!}{N!(M-N)!} p^N (1-p)^{(M-N)}$$

- The `poisson_distribution`, generating the integer number of events falling within a fixed interval of time, space, or some other variable, parametrized by the mean expected number of events μ. The probability distribution is:

$$P(N|\mu) = \frac{\mu^N}{N!} e^{-\mu}$$

- The `exponential_distribution`, generating a floating point variate according to an exponentially falling probability density, parametrized by the decay constant λ. The probability density is:

$$\rho(x|\lambda) = \lambda e^{-\lambda x}.$$

- The normal_distribution, generating a floating point variate according to a normal (or Gaussian) distribution, parametrized by the mean x_0 and standard deviation σ, and described by the probability density function:

$$\rho(x|x_0, \sigma) = \frac{1}{\sqrt{2\pi}\sigma} e^{-(x-x_0)^2/2\sigma^2}.$$

- The gamma_distribution, generating a floating point variate according to a gamma distribution, which can be expressed as:

$$\rho(x|\alpha, \beta) = \frac{1}{\Gamma(\alpha)\beta^\alpha} x^{\alpha-1} e^{-x/\beta}$$

This has some of the same features as the normal distribution, but does not extend to negative values.

- The chi_squared_distribution, generating a floating point variate according to the χ^2 distribution:

$$\rho(x|N) = \frac{1}{\Gamma(N/2)2^{N/2}} x^{N/2-1} e^{-x/2}$$

which arises frequently in statistical applications and in data analysis.

- The Cauchy_distribution, generating a floating point variate according to:

$$\rho(x|\gamma, x_0) = \frac{1}{\pi\gamma\left(1 + \left(\frac{(x-x_0)}{\gamma}\right)^2\right)}$$

is called a Cauchy distribution, also known to atomic physicists as a Lorentzian distribution and to particle and nuclear physicists as a Breit-Wigner.

This list is not exhaustive. We can very often find the distribution we need in the standard library. When that is not possible, we will need to roll our own. Before we do that we need a tool to visualize sequences of random variables, whether they are coming from off the shelf software or whether we have generated them by some other means.

7.2.4 Histograms

Visualizing probability distributions (and other things) can be very helpful (for example, when debugging). The proper way to do this is to create a histogram and plot it. In the QAT library, this can be done by creating a histogram (class Hist1D), creating a plot from it (class PlotHist1D), and adding it to the PlotView. Here is a code excerpt with the relevant parts; the whole program can be found in EXAMPLES/CH7/HIST.

```
#include "QatDataAnalysis/Hist1D.h"
#include "QatPlotting/PlotHist1D.h"
#include <random>
...
typedef std::mt19937 EngineType;
...
int main (int argc, char * * argv) {

  ...

  EngineType e;
  Hist1D histogram ("Random", 1000, -10.0, 10.0);

  std::uniform_real_distribution<double> u;
  for (int i=0;i<100000;i++) {
    double x=u(e);
    histogram.accumulate(x);
  }

  PlotHist1D pH=histogram;
  PlotView view(pH.rectHint());
  view.add(&pH);
  ...
}
```

The plot generated by this simple example is shown in Figure 7.1 (left). Now that we have a way of visualizing the data we produce, we can experiment with the random number library, changing, for example the distribution according to which random numbers are generated. The modifications required to generate according to the normal distribution are quite trivial:

```
  std::normal_distribution<double> n(3.0, 0.5);
  for (int i=0;i<100000;i++)  {
    double x = n(e);
    histogram.accumulate(x);
  }
```

Arguments to the constructor are the mean $x_0 = 3.0$ and the width $\sigma = 0.5$. Figure 7.1 (middle) shows the plot we obtain with those modifications. Figure 7.1 (right) shows an example of a gamma distribution generated using $\alpha = 3, \beta = 1$. The constructor call is:

```
gamma_distribution<double> g(3.0);
```

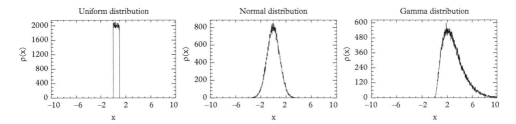

Figure 7.1 *A plot of the binned data produced by the example code, with the aid of the classes,* `Hist1D` *and* `PlotHist1D`. *On the left is a uniform distribution. In the middle, a normal distribution obtained with minor modifications described in the text. On the right is a gamma distribution, obtained again with minor modifications.*

7.2.5 Numerical methods for nonuniform variate generation

When you cannot find the distribution you need in the standard library, you'll need to generate your own. There are two widely applicable methods for sampling a distribution. The *rejection method* is simple and robust and works with either univariate PDFs or multivariate PDFs, but it is inefficient (extremely so in some cases). The *transformation method* is elegant and fast and generally preferred when applicable for univariate PDFs. If neither of these methods work then you find yourself in the domain of other tricks tailored to specific distributions; in this case we refer you to (Devroye, 1986), which is an extensive treatise on the subject.

7.2.6 The rejection method

The rejection method is also called "random throwaway" or von Neumann rejection. Suppose that one seeks to generate a random variable x according to the distribution $\rho(x) = \frac{3}{4}\left(1 - x^2\right)$ for $-1 < x < 1$. The method requires that one determine ρ_{max}, the maximum value that the PDF takes over its full range. Here $\rho_{max} = 3/4$. Next we throw a uniform variate x over the full range (from -1 to 1). Finally, we decide whether to keep the variable that was just generated. For that purpose, we throw a second random variable y generated uniformly from 0 to ρ_{max}. Keep the first variable x only if the second variable y satisfies $y < \rho(x)$, otherwise throw it away. In other words, we generate points uniformly over a rectangular region in a two-dimensional space, retaining only those points that fall beneath the desired PDF (in y) and then ignoring their y coordinate.

That this procedure samples $\rho(x)$ can easily be seen by plotting the two variables in the xy plane, applying the condition $y < \rho(x)$, and then projecting out the conditional probability onto the x-axis.

The code looks something like this:

```
EngineType e(100);
std::uniform_real_distribution<double> rx(-1, 1),ry
    (0,0.75);
```

```
Hist1D hx("X", 100, -2.0, 2.0);
Genfun::Variable    X;
Genfun::GENFUNCTION f=3/4.0*(1-X*X);
for (int i=0;i<1000000;i++) {
  double x = rx(e);
  double y = ry(e);
  if (y<f(x)) {
    hx.accumulate(x);
  }
}
```

The full program can be found in EXAMPLES/CH7/THROWAWAY1. Note that it is important that $f(x)$ never exceeds the maximum allowable value of the random variable y, in this case 3/4. On the other hand, if the maximum value of y is chosen to be too large, the distribution will be correct, but inefficient. Note also that while we have used a normalized probability density in this example, in general the method does not require that the probability density be normalized.

The technique can be trivially extended to generate joint probability distributions as well. As an example we take the probability distribution:

$$\sigma \sim \frac{1}{2-x} + (2-x) - 2(1-x^2)\cos^2\phi$$

which occurs in the Compton scattering of polarized light, as we shall see in Chapter 15. The code now requires three random variables to be generated per attempt ($\hat{x}$, $\hat{\phi}$, and the rejection variate). It is illustrated in the following code, excerpted from the complete program in EXAMPLES/CH7/THROWAWAY2. The corresponding output is shown in Figure 7.2. The code also illustrates the use of the Hist2D and PlotHist2D classes from the QAT library for making and displaying two dimensional histograms.

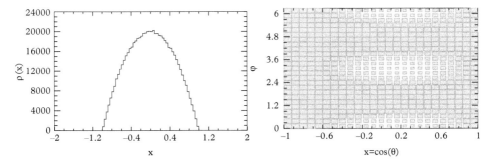

Figure 7.2 *Two plots illustrating the rejection technique. On the left is the technique applied to sample a simple 1d PDF, $x = 3/4(1-x^2)$. On the right, a plot of the Compton scattering cross section vs. the two variables $x = \cos\theta$ and ϕ, for annihilation radiation with $E_\gamma = 511$ keV. The code for the two examples is in EXAMPLES/CH7/THROWAWAY1 and EXAMPLES/CH7/THROWAWAY2.*

Generation of random variates

```cpp
#include "QatDataAnalysis/Hist2D.h"
#include "QatPlotting/PlotHist2D.h"
#include <random>
typedef std::mt19937 EngineType;
int main (int argc, char * * argv) {

  ...

  const double MAXTHROW=2.0;
  EngineType e;
  std::uniform_real_distribution<double>
    rx(-1.0,1.0),
    rphi(0.0,2*M_PI),
    ry(0.0,MAXTHROW);

  Hist2D hxy("X vs PHI",
             25, -1.0, 1.0,
             25,  0, 2.0*M_PI);


  Genfun::Variable    X(0,2),PHI(1,2);
  Genfun::Square      Square;
  Genfun::Cos         Cos;
  Genfun::GENFUNCTION f = 1/(2-X) + (2-X) - 2.0*(1-X*X)*
    Square(Cos(PHI));

  for (int i=0;i<1000000;i++) {
    Genfun::Argument arg(2);
    arg[0]=rx(e);
    arg[1]=rphi(e);
    double y = ry(e);
    if (y<f(arg)) {
      hxy.accumulate(arg[0],arg[1]);
    }
    else if (y>MAXTHROW) {
      throw std::runtime_error ("Function value exceeds max
          random no");
    }
  }
  PlotHist2D pxy=hxy;
  {
    PlotHist2D::Properties prop;
    prop.brush.setStyle(Qt::SolidPattern);
```

```
      prop.brush.setColor("darkRed");
      pxy.setProperties(prop);
    }
    PlotView view(pxy.rectHint());
    view.add(&pxy);
    ...
}
```

7.2.7 Direct sampling (or the transformation method)

The second main method for sampling distributions is called the *transformation method*, also known as direct sampling. This is applicable when the PDF that we wish to sample can be integrated and the resulting cumulative density function can be inverted. When these conditions are satisfied the distribution can be sampled with one call to a uniform variate generator and one call to a function that carries out the transformation from the uniform variate to a nonuniform variate distributed according to the desired PDF.

Let x be a random variable distributed uniformly on the interval [0,1]. The function we use to transform x to another variable y is called the *sampling function*, and we will designate it as S, so that $y = S(x)$. We wish y to be distributed according to some probability distribution $\rho(y)$, and our job is to determine the sampling function.

To do so, consider a small interval in the range of x of width dx. The probability for the variate generator to draw a value in this interval is $\rho_x(x)dx$. The interval corresponds to an interval of the transformed variable y, with width dy such that the probability will fall within the interval of $\rho_y(y)dy$. Setting these two probabilities equal we get:

$$\rho_x(x)dx = \rho_y(y)dy.$$

However $\rho_x(x) = 1$ since x is uniformly distributed so:

$$dx = \rho_y(y)dy$$

or:

$$\int_0^x dx = \int_{y_0}^y \rho_y(y)dy \equiv F(y) \qquad (7.1)$$
$$x = F(y)$$

or:

$$y = F^{-1}(x).$$

We therefore see that the sampling function is $S = F^{-1}$, which is the inverse of the *cumulative probability density* (CDF):

$$F(y) = \int_{y_0}^{y} \rho_y(y) dy.$$

In Eq. 7.1, the quantity y_0 varies from problem to problem but represents the lowest value for which the PDF $\rho(y)$ is defined. This could be $-\infty$ in certain cases, or 0 in others, depending on the nature of the PDF. To summarize the requirements of this method: the PDF must be integrable and indeed normalized; and the integral of the normalized PDF must be invertible.

Let us now take a look at several PDFs that can be generated by the transformation method. We start with the normalized PDF for a typical radioactive decay process:

$$\rho(t; \tau) = \frac{1}{\tau} e^{-t/\tau}.$$

where τ is the mean life of the decaying isotope. The cumulative distribution function is:

$$y = F(t) = \frac{1}{\tau} \int_0^t e^{-t/\tau} dt = 1 - e^{-t/\tau}$$

and the inverse is:

$$t = -\tau \ln(1 - y).$$

Thus $S(x) = F^{-1}(x) = -\tau \ln(1 - x)$ is the sampling equation. Recall that x is a uniform variate on the interval $[0,1]$.

Another useful distribution that can be generated this way is the Cauchy distribution, which also goes by the name of a Breit-Wigner (for nuclear and particle physicists) and Lorentzian (for atomic physicists). The Cauchy distribution describes unstable resonances in physics, such as a spectral line of energy x_0 and half-width γ. The PDF is:

$$\rho(x; x_0, \gamma) = \frac{1}{\pi \gamma \left(1 + \left(\frac{x - x_0}{\gamma}\right)^2\right)}.$$

The distribution is characterized by the mean x_0, and the half-width at half maximum γ. The problem can be reduced to sampling the **standard Cauchy distribution**:

$$\rho(x; x_0 = 0, \gamma = 1) = \frac{1}{\pi (1 + x^2)}$$

by rescaling and shifting the result.

The CDF for the standard Cauchy distribution is:

$$F(x) = \frac{1}{\pi} \int_{-\infty}^{x} \frac{1}{1+x^2} dx$$
$$= \frac{1}{\pi} \tan^{-1} x + \frac{1}{2}.$$

Inverting gives the sampling function:

$$S(x) = \tan\left(\pi\left(x - \frac{1}{2}\right)\right).$$

As a third example of the transformation method, we sample a symmetric triangular distribution on the interval $[-2,2]$. The distribution is:

$$\rho(x) = \frac{1}{4}(2+x) \qquad -2 < x < 0$$
$$\rho(x) = \frac{1}{4}(2-x) \qquad 0 < x < 2.$$

The distribution is piecewise continuous, but we will find a trick to generate it efficiently. The idea will be to generate a uniform variate on the interval $[-1,1]$. If the result is positive it is used to generate the negative half of the distribution $(-2 < x < 0)$, if it is negative it will be made positive and used to generate the positive half. We re-normalize the negative half of the distribution so that it integrates to unity:

$$\rho_+(x) = \frac{1}{2}(2+x) \quad -2 < x < 0,$$

and integrate to obtain the cumulative distribution function:

$$F(x) = \frac{1}{4}(x+2)^2.$$

Finally, invert to obtain the sampling equation.

$$S(x) = \sqrt{4x} - 2.$$

When x (generated from -1 to 1) is positive we return the value $S(x)$ and when it is negative we return $-S(-x)$.

Code that implements all three of the examples in this section can be found in the directory EXAMPLES/CH7/DIRECTSAMPLE. The area is set up to build a program called directsample that can be invoked with a flag -ex taking the value 1, 2, or 3, to generate the first (exponential), second (Cauchy/Breit-Wigner), or third (triangular) examples.

7.2.8 Sum of two random variables

Imagine that we have two random variates x_1 and x_2 drawn from probability distributions $\rho_1(x_1)$ and $\rho_2(x_2)$. What is the probability distribution of the sum $u = x_1 + x_2$?

To answer this question form the joint probability density for the two variables $\rho(x_1, x_2) \equiv \rho_1(x_1) \cdot \rho_2(x_2)$ and change variables:

$$(x_1, x_2) \to u, v$$
$$u = x_1 + x_2$$
$$v = x_2.$$

The Jacobian of this transformation has determinant $|\mathbf{J}| = 1$.

Now we transform the probability density as follows:

$$\rho(x_1, x_2)dx_1 dx_2 = \frac{\rho_1(x_1) \cdot \rho_2(x_2)}{|\mathbf{J}|} du dv$$
$$= \rho_1(u-v)\rho(v) \, du dv$$

so that we identify the joint probability density $\rho(u, v) = \rho_1(u-v)\rho_2(v)$. To obtain the probability for $u = x_1 + x_2$ alone, we integrate over v:

$$\rho(u) = \int \rho(u,v) dv = \int \rho_1(u-v)\rho_2(v) \, dv.$$

This latter function is called the **convolution**, denoted $(\rho_1 * \rho_2)$, of the two PDFs. Thus the sum of two random variates is distributed according to the convolution of their PDF's.

A common application of convolution is in describing the resolution of a detector or other effects that smear out the response of an instrument to a signal. Imagine that one is observing a spectral line with a spectrometer. If resolution effects can be neglected the spectrometer should record a Lorentzian lineshape, i.e., a lineshape described by a Cauchy distribution. However a thermal Doppler shift can increase or decrease the energy of a single detected photon, depending upon the velocity of the atom emitting the photon. The increase/decrease is governed by a normal distribution with

$$\sigma = \sqrt{kT/mc^2} x_0,$$

where x_0 is the natural frequency of the emission line. Thus the observed lineshape will be the convolution of a Gaussian distribution with a Cauchy distribution. This curve, having three parameters ($\Gamma = 2\gamma$, x_0 and σ), is called a ***Voigt distribution***. You can find an implementation in `QatGenericFunctions/VoigtDistribution`.

If we sum up more random variates something interesting begins to happen. The result of summing 1, 2, 3, and 4 uniform variates (from -1 to 1) is shown in Figure 7.3.

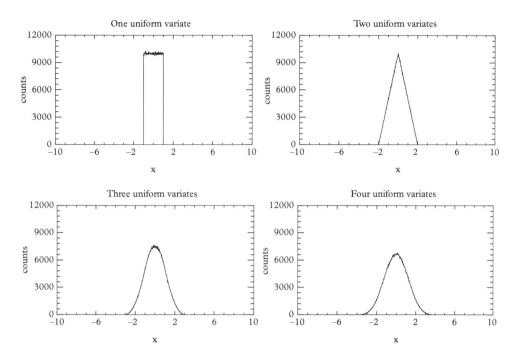

Figure 7.3 *Sum of one, two, three, and four uniform variates. The Gaussian behavior of the sum emerges as a consequence of the central limit theorem.*

The sum of a large number of random variates with identical distributions will be Gaussian according to the central limit theorem. In this example we see that reasonably Gaussian behavior begins to emerge after only a few uniform variates are summed up. You therefore might be tempted to add variates to approximate a Guassian distribution. The distribution, however, cuts off entirely at $x = \pm N$ where N is the number of uniform variates used in the sum, whereas a real Gaussian distribution has infinite extent in either direction. This might not be a problem if the statistics you intend to generate is such that fewer than one single event will be found past the cutoff point. But there are better ways to generate a Gaussian distribution.

7.2.9 The Gaussian (or normal) distribution

Because of its importance in mathematics and physics, many sources of routines for Gaussian sampling, including the C++ standard library, exist. However, the prominence of the Gaussian distribution makes a closer look worthwhile.

To start, note that the CDF is related to an error function, and this is not readily invertible, so the transformation method is awkward. Nevertheless, it *is* possible to construct a working algorithm if an accurate implementation of erf^{-1} can be obtained (see (Abramowitz and Stegun, 1972) for this). But a better algorithm, called the **Box-Müller method**, obviates the need for this labor.

The method is based on the following simple observation:

$$\frac{1}{\sqrt{2\pi}}e^{-x^2/2}\frac{1}{\sqrt{2\pi}}e^{-y^2/2}dxdy = \frac{1}{2\pi}e^{-r^2/2}r\,dr\,d\phi.$$

Because the radial dependence is simple, one can generate points in the $r - \phi$ plane and then transform to Cartesian coordinates (hence *two* Gaussian variates are created). Specifically, one generates ϕ according to a uniform distribion on $[0, 2\pi]$ and r according to $\rho_r(r) = e^{-r^2/2}r$. To do this, note that:

$$\rho_r(r)\,rdr = e^{-r^2/2}\,rdr$$
$$= e^{-z}\,dz.$$

To summarize, one has the algorithm to:

1. Generate ϕ uniformly on the interval $[0, 2\pi]$.
2. Generate z according to an exponential distribution.
3. Compute $r = \sqrt{2z}$.
4. $x = r\cos\phi$ and $y = r\sin\phi$ are independent normally distributed random variables.

The Box-Müller method is commonly available within most random number packages including those of the C++ standard library so you will never actually have to code it yourself. Use

```
std::normal_distribution
```

instead.

7.3 The multivariate normal distribution, χ^2, and correlation

Let x and y be two independent, normally distributed random variables with mean x_0 and y_0 and standard deviation σ_x and σ_y. Their joint probability density is simply the direct product of two normal distributions:

$$\rho(x,y) = \rho(x)\cdot\rho(y) = \frac{1}{\sqrt{2\pi}\sigma_x}e^{-\frac{(x-x_0)^2}{2\sigma_x^2}}\frac{1}{\sqrt{2\pi}\sigma_y}e^{-\frac{(y-y_0)^2}{2\sigma_y^2}}. \tag{7.2}$$

This is normalized such that $\int_{-\infty}^{\infty}\int_{-\infty}^{\infty}\rho(x,y)dxdy = 1$. Contours of constant $\rho(x,y)$ in the x-y plane are ellipses, with the principal axes oriented along the x and y axes. To obtain a more general expression for the distribution we will make the following definitions:

$$\vec{\alpha} \equiv (x, y)$$
$$\vec{\alpha}_0 \equiv (x_0, y_0)$$
$$\mathbf{D} \equiv \begin{pmatrix} \sigma_x^2 & 0 \\ 0 & \sigma_y^2 \end{pmatrix}.$$

These definitions permit a more compact expression for Eq. 7.2:

$$\rho(\vec{\alpha}) = \frac{1}{(2\pi)\,|\mathbf{D}|^{1/2}} e^{-(\vec{\alpha}-\vec{\alpha}_0)^T \cdot \mathbf{D}^{-1}(\vec{\alpha}-\vec{\alpha}_0)/2}.$$

Imagine that we choose to express $\vec{\alpha} - \vec{\alpha}_0$ in terms of another set of random variables:

$$\vec{\alpha} - \vec{\alpha}_0 = \mathbf{A} \cdot (\vec{\alpha}' - \vec{\alpha}_0')$$

where $\mathbf{A}$ is the matrix of a rotation in the x-y plane; recall that for rotations $\mathbf{A}^T \mathbf{A} = 1$. The Jacobian of this transformation is the matrix $\mathbf{A}$, and its determinant is unity. The rule by which we transform probability densities is the following:

$$\rho(x', y') = \rho(x, y)/||\mathbf{J}||$$

and in our case this is:

$$\rho(\vec{\alpha}') = \frac{1}{(2\pi)\,|\mathbf{D}|^{1/2}} e^{-(\vec{\alpha}'-\vec{\alpha}_0')^T \cdot (\mathbf{A}^T \mathbf{D}^{-1} \mathbf{A}) \cdot (\vec{\alpha}'-\vec{\alpha}_0')/2}.$$

The expression

$$\mathbf{A}^T \mathbf{D}^{-1} \mathbf{A} = \left(\mathbf{A}^T \mathbf{D} \mathbf{A}\right)^{-1}$$
$$= \mathbf{C}^{-1}$$

defines the *covariance matrix*, $\mathbf{C}$. This is symmetric, real, and positive-definite (because all of its eigenvalues are positive). Since $|\mathbf{C}| = |\mathbf{D}|$, we can write (dropping the primes)

$$\rho(\vec{\alpha}) = \frac{1}{(2\pi)^{N/2}\,|\mathbf{C}|^{1/2}} e^{-(\vec{\alpha}-\vec{\alpha}_0)^T \cdot \mathbf{C}^{-1}(\vec{\alpha}-\vec{\alpha}_0)/2}. \quad (7.3)$$

In the case $N = 2$ this is the usual form of a bivariate normal distribution. The generalization to an N dimensional parameter space is straightforward (as shown) and is called a multivariate normal distribution.

214   *The multivariate normal distribution, χ^2, and correlation*

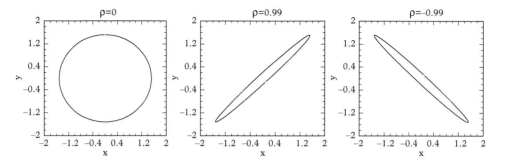

Figure 7.4 *Profiles of a bivariate normal distribution. Left: with a diagonal covariance matrix. Middle: with positive correlation. Right: with negative correlation.*

Returning to the bivariate normal distribution, since we have obtained the general expression taking the case of independent, normally-distributed random variables and rotating the two-dimensional parameters space, the profile of the PDF now resembles an ellipse whose major and minor axes do not coincide with the coordinate axes, as shown in Figure 7.4 (middle plot and right-hand plot). This means that the two variables x and y are no longer independent, but correlated. We say that the correlation is positive when a positive deviation in x is most likely to be accompanied by a positive deviation in y; and it is negative when a positive deviation in x is most likely to be accompanied by a negative deviation in y. When the random variables are uncorrelated or independent, deviations in one coordinate are unrelated to those in the other coordinate.

An explicit expression for the covariance matrix is:

$$\mathbf{C} = \begin{pmatrix} \sigma_x^2 & \sigma_{xy} \\ \sigma_{xy} & \sigma_y^2 \end{pmatrix}.$$

Since the matrix is positive-definite the determinant is greater than zero, which implies:

$$-1 < \frac{\sigma_{xy}}{\sqrt{\sigma_x^2 \sigma_y^2}} < 1.$$

The quantity $r = \sigma_{xy}/\sqrt{\sigma_x^2 \sigma_y^2}$ is called the **correlation coefficient**. When it is exactly equal to ± 1, from the above equations one can see that the covariance has a zero eigenvalue, and the elliptical profile is degenerate. In such circumstances, the value of one random variable predicts exactly the value of the other random variable and the two variables are said to be completely correlated ($r = 1$) or completely anticorrelated ($r = -1$).

Given a known multivariate normal probability distribution and a single vector of random variables $\vec{\mu}$ (a measurement) presumed to have been drawn from the distribution (either from a real or a simulated experiment), one is sometimes called upon to decide how probable the vector $\vec{\mu}$ is. In probability theory we do not discuss the probability of a single point, but rather the probability of an interval, and the way to properly frame the question is how probable is it that another vector of random variables drawn from the same distribution has an *even lower* value of the probability density.

In Eq. 7.3, the quantity $\chi^2 \equiv (\vec{\mu} - \vec{\alpha}_0)^T \cdot \mathbf{C}^{-1}(\vec{\mu} - \vec{\alpha}_0)$ appearing in the exponent, is called the *Mahalanobis distance* squared or the *standard distance* squared, or more commonly the χ^2 between $\vec{\mu}$ and $\vec{\alpha}_0$. This is like the normal Euclidian distance between a given point $\vec{\mu}$ in the N-dimensional space and the most probable point $\vec{\alpha}_0$, except it uses the inverse covariance matrix as a metric. We integrate the multivariate normal distribution over a region centered at $\vec{\alpha}_0$ and bounded by the surface $(\vec{\mu} - \vec{\alpha}_0)^T \cdot \mathbf{C}^{-1}(\vec{\mu} - \vec{\alpha}_0) = \chi^2$. This integral depends only on the value of χ^2, and on the number of dimensions, or *degrees of freedom*, N, of the parameter space. It is called the *cumulative χ^2 distribution*:

$$P(\chi^2|N) = \int_V \rho(\vec{\alpha})d^N\alpha = \int_V \frac{1}{(2\pi)^{N/2}|\mathbf{C}|^{1/2}} e^{-(\vec{\alpha}-\vec{\alpha}_0)^T \cdot \mathbf{C}^{-1}(\vec{\alpha}-\vec{\alpha}_0)/2} d^N\alpha$$

$$= \int_{V'} \frac{1}{(2\pi)^{N/2}} \prod_{i=0}^{N-1} \frac{1}{\sigma_i} e^{u_i^2/2\sigma_i^2} d^N u$$

where V is the elliptical (ellipsoidal or hyperellipsoidal) region enclosed by the contour of constant χ^2, and in the second line we transformed variables $\alpha_i \to u_i$ to independent random variables u_i by rotating the hyperellipsoid so that its principal axes coincide with the N coordinate axes; V' is the reoriented hyperellipsoid. A change of variables allows the integral to be reduced to a simple form:

$$P(\chi^2|N) = \frac{1}{\Gamma(\frac{N}{2})} \int_0^{\frac{\chi^2}{2}} t^{\frac{N}{2}-1} e^{-t} dt.$$

This function is known as the **cumulative χ^2 distribution**. Implementation of this function is often made in terms of the incomplete gamma function $P(a, x)$ (Press, 2007):

$$P(\chi^2|N) = P\left(\frac{N}{2}, \frac{\chi^2}{2}\right)$$

and in the `QatGenericFunctions` library it is implemented as `CumulativeChi Square`. The complement of the function, $Q(\chi^2|N) \equiv 1 - P(\chi^2|N)$ is most often referred to as the "probability χ^2 for N degrees of freedom". Figure 7.5 shows these distributions for $N = 1, 2 \ldots 6$ degrees of freedom.

If one has a dataset consisting of N independent random (column) vectors $\vec{x}$ i.e., $\mathcal{D} = \{\vec{x}_0, \vec{x}_1 \ldots \vec{x}_{N-1}\}$, one can also define a covariance matrix for the dataset:

$$\mathbf{C} = \langle \vec{x}\vec{x}^T \rangle - \langle \vec{x} \rangle \langle \vec{x}^T \rangle \qquad (7.4)$$

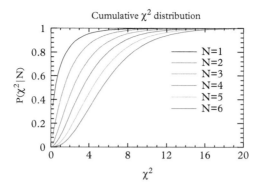

Figure 7.5 *Graphs of the cumulative χ^2 distribution $P(\chi^2|N)$ for $N = 1, 2 \ldots 6$ degrees of freedom.*

where the angle brackets denote averaging over the dataset. This definition is valid no matter what the joint PDF is that governs the random vectors x_i. If the PDF is a multivariate Gaussian distribution, then the covariance matrix for the dataset converges to the covariance matrix of the PDF in the limit $N \to \infty$. The correlation coefficient

$$r = C_{ij}/\sqrt{C_{ii}C_{jj}} \qquad (7.5)$$

can also be estimated from the data. Because of Schwartz's inequality it is bounded by $(-1, 1)$.

7.4 Monte Carlo integration

An immediate application of the stochastic methods developed in this chapter is the integration of multidimensional functions. Rewind your miserable life to your first semester of integral calculus. There you are, computing the area of some geometric shape, maybe the unit circle. Our task will be to recreate this experience with Monte Carlo methods. Since the area is known to be $A = \pi$, we can also pretend that we do not know the value of π and call this a numerical computation of π.

The unit circle is shown in Figure 7.6. The area of this circle is:

$$A = \int_{-1}^{1} dx \int_{-\sqrt{1-x^2}}^{\sqrt{1-x^2}} dy. \qquad (7.6)$$

In the Monte Carlo method, we compute this integral by throwing (over and over again) two uniform variates on the interval [-1,1]. Each pair maps onto a point on the x-y plane.

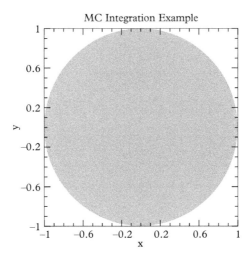

Figure 7.6 *The area area of the circle is computed by throwing two random numbers between -1 and 1. Each pair of numbers maps to a point in the $x - y$ plane. The fraction of points falling within the shaded region is computed.*

If the point falls within the unit circle (the shaded region in Figure 7.6), we count it as a "success", otherwise it is a "failure". The number of successes is denoted N and the number of failures M. The area of the circle is the area of the box (4 units) times the fraction of successes: $A = 4r$ with $r = N/(N + M)$.

Obviously this procedure works well with a large number of points and less well with, say, five or ten points. It certainly behooves us to estimate the error. The computed value of the integral, like the points themselves, is a random variable whose expectation value is π. We can use statistical methods to quantify the variance and/or standard deviation. Since M and N are independent, we propagate the statistical errors in these quantities ($\sqrt{M}$ and $\sqrt{N}$, respectively) to obtain:

$$\begin{aligned}
\sigma_r^2 &= \left(\frac{\partial r}{\partial N}\right)^2 \sigma_N^2 + \left(\frac{\partial r}{\partial M}\right)^2 \sigma_M^2 \\
&= \left(\frac{M}{(M+N)^2}\right)^2 N + \left(\frac{-N}{(M+N)^2}\right)^2 M \\
&= \frac{r(1-r)}{M+N}.
\end{aligned} \quad (7.7)$$

In other words $\sigma_r = \sqrt{r(1-r)/N_{tot}}$, where $N_{tot} = M+N$, and the error in the area is $\sigma_A = 4\sigma_r = 4\sqrt{r(1-r)/N_{tot}}$. This error decreases as the square root of the number of points, and thus as the number of function evaluations. Compared with the Trapezoid rule (error

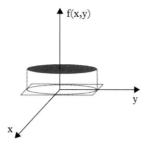

Figure 7.7 *Another view of the integration in Eq. 7.6.*

decreases as the square of the number of evaluations) or Simpson's rule (fourth power), the convergence rate is very slow. This slow convergence is disappointing, however, *it is independent of dimension*, and therefore represents a massive advantage for high dimensional problems.

Another way of viewing the quantity r is shown in Figure 7.7. Consider a function $f(x, y)$ (whose shape is like a top-hat, rising to a value of 1.0 where $x^2 + y^2 < 1$ and having the value zero everywhere else) over the square region defined by $|x| < 1$ and $|y| < 1$. The value of r is the average value of this function. The value of $f_i = f(x_i, y_i)$ for the ith point is random and equal to either 0 or 1; its mean value over the whole sample is $\bar{f} = \frac{0 \cdot M + 1 \cdot N}{M+N} = r$. The variance of r, Eq. 7.7, can be related to the variance of the function f:

$$\sigma_f^2 = \frac{N}{N+M}(1-f)^2 + \frac{M}{N+M}f^2$$
$$= f(1-f)^2 + (1-f)f^2$$
$$= f(1-f)$$

so that:

$$\sigma_r^2 = \frac{\sigma_f^2}{N_{tot}}.$$

There are two important lessons which one can see already from this simple integration exercise:

- The integrals converge as "something" divided by $\sqrt{N_{tot}}$. There is little one can do to influence this situation!
- However, the "something" (in this case σ_f) has everything to do with how much the function we integrate varies over the domain of integration. Functions with little variation will have a smaller error.

It is also useful to consider the fractional error in the integral,

$$\sigma_A/A = \frac{4\sqrt{r(1-r)/N_{tot}}}{4r}$$

$$= \frac{1}{\sqrt{N_{tot}}}\sqrt{\frac{1-r}{r}}.$$

For two dimensions (a circle) the value of r, which is $\pi/4$, is high. If, however, we were using this method to estimate the volume of a sphere in three dimensions, the value would be $\pi/6$. In higher dimensions it is even lower. The problem here is that more and more of the points are being laid down in regions where the value of the function is small. We will learn how to deal with this problem next.

7.4.1 Importance sampling

Let us drive home the last point. We seek to compute the probability that a random point falls inside a hypersphere of unit radius in a D-dimensional hypercube with sides of length 2. The answer is given by the ratio of volumes:

$$P(D) = \frac{2}{\Gamma(D/2)}\left(\frac{\sqrt{\pi}}{2}\right)^D.$$

For $D = 100$ this is $1.868 \cdot 10^{-68}$. Clearly the method we have described for evaluating integrals will fail drastically for hyperspheres of high dimensionality (Gibbs, 2006).

The resolution of the problem is to throw random points as closely as possible to the region of interest. In practice this means factoring out a function whose integral is *known* from the integrand. This function is denoted w and is called the **weight function**. Thus we set:

$$\int f(\vec{x})\,d\vec{x} = \int h(\vec{x})w(\vec{x})\,d\vec{x}.$$

The procedure requires generating abscissas according to the weight function, and thus w must be a *positive semi-definite* and *normalized* PDF.

The next step is to generate N random points according to the PDF $w(\vec{x})$. The integral is then given by:

$$\int f(\vec{x})\,d\vec{x} \approx \overline{h}$$

with error:

$$\sigma = \sqrt{\frac{\overline{h^2} - \left(\overline{h}\right)^2}{N-1}}.$$

Here:

$$\overline{h} \equiv \frac{1}{N} \sum_{i=1}^{N} h(x_i)$$

(with a similar definition for $\overline{h^2}$). The factor of $N-1$ in the expression for σ reflects the use of the unbiased estimator for the standard deviation.

There is considerable freedom in how to split the integrand into two parts $h(x)$ and $w(x)$. As discussed above, when carrying out a Monte Carlo integration it is best to arrange the splitting so that most of the variation is carried by $w(x)$ and as little as possible in $h(x)$. In practice this desire will be heavily constrained by what you can actually integrate when forming the weight function.

7.4.2 Example

We seek to evaluate the integral:

$$\int_0^\infty x^2 \, e^{-x} \, dx.$$

Possible weight functions are $\exp(-x)$ and $x \exp(-x)$ (and of course $x^2 \exp(-x)$ since the integral can be done exactly, but let us not consider this case). These options are implemented in the directory EXAMPLES/CH7/BASICMCINTEGRATE. An excerpt of the relevant code is

```
std::mt19937 e;

const unsigned int N=10000;
{
  std::exponential_distribution<double> eDist;
  double sum=0,sum2=0;
  for (int i=0;i<N;i++) {
    double x=eDist(e);
    double h=x*x;
    sum  +=   h;
    sum2 += h*h;
  }
  double hBar=sum/N,h2Bar=sum2/N, sigma=sqrt((h2Bar-hBar*
      hBar)/(N-1));
  std::cout << "Integral is " << hBar  << "+-" << sigma <<
      std::endl;
}
```

Table 7.1 *Estimating $\int_0^\infty x^2 e^{-x}\, dx$ with various weight functions.*

w	h	Integral	Error
e^{-x}	x^2	1.947	0.043
xe^{-x}	x	2.019	0.014
$x^2 e^{-x}$	1	2	–

The results from running the code are given in Table 7.1. One sees that employing a 'smoother' integrand h leads to a factor of three or four reduction in the error, as expected.

7.5 Markov chain Monte Carlo

It is often very difficult to sample a probability distribution and one must employ more powerful techniques such as the *Markov chain Monte Carlo method*. The basic idea of this method is to employ a stochastic process called a *Markov chain* (defined below) to generate random variables according to a multivariate probability distribution. This distribution, which will be denoted ρ in the following, can be either a continuous probability density or a discrete probability distribution. The first case is reasonably intuitive. In the second case, a common application is to a spin lattice, whose configuration is governed by a discrete multivariate PDF; discrete, because the spin can take a finite number of orientations (↑ or ↓), and multivariate because it describes every site on the lattice. In thermodynamic calculations the state of the lattice is governed by a Boltzmann distribution.

A *Markov chain* (named after Russian mathematician Andrei Markov, 1856–1922) is a sequence of *states* that are produced by a simple stochastic process wherein the current state is solely a function of the previous state. A state is a complete specification of a system. For example, if one is describing a spatial probability density the state consists of three position variables x, y, and z and the number of states is infinite. In the case in which the system is a lattice of four spin-1/2 ions, the configurations (↑, ↑, ↑, ↑), (↓, ↑, ↑, ↑), etc. are possible. Concentrating for simplicity on the latter case, we map the 16 states to unit vectors in a $M = 2^4 = 16$ dimensional space denoted:

$$\hat{X}_1 = (1, 0, \ldots, 0) \to (\uparrow, \uparrow, \uparrow, \uparrow),$$
$$\hat{X}_2 = (0, 1, \ldots, 0) \to (\downarrow, \uparrow, \uparrow, \uparrow),$$

etc.

A Markov chain is generated by stepping from one element of the set, $\hat{X}_i$, to another, $\hat{X}_j$, with probability p_{ij}. The p_{ij} are called *transition probabilities* and the $M \times M$ matrix **P** that is comprised of these probabilities is called the *transition matrix* (or

Markov matrix). Transition probabilities will be taken to be independent of time in our discussion and can be written as:

$$p_{ij} = \hat{X}_i^T \mathbf{P} \hat{X}_j.$$

The probability of transitioning to any state i must sum to unity, thus:

$$\sum_{j=1}^{M} p_{ij} = 1. \tag{7.8}$$

A Markov process creates a sequence of states $\hat{X}^{(t)}$ that are labelled by an index t, often called ***algorithmic time***. A particular $\hat{X}^{(t)}$ is drawn from a probability distribution $\mathbf{x}^{(t)}$

$$\mathbf{x}^{(t)T} = \mathbf{x}^{(t-1)T} \mathbf{P} = \hat{X}^{(0)T} \mathbf{P}^t. \tag{7.9}$$

Notice that the sequence of probability distributions has been initialized with a particular state. In principle, one could generate the sequence of probability distributions; in practice this is prohibitively expensive because the dimension M is typically extremely large. Instead, one generates an ensemble of sequences of states drawn from the sequence of probability distributions. The ensemble can then be used to compute expectation values of operators.

Crucially, under certain conditions Markov chains obtain a unique fixed point distribution,

$$\mathbf{w}^T \equiv \hat{X}^{(0)T} \lim_{t \to \infty} \mathbf{P}^t.$$

This result is independent of the starting state. The "conditions" mentioned are often summarized by saying that the Markov chain must be ***ergodic***. Roughly speaking this means that all states eventually appear in an infinitely long sequence and that the sequence does not get stuck in subspaces of the system. A simple way to ensure that a Markov chain is ergodic is to impose ***detailed balance***:

$$w_i p_{ij} = w_j p_{ji} \quad \text{(no sum)}, \tag{7.10}$$

where w_i are the components of $\mathbf{w}$. Summing this expression over j establishes that $\mathbf{w}^T$ forms a fixed point of the Markov process:

$$\mathbf{w}^T = \mathbf{w}^T \mathbf{P}.$$

The Markov chain Monte Carlo method is completed by equating the fixed point to the desired probability distribution, ρ, by appropriately arranging transition probabilities. The manner in which this is done is the subject of the following section.

7.5.1 The Metropolis-Hastings algorithm

The central task is to design the transition probabilities such that the desired probability density $\rho(\hat{X})$ is achieved. A number of solutions to this problem exist; here we present that due to Metropolis *et al.* (1953) and Hastings (1970).

A simple and general solution to generating this distribution is provided by the ***Metropolis-Hastings algorithm***. The algorithm generates a new state $\hat{X}'$ from an old state $\hat{X}$ according to ***proposal distribution***, $R(\hat{X} \to \hat{X}')$, which is at the discretion of the programmer. This is done using a source of random variates, either discrete or continuous, according to the nature of the problem. The algorithm is then defined by:

```
loop in algorithmic time
  generate the next state in the Markov chain according to
    R(X̂ → X̂′)
  accept the new state with probability min(1, R(X̂′→X̂)ρ(X̂′)/R(X̂→X̂′)ρ(X̂))
  if accepted
    use the new state
  else
    re-use the old state
  end if
end loop
```

It is a good idea to tune any parameters in the proposal function to obtain about a 50% acceptance rate in the state updates; acceptance rates between 30% and 70% are generally considered reasonable. A higher acceptance rate implies that the integration volume is being explored too slowly, while a lower acceptance rate wastes computational effort on rejected update attempts.

We note that if the proposal distribution is reversible:

$$R(\hat{X} \to \hat{X}') = R(\hat{X}' \to \hat{X})$$

then the acceptance probability simplifies to:

$$\min\left(1, \frac{\rho(\hat{X}')}{\rho(\hat{X})}\right).$$

In this case the algorithm is called the ***Metropolis method***. A Gaussian of fixed width is a typical example of a reversible proposal distribution.

7.5.2 Slow mixing

While the Metropolis-Hastings algorithm is simple and straightforward to implement, the algorithm is prone to several types of problems which we must identify and ameliorate.

A process that moves frequently and converges quickly to a stationary distribution is said to be ***rapidly mixing***. One that remains stuck or does not converge quickly to a stationary distribution is said to be slowly mixing. Achieving rapid mixing is the principal challenge in Markov chain Monte Carlo (MCMC) implementation. While large steps in the proposal distribution allow MCMC to explore the parameter space more fully, a strategy that is too bold will result in a low probability that any one step is accepted. If, on the other hand, a more cautious strategy with smaller steps is used, the chain takes a long time to approach the stationary distribution. Worse, an overly cautious strategy may never reach disconnected regions in the sampled probability density function, because "tunneling through" to these regions takes a prohibitively long time. Both of these points are discussed further below. Slow mixing can be sharply reduced if the proposal distribution is tailored to the desired PDF $\rho(\hat{X})$ and constructed in such a way that the proposed step can reach all of the regions in parameter space where $\rho(\hat{X})$ is non-negligible.

We illustrate this effect by considering the effect of step size in generating the electron probability distribution for a hydrogen atom. This is the square magnitude of the hydrogenic wavefunction:

$$\phi_{nlm}(r,\theta,\phi) \propto \rho^l e^{-r/n} L_{n-l-1}^{2l+1}(2r/n) Y_l^m(\theta,\phi),$$

where r is measured in units of a_0, the Bohr radius and $L_p^q(x)$ is the associated Laguerre polynomial. The normalization of the wavefunction is not required.

Distributions can be generated using the example program in:

EXAMPLES/CH7/MCMC/HYDROGEN

This program has a few command line arguments to set the quantum numbers *n*, *l*, and *m*, in addition to the number of points used in the simulation. The program uses the normal distribution for the proposal function, which is provided by std::normal_distribution in the C++ standard library. This proposal function is perfectly reversible, so we apply the Metropolis method. The width of the proposal distribution is likewise specified on the command line. You can invoke the program with:

‖ mcmc-hydrogen NPOINTS=val N=val L=val M=val fSigma=val

The optional command line arguments have default values of NPOINTS=10K, N=0, L=0, M=0, and fSigma=1.0. An animated version of the program exists too; it lives in

EXAMPLES/CH7/MCMC/HYDROGEN-ANIM

and allows you to witness the "diffusion" of probability density in Markov Chain Monte Carlo. Animated code is invoked with the command

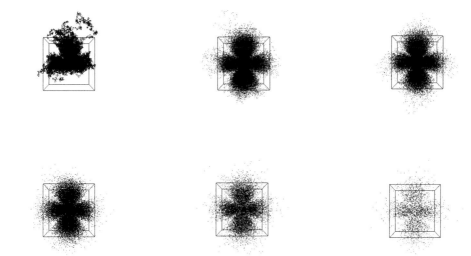

Figure 7.8 *Sampling the $n = 2, \ell = 1$ hydrogenic probability distribution. From left to right and the top to bottom we increase the width of the proposal distribution by scaling the nominal proposal distribution by : 0.1, 0.5, 1.0, 2.0, 4.0, and 8.0. Timid steps (upper left) cause the chains to move slowly through the parameter space while bold steps result in inefficiency.*

```
|| mcmc-hydrogen-anim NPOINTS=val N=val L=val M=val
```

The effect of varying the sampling distribution is shown in Figure 7.8. The sequence of plots runs from conditions which are too timid in the upper left, to too bold in the lower right. Ideal conditions lie somewhere in between. There is a clear lesson: sampling distributions should be carefully explored before full scale Monte Carlo computations are undertaken.

7.5.3 Thermalization

Markov Chain Monte Carlo needs to be run for long enough for the system to reach equilibrium and for the probability density function to be sampled enough to obtain observables to the required precision. If the system is started in a very unlikely state, a certain amount of time is required for the Markov process to move out of the tails of the distribution and into the peak. This is shown in Figure 7.9. Even if the chain is started in a likely configuration, MCMC often needs substantial algorithmic time to settle into the fixed point distribution. If the chain is terminated before this occurs, the distribution tails are over-represented.

The usual procedure for dealing with this issue is to let the Markov chain run for a certain period until the main modes of the likelihood function are found. The initial part of the chain is discarded. This is called ***thermalization*** or ***burn-in***.

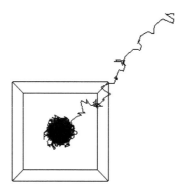

Figure 7.9 *A Markov Chain Monte Carlo computation of electron probability densities in hydrogen for the n = 0 state. The Metropolis method is used and the system is started from a very unlikely point. Its trajectory evolves inexorably towards a major mode of probability. The time taken to reach a likely configuration is called the burn-in or thermalization time and can be identified with time-series plots.*

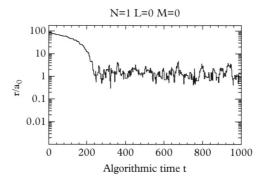

Figure 7.10 *A time-series plot of the radial coordinate in a Markov Chain computation of electron probability densities in a hydrogen atom for the n=0 state, when the computation begins from a very unlikely point. From this plot one can identify a burn-in phase of about 200 iterations.*

A time-series plot showing the value of one or more parameters vs. iteration number is usually informative. An example is shown in Figure 7.10. Evidently about 200 iterations are required for burn-in in this case. Another strategy is to initiate the Markov process with different starting states. Observables will typically have very different values until thermalization has occurred. An example of this approach is given in Figure 20.2. The figure displays a time series plot of the magnetization of an Ising spin system with "hot" and "cold" starting configurations. One sees that for $k_B T = 2.0$ around 200 Markov steps are required before the system has equilibrated.

Both of the preceding methods are simple to implement and give a good indication of how well thermalization is proceeding.

7.5.4 Autocorrelation

It is the central limit theorem that underpins the validity (and usefulness) of the Monte Carlo method for evaluating integrals. We seek the same functionality here; however, the existence of *autocorrelation* invalidates a direct application of the theorem.

Autocorrelations are best illustrated by considering a computation of the average error in a measurement of a quantity $\langle \mathcal{O} \rangle$. If one has a sequence of N such measurements, then:

$$\langle \delta \mathcal{O}^2 \rangle = \left\langle \left(\frac{1}{N} \sum_{t=1}^{N} (\mathcal{O}_t - \langle \mathcal{O} \rangle) \right)^2 \right\rangle = \frac{1}{N^2} \sum_{t=1}^{N} \left\langle (\mathcal{O}_t - \langle \mathcal{O} \rangle)^2 \right\rangle + \frac{2}{N^2} \sum_{t=1}^{N} \sum_{\ell=t+1}^{N} \left(\langle \mathcal{O}_t \mathcal{O}_\ell \rangle - \langle \mathcal{O} \rangle^2 \right) \quad (7.11)$$

where $\mathcal{O}_t$ represents the value of the operator at algorithmic time t. This can be rewritten as:

$$\langle \delta \mathcal{O}^2 \rangle = \frac{1}{N} \left(\langle \mathcal{O}^2 \rangle - \langle \mathcal{O} \rangle^2 \right) (1 + 2\tau_{\text{int}}) \quad (7.12)$$

with:

$$\tau_{\text{int}} = \sum_t \frac{\langle \mathcal{O}_0 \mathcal{O}_t \rangle - \langle \mathcal{O} \rangle^2}{\langle \mathcal{O}^2 \rangle - \langle \mathcal{O} \rangle^2}. \quad (7.13)$$

In deriving this we have used homogeneity in time of expectation values, which is true for any stationary sequence. The factor $1 + 2\tau_{\text{int}}$ is called the *statistical inefficiency* and can be disastrously large under certain conditions (to be explored in Chapter 20). The time scale, τ_{int}, is sometimes called the *integrated autocorrelation time*. It can be defined in terms of a temporal *autocorrelation function*:

$$C(t) = \frac{\langle \mathcal{O}_i \mathcal{O}_{i+t} \rangle - \langle \mathcal{O}_i \rangle^2}{\langle \mathcal{O}_i^2 \rangle - \langle \mathcal{O}_i \rangle^2} \quad (7.14)$$

Notice that this expression coincides with the summand in Eq. 7.13; and by comparison with Eq. 7.5 one can recognize it as the correlation coefficient r between a quantity at a time i and a later time $i + t$. Like the correlation coefficient, the function $C(t)$ is bounded by $(-1, 1)$ because of Schwartz's inequality. Figure 7.11 shows the autocorrelation plot for the example of the previous section.

Markov chain Monte Carlo, like ordinary Monte Carlo, can be used to evaluate integrals; when doing so, Eq. 7.12 implies that:

$$\int f(\vec{x}) w(\vec{x}) \, d\vec{x} = \bar{f} \pm \sqrt{\frac{(\overline{f^2} - (\bar{f})^2)(1 + 2\tau_{\text{int}})}{N - 1}}. \quad (7.15)$$

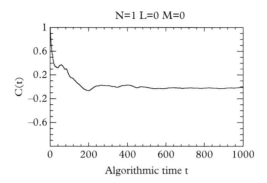

Figure 7.11 *The autocorrelation of the radial variable in a computation of electron probability densities in the hydrogen atom for the $n = 0$ state. This plot indicates that the system forgets its state after approximately 200 iterations.*

Here $\bar{f}$ is the average of a quantity $f(\vec{x})$ evaluated over the elements of the Markov chain (after thermalization) and the integrated autocorrelation is given by Eq. 7.13 with the operator $\mathcal{O}$ replaced by f. Finally, the probability density $w(\vec{x})$ is the fixed point distribution of the Markov chain.

Eq. 7.12 shows the importance of minimizing the correlation between states in the stochastic sequence. Usually the dependence decreases as the number of steps between elements in the chain increases, so a simple way to decrease autocorrelations is to skip elements in the sequence when evaluating observables.

7.5.5 Multimodality

Slow mixing is particularly aggravated when probability density functions are ***multimodal***, that is, they have separated regions of high probability. The 3-dimensional electron probability densities discussed above offer a good example. In such cases the Markov chain must "tunnel" through areas of low probability to probe other important regions in the distribution, otherwise observables will be poorly estimated.

Multimodality is best handled by carefully constructing proposal functions that transition to other nodes with an appropriate frequency. Unfortunately, no general algorithm exists and one must tune according to the problem at hand. This is reminiscent of the situation with integrals: the numericist is well advised to know her foe!

To summarize, the Metropolis-Hastings algorithm is easily implemented but requires monitoring and tuning to ensure fidelity of the computation. The main challenges are to achieve good mixing, particularly among the modes of the distribution when multiple modes exist; to ensure a proper identification of the burn-in phase; and to generate Markov chains that are long enough to detect possible anomalies. Patience, and liberal use of time series plots, autocorrelation plots, and multiple redundant computations using independent chains are recommended.

7.6 The heat bath algorithm

The *heat bath algorithm* is a type of Markov chain method that is closer to direct Monte Carlo than MCMC. Again we assume that it is impractical to evaluate the complete PDF, and that other methods must be employed. The heat bath algorithm achieves this by sampling a given multivariate distribution in terms of its conditional distribution; which, presumably, is easier to compute. This form of sampling is also known as *Gibbs sampling*.

The algorithm uses the components of a state, which we denote:

$$\mathbf{x}^{(k)} = (X_1^{(k)}, \ldots, X_M^{(k)}) \tag{7.16}$$

as before. With this notation, the Gibbs sampling algorithm is:

```
select an initial X̂⁽⁰⁾
iterate in algorithmic time, k
  iterate in components, i
    sample X_i^(k+1) from the conditional probability
       p(X_i^(k+1)|X_1^(k+1),...,X_{i-1}^(k+1),X_i^(k),...,X_M^(k))
  end
  evaluate observables
end
```

Notice that the conditional probability is evaluated with states that *have already been updated* up to the index location, $i-1$. After this location the current states are used.

Although this is the usual way that Gibbs sampling is presented, the reader will note that the ordering of state components is arbitrary and hence it is not necessary to sweep linearly through state components. In fact, an acceptable alternative is to choose the index i at random. This amounts to subsuming the index i into the algorithmic time k, yielding a "component-algorithmic time".

7.6.1 An application: Ising spin systems

Let us make the preceding, rather theoretical, discussion concrete by considering an application to compute the thermodynamic properties of a one-dimensional Ising model. The Ising model we explore (see Chapter 20 for much more) consists of a series of "spins" that take on the values $\sigma_i = \pm 1$ (representing "up" or "down") which interact with their neighbors according to the Hamiltonian:

$$H = -\sum_{i=1}^{M} \sigma_i \sigma_{i+1}. \tag{7.17}$$

The index i refers to the position of the spin in the Ising chain. The minus sign implies that the ground state of the system is the magnetized state, $+++\ldots+$ (or, of course $---\ldots-$).

We remind the reader that all thermodynamic properties can be obtained from the partition function:

$$Z = \sum_{\sigma_1,\ldots,\sigma_M} \exp[-\beta H(\sigma_1,\ldots,\sigma_M)], \qquad (7.18)$$

with $\beta = 1/(k_B T)$. Thus, for example, the energy of the system is:

$$\langle E \rangle = -\frac{\partial}{\partial \beta} \log Z.$$

Implementing the heat bath algorithm requires evaluating the conditional probability for the Ith spin:

$$p(\sigma_I | \sigma_1, \sigma_2, \ldots \sigma_M) = \frac{\exp(-\beta H(\ldots \sigma_I \ldots))}{\sum_{\sigma_I} \exp(-\beta H(\ldots \sigma_I \ldots))}. \qquad (7.19)$$

The crucial feature of this expression — the one that makes Gibbs sampling feasible — is that it simplifies drastically because of the form of the Hamiltonian, Eq. 7.17. Indeed, substituting for H reveals that *all spin-dependence cancels between numerator and denominator* except for those spins that interact with σ_I. Thus the conditional probability simplifies to:

$$p(\sigma_I | \sigma_{I-1}, \sigma_{I+1}) = \frac{\exp[\beta \sigma_I \cdot (\sigma_{I-1} + \sigma_{I+1})]}{\sum_{\sigma_I} \exp[\beta \sigma_I \cdot (\sigma_{I-1} + \sigma_{I+1})]}. \qquad (7.20)$$

This is a simple function that depends only on σ_I and:

$$\sigma_{I-1} + \sigma_{I-1} \in \{-2, 0, 2\}$$

and hence can be pre-computed and stored in an array.

This remarkable cancellation is no accident: it exists because the Ising Hamiltonian is *local*, meaning that a given degree of freedom only interacts with a small collection of other degrees of freedom.

The final step in implementing the algorithm is choosing an update scheme. The method displayed in the pseudocode of the previous section can be achieved by simply iterating through the spin index I while using σ_{I-1} (which has been updated) and σ_{I+1} (which has not been updated) to form the conditional probability. Alternatively, one can select the index I at random, as mentioned above.

7.6.2 Markov chains for quantum problems

The Ising model example makes it clear that Markov chain algorithms are powerful methods for evaluating thermodynamic observables. However, their utility is much greater; for example later chapters will use them to simulate physical processes. Perhaps surprisingly, Markov chain Monte Carlo is also an effective tool for investigating quantum mechanical problems.

An intuitive way to see the connection of quantum mechanics to stochastic processes is by considering the Schrödinger equation in imaginary time (this is also sometimes called *Euclidean time*, or *Wick rotating*). We thus let $t \to -i\tau$ and obtain:

$$i\hbar \frac{\partial}{\partial t}\psi = -\frac{\hbar^2}{2\mu}\nabla^2\psi + V(x)\psi \quad \to$$
$$\frac{\partial}{\partial \tau}\psi = \frac{\hbar}{2\mu}\nabla^2\psi - \frac{1}{\hbar}V(x)\psi.$$

The latter equation is a *diffusion equation* with diffusion coefficient $D = \hbar/(2\mu)$. The interaction term can be thought of as an additional drift that affects the diffusing particles (this is then called the *advection-diffusion equation*).

The connection to stochastic processes is provided by the physics of diffusion: namely the equation is a continuum version of random walks by microscopic particles. Thus in a very real sense quantum mechanics is equivalent to random walks, and random walks are a type of Markov chain. This equivalence will be exploited to construct a powerful stochastic method to evaluate quantum mechanical observables in Chapter 21.

Another deep connection of quantum mechanics to thermodynamics, and hence Markov chains, was noted in the 1940s by Richard Feynman (American physicist, 1918 – 1988). At that time Feynman was developing the *path integral* from a suggestion due to Dirac (Feynman, 1948). The idea is that the quantum mechanical amplitude for a particle to transit from (x_0, t_0) to (x_f, t_f) is given by a sum over all possible paths the particle can take (physical or not) weighted by the exponentiated action evaluated for that path. The mathematical expression for this is:

$$\langle x_f t_f | x_0 t_0 \rangle = \int_{x_0}^{x_f} Dx(t) \, \exp\left[\frac{i}{\hbar} \int_{t_0}^{t_f} dt L(x, \dot{x})\right], \tag{7.21}$$

where L is the particle lagrangian:

$$L = \frac{1}{2}m\dot{x}^2 - V(x)$$

and Dx is notation that summarizes the concept of "summing over paths".

Feynman noted that making the transition to Euclidean time changes the path integral to the form:

$$\int_{x_0}^{x_f} Dx(\tau) \, \exp\left[-\frac{1}{\hbar}\int_{\tau_0}^{\tau_f} d\tau \left\{\frac{1}{2}m\left(\frac{dx}{d\tau}\right)^2 + V(x)\right\}\right].$$

and that this resembles a thermodynamic partition function. The latter can be seen by approximating the path $x(\tau)$ with a set of discrete variables x_i and replacing the Euclidean temporal derivative with $(x_i - x_{i+1})/\epsilon$, where ϵ is a small time difference. The result is:

$$\langle x_f \tau_f | x_0 \tau_0 \rangle = \int \prod_i dx_i \, \exp\left[-\frac{\epsilon}{\hbar}\sum_j \left\{\frac{1}{2}m\left(\frac{x_j - x_{j+1}}{\epsilon}\right)^2 + V(x_j)\right\}\right]. \quad (7.22)$$

The expression in braces is a Hamiltonian for an unusual many-particle system and the full expression is a partition function with temperature $k_B T = \hbar/\epsilon$. Thus all the methods used to evaluate many-body thermodynamic observables can be applied to quantum problems. Finally, this equivalence applies equally well to field-theoretic path integrals. This will be exploited to perform computations in quantum field theory in Chapter 24.

7.7 Where to go from here

Monte Carlo methods form an important and ubiquitous tool in modern science. As a result they appear often, and in many guises, in the remainder of this book. Here we provide a short guide to subsequent Monte Carlo applications so that the assiduous reader can continue her education in this topic.

The heat bath algorithm is applied to classical spin systems in Chapter 20. Different initial configurations and boundary conditions are discussed and an explicit computation of autocorrelation is made. ***Cluster algorithms*** are introduced as a way to overcome the long autocorrelation times that exist near phase transitions. An alternative Monte Carlo update, called ***Glauber updating***, is discussed in exercise 17. Finally, the technique of ***reweighting***, which permits evaluation of observables at many different parameter (such as temperature) values, is explained.

A novel MCMC approach to evaluating quantum mechanical path integrals, called ***guided random walks***, is introduced in Chapter 21. This method leverages the similarity between the Schrödinger equation and diffusion once a Wick rotation has been made. The guided random walks method is extended to the case of discrete systems in Section 22.4.

The application of MCMC to quantum field theory is the subject of Chapter 24, where the topic of ***microcanonical updating*** is discussed. This method permits making large-scale changes to configurations that assist in overcoming multimodality and strong autocorrelation.

7.8 Exercises

1. Find a method for generating points uniformly:

   a) on the surface of a sphere,
   b) on the surface of a sphere in N dimensions,
   c) within the volume of a sphere,
   d) within the volume of a sphere in N dimensions.

   Use direct sampling rather than von Neumann rejection to make your method efficient.

2. A gas is in thermal equilibrium at a temperature T, thus the velocity distribution is:

$$\rho(v)\,dv = \left(\frac{m}{2\pi\tau}\right)^{3/2} 4\pi v^2 e^{-mv^2/2\tau}\,dv,$$

   where $\tau = k_B T$, m is the mass of a molecule and k_B is the Boltzmann constant. Nondimensionalize this and show that the Maxwell velocity distribution can be sampled with the aid of a gamma distribution. Make a plot of the velocity distribution together with data generated using the `std::gamma_distribution` of the standard library.

3. Repeat the previous problem, but this time generate the Maxwell distribution using Markov chain Monte Carlo. Plot the distributions and check that they agree with the theoretical line shape.

4. Repeat again, but this time also add a strong gravitational field to the system. Compute the probability densities for both position and momentum. Assume a binary mixture that is 50% argon and 50% nitrogen.

5. Generate data using the Monte Carlo method according to an exponential distribution convolved with a Gaussian distribution.

   a) Carry out the convolution analytically and plot it, using the value $\tau = 3$ and $\sigma = 2$.
   b) Generate two random variables, one drawn from an exponential distribution with $\tau = 3$ and one from a Gaussian distribution with $\sigma = 2$. Make a histogram of the sum of these two variables and compare with the analytic convolution from part (a).
   c) Use the rejection method to generate data distributed according to the analytic convolution of part (a).

6. A particle that is polarized along the z direction decays to a pair of lighter particles, of which one is visible. The angular distribution of the visible particle is given by:

$$\rho(\cos\theta) = \left(\frac{\sqrt{3}}{\sqrt{8}}P_1(\cos\theta) - \frac{\sqrt{7}}{\sqrt{6}}P_3(\cos\theta) + \frac{\sqrt{11}}{\sqrt{24}}P_5(\cos\theta) - \frac{\sqrt{15}}{\sqrt{6}}P_7(\cos\theta)\right)^2$$

Generate data according to this distribution, and plot it vs. $\cos\theta$ in a histogram overlaid with the original function.

7. Importance sampling.

   a) Use the Monte Carlo technique to estimate the volume of a hypersphere in 10 dimensions. The convergence rate is a constant divided by $\sqrt{N}$. Estimate the value of this constant.

   b) Estimate the probability density in the N-dimensional space by making a histogram of x_j where j is any index between 1 and 10. From the one dimensional distribution construct an estimate of the 10-dimensional distribution $g(x_1, x_2, x_3 \ldots x_{10})$. The estimate does not have to be extremely accurate. *Note: this is the approach employed by the well-known Monte Carlo program Vegas.*

   c) Using your estimate for $g(x)$, re-evaluate the volume of the sphere and estimate again the value of convergence rate constant.

8. A hydrogen molecule ion consists of two protons and a single electron. The Hamiltonian for the system is:

   $$H = -\frac{1}{2}\nabla^2\psi(\vec{r}) - \frac{1}{|\vec{r}+\hat{z}d/2|} - \frac{1}{|\vec{r}-\hat{z}d/2|}$$

   where d is the separation of the two protons, the unit of length is the Bohr radius, and the unit of energy is the Hartree (1 Hartree = 27.2 eV). Consider trial wavefunctions consisting of the symmetric and antisymmetric linear combination of hydrogenic ground-state wavefunctions, $\psi_{s,a}(\vec{r}) \equiv N_{s,a}(\phi_1(\vec{r}) \pm \phi_2(\vec{r}))$, where:

   $$\phi_1 \equiv \frac{1}{\sqrt{\pi}}e^{-(|\vec{r}-\hat{z}d/2|)} \qquad \phi_2 \equiv \frac{1}{\sqrt{\pi}}e^{-(|\vec{r}+\hat{z}d/2|)}.$$

   a) Compute and plot the normalization factors N_s and N_a as a function of the interatomic distance d.

   b) Compute and plot the energy of the system as a function of interatomic distance d, in both cases.

   Use Monte Carlo integration to evaluate the integrals occurring in this problem.

9. In Problem 12 of Chapter 5 you solved an integral to determine the rate of gamma radiation originating from K^{40} decays. Revisit this integration exercise, but this time framed as an exercise in simulation in which a radioactive atom decays at a random

distance and angle within the granite and the radiation is absorbed (or not) at a random location along its path. From the simulation estimate the flux of gamma rays emerging from the granite and compare to the result from Chapter 5.

10. [T] Convolution.
    Show that the convolution of two uniform distributions, each over the interval [-1,1] is a triangular function. A change of variable from x_1, x_2 to $u = x_1 + x_2$ and $w = x_1 - x_2$ will help. Project the joint probability distribution onto u by integrating with respect to w.

11. Line Broadening.
    Study the code in the directory EXAMPLES/CH7/BWDOPPLER. Use this code to make a plot of a photon emission line that has been generated with a Breit-Wigner distribution with additional Doppler broadening. The result is called the Voigt distribution.

12. In the transformation method (Section 7.2.7), a known probability density function $\rho(x)$ is sampled by transforming a uniform variate by a transformation equation. The transformation equation is obtained by integrating $\rho(x)$ and inverting the integral. If instead the transformation equation is given, one can reverse these steps to determine the probability density of a transformed uniform variate.

    Generate a uniform variate x on the interval [0,1] and transform it according to either of the transformation equations below. In each case, plot a histogram of the resulting variate; determine an analytic form for the distribution, and plot the analytic function on top of the data to check your answer:

    $$f(x) = x_{min}(x_{max}/x_{min})^x$$

    for $x_{max} > x_{min} > 0$ (you can take $x_{min} = 1$ and $x_{max} = 10$);

    $$f(x) = \sqrt{M^2 + M\Gamma \tan(M\Gamma x)}$$

    for $M > 0$ and $\Gamma > 0$ (you can take $M = 100$ and $\Gamma = 10$).

13. [T] Heat Bath and Detailed Balance.
    Establish that the heat bath algorithm obeys the condition of detailed balance.

14. Importance sampling i.
    Consider the example of Section 7.4.2 again.

    a) Modify the code of EXAMPLE/CH7/BASICMCINTEGRATE to initiate the Mersenne twister engine with a user-inputted random seed.

    b) Add a section that computes the integral by generating uniform variates over the interval $(0, 1)$. Implement this with the change of variables $u = x/(1 + x)$. Use the weight function $w = 1$. How does the error compare to the other cases? Can you explain this result?

15. **Importance Sampling ii.**
    Consider evaluating the integral:
    $$I = \int_0^1 \frac{e^x - 1}{e - 1} dx = \frac{e-2}{e-1}.$$

    a) If abscissas are drawn uniformly on $(0, 1)$ show that the error in a Monte Carlo estimate of I is given by:
    $$\frac{0.286}{\sqrt{N}}.$$
    What sample size need be employed to achieve 1% accuracy?

    b) Now draw abscissas with a weight function $w = x$. Show that the error is now
    $$\frac{0.0523}{\sqrt{N}}.$$
    Again, what sample size need be employed to achieve 1% accuracy?

    c) Repeat the previous exercise in D dimensions assuming $f = \prod_i f(x_i)$. Show that for large D the ratio of variances goes as:
    $$10^{-aD}$$
    and determine the constant a.

16. **[T] A Simple Markov Chain.**
    Consider the transition matrix
    $$P = \begin{pmatrix} 3/4 & 1/4 & 0 \\ 0 & 2/3 & 1/3 \\ 1/4 & 1/4 & 1/2 \end{pmatrix}.$$
    Find

    a) the eigenvectors and eigenvalues of P.
    b) the left eigenvectors and left eigenvalues of P.
    c) the left fixed point probability vector, **w**.
    d) $\lim_{n\to\infty} P^n$. Is the fixed point independent of the starting vector?

17. **[T] Simple properties of Markov chains.**

    a) Prove that every eigenvalue of a transition matrix is bounded by unity, $|\lambda| \leq 1$.
    b) Prove that every transition matrix has at least one eigenvalue equal to unity.

18. [T] Construct an Ising-like spin model for which the simplification necessary to implement the heat bath algorithm is *not* realized.
19. [P] Simple Harmonic Oscillator.
    We consider the evaluation of matrix elements for the one-dimensional quantum mechanical simple harmonic oscillator. The imaginary time action is given by:

    $$S[x(\tau)] = \int_{\tau_i}^{\tau_f} (\frac{1}{2}m\dot{x}^2 + \frac{1}{2}m\omega^2 x^2) \, d\tau$$

    a) Discretize time via $N\epsilon = \tau_f - \tau_i$.
    b) Nondimensionalize. You should obtain an expression like:

    $$S = \frac{1}{2} \sum_{j=0}^{N-1} [(\xi_{j+1} - \xi_j)^2 + \kappa(\xi_{j+1} + \xi_j)^2]$$

    Identify κ.
    c) Manipulate further to obtain the form:

    $$S = \sum_{j=1}^{N-1} u_j^2 - g \sum_{j=0}^{N-1} u_j u_{j+1}.$$

    What condition is required on x_0 and x_N to obtain this form? Obtain an expression for the coupling g.
    d) Implement the following Metropolis-Hastings algorithm:

    ```
    initialize u_j
    loop
       randomly select a time slice t
       propose an update u_t → u'_t = u_t + â where â is chosen
          uniformly in the range -a < â < a.
       compute the change in action   ΔS = S(u'_t) - S(u_t)
       accept u'_t with probability min(1, exp(-ΔS))
       update observables
    end loop
    ```

    e) Select a cold start ($u_j = 0$) or a hot start (random u_j). Set $\omega\epsilon = 0.25$ and $N_t = 1000$. Tune a to achieve approximately 50% update acceptance.
    f) Measure autocorrelation with $\langle u(\tau + 5)u(\tau) \rangle$. About how many updates are required to reduce autocorrelation to 10%?

g) Measure $\langle x^2 \rangle$ and compare to the analytic value. Make a plot of $\langle x^2 \rangle$ as a function of Metropolis step and compare this observable's thermalization to that obtained above.

20. [P] The traveling salesman problem.
    The ***traveling salesman problem*** is an old and famous problem in combinatorial optimization. The goal is to find the shortest path between a collection of N cities. The brute force method for solving this problem requires examining all $N!$ possible paths; something that is computationally infeasible for N as low as 20.

    You are asked to implement a ***heuristic algorithm*** that is designed to find a solution, but is not guaranteed to find the optimal solution. Our algorithm is known as ***simulated annealing***, which is implemented as follows:

    ```
    set T_i, T_f, r
    generate node coordinates
    set initial path
    T = T_i

    while T > T_f
       generate new path
       compute Δ = d(newpath) − d(oldpath)
       if Δ < 0 or exp(−Δ/T) > rnd(0,1)
          distance = distance + Δ
          path = new path
       end if
       T = r * T
    end while
    ```

    As the pseudocode indicates, the idea is to randomly explore "path space", while accepting paths with shorter total lengths according to the Metropolis algorithm. Paths that are longer are sometimes accepted according to the parameter T, which acts as a pseudo-temperature. In this way one hopes to avoid being trapped in a local minimum as the temperature is slowly lowered.

    a) Implement a class for two vectors to represent the city locations, a method to compute total distance as a function of the path and the city locations, and a way to generate new paths. Permuting cities in the path is a good way to achieve the latter. You should find results similar to those shown in Figure 7.12.
    b) Assuming that N city locations are placed randomly on a square of size $(0, 1) \times (0, 1)$, what is a reasonable starting temperature for the simulation?
    c) Explore the behavior of the algorithm as a function of the cooling rate, r. We have found that very slow cooling rates (and therefore very many iterations) are required to find reasonably reliable global minima.

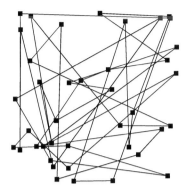 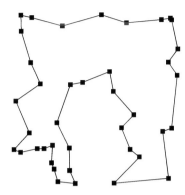

Figure 7.12 *Initial and final paths for 40 random cities.*

d) Consider making many Monte Carlo updates at each temperature. Explore the behavior of the algorithm as a function of this parameter.

e) If you have not already done so, implement an efficient way to compute Δ that leverages the fact that most inter-city distances do not change when only two cities in the path are permuted.

•••

BIBLIOGRAPHY

Abramowitz, M. and I.A. Stegun (1972). *Handbook of Mathematical Functions*. Dover.
The ATLAS Collaboration (2008). *Expected Performance of the ATLAS detector*. CERN-OPEN-2008-020.
Devroye, L. (1986). *Nonuniform Variate Generation*. Springer Verlag.
Feynman, R. P. (1948). *Space-Time Approach to Non-Relativistic Quantum Mechanics*. Reviews of Modern Physics **20**, 367.
Gibbs, W. (2006). Computation in Monder Physics. 3rd ed. World Scientific.
Hastings, W. (1970). *Monte Carlo sampling methods using Markov chains and their applications*. Biometrika **57**, 97.
Heitler, W. (1955). *The Quantum Theory of Radiation*. Oxford Clarendon Press.
Press, W., S. Teukolsky, W. Vetterling, and B. Flannery (2007). *Numerical Recipes, the Art of Scientific Computing*. 3rd ed. Cambridge University Press.
Matsumoto, M. and T. Nishimura (1998). *Mersenne Twister: A 623-dimensionally equidistributed uniform pseudo-random number generator*. ACM Transactions on Modeling and Computer Simulation: Special Issue on Uniform Random Number Generation **8**, 1.
Metropolis, N., A. Rosenbluth, M. Rosenbluth, A. Teller, and E. Teller (1953). *Equation of State Calculations by Fast Computing Machines*. J. Chem. Phys. **21**, 1087.
Watson, Geoffrey S. (1969). *Density Estimation by Orthogonal Series*. The Ann. of Math. Stat. **40** 4, 1496.

8
Percolation and universality

8.1	Site percolation	241
	8.1.1 The cluster algorithm	241
	8.1.2 Code verification	244
	8.1.3 The percolation probability	244
8.2	Fractals	249
8.3	Scaling and critical exponents	252
	8.3.1 The correlation length and the anomalous dimension	253
	8.3.2 Additional scaling laws	256
8.4	Universality and the renormalization group	258
	8.4.1 Coarse graining	260
	8.4.2 Monte Carlo renormalization group	263
8.5	Epilogue	264
8.6	Exercises	265
	Bibliography	272

Consider a field randomly populated with clumps of grass. If one clump catches fire how will the fire spread? This is a problem of *percolation* where one seeks global properties of spatially random configurations of local objects. **Bond percolation** in two dimensions provides a simple example: take a two dimensional grid of points. Connect neighboring points with probability p. What is the probability that two sides of the grid are connected by a sequence of bonds? How does the probability change as the size of the grid changes? What features do the connected regions display? The answers to these questions are surprising and reveal deep truths about the structure of physical phase transitions.

Percolation theory is also useful, having found application in a variety of physical systems including the flow of oil through porous rock, the spread of epidemic disease, electron mobility in a disordered semiconductor, switching in amorphous semiconductors, and gelation of branched macromolecules. We will consider it to be primarily a problem in statistics and will use it as a vehicle to learn computational methods for Monte Carlo configuration generation, pattern recognition, data analysis, and error estimation. Along the way we will also be exposed to the important concepts of scaling, critical exponents, universality, and fractals.

8.1 Site percolation

We shall take *site percolation* in two dimensions as our percolation paradigm. Thus consider an $L \times L$ checkerboard of squares. Color each square with probability p. We define a *cluster* to be a collection of colored squares which share a common side; these will be called nearest neighbors. Note that sites touching at a corner do not qualify as nearest neighbors. A cluster is said to *percolate* if it connects the top and bottom of the grid (other definitions of percolation are possible and yield similar results).

The first question we address is what is the probability of percolation as a function of the coloring probability p and the grid size, L? We call this $P(p, L)$. It is possible to analytically compute the percolation probability for small grids, but the problem rapidly becomes intractable since the number of configuration grows as 2^{L^d}, where d is the number of dimensions of the lattice.

Show that $P(p, 2) = p^2(2 - p^2)$.

It is clear that some sort of computational scheme is required to answer our question. Our strategy will be to set up a lattice of points ℓ_i (where $i = (i_1, i_2)$ is an ordered pair labeling the ith square) of size $L \times L$ and set $\ell_i = 1$ with probability p. We then generate an ensemble of stochastic grid *configurations* and compute the probability that percolation occurs. Some example configurations are shown in Figure 8.1.

8.1.1 The cluster algorithm

Generating configurations is simple; detecting clusters in a configuration is a more difficult task. It is easy for people to see clusters but how does one teach a computer to do it? The natural first step is to create a cluster matrix that labels each site according

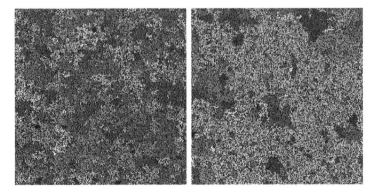

Figure 8.1 *Clusters on a 400×400 grid. On the left, the configuration was generated using $p = 0.55$, on the right, $p = 0.59$. The right-hand configuration includes a percolating cluster, which is drawn in a light purple color and extends from top to bottom.*

to the cluster of which it is an element. Let's call $\ell_i \in (0, 1)$ the lattice of occupied or unoccupied sites and c_i the cluster number associated with the site i. As above, i is a generic location label: it could be an array of integers or a mapping from such an array to a super-index. The quantity $i + 1$ then means 'the neighbors of i' as defined by the geometry of the lattice in question.

The algorithm is now reasonably simple to specify:

```
// first attempt at a cluster algorithm
cnum = 1
{c_i = 0} for all i

sweep through the lattice
  if ℓ_i and ℓ_{i+1} are occupied

    (1) if c_i = 0 and c_{i+1} = 0
          c_i = c_{i+1} = cnum
          increment cnum

    (2) else if c_i ≠ 0 and c_{i+1} = 0
          c_{i+1} = c_i

    (3) else if c_i = 0 and c_{i+1} ≠ 0
          c_i = c_{i+1}

    (4) else if c_i ≠ 0 and c_{i+1} ≠ 0
          set max(c_i, c_{i+1}) = min(c_i, c_{i+1})

  end if
end sweep
```

There are four possibilities on sweeping through the lattice.

(1) neither site belongs to a cluster: assign them to cnum and increment cnum.
(2) site i is in a cluster and $i + 1$ is not: assign site $i + 1$ the cluster number of site i.
(3) the same as (2) but reverse the roles of i and $i + 1$.
(4) both sites are elements of clusters, thus we merge the clusters by retaining the lower cluster number.

While this is pretty clear, the reader may be wondering how exactly one implements step (4). One could, for example, sweep through the lattice and replace all the instances of $\max(c_i, c_{i+1})$ with $\min(c_i, c_{i+1})$. But this turns an order L^d algorithm into an order L^{2d} algorithm, which is disastrous for large lattices. A better approach is to keep track of which clusters have been linked. One can do this with a pointer array `cluster`. Thus if clusters 3 and 7 are to be merged one simply sets `cluster[7] = 3`. It might happen that cluster 3 has already been linked so we introduce a function `root` to scan the path of cluster links to determine the 'root' cluster number

```
root(c):
  while cluster[c] ≠ 0
    c = cluster[c]
  end while
return c
```

Lastly, note that one must set `cluster[cnum] = 0` to terminate the cluster pointer tree.

We thus arrive at an efficient version of the cluster algorithm:

```
// efficient cluster algorithm
cnum = 1
{cᵢ = 0} for all i

sweep through the lattice
  if ℓᵢ and ℓᵢ₊₁ are occupied

    (1) if cᵢ = 0 and cᵢ₊₁ = 0
          cᵢ = cᵢ₊₁ = cnum
          cluster[cnum] = 0
          increment cnum

    (2) else if cᵢ ≠ 0 and cᵢ₊₁ = 0
          cᵢ₊₁ = cᵢ

    (3) else if cᵢ = 0 and cᵢ₊₁ ≠ 0
          cᵢ = cᵢ₊₁

    (4) else if cᵢ ≠ 0 and cᵢ₊₁ ≠ 0 and root(cᵢ) ≠ root(cᵢ₊₁)
          mx = max(root(cᵢ), root(cᵢ₊₁))
          mn = min(root(cᵢ), root(cᵢ₊₁))
          cluster[mx] = mn

  end if
end sweep
```

Once the lattice has been scanned one can clean house by resetting all the cluster numbers:

```
sweep through the lattice
  cᵢ = root(cᵢ)
end sweep
```

Determining whether a cluster percolates between two sets of border sites can be done simply by checking if the cluster numbers on any two sites of the borders agree:

```
pc = 0     // the percolating cluster number
percolate = false

sweep along border 1
  sweep along border 2
    if c_border 1 ≠ 0 and c_border 1 = c_border 2
      percolate = true
      pc = c_border 1
      break
    end if
  end sweep
end sweep
```

A final word: the increase in algorithm speed has been gained at the expense of memory, the array `cluster` must be of size L^d and the total memory required goes from $2L^d$ to $3L^d$. The exchange of memory for speed (or vice versa) is common in algorithm design.

8.1.2 Code verification

Debugging Monte Carlo code is notoriously difficult–how does one determine an error in a random number? In general one must compute the expectation value of a quantity and compare that to an analytically known result. In this case we have the expression for $P(p, 2)$. The comparison is shown in Figure 8.2 and indicates that we are on the right track. Numerical data were generated with eight groups of 100 configurations. Because the lattice size is only 2×2, the test is not very stringent, so some additional visual debugging may also be necessary. Notice that the points do not lie exactly on the analytic curve–as one expects for randomly generated numbers!

8.1.3 The percolation probability

It is time to answer the question posed in this section: what is the percolation probability? Certainly P runs to zero as $p \to 0$ and $P \to 1$ as $p \to 1$. We expect a monotonic curve in between. But how does the curve change as a function of the grid size? The numerical answer is provided in Figure 8.3, where we have chosen to focus on the range $p \in (0.55, 0.65)$.

> The error bars in Figure 8.3 seem to get smaller as L increases. Explain.

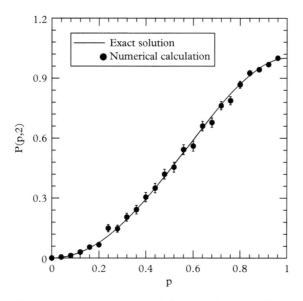

Figure 8.2 *Exact and numerical percolation probability for L = 2.*

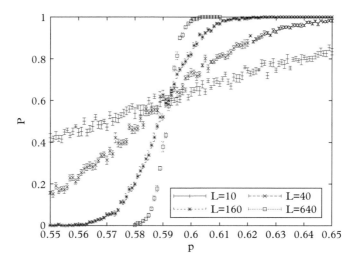

Figure 8.3 *Critical percolation probability intersections.*

246 *Site percolation*

The percolation probability rapidly gets steeper as L gets larger. In fact as the lattice size approaches infinity the percolation probability approaches a step function:

$$\lim_{L\to\infty} P(p,L) = \theta(p - p_c), \tag{8.1}$$

where p_c is the **critical probability**. In the bulk limit generating configurations with a density below p_c *never* yields a percolating cluster, while it *always* yields one above p_c. While reminiscent of a phase transition, this remarkable property of percolation has nothing to do with dynamics, it is purely statistical in nature!

Next we focus on computing the percolation probability. The reader is likely aware of fitting algorithms that are designed to adjust a parameterized function to sequences of data such as those shown in Figure 8.3. Such algorithms are common components of data analysis packages. The common ones are χ^2 (or linear least-squared) fitting, binned maximum likelihood fitting, and unbinned maximum likelihood fitting. These techniques are based at some level on the notion of likelihood, which we encounter first in the following paragraphs.

To make maximum use of the information we have on hand, we wish to incorporate all points in our estimate. We focus first on one set of points in Figure 8.3 collected with fixed L, which we will call a *sequence* and denote by $\mathcal{S}$, consisting of many sets of trials, each set with fixed p. For each value of p in the sequence, the set of trials is represented on the graph with a *central value* and an *error bar*, which together are meant to summarize all the information about result of the many trials, i.e. that there were N successes out of M trials. The probability of that happening is given by the binomial distribution:

$$Pr(N|M,f) = \frac{M!}{N!(M-N)!} f^N (1-f)^{M-N} \tag{8.2}$$

where $f = P(p, L)$. This probability predicts the number of successes given f. However, having observed N successes in M trials, it can also be re-interpreted as an *a posteriori* probability density[1] for the unknown parameter f. Accompanying our re-interpretation of Eq. 8.2 is a change of vocabulary. The same expression is now called a **likelihood**:

$$\mathcal{L}(f|N,M) = \frac{M!}{N!(M-N)!} f^N (1-f)^{M-N}. \tag{8.3}$$

This likelihood should be normalized such that $\int \mathcal{L}(f)df = 1$, as is already the case in Eq. 8.3. The function happens to be a beta distribution in the variable f. As you can check for yourself, the standard deviation of the beta distribution in the limit of large M is:

$$\sqrt{\frac{f(1-f)}{M}}$$

---

[1] The theoretical justification for this is subject to certain caveats which are discussed later in the text.

so this should be the size of the error bar. This type of error is called a ***binomial error***. The central value and the error bar are both important to include in the graph for purposes of visualization, but keep in mind that it cannot substitute for full information at a point i, namely the number of successes N_i and the number of trials M_i. In fact, it merely summarizes the lowest two moments of the binomial distribution.

We seek next a function that describes the data. To do that, we need to select a plausible functional form to construct a parametrized *fitting function*, i.e., a function which predicts the value of f given p. The curves in Figure 8.3 all have an S-shape. Such functions are called sigmoids. Several sigmoid functions that one could consider are the error function, hyperbolic tangent, or arc tangent; since the error function occurs frequently in statistics, we consider it the most likely form to describe this data, and we choose this function to work with. We obtain a parametrized fitting function from `erf` by translating it so that its midpoint occurs at p_{50}, and then scaling its width:

$$f(p|p_{50}, \sigma) = \text{erf}\left((p - p_{50})/\sigma\right) \tag{8.4}$$

This is to be understood as a function of p, whose shape is determined by parameters p_{50} and σ. Now, the likelihood for a single point at p_i with N_i successes out of M_i trials is:

$$\mathcal{L}_i(p_{50}, \sigma | N_i, M_i) = \frac{M_i!}{N_i!(M_i - N_i)!} f(p_i|p_{50}, \sigma)_i^N (1 - f(p_i|p_{50}, \sigma))^{M_i - N_i}. \tag{8.5}$$

and we can see that this is a funtion of p_{50} and σ. The likelihood for the entire data set is simply the product of the likelihood for all of the points in the sequence, i.e.:

$$\mathcal{L}(p_{50}, \sigma | \mathcal{S}) = \prod_i \mathcal{L}_i(p_{50}, \sigma) \tag{8.6}$$

This function, the likelihood function, can be considered as a joint probability distribution for the parameters p_{50} and σ. It is defined by the sequence $\mathcal{S}$. It is usually a simple matter to compute the likelihood. In this example, we first define a data structure representing the data of each point:

```
struct DataPoint {
double p;
unsigned int success;
unsigned int total;
};
```

The entire sequence is a `std::vector` of these `DataPoints`. The computation of the likelihood function $\mathcal{L}(p_{50}, \sigma | \mathcal{S})$ is given by:

```
double likelihood(double p50, double sigma, const std::vector
    <DataPoint> & sequence) {
```

```
  double lL=0;
  for (unsigned int i=0;i<sequence.size();i++) {
    double P=(1+erf((sequence[i].p-p50)/sigma))/2.0;
    double M=sequence[i].total;
    double N=sequence[i].success;
    double logProb= lgamma(M+1) - lgamma(N+1) -lgamma(M-N+1)+
        sequence[i].success*log(P) + (sequence[i].total-
        sequence[i].success)*log(1-P);
    lL+=logProb;
  }
  return exp(lL);
}
```

Here we have preferred the use of the lgamma function, computing the logarithm of the gamma function, to factorials which lead more quickly to overflow errors.

It is instructive to plot the likelihood function. To do this, we can use Markov chain Monte Carlo to generate a realization of the likelihood function, following the methods of Chapter 7. After some trial, error, and tuning, we obtain the following plots for the parameters p_{50} and σ: In this case (and many others) the likelihood function has a simple Gaussian form, so we can estimate the central value for p_{50} and σ from the mean of the histograms in Figure 8.4. For this sequence the values are $p_{50} = 0.592713 \pm 0.000048$ and $\sigma = 0.00594091 \pm 0.000061$. For the purposes of a visual cross-check we co-display the function $f(p|p_{50}, \sigma)$ with the data after fixing the parameter values to those extracted from MCMC. We obtain the plot in Figure 8.5, which indicates a very good fit to the data. Repeating this for the other three sequences at $L = 160, L = 40$, and $L = 10$ gives the results in Table 8.1.

The critical probability p_c is the limit $\lim_{L \to \infty} p_{50}$. We conclude that at a lattice size of $L = 640$, our procedure has converged to the accuracy of the fit and quote a numerical value of this constant as 0.59271 ± 0.0005. These data required 100K Monte Carlo configurations per lattice site. If one wanted to reduce the error by an order of magnitude,

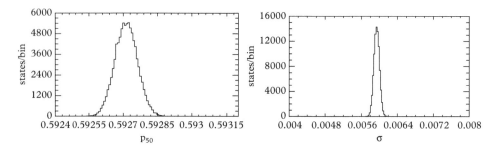

Figure 8.4 Likelihood projections in the parameters p_{50} and σ for the sequence with $L = 640$.

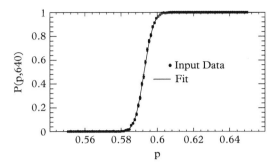

Figure 8.5 *In this plot we co-display the fitted function together with the data for a visual cross-check of the fitting procedure.*

Table 8.1 *Extracted values of p_{50} for four sequences at different lattice sizes L.*

L	p_{50}	σ
10	0.58646 ± 0.00039	0.12459±0.0016
40	0.59137 ± 0.00031	0.04616±0.0012
160	0.59277 ± 0.00008	0.01680±0.0010
640	0.59271 ± 0.00005	0.00594±0.0006

10 million configurations would be needed. For comparison, the exact value of the critical probability (to five digits) is 0.59275.

8.2 Fractals

Take another look at Figure 8.1. The percolating cluster has a ragged perimeter and is riven with holes. The cluster is reminiscent of other rough structures such as coast lines or cell phone antennas or the Sierpinski carpet (Figure 8.6). Such objects were popularized by Mandelbrot[2] who called them *fractals*. Although a strict mathematical definition of a fractal has not been established, it is generally agreed that they are **self-similar** and carry fractional dimension.

A solid material of linear size r and volume v obeys the simple scaling law:

$$v \propto r^d \qquad (8.7)$$

---

[2] Benoît Mandelbrot (1924–2010), Polish-French-American mathematician. Earlier investigations were made by Gottfried Leibniz, Karl Weierstrass, Georg Cantor, Henri Poincaré, Wacław Sierpiński, and others.

250 *Fractals*

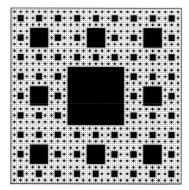

Figure 8.6 *Sierpinski carpet.*

in d dimensions. Fractals, on the other hand, are perforated with holes on all scales, which distorts the scaling relation:

$$v \propto r^D \qquad (8.8)$$

where D is in general not an integer. Let's compute the fractal dimension of the Sierpinski carpet as an example (this is also called the ***Hausdorff dimension***). The carpet is made by iteration; the first stage is a 3×3 grid with the central square empty. Thus the volume is $v_1 = 8$ (in units of the fundamental square) and the length scale is $r_1 = 3$. At the next stage each occupied square is subdivided into 3×3 grids and the central square is again empty. Thus $v_2 = 8^2$ and $r_2 = 9$. Continuing the process yields $r_n = 3^n$ and $v_n = 8^n$, thus the fractal dimension of the Sierpinski carpet is:

$$D = \frac{\log 8}{\log 3}. \qquad (8.9)$$

Similar computations can be made for any 'regular fractal'; see the exercises for an example with the Koch snowflake.

Perhaps the construction of the Sierpinksi carpet gives an idea of what 'self-similarity' means. In this case the recursive nature of the construction implies that the carpet looks precisely the same if one were to zoom in by a factor of 3. Hence the carpet is self-similar. This is pretty clear for regular fractals like the Sierpinski carpet, but what about our percolating clusters? Zooming in on a cluster yields another ragged looking cluster, but it certainly does *not* look the same. The trick here is that is *does* look the same *on average*.

Let's firm this up by computing the Hausdorff dimension of clusters. We focus on the average number of sites, v, in the largest cluster at a particular value of p. The length scale r can be defined as the mean square distance:

$$r^2 = \frac{1}{v}\sum_{i=1}^{v}||r_i - r_0||^2 \tag{8.10}$$

from the barycenter of the cluster:

$$r_0 = \frac{1}{v}\sum_{i=1}^{v} r_i \tag{8.11}$$

and r_i is the location of the ith occupied site in a given cluster. We thus seek the exponent in the relation $v \propto r^D$. It is, in fact, not necessary to spend the effort computing r because any appropriate length scale will do to determine the Hausdorff dimension. In this case it suffices to use L as a proxy for r; we therefore compute the average size (and error) of the largest cluster as a function of lattice size for $p = 0.55, 0.59275$, and 0.65. The results are presented in Figure 8.7. For $p < p_c$ clusters are rare and small and become increasingly so as the lattice size increases. Thus one expects:

$$D(p < p_c) = 0 \tag{8.12}$$

and this is what is observed. Alternatively, for $p > p_c$ clusters tend to span the entire lattice and are 'normal', thus one expects $D = d$. A fit to the data for $p = 0.65$ yields:

$$D(p > p_c) = 2.0005 \pm 0.0001, \tag{8.13}$$

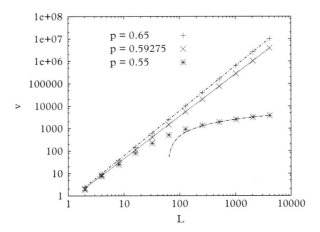

Figure 8.7 *Volume scaling for various p. The lines are fits to the form aL^b and $a\log(bL)$. Error bars are smaller than the symbols.*

reasonably close to expectations. Finally, right at the percolating threshold, clusters are ragged and spongy and one anticipates fractal behavior. In fact the fit to the data at $p = 0.59275$ yields:

$$D = 1.887 \pm 0.003, \tag{8.14}$$

reasonably close to the exact result of $D = 91/48 = 1.8958$. It appears that percolating clusters are indeed fractal! The general lattice dependence is given by:

$$v \sim \begin{cases} \log L & p < p_c \\ L^{91/48} & p = p_c \\ L^2 & p > p_c. \end{cases} \tag{8.15}$$

8.3 Scaling and critical exponents

You are familiar with the utility of reduced units (cf. Section 5.11). For example the SHO problem is naively defined in terms of two dimensionful quantities m and k:

$$m\ddot{x} + kx = 0 \tag{8.16}$$

but can be simply recast in terms of a reduced time, $\tau = t\sqrt{k/m}$ as:

$$\frac{d^2x}{d\tau^2} + x = 0. \tag{8.17}$$

Thus all temporal dependence must scale according to $\sqrt{m/k}$. Similarly the quantum mechanical problem:

$$-\frac{\hbar^2}{2m}\nabla^2\psi + \kappa x^n \psi = E\psi(x) \tag{8.18}$$

has eigenvalues that scale as $E \propto (\hbar^2 \kappa^{2/n}/m)^{n/(n+2)}$.

Does an analogous simplification occur in the percolation problem? The answer, perhaps surprisingly, is yes. It turns out, for example, that the percolation probabilities, $P(p, L)$ are really a function of one variable:

$$P(p, L) = f((p - p_c)L^{1/\nu}). \tag{8.19}$$

We verify the scaling by replotting the data of Figure 8.3 as a function of the scaling variable. The result (Figure 8.8) shows the remarkable collapse of all points to a single universal function–even for small values of L and p away from p_c.

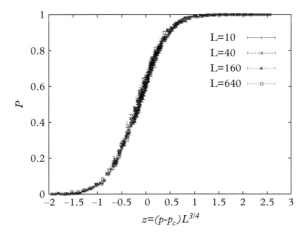

Figure 8.8 *Percolation probability scaled results. The exponent is $\nu = 4/3$; a method to obtain this is given in Section 8.4.1.*

This result can now be used to obtain an estimate of the critical percolation probability. Indeed, Eq. 8.19 implies that:

$$\left.\frac{\partial P}{\partial L}\right|_{p=p_c} = 0. \tag{8.20}$$

Thus plots of P for various L must cross–and where they cross is p_c. Consult Figure 8.3 again. It is clear that the curves (especially for larger L) are crossing at a common point. Evidently:

$$p_c \approx 0.593. \tag{8.21}$$

in agreement with the fits of Section 8.1.3.

8.3.1 The correlation length and the anomalous dimension

Why does the scaling law hold? In the example of the SHO, scaling holds because there is only one scale to the problem, $\sqrt{m/k}$, not two. The same must occur here: for large lattices there is only one relevant spatial scale. This scale is called the ***correlation length*** and is often denoted as ξ.

Roughly speaking the correlation length is the size of a typical cluster, as defined, for example, by Eq. 8.10. It is, however, traditional to define it in terms of a more fundamental quantity called the ***pair correlation function***. This function is the probability that a site at location y a distance r from an occupied site at location x belongs to the same cluster. If one sets:

$$G(x,y) = \ell_x \ell_y \delta(c_x, c_y) \qquad (8.22)$$

then the correlation function is given by:

$$g(r = ||x - y||) = \langle G(x,y) \rangle. \qquad (8.23)$$

Translation invariance implies that g can only be a function of the distance between the locations x and y.

> Argue that
> $$\sum_r g(r) = v$$
> for a given cluster.

The correlation function can now be used to obtain the correlation length via a moment:

$$\xi^2 = \frac{\sum_r r^2 g(r)}{\sum_r g(r)}. \qquad (8.24)$$

Notice that if one defines g as an average over all sites in the lattice then $g(r) = p$ for all r. This is not useful. Rather we define the correlation function as an average over all sites that are both in a single cluster. For large p most sites are filled and belong to the same cluster, thus:

$$g(r) \approx p, \; p > p_c \qquad (8.25)$$

For small p clusters are small and one expects that g drops rapidly beyond the typical cluster size. In this case:

$$g(r) \sim \frac{1}{r^{(d-1)/2}} e^{-r/\xi}, \; p < p_c, \; r \to \infty. \qquad (8.26)$$

This expression dates back a century to the work of Ornstein and Zernike (1914), who postulated that the Fourier transform of the correlation function is given by the following equation:

$$g(r) = \int \frac{d^d q}{(2\pi)^d} e^{-i\vec{q}\cdot\vec{r}} \frac{1}{q^2 + \xi^{-2}}. \tag{8.27}$$

Evaluating this integral for large r yields the form given in Eq. 8.26.

The correlation length must diverge at the critical coupling. This is because a percolating cluster is self-similar and hence looks the same (on average) at all scales. If the correlation were finite, it would provide a scale beyond which self-similarity would break down. Thus new scaling laws emerge at criticality. For example:

$$\xi \sim |p - p_c|^{-\nu} \tag{8.28}$$

where ν is a *critical exponent* to be determined. Thus at criticality the only scale in the system disappears and the correlation function must assume a power law behavior, $g(r) \sim r^{2-d}$. In two dimensions one obtains $g(r) \sim \log(r)$, which cannot be correct. Fisher (1964)[3] corrected this problem by assuming that $g(q) \sim q^{-2+\eta}$ for small q. This yields the expectation:

$$g(r) \sim \frac{1}{r^{d-2+\eta}}, \quad p = p_c, \ r \to \infty. \tag{8.29}$$

The new exponent η is termed the *anomalous dimension*.

Determining the anomalous dimension is not an easy numerical exercise (although, as always, there are good and bad ways to achieve numerical goals). We shall take a direct approach and evaluate $g(r, L, p)$ at $p = p_c$ and attempt fits of the form of Eq. 8.29. Various correlation functions (arbitrarily normalized to fit on the same graph) are shown in Figure 8.9. One immediately detects a problem: the curves appear linear (on a log scale) at large distances, which does not agree with the expected form of Eq. 8.29. The problem is that we necessarily work with finite systems, hence the correlation lengths are limited by L. In fact, because we compute at criticality, one expects $\xi \sim L$. A simple-minded guess that accommodates this reality is:

$$g(r) \sim \frac{1}{r^{d-2+\eta}} \exp(-r/\xi_L), \quad p = p_c, \ r \to \infty, \text{ finite } L. \tag{8.30}$$

Let us now examine Figure 8.9 in more detail. First, the data have been generated as an array $g(i,j)$ where $i = |x_1 - y_1|$ and $j = |x_2 - y_2|$. The correlation function must be a function of $r = \sqrt{(i^2 + j^2)}$ in the bulk limit. However for finite lattice sizes residual dependence on i and j will remain. We handle this situation (and avoid having to bin in r) by computing $g(i,j)$ and plotting g versus $\sqrt{(i^2+j^2)}$. The resulting points should fall onto a universal curve if translation invariance is good. The ragged appearance of the curves in Figure 8.9 is an indication of how close one is to the bulk limit.

---

[3] Michael E. Fisher (1931 –). British physicist renowned for his contributions to understanding critical phenomena and phase transitions.

256 Scaling and critical exponents

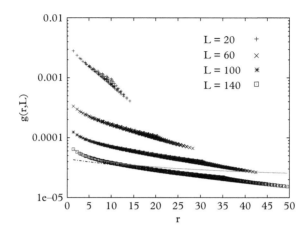

Figure 8.9 *Pair correlation function at $p = 0.59275$. Fit to $L = 140$ data for large r: dash-dot curve; fit for small r: grey line.*

Table 8.2 *Fit parameters and extrapolation for the pair correlation function at $p = 0.59275$.*

L	40	60	80	100	120	140	∞
ξ_L	14.1(2)	20.1(1)	26.8(1)	36.3(3)	38.8(1)	46.5(2)	∞
η	0.401(8)	0.349(6)	0.320(5)	0.280(5)	0.269(3)	0.262(3)	0.201(8)

The parameters η and ξ_L of Eq. 8.30 have been obtained by fitting in the small r and large r regimes respectively. Sample fits are shown for the $L = 140$ data in the figure (the dash-dot line disappears under the data points past $r = 10$); fit results are given in Table 8.2. It is clear that the correlation length is diverging with the lattice size, and indeed, a linear fit gives $\xi_L \sim 0.33(2)L$. Thus, as expected, the exponential portion of the fit Ansatz (Eq. 8.30) becomes irrelevant as the bulk limit is approached. Alternatively, the short distance behavior starts at $\eta \approx 1/2$ for small lattices but rapidly becomes smaller with increasing lattice size. A linear fit in $1/L$ yields $\eta(L = \infty) = 0.201(8)$, which is in good agreement with the analytically known value $\eta = 5/24 \approx 0.208$.

8.3.2 Additional scaling laws

It is perhaps not surprising that scaling relations exist for a variety of quantities near the critical point. For example, the fraction of sites that belong to an infinite cluster obeys the relation:

$$F \sim |p - p_c|^\beta, \qquad p > p_c, \qquad (8.31)$$

while the mean cluster size follows:

$$S \sim |p - p_c|^{-\gamma}. \tag{8.32}$$

Here S is defined as:

$$S = \frac{\sum_s s^2 n_s}{\sum_s s n_s}, \tag{8.33}$$

where n_s is the number of clusters of size s (Stauffer and Aharony, 1994). The parameter ν that appeared in the scaling law for the percolation probability (Eq. 8.19) also features in an expression for the dispersion in the percolation probabilities near criticality:

$$\langle (p_c - \langle p_c \rangle)^2 \rangle \sim L^{-2/\nu}. \tag{8.34}$$

Finally, Fisher (1964) has introduced a scaling law for the quantity n_s:

$$n_s \sim s^{-\tau} f((p - p_c) s^\sigma), \quad p \to p_c, \ s \to \infty. \tag{8.35}$$

At first sight one might think that these critical exponents are independent. But this cannot be true if more elemental quantities also scale, which is the case. The number of clusters of size s, n_s is one such quantity.[4] A variety of **scaling relations** thus exist (Levinshtein, 1976; Stanley, 1971); some of these are listed here.

$$2D = d + 2 - \eta \tag{8.36}$$

$$D = d - \beta/2 \tag{8.37}$$

$$\tau = d/D + 1 \tag{8.38}$$

$$\beta = (\tau - 2)/\sigma \tag{8.39}$$

$$\gamma = (3 - \tau)/\sigma \tag{8.40}$$

$$\nu = (\tau - 1)/(\sigma d) \tag{8.41}$$

$$d\nu = \gamma + 2\beta. \tag{8.42}$$

Some of these are merely restatements, for example Eq. 8.42 follows from Eqs 8.39, 8.40, and 8.41. Other relations follow from elementary scaling. For example the relations:

$$F + \sum_s s n_s(p_c) = p, \tag{8.43}$$

$$p = \sum_s s n_s(p), \tag{8.44}$$

---

[4] In a more familiar setting, the free energy scales near phase transitions; all thermodynamic quantities can be derived from the free energy, and hence also scale. (Widom, 1965).

and the Fisher scaling Ansatz yield:

$$F \sim \sum_s s^{1-\tau}\left(f((p-p_c)s^\sigma) - f(0)\right)$$
$$\sim (p-p_c)^{(\tau-2)/\sigma} \int dz \left(f(z) - f(0)\right) z^{-1+(2-\tau)/\sigma}. \quad (8.45)$$

In the second line we have used a standard trick to replace a sum with an integral and have made a change of variables. When combined with the definition of Eq. 8.31 one obtains the scaling relation of Eq. 8.39.

Argue that Eqs 8.43 and 8.44 are true.

Similarly:

$$S \sim \int ds\, s^{2-\tau} f((p-p_c)s^\sigma) \sim (p-p_c)^{(\tau-3)/\sigma} \int dz\, z^{-1+(3-\tau)/\sigma} f(z), \quad (8.46)$$

which, along with Eq. 8.32, yields Eq. 8.40.

Notice that the scaling relations imply that we have wasted effort (but not instructional opportunity!) in computing the anomalous dimension because it is simply related to the fractal dimension. Perhaps this is not surprising–somehow the 'fractalness' of the system should be related to the power law behavior of the pair correlation function in the percolating cluster. Alternatively, scaling is a hypothesis and our computations have confirmed that:

$$2D + \eta - d - 2 = -0.025 \pm 0.01. \quad (8.47)$$

Note that the result, despite being small, is 2.5σ removed from zero. Although it is statistically possible that this result is consistent with zero it is more likely that errors have been underestimated (this is a common occurrence!)–indeed systematic errors have been ignored in this estimate, and can be considerable.

8.4 Universality and the renormalization group

So far we have concentrated effort on site percolation in the two-dimensional square lattice. What happens in three or more dimensions? What about other types of two-dimensional lattices? In this case one would retain an array $\ell_{i,j}$ to describe the configuration but nearest neighbors would be described by a different set of "cell" vectors. For example, for the triangular lattice one could employ $\{\hat{x}, (\sqrt{3}\hat{y} + \hat{x})/2, (-\sqrt{3}\hat{y} + \hat{x})/2\}$ as a basis set. This can be implemented in code by distorting the basis so that the three neighbors are, for example, at $\{(i+1,j), (i,j+1), (i,j-1)\}$.

Once a suitable geometry has been implemented a lot of hard work of the type discussed above will yield results similar to that of Table 8.3 (see (Essam, 1983) if you are interested in how one makes the analytic computations). The notations 'cubic', 'bcc', and 'fcc' in the table refer to three-dimensional cubic, body centered cubic, and face centered cubic lattice structures.

If one were to press on and compute critical exponents and fractal dimensions for the new systems one would run into a rather remarkable result: the exponents only appear to depend on the dimensionality of the lattice, and not on the lattice type! Some critical exponents are given in Table 8.4. This stunning phenomenon, called **universality** (Fisher, 1974), is a deep result of the self-similar nature of percolation models at criticality. In brief, it is properties of the percolation cluster that determine critical exponents, and these properties are long range (or better: *all* range). However lattices differ only in their short range structure, and hence universality is a plausible outcome. Of course many quantities, such as the critical probability p_c, *do* depend on short range structure.

The loss of sensitivity to short-range properties is a familiar property of nature. One does not need, after all, to understand the details of the quantum mechanical interactions of H_2O molecules to obtain an accurate and useful macroscopic description of water. Similarly, all manners of microscopic diffusion processes can be described by the large scale effective equation[5]

$$\frac{\partial \phi}{\partial t} = \nabla \cdot (D \nabla \phi). \tag{8.48}$$

Table 8.3 *Critical probabilities for various lattices.* $s = 2 \sin \pi/18$.

Lattice	Site	Bond
hex	0.6962	$1 - s$
square	0.59275	$\frac{1}{2}$
triangular	$\frac{1}{2}$	s
diamond	0.428	0.388
cubic	0.3117	0.2492
bcc	0.245	0.1785
fcc	0.198	0.119

---

[5] Let's pursue the diffusion example a bit more. Consider Pascal's recurrence relation for the binomial coefficients:

$$P(X, T+1) = \frac{1}{2}[P(X-1, T) + P(X+1, T)]$$

Table 8.4 *Critical exponents. See Ex. 4 for Cayley trees. The singular behavior of the total number of clusters is specified by α.*

Exponent	$d=2$	$d=3$	Cayley
α	$-2/3$	-0.6	-1
β	$5/36$	0.4	1
γ	$43/18$	1.8	1
ν	$4/3$	0.9	$1/2$
σ	$36/91$	0.45	$1/2$
τ	$187/91$	2.2	$5/2$
$D(p=p_c)$	$91/48$	2.5	4
$D(p<p_c)$	1.56	2	4
$D(p>p_c)$	2	3	4

Lastly, the immense successes of thermodynamics are underpinned by a lack of sensitivity to microscopic structure. It appears that the presence of very many interacting entities serves to wash out dependence on the details of the entities themselves.

8.4.1 Coarse graining

It is useful to think of self-similarity of a percolating cluster in terms of invariance under *coarse graining*. Coarse graining is defined generically as a process of removing degrees of freedom (Sethna, 2006). For the case of percolation it is convenient to implement this idea by replacing a sublattice of size $b \times b$ with a single superlattice site. The superlattice site will be deemed to be occupied if the original sublattice percolated. Thus an $L \times L$ lattice is mapped to an $L/b \times L/b$ superlattice with an effective percolation probability called p'. It is clear that (i) the superlattice is simply another percolation problem with a different scale ξ' and probability p' (ii) the coarse graining process can be repeated yielding a sequence of superlattices.

and suppose that the triangle positions X and T map onto a lattice with cutoffs δx and δt so that $X = x/\delta x$ and $T = t/\delta t$. Let $p(x,t) = P(x/\delta x, t/\delta t)/\delta x$ then Pascal's equation is equivalent to:

$$\frac{p(x, t+\delta t) - p(x,t)}{\delta t} = D \frac{p(x-\delta x, t) + p(x+\delta x, t) - 2p(x,t)}{(\delta x)^2}$$

where $D = (\delta x)^2/(2\delta t)$. This equation is strongly dependent on the microscopic cutoffs δt and δx unless the continuum limit is approached in such a way that D is kept constant. It is in this sense that sensitivity to microscopic details are washed out in the large scale limit.

The effective correlation length is related to the original correlation length in a specific way:

$$\xi = k(p - p_c)^{-\nu} \tag{8.49}$$

$$\xi' = kb(p' - p_c)^{-\nu}. \tag{8.50}$$

(We have supplied the constant of proportionality in Eq. 8.28.) At criticality one must have $p = p'$ and $\xi = \xi'$ thus:

$$\nu = \lim_{p' \to p} \frac{\log b}{\log \left(\frac{p' - p_c}{p - p_c} \right)}, \tag{8.51}$$

which provides a useful mechanism for determining the critical exponent ν.

To see this in action let us consider the two-dimensional triangular bond lattice (with critical probability $p_c = 1/2$). We construct the superlattice by averaging three neighboring sites. The probability that this sublattice percolates is given by:

$$p' = p^3 + 3p^2(1 - p). \tag{8.52}$$

The critical probability can be estimated by imposing the criticality condition $p' = p = p_c$. Solving Eq 8.52 gives two trivial solutions $p_c = 0$ and $p_c = 1$ and the nontrivial estimate $p_c = 1/2$ (which in this case happens to be exactly right).

Expanding about $p = p_c$ and setting $b = \sqrt{3}$ then yields a simple estimate:

$$\nu \approx \frac{\log \sqrt{3}}{\log 3/2} = 1.355 \tag{8.53}$$

This is not a bad first approximation to the exact value $\nu = 4/3$.

We add two observations: (i) such successes do not come so easily in other models, (ii) the larger one can take b, the more accurate the calculation is. Unfortunately, even dealing with the relatively simple case of 3×3 sublattices on the square lattice means enumerating 2^9 configurations! How this can be done in practice is the subject of the next section.

More than forty years ago Wilson (1971)[6] suggested that the process of coarse graining configurations can be abstracted to *coarse graining theories*. It is convenient to postulate the existence of *theory space* when conceptualizing Wilson's idea. By this we mean the set of all possible theories that describe a given general situation, such as percolation, spin systems, or quantum field theory. If a generic Hamiltonian in such a set can be written as:

---

[6] Kenneth G. Wilson (1936 –). American physicist. Inventor of 'Wilsonian renormalization' for which he won the Nobel prize, the operator product expansion, and lattice field theory.

Universality and the renormalization group

$$H(\{\lambda\}) = H_0 + \lambda_1 H_1 + \lambda_2 H_2 + \ldots \tag{8.54}$$

where the H_i are interaction terms of fixed form and the λ_i are coupling constants, then theory space would consist of the set of points labelled by $(\lambda_1, \lambda_2, \ldots)$.

Wilson's idea was to study how theory space maps onto itself under coarse graining. Indeed, one can consider coarse graining to be a map $\mathcal{M}$ that effects $H' = \mathcal{M}(H)$, or more specifically,

$$H(\{\lambda'\}) = \mathcal{M}(H(\{\lambda\})). \tag{8.55}$$

One says that the theory parameters $\{\lambda'\}$ have been **renormalized** by the mapping. The mapping itself is often called a **renormalization group**, although it does not satisfy all the requirements to form a mathematical group.

Figure 8.10 illustrates what can happen under **renormalization group flow**, in which a map that corresponds to an infinitesimal coarse graining is iterated many times. Lines indicate how Hamiltonians are renormalized as the map is iterated. The arrows indicate the direction of flow, which is typically towards large scales (ie., $b > 1$)[7]. It may happen that a *fixed point* of the map exists:

$$H_\star = \mathcal{M}(H_\star), \tag{8.56}$$

where $H_\star = H(\{\lambda_\star\})$. This is indicated by the point labelled $H_\star$ in the figure. All theories that flow into this point are in the **basin of attraction** of the fixed point and are said to be in a **universality class**. We have already learned that all two-dimensional percolation models are members of one universality class.

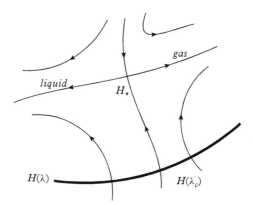

Figure 8.10 *Renormalization group flow in theory space.*

---

[7] The fancy way to say this is 'the renormalization group flows towards the infrared'.

The dark line in Figure 8.10 represents a particular theory as a coupling is varied (in our familiar percolation model the parameter would be p). At some point the model crosses a flow line that feeds into the fixed point. This point is the critical coupling λ_c.

Notice that the fixed point has flow lines that diverge from it (the fixed point is thus conditionally stable).[8] The physical behavior the theory is qualitatively different depending on which direction one follows on this line. In fact the two sides of the diverging flow line represent different phases which are separated by a phase transition. Right at the fixed point the system does not 'know' which phase it is in and experiences fluctuations at all scales between the phase options (labelled gas and liquid in the figure).

> How would you a interpret a diagram in which all flows approached a fixed point?

8.4.2 Monte Carlo renormalization group

In general critical exponents can be obtained from Eq. 8.55 by linearizing the map near the fixed point (which must be determined by other means). We call the values of the couplings after n iterations $\lambda_i^{(n)}$ and the fixed point couplings $\lambda_i^\star$. Define the distance from the fixed point:

$$\Delta \lambda_i^{(n)} = \lambda_i^{(n)} - \lambda_i^\star. \tag{8.57}$$

Then linearizing the coarse graining map near the fixed point yields:

$$\Delta \lambda_i^{(n+1)} = \sum_j T_{ij}^\star \Delta \lambda_j^{(n)} \tag{8.58}$$

where

$$T_{ij}^\star = \frac{\partial \mathcal{M}_i}{\partial \lambda_j}\bigg|_{\lambda=\lambda^\star}. \tag{8.59}$$

The eigenvalues of the linearized coarse graining map are related to critical exponents. For example, the largest eigenvalue of T is defined to be $b^{1/\nu}$. Notice that Eq. 8.51 is the one-dimensional version of this result.

Implementing the renormalization procedure in this more general case is a daunting task. One must remove (or 'integrate out') degrees of freedom in a system that can have many couplings. This can be nearly as difficult as solving the original problem, and theorists often resort to drastic approximations such as the 'ϵ-expansion' (where $d = 4 - \epsilon$) or the '$1/N$-expansion' (where N is the dimension of an internal symmetry group) (Wilson and Kogut, 1974). In addition every mapping step induces many (often

---

[8] Fixed points and their classification are discussed at depth in Chapter 13.

infinitely many) new operators and their couplings, and these must be truncated in an uncontrollable way to render any computation practical.

The great difficulties involved in implementing renormalization group ideas naturally suggest employing numerical methods. This suggestion was first made by Ma (1976), while the efficiency of the method was improved by Swendsen (1979).

The case of percolation is unusually simple: one seeks to numerically evaluate the percolation probability p' on a $L/b \times L/b$ lattice. But this is equivalent to the original problem, so there is literally no additional work to perform, only a reinterpretation of old results. Specifically p' will be given by the percolation probability computed on a lattice of size $L' = L/b$ thus:

$$p' = P(p, L'). \tag{8.60}$$

Evaluating the derivative at p_c then gives the numerator of Eq. 8.51 and hence the exponent ν.

In more complicated situations one can invert Eq. 8.58 to obtain:

$$T_{ij} = \frac{\partial \lambda_i^{(n+1)}}{\partial \lambda_j^{(n)}} \tag{8.61}$$

and use the chain rule to obtain the derivatives:

$$\frac{\partial \langle H_i^{(n+1)} \rangle}{\partial \lambda_j^{(n)}} = \sum_k \frac{\partial \lambda_k^{(n+1)}}{\partial \lambda_j^{(n)}} \cdot \frac{\partial \langle H_i^{(n+1)} \rangle}{\partial \lambda_k^{(n+1)}} \tag{8.62}$$

Finally, the derivatives of the expectation values of interaction terms can be related to correlation functions (Landau and Binder, 2000) as follows:

$$\frac{\partial \langle H_i^{(n)} \rangle}{\partial \lambda_j^{(m)}} = \langle H_i^{(n)} H_j^{(m)} \rangle - \langle H_i^{(n)} \rangle \langle H_j^{(m)} \rangle. \tag{8.63}$$

Numerical implementations of this scheme must carefully explore the dependence on the algorithmic parameters that specify the size of the system, the number of coarse graining iterations, and the truncation in the number of operators. One finds that convergence is quite fast for the two-dimensional Ising model, but that this fortunate situation does not hold for other systems (Landau and Binder, 2000).

8.5 Epilogue

Perhaps the reader is feeling a little pummelled at this point: what started as an investigation of a nearly trivial problem in statistics has turned into a whirlwind tour of some of the most profound concepts of modern physics. We stress that the presentation has not been historical, all of the general concepts–scaling, self-similarity, fractals, critical exponents, the renormalization group–were developed in different contexts. However the application of them to percolation is particularly simple and hence this chapter served as a useful introduction to these sophisticated concepts. They will be re-visited in chapters on nonlinear dynamics (Chapter 13), molecular dynamics (Chapter 18), classical and quantum spin systems (Chapters 20 and 22), and quantum field theory (24). In short, any problem with many degrees of freedom and a phase transition can benefit from the renormalization group methodology.

Computationally, we have learned a useful technique in pattern recognition, and how to apply scaling laws to extract critical couplings. Previous expertise in extrapolation and data fitting has also been leveraged.

The existence of fluctuations at all scales underpins self-similarity at a critical point. This in turn implies that a large set of theories, all of which differ at short distances, share the same singular structure at large distances. Although the concept of universality is perhaps new to the reader, it is vital to our ability to describe nature. Indeed, if the phenomena of every day life depended on the details of the microscopic quantum mechanical world that supports such phenomena we would never have been able to perceive simple physical laws.

8.6 Exercises

1. [T] Elementary Statistics.
   Consider Figure 8.2. If the error bars represent one standard deviation then approximately how many points should lie 1σ or more from the exact curve? If you saw data as in Figure 8.11, would you be suspicious? See (Volkow, 2011) for a real-world example of this problem.

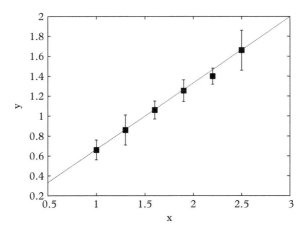

Figure 8.11 *Dubious data?*

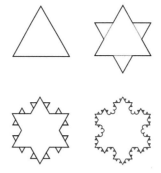

Figure 8.12 *Koch snowflake.*

Figure 8.13 *Affine leaf.*

2. [T] Koch Snowflake.
   A Koch snowflake is obtained by adding triangles to sides of triangle, as illustrated in Figure 8.12. Obtain expressions for the perimeter and area of the nth order Koch snowflake. How do they behave as n runs to infinity?

3. Affine Maps.
   Write a program to generate points according to the algorithm:

$$\left\| \begin{array}{l} \vec{x} = \vec{x}_0 \\ \texttt{iterate} \\ \quad \vec{x} = \vec{f}(\vec{x}; \hat{\mu}) \\ \texttt{end iterate} \end{array} \right.$$

Here $\hat{\mu}$ is a random variable that is regenerated at each iteration. It is often better to let the components of $\vec{x}$ update on the fly in the iteration. Obtain something like Figure 8.13 with the function:

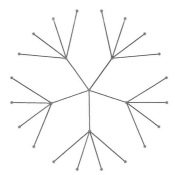

Figure 8.14 *A Cayley tree with z = 5 and two generations.*

$$f_x = \begin{cases} 0 & 3\% \\ 0.2x - 0.26y & 13\% \\ -0.15x + 0.28y & 11\% \\ 0.85x + 0.04y & 73\% \end{cases}.$$

$$f_y = \begin{cases} 0.16y & 3\% \\ -0.23x + 0.22y + 1.6 & 13\% \\ 0.26x + 0.24y + 0.44 & 11\% \\ -0.04x + 0.85y + 1.6 & 73\% \end{cases}$$

where the percentages indicate the probability of using the particular map. Is the object self-similar?

4. [T] Cayley Trees.
   A *Cayley tree* with coordination number z is a structure formed by adding z links to a node such that only one link connects two nodes as in Figure 8.14

   a) Terminate a z-tree after r generations and call the terminal sites the 'surface' of the tree. Derive an expression for the ratio of the surface sites to the total number of sites. How do you interpret the dimensionality of the tree?

   b) Occupy each site in a z-tree with probability p. A percolating cluster will be one that starts at the center and reaches the surface. Prove that:

   $$p_c = \frac{1}{z-1}.$$

   c) Prove that the probability that a given site belongs to an infinite cluster for $z = 3$ is:

   $$F = p - \frac{(1-p)^3}{p^2}.$$

   Obtain β. See (Flory, 1941) if you are having trouble.

d) Obtain the mean cluster size, defined as the average number of sites in a cluster containing the origin, for $z = 3$:

$$S = \frac{1+p}{1-2p}.$$

Obtain γ.

5. **[T] One-dimensional Percolation.**
Consider site percolation in a one-dimensional chain of squares. Show the following

   a) The critical probability is:

   $$p_c = 1.$$

   b) The probability that a site in the lattice is the left-hand end of a cluster of size s is:[9]

   $$n_s = p^s(1-p)^2.$$

   c) The pair correlation function obeys:

   $$g(r) = p^r.$$

   d) Use $g(r)$ to obtain ν.
   e) Lastly, prove that the average cluster size obeys:

   $$S = \frac{p_c + p}{p_c - p}.$$

6. **Binned Likelihood Fitting.**
In the directory DATA/CH8/EX6 you will find a data file containing measurements obtained in a simulation of a one-dimensional percolation problem. Each row of the file contains, first, the cluster size s, and second, the number of clusters found to have size s in the simulation. The physicist who simulated the data forgot which value of p was used to generate those data. Construct a likelihood function $\mathcal{L}(p)$ for the data sample based upon the Poisson distribution:

---

[9] this is called the *cluster number*, which is perhaps a misleading term.

$$P(N|\lambda) = e^{-\lambda}\frac{\lambda^N}{N!}.$$

At what value of p does your $\mathcal{L}(p)$ take its maximum value? Over what distance does it fall by a factor $e^{-1/2}$?

7. 1d Clusters.
   Write your own algorithm to generate and cluster configurations in 1 dimension. Use the algorithm developed in the previous exercise to re-extract from the simulation the value of the probability p. How does the performance of your algorithm depend upon the lattice size L?

8. [T] Critical Exponent ν.
   Compute analytically the percolation probability $P(p, 3)$ on a 3×3 lattice as a function of p. From this, estimate the critical probability on a 9×9 lattice, a 27×27 lattice, and an infinite lattice. Then, estimate the critical exponent ν.

9. Parallel Cluster Recognition Algorithm.
   Design a parallel cluster recognition algorithm. Write your answer in pseudocode or, better, implement and test it. Under what conditions is it less efficient than the serial cluster recognition algorithm?

10. Universality.
    Implement code to analyze the *bond percolation* model on a two-dimensional square lattice. This model joins neighboring sites with probability p. Percolation is defined when two sides of a lattice are continuously connected by occupied bonds.

    a) Determine p_c and show that is different from that of the 2d site percolation model.
    b) Implement viable schemes to compute the critical exponents ν, β, and γ.
    c) Verify that the site and bond models belong to the same universality class.

11. Monte Carlo Renormalization Group for Percolation.
    Consider the square site percolation model.

    a) Attempt to obtain an accurate estimate of the critical probability p_c by using the renormalization group equation $p' = P(p, L')$ at criticality.
    b) The deviation of this solution from the exact result scales as $L^{-1/\nu}$. Verify this form and use it to estimate ν.

12. Cluster Growth Algorithm.
    An alternative approach to analyzing clusters is to generate them from a single occupied site by visiting all neighboring sites and marking them with probability p. In this way the system automatically percolates on a lattice up to the largest linear size of the cluster. Iteration stops when no neighbouring sites have been marked. Implement this algorithm in your favorite language and test it by computing the anomalous dimension of the percolation cluster for a square lattice.

13. **One-dimensional Ising Model.**

    The one-dimensional Ising model is a traditional instructional tool for statistical mechanics. Here we compare the exact solution to a renormalization group computation of the partition function.

    a) Obtain the exact partition function for the one-dimensional chain with periodic boundary conditions,

    $$Z = \sum_{\{\sigma_1, \sigma_2, \ldots, \sigma_N\}} \exp\left(K \sum_i \sigma_i \sigma_{i+1}\right).$$

    Here σ is a classical spin variable, $\sigma = +1, -1$. *Hint:* use the transfer matrix method where $Z = tr V^N$, with:

    $$V_{\sigma\sigma'} = \exp(K\sigma\sigma') = \begin{pmatrix} \exp(K) & \exp(-K) \\ \exp(-K) & \exp(K) \end{pmatrix}.$$

    Diagonalize V and obtain:

    $$F = \lim_{N\to\infty} \frac{1}{N} \log(Z) = \ln 2 \cosh K.$$

    b) Coarse grain by summing over the even spins. For example, summing over σ_2 gives a factor:

    $$\exp(K(\sigma_1 + \sigma_3)) + \exp(-K(\sigma_1 + \sigma_3)).$$

    This needs to be rewritten in a form that mimics the original Hamiltonian. Fortunately this is easily achieved via:

    $$\exp(K(\sigma_1 + \sigma_3)) + \exp(-K(\sigma_1 + \sigma_3)) = f(k) \exp(K' \sigma_1 \sigma_3)$$

    where $f(K) = 2\sqrt{\cosh(2K)}$ and $K' = \frac{1}{2} \ln \cosh 2K$. The end result is a partition function of precisely the same form as the original but with one half the degrees of freedom and a renormalized coupling. Specifically, for large N:

    $$Z(K, N) = f(K)^{N/2} Z(K', N/2)$$

    and hence:

    $$F(K) = \frac{1}{2} \ln f(K) + \frac{1}{2} F(K').$$

    Invert the expression for $f(K)$ to obtain $K = f^{-1}(K')$. Notice that these equations iterate towards larger coupling K. Choosing a small initial value for K' then permits an estimate of the initial value for $F \approx \ln 2$ and estimates of F for larger K.

Iterate from small K' and obtain a series of estimates for F. Compare this to the exact values. Remarkably, the error in the initial estimate for F actually gets smaller as the iteration proceeds. Notice that there is no fixed point in the iteration so no phase transition is predicted.

14. **Two-dimensional Ising Model.**
Apply the renormalization group to the two-dimensional Ising model. Take the interaction to be $\sum_{(i,j)} \sigma_i \sigma_j$ where the indices run over nearest neighbors on a square lattice. As in the previous question, sum over the odd sublattice to obtain a new expression for the partition function:

$$Z(K, N) = f(K)^{N/2} \sum_{\text{even}} \exp(H_1 + H_2 + H_3)$$

where:

$$f(K) = 2 \cosh^{1/2}(2K) \cosh^{1/8}(4K)$$

and:

$$H_1 = \frac{1}{4} \ln \cosh(4K) \sum_{nn} \sigma_i \sigma_j \quad (8.64)$$

$$H_2 = \frac{1}{8} \ln \cosh(4K) \sum_{nnn} \sigma_i \sigma_j \quad (8.65)$$

$$H_3 = \left(\frac{1}{8} \ln \cosh(4K) - \frac{1}{2} \ln \cosh(2K)\right) \sum_{\square} \sigma_i \sigma_j \sigma_k \sigma_\ell.$$

Here *nn* means 'nearest neighbor'–those sites $\sqrt{2}$ in distance from each other; *nnn* means 'next nearest neighbor' and refers to sites a distance 2 apart. The last term is a sum over sites that form a square around the former site.

Unlike the one-dimensional case, coarse graining generates new terms in the effective Hamiltonian which make mapping to the original form ambiguous. One approach would be to simply ignore the new terms H_2 and H_3. Then:

$$K' = \frac{1}{4} \ln \cosh(4K) \quad (*)$$

and:

$$F(K') = 2F(K) - \ln f(K).$$

Verify that no phase transition occurs with this truncation. Replace Eq. (∗) with:

$$K' = \frac{3}{8} \ln \cosh(4K).$$

What happens in this case? You should find a fixed point in the transformation. Locate the fixed point and compare to the exact result $K_c = \frac{1}{2}\sinh^{-1}(1)$. The new coefficient in the expression for K' is an attempt to incorporate H_2 in the computation.

15. The Central Limit Theorem and the Gaussian Fixed Point.

    The central limit theorem is a general statement about sums of variates that is (largely) independent of the details of the initial probability distribution. This is reminiscent of renormalization group arguments, and in fact one can derive the central limit theorem by considering a map from the space of probability functions onto itself.

    a) Coarse grain by combining two variates. Show that the probability distribution for the sum of two numbers drawn from the distribution ρ is:

    $$C(x) = \int dy\, \rho(x-y)\rho(y).$$

    b) Rescale C so that width of the coarse grained distribution remains fixed (take $\mu = 0$). This means multiplying the argument of ρ by $\sqrt{2}$ and normalizing. Thus the renormalization map is:

    $$\mathcal{M}[\rho] = \sqrt{2} \int dy\, \rho(\sqrt{2}x - y)\rho(y).$$

    Show that a Gaussian is a fixed point of this map. Studying the eigenvalues of the linearization of the map near the fixed point yields information on how the Gaussian distribution is approached as the number of variates becomes large.

BIBLIOGRAPHY

Essam, J. W, D. S. Gaunt, and A. J. Guttmann (1978). *Percolation theory at the critical dimension.* J. Phys. A **11**, 1983.

Fisher, M. E. (1964). *Correlation Functions and the Critical Region of Simple Fluids.* J. Math. Phys. **5**, 944.

Fisher, M. E. (1974). *The Renormalization Group in the Theory of Critical Behavior.* Rev. Mod. Phys. **46**, 597.

Flory, P. J. (1941). *Molecular Size Distribution in Three Dimensional Polymers.* Am. Chem. Soc. **63**, 3091.

Landau, D. P. and K. Binder (2000). *A Guide to Monte Carlo Simulations in Statistical Physics.* Cambridge University Press.

Last, B. J. and D. J. Thouless (1971). *Percolation Theory and Electrical Conductivity.* Phys. Rev. Lett. **27**, 1719.

Levinshtein, M. E. *et al.* (1976). *The Relation Between the Critical Exponents of Percolation Theory.* Sov. Phys. JETP, **42**, 197.

Ma, S. -K. (1976). *Renormalization Group by Monte Carlo Methods.* Phys. Rev. Lett. **37**, 461.

Mandelbrot, B. (1967). *How Long Is the Coast of Britain? Statistical Self-Similarity and Fractional Dimension.* Science, **156**, 636.

Ornstein, L. S. and F. Zernike (1914). *Accidental Deviations of Density and Opalescence at the Critical Point of a Single Substance.* Proc. Acad. Sci. Amsterdam, **17**, 793.

Sethna, J. P. (2006). *Statistical Mechanics.* Oxford University Press.

Stanley, H. E. (1971). *Introduction to Phase Transistions and Critical Phenomena.* Oxford University Press.

Stauffer, D. and A. Aharony (1994). *Introduction to Percolation Theory.* CRC Press.

Swendsen, R. H. (1979). *Monte Carlo Renormalization Group.* Phys. Rev. Lett. **42**, 859.

Volkow, N. D., D. Tomasi, G. J. Wang, P. Vaska, J. S. Fowler, F. Telang, D. Alexoff, J. Logan, and C. Wong (2011). *Effects of cell phone radiofrequency signal exposure on brain glucose.* JAMA **305**, 808.

Widom, B. (1965). *Equation of State in the Neighborhood of the Critical Point.* J. Chem. Phys. **43**, 3898.

Wilson, K. G. (1971). *Renormalization Group and Critical Phenomena. I.* Phys. Rev. B **4** 3174, 3184.

Wilson, K. G. and J. B. Kogut (1974). *The Renormalization group and the epsilon expansion.* Phys. Rept. **12**, 75.

9
Parallel computing

9.1	High performance computing	274
9.2	Parallel computing architecture	278
9.3	Parallel computing paradigms	279
	9.3.1 MPI	279
	9.3.2 openMP	285
	9.3.3 C++11 concurrency library	289
9.4	Parallel coding	299
9.5	Forking subprocesses	300
9.6	Interprocess communication and sockets	302
9.7	Exercises	308
	Bibliography	310

The topic of parallel computing tends to be neglected or added as an afterthought in computational text books. We have chosen to present it relatively early because parallel computing is no longer an esoteric subject only meant for power users. In fact, parallel computing has become a commonplace experience in modern life: almost all current cell phones leverage parallel computing to enhance battery life, and massively parallel cloud computing is utilized every time someone seeks driving directions or asks Siri a question.

This chapter begins with an overview of the terminology encountered in the field of high performance computing. This leads to an examination of methods to measure performance and a brief look at Moore's Law. We then introduce different schemes used to implement parallel computing in hardware. Of course the main goal is to learn how to leverage this hardware, which is the objective of this chapter. We thus introduce several popular parallel computing protocols: MPI, openMP, and the C++11 concurrency library. Finally, parallel computing implies inter-computer communication; accordingly, the chapter closes with an examination of C++ sockets.

9.1 High performance computing

Modern computers are complex devices with structure over many length scales[1]. At the largest scale typical supercomputers are comprised of many **nodes**. A node is a

---

[1] In fact the scales range over an astounding ten orders of magnitude.

physical unit with its own processors, memory, and I/O bus. A *processor* is an integrated circuit chip whose primary role is to perform operations on data. The operations can include arithmetic computations (performed by an *arithmetic logic unit*) or memory manipulation. Processors often contain several *central processing units*, or "CPUs". When a single chip contains several CPUs, these are referred to as *cores*. Finally, a core may support multiple *threads*, which are independent instruction streams that share memory and other logical units in the chip. Threads are distinct from *processes* that demand more resources, such as memory, of the hardware.

It is possible that a core implements *pipelining*, where instructions can be executed in parallel. This is called *instruction-parallelism* and, because it is transparent to the user, is typically not counted towards "parallelism" of the device.

Characteristics of the world's largest supercomputers are given in Table 9.1. These machines contain millions of cores and petabytes ($2^{50} \approx 10^{15}$ bytes) of memory. Their computational capabilities are measured in *flops* (or flop/s), which is a floating point operation per second. Typical calculators operate at around 10 flops, whereas the six cores of an Apple A9 can maintain 173 Gflops. Finally, the last column of the table gives the power requirements of the respective machines. Notice that they are quite large! The US government is actively pursuing "exascale supercomputers" capable of exaflops of sustained computations. If power requirements are naively scaled up from those of Table 9.1, tens of gigawatts (i.e., several large scale nuclear reactors) would be needed to power a single computer. Clearly energy efficiency is a primary concern in future supercomputer development.

At the other end of the computational spectrum, my laptop has an Intel Core i7 processor with two cores, each of which is capable of running two threads. The most familiar computer is the cell phone, which owes its existence to the miniaturization revolution. A significant challenge with cell phones is powering their vast computational capabilities. One approach is to employ processors with multiple low power cores. When the phone is idle only a single core need run; additional cores are activated for

Table 9.1 *World's leading supercomputers. The theoretical peak capacity is listed.*

Machine	Location	Cores (M)	Peak capacity (Pflop/s)	Memory (PB)	Power (MW)
Sunway TaihuLight	National Supercomputing Center, Wuxi, China	10.65	125	1.3	15.4
Tianhe-2	National Super Computer Center, Guangzhou, China	3.1	55	1.0	17.8
Titan Cray XK7	Oak Ridge National Laboratory, TN, USA	0.56	27	0.71	8.2
Sequoia BlueGene/Q	Lawrence Livermore National Laboratory, CA, USA	1.57	20	1.6	7.9

Table 9.2 *Cell phone SoC characteristics.*

Model	Introduced	Chip	Speed	Word size	Cores	Feature size
Galaxy S7	2016	snapdragon 820	1.6 GHz	64 bits	4	14 nm
iPhone 6S	2015	A7	1.4 GHz	64 bits	2	16 nm
HTC One M8	2014	snapdragon 801	2.45 GHz	32 bits	4	28 nm

computationally demanding tasks such as navigation or searching for Japanese cartoon monsters.

Table 9.2 lists some recent cell phones and characteristics of their primary computing engines. These are often called *system on a chip* (SoC) devices because the chips incorporate CPUs; *graphical processing units* (GPUs) which are optimized to perform the computations required for graphics visualization; power management hardware; sensors (motion, magnetic field, orientation); and devices in support of communication protocols such as wifi and bluetooth. Each of the cell phones listed below encompass approximately one hundred times the memory and computational power of a typical supercomputer of the 1980s.

Table 9.2 has a column labelled "feature size". This refers to the size, typically measured in nanometers, of integrated circuit features such as a transistor gate length or the distance between wires. Historically, the bulk of the advances in computational power has arisen due to rapidly decreasing feature size. This carries three benefits: lower power consumption, increased speed, and smaller chip size. These favorable characteristics have underpinned *Moore's law* (named after American businessman, Gordon Moore, 1929–), which states that the number of components in an integrated circuit double every year[2]. Nevertheless, the specific reason for performance increases has changed with time: from 1960 to 1986 this was due to increasing word size, from 1986 to 2004 decreasing feature size was used to leverage chip clock speed increases, and since 2004, a steadily increasing number of cores has pushed the performance boundary.

The right panel of Figure 9.1 shows the approximate validity of Moore's law over four decades; the left panel makes clear the incredible decrease in the cost of computing since 1960. Recent reduction in feature size is shown in Figure 9.2.

Recently the ability to shrink feature size has slowed and the industry has turned to increasing core count to keep Moore's law afloat. Certainly, the exponential behavior seen in Figure 9.1 cannot be maintained indefinitely. At a fundamental level chip features are already approaching the limit where quantum effects will ruin chip reliability. And the problem with power requirements has already been mentioned. One expects that ingenuity will carry the stupendous advances of the previous 50 years on for some time; however, it all must end some time. This is a sobering thought as it implies that there will be limits to what can be computed–it may never be possible to predict weather a year in advance or to compute the mass of the $X(3872)$ meson to four digits accuracy.

---

[2] The doubling period has been revised several times since the original observation.

Parallel computing 277

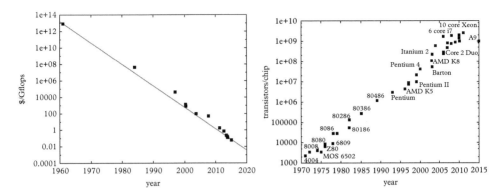

Figure 9.1 *Left: inflation adjusted dollars per gflops by year. The line is an exponential with coefficient of -0.6 year^{-1}. Right: Transistors per chip. The coefficient of the exponential growth rate is approximately 0.35 year^{-1}.*

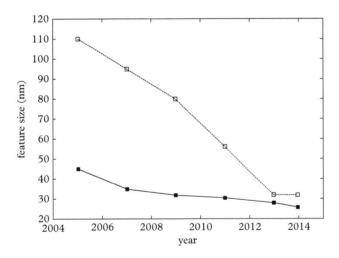

Figure 9.2 *Chip feature size. Transistor gate length (solid), one half the distance between wires on the first metal layer of the chip (open).*

9.2 Parallel computing architecture

A CPU operates on data according to instructions. Thus the CPU must deal with data streams and instruction streams. It is possible to make either of these parallel, depending on the hardware implementation chosen. In 1966, Michael Flynn (American computer scientist, 1934 –) proposed *Flynn's taxonomy* which labels the four possibilities for parallelism. These are:

SISD Single instruction stream; single data stream.

SIMD Single instruction stream; multiple data stream.

MISD Multiple instruction stream; single data stream.

MIMD Multiple instruction stream; multiple data stream.

All four types of architecture exist. SISD machines are not parallel and are chiefly older PCs and mainframe computers. SIMD computers execute a common code against multiple data streams. This is a useful paradigm for GPUs. MISD computers execute several instruction streams which act on identical data. The most common type of supercomputer implements the MIMD architecture. In this case independent instruction streams can access independent data streams. Data can be stored in a single common memory or it can be distributed across nodes.

Parallel computers must communicate information between cores and nodes. This often forms a large portion of computational effort and it is therefore important to design algorithms that reflect the physical constraints of the computer to be used. Some terminology associated with this problem is as follows.

bandwidth The maximum rate at which data can be sent.

latency The time required to send a length zero message (thus this is the communication overhead).

granularity The ratio of computational demands to communication demands of a program. Coarse granularity calls for few communication requests while fine granularity means many communications are required per unit computation.

embarrassingly parallel Embarrassingly parallel code requires no (or nearly no) communication between subprocesses.

Another important concept is *synchronization*. Synchronization refers to the timing in which tasks are performed; for example, a thread may require input from another thread before it can continue its computation. Alternatively, two threads may modify a single variable at the same time, leading to unpredictable outcomes. A parallel programmer must be aware of these situations and construct code accordingly.

Scalability refers to how performance varies as the number of CPUs or threads is increased. This is rarely linear because more cores means greater communication overhead. ***Load balancing*** is another important concept associated with parallel computing. It may occur that some threads carry most of the computational effect while others lie idle. This is clearly an inefficient situation and astute programmers will work to maintain appropriate load balance in their code.

Unlike SISD coding, it is apparent that machine architecture affects parallel coding design in important ways. This added layer of complexity cannot be avoided by the researcher. However, scientific parallel coding is often fairly simple, and even novices can make good use of the power of multicore machines.

9.3 Parallel computing paradigms

A variety of methods for achieving parallelism exist: the main paradigms in parallel coding are ***threads***, ***message passing***, and the ***data parallel model***.

The data parallel model achieves parallelization by independently manipulating portions of a large data set. For example, an array can be divided into N portions and a task assigned to each portion with each task performing the same operations. This paradigm is implemented in Fortran 95 via extensions called ***Coarray Fortran***. ***Global Arrays*** is a public domain library with Fortran 77 and C bindings.

The message passing model separates tasks and their associated memory. Tasks and memory can reside in the same node or across multiple machines. Tasks exchange information by sending and receiving messages. Message passing has been employed since the 1980s, and, starting in the 1990s, has been standardized via the ***message passing interface*** (MPI). Basic features of the MPI are discussed below.

The final major paradigm for parallel computing is the thread model. Threads are usually implemented as independent streams of instructions within a controlling ***heavy weight*** program. This program acquires all of the resources necessary to run the code. The heavy weight program spawns threads as subtasks that can have local data but also share the resources of the heavy weight program. This carries the benefit of avoiding data replication. Because of this the subtasks are called ***light weight***. Threads communicate via shared memory, which raises synchronization issues of which the programmer needs to be acutely aware. The two most common threading implementations are ***posix threads*** (also called ***pthreads***) and ***openMP***. Both of these will be discussed below.

We note that it is also possible (and useful) to consider ***hybrid models*** wherein the previous models are combined. For example, threads can be used to perform computationally intensive work on local nodes and MPI can be used for internode communication. This paradigm lends itself well to common supercomputer architecture that consists of clusters of multi-core nodes.

9.3.1 MPI

Implementation of the ***message passing interface*** (MPI) started in 1992 as a (successful) attempt to standardize the message passing paradigm. The most recent implementation was made in 2015 (MPI-3.1) and includes specifications for C, C++, and Fortran90. MPI consists of several hundred routines that manage tasks and the communication between tasks which run on one or more computers.

The following `Hello World` code illustrates the use of a few of these routines. The routine `MPI_Init` must be called once before any parallel tasks are to be undertaken. This routine establishes a series of tasks (to be run in parallel, of course) according to specifications made in your local job submission scheme. Each task has a *rank* associated with it that can be accessed with the routine `MPI_Comm_rank`. In the following, we refer to these as `taskids`. Notice that most of the MPI routines reference `MPI_COMM_WORLD`; this is a predefined communicator that accesses all the MPI tasks available to the program.

```
#include "mpi.h"
#include <iostream>

int main(int argc, char *argv[]) {
 using namespace std;
 int numtasks, taskid, len, rc;
 char hostname[MPI_MAX_PROCESSOR_NAME];

 MPI_Init(&argc,&argv);                           // initialize
                                                  //    MPI
 MPI_Comm_size(MPI_COMM_WORLD,&numtasks);         // get number
                                                  //    of tasks
 MPI_Comm_rank(MPI_COMM_WORLD,&taskid);           // get my rank
                                                  //    (taskid)
 MPI_Get_processor_name(hostname, &len);          // obtain
                                                  //    hostname

 cout << " Hello World from task " << taskid << endl;
 cout << " hostname: " << hostname << " number of tasks: 
    " << numtasks << endl;

 MPI_Finalize();                                  // clean up
 return 0;
}
```

This code was compiled with the wrapper `mpic++` which implements `openmpi` for the `gcc` compiler. The code can be executed on one or many machines using the `runmpi` command; alternately, and depending on the configuration of the host machine or cluster, the code can be submitted under a batch system. Running the code with two nodes and 20 tasks per node gave the following output:

```
Hello World from task 6
hostname: n197.mpi.sam.pitt.edu number of tasks: 40
```

```
Hello World from task 24
hostname: n198.mpi.sam.pitt.edu number of tasks: 40
Hello World from task 31
hostname: n198.mpi.sam.pitt.edu number of tasks: 40
...
```

Tasks 0 to 19 have been assigned to node `n197.mpi.sam.pitt.edu` while tasks 20 to 40 have been assigned to `n198.mpi.sam.pitt.edu`. Notice also that the order in which the tasks are reported is random. This occurs because many tasks are attempting to access a single resource (the output file) at the same time, and is an issue with synchronization. Indeed, we have often seen cases where letters from each output stream are intermingled in the output file. In general, I/O is problematic in parallel computing and one is well-advised to minimize it, or to only have a "master task" deal with it.

Embarrassingly parallel example

The `hello world` application did not need intertask communication. We now consider an embarrassingly parallelizable problem that it often encountered by the numericist, namely Monte Carlo computation. Stochastic calculations rely on executing independent actions very many times–clearly it is useful to divide these among processors. Communication is required to initiate the tasks and to combine their results after the computations are complete.

Our application will be a computation of the integral $\int_0^\infty x^2 \exp(-x)\, dx$ via direct Monte Carlo. The reader is directed to Section 7.4.2 to see the serial solution to this problem. Most of the code for the parallel implementation is given below. The new element in this example is `MPI_Reduce`, which collects the Monte Carlo sums from each task and combines them (into `fsum`). The communication necessary to do this is implemented via *send* and *receive buffers* whose jobs are to store information being generated by tasks and receive that information for further use.

```
int main (int argc, char *argv[]) {
  int taskid,numtasks;
  int seed = 12341;
  const unsigned int N=10000000;
  const unsigned int MASTER=0;
  MPI_Init(&argc,&argv);
  MPI_Comm_size(MPI_COMM_WORLD,&numtasks);
  MPI_Comm_rank(MPI_COMM_WORLD,&taskid);
  const unsigned int Npertask=N/numtasks;

  std::mt19937 e(seed+taskid);            // assign a unique
                                          //      seed to each task
  std::exponential_distribution<double> eDist;
  double sum=0,sum2=0;
  for (unsigned int i=0;i<Npertask;i++) {
```

```
      double x=eDist(e);
      double f=x*x;
      sum  +=   f;
      sum2 += f*f;
   }

   double fsum,f2sum;
   int rc = MPI_Reduce(&sum, &fsum, 1, MPI_DOUBLE, MPI_SUM,
                  MASTER, MPI_COMM_WORLD);
                  // fsum is the receive buffer
   int rc2 = MPI_Reduce(&sum2, &f2sum, 1, MPI_DOUBLE, MPI_SUM,
                  MASTER, MPI_COMM_WORLD);
   if ((rc != MPI_SUCCESS) || (rc2 != MPI_SUCCESS))
       std::cout << " MPI_reduce failure " << std::endl;

   // MASTER computes average and sigma
   if (taskid == MASTER) {
      fsum /= N;
      f2sum /= N;
      double sigma=sqrt((f2sum-fsum*fsum)/N);
      std::cout << " numtasks: " << numtasks << std::endl;
      std::cout << " fbar = " << fsum << " +/- " << sigma <<
         std::endl;
   }
   MPI_Finalize();
   return 0;
}
```

The specification of this routine is:

```
int MPI_Reduce(
    void* send_buffer,
    void* recv_buffer,
    int count,
    MPI_Datatype datatype,
    MPI_Op op,
    int master,
    MPI_Comm communicator)
```

where `send_buffer` is the variable to be sent, in this case `sum`, and `recv_buffer` is the receiving variable (`fsum`). The latter can be an array of size `count`; in our case we combine variables into a scalar so `count=1`. Note that `revc_buffer` is only known to the task with rank `master`. `MPI_Datatype` is one of the predefined MPI datatypes; in this case `MPI_DOUBLE`. A subset of MPI datatypes is given in Table 9.3. The type of

Table 9.3 *Some MPI datatypes.*

MPI_datatype	C datatype
MPI_CHAR	signed char
MPI_INT	signed int
MPI_FLOAT	float
MPI_DOUBLE	double
MPI_C_BOOL	_Bool
MPI_C_COMPLEX	float _Complex

Table 9.4 *MPI operations.*

MPI_Op	Function
MPI_MAX	maximum element
MPI_MIN	minimum element
MPI_SUM	element sum
MPI_PROD	element product
MPI_LAND	logical and the elements
MPI_LOR	logical or across the elements
MPI_BAND	bitwise and across the bits of the elements
MPI_BOR	bitwise or across the bits of the elements
MPI_MAXLOC	maximum value and the rank of the process that owns it
MPI_MINLOC	minimum value and the rank of the process that owns it

operation to perform on the data is specified by `MPI_Op`. Possible values for this variable are listed in Table 9.4. Finally, the MPI world communicator (`MPI_COMM_WORLD`) is specified by the last variable.

The final point concerning the code is that a single task, denoted `MASTER`, has been assigned to manipulate the received data and to print out this information. If this procedure were not followed, `numtasks` integrals would be printed and most of them would be wrong.

The routine `MPI_Reduce` is termed ***thread safe*** because all communications are executed in a manner that prevents overwriting memory in an uncontrolled fashion. The importance of this can be seen when considering task-to-task communication.

This is typically achieved with paired *send* and *receive* routines. Thus, for example, the pseudocode

```
MPI_Comm_Rank(... currentTask ...);

if (currentTask == sendTask)
  MPI_generic_Send(&sendBuffer, ..., destTask,...);

if (currentTask == destTask)
  MPI_generic_Recv(&recvBuffer, ..., sendTask, ...);
```

sends information from `sendTask` to `destTask`. Problems can occur if this code is iterated and the send operation has not completed, in this case the send buffer will be overwritten and unexpected results will ensue. This behavior can be avoided by using **blocking** routines that do not permit program execution to continue until the send buffer has cleared. Similarly, the receiving routine can also block until it has processed its receive buffer. This can slow program execution, thus **nonblocking** routines are also provided. The MPI implementations of the blocking routine is:

```
int MPI_Send(
    void *buf,
    int count,
    MPI_Datatype datatype,
    int dest,
    int tag,
    MPI_Comm comm)
```

The new element in this implementation is the `tag` which is a non-negative integer assigned by the coder to identify a message.

The nonblocking routine is specified as:

```
int MPI_Isend(
    void *buf,
    int count,
    MPI_Datatype datatype,
    int dest,
    int tag,
    MPI_Comm comm,
    MPI_Request *request)
```

The new element in this case is `request`. Because non-blocking operations can return before the requested system buffer space is obtained, the system issues a unique "request number" which can be used by the coder to confirm completion of the operation.

Although we have only explored a small part of the MPI implementation, the reader should be able assess whether MPI is a useful tool for her computational problem. For more information consult the excellent online documentation or the sources listed in the bibliography.

9.3.2 openMP

The openMP (open multi-processing) initiative is a joint effort of several hardware and software companies that seek to standardize thread-based parallel computing. The central idea is to use compiler directives to permit a master thread to fork a number of slave threads which run concurrently. The system divides the computational effort among threads and allocates threads to different processors. Once the parallel section of code completes, the threads synchronize and control is returned to the master thread (this is called a *join*).

These ideas are illustrated with a simple "hello world" program.

```
#include <iostream>
#include "omp.h"

int main(void) {
  #pragma omp parallel
  {
    int numthreads = omp_get_num_threads();
    int id =  omp_get_thread_num();
    std::cout << "Hello, world " << id << " [" << numthreads
        << "] " <<std::endl;
  } // end of scope; returns to serial execution
  return 0;
}
```

The compiler directive `pragma omp parallel` instructs the compiler to fork threads, which join after the pragma goes out of scope. As can be seen openMP routines are available to obtain the number of threads and the current thread number.

The number of threads can be set with an environment variable:

```
>export OMP_NUM_THREADS = <n>
```

or in-code with the line (which should precede the pragma):

```
omp_set_num_threads(<n>)
```

The number of supported threads can be queried with the unix command

```
>ulimit -a
```

the maximum possible number of threads with

```
>ulimit -Hu
```

and the number can be set with

```
>ulimit -a <n>
```

In general many more threads are available than cores to run them. This is useful for software applications in which concurrency is valuable (for example, a cell phone app can run communication procedures in one thread and monitor user input in another), but this is generally not useful in scientific applications where dividing intensive computations is the desired effect (*divide et impera*). In this case the maximum number of threads should be set to the number of cores available.

The `hello world` code was compiled with the command

```
g++ -fopenmp -o helloWorld helloWorld.cpp
```

The output was

```
Hello, world 6 [24]
Hello, world 1 [24]
10 [24]
Hello, world 0 [24]
Hello, world 15 [24Hello, world 22 [24] ]

9 [24]
Hello, world 21 [24]
Hello, world 13 [24]
...
```

As can be seen, a total of 24 threads were created. As with MPI, the threads reported to the output in random order, although in this case more severe synchronization issues are apparent. This is no surprise: `std::iostream` is not thread-safe.

The problem can fixed by invoking several other features of openMP as illustrated in the following code snippet.

```
int main(void) {
  int th_id, nthreads;
  #pragma omp parallel private(th_id) shared(nthreads)
  {
    th_id = omp_get_thread_num();
    #pragma omp critical
```

```
    {
      std::cout << " Hello World from thread " << th_id <<
        std::endl;
    }
    #pragma omp barrier          // wait until all threads
                                 //      reach here    [1]
    #pragma omp master           // only the master thread
                                 //      runs this section
    {
      nthreads = omp_get_num_threads();
      std::cout << " There are " << nthreads << " threads "
        << std::endl;
    }
  }
  return 0;
}
```

The keyword `critical` is used to specify a region where only one thread can be run at a time. This avoids the problem of multiple threads accessing the output stream at the same time. The `barrier` directive imposes synchronization on threads. In particular, threads will stop execution once they reach this statement; once all threads reach this point, parallel execution will resume. Finally, the `omp master` directive specifies a region of code that is only executed by the master thread. All of the other threads skip this region.

The other new elements in this code snippet are *scoping directives*, `private` and `shared`. The default scoping behavior is to make variables declared outside of the parallel region shared (i.e., known to all threads), while variables declared within the parallel region are private, which means that a local copy is kept by each thread. Other threads do not have access to private variables. The `shared` and `private` keywords can be used to override this behavior in way that we hope is clear.

Another useful openMP construct permits parallelization that is specific to loops wherein the loop is shared amongst threads. The syntax is:

```
#pragma omp parallel for
for (i=0;i<100;++i) {
  A(i) = ...               // work will be divided amongst
                           //     threads
}
```

Scheduling

This example raises the issue of *scheduling*, which is the strategy employed for sharing the work load represented by the loop. Two common strategies are *static scheduling* and *dynamic scheduling*. The former approach divides loops into *chunks* which are then assigned to threads. With dynamic scheduling the loops are divided into chunks and chunks are assigned to threads as the threads become available. The latter approach is faster in situations when threads can take greatly varying times to perform their tasks; however, additional computational overhead is incurred. Both options can be set in the `omp parallel for` command (details can be found in the bibliography).

Embarrassingly parallel example

As discussed in Section 9.3.1, Monte Carlo computations are a canonical example of embarrassingly parallel applications. We illustrate this, again, by evaluating the integral, $\int_0^\infty x^2 \exp(-x)\, dx$ with the direct Monte Carlo method. The problem is solved with the following code

```
int main (void) {
  int taskid=0,numtasks=1,Npertask;
  int seed = 12341;
  const unsigned int N=10000000;

  double sum=0.0,sum2=0.0;
  #pragma omp parallel reduction(+:sum,sum2)
  {
    numtasks = omp_get_num_threads();
    Npertask=N/numtasks;
    taskid = omp_get_thread_num();
    std::mt19937 e(seed+taskid); // assign a unique seed to
                                      each task

    std::exponential_distribution<double> eDist;
    for (unsigned int i=0;i<Npertask;i++) {
      double x=eDist(e);
      double f=x*x;
      sum  +=   f;
      sum2 += f*f;
    }
  }  // [1]
  #pragma omp master
  {
    int nn = Npertask*numtasks;
    sum /= nn;
    sum2 /= nn;
    double sigma=sqrt((sum2-sum*sum)/nn);
    std::cout << " fbar = " << sum << " +/- " << sigma <<std
       ::endl;
  }
     // [2]
  return 0;
}
```

The only new element of this code is the statement

```
#pragma omp parallel reduction(+:sum,sum2)
```

The phrase `reduction(op:var)` tells the compiler to perform the operation `op` on the variable `var` across threads. In the above example, this means that the sum of all `sum`s and `sum2`s is performed and made available at location [1]. Variables must be scalars and declared shared. Operators can be any of +, -, *, /, &, ˆ, |, &&, or ||.

openMP environment variables

All threading environments must provide local memory for each thread, which can be problematic for applications with extensive memory requirements. In fact openMP creates separate data stacks for every worker thread to store copies of private variables (the master thread uses the regular stack). The size of these stacks is not defined in the openMP standards and can vary substantially. The default stack size for g++ is a paltry 2 MB–once this limit is breached undefined program behavior results, although most compilers throw a segmentation fault. Fortunately, it is relatively easy to overcome this restriction, either memory can be assigned on the heap and manually divided amongst threads, or the default stack size can be changed with an environment variable. For example:

```
export OMP_STACKSIZE=1GB
```

Other openMP environment variables permit controlling the number of threads, the way that loop iterations are divided, binding threads to processors, setting thread wait policy, and enabling dynamic threads.

9.3.3 C++11 concurrency library

The C++ concurrency library will be the final system we consider for implementing parallelism. This is likely the system that will be most accessible to readers since it is part of the C++ standard and is implemented in clang or newer versions of gcc. The standard achieves parallelism with threads, hence all the caveats associated with openMP also apply here; in particular, the user needs to carefully maintain thread safety. In our experience, threads can be much faster than MPI, but they are only useful for implementing concurrency on a single computer.

All threads have access to the same global shared memory and their own private data. Programmers are responsible for synchronizing access (protecting) to globally shared data.

Threads are created with a declaration (look at the code below). Notice that the thread is associated with a function (in this case `f`) and that a value can be passed to that function. The idea is that many functions (or single functions of different variables) can run concurrently. The `join` member method of `thread` *must* be called some time before exiting `main`. It's function is to block execution until all threads finish. As with

Table 9.5 *Timing tests with eight threads. Times with the -O3 compile optimization are also listed.*

Threads	-O3	Time	Speed up
1	no	2.062s	1
1	yes	0.733s	2.81
8	no	0.341s	6.05
8	yes	0.114s	18.1

all previous examples of this type, the ouput is a scrambled version of the numbers 400 and 100 (recall that `iostream` is not thread-safe).

```
#include <thread>
#include <iostream>

void f(int i) {
 int j = i*i;
 std::cout << j << std::endl;
}

int main() {
    std::thread t1(f,20);
    std::thread t2(f,10);

    t1.join();
    t2.join();
}
```

Does this really work? Run times for adding $8 \cdot 10^8$ integers are shown in Table 9.5. These were generated on a quad-core Intel Mac with eight threads. As we have mentioned, running more threads than physically supported wastes overhead and is likely not useful in a numerically intensive application.

At this point it is not difficult to assemble a `Hello World` program. One solution is:

```
#include <thread>
#include <iostream>
#include <string>

void hw(std::string const& message, int i) {
    std::cout << message << " " << i << std::endl;
}
```

```
int main() {
   unsigned int nthreads = std::thread::hardware_concurrency
       ();
   std::cout << " threads/cores available = " << nthreads <<
       std::endl;

   std::thread *th = NULL;
   th = new std::thread[nthreads];
   for (int i=0;i<nthreads;i++) {
     th[i] = std::thread(hw,"hello world!",i);
   }
   for (int i=0;i<nthreads;i++) {
     th[i].join();
   }
   delete [] th;
   return 0;
}
```

The only new element is creating an array of threads of length `nthreads`, where

`std::thread::hardware_concurrency()`

is used to determine the number of threads that are available.

This code is compiled with the command:

`g++ -std=c++11 -pthread -o helloWorld helloWorld.cpp`

Some systems do not need the `-pthread` option. As is usual, the code output is a mess, similar to that shown on page 286.

One can use more modern constructs such as the `vector` class template to achieve the same thing, thereby accruing all of the benefits associated with that class (dynamical memory allocation and automatic garbage collection among them).

```
#include <thread>
#include <iostream>
#include <string>
#include <vector>

void hw(std::string const& message, int i) {
   std::cout << message << " " << i << std::endl;
}

int main() {
   unsigned int nthreads = std::thread::hardware_concurrency
       ();
```

```cpp
    std::cout << " threads/cores available = " << nthreads <<
        std::endl;

    std::vector<std::thread> th;
    for (int i=0;i<nthreads;i++) {
      th.push_back(std::thread(hw,"hello world!",i));
    }
    for (auto &t : th) {
      t.join();
    }
    th.clear();
    return 0;
}
```

There is yet another way to say hello to the world that uses the `future` class template. This is discussed below.

Synchronization: mutex and atomic

Having multiple threads all attempting to update a global variable is a clear recipe for chaos. Consider the following code, where many threads increment the global variable sh. If there are 24 threads, one might naively hope that the output is 24, but two threads may access sh when its value is 12 and leave it with a value of 13. Thus, in general, the printed output is a number less than 24 (our tests in this situation commonly gave 24, 23, and 22 as results).

```cpp
#include <thread>
#include <iostream>
#include <vector>

int sh=0;

void incr(void) {
  sh++;
}

int main() {
    unsigned int nthreads = std::thread::hardware_concurrency
        ();
    std::cout << " threads/cores available = " << nthreads <<
        std::endl;

    std::vector<std::thread> th;
    for (int i=0;i<nthreads;i++) {
      th.push_back(std::thread(incr));
```

```
    }
    for (auto &t : th) {
      t.join();
    }
    std::cout << "got " << sh << std::endl;
    th.clear();
    return 0;
}
```

None of this is a surprise. Like MPI and openMP, the thread class provides methods to deal with synchronization. One such method is called a *mutex* ("mutual exclusion"), whose purpose is to block thread execution while a code segment is owned by another thread. Usage is controlled with `lock` and `unlock` methods:

```
std::mutex mx;
int sh;            // shared data
 ...
mx.lock();
  sh++;            // manipulate shared data
mx.unlock();
 ...
```

Only one thread can execute in the critical region between the lock and unlock commands. Any other thread that attempts to execute statements in this region will be blocked until the owning thread executes `mx.unlock()`.

Resolving the conflict in our incrementing code is now simple: one simply declares a mutex and surrounds the offending statements with lock and unlock statements. All other code remains unchanged. This version of the program will always produce 24 as its output.

```
 ...
#include <mutex>

int sh=0;
std::mutex mx;

void incr(void) {
  mx.lock();
    sh++;
  mx.unlock();
}
 ...
```

294 *Parallel computing paradigms*

We remark that this seemingly benign construct can be dangerous at times. For example, if an exception is thrown before an `unlock` statement, execution will block with no escape possible.

The C++11 Concurrency Library has implemented another way to resolve synchronization issues via the *atomic* class template. The class permits declaring atomic type variables whose access is guaranteed to be thread-safe. Coding is similar to mutexes, but locking and unlocking are handled internally.

```
...
#include <atomic>

std::atomic<int> sh (0);     // initialize to 0

void incr(void) {
  sh++;
}
...
```

Again, running this code always yields 24 as the output.

While all this is fine, we wish to stress that a parallel numericist should generally avoid these techniques. The issue is that using atomic data types is blocking, and hence antithetical to parallel computing. It is far preferable to avoid relying on global variables in the first place. One way to do this is shown in the next section, where we examine a simple Monte Carlo integral for the third time.

Embarrassingly parallel example

Once again, our application will be a computation of the integral $\int_0^\infty x^2 \exp(-x)\, dx$ via direct Monte Carlo. As before, 10^7 integrand evaluations are divided amongst `nthreads` tasks. Each thread is given its own random number engine with a unique seed. Threads run independently while summing the integrand in a global variable that is unique to the thread (we implement this with an array `vector<double> sum`). In this way all synchronization issues are avoided. Once the threads exit the code combines the results from each thread to produce the final estimate for the integral.

```
...
typedef mt19937_64 MersenneTwister;
vector<thread> th;
vector<MersenneTwister> MT;
vector<double> sum, sum2;
const unsigned int N=10000000;
unsigned int nthreads = thread::hardware_concurrency();
unsigned int Npertask = N/nthreads;

void run_th(int th_id) {
```

```
    exponential_distribution<double> eDist;
    for (int i=0; i<Npertask;i++) {
      double x=eDist(MT[th_id]); // access exp distn with
          different engines
      double f=x*x;
      sum.at(th_id)    +=   f;
      sum2.at(th_id)   += f*f;
    }
  }

  int main(void) {
    unsigned long int seed;
    cout << " enter seed " << endl;
    cin >> seed;

    for (int th_id=0;th_id<nthreads;th_id++) {
      sum.push_back(0.0);
      sum2.push_back(0.0);
      MT.push_back(MersenneTwister(seed+th_id));
      // create unique engines
    }
    for (int th_id=0;th_id<nthreads;th_id++) {
      th.push_back(thread(run_th,th_id));        // [1]
    }
    for (auto &t : th) {   // wait for threads to exit
      t.join();
    }
    int nn = Npertask*nthreads;          //  analyze  results
    double fsum=0.0,f2sum=0.0;
    for (int th_id=0;th_id<nthreads;th_id++) {
      fsum += sum[th_id];
      f2sum += sum2[th_id];
    }
    fsum /= nn;
    f2sum /= nn;
    double sigma=sqrt((f2sum-fsum*fsum)/nn);
    cout << " fbar = " << fsum << " +/- " << sigma << endl;

    return 0;
  }
```

The reader will notice that this code is somewhat more elaborate than the analogous openMP program. In that case all the effort in creating the threads and in rendering the sums thread-safe was handled by the compiler.

Creating thread-safe code can be subtle, sometimes extremely so. The reader is directed to Ex. 10 to see a particularly brutal example that involves the code just presented.

Future and promise

The C++ standard library provides an alternative mechanism to execute parallel code with the `future` class template. This is implemented with *future* and *promise* objects. Future and `promise` enable the transfer of data between tasks without implementing a lock. In particular, a task places data into a `promise` when the task wants to return data to the thread that created it. The system then makes that data appear in the `future` object that is associated with the promise.

The central mechanism for our purposes is the function

```
async(launch::async, fn, args)
```

which calls a function `fn(args)` and returns without waiting for its execution to complete. The value eventually returned by `fn` can be accessed by a *future* object that is returned by `async`. These objects are of type

```
future<type>
```

and can be accessed with `future::get`. Finally, the phrase `launch::async` is a directive that instructs the compiler to create a new thread which calls `fn`.

The concepts are illustrated in the following code, which implements the third variant of `Hello World` promised above.

```
#include <iostream>
#include <future>
#include <vector>
#include <string>

void hw(string const& message, int i) {
  std::cout << message << " " << i << std::endl;
}

int main() {
   unsigned int nthreads = std::thread::hardware_concurrency
       ();
   std::cout << " threads/cores available = " << nthreads <<
       std::endl;
   std::vector<future<void>> th;
   for (int i=0;i<nthreads;i++) {
```

```
      th.push_back(async(launch::async,&hw,"hello world!",i));
   }
   th.clear();
   return 0;
}
```

An advantage of `async` is that it has a simple mechanism by which threads can pass data back to `main`. This is illustrated in the following example:

```
#include <iostream>
#include <future>
#include <vector>

using namespace std;

int square(int x) {
   return x * x;
}

int main() {
   unsigned int nthreads = thread::hardware_concurrency();
   vector<future<int>> fut;
   for (int i=0;i<nthreads;i++) {
      fut.push_back(async(launch::async,&square,i));    // [1]
   }
   int sum=0;
   for (auto &f : fut) {
      sum += f.get();                                   // [2]
   }
   cout << " the sum is " << sum << endl;
   return 0;
}
```

Here `nthreads` threads are created in line [1], each thread computes the square of its thread number and this is returned as a future object. Line [2] serves to obtain the values of these objects. Importantly, the `get` function blocks until the return value becomes available, hence the sum being computed is thread-safe.

We now consider our favorite embarrassingly parallel application yet again; this time with the future-promise paradigm.

```
...
typedef mt19937_64 MersenneTwister;
vector<MersenneTwister> MT;
const unsigned int N=10000000;
```

```cpp
unsigned int nthreads = thread::hardware_concurrency();
unsigned int Npertask = N/nthreads;
struct vevs {
    double mu;
    double musq;
};

vevs run_th(int th_id) {
    exponential_distribution<double> eDist;
    vevs v = {0.0,0.0};
    for (int i=0; i<Npertask;i++) {
        double x=eDist(MT[th_id]);      // access uniform distn
                                        //        with different engines
        double f=x*x;
        v.mu   += f;
        v.musq += f*f;
    }
    return v;
}

int main(void) {
    unsigned long int seed;
    cout << " enter seed " << endl;
    cin >> seed;

    vector<future<vevs>> fut_sum;
    for (int th_id=0;th_id<nthreads;th_id++) {
        MT.push_back(MersenneTwister(seed+th_id));
    }
    for (int th_id=0;th_id<nthreads;th_id++) {
        fut_sum.push_back(async(launch::async,run_th,th_id));
    }
    int nn = Npertask*nthreads;
    double fsum=0.0,f2sum=0.0;
    for (auto &fs : fut_sum) {
        vevs v = fs.get();
        fsum  += v.mu;
        f2sum += v.musq;
    }
    fsum  /= nn;
    f2sum /= nn;
```

```
    double sigma=sqrt((f2sum-fsum*fsum)/nn);
    cout << " numtasks: " << nthreads << " [" << nn << "]" <<
        endl;
    cout << " fbar = " << fsum << " +/- " << sigma << endl;

    return 0;
}
```

The main difference from the threaded implementation are the use of `async` to launch threads and the use of `sync::get` to block and obtain results. We have also used a `struc` to clean up the passing of data between threads and the master thread.

9.4 Parallel coding

The previous sections have presented several parallel methods for evaluating an integral with direct Monte Carlo. Variations of this code can be immediately applied to any problem that is embarrassingly parallel. The issue of non-embarrassing parallel problems remains to be addressed. Unfortunately, no general method exists–every problem will have a different strategy for (parallel) solution, and these will even vary depending on the machine architecture and programming paradigm.

To illustrate we consider creating a parallel solution to the percolation problem of Chapter 8. Certainly, generating the percolation configurations is embarrassingly parallel and the solutions presented above apply. However, cluster recognition is nontrivial. One way to proceed would be to divide the grid into subgrids (say nine of them) and run the cluster recognition algorithm in a thread which has been assigned one subgrid. Again, this step is easy to implement because it is embarrassing. The next step would be to merge clusters by traversing the boundaries between the subgrids and equating cluster identification numbers when sub-clusters "touch". This latter step is easiest to implement in a serial fashion; however, it can be made parallel by assigning boundaries to different threads to join neighbouring subgrids, and iterating the joins until the entire grid has been merged.

Molecular dynamics (the subject of Chapter 18) is another area that can benefit from parallel code. In this case one might divide the N particles being considered into groups of size M and have a thread manage their interactions. Collisions between particles in different groups would have to be signalled by inter-thread communication. Alternatively, one could divide the spatial volume into subvolumes which are managed by a thread. Once molecules pass from one subvolume to another, inter-thread communication would be required to reassign the particles.

The reader will appreciate that coding parallel algorithm is not a simple task and there will be many opportunities to destroy thread-safety!

9.5 Forking subprocesses

Multithreading is a decent way of parallelizing a calculation if your goal is to use all of the CPUs on a given processor. However, in order to multithread a computation, *all* of the code that runs in the thread needs to be thread-safe. This includes code that you run indirectly, over which you may have no control. In this unhappy situation, forking subprocesses may be a viable alternative to multithreading.

With a `fork` system call a parent process clones itself, and both parent and daughter processes pick up from where they left off. Processes, however, unlike threads, have completely independent memory spaces, so write operations to variables (local or global) that occur within the daughter processes affect neither other daughter processes nor the parent process. Initially, the memory space of the daughter process is an exact clone of the parents. If the computation requires a lot of memory, this strategy may be a poor choice because the memory of the parent is duplicated as many times as daughter processes are created. The advantage of the approach is that it is applicable even if the code is not thread-safe, such as the code that your ignorant colleagues are likely to have written, probably years ago.

Parent and daughter processes, picking up from the same place after a `fork` system call, follow different paths of execution. Continuing from that point, each of the two processes knows its identity from the return value of the `fork` call. In the parent process, `fork` returns the process id for the subprocess which was spawned. In its clone, the daughter process, the call to `fork` returns 0. Here is an example:

```
#include <unistd.h>
  pid_t child = fork();

  if (child != 0) {
    std::cout << "HELLO from the master" << std::endl;
    ...
  }
  else {
    std::cout << "HELLO from the slave " << std::endl;
    ...
  }
```

The unix command `ps xf` shows the process tree, and if you execute this command while both parent and slave are executing you should be able to identify both processes from the output. In most circumstances, the parent process should wait for the child process to finish. This is accomplished by invoking the `wait` (or `waidpid`) system call; if you fail to do so you will create a *zombie* process which will not die until you kill it with the `kill` command. Killing zombies might seem like fun at first, but you will tire of it ultimately. Your child processes can otherwise be made to die when they finish executing by making the following invocation prior to forking:

```
signal(SIGCHLD, SIG_IGN);
```

This tells the parent process to ignore the `SIGCHLD` signal, which is raised whenever a child process terminates. In fact, the signal is already ignored by default, but when you explicitly set its disposition to `SIG_IGN`, it has the side affect of allowing the child processes to die when they terminate. In other words, no zombies.

One last issue: how does one harvest results from the child process? What is needed is a communication channel from the daughter process to the parent process. In unix this is called a pipe. The `pipe` system call fills an array of two integers, one reading and one for writing. Here is an example in which data is transmitted in one direction, from child to parent; the complete code and project file can be found in `CH9/EXAMPLES/PIPE`.

```
//
// Pipe: a simple program to pass some data (a string) from a
       child
// to a parent.  The parent prints it to the terminal.
//
int main() {

  // Don't wait for child to finish.
  signal(SIGCHLD,SIG_IGN);

  // Open a pipe:
  int pipefd[2];
  pipe(pipefd);

  // For subprocess
  pid_t child = fork();


  if (child == 0) {     // Child writes string to the pipe:
    close(pipefd[0]);   // Close unused read end

    std::string blurb="Ceci n'est pas une pipe";
    size_t bsize=blurb.size();

    write(pipefd[1], &bsize, sizeof(bsize));
    // write the data size
    write(pipefd[1], blurb.c_str(), blurb.size());
    // write the date
    close(pipefd[1]);
    // close the write end

  } else {              // Parent reads string from pipe
```

```
        close(pipefd[1]);    // Close unused write end

        size_t bsize;
        read(pipefd[0], &bsize, sizeof(bsize)); // Read the
                                                //   string size

        char buf[bsize];                        // Allocate space
        read(pipefd[0], buf, bsize);            // Read the
                                                //   string

        std::cout << buf << std::endl;          // Print the
                                                //   string
        close(pipefd[0]);                       // Close the read
                                                //   end

    }
}
```

9.6 Interprocess communication and sockets

A client/server architecture is yet another way to achieve parallelism, though it is interesting in its own right. There are many examples of systems with a client/server architecture, from web servers to email servers. But servers can handle requests for heavy scientific calculation as well.

We can see how parallelism comes about in these architectures by considering an email server. You probably connect to a central Exchange server or IMAP server at your university, laboratory, or place of employment from your personal computer, laptop, or smartphone. You may make a request to search for a particular email. This search will be executed on the server, rather than on your local machine, whatever it may be. But at the same time, in a large organization, many other clients are executing similar requests. These are carried out in parallel, typically within different processes on the server.

The way in which this works is described succinctly in Tougher (2002). A server process opens a numbered port and listens for requests. The client, possibly executing on a different computer, connects to the host using the ip address and the port number. If the connection is successful, a two-way communication channel is opened between the two machines. System calls `send` and `recv` are used then to transfer data between the two machines. Two classes, `ClientSocket` and `ServerSocket`, based upon Tougher (2002), are included with the `QatDataAnalysis` library and facilitate client/server communication, hiding the details. A basic example of their use can be found in our examples area. The code for the client is shown here:

```cpp
#include "QatDataAnalysis/ClientSocket.h"
#include <stdexcept>
#include <iostream>
#include <string>
#include <sstream>
#include <stdexcept>


enum Command {GET_VERSION_INFO,
              CALCULATE,
              HANGUP_SERVER};

int main ( int argc, char * argv[] )
{
  using namespace std;
  //
  string usage = string ("usage") + argv[0] + "[-h host/
      localhost] [-p port/7000]";
  //
  string host            = "localhost";
  unsigned int port      = 70000;

  for (int i=1; i<argc;i++) {
    if (string(argv[i])=="-h") {
      i++;
      host=argv[i];
    }
    else if (string(argv[i])=="-p") {
      i++;
      istringstream stream(argv[i]);
      if (!(stream>>port)) {
        throw runtime_error("Cannot parse port");
      }
    }
    else {
      cerr << usage << endl;
      return 1;
    }
  }
  try {

    ClientSocket clientSocket(host,port);
    unsigned int    version, concurrency, pid;
```

```
    clientSocket << GET_VERSION_INFO; // Sample size and
                                      version;
    clientSocket >> version;
    clientSocket >> concurrency;
    clientSocket >> pid;
    cout << "Client Connected to host "
         << host
         << " port "
         << port
         << " server version "
         << version
         << " hardware concurrency "
         << concurrency
         << " pid "
         << pid
         << endl;

    clientSocket << CALCULATE;
    clientSocket << HANGUP_SERVER;
  }
  catch (const exception & e ) {
    cout << "Exception caught: " << e.what() << endl;
  }

  return 0;
}
```

This code can be found in EXAMPLES/CH9/CLIENT. The simplest server code can be found in EXAMPLES/CH9/SERVER, but the example there has a fundamental drawback: it can only handle one request at a time, and blocks for the entire time it is connected to a client. This is usually not the desired behavior: imagine what would happen if webservers behaved like this! A slightly more sophisticated version of this code can be found in EXAMPLES/CH9/MULTISERVER. This version forks a subprocess every time a client connects; the subprocess handles all communication with the client. The multiserver can be monitored with the unix command ps xf. With five active connections that command will show something like this:

```
18771 pts/1    Ss    0:01       \_ bash
20504 pts/1    S+    0:00       |   \_ ../local/bin/multiServer
20507 pts/1    S+    0:00       |       \_ ../local/bin/multiServer
20509 pts/1    S+    0:00       |       \_ ../local/bin/multiServer
20512 pts/1    S+    0:00       |       \_ ../local/bin/multiServer
20514 pts/1    S+    0:00       |       \_ ../local/bin/multiServer
20516 pts/1    S+    0:00       |       \_ ../local/bin/multiServer
```

The (multi)server code is:

```
#include "QatDataAnalysis/ServerSocket.h"
#include <stdexcept>
#include <sstream>
#include <thread>
#include <signal.h>

enum Command {GET_VERSION_INFO,
              CALCULATE,
              HANGUP_SERVER};

using namespace std;


// Version this server.  The client can then check for
//    compatibility.
const unsigned int version=0;


int main(int argc, char ** argv ) {

  // Usage statement:
  string usage = string ("usage") + argv[0] + "[-p port]";

  // Default port is 70000
  unsigned int port  = 70000;

  // Parse the command line:
  for (int i=1; i<argc;i++) {
    if (string(argv[i])=="-p") {
      i++;
      istringstream stream(argv[i]);
      if (!(stream>>port)) {
        throw runtime_error("Cannot parse port");
      }
    }
    else {
      cerr << usage << endl;
    }
  }
  //
```

```cpp
  // This call prevents child processes from living on as
     Zombies:
  //
  signal(SIGCHLD,SIG_IGN);
  //
  try {
    //
    // Create the socket:
    //
    ServerSocket *server = new ServerSocket(port);
    while (true) {

      cout << "Server ready, waiting.." << endl;
      ServerSocket newSocket;
      server->accept(newSocket);
      pid_t child = fork();

      if (child == 0) {

        delete server;
        server=NULL;

        while (true) {
          //
          // Read the command:
          //
          Command command;
          newSocket >> command;
          //
          // Execute the command:
          //
          if (command==HANGUP_SERVER) {
            std::cout << getpid() << " HANGUP" << std::endl;
            break;
          }
          else if (command==GET_VERSION_INFO) {
            std::cout << getpid() << " VERSION" << std::endl;
            newSocket << version;
            newSocket << std::thread::hardware_concurrency();
            newSocket << getpid();
          }
          else if (command==CALCULATE) {
```

```
              std::cout << getpid() << " CALCULATE" << std::
                  endl;
            }
          }
        }
      if (!server) break;
    }
  }
  catch (exception & e) {
    cerr << "Exception was caught: " << e.what() << "\n
       Exiting" << endl;
  }
  return 0;
}
```

The server is invoked with the command:

`multiServer -p 7000`

(the server will listen on port 7000) and the client with

`simpleClient -h localhost -p 7000`

Output from a single client is

```
Client Connected to host localhost port 70000 server
           version 0 hardware concurrency  4 pid 20509
```

While the server output, which might typically be directed to a log file, is:

```
Server ready, waiting..
Server ready, waiting..
20507 VERSION
Server ready, waiting..
20509 VERSION
Server ready, waiting..
20512 VERSION
Server ready, waiting..
20514 VERSION
Server ready, waiting..
20516 VERSION
20507 CALCULATE
```

```
20507 HANGUP
20509 CALCULATE
20509 HANGUP
20512 CALCULATE
20512 HANGUP
20514 CALCULATE
20514 HANGUP
20516 CALCULATE
20516 HANGUP
```

A few final remarks may be in order. We have demonstrated a system with one server connecting to N clients, but it is also possible for each client to open multiple connections, causing multiple independent subprocesses to be spawned on the server side, and thus achieving a high degree of parallelism.

Typically, many services will run on any given processor, and each service listens on its own dedicated port. The lower number ports are assigned to common linux services, such as telnet (23), ssh (22), ftp (21), irc(994). Those ports with number 1024 and greater are unassigned and may be used for custom services, provided they do not clash with other services.

Communication between client and server on a single host usually poses no problems, but host-to-host communication may be blocked by firewall settings. Another issue arises when a single "head" node handles all the communication for a cluster, which communicates only through the head node. In such cases, specific ports on the hidden machines may be "forwarded" to the outside world. The configuration of services on a server machine and/or cluster can lead such issues of system and network administration; your degree of success in solving these may depend upon the quality of your relationship with the local administrators, so be nice to them.

9.7 Exercises

1. Construct Table 9.5 for your computer and your compiler using `std::thread`.
2. Implement the following code in place of that described in Section 9.3.3:

   ```
   for (int i=0;i<nthreads;i++) {
     th.push_back(std::thread(sum,i));
     th[i].join();
   }
   ```

   How do your timing tests differ? Explain the results.
3. Discuss the impact of I/O on parallel program performance.

4. Implement the following code in an environment that supports openMP.

    ```
    #include <iostream>
    #include "omp.h"
    int main(int argc, char **argv) {
      int N;
      std::cout << " enter number of nodes " << std::endl;
      std::cin >> N;
      omp_set_num_threads(N);
      #pragma omp parallel for
      for (int i = 0; i < 1000000000; i++) {
        double x = 2.0 * i;
      }
      return 0;
    }
    ```

    Run `time ./<fn>` for various N (depending on your system) and interpret the results.

5. openMP reduce.

    a) Implement the code for parallel direct Monte Carlo on page 288 and run it.
    b) Now move the scope brace from location [1] to [2] and rerun. Explain the difference in behavior.

6. Implement a matrix multiply $\mathbf{M}\vec{x}$ in parallel using openMP directives.

7. Remove the line labelled [1] in the Hello World code on page 286 and run the code. Explain the resulting behavior.

8. Dot Product.

    a) Write a code snippet that performs dot products with openMP. Use dynamic scheduling with chunk size, `chunk`.
    b) Why might reduction not be associative for real numbers?

9. Thread and Mutex.

    a) Write code using `thread` that prints out 40 characters (in a row) in a thread. Create several threads with different characters and confirm that the characters are interleaved.
    b) Repeat this exercise with a `mutex` and confirm that the lines are unmixed. What sets the order of the lines?

10. Threaded Coding is Hard.

   a) Implement the threaded parallel code on page 294 on your system. Run it several times with the same seed and note the behavior.

   b) Now place the thread creation statement (labelled [1]) into the loop above. Run this code several times with the same seed. Note that the behavior is different! This seemingly benign change has introduced a bug. What is it? (Hint: how does vector manage memory?)

11. [P] Rainbows.
    A rainbow is produced by white light refracting into a spherical droplet, reflecting from the back surface, and refracting back into the air (for which we approximate the index of refraction to be $n = 1$). The index of refraction of water varies with wavelength (see Chapter 4, Exercise 7). Simulate the formation of a rainbow. Start by randomly choosing the wavelength of a photon between 400-700 nm. Let it strike a raindrop at a fixed direction and a random position upon the spherical surface, and follow the trajectory through the raindrop until in re-emerges after a single reflection.

   a) Plot the angle between the incident beam and the scattered beam for light of various frequencies. You should obtain a sharp peak near 40 degrees.

   b) Generate a color image of the rainbow, or a segment thereof.

   c) Choose a convenient paradigm for parallelizing your code and implement it.

   d) Get your hands on a machine with as much hardware concurrency as possible and run your procedure, following at least 25 million photons.

You will have created a simulation of rainbow formation which captures some of the main features of the rainbow, but leaves out many details. The degree of reflection and refraction, governed by *Fresnel's equations*, vary according to the light wavelength, angle of incidence, and polarization. The resultant rainbow is therefore polarized! Thin film interference can also play a role. Multiple reflections within the rainbow can generate secondary rainbows. Consider expanding your code to explore these features.

BIBLIOGRAPHY

Tougher, R. (2002). *Linux Socket Programming in C++*. Linux Journal, http://tldp.org/LDP/LG/issue74/tougher.html.
For further information consult the following.

Pthreads Programming by B. Nichols *et al.* O'Reilly and Associates.
Programming With POSIX Threads by D. Butenhof. Addison Wesley.

Programming With Threads by S. Kleiman *et al.* Prentice-Hall.
Designing and Building Parallel Programs by Ian Foster http://www.mcs.anl.gov/itf/dbpp/.

See https://computing.llnl.gov/tutorials/mpi/#AppendixA for a list of MPI routines.
See https://www.open-mpi.org for more on mpi.
See http://openmp.org/wp/ for more on openmp.
See http://www.cplusplus.com/reference/multithreading/ for multithreading in C++.

10

Graphics for physicists

10.1	Graphics engines	312
	10.1.1 3d libraries and software	313
	10.1.2 Generating graphics	314
	10.1.3 The `Open Inventor/Coin3d` toolkit	315
10.2	First steps in a 3d world–3d visualization	316
	10.2.1 The basic skeleton of a 3d application	316
	10.2.2 A three-dimensional greeting to the world	318
	10.2.3 A colorful spherical world	321
	10.2.4 Deleting nodes in the scene graph	322
10.3	Finding patterns–Testing random number generators	322
10.4	Describing nature's shapes–fractals	326
	10.4.1 Shared nodes	331
10.5	Animations	335
	10.5.1 `Coin` engines: a rotating world	335
	10.5.2 `Coin` sensors: an orbiting planet	338
10.6	The `Inventor` system	339
10.7	Exercises	340
	Bibliography	342

Nature exists in three dimensions. Effectively representing phenomena like applied forces and vector fields, moving objects, colliding galaxies, spreading micro-organisms, interacting particles, and travelling light rays in two dimensions is a difficult task. Data from experiments and simulations can be often better explored and understood if visualized with an interactive three-dimensional representation. The goal of this chapter is to develop important skills in order to visualize data from simulated and actual experiments. Our first application will come in the following chapter, in which we compute the trajectories of dynamical systems. We use open source object modeling software called `Open Inventor/Coin3D`.

10.1 Graphics engines

The rules for representing a three-dimensional world on a two-dimensional screen were first codified in the early 1400s by the Italian architect Filippo Brunelleschi, who observed

Figure 10.1 *The machine used by Albrecht Dürer to help draw objects in perspective (adapted from https://commons.wikimedia.org/wiki/File:Dürer_Stich_aus_Anweisung_2.jpg. Public domain)*

that when using a fixed single point of view, parallel lines in a scene appear to converge to a single point in the distance called the "vanishing point". He used a double-mirror device to draw the famed Florentine baptistry in a realistic way, thereby inventing "linear perspective"–a technique that has been explored and enhanced by artists and architects throughout the following centuries (Figure 10.1). These rules remain the foundation of modern 3d graphics systems. Nowadays, modern computer graphics software packages use virtual cameras to render three-dimensional images on the two-dimensional screen of the computer monitor. This can be achieved in different ways: Figure 10.2 shows the two most common camera configurations, namely *perspective* and *orthographic* cameras.

10.1.1 3d libraries and software

There are two major approaches to 3d computer graphics. The first approach uses **real-time rendering** to produce images quickly. This is useful when user interactivity is desired. The second approach, called **deferred rendering**, uses many hours on thousands of cores to produce photorealistic images to compose the frames for animated films. Real-time rendering is more important for scientific applications and will be the focus of our attention. Real time rendering relies on APIs known as **graphics engines**, which produce images on demand at upwards of 32 frames/second (the frame rate of motion pictures).

314 *Graphics engines*

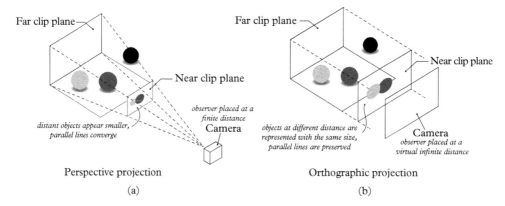

Figure 10.2 *Modern computer graphics software packages use virtual cameras to get correct three-dimensional images on the flat 2d screen of the monitor. Different cameras project 3d objects on the view plane in different ways: on the left, a* perspective camera, *which makes distant objects appear smaller, and which is the natural view for human beings; on the right, an* orthographic camera, *used for technical purposes, which makes equal distances appear the same in the whole image.*

What follows is only a very brief introduction. More information can be found in specialized books like Hughes *et al.* (2013) and Akenine-Moller (2008)[1].

10.1.2 Generating graphics

Real-time rendering of 3d objects onto a two-dimensional screen is accomplished in several steps, requiring both a graphics engine, which is a software component, and a *graphics pipeline*, which is a hardware component that resides in the primary CPU or in a dedicated graphics processing unit (GPU).

We start by assuming that there exists, in software, a three-dimensional description of the geometry to be rendered. We call this a *scene*. The graphics engine decomposes a scene into simpler geometrical objects (such as triangles) that can be handled by the graphics hardware, and transforms them into *world coordinates*. The following steps then occur in the graphics pipeline. World coordinates are transformed into *camera coordinates*, according to whether an orthographic or a perspective camera is used. The effect of light sources is applied. The 3d scene is culled to remove geometric primitives lying outside of the cone of view, and the entire scene is mapped onto the screen. In a step known as *rasterization*, RGB color values are assigned to each pixel on the screen, and finally, a step called *texturing* may then apply external images or patterns to the shapes. The whole chain can be run many times per second, and some pieces of the chain such as lighting may even require multiple passes.

---

[1] The authors of (Akenine-Moller, 2008) also maintain an annotated and updated on-line version of the bibliography on the book's website, where a list of very useful books on computer graphics and a review of free resources available on the web can be found.

10.1.3 The `Open Inventor/Coin3d` toolkit

A graphics engine is an API designed to carry out the translation from high-level objects including shapes, transformations, lights, and cameras into triangles, so that the programmer does not have to care about these processing steps. The graphics engine also handles events such as mouse clicks and key strokes, which are typically important for the selection of objects displayed on the screen. In this text our examples and exercises are carried out using the free `Coin`[2] implementation of the Open Inventor class library. `Coin` uses the ubiquitous OpenGL API for object rendering. OpenGL is one of two low-level graphics APIs. The other is Direct3d, a proprietary standard of Microsoft Corporation that is historically linked to game development. OpenGL, being an open standard, can run on virtually any platform.

The development of Open Inventor started in 1988 as a proprietary software package written by SGI (Silicon Graphics International Corp, 2009) that implemented a high level approach to the creation of 3d graphics. In parallel, a "clone" named `Coin` was independently created by Systems in Motion (SIM) (Kongsberg, 2014). `Coin` reimplemented and optimized all Open Inventor classes and functions while adding some functionality. Both Open Inventor and `Coin` were released open-source and they continue to be used in a number of projects, both in industry and in academe. The original Open Inventor toolkit has been taken over by software companies, first Template Graphics (TGS), and more recently FEI (part of Thermo Fisher Scientific) and packaged into commercial products. These are used in the automotive, oil and gas, and medical industries. See FEI (2015) for more details.

In this book we will use the `Coin` toolkit, which is currently the most popular of the two open-source implementations. Moreover, the documentation is excellent and available for all of the classes, numbering in the hundreds; and a very good pedagogical introduction to graphics programming with `Coin/Inventor` is available (Wernecke, 1994a). The basic ideas behind the toolkit have migrated to other C++ toolkits such as (Osfield, 2015), or to other computer languages, such as Java in the form of Java3d (Selman, 2002). Thus the reader will able to use the concepts acquired here in other contexts or with other technologies as well.

`Coin/Open Inventor` (in the following "`Coin`" only) is a C++ class library containing *shape nodes* that represent primitive constructed solids (like Sphere and Cone), objects called *property nodes* to store parameters and settings (like Color and Material), objects to be used to set the 3d scene (like Light and Camera), called *light nodes* and *camera nodes* respectively[3]. Figure 10.3 shows an example of the icons used by Open Inventor and `Coin` to graphically representing those nodes. We will see other nodes later on.

---

[2] The `Coin` graphics engine is part of the `Coin3d` framework, which contains some other libraries. Very often `Coin` and `Coin3d` are used as synonyms.

[3] A complete list of all nodes can be found in Wernecke (1994b) and documentation for all the nodes and classes can be found in the *Open Inventor C++ Reference Manual* (Open Inventor Architecture Group, 1994).

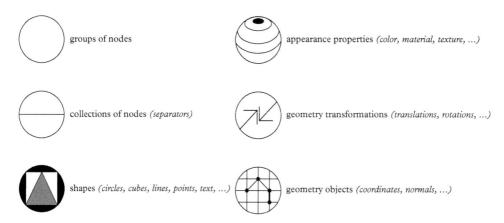

Figure 10.3 *An example of* Open Inventor/Coin *node icons.* Open Inventor *and* Coin *use icons to represent the different nodes which define a shape, like a circle, or a parameter, like a color. Some examples are displayed here.*

All objects are stored in a *scene graph*, a hierarchical tree of objects linked together in a manner similar to a file directory tree. These objects are called *scene objects*, and the scene graph describes a *scene*, which is a collection of objects to be rendered onto the screen. When an image of the scene has to be generated, the objects are rendered.

The order and the structure of the scene graph affect the final image; this will be discussed more later. For complicated scenes, it may be a good idea to sketch the scene graph before implementing it in code.

10.2 First steps in a 3d world–3d visualization

This section contains some simple examples of how to create a scene and to visualize it in a window that gives the user control over the camera. The final goal is building a complete program to take and visualize plain data from a simulation program in an interactive three-dimensional scene. We start with a simple example that establishes the basic structure of a 3d application by creating a 3d scene and then adding three-dimensional text to it.

10.2.1 The basic skeleton of a 3d application

The main ingredients in the example to appear in this chapter are the Coin and SoQt APIs. The SoQt library binds the graphics library Coin to the windowing system *Qt*. Our first example creates the window shown in Figure 10.4. The scene is empty in this example, but a large amount of infrastructure for user interaction such as selection and zoom is evident from the figure. Listing 10.1 shows the code, which is the basic skeleton of all graphics applications that use Coin and SoQt. It can be found in EXAMPLES/CH10/SKELETON.

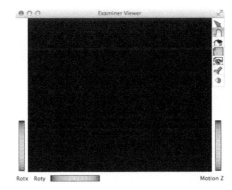

Figure 10.4 *The basic user interface to display 3d content, using the* `Coin` *and* `SoQt` *packages.*

Listing 10.1: The basic `Coin` graphical user interface

```
// SoQtSkeleton.cpp
//
// SoQt includes
#include <Inventor/Qt/SoQt.h>
#include <Inventor/Qt/viewers/SoQtExaminerViewer.h>
// Coin includes
#include <Inventor/nodes/SoSeparator.h>

int main(int argc, char ** argv)
{
  // Initialize the Qt system:
  QApplication app(argc, argv);

  // Make a main window:
  QWidget mainwindow;
  mainwindow.resize(400,400);

  // Initialize SoQt
  SoQt::init(&mainwindow);

  // The root of a scene graph
  SoSeparator *root = new SoSeparator;
  root->ref();

  // Initialize an examiner viewer:
  SoQtExaminerViewer * eviewer = new SoQtExaminerViewer
      (&mainwindow);
  eviewer->setSceneGraph(root);
```

```
29      eviewer->show();
30
31      // Pop up the main window.
32      SoQt::show(&mainwindow);
33
34      // Loop until exit.
35      SoQt::mainLoop();
36
37      // Clean up resources.
38      delete eviewer;
39      root->unref();
40
41      // exit.
42      return 0;
43  }
```

The program initializes the *Qt* windowing system, creates a top-level window with a size of 400 by 400 pixels, then initializes the `SoQt` system which is the "glue" between the `Coin` graphics API and *Qt*. A `SoSeparator` is an empty scene; empty, at least, until we start to populate it with shapes and properties. Finally, the `SoQtExaminerViewer` is an object that provides for both scene rendering and a high level of user interaction.

Memory management is facilitated by **reference counting**: the `ref()` method increases the objects **reference count** (here, from zero to one) while `unref()` decreases it (from one to zero). When the reference count drops to zero, the object is deleted. The usefulness of this system will become apparent later. The program blocks in the call to `SoQt::mainLoop`, during which all user interaction is handled, until the user closes the main window or activates a Quit button.

The compilation of this kind of program is controlled by a qt project file, an example of which can be found in

EXAMPLES/CH10/SKELETON/src/qt.pro.

10.2.2 A three-dimensional greeting to the world

The next step is to add graphical content to the empty scene. This content can be built from a set of predefined objects in the `Coin` library called *nodes*. These include 3d shapes, like `SoCube`, `SoCone` or `SoLight`, as well as objects to group these together like `SoGroup`. The prefix "So" means *Scene Object*.[4].

Out of respect for tradition, the first content that we shall add to the scene is the classic message, "Hello world!", which will be implemented with an `SoText3` object. We color it yellow by combining one unit of red with one unit of blue and no green, using the

---

[4] `Coin` on-line documentation can be found at (Kongsberg, 2014) or (CCNLab, 2015).

SoBaseColor node. Listing 10.2 shows the relevant lines of code to be added to that above. You can find the whole working code here:

EXAMPLES/CH10/HELLO_WORLD.

Figure 10.5 shows the final structure of our simple scene graph and Figure 10.6 shows the resulting image. The order in which the nodes are added to the scene graph is important: here, for example, the SoBaseColor node is added to the scene graph **before** the SoText3 node, because the SoBaseColor node, as many other Coin nodes, affects all the objects which *follow* that node in the node tree. For example, when we add a yellow SoBaseColor node to the scene graph, all subsequent nodes that have a color property will be colored yellow, while nodes added before will be not affected. When a SoBaseColor node is added to the scene graph, it replaces any other color that was set before. Note that other types of nodes—like those describing geometrical transformations—accumulate their effect when added to the same root node.

Listing 10.2: The 3d "Hello World!" program in Coin: adding a three-dimensional text to the scene graph

```
// Coin includes
#include <Inventor/nodes/SoBaseColor.h>
#include <Inventor/nodes/SoText3.h>
#include <Inventor/nodes/SoSeparator.h>

  ...

// Set the main node for the "scene graph"
SoSeparator *root = new SoSeparator;
root->ref();

// Set the color for the text (in RGB mode)
SoBaseColor *color = new SoBaseColor;
color->rgb = SbColor(1, 1, 0); // Yellow
root->addChild(color);

// Set the node for the 3d text, set the visible surfaces
// and add actual text to it
SoText3 * text3d = new SoText3();
text3d->string.setValue("Hello World!");
text3d->parts.setValue("ALL");
root->addChild(text3d);
```

In this example the SoSeparator functions as a group node, SoGroup, from which it inherits. Group nodes can be used to group scene objects into more complex structures, and in addition they can be used to group together different properties; for example

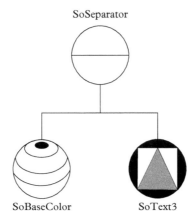

Figure 10.5 *The scene graph for the 3d "Hello World!" example.*

Figure 10.6 *A first basic 3d visualization, which displays a three-dimensional greeting to the world.*

one might bundle a draw style together with a color. The `SoSeparator` differs from the `SoGroup` in one important way: property nodes in the `SoSeparator` can only affect children below the separator, whereas those under the `SoGroup` affect other nodes encountered in the top-to-bottom and left-to-right traversal of the scene graph, which occurs during rendering.

If you compile and run this example you will see that the controls on the `SoQtExaminerViewer` window allow you to zoom and rotate the scene. You can also use the mouse to zoom and rotate. Be aware that the top two buttons in the toolbox on the right toggle between two interaction modes; the default interaction mode ("the hand") is used for viewpoint control, while the other interaction mode ("the arrow") is for object selection. The seventh and last button in the toolbox (the red/yellow cube) is also noteworthy: it toggles between two different visualization modes: "perspective" (the default mode) or "orthographic" (see Figure 10.2). More information on the user interface can be found in the manual pages.

10.2.3 A colorful spherical world

Our second basic example will show how to use ***textures***. Texturing an object means applying a flat image (the *texture*) onto that 3d object. In computer graphics texturing is used to mimic a surface's texture or material. Here we will use it to build a model of the Earth.

Listing 10.3 shows the code of our simple example; this codes constructs a simple scene graph shown in Figure 10.7. The node SoTexture2 is a property node that affects subsequent shapes. It is placed under a separator, and before a sphere, and reads a texture file containing the image in Figure 10.8(a). The outcome of our basic example is a 3d model of the Earth, shown in Figure 10.8(b).

Listing 10.3: How to apply a texture to a shape

```
// A node for the whole Earth object
SoSeparator *earth = new SoSeparator;

// A node for the texture
SoTexture2 *texture_earth = new SoTexture2;

// Name of texture file
texture_earth->filename = "world32k.jpg";

// Add texture to group node
earth->addChild(texture_earth);

// Add sphere to group node
// => Draw texture on surface of sphere
earth->addChild(new SoSphere);
```

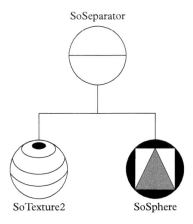

Figure 10.7 *Graph of a textured sphere (SoTexture2+SoSphere) used to represent a colorful Earth.*

(a) A world image to be used as a texture

(b) Creating a 3d rotating Earth applying a texture to a sphere

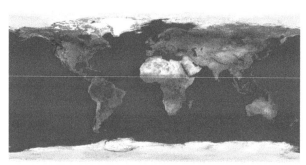
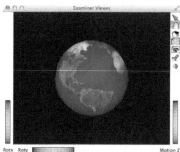

Figure 10.8 *(a), the flat image of the Earth that will be used as texture; (b), the resulting 3d rotating Earth, obtained applying the texture (a) to a 3d sphere.*

10.2.4 Deleting nodes in the scene graph

We return now to the issue of node deletion. Recall from Chapter 6 that when memory is allocated using `operator new` it should be released with `operator delete`. In Coin, `operator delete` is in fact private and the scene objects take care of deleting themselves when the reference count reaches zero.

Reference counts may be incremented explicitly, but they are also incremented when parent-child links are established. When these links are broken, for example when a parent is destroyed, the reference count is decremented. When an explicit `unref()` method is invoked on the root node, the root node destroys itself, decrementing the reference count of its children, which will also delete themselves if their reference count drops to zero. This is more sophisticated than a simple recursive delete of children, and with good reason: it is necessary if any of scene objects are shared, i.e., if the same scene object appears in different portions of a complex scene graph (see Section 10.4.1). In our example, one call to `unref()` on the root node suffices to free all of the memory allocated to scene objects, and the program does not leak memory.

10.3 Finding patterns–Testing random number generators

In Chapter 7, we discussed the high-quality Mersenne twister engine, a pseudorandom number generator (PNG) which we highly recommended. We now discuss (but do not recommend) the RANDU PNG, for reasons that will become apparent after we implement some data visualization.

RANDU is a *linear congruential generator* (or LCG), which generates sequences of pseudorandom numbers according to:

$$X_{n+1} = (aX_n + c) \pmod{m}, \tag{10.1}$$

where *a* is an integer called the *multiplier*, *c* is the integer *increment*, *m* is a large prime integer called the *modulus*, and X_0 is the *seed*. The parameters characterize the behavior of the generator and their particular values are very important (Park, 1988). If the parameter *c* is equal to 0, then the generator is usually called a *multiplicative congruential generator* (or MCG) (or *Lehmer random number generator*, or *Park-Miller random number generator*). RANDU is an example of an MCG since it is defined as:

$$X_{n+1} = 65539 \, X_n \cdot \mathrm{mod} \, 2^{31}.$$

These parameters were chosen because they were easy to handle by older machine architectures. The non-prime *modulus* was chosen for simplicity in computing the basic CPU-operation MOD, and the *multiplier* was picked because of its simple binary representation.

Today, no LCG PNG is considered good enough for serious numerical work like Monte Carlo simulation, let alone more demanding applications like cryptography. In fact RANDU is considered an example of an LCG with a particularly tragic choice of defining parameters. Why this is so is established by the code in the directory

EXAMPLES/CH10/RANDOM_NUMBERS

We can make an LCG identical to the RANDU generator using template class

std::linear_congruential_engine

as in listing 10.4.

Listing 10.4: Instantiation of an LCG of type RANDU using std::linear_congruential_engine

```
1  // Instantiate a random number generator identical to the
     deficient RANDU
2  linear_congruential_engine<uint_fast32_t,      65539,    0,0
     X80000000> randu(1);
```

We use this generator to create a triplet of points, each point on the interval [0,1], so that the points fill the interior of a cube in three dimensions. The three axes are colored red for the *x* axis, green for the *y*, and blue for the *z* (this is a common convention). The axes are represented by the scene (sub)graph in Figure 10.9. Each axis is represented by an SoLineSet, however the coordinates of the line are determined by a property node, SoCoordinate3 inserted before the SoLineSet. The SoCoordinate3 stores three-vector coordinates (i.e., (x,y,z) triplets) in an array. In Listing 10.5, all three axes are built by instantiating the scene objects, setting their properties, and assembling them into a scene graph in order to obtain the more complete subgraph of Figure 10.9. Properties must always be added before shapes. In Figure 10.10, a single SoDrawStyle

324 *Finding patterns–Testing random number generators*

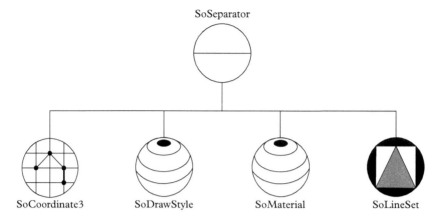

Figure 10.9 *The sub-graph which defines a single axis.*

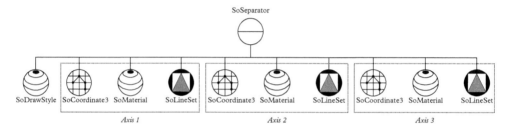

Figure 10.10 *The graph that defines the three axes.*

node suffices to affect the draw style of all subsequent scene objects in the graph, until a new SoDrawStyle node is encountered. The order of the SoCoordinate3 and the SoMaterial nodes is not important as long as they are inserted into the graph before the node which uses them, i.e., the SoLineSet. Each new property node like SoLineSet and SoMaterial changes the previous property settings and makes them default. The SoSeparator, like SoGroup, is a group node, but it also serves a special purpose: it prevents property nodes rooted beneath the SoSeparator from propagating to nodes above the SoSeparator.

Listing 10.5: Defining the three X,Y and Z axes

```
1   SoSeparator *aSep=new SoSeparator();
2   root->addChild(aSep);
3
4   // a loop to create the three axes
5   for (int i=0;i<3;i++) {
6
```

```cpp
        // coordinates for the starting and ending points of the
            axes
        SoCoordinate3 *arrow = new SoCoordinate3();
        arrow->point.set1Value(0,0,0,0);

        // (1,1,0,0), (1,0,1,0) or (1,0,0,1)
        arrow->point.set1Value(1,i==0,i==1,i==2);
        aSep->addChild(arrow);

        // drawing style
        SoDrawStyle *ds= new SoDrawStyle();
        ds->style=SoDrawStyle::LINES;
        ds->lineWidth=3;
        aSep->addChild(ds);

        // material/color for the axes
        SoMaterial *aMaterial = new SoMaterial();
        // (1,0,0), (0,1,0) or (0,0,1)
        aMaterial->diffuseColor.setValue(i==0,i==1,i==2);
        aSep->addChild(aMaterial);

        // a LineSet, which uses the Coordinates to draw the axes
            lines
        SoLineSet *line = new SoLineSet();
        line->numVertices=2;
        aSep->addChild(line);
    }
```

Next we add the randomly generated points to the scene. We will render them as white dots. For this we need both a new `SoCoordinate3` to define the position of each dot and an `SoPointSet` to interpret these positions as dots. These steps are shown in Listing 10.6.

If you want to explore this example more, the code can be found in

EXAMPLES/CH10/RANDOM_NUMBERS.

This code produced the graphical output in Figure 10.11. From our visualization we can easily see that RANDU badly fails the *spectral test*: random numbers are not generated evenly in the whole space; instead, they are distributed over planes in the 3d space.

Listing 10.6: Drawing the random points in the 3d scene

```cpp
    // Adding the Coordinates and the PointSet to the root
        separator
```

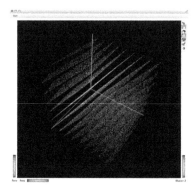

Figure 10.11 *Visualizing a spectral test of pseudo-random numbers.*

```
2   SoCoordinate3 *coordinates = new SoCoordinate3();
3   root->addChild(coordinates);
4   SoPointSet *pointSet =new SoPointSet;
5   root->addChild(pointSet);
6
7   // Fill the coordinates with uniform random numbers:
8   for (int i=0;i<n_points;i++) {
9     // generate new random coordinates
10    double X = double(randu())/randu.max();
11    double Y = double(randu())/randu.max();
12    double Z = double(randu())/randu.max();
13    coordinates->point.set1Value(i,X,Y,Z);
14    pointSet->numPoints=i+1;
15  }
```

10.4 Describing nature's shapes–fractals

Nature cannot always be represented by basic geometrical shapes and rules; natural phenomena like thunder, or the growth of a cell culture, or even the shape of a tree are difficult to represent using basic Euclidean entities. Instead they are represented by fractal geometry. In the following example we will see how to generate and visualize fractal shapes, using `Coin SoLineSet`, while encountering *geometry transformations* and employing *recursion*, using space transformations to draw 3d objects in more complex scenes.

Let us start with a classical example: drawing a tree-like shape. The branches of trees follow some basic rules: from each branch, a number of smaller sub-branches are generated, which are similar to the main branch. We observe the same behavior at each level of the tree, decreasing always in size, starting from the trunk. Thus a tree can be generated easily with a *recursive function*—i.e. a function that calls itself—starting from the trunk and generating a number of smaller sub-branches for each step. While a true fractal would be infinite, our representation will only display a finite portion of the sequence.

A tree can be represented by a line, the trunk, with two smaller lines emerging from its tip, the branches. This structure is repeated to build an entire tree. With each step a few parameters, like the size of the branches, will change. Thus the tree is ***self-similar***, which is a common feature of fractals (see Chapter 8 for much more on this).

Two functions are required: first, a function to draw a single line on the screen, and second, a function to draw a main branch followed by two sub-branches, the latter being recursive. A full working version of the example code can be found in the directory:

EXAMPLES/CH10/FRACTALS/DrawFractalTree

Listing 10.7 shows the function `drawLine` which draws a single line segment. It takes two pairs of (x, y) values, which set the endpoints of the line segment in a two-dimensional space, and a pointer to a `SoGroup`. The function constructs the line segment according to methods which are now familiar to the reader, under an `SoSeparator`, and adds the `SoSeparator` to the `SoGroup`. Figure 10.12 shows the scene graph representing the line segment.

Listing 10.7: Drawing the basic line for the fractal tree-like shape

```
void drawLine(float x1, float y1, float x2, float y2,
    SoGroup* group)
{
    // Declare a new Separator to contain the line
    SoSeparator *sep = new SoSeparator;

    // Declaring a set of 3d coordinates.
    SoCoordinate3* points = new SoCoordinate3;

    //Using the two points as coordinates
    points->point.set1Value(0, x1, y1, 0);
    points->point.set1Value(1, x2, y2, 0);

    // Declare a new LineSet, to draw our line
    SoLineSet *pLineSet = new SoLineSet;

    // Build the line object, adding to the Separator node...
    // ...first, the two points representing
    // the start and the end of the line...
    sep->addChild(points);
    // ...then, the line, which will be
    // drawn between the two points
```

328 *Describing nature's shapes–fractals*

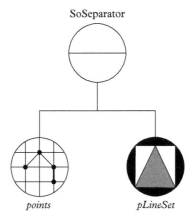

Figure 10.12 *Graph of the single branch used in the tree-like shape.*

```
22      sep->addChild(pLineSet);
23
24      // Add the line to the scene graph
25      group->addChild(sep);
26  }
```

Listing 10.8 shows an excerpt of the `drawTree` function that draws the simplest basic tree: a main branch with two sub-branches. Its input parameters include: the coordinates of the starting point, the length of the main branch, the angle under which the two sub-branches are generated from the tip of the main branch, the width of the line used to draw the branches, and a pointer to an `SoGroup` to which the generated branches are added.

A recursive function is a function that calls itself. It cannot do this indefinitely; instead the recursion must stop at the ***base case***. In our example, since we will arrange for the branches to become smaller and smaller, we define the base case as the case in which the branches are smaller than one unit. When this condition is detected, the recursion stops. This is achieved in the recursive function, `drawTree`, by wrapping the implementation in a conditional statement as follows:

```
void drawTree(float x, float y, float len, float angle,
    float stroke, SoGroup* group)
{
  if (len >= 1.) {
    ...
  }
}
```

The function also sets a drawing style for the lines, creates a single line segment to represent the main branch (using the `DrawLine` function) and creates two new

SoSeparators to hold the sub-branches. The result is the scene graph shown in Figure 10.13.

Listing 10.8: Drawing the recursive basic module of the fractal tree-like shape

```
        // First, set up the line style...
        SoDrawStyle* style = new SoDrawStyle();
        style->lineWidth = std::max(0.5,stroke/2.0);
        group->addChild(style);

        // Draw the "mother branch" line in the current
            separator
        drawLine(x, y, x, y+len, group);

        // Then, create two new separators,
        // to host the transformations and the "children
            branches"...
        SoSeparator* trSep1 = new SoSeparator;
        SoSeparator* trSep2 = new SoSeparator;
        // ... and add them to the "mother" Separator
        group->addChild(trSep1);
        group->addChild(trSep2);
```

Two sub-branches are to be drawn in the *x-y* plane at a fixed angle from the main branch, so we insert a geometry transformations into the scene graph for each

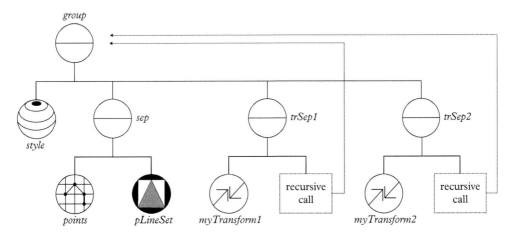

Figure 10.13 *Graph of the base* tree element *used to draw the fractal tree-like figure. The base element consists of a main branch, followed by two rotated sub-branches.*

sub-branch. The class `SoTransform` can be used for this; it implements the most general kind of transforms, which includes rotations, translations, scale operations, and all combinations of these. In order to build this rotation, we must use two operations: first we set the center of rotation at the tip of the main branch, and second we set the rotation axis and the rotation angle. Transformations in `Coin` are additive, which means that their effects accumulate within the same `SoSeparator`. The transformation we set will be applied to any object subsequently added to the separator.

Listing 10.9: Drawing the recursive basic module of the fractal tree-like shape *(continues from previous listing)*

```
16        // length of sub branches = 2/3 of the main branch
17        float branch_len = len * 0.66;
18        // set the stroke width of the sub branch
19        stroke = stroke * 0.66;
20
21        // Recursive calls generate each sub-branch:
22        drawTree(x, y+len, branch_len, angle, stroke, trSep1);
23        drawTree(x, y+len, branch_len, -angle, stroke, trSep2);
```

Sub-branches in a real tree are smaller and thinner than the main branch, so we set a shorter length and a smaller stroke weight for the sub-branches (Listing 10.9). As a last step, we create the two sub-branches in a recursive call to `drawTree`. Figure 10.13 shows the resulting graph for a single "tree element", consisting of a main branch plus two rotated sub-branches.

The whole tree can now be generated by a single call to `drawTree` from `main`. At each iteration, two new branches are recursively generated from the tip of the mother branch. These new branches will be drawn with a line weight equal to half of the line width of the mother branch and with a fraction of the length of the mother branch.

Figure 10.14(a) shows the output of our example code: a fractal-like tree shape has been drawn. We can further develop this example to draw a random forest of fractal-trees by using random angles while constructing the trees, as shown in Figure 10.14(b); the interested reader can find example code in the directory

EXAMPLES/CH10/FRACTALS/DrawRandomForest

We can extend and generalize this example to draw the *Cayley Tree*, which was introduced in Chapter 8. Recall that Cayley trees have all nodes connected to a number of neighbors, z, where z is called the *coordination number*; moreover, all nodes are arranged around a *root node* (the origin of the graph) in n shells, where n is called the *generation number*. Figure 10.15 shows a Cayley tree generated with $z = 3$ (each node is connected to exactly three neighbors) and $n = 10$ (ten shells of nodes have been computed and drawn). The tree shape from our previous example (Figure 10.14(a)) is a special case of Cayley Tree where the root node is connected only to one neighbour, as is the case for

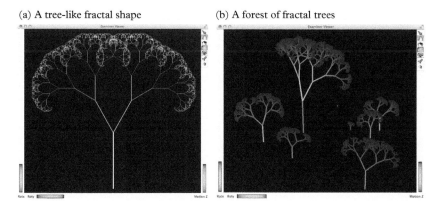

Figure 10.14 *(a), A tree-shape created by using basic fractal rules; (b), A forest generated using the previous fractal tree-shape.*

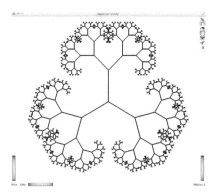

Figure 10.15 *Generation and visualization of a Cayley tree with* coordination number $z = 3$ *and* generation number $n = 10$.

ending nodes. The reader can find the code to generate and visualize a *n*-Cayley tree in the directory

EXAMPLES/CH10/FRACTALS/CayleyTree

10.4.1 Shared nodes

So far we have built 3d objects and scene graphs by newing a separate instance of every component of the scene; an alternative is to share instances representing identical components, such as a branch of a Cayley tree. A ***shared node*** is instantiated only once, and then its reference is used multiple times in the scene graph. Shared nodes are often used to optimize the memory usage of the scene and permit more efficient code. We demonstrate their use by modifying the Cayley Tree example. A complete version of the example can be found in the directory

EXAMPLES/CH10/FRACTALS/CayleyTree_sharedNodes

In our example we recursively draw tree branches using a straight line. The line is built with only two nodes: a `SoCoordinate3` to define the space coordinates of its ending points and a `SoLineSet` to actually draw the line segment between them. The branch assembly is simple, but it is created many times while building the scene: three new line segments are built for each generation, and the only differences among the segments are the coordinates of the ending points and the line length. We define a basic line segment of unit length, then use it in the graph together with an `SoTransform` that rescales the line segment, translates to the starting point coordinates, and rotates it into position.

Listing 10.10 shows the code that builds the basic line and stores it in an `SoGroup` node with global scope[5]. This function will be called by the `main()` function, once only, to initialize the shared node.

Listing 10.10: Building a node that is used as *shared node*

```
1  // a global pointer for a shared node
2  SoGroup* branch = 0;
3
4  // build the shared node once
5  void buildBasicLine()
6  {
7    // Construct parts for single basic branch
8    SoCoordinate3* points = new SoCoordinate3;
9    // Setting two points: the start and the end of the line
10   const SbVec3f p1 =   SbVec3f(0, 0, 0);
11   const SbVec3f p2 =   SbVec3f(0, 1, 0);
12   //Using the two points as coordinates
13   points->point.set1Value(0, p1);
14   points->point.set1Value(1, p2);
15   // Declare a new LineSet, to draw our line
16   SoLineSet *pLineSet = new SoLineSet;
17
18   // Put branch parts together
19   branch = new SoGroup;
```

---

[5] Global variables are not preferred in serious designs because they are completely un-encapsulated. In a better design, a class is declared and a class variable is used. We use a global variable here to simplify the example.

```
20      branch->addChild(points);
21      branch->addChild(pLineSet);
22    }
```

Listing 10.11 shows the `drawLine` function modified to use the new shared node. This function is called by the recursive function that draws the whole tree. It instantiates a geometry transformation, translates the basic line to the appropriate location, scales it along the y axis to set the length of the branch, and incorporates it into the scene graph. Thus, instead of creating z line segments for n generations, we create only one line object and we reuse it all over the scene graph, which is shown in Figure 10.16.

Comparing Figure 10.13 to Figure 10.16, one sees that the latter has been implemented with fewer nodes. While the Cayley tree appears simple, you will find that a tree with many generations can consume large amounts of memory—the number of branches grows exponentially with the number of generations. Thus, node sharing is an important technique.

Listing 10.11: Using a *shared node*

```
23    // draw the basic line
24    void drawLine(float x, float y, float len, SoGroup* group)
25    {
26      // Declare a new Separator to contain the line
27      SoSeparator *sep = new SoSeparator;
28
29      // A transformation for the main branch
30      SoTransform *mainBranchTransform = new SoTransform;
31      // translate the main branch to (x,y,0)
32      mainBranchTransform->translation = SbVec3f(x, y, 0);
33      // scale the branch line over the y axis, to match 'len'
            length
34      mainBranchTransform->scaleFactor.setValue(1, len, 1);
35
36      // Draw the "mother branch" line in the current separator
37      sep->addChild(mainBranchTransform);
38      sep->addChild(branch); // here we use a shared node!!
39
40      // Add the line to the scene graph
41      group->addChild(sep);
42    }
```

We can, in fact, share even more nodes for additional optimization. The Cayley tree's self-similarity permits one to share entire tree segments, and at every level. Only z copies of the first generation need to be instantiated. Then, only z copies of the second

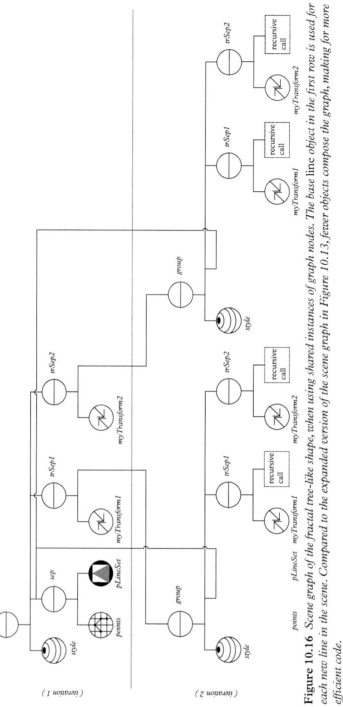

Figure 10.16 *Scene graph of the fractal tree-like shape, when using shared instances of graph nodes. The base line object in the first row is used for each new line in the scene. Compared to the expanded version of the scene graph in Figure 10.13, fewer objects compose the graph, making for more efficient code.*

generation need to be instantiated. The self-similarity allows for a massive sharing of nodes. Generating deep Cayley trees would be an intractable problem without the use of shared nodes. See Exercise 7 for more on this point.

10.5 Animations

Until now we have only built static objects within a static scene: users are able to move the camera around to see the scene from different angles, but the objects themselves are not moving relative to their positions in the scene. Sometimes animating a scene is desirable; fortunately, Coin has functionality for animation, which we describe briefly here.

In Coin, member data are generally stored in *fields*, which are like ordinary member data except that a notification mechanism triggers re-rendering of the whole scene whenever a field is modified. Therefore in order to animate a scene, we have only to change the field of a scene object. If the field is part of a transformation, the transformation is modified and affected objects move. If the field is part of a shape, then the shape changes. Moreover, fields can be connected together.

Animation can be implemented in two ways: fields can be connected to a Coin *engine*, which puts out a stream of changing numbers, or use can be made of a SoTimerSensor. In the second approach one arranges for a callback function to be repeatedly invoked during the program's main loop. The callback function is used to change the value of any field in the scene. The first mechanism is convenient for simple animations, while the second is more powerful and general.

10.5.1 Coin engines: a rotating world

Programming animations with a Coin engine involves two steps: first one must build an *engine*, then link the engine output to one of the fields of an object in the scene.

To illustrate a basic animation we shall modify the example code described in Section 10.2.3 to make a rotating Earth. Listing 10.12 shows the relevant part of the new code, which creates an SoSeparator containing an SoRotationXYZ and adds to that a subgraph containing the Earth, which contains a textured sphere. The reader can find a full working version in the directory

EXAMPLES/CH10/ANIMATIONS/ROTATING_EARTH

An SoRotationXYZ is used for the rotation, and the "Y" axis as is set as the rotation axis[6] We place the Earth object (whose code has been moved to a separate function for convenience) and add it after the SoRotationXYZ. We now have an Earth object preceded by a transformation node; this allows one to set the position of the Earth object by changing the attributes of the transformation node. In particular, we can change its rotation angle at a given rate to let the Earth rotate.

---

[6] In Coin the y axis is along the vertical direction, the x axis is to the right, and the z axis is out of the screen.

336 *Animations*

Listing 10.12: A transformation node to rotate the Earth

```
// A separator for the Earth
SoSeparator *earth = new SoSeparator;

// Create a rotation node
// and add it to the local 'earth' separator
// *before* adding the Earth shape object
SoRotationXYZ *earthRotation = new SoRotationXYZ;
earthRotation->axis.setValue("Y"); // set the rotation axis
earth->addChild(earthRotation);

// Create the Earth object
// and add it to the local 'earth' separator
earth->addChild( drawEarth() );
```

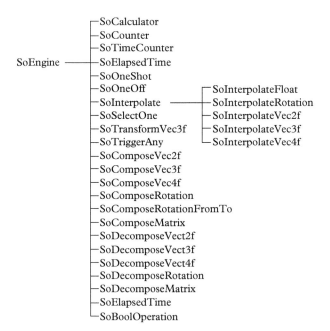

Figure 10.17 *Class tree of* Coin *"engine" nodes.*

To animate the Earth one must create a counter object to create the time steps and to trigger the animation. For that, we make use of `SoTimeCounter`, one of the *engine nodes* provided by Coin. There are many engines, each of them with different characteristics; Figure 10.17 shows their class tree. The `SoTimeCounter` node is a programmable integer counter, which can be customized through changing, for example, its max, step and frequency attributes. The counter also has an attribute called output whose value is updated according to the settings. Listing 10.13 shows how the counter node has been set for our example: as we want a complete rotation of the Earth sphere, we set max to 360 and step to 1 in order to mimic a rotation of 360 degrees, composed of steps of 1 degree; also, we set the frequency value to 0.03, which means that the counter performs 0.03 complete cycles (i.e. from 0 to 360) per second.

Listing 10.13: Setting the counter node which triggers the animation
(continues from previous listing)

```
14    SoTimeCounter *counterEngine = new SoTimeCounter;
15    counterEngine->max=360;      // set max steps
16    counterEngine->step=1;       // set step value
17    counterEngine->frequency=0.03; // set the frequency
```

At the moment we have an object, the Earth, which can be rotated by changing the value of the rotation angle of its geometry transformation; and a counter that counts from zero to 360, in degrees. The rotation transformation of the Earth handles angles in radians, so the values from the counter must be converted. To do this we make use of another Coin engine node, the SoCalculator. This node features 8 scalar inputs, named from a to h, and 4 scalar outputs (besides other attributes) named from oa to od; the values of the input attributes are used in custom formulas provided by the programmer and the outcome placed in the output attributes. Listing 10.14 shows the code. After having instantiated our calculator object, we connect the output attribute of the counter—which is an integer between 0 and 360—to the input a of the calculator; after that, the conversion formula is set using the values of the input attribute a and the output attribute oa. For each time step the counter produces a value in degrees, which is then converted to radians.

Listing 10.14: An engine node to convert from degrees to radians
(continues from previous listing)

```
18    SoCalculator *converterDegRad = new SoCalculator;
19    converterDegRad->a.connectFrom( &counterEngine->output );
          // connect 'output' of the counter engine to 'a'
20    converterDegRad->expression.set1Value(0,"oa=a/(2*M_PI)");
          // set the (Deg->Rad) conversion formula
```

The next step is to rotate the Earth. In listing 10.15 the output of the calculator is first connected to the angle attribute of the rotation node; then the whole earth separator is added to the main scene graph. From now on, for each time step, the Earth is rotated by an angle of 1 degree, resulting in an animated rotating globe.

Listing 10.15: Connecting the angles to the rotation transformation input *(continues from previous listing)*

```
21    // Connect the converter output 'oa' to the rotation
         'angle'
22    earthRotation->angle.connectFrom( &converterDegRad->oa );
23
24    // Add the Earth node to the root node
25    root->addChild( earth );
```

10.5.2 `Coin` sensors: an orbiting planet

In the last example we used the `Coin SoCalculator` to convert degrees to radians. If the calculations become difficult, as in Exercise 5, it is more straightforward and less cumbersome to carry them out in more standard code rather than with engines connected to fields. This is done by modifying the value of a field of an object in the scene. In the directory

EXAMPLES/CH10/ANIMATIONS/ORBITS

the reader will find code that animates the motion of the Earth around the sun. In this example a function called `callback` modifies an `SoTranslation` which controls the position of the earth in its orbit:

```
SoTranslation *planetTranslation=NULL;
double tPhi=0.0;
void callback(void *, SoSensor *) {
  static const double R=10;
  planetTranslation->translation.setValue(R*cos(tPhi), R*sin(
      tPhi), 0);
  tPhi+=0.01;
}
```

The variable `planetTranslation` has global scope, though more heavily engineered designs are possible. Two arguments to this function are anonymous and are not used. They are necessary; however, because it's important that this function has a specific signature: a void function taking a `void *` and an `SoSensor *`.

`SoSensors` (class tree diagram in Figure 10.18) are scene objects that detect a specific condition and invoke a callback function when the condition is detected. The `SoTimerSensor` invokes the callback at specific programmable intervals. In the following lines of code the callback function is scheduled and the update interval is set to 10 ms:

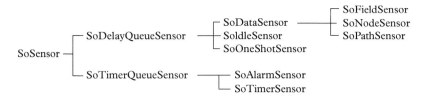

Figure 10.18 *The class tree of* `SoSensor` *classes. Among them we can find the classes* `SoTimerSensor` *used in the orbiting earth example.*

```
// Schedule the update right here:
  SoTimerSensor * timerSensor = new SoTimerSensor;
  timerSensor->setFunction(callback);
  timerSensor->setInterval(0.01);
  timerSensor->schedule();
```

Note that the somewhat involved mechanism is necessary because the updates to the position of the Earth must happen *in the* `SoQt::mainLoop()` *function, during which the program blocks*. Callbacks are used in such contexts, among others, and are particularly common in programs with graphical user interfaces. Another context in which callbacks are used is to take action in response to the selection of objects (see the documentation for `SoSelection`).

In this example the callback is repeatedly invoked and the position of the planet updated. This occurs in the main loop while `Coin` renders the scene and handles user interaction. This mechanism is preferred, particularly when the position, shape, size, or attributes of several objects must be updated. Exercise 5 illustrates this with an application that requires updating the positions of one thousand molecules.

10.6 The `Inventor` system

In this chapter we have briefly explored a small fraction of the classes of the `Coin` library. Nevertheless, we have seen that it is possible to build reasonably powerful visualizations. The `Inventor` class list contains many more classes than those introduced here and the reader is invited to experiment with them in order to get a feeling of their potential in terms of both graphics and user interaction. The most useful reference is Wernecke (1994a). With the exception of `SoSensors`, all of the scene objects have a common `base class`, `SoNode`, from which they inherit. A class tree diagram for the majority of our scene objects is shown in Figure 10.19.

The practical consequences of the inheritance relationship have hopefully become apparent by now. The primary advantage is the ability to use any specific type of `SoNode`, representing a cone, a sphere, a material, a translation, etc., wherever an `SoNode` is required; for example, when assembling such nodes under a `SoGroup` or an `SoSeparator`. This feature is called *polymorphism*. A second advantage of

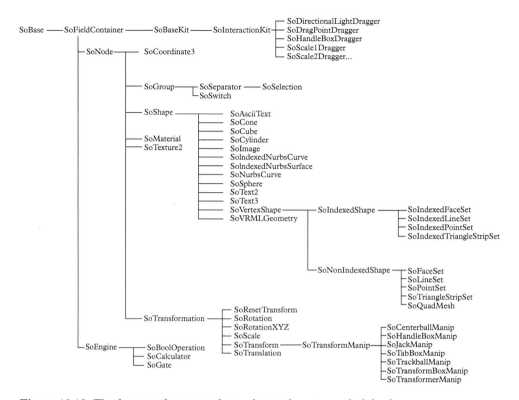

Figure 10.19 *The class tree of* SoNode *classes. Among them we can find the classes* SoCoordinate3 *and* SoTexture *that have been used in the examples described in this chapter.*

polymorphism, when properly executed, is a mechanism for acheiving flexibility and extensibility of a software system; indeed, in the Inventor Toolmaker (Wernecke, 1994b), one can find guidance on extending the Inventor system with custom SoShapes and other objects in Inventor. A detailed discussion of inheritance and polymorphism is deferred to Chapter 12.

10.7 Exercises

1. Using SoCone, draw a semi-transparent cone, and using SoFaceSet draw a plane which intersects the cone. Do this in three ways, in order to generate the three conic sections: parabola, ellipse, hyperbola. Calculate these curves and co-display them with the cone using SoLineSet to check the calculation.
2. In Chapter 3, Exercise 13, you solved the problem of the motion of five massive balls connected by springs of different strength. Revisit this problem, this time animating the solution.

3. Use a SoLineSet to animate standing waves of various frequencies on a string.
4. Drum Heads.
   In two dimensions, the wave equation is
   $$\left[\nabla^2 - \frac{1}{v^2}\frac{\partial^2}{\partial t^2}\right]u(x,y) = 0 \tag{10.2}$$
   where $\nabla^2 = \frac{\partial^2}{\partial x^2} + \frac{\partial^2}{\partial y^2}$. The equation describes waves on a membrane, such as a drum head, where boundary conditions constrain $u(x,y) = 0$ at $x^2 + y^2 = a^2$. Solve the wave equation for this set of boundary conditions and animate standing wave solutions for the first few modes of vibration. Use SoFaceSet to describe the membrane.
5. [P] Gas in a Piston.
   Simulate a confined volume of an ideal gas expanding against a piston. Keep track of about 1000 molecules. Start the molecules at random positions and random velocities distributed according to a Maxwell-Boltzman distribution. Each molecule collides with fixed cylindrical walls, and one fixed "bottom" wall, but the top wall is a piston free to move in the vertical direction. Collisions between gas molecules and the piston are elastic. Keep track of the kinetic energy of the gas as well as that of the piston.
6. A Solar System.
   It is well known that planets move in ellipses around the sun. The speed at which they move is less familiar, but can be obtained either by numerical inversion of *Kepler's equation* (see Goldstein, 2001), or by the application of Kepler's law of orbits, law of areas, and law of periods. Animate a toy solar system with two planets orbiting a massive sun. The orbits of the two planets should have significant eccentricity.
7. Execute the Cayley tree example EXAMPLES/CH10/FRACTALS/CayleyTree and determine the maximum number of generations that can be visualized without the use of shared nodes. Then, optimize the code very heavily by sharing as many branches and subtrees as is possible, and determine the maximum number of nodes again.
8. Visualize a Cayley tree with a coordinate number $z = 4$. The tree should develop symmetrically in all three dimensions, rather than being confined to a plane as with the example in Section 10.4.1. Use shared nodes as in the preceding problem to generate a deep tree.
9. From the depths of your subconscious mind, dream up a scene and capture it in an animated or still scene using Coin. You may choose any topic you like: from the world of science, the world of technology, the world of art, politics, sports, religion, exoticism, cuisine culture, or whatever goes on inside your head.
10. [P] Now we examine the outside of your head. Obtain a *profile gauge* from a hardware store (for about US $10). This is a device that can be pushed against any surface

and retains the shape of the surface. Prefer the plastic to the metal variety since the latter is too stiff and may cause pain or injury. Use the device to measure the contours of your head, and transfer them to a graphics program. You can do this with an `SoFaceSet`. Verify graphically that the shape looks right, then determine the volume of your head using Monte Carlo integration.

BIBLIOGRAPHY

H. Goldstein, C. P. Poole, and J. Safko (2001). *Classical Mechanics*. 3rd ed. Addison-Wesley.

Hughes, J., A. van Dam, M. McGuire, D. Sklar, J. Foley, S. Feiner, and K. Akeley (2013). *Computer graphics: principles and practice*. 3rd ed. Addison-Wesley, `cgpp.net`.

Akenine-Moller, T., E. Haines, and N. Hoffman (2008). *Real-time rendering*. 3rd ed. A K Peters.

Kongsberg (2014). *Coin3D* bitbucket.org/Coin3D/coin/wiki/Home.

FEI (2015). *OpenInventor*, www.openinventor.com.

Open Inventor Architecture Group (1994). *Open Inventor C++ Reference Manual, The Official Reference Document for Open Inventor, Release 2*. Addison-Wesley.

Osfield, R. (2015). *Openscenegraph*. www.openscenegraph.org.

Selman, D. (2002). *Java 3D programming*. Manning Publications.

Silicon Graphics International Corp. (2009). *Open Inventor*. oss.sgi.com/projects/inventor.

Wernecke, J. (1994a). *The Inventor Mentor*. Addison Wesley.

Wernecke, J. (1994b). *Open Inventor Nodes Quick Reference*. techpubs.sgi.com/library/manuals/2000/007-2469-001/pdf/007-2469-001.pdf.

Wernecke, J. (1994c). *The Inventor Toolmaker, Extending Open Inventor, Release 2*. 2nd ed. Addison-Wesley.

CCNLab (2015). *Coin3D*. grey.colorado.edu/emergent/index.php/Coin3d.

Qt (2015). *Qt*. www.qt.io.

Park, S. K. and K. W. Miller (1988). *Random Number Generators: Good Ones Are Hard to Find*. Commun. ACM **31** 10, 1192.

11
Ordinary differential equations

11.1 Introduction	344
11.2 Example applications	345
11.2.1 Projectile motion with air resistance	345
11.2.2 Motion of a charged particle in a magnetic field	346
11.2.3 Simple nonlinear systems: the Lorenz model	347
11.2.4 The Lagrangian formulation of classical mechanics	348
11.2.5 The Hamiltonian formulation of classical mechanics	350
11.2.6 The Schrödinger equation	351
11.3 A high-level look at an ODE solver	353
11.3.1 A simple integrator class	354
11.3.2 Example: the harmonic oscillator	357
11.4 Numerical methods for integrating ordinary differential equations	358
11.4.1 The Euler method	359
11.4.2 The midpoint method	361
11.4.3 The trapezoid method	361
11.4.4 The 4th order Runge-Kutta method	361
11.4.5 Properties of nth order Runge-Kutta methods	362
11.5 Automated solution of classical problems	368
11.5.1 Taking partial derivatives with GENFUNCTIONs	368
11.5.2 Computing *and* solving the equations of motion	370
11.5.3 A classical Hamiltonian solver	371
11.6 Adaptive step size control	373
11.6.1 Step doubling	375
11.6.2 Embedded Runge Kutta methods	379
11.7 Symplectic integration schemes	383
11.7.1 Symplectic transformations	386
11.8 Symplectic integrators of first and higher order	390
11.9 Algorithmic inadequacies	393
11.9.1 Stability	393
11.9.2 Solution mixing	394
11.9.3 Multiscale problems	395
11.10 Exercises	396
Bibliography	401

Applied Computational Physics. Joseph F. Boudreau and Eric S. Swanson, Oxford University Press (2018).
© Joseph F. Boudreau and Eric S. Swanson. DOI:10.1093/oso/9780198708636.001.0001

11.1 Introduction

Ordinary differential equations (ODEs) arise frequently in classical and quantum physics. An ordinary differential equation involves one independent variable which we call t, and one or more dependent variables which we might call x in the case of a single dependent variable. These equations contain derivatives of x up to the *order* of the ODE. A familiar example is the equation governing the damped harmonic oscillator:

$$\frac{d^2x}{dt^2} + 2\omega_0\zeta\frac{dx}{dt} + \omega_0^2 x = 0 \qquad (11.1)$$

where ω_0 is the undamped angular frequency of the oscillator and ζ is the damping ratio. This is a second order ordinary differential equation, but it can be reduced to two coupled first-order differential equations:

$$\frac{dx}{dt} = v$$
$$\frac{dv}{dt} = -2\omega_0\zeta v - \omega_0^2 x. \qquad (11.2)$$

A similar reduction can be carried out on any set of nth-order differential equations. Eq. 11.2 is called **autonomous** because it does not contain the independent variable t on the right hand side. By contrast, if we add a harmonic driving force to the oscillator, we obtain a different set:

$$\frac{dx}{dt} = v$$
$$\frac{dv}{dt} = -2\omega_0\zeta v - \omega_0^2 x + \frac{F}{m}\cos(\omega t) \qquad (11.3)$$

which is called **nonautonomous** because the right hand side contains the independent variable t explicitly.

A nonautonomous set of equations can be trivially transformed into an autonomous set by making the substitution $t \to T$ and adding the equation:

$$\frac{dT}{dt} = 1$$

to the set; for example Eq. 11.3 would become:

$$\frac{dx}{dt} = v$$
$$\frac{dv}{dt} = -2\omega_0\zeta\frac{dx}{dt} - \omega_0^2 x + \frac{F}{m}\cos(\omega T)$$
$$\frac{dT}{dt} = 1; \tag{11.4}$$

and consequently an integration procedure designed for autonomous ODEs suffices to integrate nonautonomous ODEs as well.

Finding a solution to a coupled set of ordinary differential equations usually amounts to finding the time-dependence of the dependent variables, v and x in this case. For an nth-order ODE, n constants of integration are required. These are usually the starting values of the n variables, $x_0 \equiv x(0)$ and $v_0 \equiv v(0)$ in our example. For a system of particles evolving according to the laws of classical dynamics the starting values could be the initial values of the coordinates and the velocities of each particle in the system; or, taking the more advanced view of Lagrangian mechanics, the *generalized* coordinates and velocities. This type of problem is called an ***initial value*** problem. It is rare to find initial value problems with a closed-form solution, thus computational methods become vital. Problems that do possess closed-form solutions can be useful to check the accuracy of the numerical integration.

One view of ordinary differential equations is that they define a family of curves in N dimensions, N being the number of independent variables. The curves are parametrized by the starting values of the coordinates and the velocities, and perhaps by other constants, e.g., in Eq. 11.2, the natural frequency ω_0. Object-oriented techniques make it possible to endow objects with practically any type of interface, and ODEs are no exception. We will demonstrate in this chapter a set of classes that can be used to define functions through the differential equations that govern their time development. The reward will be a mechanism to produce functions that can be used just like any other function, like `sin` or `exp`, for plotting or data modeling. The techniques depend on object-oriented features of the C++ programming language that will be developed later in the chapter *Polymorphism* and are not fully explained here. The goal here is to demonstrate the power of these techniques and to whet the reader's appetite for developments to come in later chapters.

Before developing a computational solution to ODEs, we give a few examples of familiar situations in which they arise in classical and quantum mechanics.

11.2 Example applications

11.2.1 Projectile motion with air resistance

Air resistance may be modeled as a force proportional to a power of the velocity of a body moving through a stationary fluid: $\vec{F} = -\gamma v^2 \hat{v}$, where γ is a constant whose value depends on the shape of the object and the viscosity of the fluid.[1] With a force of $F_g = -mg\hat{x}_1$, and defining the horizontal direction as x_0 and the vertical direction as x_1, the equation of motion of a projectile moving through viscous air is:

$$m\frac{d^2\vec{x}}{dt^2} = -mg\hat{x}_1 - \gamma v^2 \hat{v}$$

which is actually two second-order differential equations. Reducing these two equations to four first-order differential equations gives:

$$\begin{aligned}\frac{dx_0}{dt} &= v_0 \\ \frac{dx_1}{dt} &= v_1 \\ \frac{dv_0}{dt} &= -\frac{\gamma}{m}v_0\sqrt{v_0^2 + v_1^2} \\ \frac{dv_1}{dt} &= -g - \frac{\gamma}{m}v_1\sqrt{v_0^2 + v_1^2}.\end{aligned} \qquad (11.5)$$

In this set of coupled equations, nonlinear terms involving the dependent variables v_i for $i \in \{0, 1\}$ occur on the right hand side, making this a system of *nonlinear* differential equations. Nonlinear ODEs almost never have a closed-form solution, and computational methods play a dominant role in understanding the solutions.

11.2.2 Motion of a charged particle in a magnetic field

Figure 11.1 shows a system of elementary particles after a proton-antiproton collision, surrounded by a simple silicon tracking system. In actual particle detectors the momentum analysis of these particles is an important ingredient in the study of such collisions and is carried out with a magnetic field. Exceptionally uniform magnetic fields are sometimes achieved, in which case the trajectory of a charged particle is known to be a helix. When the fields are not uniform, trajectories are determined by the Lorentz force law. Since particles are relativistic in high-energy particle-antiparticle collisions we will use the relativistic form of the law (in CGS-Gaussian units), which is:

$$\frac{d\vec{p}}{d\tau} = \frac{q}{m}\left(\vec{p} \times \vec{B}(\vec{r})\right)$$

---

[1] The effects of fluid flow around objects are characterized by a dimensionless number known as the Reynolds number $Re = \rho vL/\eta$ where ρ is the density of the fluid, η is its viscosity, L is a characteristic length of the object, and v is its speed through the fluid. The quadratic drag described here is a good description in situations with high Reynolds number, $Re > 1000$. A baseball thrown at 100 mph has $Re = 2.3 \times 10^5$.

Figure 11.1 *Trajectories of charged particles in a (fictitious) particle detection system.*

where $\vec{p} = mc\vec{\beta}\gamma$ is the relativistic momentum and τ is the proper time of the particle, $\tau = t/\gamma$. It may be more useful to parametrize the momentum as a function of path length $s = vt = c\beta\gamma\tau$,

$$\frac{d\vec{p}}{ds} = \frac{d\vec{p}}{d\tau} \cdot \frac{d\tau}{ds} = \frac{q}{mc\beta\gamma}\left(\vec{p} \times \vec{B}(\vec{r})\right).$$

From the definition of momentum we have $\vec{p} = m\gamma \vec{v} = m\gamma \frac{d\vec{r}}{dt}$, again writing this in terms of the path length covered gives:

$$\frac{d\vec{r}}{ds} = \frac{d\vec{r}}{dt} \cdot \frac{dt}{ds} = \frac{1}{m\gamma}\vec{p} \cdot \frac{1}{v} = \frac{1}{mc\beta\gamma}\vec{p}.$$

In summary we have six first-order autonomous ordinary differential equations describing the path of a particle in a magnetic field, namely:

$$\frac{d\vec{r}}{ds} = \frac{1}{mc\beta\gamma}\vec{p}$$
$$\frac{d\vec{p}}{ds} = \frac{q}{mc\beta\gamma}\left(\vec{p} \times \vec{B}(\vec{r})\right).$$

11.2.3 Simple nonlinear systems: the Lorenz model

A simple model for convection in the atmosphere, introduced by Edward Lorenz in 1963, is described by the differential equations:

$$\frac{dx}{dt} = \sigma(y - x) \tag{11.6}$$

$$\frac{dy}{dt} = x(\rho - z) - y \tag{11.7}$$

$$\frac{dz}{dt} = xy - \beta z. \tag{11.8}$$

This is another example of a system of nonlinear differential equations, owing to the presence of the terms xy and xz on the right hand side. Such problems are treated in detail in the following chapter. The parameters σ, ρ, and β are control parameters. For certain values of the control parameters the Lorenz model exhibits chaotic behavior, with a space-filling quality of the trajectories and a rapid divergence of nearby trajectories—a feature that is discussed extensively in the next chapter.

11.2.4 The Lagrangian formulation of classical mechanics

The Lagrangian formulation of classical mechanics is an approach that every advanced undergraduate and graduate student learns (Goldstein 2001; Thornton, 2004). It provides a simple, uniform framework for solving mechanics problems in terms of generalized coordinates that specify the configuration of the mechanical system. Many of these problems would not be tractable with other methods. One identifies the generalized coordinates $q_i, i = 0, 1, 2..n-1$ of the system, which are any set of independent variables fully characterizing the configuration of the system. Then, one expresses the potential energy and the kinetic energy in terms of the generalized coordinates q_i and their time derivatives $\dot{q}_i \equiv dq/dt$. We denote the potential energy $V = V(q_i)$ and the kinetic energy $T = T(q_i, \dot{q}_i)$. The Lagrangian is then $L = T - V$. The n equations of motion follow from the Euler-Lagrange equations:

$$\frac{\partial L}{\partial q_i} - \frac{d}{dt}\frac{\partial L}{\partial \dot{q}_i} = 0. \tag{11.9}$$

While partial derivatives appear in the Euler-Lagrange equations, the equations they generate are second-order differential equations[2].

For the one-dimensional harmonic oscillator, for example, $V = \frac{1}{2}kx^2$, $T = \frac{1}{2}m\dot{x}^2$, so $L = \frac{1}{2}m\dot{x}^2 - \frac{1}{2}kx^2$, and the Euler-Lagrange equation generates a single equation of motion:

$$m\ddot{x} + kx = 0.$$

---

[2] Usually the potential energy will be expressed only in terms of the coordinates $V = V(q_i)$, but a velocity-dependent potential can also be introduced. The case of a charged particle in a magnetic field is particularly important. Let the generalized coordinates of the system be the particle's position $\vec{r} \equiv \{x, y, z\}$; its derivative $\vec{v} \equiv d\vec{r}/dt$; the scalar potential is ϕ and the magnetic vector potential is $\vec{A}$. In this case, the velocity-dependent magnetic forces can be incorporated by replacing the normal potential energy $-q\phi$ with an interaction term $q(\vec{v} \cdot \vec{A} - \phi)$.

While it is easy to obtain second order differential equations in the Lagrangian formulation it is usually highly nontrivial to solve them. We take the example of the heavy symmetric top in a gravitational field with one point fixed. The mass of the top is M, the center of gravity is a distance l from the fixed point along the symmetry axis, and the inertia tensor is characterized by only two values I_1 and I_3 since the symmetry imposes $I_1 = I_2$. We sketch here the key points of the solution developed in Goldstein (2001). The generalized coordinates are the Euler angles θ, ϕ, and ψ and the Lagrangian for the system is:

$$L = \frac{1}{2}I_1(\dot{\theta}^2 + \dot{\phi}^2 \sin^2\theta) + \frac{1}{2}I_3(\dot{\phi}\cos\theta + \dot{\psi})^2 - Mgl\cos\theta. \qquad (11.10)$$

One obtains three equations of motion; the first is:

$$I_3(\dot{\phi}\cos\theta + \dot{\psi}) = const$$
$$= I_1 a, \qquad (11.11)$$

which effectively defines a new constant a; the second is:

$$(I_1 \sin^2\theta + I_3 \cos^2\theta)\dot{\phi} + I_3 \dot{\psi}\cos\theta = const$$
$$= I_1 b, \qquad (11.12)$$

which defines the constant b; and, using the conserved energy instead of the third Euler-Lagrange equation,

$$\frac{1}{2}I_1(\dot{\theta}^2 + \dot{\phi}^2 \sin^2\theta) + \frac{1}{2}I_3(\dot{\phi}\cos\theta + \dot{\psi})^2 + Mgl\cos\theta = const$$
$$= E. \qquad (11.13)$$

From Eqs. 11.11 and 11.12 we obtain:

$$\dot{\phi} = \frac{b - a\cos\theta}{\sin^2\theta} \qquad (11.14)$$

$$\dot{\psi} = \frac{I_1 a}{I_3} - \cos\theta \left(\frac{b - a\cos\theta}{\sin^2\theta}\right). \qquad (11.15)$$

Define two more constants:

$$\alpha \equiv \frac{2E}{I_1} - \frac{I_1}{I_3}a^2 \qquad \beta = \frac{2Mgl}{I_1}$$

and write Eq. 11.13 as:

$$\alpha = \dot{\theta}^2 + \left(\frac{b - a\cos\theta}{\sin\theta}\right)^2 + \beta\cos\theta. \tag{11.16}$$

Finally, with the substitution $u = \cos\theta$, Eqs. 11.14, 11.15, and 11.16 take the form:

$$\frac{du}{dt} = \sqrt{(1-u^2)(\alpha - \beta u) - (b - au)^2}$$
$$\frac{d\phi}{dt} = \frac{b - au}{1 - u^2}$$
$$\frac{d\psi}{dt} = \frac{I_1 a}{I_3} - u\left(\frac{b - au}{1 - u^2}\right).$$

You may be wondering why we are left with only three differential equations instead of six. The answer is that we have already found three "first integrals" of the motion, which were used to define the constants E, a, and b.

While you may not be familiar with this particular problem, the essential observation is that the Euler-Lagrange equations produce a set of equations that typically require numerical techniques to produce useful predictions.

11.2.5 The Hamiltonian formulation of classical mechanics

An alternate way of formulating classical mechanics is in terms of the Hamiltonian. The Hamiltonian method is useful because the equations of motion, namely *Hamilton's equations*, are a set of first order differential equations. A Hamiltonian can be written as:

$$H(p_i, q_i) = T(p_i, q_i) + V(q_i), \tag{11.17}$$

where

$$p_i = \frac{\partial L}{\partial \dot{q}_i} \tag{11.18}$$

are the "generalized momenta". Hamilton's equations are:

$$\dot{p}_i = -\frac{\partial H}{\partial q_i} \qquad \dot{q}_i = \frac{\partial H}{\partial p_i}. \tag{11.19}$$

For example, the harmonic oscillator has the Hamiltonian:

$$H = \frac{p^2}{2m} + \frac{1}{2}kx^2, \tag{11.20}$$

so Hamilton's equations read:

$$\frac{dp}{dt} = -kx$$
$$\frac{dx}{dt} = \frac{p}{m}, \quad (11.21)$$

which have the desired first order form.

11.2.6 The Schrödinger equation

In quantum mechanics, a particle is not described by a position or momentum, but rather a wavefunction, which is a complex function of spatial coordinates and time. All dynamical variables of the particle can take on a spectrum of values, which are realized with a probability determined by the wavefunction. The time development of the wavefunction itself is governed by the quantum mechanical Hamiltonian operator:

$$H\psi = \left[-\frac{\hbar^2}{2m}\nabla^2 + V(\vec{r}) \right] \psi(\vec{r}, t) \quad (11.22)$$

such that:

$$i\hbar \frac{\partial}{\partial t}\psi(\vec{r}, t) = H\psi(\vec{r}, t). \quad (11.23)$$

This is called the time-dependent Schrödinger equation. By substituting a trial solution $\psi(\vec{r}, t) = \phi_i(\vec{r})e^{-iE_i t/\hbar}$ one accomplishes a separation of variables, removing the time dependence and obtaining the time-independent Schrödinger equation:

$$H\phi_i(\vec{r}) = E_i \phi_i(\vec{r}) \quad (11.24)$$

or:

$$\left[-\frac{\hbar^2}{2m}\nabla^2 + V(\vec{r}) \right] \phi_i(\vec{r}) = E_i \phi_i(\vec{r}). \quad (11.25)$$

This is still a partial differential equation; however, in many situations the potential is central, $V(\vec{r}) = V(r)$, and a further separation can be accomplished. In this case the solution breaks down into a universal angular part times a radial part:

$$\phi_{n,l,m}(\vec{r}) = \frac{1}{r} U_{n,l}(r)\, Y_l^m(\theta, \phi). \quad (11.26)$$

Example applications

The function $U_{n,l}$ is the solution to the effective one-dimensional problem,

$$\left[-\frac{\hbar^2}{2m}\frac{d^2}{dr^2} + V(r) + \frac{\hbar^2 l(l+1)}{2mr^2}\right] U_{n,l}(r) = E_{n,l} U_{n,l}(r) \qquad (11.27)$$

which we may write as:

$$\frac{d^2 U_{n,l}(r)}{dr^2} = \left[\frac{2m}{\hbar^2}(V(r) - E_{n,l}) + \frac{l(l+1)}{r^2}\right] U_{n,l}(r) \qquad (11.28)$$

or:

$$\frac{dU_{n,l}(r)}{dr} = U'_{n,l}(r)$$
$$\frac{dU'_{n,l}(r)}{dr} = \left[\frac{2m}{\hbar^2}(V(r) - E_{n,l}) + \frac{l(l+1)}{r^2}\right] U_{n,l}(r), \qquad (11.29)$$

which again has the desired form of a set of coupled first-order differential equations. Moreover, the equations are linear.

This last point merits a small digression. Linear ordinary differential equations are those which, when put into the canonical form:

$$\frac{d\vec{y}(t)}{dt} = f(t, \vec{y}(t))$$

can be written as:

$$\frac{d\vec{y}(t)}{dt} = \mathbf{A}(t) \cdot \vec{y}(t) + \vec{\phi}(t). \qquad (11.30)$$

Eq. 11.30 is said to be an ***inhomogeneous*** set of equations. The corresponding homogeneous set of equations is:

$$\frac{d\vec{y}(t)}{dt} = \mathbf{A}(t) \cdot \vec{y}(t). \qquad (11.31)$$

Linear ordinary differential equations satisfy a *superposition principle*. Given any two solutions $\vec{y}_1(t)$ and $\vec{y}_2(t)$, a linear combination $\alpha \vec{y}_1(t) + \beta \vec{y}_2(t)$ is also a solution. To any solution of the inhomogeneous set of equations we can add an arbitrary superposition of solutions to the homogeneous set of equations; this result is also a solution to the inhomogeneous set of equations. The vector space of solutions to a second-order linear homogeneous equation such as Eq. 11.28, or two coupled first-order linear differential equations such as Eq. 11.29 has a dimensionality of two – i.e., we will, in general, have *two* linearly independent solutions.

Eq 11.29 has the form of a second-order linear homogeneous equation that lends itself readily to the treatment we are discussing. However the boundary conditions can add a layer of complexity to the problem. For example, bound state solutions are constrained at two spatially separated points rather than at the origin. Thus for a particle in a spherical box centered at the origin with radius a the wavefunction must be zero at $r = a$ and finite at $r = 0$, which occurs only for quantized values of the energy E. This is a *two-point boundary value problem*. However for positive-energy scattering states the radial wavefunctions $U_{n,l}$ can be obtained by starting the function with an arbitrary value of $U_{n,l}(r_0)$ and $U'_{n,l}(r_0)$ at some point r_0 and evolving forward (or backward) with r. In these circumstances we have an initial value problem, and the techniques we discuss in this chapter are applicable. There are usually two independent solutions to the radial Schrödinger equation; they can be obtained by taking the same starting value for $U_{n,l}(r_0)$; setting $U'_{n,l}(r_0) = 0$ in one case and $U'_{n,l}(r_0) \neq 0$ in the other case. Once these functions are obtained, one can construct the general solution by superposition.

11.3 A high-level look at an ODE solver

All of the problems in the previous section can be expressed as a set of first-order ordinary differential equations of form:

$$\frac{d\vec{y}(x)}{dx} = \vec{f}(x, \vec{y}) \tag{11.32}$$

subject to constraints $\vec{y}(x_0) = \vec{y}_0$. Our first concern is to provide accurate solutions to the problem at hand; as we will see, a fair amount of numerical sophistication is required. As a result, we also concern ourselves with another objective: to hide the computational complexity behind an interface that exposes only the features of the problem which are important to the user. There are many imaginative ways to formulate an interface that does the job; here we describe one way to do it. The resulting code is part of QatGenericFunctions.

We take the view that the functions $\vec{f}(x, \vec{y}(x))$ together with the initial conditions $\vec{y}_0$ define a set of functions $\vec{y}(x)$. The design of our "integrator" is founded on the notion that it is *merely a procedure to manufacture these functions*.

To let this idea sink in we're going to interrupt the typical narrative in which numerical algorithms are discussed in great detail, and instead give an example of an integrator incorporating this notion into its design. The neglect of mathematical detail will be quickly remedied, since we will discuss the important algorithms in Section 11.4. Some of the object-oriented programming practices that enable this type of design have already been discussed in previous chapters, others will be discussed later (see Chapter 12).

The functions are implemented as functors, specifically, as another type of function in the Genfun namespace[3]. Because nonautonomous differential equations can be trivially

---

[3] The classes are introduced and discussed in great detail in Chapter 3.

transformed into autonomous differential equations (see Section 11.1), it is sufficient to write an integrator that works with autonomous differential equations, i.e., those with no explicit dependence on the independent variable so that $\vec{f}(x,\vec{y}(x)) = \vec{f}(\vec{y}(x))$ The equations $\vec{f}(\vec{y}(x))$ may have control parameters, for example, σ, ρ, and β in Eq. 11.8. We call these parameters $\vec{\alpha}$ and rewrite the functions as $\vec{f}(\vec{y}(x)) = \vec{f}(\vec{y}(x); \vec{\alpha})$ in order to show explicitly their dependence upon the control parameters. Any parameter of the functions $\vec{f}$ is automatically a parameter of the functions $\vec{y}(x)$, as are the starting values $\vec{y}_0$; accordingly we may rewrite the solution as $\vec{y} = \vec{y}(x; \vec{\alpha}, \vec{y}_0)$.

We discuss an integrator that returns this set of functions *and* gives access to their controlling parameters $\vec{\alpha}$ and $\vec{y}_0$. An advantage of this viewpoint is that the functions become useful in the context of applications such data modeling, where functions change their shape in response to changes of a parameter. Because the functions $\vec{y}(x)$ are defined as the solution to an ODE, a large amount of integration takes place "under the hood" when the function is evaluated. In order to avoid repeating the calculation, the results are cached. When a second evaluation is requested, the integration up to the requested point can be carried out starting from a nearby, previously-evaluated point. Of course when any parameter is changed the entire cache is invalid and must be cleared.

It should be clear that we are talking about a powerful but somewhat heavily engineered piece of code. The first step in this engineering process is to decide what the user should see, not how an object or objects should look internally. Therefore in the next section we describe a simple program that uses the integrator to manufacture the required functions. In software engineering one typically thinks the use cases through before actually writing any code. The entities (classes) that represent the key abstractions in the design are determined, then the protocol for interacting with the classes, namely the class interface as expressed in the class declaration, is fixed. In the case of an integrator one requires a mechanism to specify the differential equations to be integrated and the starting values, a mechanism for returning the solution(s) to these equation(s), and access to adjustable parameters. Additionally, management of a cache of previously-calculated information to speed up future function evaluations is required; the cache is shared information common to all of the functions but completely hidden from the user. Care needs to be given to memory management so that the cache is deleted when the last client object is destroyed.

The basic usage of our class is described with an example in the next section.

11.3.1 A simple integrator class

The example we discuss in this section integrates the equations of motion of a projectile in the presence of air resistance. The example code is included in the EXAMPLES/ CH11/PROJECTILE area. It uses an integrator class from `QatGenericFunctions` called (RKIntegrator). This integrator employs a widely-used technique for integrating differential equations called the ***Runge-Kutta method***—actually a family of related methods discussed in Section 11.4. The `RKIntegrator` is capable of handling autonomous first-order differential equations in the form of Eq. 11.32.

Our example code `projectile.cpp` first creates an instance of the integrator:

```
#include "QatGenericFunctions/RKIntegrator.h"
using namespace Genfun;
.
.
.
RKIntegrator integrator;
```

And then defines the functions $f_i(y_0, y_1 \ldots)$ (see Eqs. 11.5):

```
  Variable X(0,4),Y(1,4),VX(2,4),VY(3,4);
  double gamma=0.02;
  double m     =0.5; // kg
  double g     =9.8; // m/s/s

  GENFUNCTION DXDT  = VX;
  GENFUNCTION DYDT  = VY;
  GENFUNCTION DVXDT = -gamma/m*VX*Sqrt()(VX*VX+VY*VY);
  GENFUNCTION DVYDT = -g -gamma/m*VY*Sqrt()(VX*VX+VY*VY);
```

Notice that we are calling a new form of the constructor for Genfun::Variable in the first line. The constructor with two integer arguments (1,4) creates a variable that is a component of a four dimensional vector with index value 1 (the indices in this case run from 0-3).

Our next step will be to add these differential equations to the integrator. The integrator then takes these differential equations *plus a starting value for each variable being integrated* and uses it to define the function.

```
  integrator.addDiffEquation(&DXDT,  "X", 0);
  integrator.addDiffEquation(&DYDT,  "Y", 0);
  integrator.addDiffEquation(&DVXDT, "VX",50);
  integrator.addDiffEquation(&DVYDT, "VY",50);
```

These lines tell the integrator that the quantity X governed by the differential equation DXDT (defined above) has a starting value of 0. The quantity Y starts at 0, too. The quantities VX and VY, representing the *x*- and *y*-components of the projectile's velocity, both start at 50 m/s.

That's all the integrator needs to know. The fully integrated trajectory comes back from the integrator like this:

```
GENFUNCTION  x = *integrator.getFunction(X);
GENFUNCTION  y = *integrator.getFunction(Y);
```

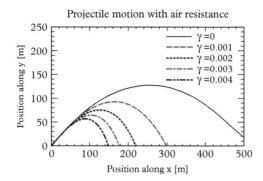

Figure 11.2 *The orbit of a projectile in the presence of air resistance for various values of the parameter γ.*

The resulting functions[4] can be used like any other QatGenericFunctions functor. In the example program projectile.cpp we have used these functions to make a plot of an orbit. The actual work is done when the function is evaluated, since then the integration is carried out, from $t = 0$ up to the specified value of the independent variable t. The plot in Figure 11.2 shows y vs x, in other words, the orbit of the projectile. Since the functions x and y are functions of t, we need to do a little work to get the orbit equations. In Figure 11.2, we have evaluated $x(t)$ and $y(t)$ for various values of t and then used the CubicSplineInterpolator (Chapter 4) to interpolate between the points.

One can do a little more with the integrator than has been described. For one thing, we know that we can regard the starting value of each variable as a parameter that governs the shape of the function. The function RKIntegrator::addDiffEquation method has the following signature

```
Parameter * addDiffEquation (
      const AbsFunction   * diffEquation,
      const std::string & variableName="anon",
      double defStartingValue=0.0,
      double startingValueMin=0.0,
      double startingValueMax=0.0);
```

so in the above example we have been throwing away the return value and taking most of the defaults. If we wanted to, we could have set a range for the starting value and

---

[4] You may be wondering about the memory management. The integrator allocates memory for the functions and cleans these functions up when the integrator is destroyed. The shared cache is referenced by all of the functions and any copies that may exist. It is freed when the last referencing function, original or copy, is destroyed.

held onto the parameter returned by the function. This would have allowed us to use the starting value of the integrated quantity as a parameter to the function returned by the integrator[5]. Another method is also available,

```
Parameter * createControlParameter (
        const std::string & variableName="anon",
        double defStartingValue=0.0,
        double startingValueMin=0.0,
        double startingValueMax=0.0);
```

which creates a control parameter that can be used in defining the differential equations governing the evolution of the variables[6]. Thus, we achieve the goal of using the ordinary differential equations to define a set of functions parametrized by their starting values and other control parameters which appear in the differential equations.

11.3.2 Example: the harmonic oscillator

The harmonic oscillator in one dimension furnishes a simple example with a familiar solution. We start with the Hamiltonian of the system, Eq. 11.20, and the first-order differential equations that can be derived from the Hamiltonian, Eqs. 11.21. These equations are integrated in an example which you can find in EXAMPLES/CH11/HARMONIC; the program is called harmonic.cpp. Here are some excerpts:

```
//
// Harmonic oscillator:
//

double K=1.0; // Newtons/m
double M=1.0; // Kg
Variable X(0,2),P(1,2);
GENFUNCTION DXDT=P/M;
GENFUNCTION DPDT=-K*X;
//
// Integrate the equations of motion:
//
RKIntegrator integrator;

integrator.addDiffEquation(&DXDT,"X",  0);
integrator.addDiffEquation(&DPDT,"P",1.0);
```

---

[5] Parameter is described in a previous chapter. The memory allocated for this operation is part of the cache and is released when the cache is destroyed.

[6] The memory is released when the cache is destroyed.

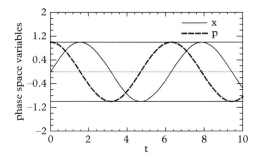

Figure 11.3 *Numerical solution to the harmonic oscillator problem, integrated up to t = 10. This problem is discussed further in Section 11.7.*

```
GENFUNCTION x=*integrator.getFunction(X);
GENFUNCTION p=*integrator.getFunction(P);
```

and then one can use the functions x and p in the usual way, for example plotting them (vs. time) as in Figure 11.3. We return to this problem in Section 11.7 where it will be used to illustrate problems with simpler algorithms.

11.4 Numerical methods for integrating ordinary differential equations

We now turn to numerical methods for solving ordinary differential equations of the form:

$$\frac{d\vec{y}(x)}{dx} = \vec{f}(x, \vec{y}(x)) \tag{11.33}$$

subject to the initial-value boundary conditions:

$$\vec{y}(x_0) = \vec{y}_0 \tag{11.34}$$

where $\vec{y}_0$ is a vector of known starting values. Solutions to this set of equations exist and are unique under a set of general conditions. In the theory of ordinary differential equations the most important characteristic of the functions $\vec{f}(x, \vec{y}(x))$ is a notion of continuity called **Lipshitz continuity**, which is stronger than the usual type of continuity and implies that the magnitude of the derivative remains bounded. A function of one variable $g(x)$ is Lipshitz continuous if there exists a constant $L > 0$ (called the Lipshitz constant) such that $||g(x) - g(y)|| < L||x - y||$ for any real numbers x and y. Most of the theorems in the numerical analysis of ordinary differential equations are predicated on the requirement that $\vec{f}(x, \vec{y}(x))$ in Eq. 11.33 satisfy a Lipshitz condition on their second variable:

$$||\vec{f}(x, \vec{y}(x)) - \vec{f}(x, \vec{z}(x))|| \leq L||\vec{y}(x) - \vec{z}(x)|| \qquad (11.35)$$

for all values of x.

Numerical integration of Eq. 11.32 or 11.33 boils down to this: taking a presumed starting point for all of the variables and using the equations to determine how they evolve at each subsequent point in time. Temporal evolution is accomplished on a mesh with (possible variable) spacing h.

A good integration algorithm has two components: a method for evolving the equation at each step h, and a method for choosing (and even changing when required) the step size h. The most important feature of the stepping algorithm is its *order*. Let the numerical solution at the ith mesh point be designated as y_i. An algorithm is said to be of order p if the difference between the estimated solution $\vec{y}_i$ at the ith step and the true solution $\vec{y}(x_0 + ih)$ satifies $||\vec{y}_i - \vec{y}(x_0 + ih)|| = O(h^{p+1})$.

This is referred to as the *local error*. Usually, one is more concerned with the *global error* accumulated over all of the steps within a fixed interval of integration because error can accumulate during temporal evolution.

In the following sections we introduce several techniques for integrating ordinary differential equations. Like the integration formulae of Chapter 5, these methods are expressed in simple formulae with different accuracy and stability properties. The methods have been extensively investigated and are well understood. The analysis of all but the simplest methods is however quite complicated, so we will summarize the important results and refer the interested reader to the references (Butcher, 2003; Hairer, 2008; Hairer, 2002). All the methods we discuss are convergent: the global error approaches zero as $h \to 0$, but the order of the convergence varies greatly from method to method. The main theoretical requirement is that $\vec{f}(x, \vec{y})$ is continuous in the first variable and Lipshitz continuous in the second.

11.4.1 The Euler method

The simplest method, called Euler's method, dates from 1768 and steps according to the formula:

$$\vec{y}_{n+1} = \vec{y}_n + h\vec{f}(x, \vec{y}_n(x)). \qquad (11.36)$$

A great deal is known about the numerical properties of this algorithm when it is applied to a set of first-order differential equations (for which $\vec{f}(x, \vec{y}(x))$ satisfies the conditions of the previous paragraph). It is known, for example, that the method is first-order. This can be demonstrated by considering the Taylor series expansion:

$$\begin{aligned}
\vec{y}(x_0 + h) &= \vec{y}(x_0) + h\vec{y}'(x_0) + O(h^2) \\
&= \vec{y}(x_0) + h\vec{f}(x_0, \vec{y}(x_0)) + O(h^2) \\
&= \vec{y}_1 + O(h^2)
\end{aligned} \qquad (11.37)$$

implying:

$$\|\vec{y}_1 - \vec{y}(x_0 + h)\| = O(h^2) \tag{11.38}$$

A more detailed treatment (Butcher, 2003) shows that the error in any component of $\vec{y}$ can be written[7] as the corresponding component of:

$$\frac{1}{2}\vec{y}''(x_0) \cdot h^2 + O(h^3). \tag{11.39}$$

These statements are concerned with the local error. When one wishes to integrate the differential equation over some fixed interval in x, the global error decreases as h rather than h^2. The reason for this is clear: while the error committed in each step decreases as h^2, the number of steps grows as h. The first-order convergence of Eq. 11.36 as $h \to 0$ was proven by Cauchy in 1824 and constitutes the mathematical foundation on which these computational algorithms are based.

When Euler's method is applied to a function of x that does not depend on y it reduces to a classical integration formula. Let there be only one dependent variable y and assume $f(x, y) = f(x)$ so that we are solving:

$$\frac{dy}{dx} = f(x)$$

starting from $x = a$ and integrating over n steps of step size $h = (b - a)/n$, we obtain:

$$y(a) = 0$$
$$y(a + h) = hf(a)$$
$$y(a + 2h) = hf(a) + hf(a + h)$$
$$\cdot$$
$$\cdot$$
$$y(a + nh) = h(f(a) + f(a + h) + \ldots f(a + (n-1)h)$$

The last expression is a quadrature formula (the rectangle rule) for:

$$\int_a^{a+nh} \frac{dy}{dx} dx$$

that one does not usually discuss because of its poor relative error, which is $O(h)$ (compare to the trapezoid rule, $O(h^2)$ or Simpson's rule of order $O(h^4)$). The abscissas of this quadrature formula are the starting points of each interval. The reason for the poor

---

[7] The proof follows from considering the truncated Taylor series with remainder (Lagrange form), or by the method of Butcher (2004).

behavior is the one-sided nature of the formula: only the derivative at the beginning of the interval is used. The following subsections develop the topic of Runge-Kutta methods; which generalize and improve the Euler method.

11.4.2 The midpoint method

An immediate improvement is to evaluate the derivative at the midpoint of the interval $[0, h]$; in that case we use the slope at x_0 to evaluate an approximation to $y(x_0 + h/2)$, then at this mid-point we re-evaluate the slope and use it to take a step over the full interval h; thus there are two evaluations of the slope per step. The procedure can be summarized in the formula:

$$\vec{k}_1 = \vec{f}(x_0, \vec{y}_0)$$
$$\vec{k}_2 = \vec{f}\left(x_0 + \frac{h}{2}, \vec{y}_0 + \frac{h}{2}\vec{k}_1\right)$$
$$\vec{y}_1 = \vec{y}_0 + h\vec{k}_2. \qquad (11.40)$$

The midpoint method has order 2. In the case of a single variable and $f(x,y) = f(x)$, the midpoint method reduces to the midpoint quadrature rule, an example of an open Newton-Cotes quadrature rule.

11.4.3 The trapezoid method

Another second order numerical integration method can be obtained with a slight rearrangement of the formula in Section 11.4.2. This method is summarized in the following way:

$$\vec{k}_1 = \vec{f}(x_0, \vec{y}_0) \qquad (11.41)$$
$$\vec{k}_2 = \vec{f}(x_0 + h, \vec{y}_0 + h\vec{k}_1)$$
$$\vec{y}_1 = \vec{y}_0 + \frac{h}{2}(\vec{k}_1 + \vec{k}_2) \qquad (11.42)$$

The formula is based on the trapezoid rule in the case of a single variable and $f(x,y) = f(x)$.

11.4.4 The 4th order Runge-Kutta method

A fourth order (error going like h^5) formula is nearly as simple but at the price of needing four evaluations of the derivative per step:

$$\vec{k}_1 = h\vec{f}(x_0, \vec{y}_0)$$
$$\vec{k}_2 = h\vec{f}\left(x_0 + \frac{h}{2}, \vec{y}_0 + h\frac{\vec{k}_1}{2}\right)$$

$$\vec{k}_3 = h\vec{f}\left(x_0 + \frac{h}{2}, \vec{y}_0 + h\frac{\vec{k}_2}{2}\right)$$

$$\vec{k}_4 = h\vec{f}(x_0 + h, \vec{y}_0 + h\vec{k}_3)$$

$$\vec{y}_1 = \vec{y}_0 + \frac{\vec{k}_1}{6} + \frac{\vec{k}_2}{3} + \frac{\vec{k}_3}{3} + \frac{\vec{k}_4}{6}. \tag{11.43}$$

Each piece of the final calculation is an estimate of the slope of $\vec{y}(x)$, and each piece has its own leading error. The final result can be viewed as an admixture of all of these estimates, arranged in such a way that the leading errors cancel. The coefficients required for the cancellation were found by kutta (1901).

11.4.5 Properties of nth order Runge-Kutta methods

A generic Runge-Kutta method is characterized by a multistep formula involving coefficients c_i, a_{ij}, and b_i such that:

$$\vec{k}_1 = \vec{f}(x_0 + c_1 h, \vec{y}_0) \tag{11.44}$$

$$\vec{k}_2 = \vec{f}(x_0 + c_2 h, \vec{y}_0 + h a_{21} \vec{k}_1) \tag{11.45}$$

$$\vec{k}_3 = \vec{f}(x_0 + c_3 h, \vec{y}_0 + h a_{32} \vec{k}_2 + h a_{31} \vec{k}_1) \tag{11.46}$$

$$\ldots$$

$$\vec{k}_s = \vec{f}(x_0 + c_s h, \vec{y}_0 + h a_{ss-1} \vec{k}_{s-1}$$
$$+ h a_{ss-1} \vec{k}_{s-2} \ldots + h a_{s1} \vec{k}_1) \tag{11.47}$$

$$\vec{y}_1 = \vec{y}_0 + h \sum_i b_i \vec{k}_i \tag{11.48}$$

Nowadays, the construction of nth order Runge-Kutta methods is largely a mechanical procedure. While the procedure itself is not easy to understand[8], its consequences are simple: a host of higher-order variants is known and fully characterized. This being the case, the information you'll need in order to implement a stepping routine is: what are the coefficients? What order do we obtain with them?

The first question can be succinctly addressed by summarizing the coefficients in a figure known as a **Butcher tableau** (Butcher, 2003) with the following form:

$\vec{c}$	A
	b^T

---

[8] The procedure for "discovering" higher order Runge-Kutta techniques starts with the creation of a graphical representations called a **rooted tree**, which is a simply-connected, acyclic graph with a fixed number of nodes corresponding to the desired order of the method. From the rooted trees one obtains a set of equations called order conditions on the coefficients a_{ij}, b_i and c_i. Several sets of coefficients may satisfy the order conditions (Butcher, 2003).

where $\vec{b}$ and $\vec{c}$ are column-vectors of coefficients and $\mathbf{A}$ is a matrix of coefficients (for the algorithms considered here, it is lower-diagonal).

We will address the second question by grouping the tableau according to the order of the method so obtained. Here are a few of them:

First order methods

$$\begin{array}{c|c} 0 & \\ \hline & 1 \end{array} \quad \text{Euler method}$$

Second order methods

$$\begin{array}{c|cc} 0 & & \\ 1 & 1 & \\ \hline & \frac{1}{2} & \frac{1}{2} \end{array} \quad \text{trapezoidal rule method}$$

$$\begin{array}{c|cc} 0 & & \\ \frac{1}{2} & \frac{1}{2} & \\ \hline & 0 & 1 \end{array} \quad \text{midpoint method}$$

Third order methods

Two possible third-order methods exist:

$$\begin{array}{c|ccc} 0 & & & \\ \frac{2}{3} & \frac{2}{3} & & \\ \frac{2}{3} & \frac{1}{3} & \frac{1}{3} & \\ \hline & \frac{1}{4} & 0 & \frac{3}{4} \end{array} \quad \text{RK31}$$

$$\begin{array}{c|ccc} 0 & & & \\ \frac{1}{2} & \frac{1}{2} & & \\ 1 & -1 & 2 & \\ \hline & \frac{1}{6} & \frac{2}{3} & \frac{1}{6} \end{array} \quad \text{RK32}$$

Fourth order methods

There are many coefficient sets (or tableaux) which satisfy the order conditions for a fourth-order Runge-Kutta Integrator. The first example is the Kutta's formula, Eq. 11.43, summarized as:

$$\begin{array}{c|cccc} 0 & & & & \\ \frac{1}{2} & \frac{1}{2} & & & \\ \frac{1}{2} & 0 & \frac{1}{2} & & \\ 1 & 0 & 0 & 1 & \\ \hline & \frac{1}{6} & \frac{1}{3} & \frac{1}{3} & \frac{1}{6} \end{array} \quad \text{RK4 or classical Runge-Kutta method}$$

364 *Numerical methods for integrating ordinary differential equations*

The second example is a tableau that gives Simpson's 3/8 rule in the special case of $f(x,y(x)) = f(x)$:

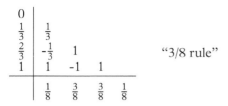

"3/8 rule"

A Class for Butcher Tableaux

When building an integrator, why not arrange for maximum flexibility in the choice of Runge-Kutta algorithm? Since the stepping algorithm is completely specified by the Butcher tableau (and the order that it achieves), an easy way to do this is to invent a class for such tableau; the class can then be used as input to the integrator in order to steer the stepping component. A generic Butcher tableau has been encapsulated in the class `ButcherTableau` (namespace `Genfun`). This class acts as an empty data structure that can be filled with the elements ($\mathbf{A}, \vec{b},$ and $\vec{c}$) of your favorite tableau. It has the following public interface:

```
class ButcherTableau {

public:

  // Constructor:
  inline ButcherTableau(const std::string &name,
                        unsigned int order);

  // Returns the name:
  inline const std::string & name() const;

  // Returns the order:
  inline const unsigned int order() const;

  // Returns the number of steps:
  inline const unsigned int nSteps() const;

  // Write access to elements:
  inline double & A(unsigned int i, unsigned int j);
  inline double & b(unsigned int i);
  inline double & c(unsigned int i);

  // Read access to elements (inline for speed)
  inline const double & A(unsigned int i,
                          unsigned int j) const;
```

```
    inline const double & b(unsigned int i) const;
    inline const double & c(unsigned int i) const;
}
```

Generally it will not be necessary for you to use this class directly since the header file defines the following subclasses which preloaded coefficients:

- EulerTableau
- MidpointTableau
- TrapezoidTableau
- RK31Tableau
- RK32Tableau
- ClassicalRungeKuttaTableau
- ThreeEighthsRuleTableau

Fixed-step size Runge Kutta integration

In Section 11.3.1 we described a class for carrying out integration of functions. By allowing you to specify an alternate "stepper" the class allows you to work with any of the above integration schemes. By default it chooses a sophisticated stepper, of the kind we will describe in Section 11.6, but if you need to you can change this in the following way. First, you need to instantiate a Butcher tableau to describe the method you will apply. If you choose one of the preloaded tableau you can do this with:

```
#include "QatGenericFunctions/ButcherTableau.h"
ClassicalRungeKuttaTableau t;
```

Then you instantiate a `SimpleRKStepper` with the following code:

```
#include "QatGenericFunctions/SimpleRKStepper.h"
SimpleRKStepper stepper(t, 0.001);
```

where the second argument is the fixed step size. Finally, you can use your stepper within the integrator by changing the way the integrator is instantiated:

```
RKIntegrator integrator(&stepper);
```

and proceed as normal, as described in Section 11.3.1. The interface to `SimpleRKStepper` is here:

```
  class SimpleRKStepper:public RKIntegrator::RKStepper{

  public:
```

```cpp
    // Constructor:
    SimpleRKStepper(const ButcherTableau & tableau,
                    double stepsize);

    // Destructor:
    virtual ~SimpleRKStepper();

    // Take a step:
    virtual void step (const RKIntegrator::RKData      *
                          data,         // functions
                       const RKIntegrator::RKData::Data &
                          sdata,        // start point
                       RKIntegrator::RKData::Data       &
                          ddata,        // end point
                       double
                          timeLimit     // time limit
                       ) const ;
    // Clone:
    virtual SimpleRKStepper *clone() const;

private:

  ButcherTableau tableau;
  double stepsize;
};
```

The main action routine is called `step`, shown below, which operates on two classes;
`RKIntegrator::RKData`,
which among other things holds the functions to be integrated and a cache of
`RKIntegrator::RKData::Data`
(the second class) which contains information about the solution at each mesh point:

```cpp
class RKIntegrator {
  class RKData {
    struct Data{

      std::vector<double>         variable;
        // solution
      mutable std::vector<double> firstDerivative;
        // it's first derivative
      double time;                // time
      ...
    }
  }
};
```

The actual work is done within the method `step(...)`, which contains the key numerical steps, carried out with the aid of a `ButcherTableau`, which is member data of the `SimpleRKStepper` class. The method has two modes of operation: it can integrate on a regular mesh point, or it can take a time-limited step to interpolate between mesh points; in either case it implements Eqs. 11.44–11.48

```
void SimpleRKStepper::step(const RKIntegrator::RKData
                                 * data,
                           const RKIntegrator::RKData::Data
                                 & s,
                           RKIntegrator::RKData::Data
                                 & d,
                           double
                                 timeLimit ) const
                           {
  const double h = timeLimit==0 ? stepsize : timeLimit -
     s.time;
  if (h<=0) throw std::runtime_error ("SimpleRKStepper:
     negative stepsize");
  const unsigned int nvar = s.variable.size();
  // Compute all of the k's..:
  //
  std::vector<std::vector<double> >k(tableau.nSteps());
  for (unsigned int i=0;i<tableau.nSteps();i++) {
    k[i].resize(nvar,0);
    Argument arg(nvar);
    for (unsigned int v=0;v<nvar;v++) arg[v]=s.variable[v];
    for (unsigned int j=0;j<i;j++) {
      for (unsigned int v=0;v<nvar;v++) {
        arg[v] += h*tableau.A(i,j)*k[j][v];
      }
    }
    for (unsigned int v=0;v<nvar;v++) {
      k[i][v]=(*data->_diffEqn[v])(arg);
    }
  }
  //
  // Final result.
  //
  for (unsigned int v=0;v<nvar;v++) d.firstDerivative[v] =
     0;
  for (unsigned int i=0;i<tableau.nSteps();i++) {
    for (unsigned int v=0;v<nvar;v++) {
      d.firstDerivative[v] += tableau.b(i)*k[i][v];
    }
```

```
    }
    for (unsigned int v=0;v<nvar;v++)    {
      d.variable[v] =s.variable[v]+h*d.firstDerivative[v];
    }
    d.time = timeLimit==0 ? s.time + h : timeLimit;

}
```

Note that simple fixed-step size Runge-Kutta integrators solve one problem for us but they create another: how to set the step size? One answer is to repeat the solution several times and to decrease the step size; comparing the value of the function(s) at one or more points and observing the order of the convergence. Then you can stop decreasing the step size when the changes are below the desired tolerance. Notice however that in the view we are taking of our solutions—they are just a set of functions parametrized by the differential equations—a large amount of variation in the step size can be expected to occur as we vary the parameters of the function, and even the argument of the function. If our functions are really going to do their job over the full range of parameters and arguments, we need an automatic way of setting the step size. This can be achieved by estimating the error at each step of the integration and adjusting the step size accordingly. We will return to this in Section 11.6. The default stepping behavior of our `RKIntegrator` is based on this type of method. For now, however, we are going to return to physics for a while.

11.5 Automated solution of classical problems

The Hamiltonian approach to classical dynamics requires taking a series of partial derivatives to set up the equations of motion. Is it possible to automate this procedure? The answer is "yes"!—it is possible to take derivatives symbolically in `QatGenericFunctions`. This capability does not depend on numerical tricks but rather on object-oriented programming tricks, and will produce for us something quite powerful: a technique that allows one to solve a broad class of problems in classical mechanics on the computer, which requires only the Hamiltonian as input. We will restrict ourselves here to Hamiltonians that do not depend explicitly on time, since these always generate autonomous first-order differential equations. In addition, they have the property that the Hamiltonian function is conserved, i.e.:

$$\frac{dH(q_i, p_i)}{dt} = 0. \tag{11.49}$$

11.5.1 Taking partial derivatives with GENFUNCTIONs

The functors in `QatGenericFunctions` can be functions of more than one variable in which case the variables are agglomerated with a class called `Genfun::Argument`.

Specifically, the function call operator() takes a Genfun::Argument as input. For example a function *G* can be made dependent on two variables, *x*(0) and *x*(1), as follows:

```
Genfun::Argument x(2);
Genfun::BivariateGaussian G;
std::cout << G.dimensionality() << std::endl; //
    output: ''2''

x[0]=0.5;
x[1]=1.2;

double y = G(x);
```

Each functor has a method called dimensionality() that returns the number of arguments required by the function. The bivariate Gaussian distribution is intrinsically a function of two variables $\vec{x} = (x_0, x_1)$. Other functions can be made by associating the *n* components of the variable list with a name and then combining the variables in algebraic expressions like:

```
Variable X(0,2),P(1,2);
GENFUNCTION DXDT=P/M;
GENFUNCTION DPDT=-K*X;
```

which allows us to make functions of two variables (such as DXDT and DPDT) where we symbolically refer to the first variable as X and the other as P. Now we will use this to define the Hamiltonian, which is itself a function of the variables X and P:

```
Variable X(0,2),P(1,2);
GENFUNCTION Hamiltonian=P*P/2.0/M + 1/2.0*X*X;
```

It is finally time here to reveal the secret of how QatGenericFunctions work. When doing any type of arithmetic on a QatGenericFunction, the result of the operation is an object which keeps track of the operation to be performed as well as the operand(s). The operation can be addition, subtraction, multiplication, division, negation, or composition, and operands can be functions, parameters, or constants. Repeated operations build up what is called a **binary expression tree**. The actual evaluation of the function is deferred until the function is invoked, at which point the expression tree is traversed. A limited amount of symbolic manipulation is available within the QatGenericFunctions package. Among this functionality is the ability to calculate a derivative (or a partial derivative in case of a function of two dimensions). Here for example we compute and evaluate the derivative of a sine function:

```
Genfun::Sin Sin;
Genfun::GENFUNCTION f=Sin.derivative();

double x = 1.0;
double y = f(x);
```

which assigns to the variable y the value of cos(1.0). Each primitive function in the library can compute its derivative (a numerical derivative is taken in case the class does not provide any overrider). Each operation node in the binary expression tree knows how to compute its derivatives: the sum rule for the sum of functions, the product rule for the product of functions, the chain rule for the composition of functions, and so forth. The operation nodes call upon the functions that comprise their operand list during the execution of this task. The end result is one binary expression tree representing a function can produce another binary expression tree representing its derivative *independently of what the function actually is*.

The same idea applies to functions of more than one variable with a slight modification, namely the partial derivative with respect to the ith element of the argument list can be taken. In this example the partial derivative with respect to the 0th variable (X) is taken:

```
Genfun::GENFUNCTION f = Hamiltonian.partial(0);
```

The partial derivative that is returned is also a function of two variables. A more readable version of the same statement can also be written:

```
Genfun::GENFUNCTION f = Hamiltonian.partial(X.index());
```

or even better:

```
Genfun::GENFUNCTION f = Hamiltonian.partial(X);
```

We will let you work with these functions for a little while longer and in the chapter *Polymorphism* we will show you how they are constructed—and how to extend them to other functions that you may wish to implement.

11.5.2 Computing *and* solving the equations of motion

We put it all together in the following sample code for the harmonic oscillator.

```
//
// Harmonic oscillator:
//

double K=1.0;  // Newtons/m
```

```
double M=1.0; // Kg
Variable X(0,2),P(1,2);
GENFUNCTION Hamiltonian = 1/2.0/M*P*P + 1/2.0*K*X*X;
GENFUNCTION DXDT=+Hamiltonian.partial(P);
GENFUNCTION DPDT=-Hamiltonian.partial(X);
//
// Integrate the equations of motion:
//
RKIntegrator integrator;

integrator.addDiffEquation(&DXDT,"X",  0);
integrator.addDiffEquation(&DPDT,"P",1.0);

GENFUNCTION x=*integrator.getFunction(X);
GENFUNCTION p=*integrator.getFunction(P);
```

The integrator works as before, except that here we specify the Hamiltonian rather than the equations of motion deriving from the Hamiltonian. This is, of course, very convenient. One could for example add a small perturbing potential to the Hamiltonian and investigate the resulting anharmonicities.

11.5.3 A classical Hamiltonian solver

The preceding ideas are encapsulated in the class `RungeKuttaSolver` (namespace `Genfun`). The solver is a fairly heavily engineered piece of software, which, since the noble discipline of software engineering is not the focus of this book, we will not describe here.

It uses the ideas of this section to solve problems in classical mechanics. The following example is provided as an illustration in the area `EXAMPLES/CH11/HAMILTONIAN`. Using the solver, it computes the orbits of a particle moving in two dimensions under a force, central or otherwise, though in the illustration we have chosen a central force. The program uses the class `Classical::PhaseSpace` to set up the phase space variables, then allows the user to construct a Hamiltonian out of the phase space variables and calls `Classical::RungeKuttaSolver` to obtain solutions. In this example both the time development and the orbit are then plotted with the usual tools. Here is the code (in `hamiltonian.cpp`)

```
#include "QatGenericFunctions/RungeKuttaClassicalSolver.h"
#include "QatGenericFunctions/PhaseSpace.h"
#include "QatGenericFunctions/Variable.h"
#include "QatGenericFunctions/Power.h"
#include "orbit.h"
#include "plot.h"
using namespace Genfun;
```

```cpp
int main (int argc, char **argv) {

  // r^N potential

  int N=4;

  // Phase Space:

  Classical::PhaseSpace phaseSpace(2);

  const Classical::PhaseSpace::Component
   & q=phaseSpace.coordinates(),
   & p=phaseSpace.momenta();

  // The Hamiltonian:

  GENFUNCTION H = (p[0]*p[0]+p[1]*p[1])/2.0 + Power(N/2)
      (q[0]*q[0]+q[1]*q[1]);

  // Starting value of q & p

  phaseSpace.start(q[0],0.0);
  phaseSpace.start(q[1],1.0);
  phaseSpace.start(p[0],1.0);
  phaseSpace.start(p[1],1.0);

  Classical::RungeKuttaSolver solver(H, phaseSpace);

  //-------------------------------------------------------

  QApplication    app(argc,argv);
  orbit(app,
        solver.equationOf(q[0]),  // We plot the orbit
        solver.equationOf(q[1]));
  plot (app,
        solver.equationOf(q[0]),  // We plot together q0
                                  //            and p0
        solver.equationOf(p[0]));
  plot (app,
        solver.equationOf(q[1]),  // We plot together q1
                                  //            and p1
        solver.equationOf(p[1]));
```

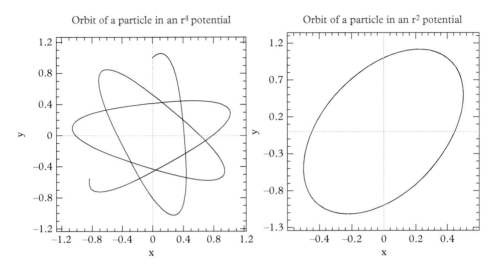

Figure 11.4 *The figure on the left shows the orbit of a particle under the influence of an r^4 potential. This is generated by the program EXAMPLES/CH11/HAMILTONIAN, right out of the box. On the right, the potential is modified to an r^2 central potential, which has a familiar elliptical orbit.*

```
  plot (app,
        solver.energy());        // We plot the energy
}
```

Figure 11.4 shows the orbit for this example (on the left) and also the orbit for a slightly modified Hamiltonian, where the r^4 potential is changed to an r^2 potential (i.e., a two dimensional simple harmonic oscillator). The program is a good starting point for solving other problems in classical mechanics.

11.6 Adaptive step size control

Numerical solutions to ordinary differential equations with fixed step size have two drawbacks; first, they require experimentation to determine an appropriate value of the step size; and second the step size will usually be smaller than required in some regions of integration when it is adjusted to the appropriate value in others. In the Euler method, for example, the error committed at each step is $1/2h^2 y''$, where y'' is evaluated at some unknown point in the interval $[x_i, x_i + h]$; so the error is proportional to the second derivative. A similar analysis can be conducted for all of the methods in the Runge-Kutta family.

The solution to this problem is to add *adaptive step size* (Butcher, 2003; Press, 2007) adjustment to the algorithm. This procedure requires, first, an estimate of the error committed at each step of the integration, and second, knowledge of the order p of the integration. Typically the error will be estimated by integrating the function(s) in two different ways at each step, either by *step doubling* (Section 11.6.1) or by the so-called *embedded Runge-Kutta* methods (Section 11.6.2). Each of these techniques individually provides two estimates of the integral, one which is propagated forward and one which is taken as a control. Then, having obtained two different results $\vec{y}_1$ and $\vec{y}_2$, we take the magnitude of the difference, $\delta = ||\vec{y}_1 - \vec{y}_2||$ as an estimate of the error. Sometimes the worst case deviation is used instead of the magnitude of the difference. In later sections we will examine the step-doubling technique and embedded Runge-Kutta methods; here we show how the error estimate is used to implement adaptive step size adjustment.

The idea is to fix the size of the error committed at each step to a user-specified tolerance T. This tolerance does *not* directly translate to the desired accuracy in global error; its purpose is to maintain the local error at each step of the integration at an approximately constant value. Let δ be the error committed at each step; since the stepping method is order p we can write that:

$$\delta = A h^{p+1}$$

where A is the constant in the leading term of the local error. This expression can be used to adjust the estimated error at each step to the tolerance by growing or shrinking the size of the step. Let h_T be the step size that gives the *desired* error, i.e. the tolerance T:

$$T = A h_T^{p+1}.$$

Dividing the two equations,

$$\frac{T}{\delta} = \left(\frac{h_T}{h}\right)^{p+1}.$$

or

$$h_T = h \left(\frac{T}{\delta}\right)^{\frac{1}{p+1}}. \tag{11.50}$$

This prescription can be used at each step. Certain practical considerations lead to simple modifications to the basic procedure.

In practice a step size that is too small will be adjusted up (for the next step), while a step size that is too large will be adjusted down, *and* the step will have to be repeated. To avoid this type of failure in the following step, any upward adjustment of the step size is moderated by introducing a safety factor S with a typical value of 0.9, so that the ratio $r = h_T/h$ in Eq. 11.50 is modified to:

$$r = S \left(\frac{T}{\delta}\right)^{\frac{1}{p+1}} \tag{11.51}$$

In addition, it is common practice to limit this factor to an interval $[r_{min}, r_{max}]$, such as $[0.5, 2.0]$. When the maximum upward adjustment r_{max} is reached, the observed error δ_{max} is:

$$\delta_{max} = T \left(\frac{S}{r_{max}} \right)^{p+1}. \qquad (11.52)$$

The error is tested against this cutoff value; if it is less than the cutoff, then the ratio $r = R$ is applied in the step adjustment; and otherwise Eq. 11.51 is used. One must be wary to avoid division by zero in this equation; this can be accomplished by adding a small regularizing positive number to the denominator.

11.6.1 Step doubling

Let's compare the value of a function at $x_0 + h$ with the numerical value y_1 from a discretized calculation of order p:

$$\begin{aligned} y(x_0 + h) &= y_1 + O(h^{p+1}) \\ &= y_1 + Ah^{p+1} + O(h^{p+2}). \end{aligned}$$

We can also take the result as two half-steps; we call the result y_2 which is related to $y(x_0 + h)$ in the following way.

$$\begin{aligned} y(x_0 + h) &= y_2 + 2A \cdot (h/2)^{p+1} + O(h^{p+2}) \\ &= y_2 + 2^{-p}Ah^{p+1} + O(h^{p+2}). \end{aligned}$$

Obviously y_2 is a better approximation to the solution than y_1; the error term $2^{-p}Ah^{p+1}$ can be obtained (to order $p - 1$) from the two equations:

$$2^{-p}Ah^{p+1} = \frac{y_1 - y_2}{2^p - 1}. \qquad (11.53)$$

By correcting y_2 for this error,

$$y_2 \to y_2 - \frac{y_1 - y_2}{2^p - 1}, \qquad (11.54)$$

we are effectively extrapolating to a smaller step size, and gaining in precision. However, the most valuable aspect of the step-doubling algorithm is that it generates an estimate of its error at each step, namely:

$$\delta = y_1 - y_2 \qquad (11.55)$$

Adaptive step size control

which can thus be used directly with the adaptive step size control algorithm described in the previous paragraph. At the end of Section 11.4.5, we described the `SimpleRKStepper` class which, together with one or more `ButcherTableau` classes (for example `EulerTableau`, `MidpointTableau`, `TrapezoidTableau`) could replace the default stepper of `RKIntegrator` with fixed-step size explicit Runge-Kutta integration according to one of the stepping algorithms. Here we describe how you can do the same, except with adaptive step size.

As usual you will choose the underlying stepping algorithm by instantiating a `ButcherTableau` of your choice, for example:

```
#include "QatGenericFunctions/ButcherTableau.h"
ThreeEightsRuleTableau t;
```

The `StepDoublingRKStepper` is a type of error-estimating stepper that implements step doubling; you can instantiate it like this:

```
#include "QatGenericFunctions/StepDoublingRKStepper.h"
StepDoublingRKStepper subStepper(t);
```

The `StepDoublingRKStepper` is not used directly in the `RKIntegrator`, instead it is used internally by the `AdaptiveRKStepper`:

```
#include "QatGenericFunctions/AdaptiveRKStepper.h"
AdaptiveRKStepper stepper(&subStepper);
```

and that can then be used in `RKIntegrator`:

```
RKIntegrator integrator(&stepper);
```

This gives a great deal of flexibility in combining the step-doubling technique. In addition, a set of modifiers can be used to change the algorithmic parameters. For example:

```
stepper.initialStepsize()=0.02;
stepper.tolerance()=2.0E-6;
```

The constructor in `AdaptiveRKStepper` does not need any arguments since it uses the embedded RK algorithm (described in the next section) by default. An excerpt from the header file of `AdaptiveRKStepper` is shown below (not shown are the signatures for private functions and those functions which are not meant to be called by the user).

```
class AdaptiveRKStepper:public RKIntegrator::RKStepper {

public:
```

```cpp
class EEStepper;

// Constructor. A default stepper is created if none
// is specified--it is an Embedded Runge Kutta stepper
// using a Cash-Karp extended Butcher Tableau, which
// is fourth-order.
AdaptiveRKStepper(const EEStepper *eeStepper=NULL);

// Copy constructor:
AdaptiveRKStepper(const AdaptiveRKStepper & right);

// Destructor:
virtual ~AdaptiveRKStepper();

// Clone
virtual AdaptiveRKStepper *clone() const;

// Accessors and modifiers to algorithmic parameters.
// Roughly speaking these are ordered according to
// importance: the user will often wish to modify the
// tolerance and the starting step size, but rarely
   should
// need to touch any of the others.

// The tolerance:
double & tolerance();
const double & tolerance() const;

// The starting step size:
double & startingStepsize();
const double & startingStepsize() const;

// The safety factor.  Step size increases are moderated
// by this factor:
double & safetyFactor();
const double & safetyFactor() const;

// The minimum amount by which a step size is decreased:
double & rmin();
const double & rmin() const;

// The maximum amount by which a step size is increased:
double & rmax();
```

378   *Adaptive step size control*

```
    const double & rmax() const;
}
```

The `AdaptiveRKStepper`'s main routine is called `step`, and is very similiar to `SimpleRKStepper::step`, except that it uses an "error estimating" stepper, `EEStepper`, at each meshpoint, and adjusts the step up or down accordingly.

```
void AdaptiveRKStepper::step(const RKIntegrator::RKData
                                  * data,
                             const RKIntegrator::RKData::
                                  Data & s,
                             RKIntegrator::RKData::Data
                                  & d,
                             double
                                  timeLimit) const
{
  //
  // Adaptive step size control
  //
  if (s.time==0.0) {
    stepsize=sStepsize;
  }
  const unsigned int p = eeStepper->order();
     // Order of the stepper
  const double deltaMax = T*pow(S/Rmax, p+1);
     // Maximum error 4 adjustment.
  const double TINY     = 1.0E-30;
     // Denominator regularization
  double hnext;
  //
  // Time limited step ?
  //
  d.time= timeLimit==0? s.time+stepsize : timeLimit;

  //------------------------------------//
  // Take one step, from s to d:        //
  //------------------------------------//
  double h = d.time-s.time;
  while (1) {
    std::vector<double> errors;
    eeStepper->step(data, s, d, errors);
    if (timeLimit!=0.0) return;

    // Take absolute value:
```

```
      for (size_t e=0;e<errors.size();e++) errors[e] =
        fabs(errors[e]);

      // Select the largest:
      double delta = (*std::max_element(errors.begin(),
        errors.end()));
      if (delta > 1) {
        //
        // Bail out and try a smaller step.
        //
        h = std::max(S*h*pow(T/(delta + TINY), 1.0/(p+1)),
          Rmin*h);
        if  (!(((float) s.time+h - (float) s.time) > 0) ) {
          std::cerr << "Warning, RK Integrator step underflow
              " << std::endl;
        }
        d.time = s.time+h;
        hnext=h;
        continue;
      }
      else {
        if (delta > deltaMax) {
          hnext = S*h*pow(T/(delta + TINY),1.0/(p+1));
        }
        else {
          hnext = Rmax*h;
        }
      }
      break;
    }
    stepsize=hnext;
    return;
}
```

11.6.2 Embedded Runge Kutta methods

Is it possible for a Runge-Kutta method to build up a set of k's (slope estimates at various points in the interval, Eq. 11.44–11.47) and then combine them as in Eq 11.48, but forming *two* estimates of the $y(x_0 + h)$, each having its own order? The answer is yes. Fehlberg (1968) found two tableaux with the same values of $\vec{A}$ and $\vec{c}$ but different values of $\vec{b}$, i.e. those constants used for the final computation. One of these gives fourth-order accuracy and the other gives fifth-order accuracy. We extend, our tableaux accordingly, so as to express succinctly these so-called embedded Runge-Kutta algorithms:

$$\begin{array}{c|c} \vec{c} & \mathbf{A} \\ \hline & b^T \\ & \hat{b}^T \end{array}$$

where the vector $\vec{b}$ is the usual output vector, and $\hat{b}$ is the vector used in an alternate computation used only for the error estimate. Typically the order of these algorithms differs by one unit, and, while there is some choice about whether the result of the higher- or lower-order algorithm is propagated forward to the next step, the higher-order algorithm is usually chosen. From Eq. 11.48 we see that the error is:

$$\vec{\delta} = \sum_i (b_i - \hat{b}_i)\vec{k}_i. \tag{11.56}$$

We list here a few of the extended Butcher tableaux for common embedded algorithms:

The "Heun-Euler" method:

$$\begin{array}{c|cc} 0 & & \\ 1 & 1 & \\ \hline & 1/2 & 1/2 \quad \text{(order 2)} \\ & 1 & 0 \quad \text{(order 1)} \end{array}$$

The "Bogacki-Shampine method":

$$\begin{array}{c|cccc} 0 & & & & \\ 1/2 & 1/2 & & & \\ 3/4 & 0 & 3/4 & & \\ 1 & 2/9 & 1/3 & 4/9 & \\ \hline & 2/9 & 1/3 & 4/9 & 0 \quad \text{(order 3)} \\ & 7/24 & 1/4 & 1/3 & 1/8 \quad \text{(order 2)} \end{array}$$

One of the formulae appearing in Fehlberg's original work (Fehlberg 1968), a fifth-order method with an embedded fourth order method which Fehlberg called RK4(5) is given below. Fehlberg also described embedded methods with orders as high as 9 (Fehlberg 1968); the Butcher tableaux used in these works are in a format similar to those used here. Unlike the other methods summarized in this section, Fehlberg's original formulation propagated the lower-order method rather than the higher-order method, so in this instance our tableau adheres to the original formulation:

$$\begin{array}{c|cccccc} 0 & & & & & & \\ 1/4 & 1/4 & & & & & \\ 3/8 & 3/32 & 9/32 & & & & \\ 12/13 & 1932/2197 & -7200/2197 & 7296/2197 & & & \\ 1 & 439/216 & -8 & 3680/513 & -845/4104 & & \\ 1/2 & -8/27 & 2 & -3544/2565 & 1859/4104 & -11/40 & \\ \hline & 25/216 & 0 & 1408/2565 & 2197/4104 & -1/5 & 0 \quad \text{(order 4)} \\ & 16/135 & 0 & 6656/12825 & 28561/56430 & -9/50 & 2/55 \quad \text{(order 5)} \end{array}$$

The "Cash-Karp" method (Cash 1990) is favored by the authors of the very influential *Numerical Recipes* series (Press, 2007):

0						
1/5	1/5					
3/10	3/40	9/40				
3/5	3/10	−9/10	6/5			
1	−11/54	5/2	−70/27	35/27		
7/8	1631/55296	175/512	575/13824	44275/110592	253/4096	
	37/378	0	250/621	125/594	0	512/1771 (order 5)
	2825/27648	0	18575/48384	13525/55296	277/14336	1/4 (order 4)

Many more algorithms and their tableau can be found in the literature; see for example (Butcher, 2003). A class `ExtendedButcherTableau` with an interface similar to that of `ButcherTableau` has been included in our implementation:

```
class ExtendedButcherTableau {

public:

  // Constructor:
  inline ExtendedButcherTableau(const std::string &name,
                                unsigned int order,
                                unsigned int orderHat);

  // Returns the name:
  inline const std::string & name() const;

  // Returns the order of the main formula
  inline const unsigned int order() const;

  // Returns the order of the controlling formula
  inline const unsigned int orderHat() const;

  // Returns the number of steps:
  inline const unsigned int nSteps() const;

  // Write access to elements:
  inline double & A(unsigned int i, unsigned int j);
  inline double & b(unsigned int i);
  inline double & bHat(unsigned int i);
  inline double & c(unsigned int i);

  // Read access to elements (inline for speed)
  inline const double & A(unsigned int i, unsigned int j)
     const;
  inline const double & b(unsigned int i) const;
```

382 *Adaptive step size control*

```
      inline const double & bHat(unsigned int i) const;
      inline const double & c(unsigned int i) const;
      ...
};
```

The following classes which preload the appropriate values:

- HeunEulerXtTableau
- BogackiShampineXtTableau
- FehlbergRK45F2XtTableau
- CashKarpXtTableau

Now, to use these methods with an `AdaptiveRKStepper` you proceed as follows. First, instantiate an extended Butcher tableau:

```
#include "QatGenericFunctions/ExtendedButcherTableau.h"
HeunEulerXtTableau t;
```

```
#include "QatGenericFunctions/EmbeddedRKStepper.h"
EmbeddedRKStepper subStepper(t);
```

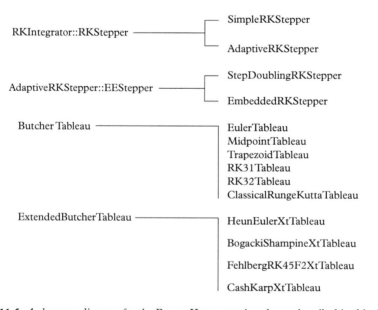

Figure 11.5 *A class tree diagram for the Runge-Kutta stepping classes described in this chapter.*

If you don't specify any tableau in the constructor, Cash-Karp is taken by default. Use this together with the `AdaptiveRKStepper`, which automatically adjusts the step size:

```
#include "QatGenericFunctions/AdaptiveRKStepper.h"
AdaptiveRKStepper stepper(subStepper);
```

which can then be used in `RKIntegrator`:

```
RKIntegrator integrator(&stepper);
```

One can now use the integrator as previously outlined.

We end this discussion on Runge-Kutta integrators with a class tree diagram showing the relationship between various classes discussed in this section (Figure 11.5). This tree can be extended if need be; details will be given in Chapter 12.

11.7 Symplectic integration schemes

When applied to classical problems, Runge-Kutta integration techniques often exhibit an inconvenient property known as *energy drift*. This is illustrated in Figure 11.3, which shows the early solution to the simple harmonic oscillator, and Figure 11.6, which shows the same solution for very long time. It is evident that the amplitude of the oscillations has grown (the horizontal lines indicate the amplitude of the oscillations near $t = 0$), and therefore the energy has increased. Since we know we are dealing with a conservative system, the effect is obviously an artifact of our numerical integration technique. This phenomenon is present in both lower-order and higher-order Runge-Kutta methods.

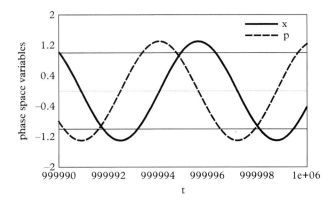

Figure 11.6 *The long term behavior of the harmonic oscillator integrated numerically with Runge-Kutta integration, at around $t = 10^6$ (approximately 160k cycles). The plot reveals an obvious drift in the energy of the system.*

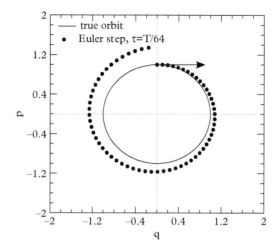

Figure 11.7 *A phase space diagram of motion in the harmonic oscillator problem. The arrow shows the direction of a typical step in a first-order algorithm. The step always brings us to a point outside the true trajectory resulting in an outward spiral of the phase space trajectory and a secular increase in energy.*

The source of the problem is easily visualized in the phase-space diagram of Figure 11.7. Whenever we take steps based upon the derivatives of q and p, we are moving along the tangent line of the path through phase space. For the harmonic oscillator problem our approximation will invariably take us to a point that lies outside the actual path of the system—and one with a higher energy. Let us calculate the size of the effect for the Euler method with a step size of τ:

$$q' = q + \tau \frac{\partial H}{\partial p} = q + \frac{p}{m}\tau$$
$$p' = p - \tau \frac{\partial H}{\partial q} = p - kq\tau$$

and after one step the energy E' is:

$$E' = \frac{1}{2m}p'^2 + \frac{1}{2}kq'^2$$
$$= \left(1 + \tau^2 \omega^2\right) E$$

where E is the energy before the step. We see that the energy increases at each step quadratically with the step size τ.

Instead of following the Euler algorithm, we try another way of advancing through phase space that does not follow the tangent exactly. At each step we apply two small linear transformations $\mathbf{D} = \mathbf{D}_2 \cdot \mathbf{D}_1$ where:

$$\mathbf{D}_2 = \begin{pmatrix} 1 & 0 \\ -\frac{\tau}{q}\frac{\partial H}{\partial q} & 1 \end{pmatrix} \quad \mathbf{D}_1 = \begin{pmatrix} 1 & \frac{\tau}{p}\frac{\partial H}{\partial p} \\ 0 & 1 \end{pmatrix}. \quad (11.57)$$

This form approximates (to order τ^2) the first-order Euler step whenever the Hamiltonian separates, i.e., whenever one can write $H(p, q) = T(p) + V(q)$. For the harmonic oscillator these become:

$$\mathbf{D}_2 = \begin{pmatrix} 1 & 0 \\ -\tau k & 1 \end{pmatrix} \quad \mathbf{D}_1 = \begin{pmatrix} 1 & \frac{\tau}{m} \\ 0 & 1 \end{pmatrix} \quad (11.58)$$

and now after one step we estimate q' and p' to be:

$$q' = q + \frac{p}{m}\tau$$
$$p' = p - kq'\tau$$
$$= p - kq\tau - \frac{kp\tau^2}{m}.$$

The energy is estimated to be:

$$E' = \left(1 + \tau^2\omega^2\right) E - \tau^2\omega^2 \frac{p^2}{m} + O(\tau^3).$$

Is this any better than before? The energy is still not conserved by the integration scheme except on average and to order τ^2 since $\langle \frac{p^2}{2m}\rangle = \frac{E}{2}$ for the harmonic oscillator. Thus $\langle E'\rangle = E + O(\tau^3)$. To get a better picture of what is going on we choose a somewhat large step size ($\tau = T/16$ and $\tau = T/64$, where $T = 2\pi/\omega$) and trace the orbit of a point in phase space under a sequence of the transformations described by Eq. 11.58.

The estimated orbit does not approximate closely the actual orbit because we have deliberately chosen a large step size τ to illustrate the interesting features of the approximation. The true trajectories shown in Figure 11.8 start at the top of the plot and move clockwise around the figure. The transformations in Eq. 11.58 applied repeatedly to the point in the phase space move the system around in a distorted path. The orbit of the point under the transformation (shown as points in Figure 11.8) is sometimes inside the true orbit where we estimate a lower energy, sometimes outside, where we estimate a higher energy. After one full cycle the point appears to comes around nearly to its original position in phase space. By choosing a smaller step size the deviation of this calculated orbit from the true particle's orbit becomes smaller and can be made very

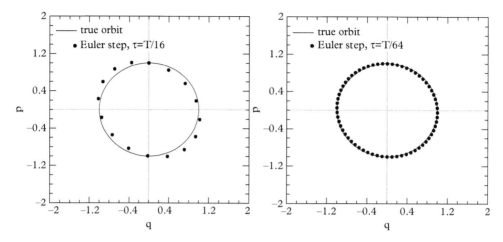

Figure 11.8 *Two plots comparing the phase space trajectory of the simple harmonic oscillator with points along the trajectory computed with the transformation of Eq. 11.58, for two different values of the step size, $T/16$ (left), and $T/64$ (right).*

accurate—while conserving the energy. In the next section we will see where this behavior comes from.

11.7.1 Symplectic transformations

Let us take the N canonical coordinates q_i and the N canonical momenta p_i that comprise a $2N$-dimensional phase space and make a vector $\vec{\eta} = (q_0, q_1, q_2 \ldots q_N; p_0, p_1, p_2 \ldots p_N)$. Define a $2N \times 2N$ block diagonal matrix:

$$\mathbf{J} = \begin{pmatrix} 0 & \mathbf{I} \\ -\mathbf{I} & 0 \end{pmatrix}. \tag{11.59}$$

This matrix defines a group of transformations on the $2N$ phase space variables in the following way. Define a transformation from one set of phase space variables $\vec{\eta}$ to a new set $\vec{\eta}'$:

$$\vec{\eta} \to \vec{\eta}'(\vec{\eta})$$

which is continuous and differentiable and let $\mathbf{M}$ be the Jacobian matrix of the transformation:

$$(\mathbf{M})_{ij} = \frac{\partial \eta'_i}{\partial \eta_j}.$$

The transformation (and the matrix) is said to be **symplectic** if:

$$\mathbf{M}^T \mathbf{J} \mathbf{M} = \mathbf{J}. \tag{11.60}$$

The set of symplectic transformations satisfies all of the conditions of a mathematical group, which is known as the **symplectic group**. In particular it satisfies the condition of closure, which means the the composition of any two symplectic transformations is again a symplectic transformation. It can easily be verified that the matrices $\mathbf{D}_1$ and $\mathbf{D}_2$, each defining half of the modified step $\mathbf{D} = \mathbf{D}_2 \cdot \mathbf{D}_1$ (step size τ, Eq. 11.57 and Eq. 11.58 above) are symplectic, and therefore so is their composition—the Jacobian matrix $\mathbf{D}$ of the combined transformation itself.

Symplectic transformations play a special role in classical mechanics because of their connection to **canonical transformations**, which are transformations of the phase space that preserve the form of Hamilton's equations. All canonical transformations can be represented as symplectic matrices, and all symplectic matrices represent transformations that are canonical.

The initial state of a classical mechanical system can be represented by a point in phase space. The evolution of the system with time is the phase space trajectory, governed by the Hamiltonian of the system. Between the initial time $t_0 = 0$ and a future time $t = \tau$, the time development can be represented as a matrix that is symplectic, and that depends on the Hamiltonian in the following way. Hamilton's equations in symplectic notation read:

$$\frac{d\vec{\eta}}{dt} = \mathbf{J} \frac{\partial H}{\partial \vec{\eta}} \equiv \mathbf{G}(\vec{\eta}) \cdot \vec{\eta} \tag{11.61}$$

(a definition of the operator $\mathbf{G}(\vec{\eta})$) which follows from Hamilton's equations. If the system is started from a point $\vec{\eta}_0$ in the phase space at $t_0 = 0$, then at a later time τ it will have the value:

$$\vec{\eta} = \exp(\mathbf{G}\tau)\vec{\eta}_0.$$

The matrix $(\mathbf{G}\tau)$ is easily seen to be symplectic (Goldstein 2001) and hence $\mathbf{U}_\tau \equiv \exp(\mathbf{G}\tau)$, called the **time development operator**, is also symplectic.

We can now state why symplectic integrators work. At each integration step the phase space variables are transformed by a symplectic map, which corresponds to a Hamiltonian. While this Hamiltonian is not the Hamiltonian of the problem at hand, it differs by only a slight perturbation. The perturbation gets smaller as the step size is decreased, approaching the desired Hamiltonian as the step size goes to zero.

We will illustrate this by examining a first-order symplectic scheme for integrating the harmonic oscillator problem[9]. Our goal will be to determine the effective Hamiltonian

---

[9] This scheme is also called the **symplectic Euler method** or the **Euler-Cromer method**.

that is being solved by the symplectic scheme by identifying the operator **D** with the operator $\mathbf{U}_\tau$ and then inverting Eq. 11.61. In other words we will:

1. Take the logarithm of the matrix which represents a step through a finite interval τ to obtain $\mathbf{G}_0 \tau$.
2. Use $\mathbf{J} \frac{\partial H}{\partial \vec{\eta}} = \mathbf{G}_0 \vec{\eta}$ to identify a set of differential equations for the Hamiltonian $H(\vec{\eta})$.
3. Integrate those equations.

Perturbed Hamiltonian of the harmonic oscillator

The operator **D** we are seeking is obtained from Eq. 11.58:

$$\mathbf{D} = \mathbf{D}_2 \cdot \mathbf{D}_1 = \begin{pmatrix} 1 & 0 \\ -hk & 1 \end{pmatrix} \cdot \begin{pmatrix} 1 & \frac{h}{m} \\ 0 & 1 \end{pmatrix} = \begin{pmatrix} 1 & \frac{h}{m} \\ -hk & (1 - \frac{k}{m} h^2) \end{pmatrix}$$

which we will write as:

$$\mathbf{D} = \begin{pmatrix} 1 & \frac{\tau}{m} \\ -\tau \omega^2 m & (1 - \omega^2 \tau^2) \end{pmatrix}$$

The first step is to find the logarithm of this matrix.

> Show that as $\tau \to 0$
> $$\ln(\mathbf{D}) = \begin{pmatrix} \omega^2 \tau^2 / 2 & \tau/m \\ -m\omega^2 \tau & -\omega^2 \tau^2/2 \end{pmatrix} + O(\tau^3).$$

From $\mathbf{G} = \ln(\mathbf{D})/\tau$ we have:

$$\mathbf{G}\vec{\eta} = \begin{pmatrix} \omega^2 \tau/2 & 1/m \\ -m\omega^2 & -\omega^2 \tau/2 \end{pmatrix} \cdot \begin{pmatrix} q \\ p \end{pmatrix}$$
$$= \begin{pmatrix} \omega^2 \tau q/2 & p/m \\ -mq\omega^2 & -\omega^2 \tau p/2 \end{pmatrix} = \frac{d\vec{\eta}}{dt} = \begin{pmatrix} \frac{dq}{dt} \\ \frac{dp}{dt} \end{pmatrix} = \begin{pmatrix} \frac{\partial H}{\partial p} \\ -\frac{\partial H}{\partial q} \end{pmatrix}.$$

We can now integrate the following equations:

$$\frac{\partial H}{\partial p} = \frac{p}{m} + \frac{1}{2}\omega^2 \tau q$$
$$\frac{\partial H}{\partial q} = mq\omega^2 + \frac{1}{2}\omega^2 \tau p$$

to obtain:

$$H = \frac{p^2}{2m} + \frac{1}{2}m\omega^2 q^2 + \frac{\tau\omega^2 qp}{2}. \qquad (11.62)$$

This is not quite the Hamiltonian of the problem we set out to solve (the harmonic oscillator), but it is close, differing only by a "perturbation" term $\tau\omega^2 qp/2$, which is first-order in the step size τ. We see that our symplectic Euler algorithm has been solving some problem exactly, but not the harmonic oscillator. In Figure 11.9 we show that the approximate solution (points) follows the exact phase space trajectory of this alternate problem. In Figure 11.10 we show the long-term behavior of a symplectic Euler method, which can be compared to the plot in Figure 11.3. The code that produces this plot can be found in EXAMPLES/CH11/SYMPLECTIC.

Of course it is not necessary to determine the effective symplectic Hamiltonian to integrate the equations of motion. In general there is always a Hamiltonian that describes the orbit of our approximate solutions in a $2N$ dimensional phase space that differs from the actual Hamiltonian at order τ. For proof we refer the reader to Yoshida (1992).

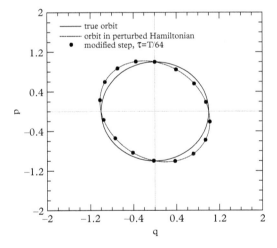

Figure 11.9 *As in Figure 11.8 (left), we compare the approximate solution generated by a sequence of steps generated by our symplectic stepper ($\tau = T/16$) with the true trajectory (solid line) but also with the trajectory of the associated perturbed Hamiltonian (dashed line). The approximate solution to the harmonic oscillator problem which we have generated is the exact solution to another problem having a slightly perturbed Hamiltonian.*

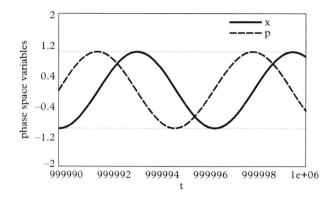

Figure 11.10 *Long-term behavior of a symplectic Euler method, to be compared with the plot of Figure 11.6. The solution shows no secular increase of the energy.*

11.8 Symplectic integrators of first and higher order

The treatment we have applied to the harmonic oscillator is an example of a *symplectic integration algorithm*. These are a family of integrators that are not in the Runge-Kutta family, and are *not* based on Eqs. 11.44–11.48. These integrators are based on symplectic transformations at each step and approach the true solution at each step at a given order in the step size τ.

The techniques we describe in this section apply to systems evolving in a $2N$ dimensional phase space, governed by a Hamiltonian:

$$H = H(q_1, q_2, \ldots q_N; p_1, p_2, \ldots p_N). \tag{11.63}$$

We restrict ourselves to cases where the Hamiltonian does not depend explicitly on time; thus:

$$\frac{dH}{dt} = \frac{\partial H}{\partial t} = 0 \tag{11.64}$$

and the system is conservative. We will further restrict ourselves to separable Hamiltonians, i.e., those which can expressed as:

$$H(\vec{\eta}) = T(\vec{p}) + V(\vec{q}) \tag{11.65}$$

where $T(\vec{p})$ is the kinetic energy, $V(\vec{q})$ is the potential energy, and:

$$\vec{\eta} = (q_0, q_1, q_2 \ldots q_N; p_0, p_1, p_2 \ldots p_N). \tag{11.66}$$

The harmonic oscillator in one dimension, as well as any central force problem, are clearly separable. The torque-free motion of a rigid body is also separable (Piña 1997), though not in generalized coordinates which are widely familiar.

Hamilton's equation in symplectic form are:

$$\frac{d\vec{\eta}}{dt} = \mathbf{J}\frac{\partial H}{\partial \vec{\eta}}. \tag{11.67}$$

A solution to these equations can be formulated in terms of the Poisson bracket. In standard notation the Poisson bracket is:

$$\{u, v\} \equiv \frac{\partial u}{\partial q_i}\frac{\partial v}{\partial p_i} - \frac{\partial u}{\partial p_i}\frac{\partial v}{\partial q_i}. \tag{11.68}$$

The temporal development of any dynamical variable not explicitly dependent on the time can be expressed in terms of the Poisson bracket:

$$\begin{aligned}\frac{du}{dt} &= \frac{\partial u}{\partial q_i}\frac{dq_i}{dt} + \frac{\partial u}{\partial p_i}\frac{dp_i}{dt} \\ &= \frac{\partial u}{\partial q_i}\frac{\partial H}{\partial p_i} - \frac{\partial u}{\partial p_i}\frac{\partial H}{\partial q_i} \\ &= \{u, H\}.\end{aligned}$$

This applies to the canonical coordinates η as well:

$$\frac{d\vec{\eta}}{dt} = \{\vec{\eta}, H\} \equiv D_H \vec{\eta}. \tag{11.69}$$

Formally the equation can be integrated from $t = 0$ to $t = \tau$:

$$\vec{\eta} = \exp(\tau D_H)\vec{\eta}_0. \tag{11.70}$$

This solution has no practical value since the right hand side cannot be computed. However an approximation due to Neri (1988) can be found:

$$\vec{\eta} = \prod_{i=1}^{k} \exp(a_i \tau D_T) \exp(b_i \tau D_V) \vec{\eta}_0 + \mathcal{O}(\tau^{k+1}), \tag{11.71}$$

where:

$$D_T \equiv \{\cdot, T(\vec{p})\} \qquad D_V \equiv \{\cdot, V(\vec{q})\}, \tag{11.72}$$

and where the a_i's and b_i's obey:

$$\sum_i a_i = \sum_i b_i = 1. \quad (11.73)$$

The transformation can be practically calculated by applying sequentially two transformations at each step i in the expansion, Eq. 11.71. First the phase space variables are updated according to:

$$\begin{pmatrix} \vec{q} \\ \vec{p} \end{pmatrix} = \begin{pmatrix} \vec{q} + a_i \tau \frac{\partial T}{\partial \vec{p}} \\ \vec{p} \end{pmatrix}; \quad (11.74)$$

and then the new values are again updated according to:

$$\begin{pmatrix} \vec{q} \\ \vec{p} \end{pmatrix} = \begin{pmatrix} \vec{q} \\ \vec{p} - b_i \tau \frac{\partial V}{\partial \vec{q}} \end{pmatrix}. \quad (11.75)$$

Verify that

$$\exp(a_i \tau D_T) \begin{pmatrix} \vec{q} \\ \vec{p} \end{pmatrix} = \begin{pmatrix} \vec{q} + a_i \tau \frac{\partial T}{\partial \vec{p}} \\ \vec{p} \end{pmatrix} \quad (11.76)$$

and

$$\exp(b_i \tau D_V) \begin{pmatrix} \vec{q} \\ \vec{p} \end{pmatrix} = \begin{pmatrix} \vec{q} \\ \vec{p} - b_i \tau \frac{\partial V}{\partial \vec{q}} \end{pmatrix} \quad (11.77)$$

Why are these expressions calculable while Eq. 11.70 is not?

The constants a_i and b_i remain to be specified. It takes considerable ingenuity to find these constants such that the leading errors cancel to some fixed order, and their derivation is beyond the scope of this book. The simplest method, known as the symplectic Euler equation (or Euler-Cromer method) is order $k = 1$; and has $a_1 = b_1 = 1$. The Verlet method (Verlet, 1967) is order $k = 2$ and has $a_1 = a_2 = 1/2$, $b_1 = 1$, and $b_2 = 0$. A third-order integrator with:

$$\begin{array}{lll} a_1 = 2/3 & a_2 = -2/3 & a_3 = 1 \\ b_1 = 7/24 & b_2 = 3/4 & b_3 = -1/24 \end{array}$$

was derived by Ruth (1983). Neri (1988) derives a fourth-order integrator with

$$a_1 = a_4 = \frac{1}{2(2 - 2^{1/3})} \qquad a_2 = a_3 = \frac{1 - 2^{1/3}}{2(2 - 2^{1/3})}$$

$$b_1 = b_4 = \frac{1}{(2 - 2^{1/3})} \qquad b_2 = -\frac{2^{1/3}}{(2 - 2^{1/3})} \qquad b_3 = 0.$$

More symplectic integrators with orders up to 6 and 8 are given in (Yoshida, 1990), together with an elegant technique for deriving the constants.

You may have noticed that the transformation equations, Eqs. 11.74 and 11.75, are expressed in terms of derivatives of the kinetic energy $T(\vec{p})$ and the potential energy $V(\vec{q})$. This feature can be combined with an automatic derivative system such as we have discussed in Section 11.5.2 to produce an integrator that can apply symplectic integration techniques to *any* separable, conservative classical system (of reasonable size) along the same lines as the Hamiltonian solver discussed in Section 11.5.3. We do not go into detail here to avoid repetition of those sections; instead the topic is explored in the exercises.

11.9 Algorithmic inadequacies

We have discussed two important properties of algorithms for solving differential equations: **accuracy** and **fidelity**. The former is characterized by the order of the discretization, while the latter refers to the ability of the discretization to retain global features of the differential equation. An example of such a feature that we have already discussed extensively is conservation of energy. Here we discuss other issues that can arise when solving differential equations.

11.9.1 Stability

In general it is not possible to choose algorithmic parameters (such as the step size) completely at will. This can be simply illustrated by considering the equation $y'(x) = -ky(x)$ with solution $y \propto \exp(-kx)$. The Euler approximation to this equation is:

$$y_{n+1} = (1 - hk)y_n \qquad (11.78)$$

and has solution:

$$y_n = (1 - hk)^n y_0. \qquad (11.79)$$

Thus obtaining $y(x) \to 0$ for large x can only be achieved if:

$$|1 - hk| < 1 \qquad (11.80)$$

or $h < 2/k$. It is not possible to obtain a reasonable solution to the Euler difference equation if too large a grid spacing is employed. Hence the Euler algorithm is termed *conditionally stable*.

The problem can be avoided by using a sufficiently small step size or by employing a more robust *implicit solution method*. For example, if one seeks to solve the stability problem for an equation like $y' = f(x, y)$ one could discretize as:

$$y_{n+1} = y_n + \frac{h}{2} \left(f(x_n, y_n) + f(x_{n+1}, y_{n+1}) \right). \tag{11.81}$$

The utility of this discretization can be seen by considering the simple exponential problem $f = -ky$. In this case one can solve for y_{n+1}:

$$y_{n+1} = \frac{1 - hk}{1 + hk} y_n, \tag{11.82}$$

which implies that the solution approaches zero if:

$$\left| \frac{1 - hk}{1 + hk} \right| < 1. \tag{11.83}$$

Since this is true for all step sizes, the implicit discretization of Eq. 11.81 is termed *unconditionally stable*. Although this feature is desirable, the form of Eq. 11.81 makes it clear that implementing it can be difficult since, in general, a nonlinear equation must be solved at each time step. Furthermore, for the relatively simple equations generated by typical explicit discretization schemes, it is not too computationally expensive to choose step sizes sufficiently small to avoid stability problems.[10]

11.9.2 Solution mixing

Inaccuracies can arise in solving differential equations when numerical errors induce spurious behavior in the solution. Consider, for example, the equation:

$$y'' - k^2 y(x) = 0 \tag{11.84}$$

with boundary condition $y(0) = 1$ and $y'(0) = -k$. The general solution is:

$$y = c_1 \exp(-kx) + c_2 \exp(kx) \tag{11.85}$$

---

[10] We shall see that similar problems (and solutions) are present in the—much more expensive – algorithms for partial differential equations.

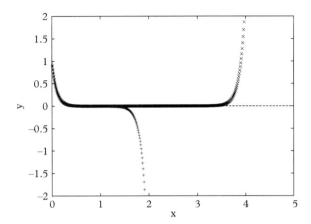

Figure 11.11 *Numerical solution to $y'' = 100y$ in single precision (pluses) and double precision (crosses).*

and the boundary conditions pick out the specific solution $y = \exp(-kx)$. As a solution algorithm is iterated, it necessarily 'picks up' a small contribution from the undesired component proportional to $\exp(kx)$ and this eventually dominates the exact solution.

There is no good solution to this problem other than exercising care when one suspects that spurious behavior is due to roundoff error. In such cases it would be prudent to examine the stability of the solution under increases in numerical precision. This is illustrated in Figure 11.11 for the case $k = 10$ with the Euler method with a step size of $h = 0.01$. The appearance of the spurious solution is apparent at large x, as is the delayed appearance when double precision reals are employed.

11.9.3 Multiscale problems

Real world problems are typically driven by physics on many scales that interact in complex ways. For example, weather systems depend on spatial scales ranging from meters to thousands of kilometers (describing, for example, water vapor exchange at the surface of a wavy ocean to hurricanes). Chemical reactions that can depend on temporal scales that differ by many orders of magnitude provide another example.

Multiscale problems introduce efficiency issues that are related to stability. Consider the two-scale problem defined by:

$$y'' + (k_0 + k_1)y' + k_1 k_0 y = 0 \tag{11.86}$$

with $y(0) = 2$ and $y'(0) = -k_0 - k_1$. The solution decays with two scales given by $1/k_0$ and $1/k_1$:

$$y = \exp(-k_0 x) + \exp(-k_1 x). \tag{11.87}$$

If $k_1 \gg k_0$ and one is employing a simple Euler-type algorithm, then stability requires that $h < 2/k_1$. In this case the transient term $\exp(-k_1 x)$ will be reliably reproduced. The problem is that for x larger than $1/k_1$ the solution is dominated by the first term, and it is grossly inefficient to solve a problem with a scale of $1/k_0$ when using a step size $h \sim 1/k_1 \ll 1/k_0$.

Two approaches to solving *stiff equations* such as this are:

(i) to employ adaptive step size routines that are capable of stepping with size $1/k_1$ in the transient regime and $1/k_0$ in the slowly varying regime;

(ii) to employ implicit methods that do not place strict requirements on step sizes, as discussed in Section 11.9.1.

11.10 Exercises

1. [T] Solution Mixing.
   Demonstrate the solution mixing discussed in Section 11.9.2 by considering the Euler solution to Eq. 11.84. Show that iteration of the initial condition $(y, y')_0 = (1, -k)$ smoothly approaches zero if one is in the stable region. Then show that a small admixture, say $(y, y')_0 = (1, -k + \delta)$ induces a growing solution.

2. [T] Symplectic Algorithms.
   Verify that the transformations of Eq. 11.76 and 11.77 are symplectic. Show that this implies that the full time evolution of the approximation of Eq. 11.71 is symplectic and exactly conserves some quantity, which differs from the Hamiltonian by terms of $\mathcal{O}(\tau^{k+1})$.

3. [T] Simple Pendulum.
   A simple pendulum consists of a length a of massless string attached to a bob of mass m. Choose, as the canonical coordinate, the angle θ between the pendulum's string and the vertical direction, and show that with these coordinates the Hamiltonian is:

   $$H = \frac{l^2}{2ma^2} - mga \cos\theta$$
   $$= E \left(\frac{1}{2}\left(\frac{l}{\lambda}\right)^2 - \cos\theta \right) \qquad (11.88)$$

   where l is the angular momentum of the pendulum, $E \equiv mga$ is a characteristic energy for the system, and $\lambda \equiv ma^{3/2}g^{1/2}$ is a characteristic angular momentum. Then plot the two-dimensional phase space trajectory for the system when the pendulum is started from $\theta = 0$ and

   a) $l_0 = \lambda/4$
   b) $l_0 = \lambda/2$

c) $l_0 = \lambda$
d) $l_0 = \sqrt{2}\lambda$
e) $l_0 = 2\lambda$
f) $l_0 = 3\lambda$.

4. **Spherical Pendulum.**
   A *spherical pendulum* is a simple pendulum that is free to move in a manner consistent with the constraint provided by the string. The pendulum is released from a position θ_0 with an angular velocity of magnitude $l = \lambda$ (see the previous problem), and a direction perpendicular both to the string and to the bob's displacement from the equilibrium point. Using spherical coordinates where θ is the angle between string and the vertical direction, and ϕ is the azimuthal angle, taking $\phi_0 = 0$, integrate the equations of motion numerically. Project the orbit into the horizontal plane (i.e., plot $\sin\theta \cos\phi$ vs. $\sin\theta \sin\phi$) – notice that this takes the form of a precessing ellipse. Compute the precession rate together with an estimate of its error (from numerical accuracy).

5. **Spinning Top.**
   Integrate the equation of the spinning top, Eq. 11.17. Plot the three variables u, ψ, and ϕ vs. time. If you prefer you can display your solution with a graphical animation.

6. **Spinning Top ii.**
   As a variation of the previous problem determine the Hamiltonian from the Lagrangian of the heavy symmetric top. Using the techniques of Section 11.5.3, express this Hamiltonian using the `PhaseSpace` class of the `QatGeneric Functions` class library and let the solver integrate the equations. Display, or animate, the solution.

7. **Restricted Three Body Problem.**
   In the celebrated *restricted three body problem*, three bodies move under their mutual gravitational attraction, but one of the bodies is far lighter than the other two, such that its influence on the heavier bodies can be neglected. A small planet moving in the gravitational field of two massive binary stars is a good example. Derive, and integrate numerically, the equations of motion for the restricted three body problem. Animate the solution. Try a set of initial conditions in which all three objects a) remain coplanar, and b) are acoplanar.

8. **Energy Conservation.**
   For the r^2 and r^4 central potential problems of Figure 11.4, plot the long-term behavior of the total energy when integrated with a fixed-order Euler method and compare with the symplectic Euler method with the same step size. Use a nondimensionalized Hamiltonian in two dimensions:

$$H = p^2/2 + r^N \qquad N = 2 \text{ or } 4 \qquad (11.89)$$

and start the motion from $\vec{r} = (0, 1)$ and $\vec{p} = (1, 1)$.

9. [P] Separable Hamiltonians.
   Write and test a symplectic integrator that will integrate any separable Hamiltonian, along the lines of Section 11.5.2. Use the automatic derivative system of `QatGenericFunctions`.

10. Damped Harmonic Oscillator.
    The damped harmonic oscillator, Eq. 11.2 can be rewritten as:

    $$\frac{dx}{d\tau} = \eta$$
    $$\frac{d\eta}{d\tau} = -2\zeta\eta - x \qquad (11.90)$$

    with the substitution $\eta \equiv v/\omega_0$ and $\tau = \omega_0 t$.

    a) Obtain the exact analytic solution for this equation in the case $\zeta = 1/2$.
    b) For each of the following numerical solutions, make a logarithmic plot of the deviation between the numerical solution and the true solution at the point $\tau = 10$ vs. the step size h: Euler, Midpoint, Trapezoid, Classical Runge-Kutta.
    c) Combine each of the methods in part (b) with adaptive step size control and make the same plot vs. the user-specified tolerance.

11. Radioactive Decay.
    An initially pure sample containing N_0 atoms of $^{213}_{83}\text{Bi}$ atoms decays according to:

    $$^{213}_{83}Bi \rightarrow\, ^{209}_{81}Tl + \alpha$$

    with a half-life of 45.6 minutes. This is followed by the decay:

    $$^{209}_{81}Tl \rightarrow\, ^{209}_{82}Pb + \beta$$

    with a half-life of 132 seconds, then:

    $$^{209}_{82}Pb \rightarrow\, ^{209}_{83}Bi + \beta$$

    with a half-life of 195 minutes. The nuclide $^{209}_{83}Bi$ has a half-life so long (1.9×10^{19} years) that it can be considered stable.
    Obtain curves for the populations of $^{213}_{83}Bi$, $^{209}_{81}Tl$, $^{209}_{82}Pb$, and $^{209}_{83}Bi$. First derive a set of coupled first-order differential equations for N_i, $i = 1, 2, 3, 4$ for each of the four species; solve these numerically and then plot N_i/N_0.

12. [P] Nucleon Scattering.
    A naive model for the interaction of nucleons in a nucleus is given by the Woods-Saxon potential:

    $$V(r) = \frac{V_0}{1 + e^{(r-R)/a}}$$

    where $R = r_0 A^{1/3}$, A is the atomic mass number, and $r_0 = 1.25$ fm, $V_0 = 50$ MeV, and $a = 0.5$ fm.

    For the unbound (scattering) states, determine the radial wavefunctions $U_{k,l}$ corresponding for neutron scattering from $^{12}_{6}C$ with energy $E_k = \frac{\hbar^2 k^2}{2m}$ when $k = \frac{1}{R}$, $k = \frac{1}{2R}$, and $k = \frac{2}{R}$. Do this for $l = 0, l = 1$, and $l = 2$.

    Start at $r = 0$; there the function is constrained to have $U_{k,l}(r = 0) = 0$. The slope $U'_{k,l}(r = 0)$ can be any arbitrary positive number – in the end the normalization procedure will readjust the slope anyway—so take $U_{k,l}(r = 0) = 1$. Integrate $|U_{k,l}|^2$ to determine a normalizing factor and normalize all of your plots.

13. Muon Detection.
    A crude model for a muon detection system at the LHC assumes an empty drift space measuring about 4.7 m that is surrounded by a toroidal magnetic field. The field is approximated to be that of an ideal toroid with an inner radius of 4.25 m and an outer radius of 11 m. Lastly, the magnetic field at the inner radius is 4.7 T.

    Consider a charged particle that enters the magnet system in the plane perpendicular to the axis of the toroid.

    a) Determine numerically the amount by which the particle bends when the particle's momentum is 50 GeV/c; 100 GeV/c; 500 GeV/c; 1000 GeV/c.

    b) A muon tracking system, roughly speaking, performs a momentum analysis on charged particles passing through this field. Tracking algorithms require a prediction of the bending through the field. Suppose that the experimental resolution of the tracking detectors located just outside the toroid field at 11m is 80 μm. It is desirable to keep the error in the predicted impact point to within 1% of this value. For each of the above listed track momenta, determine the smallest reasonable step size appropriate for this calculation using one of the following fixed step size algorithms

        i. The Euler Method
        ii. The Midpoint Method
        iii. The Classical Runge-Kutta Method.

14. Double Pendulum.
    The double pendulum consists of a simple pendulum (mass m_1, string length l_1) connected by a massless rigid rod to another simple pendulum (mass m_2, string length l_2).

a) Determine the Lagrangian for this system. Use the signed angles made by each string θ_1 and θ_2 as generalized coordinates.

b) Obtain the Hamiltonian.

c) From the Hamiltonian, obtain four first-order differential equations and identify them as "linear" or "non-linear".

d) Numerically integrate the equations of motion over a long period in order to obtain the orbit, i.e. θ_2 vs. θ_1. Plot the orbit for two values of the starting parameters:

i. $\theta_1 = \pi/12, \theta_2 = \pi/12$
ii. $\theta_1 = \pi, \theta_2 = \pi$.

15. The SEIR Disease Model.

The **SEIR disease model** simulates the spread of infectious disease through a population. This model divides the population into the categories *susceptible, exposed, infectious,* and *recovered* (hence the name). The rate at which the categories change size is shown schematically in Figure 11.12. The incoming arrow represents population increase due to births at the rate b. The outgoing arrows represent deaths in each group, specified by the rate d (Rock et al., 2014).

a) Write the four rate equations that correspond to the schematic (neglect the dashed line). This nearly trivial system is made more interesting when it is realized that the *force of infection* λ is proportional to the infected population:

$$\lambda = \beta \frac{I}{N}.$$

The denominator is the total population, $N = S + I + E + R$.

b) If $b = d$ is the population static?

c) What does the model assume about disease mortality?

d) Solve the resulting SEIR equations with the following input parameters: $\beta = 0.2$, $\gamma = 0.1$, $b = 10^{-4}$, $d = 10^{-4}$, $\alpha = 0.1$, and $\nu = 10^{-3}$. Initial conditions are $I_0 = 10^{-6}$, $N = 1$, and $S_0 = N - I_0$. Plot the infected population ratio on a log scale as a function of time assuming that the constants carry units of people/day where appropriate. Integrate for several thousand days.

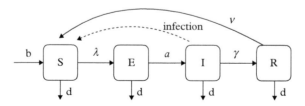

Figure 11.12 *Populations dynamics according to the SEIR model.*

BIBLIOGRAPHY

Butcher, J. C. (2003). *Numerical Methods for Ordinary Differential Equations.* John Wiley and Sons.
Cash, J. and A. Karp (1990). *A variable order Runge-Kutta method for initial value problems with rapidly varying right-hand sides.* ACM Transactions on Mathematical Software **16**, 201.
Fehlberg, E. (1968). *Classical fifth, sixth, seventh and eighth-order Runge-Kutta formulas with step size control.* NASA Technical Report 287.
Fehlberg, E. (1969). *Low-order classical Runge-Kutta formulas with step size control and their application to some heat transfer problems.* NASA Technical Report 315.
Fehlberg, E. (1970). *Klassische Runge-Kutta-Formeln vierter und niedrigerer Ordnung mit Schrittweiten-Kontrolle und ihre Anwendung auf Wärmeleitungsprobleme.* Computing Arch. Elektron. Rechnen. **6**, 61.
Goldstein, H., C. P. Poole, and J. Safko (2001). *Classical Mechanics.* 3rd ed. Addison-Wesley.
Hairer, E., S. P. Nørset, and G. Waner (2008). *Solving Ordinary Differential Equations I: Nonstiff Problems.* Springer-Verlag.
Hairer, E. and G. Waner (2002). *Solving Ordinary Differential Equations II: Stiff and Differential-Algebraic Problems.* Second Revised Edition, Springer-Verlag.
Kutta W. (1901). *Beitrag zur naeherungsweisen Integration totaler Differentialgleichungen.* Z. Math Phys. **46**, 435–453.
Neri, F. (1988). *Lie Algebras and Canonical Integration,* University of Maryland Preprint.
Piña, E. (1997). *Solution by the separation method of the motion of a rigid body with no forces.* Revista Mexicana de Física **43**, 205.
Press, W., S. Teukolsky, W. Vetterling and B. Flannery (2007). *Numerical Recipes, the Art of Scientific Computing, 3rd edition.* Cambridge University Press.
Rock, K., S. Brand, J. Moir, and M. J. Keeling (2014). *Dynamics of Infections Diseases.* Rep. Prog. Phys. 77, 026602.
Ruth, R. (1983). *A Canonical Integration Technique.* Nuclear Science, IEEE Trans. on NS-30, No. 4 (4), 2669
Thornton S. T. and J. B. Marion (2004). *Classical Dynamics of Particles and Systems.* 5th ed. Thompson.
Verlet, L. (1967). *Computer experiments on classical fluids. I. Thermodynamical properties of Lennard-Jones molecules.* Phys. Rev. **159**, 98.
Yoshida, H. (1990). *Construction of higher order symplectic integrators.* Phys. Lett. A **150**, 5–7, 262.
Yoshida, H. (1992). *Symplectic Integrators for Hamiltonian Systems: Basic Theory,* in *Chaos, Resonance, and Collective Dynamical Phenomena in the Solar System: Proceedings of the 152nd Symposium of the International Astronomical Union.* International Astronomical Union. Symposium no. 152, Angra dos Reis, Brazil ed. Sylvio Ferraz-Mello, Kluwer Academic Publishers.

12
Polymorphism

12.1	Example: output streams	404
12.2	Inheritance	406
12.3	Constructors and destructors	407
12.4	Virtual functions	409
12.5	Virtual destructors	413
12.6	Pure virtual functions and abstract base classes	415
12.7	Real example: extending the GenericFunctions package	416
12.8	Object-oriented analysis and design	421
12.9	Exercises	421
	Bibliography	423

One of the more useful features of modern programming languages, for both scientific and nonscientific applications, is the ability to write functions that accept as their input *entire categories* of classes having some degree of commonality. Broadly speaking there are two mechanisms in C++ that can be employed to achieve this feature: templates and **polymorphism**. Both have already been encountered in previous chapters and both are important tools that can provide major benefits when employed within the appropriate context.

The reader is already familiar with templates, of which `std::vector` is the archetype. Templates are examined in more detail in Chapter 17. The reader has also seen class tree diagrams (e.g. Figures 5.4, 5.7, 6.7, 10.19, and 11.5), which indicate how certain classes representing a specific type of object can be substituted for other classes in routines that require a more general type of object. A partial list of previously encountered examples is

- the std::ostream classes (`std::cout, std::ofstream, std::ostringstream`),
- the `Genfun::AbsFunction` classes of Section 3.8,
- Graphics classes (like `SoNode`) from the Open Inventor library,
- Classes used in the integration of ordinary differential equations, `ButcherTableau` and `ExtendedButcherTableau` classes, among others.

Polymorphism, the mechanism by which objects of different types share a common interface, is the topic of this chapter. We hope that the previous exposure has amply illustrated the utility and power of this technique and given the reader a clear idea of the flexibility that it engenders. In this chapter we will focus on how one achieves polymorphism in practice.

But first, a remark about when one uses templates vs. polymorphism. The answer to this question depends to a large extent on the degree of commonality in the classes. When little is required of the class, templates may be appropriate; indeed `std::vector` and other collection classes require little more than a default constructor and a copy constructor. Since many datatypes (including built-in datatypes) will satisfy such a simple set of conditions, templates are the method of choice. Another example is `std::complex`, which is used to complexify different types of floating point datatypes. It requires a little more in the way of specific behavior than `std::vector` from the datatypes it complexifies: they must, for example, implement essential operations like `+,-,/,*`.

If the requirements are more complicated, polymorphism is the technique of choice. Polymorphism organizes classes having specific common behavior into a set of more general classes. The ability to provide a derivative, for example, is a common feature of all functions of a real variable, though each function may implement that differently: a sine function would return a cosine, a sinh function would return a cosh function, etc. In polymorphism the classes are organized hierarchically, the hierarchy guarantees that the *subclasses* obey a certain protocol that is explicit in the definition of the *superclass*. It also guarantees that all interactions possible for the superclass are defined and implemented in the subclasses.

Since the set of interactions defines in some sense the abstraction represented by the class, one can put another spin on polymorphism:

- It is the way in which a programmer can abstract common features of related classes into generalizations.
- It is also the way that generalizations are refined and made more specific.

Among its benefits are that it:

- Allows one to save vast amounts of programming time by writing functions accepting whole categories of classes.
- Provides flexibility by allowing these functions to work on classes that have not yet been invented.
- Provides, when skillfully executed, an intuitive programming interface and a manageable way to organize complicated coding projects.

C++ and other languages supporting the Object Oriented programming paradigm provide mechanisms for implementing polymorphism in C++ classes. These mechanisms allow one to override the behavior of the superclass when necessary, or use the

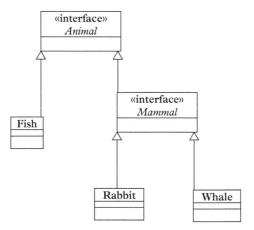

Figure 12.1 *A simplified taxonomy of animals.*

default behavior if appropriate. They can be easily applied when the designer has a clear idea about the fundamental abstractions that are represented in the classes and how they are related. We will refer to a primitive taxonomy of certain animals, shown in Figure 12.1, to illustrate how abstractions are classified.

Often the main challenge in implementing a complicated project is defining the abstractions required for the project and organizing them into a conceptual hierarchy. Figure 12.1 is intuitive to most students, but of course it was a major intellectual feat at the time of Carolus Linnaeus (1707–1778) who invented the classification scheme; a feat which indeed has something in common with software design. The process by which one develops code (either a specific application or an application library) usually follows several steps. First, the key abstractions needed for the project are identified. Then, the commonality in these abstractions is discovered. The abstractions are encapsulated in a set of classes and the classes organized in a hierarchy (like chimeric animals that exist only in the imagination). Then the software designer begins to bring these to life: the protocol for the full set of classes is refined; usually the interfaces (header files) are created before implementing any actual functions. When the design of the project is complete, the implementation can finally start. The design can be reviewed and updated as real-life complications arise during implementation. Finally one proceeds to debug, test, and tune the code.

This chapter describes how to implement a known design. Some of our illustrative examples are based upon the animal kingdom, with a "design" known well to every reader.

12.1 Example: output streams

Make a small change to your first C++ program `helloWorld.cpp` and you can begin to see what polymorphism is all about. The original program looks like this:

```
#include <iostream>
main() {
  std::cout << "Hello, world" << std::endl;
}
```

You can modify this in the following way:

```
#include <fstream>
main() {
  std::ofstream file("output.txt");
  file << "Hello, World" << std::endl;
}
```

The first version uses the free subroutine `std::ostream& operator<< (std::ostream&, const char*)` to echo a character string to the terminal. The second version uses exactly the same subroutine. There is no need to write a separate subroutine for the `std::ofstream`, because a `std::ofstream` is a `std::ostream`.

> Exercise: write a simple class which writes itself to an std::ostream using the left shift operator, or reuse one of your existing classes (eg. stack) after endowing it with the left shift operator. Test that the same routine works with `std::cout` and with `std::ofstream`.

There are several ways to represent the "is-a" relationship graphically. One of the simplest is the class tree diagram. UML, or *universal modeling language*, notation is another way to describe code graphically, many programmers use UML to visualize the relationships between classes, and a few even use graphical tools to program. Figures 12.1 and 12.2 are UML diagrams, generated with the StarUML^{TM} program.

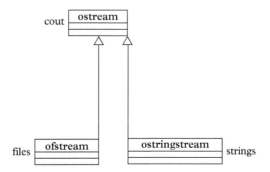

Figure 12.2 *Class diagram showing the relationship between* `std::ofstream`, `std::ostringstream`, *and* `std::ostream` *classes.*

12.2 Inheritance

A mammal is an animal. A rabbit is a mammal. ***Inheritance*** is used to establish an "is-a" relationship between the ***superclass*** or ***base class*** (Mammal for instance) and the ***subclass*** or ***derived class*** (Rabbit). Once an inheritance relation is established, the subclass can be used anywhere the superclass is required. To establish this relationship, one declares the subclass like this:

```
class Rabbit:public Mammal {
...
};
```

This is called ***public inheritance*** to distinguish from ***private inheritance***, which is rarely used (by most designers, anyway) and will not be discussed in this book. This means that one class (Mammal) generalizes another class (Rabbit); that a Rabbit will do anything that a mammal does, lactating for example, and maybe more.

The inheritance relationship is often misused. At first glance it may appear to be a coding shortcut, a way of reusing an implementation that has already been written. That is not the primary purpose of inheritance. Inheritance is used to establish an "is-a" relationship between two classes and enables polymorphism in client code. If this is not your purpose, do not use inheritance; you can re-use a class within another class in other ways, either by simply using it or by holding an instance as private member data. These are called "using" and "has" relationships, respectively.

Imagine that we are designing a class library to describe rotations and their actions on vectors in a three dimensional Euclidean space. We already have a class for matrices, implementing all of the operations of matrices. The question is:

should `Rotation` *inherit from* `Matrix`?

To answer this, ask first whether a rotation *is* a matrix. You can see that while matrices may be added and subtracted, rotations cannot—only the operation of composition applies to rotations. So a `Rotation` is not a `Matrix` and should not inherit from one. In a better design, a `Rotation` *has* a `Matrix`. Thus a `Rotation` class would be implemented as:

```
class Rotation {
public:
...
private:

  Matrix rep;

};
```

12.3 Constructors and destructors

The memory required for an instance of a subclass is all of the memory for the superclass member data, plus the additional memory of the subclass's member data. In the constructor call, this memory is allocated and should be properly initialized. The superclass will be initialized first, with a call to a constructor, then the subclass. Since the superclass may have more than one constructor, the superclass initializes the subclass within its own constructor. Let's look in detail at a part of the hierarchy `Animal > Mammal > Whale`. The interface to `Animal` has a constructor used to name the `Animal`, a few methods representing the `Animal`'s usual duties, and private member data to store the name of the animal.

```cpp
#include <string>
class Animal {
 public:

   // constructor:
   Animal (const std::string & name);

   // get the name:
   const std::string  & getName() const;

   // daily activities:
   void wakeUp();
   void eat();
   void sleep();
   void haveTypicalDay();


 private:
   std::string  _name;
};
```

The class `Mammal` inherits from `Animal`; here is it's interface:

```cpp
#include "ZOO/Animal.h"
class Mammal: public Animal {

  public:

    // Constructor:
    Mammal(const std::string & name);
```

```
    // What is a mammal after all?
    void lactate();
};
```

We can see that a `Mammal` does everything that a generic animal does, and *in addition* lactates. A `Whale` is a sea creature which—despite appearances—is not a fish but a `Mammal` that does everything that `Mammal` does, and therefore by extension everything that an `Animal` does. It also does its own thing:

```
#include "ZOO/Mammal.h"
class Whale: public Mammal {
 public:

   // Constructor:
   Whale(const std::string & name);

   // Exhale through the blow-hole:
   void spout();

};
```

We are now going to concern ourselves with the constructor calls. The constructor to `Whale` calls the constructor to `Mammal`, which calls the constructor to `Animal`. All of these require a name, which is ultimately going to be held in the private member data of `Animal` and be made accessible through the interface to `Animal`. Providing a name in the constructor call ensures that the Animal is properly initialized and ready for action. Constructor calls to base classes must be explicit in the derived classes. In the class `Whale` the constructor is implemented as follows:

```
Whale::Whale(const std::string & name):Mammal(name) {
}
```

i.e., using the initialization syntax. In the class `Mammal` it is similar:

```
Mammal::Mammal(const std::string & name):Animal(name) {
}
```

When the base class constructor requires arguments, this syntax is the mechanism by which the arguments are passed to the base class. When the base class has more than one overloaded constructor, it provides a way to select the proper one. Explicit calls to the base class constructors are required in copy constructors as well as ordinary constructors, when they exist.

When an object is destroyed (by going out of scope or by an explicit call to `delete`), the sequence of events we have just described is reversed. First the destructor of the most derived class is called, then destructor of the next most derived class is called, and so forth down to the destructor in the base class. All actions, such as memory clean-up, should be carried out in the destructor code. Since each class has only one destructor, never taking any arguments and therefore never overloaded (as constructors are), the compiler itself insures that the destructors are called when an object goes out of scope or is deleted through an explicit `delete` call. Certain pitfalls; however, may arise, their origin and cure is discussed in the following sections.

12.4 Virtual functions

It may occur to you that while all animals eat and sleep, they may all happen to carry out these activities in unique ways. A programmer seeking to capture their complex behavior needs to do *more* than keep adding to the list of behaviors as derived classes are added to the hierarchy, (s)he also needs to customize some of the behavior. If we admit that all animals eat and sleep, then we should at least allow that these activities can be carried out differently for each species. Whales, for example, can not simply doze off for the night, since they need to breathe every thirty minutes or so, so they *take short snoozes*, punctuated by occasional bouts of spouting. In addition, whales *feast on krill*, which they round up and engulf in large numbers. To deal with this challenge, C++ introduces ***virtual functions***, which are essentially functions that are intended to be overridden in derived classes.

The following piece of code (EXAMPLES/POLYMORPHIC-ZOO/src/zookeeper.cpp) illustrates an attempt to manage three animals, Stuart the Whale, Lisa the Rabbit, and Bob the Fish.

```cpp
#include "ZOO/Whale.h"
#include "ZOO/Fish.h"
#include "ZOO/Rabbit.h"
#include <vector>

void typicalDayAtTheZoo(const std::vector<Animal *> & aVect)
  {
  for (int i=0;i<aVect.size();i++) {
    aVect[i]->haveTypicalDay();
  }
}

int main() {
  std::vector<Animal *> aVect;
  aVect.push_back(new Whale("stuart"));
  aVect.push_back(new Rabbit("lisa"));
```

```
  aVect.push_back(new Fish("bob"));

  typicalDayAtTheZoo(aVect);

  return 0;
}
```

The program loads up an `std::vector` with pointers to these animals and then sends them the message `haveTypicalDay()`, defined in the class `Animal` as follows:

```
void Animal::haveTypicalDay() {
  eat();
  sleep();
}
```

For the moment we are going to redesign our code to allow "custom" versions of eat and sleep to be defined for animals, and then customize these versions in the case of the whale. There are two steps:

1. Define the functions `eat` and `sleep` to be *virtual* in the base class, `Animal`.
2. Override those functions in the derived class, `Whale`.

To declare a function virtual, simply add the keyword `virtual` to the declaration (but not the definition) of the function; for instance, change these declarations:

```
void eat();
void sleep();
```

to

```
virtual void eat();
virtual void sleep();
```

When a function is declared virtual in the base class, it means that subclasses may override the implementation of that function with their own implementation. While nothing will stop you from overriding a common function, it is almost always a mistake to do so, particularly when you are aiming to achieve polymorphism in your class design. We can go ahead now and modify the declarations of `Animal`, `Mammal`, and `Whale` so that they now have the following interfaces:

```
class Animal {

  public:
```

```cpp
  // constructor:
  Animal (const std::string & name);

  // get the name:
  const std::string  & getName() const;

  // daily activities:
  virtual void wakeUp();
  virtual void eat();
  virtual void sleep();
  void haveTypicalDay();

  private:

  std::string  _name;
};

class Mammal: public Animal {
 public:

  // Constructor:
  Mammal(const std::string & name);

  // Lactate
  void lactate();
};

class Whale: public Mammal {
 public:

  // constructor:
  Whale(const std::string & name);

  // daily activites:
  void spout();
  virtual void eat();
  virtual void sleep();

  void haveTypicalDay();
};
```

In the class `Whale`, the methods `eat` and `sleep` are again defined as virtual functions. This is in fact not necessary since they inherit their virtuality from the base class, however it is useful to include this with the definition so that the function type is apparent to anyone reading the header file.

Now we will implement the four functions `Animal::eat`, `Animal::sleep`, `Whale::eat`, and `Whale::sleep`:

```
void Animal::eat() {
  std::cout << "Animal " << _name << " eating"
      << std::endl;
}
void Animal::sleep() {
  std::cout << "Animal " << _name << " sleeping"
      << std::endl;
}
void Whale::eat() {
  std::cout << "Whale " << getName() << " FEASTING ON KRILL"
      << std::endl;
}
void Whale::sleep() {
  std::cout << "Whale " << getName() << " TAKING A SHORT
      SNOOZE" << std::endl;
}
```

Executing the example program, zookeeper.cpp, gives the following output, which shows that we have successfully overridden the methods `eat` and `sleep` for the whale, and kept the default methods for the other other animals:

```
Whale stuart FEASTING ON KRILL
Whale stuart TAKING A SHORT SNOOZE
Animal lisa eating
Animal lisa sleeping
Animal bob eating
Animal bob sleeping
```

The main distinction between virtual functions and common functions is (Meyers, 2015):

- In a common function the subclass inherits both the interface and the implementation.
- In a virtual function the subclass inherits the interfaces and may inherit the implementation.

An object can be allocated on the heap, like this:

```
Animal *a =new Whale("stuart");
```

The object pointed to by `a` is said to have three datatypes (`Animal`, `Mammal`, and `Whale`). The same holds when an object constructed on the stack is handled through its base class, as follows:

```
const Animal & a  = Whale("stuart");
```

`Animal` is referred to as the *static type* of the object, while `Whale` is the *dynamic type* of the object.

Think back to the various examples in which polymorphism has been used so far in this book, and you will see that the normal behavior that one should expect when a message is sent to an object is that the function should be resolved according to the dynamic type. A message is sent to an `Animal`, but it is received by a `Whale`. The client code is then able to handle the object(s) by pointer (or reference) to the base class. This is the mechanism by which polymorphism is achieved.

If on the other hand the function `eat` is a common function rather than a virtual function, this mechanism is not enabled, and the function is resolved according to the static type—usually this situation is unintended and produces illogical behavior. Therefore you should (essentially) never override a common function. If you are writing a C++ class library, you should also anticipate two kinds of clients; first, those who will simply use your classes (as we have been doing in most of the examples), and second, those who will extend them by further subclassing (Meyers, 2015). If your classes are designed for extension, then all of the functions that may be overridden in derived classes should be declared virtual.

It is important to master inheritance and polymorphism when writing any project of significant size and scope, and virtual functions are an important part of implementing polymorphism properly. It is very common to observe neophytes leaving out virtual functions from their design, simply because they have not taken the time to understand them.

12.5 Virtual destructors

Recall that there are two ways to allocate objects, "on the stack" and "on the heap". Object allocation on the stack is done simply by initializing the object with a call to one of the available constructors. Object allocation on the heap is achieved with `operator new`. An object allocated on the stack is destroyed when it goes out of scope; when that happens the destructor for the most-derived class (e.g. `Whale`) is called first, then the next-to-most derived class (e.g. `Mammal`), all the way back to the base class (e.g. `Animal`). The destructor call gives the doomed object the opportunity to carry out any actions that need to be taken at the end of its life-cycle, such as cleaning up any memory that it has allocated.

A problem can arise when a subclass is allocated on the heap. Imagine that the class `Whale` contains a pointer to some some memory which it allocates each time it feasts on krill:

```cpp
class Whale {

  public:
...
  // destructor:
  ~Whale() { delete [] someMemory;}

  private:

  // some memory allocated by the whale
  int *someMemory;
```

The memory is to be cleaned up eventually in the destructor call. Also, imagine that a whale has been allocated on the heap:

```cpp
Animal *a = new Whale("stuart");
```

When we call `delete a` (unless further measures are taken as discussed below) the destructor call is resolved according to the static type, `Animal`. The destructor in `Whale` should have been called *first*, but it is never invoked at all, and the memory pointed to by the pointer `someMemory` is never freed: we have created a memory leak.

However, the destructor like any member function may be declared to be virtual in the base class, and this brings an easy solution to the problem. By declaring a virtual destructor

```cpp
class Animal {

public:

    ...
    virtual ~Animal();
...
};
```

we ensure that the destructor is resolved according to the dynamic type, `Whale`. We do not need to take any further action to insure that destructors for `Mammal` and `Animal` are subsequently called, since C++ does that automatically. The takeaway message from this discussion is the simple rule:

- Declare destructors to be virtual in base classes (Meyers, 2015).

There is no need to declare destructors virtual in subclasses. The virtuality propagates, anyway. However it does no harm and is good practice, since the virtuality is then manifest in the header file.

Since C++11, two specifiers `final` and `override` are available; final "devirtualizes" a virtual member function, such that it cannot be further overridden, and `override` causes the compiler to check that the overridden function is indeed virtual (and otherwise complain). The use of these specifiers is optional. For more information, see Meyers (2015).

12.6 Pure virtual functions and abstract base classes

In our discussion of virtual functions (Section 12.4), we identified two possibilities for member functions: for common functions, the subclass inherits the interface and the implementation, while for virtual functions the subclass inherits the interface and may inherit (or override) the implementation.

There is also a third possibility: by declaring a ***pure virtual function*** in the base class, we can ensure that the subclasses *must* provide an implementation. Why is this useful? Well, consider the case of an animal again. What is a "typical day"? It may occur to you that this is an absurd notion; that in animals ranging from intestinal bacteria to birds, fish, tubeworms, land mammals, and finally water mammals, no generic typical day can possibly be conceived. Jumping to the more relevant example of functions of one variable, we can say that functions should be able to return their value (given an argument), but that there is no such thing as a typical value returned by a function.

The way to implement a pure virtual function is to declare it in the following way:

```
class Animal {
  virtual void haveTypicalDay()=0;    // note the syntax "=0"
};
```

With this declaration no function definition is required in the base class.

Any base class that defines such a function becomes, as a result, an ***abstract base class*** (ABC). An ABC has the property, which may seem bizzare at first, that no instances of the class can ever be created, i.e., the compiler insures that one cannot simply write:

```
Animal a;
```

In fact is is not so illogical: the class does not know how to handle the message `haveTypicalDay()`. On the other hand, subclasses (at least those which do provide a definition of `haveTypicalDay()`) can readily be instantiated, and they can be handled through pointers or references to the base class, i.e., as `Animal`s.

Consider again the following function which dispatches messages to a collection of Animals:

```
void typicalDayAtTheZoo( const std::vector<Animal *> & aVect)
    {
  for (int i=0;i<aVect.size();i++) {
    aVect[i]->haveTypicalDay();
  }
}
```

In Chapter 2 we drew an analogy between a properly encapsulated C++ class and an integrated circuit, which exposes its interface while hiding its implementation. We can draw another analogy between an abstract base class and a socket which is designed to hold specific types of integrated circuits (Figure 12.3). Into this socket one can place an entire family of circuits, providing that they adhere to the same protocol. The protocol puts a number of restrictions on the sockets, such as number of pins, the location of the pins, the assignment of pins to power input, ground, signal input, signal output, perhaps clock input as well. Whenever a device that adheres to the protocol is inserted into the socket the device handles the request in its own manner.

Abstract base classes function as a computer-science equivalent to a socket. They completely specify a protocol. An extreme example is a *pure interface class*, which only contains pure virtual functions and no member data or implementations. Such a class exists only to specify a protocol, i.e., an interface. Functions can then be written to the interface, rather than to the implementation (Meyers, 2015): the interface class serves as a kind of socket into which concrete implementations may be plugged. This permits efficiency and flexibility, since it allows for additional concrete instances of the basic generalization to be added at a later time.

The main difference between abstract base classes and pure interface classes is that pure interface classes provide no services while in general abstract base classes provide some services. The choice of which type of class to use is a software design decision.

12.7 Real example: extending the GenericFunctions package

To show that these principles apply beyond our fictional little world consisting of spouting whales, sleeping rabbits, and lactating mammals, we now turn our attention to a more realistic application.

The `QatGenericFunctions` library lacks certain trigonometric functions, namely $\sec x$, $\csc x$, and $\cot x$. We seek to provide classes for each of these functions that are styled after the function $\sin x$.

The base class of all functions in the `GenericFunction` library is `Genfun::AbsFunction`. It provides a number of virtual functions—these are functions which we *may* overload—and a number of pure virtual functions, functions which we *must* overload. The pure virtual functions are:

Figure 12.3 *In Chapter 2 we drew an analogy between C++ classes and integrated circuits, where a clear separation occurs between the inner workings of the circuit and the interface. In this way of thinking we can identify abstract base classes with sockets. By writing code to the interface of an abstract base class we are creating a type of socket into which we can "plug" any subclass.*

```
virtual double operator() (double argument)
const=0;
virtual double operator() (const Argument &argument)
const=0;
```

The first is straightforward: it evaluates the function at the value given by the argument, and returns its value. The second is foreseen for functions of more than one variable. For functions of one variable, the second function can be implemented in terms of the first function.

The virtual functions are those that we may override. Here is a list:

```
virtual ~AbsFunction();
virtual unsigned int dimensionality() const; // returns 1;
virtual bool hasAnalyticDerivative() const {return false;}
virtual Derivative partial(unsigned int) const;
```

418 *Real example: extending the GenericFunctions package*

If you look at the header file `QatGenericFunctions/AbsFunction.h`, you will see that other virtual functions also appear, for example `virtual AbsFunction * clone() const=0` which happens to allocate a new `AbsFunction` on the heap. This function is what is known in the jargon as *boilerplate*; a piece of code which is rewritten again and again in nearly orginal form. To save the subclasser (you) from the routine work of rewriting this boilerplate each time you derive a class, two macros are avaiable to do it automatically. The macro `FUNCTION_OBJECT_DEF` should be included in the header file, with the class name (e.g. `Sec`, `Csc`, `Cot`) as an argument. The macro `FUNCTION_OBJECT_IMP`, also taking the class name as an argument, should appear in the implementation (`.cc` or `.cpp`) file. For `GenericFunctions`, this suffices to ensure that a certain number of standard virtual methods are automatically overridden in the derived classes.

Now returning to the four virtual methods listed above, a custom destructor is, strictly speaking, only required if the function dynamically allocates memory, but we generally provide one anyway, as a matter of course, even if it is empty. The function `dimensionality()` returns the number of arguments to the function, and is one by default. The default implementation will be fine for the new classes `Sec`, `Csc`, and `Cot`, so we do not override it.

The virtual function `partial (unsigned int)` takes the partial derivative of the function (or the total derivative in case it is a function of one variable). `partial` returns a `Derivative` object, which is another `AbsFunction` in the `GenericFunction` library used explicitly for this purpose. If functions of more than one dimension are used, the argument to differentiate is the index of the variable with respect to which the partial derivative is to be taken. In case we are dealing with a function of only one dimension, `partial` should check that the index is zero. The default implementation will return a class computing the derivative numerically from the orginal function. When possible, it is highly desirable to return an analytic form of the derivative. In that case:

- override the function `virtual bool hasAnalyticDerivative() const` so that it returns true.
- override the function `virtual Derivative partial(unsigned int) const` so that it provides the analytic derivative.

To complete the example, we give below the declaration and implementation of `Genfun::Csc`. The header file defines, in addition to its own version of the virtual functions discussed above, a constructor and copy constructor. Assignment of a `Csc` object has been disabled by a trick—we make the assignment operator private and do not provide any implementation. The reason for making it illegal is, first, it is of little value to assign one `Csc` object to another, and second, while assigning an arbitrary function to another would indeed be useful, it is hardly a feasible possibility. So we simply rule it out.

```
#ifndef Csc_h
#define Csc_h 1
```

```cpp
#include "QatGenericFunctions/AbsFunction.h"
namespace Genfun {
  class Csc : public AbsFunction {

    FUNCTION_OBJECT_DEF(Csc)

  public:

    // Constructor
    Csc();

    // Destructor
    virtual ~Csc();

    // Copy constructor
    Csc(const Csc &right);

    // Retreive function value
    virtual double operator ()(double argument) const;
    virtual double operator ()(const Argument & a) const
        {return operator() (a[0 ]);}

    // Derivative.
    Derivative partial (unsigned int) const;

    // Does this function have an analytic derivative?
    virtual bool hasAnalyticDerivative() const {return true;}

    private:

    // It is illegal to assign a Csc
    const Csc & operator=(const Csc &right);
  };
} // namespace Genfun
#endif
```

The implementation is fairly straightforward:

```cpp
#include "Csc.h"
#include "QatGenericFunctions/Tan.h"
#include <cassert>
#include <cmath>
namespace Genfun {
```

```
    FUNCTION_OBJECT_IMP(Csc)

    Csc::Csc() {}

    Csc::~Csc() { }

    Csc::Csc(const Csc & right) : AbsFunction(right) {  }

    double Csc::operator() (double x) const {    return
       1.0/sin(x); }

    Derivative Csc::partial(unsigned int index) const {
       assert(index==0);
       const AbsFunction & fPrime = -Csc()/Tan();
       return Derivative(& fPrime);
    }

} // namespace Genfun
```

The function needs to be compiled and then linked to the client code. It will then behave like any other `GenericFunction`. You can plot it, use it in expressions with other functions, integrate it, etc.

Incidentally what exactly is this data type `Genfun::GENFUNCTION` that we have been using for some time now? The function we have just introduced derives from `Genfun::AbsFunction`. It turns out that many kinds of manipulations like addition, subtraction, multiplication, division, and composition return various different subclasses of `Genfun::AbsFunction`. It's possible to handle all of these through the interface to the base class, namely `Genfun::AbsFunction`, but the syntax for that is not often used and little known. For addition it is:

```
const Genfun::AbsFunction & h = f+g;
```

Normally the result of the addition operator returns a reference to a temporary and the temporary disappears when the operation is complete. However, there is a special clause in the C++ standard stipulating that the temporary is to be destroyed when the reference to it, h, goes out of scope. The unnatural look has been somewhat improved by the type definition:

```
typedef const AbsFunction & GENFUNCTION;
```

allowing one to write instead

```
Genfun::GENFUNCTION h = f+g;
```

Object oriented software libraries (such as `QatGenericFunctions`) can typically be used in two ways; the classes they contain can be created within programs and used, or they can be extended. The above example illustrates how to extend `QatGenericFunctions`. Should you ever need to extend the Open Inventor Toolkit, detailed instructions can be found in a volume called *The Inventor Toolmaker* (Wernecke, 1994).

12.8 Object-oriented analysis and design

Decisions, decisions, decisions! How does one possibly get around to actual problem solving with such a complicated set of rules and an impossibly large set of possibilities to choose from?

In any type of algorithm, numerical or otherwise, a certain investment in planning is always made at the outset of the problem-solving cycle. You break the problem up into a series of steps and define the data structures appropriate for storing the result of each step, including input, solution, iteration to convergence (if applicable), and output. Inevitably, we iterate the design process as our real-life experience shows us the weak spots. This is essentially the entire process in procedural programming.

In object-oriented programming there are other dimensions to code planning that are equally critical, since the code will now consist of objects which can handle a variety of tasks in an autonomous way. When the coding is to be carried out with an existing set of classes specifically intended for the relevant type of problem, life is simple. Often, though, *you* will have to design not only the procedures that apply to the problem but also the classes that you use within those procedures.

In these circumstances a variety of issues arise: what are the basic tasks? Which classes need to be invented to carry them out? What other classes are needed by these classes? How should the classes interact with one another? How can common functionality be identified and generalizations extracted? The questions that arise are not unique to numerical and/or scientific computing, and entire books have been written on the topic of object-oriented analysis and design, or OOAD. As soon as you find yourself spending a lot of time pondering this type of question, we suggest taking a week or so out of your busy schedule and reading a book like Booch (2007), which focuses on how to go about answering them. The topic is vast, and we will not address it further.

12.9 Exercises

1. Subclassing.
   Revisit examples 7, 8 and/or 14 of Chapter 3, this time subclassing `Genfun::AbsFunction` to eliminate any need for global variables.
2. Quadrature.
   Write and test your own classical quadrature rule, inheriting from `Genfun::QuadratureRule` and test it on the functions in Chapter 5, Exercise 1. Your quadrature rule does not need to have a specific order-of-convergence, but the weights

should sum to unity. Try making both an open and a closed quadrature rule. Make a plot of the convergence rate.

3. Special Functions.
   Choose a special function that is not implemented in QatGenericFunctions. Write your own class which extends Genfun::AbsFunction and implements this function. Plot the function and its derivative. You may, if you wish, take the implementation from the gnu scientific library and just write a wrapper.

4. Inherited Geometry.
   Write an abstract base class for simple geometrical shapes, and concrete implementations of sphere, cube, tetragon, and cylinder. These objects should all respond to a method that takes a vector in three dimensions as input and returns a boolean value indicating whether the vector lies inside the volume of the object. Choose your own conventions for the placement of the origin and orientation of the shape. Then, write a Monte Carlo integrator that handles each of the subclasses through the interface of the base class. Test it with the objects you have implemented.

5. [P] Plugins.
   It is possible to add functionality to an existing program without even recompiling the program using *plugins*, which are becoming ubiquitous in modern commercial software. The procedure to dynamically load a plugin describing a cone is as follows:

   - Inherit a subclass (call it Cone) inheriting from a superclass (call it SimpleGeoObject).
   - Write a function called

     ```
     extern "C" Cone *createCone() {
       return new Cone();
     }
     ```

     The extern "C" qualifier turns off name-mangling in C++.

   - Compile the function and the class definition into a shared object file, call it libCone.so, and put that shared object file in a directory which you dedicate to plugins.

   - Write another program to load and use the plugins. For this, you will need to scan your directory, and when you find a file named libCone.so, arrange to dynamically load the library, call the routine createCone(), and cast the return value to a Cone *. To do this you will need to use the dlopen, dlsym, and dlclose system calls. Documentation is in the unix manual pages.

   Carry out this procedure for an new Cone, and then revisit the previous exercise such that all of the geometrical shapes are loaded from plugins prior to volume integration.

BIBLIOGRAPHY

Booch, G. (2007). *Object Oriented Analysis and Design.* 3rd ed., Benjamin Cummings.
Meyers, S. (2015). *Effective Modern C++.* O'Reilly.
Wernecke, J. (1994). *The Inventor Toolmaker.* Addison-Wesley.

13
Nonlinear dynamics and chaos

13.1	Introduction	424
13.2	Nonlinear ordinary differential equations	426
13.3	Iterative maps	429
	13.3.1 The logistic map	430
	13.3.2 The Hénon map	435
	13.3.3 The quadratic map	436
13.4	The nonlinear oscillator	438
	13.4.1 The Lyapunov exponent	440
13.5	Hamiltonian systems	443
	13.5.1 The KAM theorem	444
	13.5.2 The Hénon-Heiles model	445
	13.5.3 Billiard models	447
13.6	Epilogue	449
13.7	Exercises	450
Bibliography		453

In the following sections we will use some of the techniques of the past chapters to explore simple systems that exhibit complex behavior. After categorizing some of the possible behaviors of second order nonlinear equations we examine the logistic map, which is a seemingly trivial one dimensional iterative mapping of the interval (0, 1) onto itself. Slight generalizations are also considered. The study of topics such as these lead directly to the concepts of attractors, fixed points, limit cycles, Lyapunov exponents, and period doubling. The same features are revealed again in a study of the damped, driven nonlinear pendulum. The chapter closes with a discussion of chaos in conservative systems and the KAM theorem. The FPU problem, the van der Pol and Duffing oscillators, and many other iterative maps will be visited in the exercises.

13.1 Introduction

A vague idea of 'chaos' or 'complexity' has enjoyed a burgeoning vogue in recent popular culture, propelled in part by compelling visual representations of objects such as the Mandelbrot set or the Lorenz attractor. There is no doubt that the advent of computers has advanced the field, since it was essentially impossible to solve the intricate and nonlinear equations that often seem to underpin chaotic processes. Briefly and informally: chaotic dynamics can be thought of as motion that appears random. Alternatively, chaos is motion that depends sensitively on initial conditions. Although the realization that such behavior is possible came as something of a shock, it has more recently been recognized that it is a pervasive and important facet of physical reality.

It is possible to push the genesis of the field right back to Newton's *Principia* without too much overstatement. Newton fully recognized that Jupiter must exert a noticeable and perhaps important influence on Earth's orbit. This is the famed *three body problem* that tormented theoretical astronomers for three centuries. Newton's concern was that perturbations caused by other bodies in the solar system could cause Earth's orbit to be unstable. A standard perturbative approach to the three body problem suffers from a 'small denominator' problem wherein putatively small orbital corrections can become large due to small denominators. Although many of history's eminent astronomer-mathematicians tackled the problem, none were able to evade it. Eventually King Oscar of Sweden and Norway offered a prize to anyone who could resolve the issue. The challenge was taken up, and won, by Poincaré in 1887.[1] But Poincaré realized that his analysis contained a deep and damaging error just as he was about to publish his winning essay. It was in correcting this error that Poincaré discovered mathematical chaos (Barrow-Green, 1996).

A new facet of chaos theory was revealed in the early 20th century in the work of French mathematicians Pierre Fatou and Gaston Julia. These men investigated 'holomorphic dynamics' dealing with iteration of analytic functions and introduced the concept of sensitivity to initial conditions that was to figure so prominently in subsequent developments. At the same time a diverse group of physicists and engineers explored chaotic behavior in nonlinear oscillators. The eminent physicist Hermann Helmholtz added a nonlinear term to the equation of motion for a harmonically forced undamped oscillator to serve as a model for the eardrum (Helmholtz, 1895). Some decades after this, the engineer Georg Duffing considered a forced anharmonic oscillator as a model of vibrations in industrial machinery (Duffing, 1918). The exploitation of new fields of science merely served to illustrate the ubiquity of nonlinear dynamics, for example, Balthasar van der Pol discovered chaotic behavior in electric circuits containing vacuum tubes (van der Pol and van der Mark, 1927).

It is perhaps not surprising that one of the earliest uses of computers was in an investigation of chaotic dynamics, even if it was accidental. The availability of newly built computers to the scientists of the famed Manhatten Project led Fermi to ponder using one to examine the issue of thermalization.[2] Fermi, John Pasta, and Stanisław Ulam

---

[1] Henri Poincaré (1854–1912) was a French mathematician, theoretical physicist, engineer, and philosopher of science who made seminal contributions to an astonishing array of topics, including the discovery of deterministic chaos.

[2] Enrico Fermi (1901–1954), Italian-American physicist known for building the first nuclear reactor and for extensive contributions to quantum theory, statistical mechanics, and nuclear physics.

(FPU) were eventually able to use the Los Alamos machine MANIAC-I to tackle the issue in 1953. The model FPU tackled was a system of one-dimensional linear oscillators with an additional nonlinear interaction. In the absence of the nonlinearity the vibrational modes of the system completely decouple from each other. However the nonlinear term couples modes and Fermi expected energy in one mode to gradually transfer to all other modes, in keeping with traditional concepts of ergodicity and equipartition. In fact, this is precisely what happened over short time scales, but running the simulations longer revealed, almost surreally, that the energy gradually left all the higher modes and returned to the original mode. It was several decades before this behavior was understood (Ford, 1992).

At about the same time, population biologists studied simple difference equations as models for a variety of population dynamics (May, 1987). But the biologists did not focus their attention on the chaotic solutions they were finding, perhaps out of distrust of their (mechanical) calculations. Thus no effort was made to search for generic features in the 'noise' of chaotic dynamics. Again, computers were required before such searches could be reasonably undertaken .

Attitudes had changed by the 1960s and academe was ready for developments that emerged from computational meteorology. In 1961, Edward Lorenz[3] was investigating the behavior of a numerical model of the atmosphere when he input a three digit truncation of a six digit initial condition. To his surprise the results were entirely different. Subsequent investigation led him believe that this was an actual and quite general occurrence, which came to be known as the *butterfly effect*.[4] Since this time the field has exploded, with applications found in dozens of fields and throughout nature.

13.2 Nonlinear ordinary differential equations

Much of physics can be described with second order differential equations. We therefore begin our study of nonlinear differential equations by considering the general coupled system

$$\dot{x} = P(x, y)$$
$$\dot{y} = Q(x, y). \tag{13.1}$$

A visualization of the solution to this equation can be made by making the parametric plot of the *trajectories* $(x(t), y(t))$. Notice that trajectories can never cross since this implies that it is possible for one initial condition to have multiple solutions.

Unfortunately, little can be done analytically (in general) with nonlinear equations, but some understanding may be gained by linearizing the system and determining solutions

---

[3] Edward Norton Lorenz (1917–2008) was an American mathematician and meteorologist, and a pioneer of chaos theory.
[4] For a brief history of the idea in science and in fiction see Hilborn (2004).

to the equations $P(a, b) = Q(a, b) = 0$. These ***fixed points***[5] define places where the solution becomes stationary. The behavior near the critical points can be obtained by expanding about (a, b):

$$\frac{d}{dt}\begin{pmatrix} x \\ y \end{pmatrix} = \begin{pmatrix} P_x(a, b) & P_y(a, b) \\ Q_x(a, b) & Q_y(a, b) \end{pmatrix} \begin{pmatrix} x \\ y \end{pmatrix}, \tag{13.2}$$

where $P_x \equiv \partial P/\partial x$, etc. This equation is most easily solved by diagonalizing. If the eigenvalues are denoted λ_1 and λ_2 then x and y are linear combinations of $\exp(\lambda_1 t)$ and $\exp(\lambda_2 t)$. Linearized solutions are categorized according to the nature of the eigenvalues: if they both have negative real parts, then the fixed point is ***stable***. If both have positive real points the fixed point is ***unstable***. A positive and negative eigenvalue give rise to a ***saddle point***. If the eigenvalues are purely imaginary then the trajectories are ellipses around the fixed point, while non-zero real parts will lead to ***spirals***. The ***separatrix*** divides the regions between the fixed points. As an example, consider the simple pendulum defined by (g is the acceleration due to gravity while ℓ is the length of the pendulum):

$$\ddot{\theta} + \frac{g}{\ell}\sin\theta = 0. \tag{13.3}$$

The coupled equations, $\dot{\theta} = \omega$ and $\dot{\omega} = -\frac{g}{\ell}\sin\theta$ have fixed points at $(\theta, \omega) = (n\pi, 0)$. Linearizing about even values of n yields eigenvalues $\lambda = \pm i\sqrt{g/\ell}$, hence elliptical trajectories surround the even fixed points. Alternatively, one obtains:

$$\frac{\dot{\omega}}{\dot{\theta}} \approx -\frac{g}{\ell}\frac{\theta}{\omega} \Rightarrow \omega d\omega + \frac{g}{\ell}\theta d\theta = 0 \Rightarrow \omega^2 + \frac{g}{\ell}\theta^2 = \text{const.} \tag{13.4}$$

Linearizing about the odd fixed points yields $\lambda = \pm\sqrt{g/\ell}$ and hence a saddle point behavior. In more detail:

$$\frac{\dot{\omega}}{\dot{\theta}} \approx \frac{g}{\ell}\frac{\theta}{\omega} \Rightarrow \omega d\omega - \frac{g}{\ell}\theta d\theta = 0 \Rightarrow \omega^2 - \frac{g}{\ell}\theta^2 = \text{const.} \tag{13.5}$$

Putting this information together yields the ***phase portrait*** shown in Figure 13.1.

Interesting things can happen as one moves away from the fixed points. For example, it is possible for trajectories to converge to a ***limit cycle*** regardless of their starting point within a ***basin of attraction***. A well known example is provided by the ***van der Pol oscillator*** (mentioned in the introduction). This is defined by the equation:

$$\ddot{x} - \mu(1 - x^2)\dot{x} + x = 0. \tag{13.6}$$

---

[5] Fixed points are also called equilibrium, critical, or singular points.

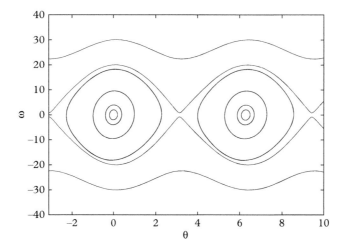

Figure 13.1 *Phase portrait for the simple pendulum.*

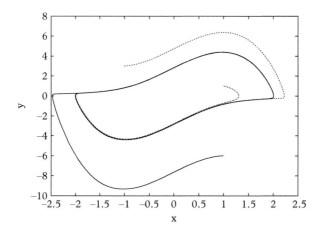

Figure 13.2 *Trajectories for the van der Pol equation, showing a limit cycle attractor. ($\mu = 2.5$).*

If one sets $\dot{y} = \mu(1 - x^2)y - x$ and $\dot{x} = y$ then the limit cycle shown in Figure 13.2 is obtained. Notice that since a limit cycle is closed and since trajectories cannot cross, a limit cycle must enclose a fixed point.

The question of the stability of a limit cycle remains. In principle one can consider linearizing around the limit cycle manifold but this is somewhat awkward, and is not necessary since Poincaré has suggested a simpler method. Cut the limit cycle along any line (say the $+y$-axis)[6] and keep track of the position of the intersection with the

---

[6] This line is called a **Poincaré section**.

trajectory in question as a function of intersection number. If the ith intersection is called P_i then these are related by some function f that propagates the intersection forward in time until the next intersection: $P_{i+1} = f(P_i)$. This is an example of an *iterative map* that will be the subject of the next section. If the trajectory *is* the limit cycle one must have $P_\star = f(P_\star)$, where $P_\star$ is a one-dimensional example of a fixed point. For trajectories that pass close to $P_\star$ one can expand on the Poincaré section and obtain:

$$(P_{i+1} - P_\star) \approx r(P_i - P_\star) \tag{13.7}$$

where:

$$r = \left.\frac{df}{dP}\right|_\star . \tag{13.8}$$

Iterating this equation reveals that trajectories are repelled from the limit cycle if $r > 1$ and are attracted to it if $r < 1$. A graph of the points P_i is called a ***Poincaré map***. Finally, a variant Poincaré map that is useful for simplifying chaotic phase space diagrams is obtained by plotting phase space points at certain 'stroboscopic' times, typically chosen to be an integral multiple of the frequency of the driving force (if present). An example of such a map is given in Section 13.4.

> Prove that r cannot be negative. Thus one can write $r = \exp(\lambda)$ where λ is real. This form of r will be re-visited below.

13.3 Iterative maps

The numerical methods presented in Chapter 11 center on the idea that differential equations can be thought of as a mapping from one configuration to another that is iterated in time. It is a small step to imagine this procedure without the backdrop of differential calculus: one simply considers the general mapping:

$$\vec{x}_{n+1} = \vec{f}(\vec{x}_n) \tag{13.9}$$

As one might expect, an enormous variety of iterative maps have found application in many areas. A famous example comes from population studies initiated by Pierre-François Verhulst in 1838. Verhulst was interested in the Malthusian problem of population growth with limited resources. His equation for a population $P(t)$ is:

$$\dot{P} = rP - \frac{r}{K}P^2 \tag{13.10}$$

where r is the reproduction rate and K is called the ***carrying capacity***. The terminology is understood by determining the condition for a stable population, namely $P(\infty) = K$.

Performing an Euler discretization, defining $\mu = 1 + hr$ and setting $P_n = (K\mu)/(hr)x_n$ gives the *logistic map*:

$$x_{n+1} = \mu x_n (1 - x_n). \tag{13.11}$$

This simple nonlinear iterative map contains a wealth of features which we will explore now.

13.3.1 The logistic map

Let's start by examining the sequence of x_n produced for a variety of couplings, μ. As with differential equations, these sequences are called *trajectories*. Trajectories for different starting positions and $\mu = 0.5$ are shown in Figure 13.3. It is pretty clear that the trajectories approach zero regardless of the starting point. Again, in analogy to the continuum case, the value to which the trajectories converge is called a *fixed point* of the trajectory (denoted $x_\star$ in the following). In this case $x_\star = 0$. The existence of a fixed point means $x_n = x_{n+1} = x_\star$ for sufficiently large n and hence:

$$x_\star = 0 \text{ or } x_\star = 1 - \frac{1}{\mu}. \tag{13.12}$$

Note that this behavior is the antithesis of ergodic dynamics—the system's phase space is *not* being uniformly sampled. The logistic map is *not* a Hamiltonian system. The subset of possible states that the system evolves to under iteration (in 'time') is called an **attractor**. In this case the attractor is two (or as we shall see, one) real numbers. In more complicated

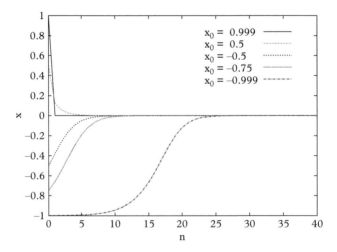

Figure 13.3 *Logistic map trajectories for $\mu = 0.5$.*

situations an attractor can have a rich geometry, and need not even have a well-defined dimensionality (we refer to the fractals of Chapter 8).

> Prove that trajectories are confined to the region:
>
> $$(f(\mu/4), \mu/4).$$

Figure 13.3 clearly agrees with the prediction $x_\star = 0$, but where is the second solution (Eq. 13.12)? A hint is provided by the large number of steps that is required for the trajectory that starts at $x_0 = -0.999$ to converge to zero. Indeed, starting at $x_0 = -1$ would yield a trajectory that remains at $x_n = -1$ for all n (recall that these statements are for the case $\mu = 0.5$).

What is going on is revealed with a one-dimensional version of the stability analysis performed in Section 13.2. Consider the general map $x_{n+1} = f(x_n)$ in the neighborhood of a fixed point. Setting $x_n = x_\star + \delta_n$ and expanding yields:

$$\delta_{n+1} = f'(x_\star)\delta_n \tag{13.13}$$

from which we conclude that the trajectory moves towards $x_\star$ if:

$$|f'(x_\star)| < 1 \tag{13.14}$$

and away from it if the magnitude of the derivative is greater than unity. The former situation refers to a **stable fixed point** wheres the latter is an **unstable fixed point**. Thus $x_\star = 0$ is stable as long as $|\mu| < 1$. Alternatively, at the nonzero fixed point one has $f'(x_\star) = 2 - \mu$ and hence this fixed point is stable when $\mu \in (1, 3)$ and it is unstable everywhere else.

So far we have established that all trajectories will approach a single value at $x_\star = 0$ if $\mu < 1$ and $x_\star = 1 - 1/\mu$ if $1 < \mu < 3$. What happens when μ is larger than 3? The answer is shown in Figure 13.4 where we see a new type of behavior when $\mu = 3.1$ or 3.5. After a short transient period, the trajectories oscillate between two ($\mu = 3.1$) or four ($\mu = 3.5$) values of x. These are called *period-two* or *period-four* trajectories of the logistic map.

What is happening can be illustrated by plotting $f(f(x))$ as shown in Figure 13.5. We see that $f(f(x)) = x$ at three points. Two of these are the upper and lower points seen in Figure 13.4 and are stable. The central point is unstable and does not feature in the trajectory. The difference between the period-two and -four trajectories is illustrated by plotting $f(f(f(f(x))))$, where it is revealed that the trajectory for $\mu = 3.1$ only has two stable fixed points, while that for $\mu = 3.5$ has four.

> Show that trajectories are period two for $3 < \mu < 1 + \sqrt{6}$.

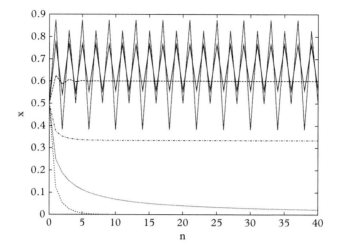

Figure 13.4 *Logistic map trajectories for $\mu = 3.5, 3.1, 2.5, 1.5, 1.0$ and 0.5.*

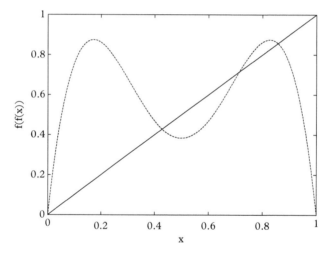

Figure 13.5 *The iterated logistic map $f(f(x))$ for $\mu = 3.5$.*

So far we know that there is one fixed point when $\mu < 3$, two for $3 < \mu < 1 + \sqrt{6}$, and four for higher coupling. What happens in general? The answer can be determined numerically by evaluating long trajectories and plotting the resulting fixed points. For those who have not seen it before, the result is a stunning verification of the onset of complexity in a seemingly trivial system (Figure 13.6). We see a general pattern of **period doubling** where the trajectory period visibly doubles at certain couplings and then degenerates into a richly patterned and complex region above a critical coupling of $\mu_\star \approx 3.56$. This is termed the **period doubling approach to chaos**, and the graph

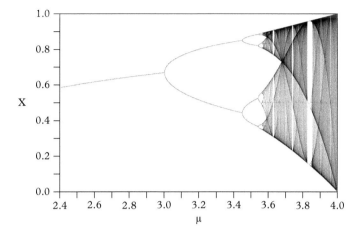

Figure 13.6 *Bifurcation diagram for the logistic map.*

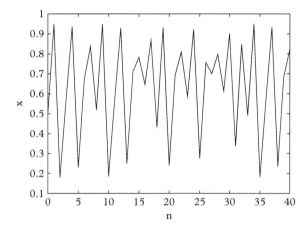

Figure 13.7 *A trajectory of the logistic map in the chaotic regime ($\mu = 3.8$).*

is called a ***bifurcation plot***. That large coupling is indeed "chaotic" is illustrated in Figure 13.7 where one sees seemingly random behavior[7] in a trajectory at $\mu = 3.8$

The rich structure seen in the bifurcation plot can be better understood by considering a quantity called the ***invariant density***. The density, $\rho(x)$, is defined as the distribution that reproduces the average of some function of x_n over a trajectory:

---

[7] At this point the reader is encouraged to ponder the relationship of this model to the linear congruence method for generating pseudo-random numbers in Chapter 10.

434 *Iterative maps*

$$\int_0^1 dx\, g(x)\rho(x) \equiv \frac{1}{N}\sum_n^N g(x_n). \tag{13.15}$$

Thus $\rho(x)dx$ is the probability that a point on a long trajectory is in the interval $(x, x+dx)$. For $\mu < 3$ we then have $\rho(x) = \delta(x_\star)$ and with period n orbits the density is given by a sum over n delta functions. Things get more complicated in the chaotic regime. For example, at $\mu = 4$ the interval $(0, 1)$ is filled with probability density $\rho \propto (x(1-x))^{-1/2}$.

A more interesting example, obtained by making a histogram of a long trajectory for $\mu = 3.8$, is shown in Figure 13.8. Notice that the density is riven with cusp features. It is these cusps that show up as boundary lines in the bifurcation plot (Figure 13.6). Intersections of boundary lines signify the existence of a fixed point or periodic orbit at the intersection point (Jensen and Myers, 1977).

The bifurcation plot (Figure 13.6) indicates that period doublings occur at specific values of the coupling. We call the nth such coupling $\mu_\star(n)$. Evidently $\mu_\star(1) \approx 3.0$, $\mu_\star(2) \approx 3.5$; furthermore the spacing between critical couplings gets smaller as the chaotic regime is approached. Is it possible that the difference of critical couplings approaches a fixed number? Surprisingly, the answer is yes. If one forms the ratio:

$$\Delta_n = \frac{\mu_\star(n-1) - \mu_\star(n-2)}{\mu_\star(n) - \mu_\star(n-1)} \tag{13.16}$$

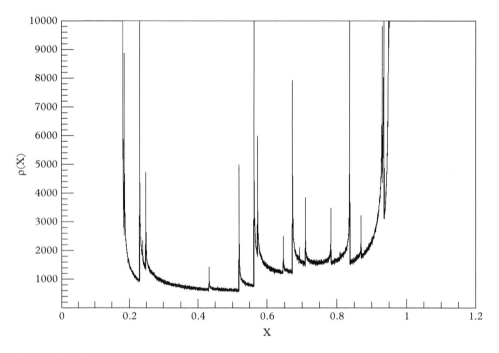

Figure 13.8 *Invariant density for logistic map for $\mu = 3.8$.*

Table 13.1 *Feigenbaum ratios.*

n	Period	$\mu_\star$	Δ_n
1	2	3	–
2	4	3.4494897	–
3	8	3.5440903	4.7514
4	16	3.5644073	4.6562
5	32	3.5687594	4.6683
6	64	3.5696916	4.6686
7	128	3.5698913	4.6692
8	256	3.5699340	4.6694

and considers it for large n one discovers (see Table 13.1) that it approaches a fixed constant:

$$\lim_{n\to\infty} \Delta_n = 4.669201\ldots. \tag{13.17}$$

This number is called the **Feigenbaum constant**, after its discoverer, Mitchell Feigenbaum (1978). Remarkably, all one-dimensional iterative maps with a single quadratic maximum bifurcate at the same asymptotic rate. This is an example of *universality*, where certain features of classes of models are shared. We have previously seen similar behavior in percolation models, in which certain critical exponents are independent of the geometry of the underlying percolation network.[8]

Use the table entries and the methods of Section 4.4 to obtain an accurate estimate of the critical coupling, $\mu_\star(\infty)$.

13.3.2 The Hénon map

The logistic map can be thought of as a discrete analogue of a nonlinear first order ordinary differential equation. What happens with higher order differential equations? As you know, an Nth order equation can be written as a system of N first order equations. Thus our analogue is found in an N-dimensional iterative map:

$$\vec{x}_{n+1} = \vec{f}(\vec{x}_n). \tag{13.18}$$

---

[8] More examples of universality are discussed in Chapter 8.

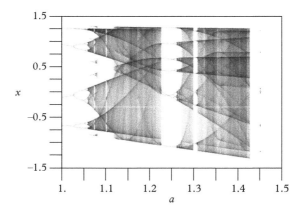

Figure 13.9 *Bifurcation plot for the Hénon model with $b = 0.3$. (Figure courtesy of Jordan Pierce).*

A famous example of a two-dimensional map was created by Michel Hénon as a simplified model of a Poincaré section of the Lorenz model. In contrast with the logistic map, attractors of the **Hénon map** cannot be described in a simple fashion.

The Hénon map is defined by:

$$\begin{aligned} x_{n+1} &= 1 - ax_n^2 + y_n \\ y_{n+1} &= bx_n. \end{aligned} \qquad (13.19)$$

Note that this is equivalent to $x_{n+1} = 1 - ax_n^2 + bx_{n-1}$. This map creates an intricate bifurcation plot, as shown in Figure 13.9.

The attractors of the map depend on the parameters; choosing $a = 1.4$ and $b = 0.3$ yields chaotic motion that is similar to that of Figure 13.7. Similar behavior is exhibited by y_n. An interesting simplification of the trajectory is obtained if one plots the phase diagram (interpreting y as a speed) for the chaotic motion. The result, shown in Figure 13.10 is an attractor of compelling simplicity. However, closer inspection shows that the attractor is riddled with gaps in random places, revealing that it is fractal. Fractal attractors are called **strange attractors** to distinguish them from the simpler geometrical sets that are often seen.

13.3.3 The quadratic map

A variant of the Hénon and logistic maps is simply given by:

$$z_{n+1} = f(z_n) \equiv z_n^2 + c \qquad (13.20)$$

where z and c are complex numbers. A simple example is provided by the case $c = 0$. Setting $z = r \exp(i\theta)$ then reveals that the initial point rotates and either moves away from the unit circle to infinity if $|z_0| > 1$ or towards the origin if $|z_0| < 1$. Alternatively, a

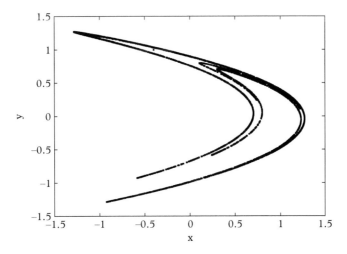

Figure 13.10 *Strange attractor for the Hénon map;* $a = 1.4, b = 0.3$.

trajectory that starts at $|z_0| = 1$ remains on the unit circle. The unit circle is an example of a *Julia set*, whereas the other points in the complex plane form two **Fatou sets**. The Fatou set is informally defined as the union of the domains of points that iterate to a given termination set; the Julia set is the complement. Thus the Julia set consists of the values of z_0 in which small perturbations can lead to large differences in the behavior of the trajectories. This matches our intuitive definition of chaoticity. In this case, two nearby points that start on the unit circle can swing in wildly different trajectories.

The following properties of Julia and Fatou sets are perhaps intuitive:

(i) they are complements of each other
(ii) they are invariant under the mapping: $f(\mathcal{J}) = \mathcal{J}$ and $f(F) = F$
(iii) trajectories within a Julia set are chaotic.

A numerical determination of the Julia and Fatou sets can be tricky, fortunately another theorem provides a simpler method: if f is a polynomial, then the Julia set is the boundary of the *finite Fatou set*; namely the set of z_0 with bounded trajectories. Thus an approximation to the Julia set can be made simply by iterating the quadratic map a certain number of times, N, and declaring an initial point in the set if $|z_N| < R$ where R is some scale chosen by the programmer. An example obtained for $c = 0.4 - i0.231$ and $R = 10$ is displayed in Figure 13.11. The white region indicates the set $\{z_0\}$ that iterate far from the origin. The central dark region remains bounded under iteration and forms the finite Fatou set, whose boundary is the Julia set. Notice that the finite Fatou set is apparently fractal in nature.

It is interesting to consider the complementary problem to determining the Julia set for a given map, namely determine the set of parameters c for which iterations of $z_0 = 0$

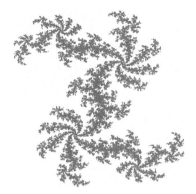

Figure 13.11 *The finite Fatou region for the quadratic map (c = 0.4 − i0.231).*

remain bounded. This is the **Mandelbrot set** famously associated with fractals, self-similarity, and complexity.

13.4 The nonlinear oscillator

Iterative maps provide a fascinating introduction to the concepts of chaotic dynamics, but they typically do not represent continuum systems. Can chaos be observed in physically relevant systems? The answer is yes—in fact many of the features we have already explored will appear again.

Perhaps the simplest chaotic physical system is represented by the damped, driven pendulum. The equation describing the system is:

$$\ddot{\theta} = -\frac{g}{\ell}\sin\theta - b\dot{\theta} + F\cos(\Omega t). \qquad (13.21)$$

Here we consider a simple pendulum of length ℓ in a uniform gravitational field, with no stretching, and no other external forces. The driving force is taken to be periodic with period Ω and strength F and the friction is modeled with the term $-b\dot{\theta}$.

We choose units such that $\omega^2 = g/\ell = 1$, set $b = 0.5$ and $\Omega = 2/3$ and integrate numerically. Because we are interested in long-time solutions, numerical accuracy is important and the Euler method is inappropriate. The trajectories for $F = 0$, $F = 0.5$, and $F = 1.2$ shown in Figure 13.12 have been obtained with the Euler-Cromer method.[9] The first two trajectories have been offset so that the plot is less cluttered. The solution for $F = 0$ displays **damping** with an initial **transient**, as is expected for a system with a strong frictional force and no external forcing. The second solution ($F = 0.5$) again displays a transient for $t < 10$ and then settles into simple periodic motion. Finally, increasing F beyond a critical value[10] of $F_\star(\infty) \approx 1.15$ yields the chaotic motion seen in

---

[9] The Euler-Cromer method is discussed in Section 11.7.
[10] The bifurcation plot for the damped oscillator actually alternates between chaotic and periodic regions as F varies.

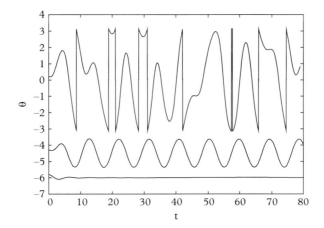

Figure 13.12 *The damped driven pendulum for $b = 0.5, F = 0, F = 0.5$ and $F = 1.2$.*

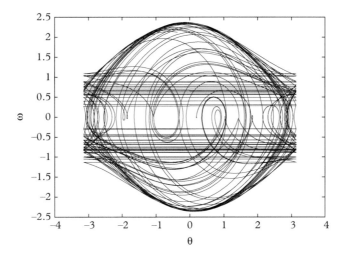

Figure 13.13 *Phase diagram for the damped, driven pendulum in the chaotic regime ($F = 1.2$).*

the top trace. Note that the frequent excursions to $\theta = \pm\pi$ indicate that the pendulum is swinging in complete circles many times during this trajectory. The pendulum also reverses direction 16 times during its evolution.

Some order can be brought to the chaos by considering the phase diagram for the different trajectories. As expected, the phase diagram for the case $F = 0.5$ is a circle in the $(\theta, \omega = \dot{\theta})$ plane (after an initial transient). However the phase diagram for the chaotic case is quite different (Figure 13.13). As with the Hénon attractor, the pattern is certainly not simple, but it is also certainly not random either.

It is possible to simplify the representation of the trajectories further by plotting the phase diagram at times that are integral multiples of the driving frequency ($t_n = 2\pi n/\Omega$).

Figure 13.14 *Strange attractor for the damped, driven pendulum in the chaotic regime ($F = 1.15$).*

The resulting plot is a Poincaré map, mentioned in Section 13.2. Of course, the periodic trajectory will appear as a single point representing the value of θ and ω at t_n (this is because the system oscillates at the driving frequency). Alternatively, the chaotic trajectory will be some subset of the full trajectory of Figure 13.13. The result, seen in Figure 13.14 as a polar plot with θ mapped to the angular variable and ω mapped to the radial variable, is reminiscent of the Hénon phase space plot and is a strange attractor.

13.4.1 The Lyapunov exponent

As we have learned in the case of the quadratic map, the Julia set can be thought of as the set of points that do not evolve smoothly with respect to each other under iteration. It is possible to formalize this concept with the aid of the ***Lyapunov exponent***.[11] The idea is to trace two nearby trajectories over time keeping track of the distance between them. If the distance is denoted $\delta\theta(t)$ then one has:

$$\delta\theta(t) = e^{\lambda t}\,\delta\theta(0) \qquad (13.22)$$

where λ is the Lyapunov exponent. Clearly, if the exponent is greater than zero, two trajectories that were initially close can be driven very far apart under system evolution.

Notice that the factor r that appeared in the discussion of the stability of limit cycles can be interpreted as the exponential of a Lyapunov exponent. In fact, with a

---

[11] Named after the Russian mathematician, Aleksandr Mikhailovich Lyapunov (1857–1918), who worked on stability of dynamical systems.

one-dimensional differential equation of the form $\dot{x} = g(x)$ expanding nearby trajectories, $x_1(t)$ and $x_0(t)$ yields[12]

$$\lambda = \frac{dg}{dx}\bigg|_{x_0}. \qquad (13.23)$$

In a D-dimensional state (phase) space there are D Lyapunov exponents given by the eigenvalues of the Jacobean matrix at the point of interest.

Notice that in general Lyapunov exponents depend on the starting position and vary in time. It is thus useful to recast the definition in terms of an average over a trajectory:

$$\lambda = \lim_{N \to \infty} \frac{1}{N} \sum_{n=0}^{N-1} \log |f'(x_n)|. \qquad (13.24)$$

(This expression is for an iterative map.) In this way the Lyapunov exponent has been conveniently divorced from initial points. The large 'time' limit in this expression is important: unstable fixed points initiate trajectories that also diverge exponentially, but only for a short time. Unfortunately, the definition of 'short' is difficult to quantify, and appropriate care must be taken by the practitioner.

> Show that Eq. 13.24 follows from $x_{n+1} = f(x_n)$. See (Earnshaw, 1993) if you are having trouble.

We are finally in a position to define chaos. If 'chaotic' is taken to mean 'extreme dependence on initial conditions', then chaos can be quantified as follows:

> A choatic system has at least one positive average Lyapunov exponent.

All continuous chaotic systems must contain at least three **degrees of freedom**[13]. Indeed, the nonlinear oscillator being studied here has three degrees of freedom (t, θ, ω) because it is nonautonomous (see Section 11.1). You can convince yourself that *three* degrees of freedom are required for chaos by considering a rank two equation. In this case if the trajectories are bounded they must either be attracted to a fixed point or be trapped by a limit cycle—there is no possibility for chaotic behavior to develop. Thus one must consider the spectrum of at least three Lyapunov exponents. If the system is dissipative then the sum of all the Lyapunov exponents must be negative (so that trajectories collapse to a fixed point). For systems without fixed points, one of the Lyapunov exponents must be negative (Haken, 1983). We are thus able to generalize the classification scheme given

---

[12] The differential equation $\dot{x} = g(x)$ is quite different from the iterative equation $x_{n+1} = f(x_n)$, hence the difference in the expressions for the Lyapunov exponents.

[13] Unfortunately the term 'degrees of freedom' is used differently in different fields. Here we mean the number of variables needed to define the coupled first order description of a system.

Table 13.2 *Lyapunov spectra and attractors for dissipative systems.*

Spectrum signs	Attractor
$(-,-,-)$	fixed point
$(0,-,-)$	limit cycle
$(0,0,-)$	quasi-periodic torus
$(+,0,-)$	chaotic

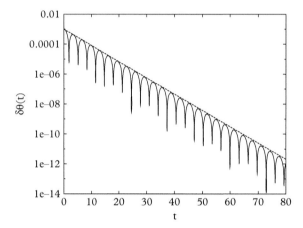

Figure 13.15 *Evolution of nearby trajectories for $F = 0.5$. The dips correspond to crossings of the trajectories.*

in Section 13.2 for two-dimensional systems to dissipative three-dimensional systems as shown in Table 13.2. The 'quasi-periodic torus' mentioned in the table will be discussed in Section 13.5.

Let us return to the nonlinear oscillator and examine its Lyapunov exponent in the θ coordinate. We employ the 'naive' definition of Eq. 13.22 and simply plot the deviation of two nearby trajectories, $\delta\theta(t) = |\theta_1(t) - \theta_2(t)|$, as a function of time. One might expect that in the periodic regime $\delta\theta$ stays constant, however this is not the case—nearby trajectories are rapidly driven toward each other. This is illustrated in Figure 13.15, which was generated for the periodic case $F = 0.5$ and an initial separation of $\delta\theta(0) = 0.001$. A fit to a simple exponential is shown as a dotted line and yields $\lambda = -0.35$.

The behavior of $\delta\theta$, shown in Figure 13.16, is completely different in the chaotic regime. Notice that the maximum separation (in our convention) is $\delta\theta = \pi$ so the separation must eventually saturate. This makes extracting the Lyapunov exponent difficult. A rough fit is shown as the dotted line and yields $\lambda \approx 0.15$.

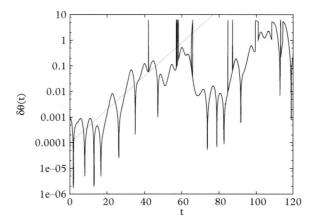

Figure 13.16 *Evolution of nearby trajectories for $F = 1.2$.*

It is evident that chaotic systems possess fractal properties, as we have seen in percolation. Are other aspects of percolation, such as self-similarity and the existence of critical scaling, manifested in chaotic systems? The answer is yes—see the exercises for an examination of self-similarity in the logistic map. Critical scaling can also be demonstrated for systems exhibiting the period doubling route to chaos. For example, in the logistic map it is found that just above μ_c the Lyapunov exponent behaves as:

$$\lambda(\mu) = \lambda_0(\mu - \mu_c)^\tau, \qquad (13.25)$$

where τ is a critical exponent with a value of approximately 0.4498.

13.5 Hamiltonian systems

The continuous systems studied so far have required nonlinearity, dissipation, and forcing to see chaotic behavior. What happens with systems that conserve energy such as the three-body or FPU problems? Such systems are called *conservative systems* or *Hamiltonian systems*.[14]

Recall that Liouville's theorem says that the phase space density is a constant of the motion:

$$\frac{d\rho}{dt} = \{\rho, H\} + \frac{\partial \rho}{\partial t} = 0. \qquad (13.26)$$

---

[14] Undergraduate physics students might be under the impression that most physics systems of interest are conservative. However, this is an idealization that is rarely achieved in actuality. All models must truncate degrees of freedom somewhere. The neglected degrees of freedom form a heat bath that influences energy flow in the subsystem of interest.

Hamiltonian systems

Thus volumes in phase space cannot change size and chaotic behavior cannot arise in the phase space density. Alternatively, the temporal evolution of ρ is linear and hence cannot be chaotic. This implies that averages over trajectories cannot be chaotic (since these can be used to define the phase space density)—only individual trajectories can show chaotic behavior.

> Show that $d\rho/dt = -b\rho$ for the nonlinear oscillator of Section 13.4.

If a Hamiltonian with N degrees of freedom[15] has N constants of motion it is said to be *integrable*. This class of Hamiltonians will form a central concept in the following discussion. If these constants of motion are called $\mathcal{J}_i(q,p)$ then it is possible to find a canonical transformation such that $H = H(\mathcal{J})$ and hence $\dot{\mathcal{J}}_i = 0$. We note that:

(i) all Hamiltonians with one degree of freedom are integrable since the energy is always a constant of the motion. (However, H must be an analytic function of q and p);
(ii) if Hamilton's equations are linear in q and p, the Hamiltonian is integrable;
(iii) if a Hamiltonian can be separated into systems with one degree of freedom it is integrable (Helleman, 1980).

An integrable system with N degrees of freedom generates trajectories that reside on an N-dimensional torus in the $2N$-dimensional phase space. The trajectories traverse the tori with characteristic frequencies, ω_i, in each of the tori dimensions. If these frequencies are incommensurate (not simple multiples of each other) then the trajectory eventually visits all parts of the torus surface and the system is said to be *ergodic*. The trajectories in this case are called *quasi-periodic* and the tori are called *KAM tori*. Alternatively, tori with frequencies that are simple multiples of each other (i.e., there exist integers such that $\sum n_i \omega_i = 0$) are called *resonant tori*. Finally, since integrable systems display only periodic or quasi-periodic trajectories, they are not chaotic.

> Show that the Lyapunov spectrum can be determined by integrating the eigenvalues of the system's Jacobean matrix (defined by generalizing Eq. 13.23) down a trajectory.

13.5.1 The KAM theorem

We now consider systems that are 'nearly' integrable. For the moment we restrict attention to systems with $N = 2$. If the system were integrable, trajectories would be restricted to a 2-dimensional torus in a four-dimensional space. If the system is nonintegrable, the trajectories can move in a three-dimensional region of phase space, and hence chaotic

---
[15] Here N is the total number of spatial coordinates for the system.

behavior is possible. Now add a small perturbation to the integrable system that destroys integrability. In the right coordinates we can write the new Hamiltonian as:

$$H'(p,q) = H(p) + \epsilon V(p,q). \tag{13.27}$$

Poincaré was the first to show that resonant tori are destroyed by arbitrarily small perturbations (Pöschel, 2001).[16] Since resonant tori form a dense set, the future stability of the solar system looked rather grim!

However, the future of humanity was saved by Kolmogorov in 1954, when he showed that the majority of tori survive the perturbation. This celebrated result is now called the *KAM theorem*.[17]

The KAM theorem states that if the system is subjected to a weak nonlinear perturbation, some of the KAM tori are deformed and survive, while others are destroyed. As stated above, resonant tori are destroyed, but tori with irrational ratios of tori frequencies survive. More rigorously, only **strongly nonresonant** tori—those with frequencies that obey:

$$\sum_i n_i \omega_i \geq \frac{\alpha}{\left(\sum_i |n_i|\right)^\tau} \tag{13.28}$$

for some α and τ and for all $\{n\} \in Z$ will survive the perturbation. As the strength of the perturbation is increased more tori dissolve, with those with frequency ratios closest to a rational number going first.

In these systems chaotic behavior is associated with the breakup of resonant tori. However for $N = 2$ the associated chaotic trajectories are constrained to small regions in phase space by the surviving KAM tori. As the perturbation grows larger more KAM tori dissolve and the region of phase space that is chaotic begins to dominate. Notice that if $N > 2$ KAM tori are no longer capable of constraining chaotic trajectories and they can diffuse throughout phase space.

13.5.2 The Hénon-Heiles model

The preceding discussion has been rather abstract; here we seek to flesh it out with a simple model invented by Hénon and Heiles (1964) to describe the motion of a star in a galaxy. As (minimally) required for chaotic behavior, the Hamiltonian has $N = 2$ and is nonintegrable:

$$H = \frac{1}{2}(p_x^2 + p_y^2) + \frac{1}{2}(x^2 + y^2) + x^2 y - \frac{1}{3}y^3. \tag{13.29}$$

---

[16] This is the famous incident referred to in the introduction.
[17] Named after Soviet mathematicians Andrei Nikolaevich Kolmogorov(1903 – 1987) and Vladimir Igorevich Arnold (1937 – 2010) and the German-American mathematician Jürgen Kurt Moser (1928 – 1999).

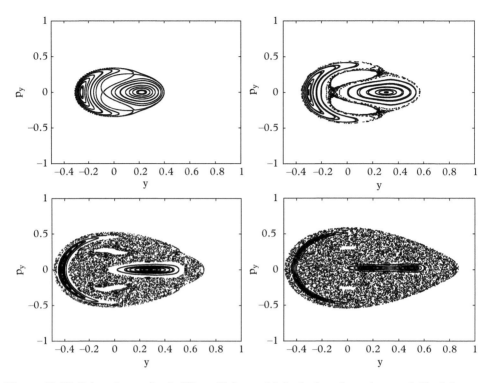

Figure 13.17 *Poincaré maps for the Hénon-Heiles model obtained on the section $x = 0$. Top left: $E = 0.06$, top right: $E = 0.10$, lower left: $E = 0.14$, lower right: $E = 0.16$.*

The potential has a local minimum at $(x, y) = (0, 0)$. Notice that the system is unbound in the directions $(x, y) \to (0, \infty)$ and $(x, y) \to (\pm\infty, -\infty)$ and that it has fixed points at $(x, y) = (0, 0), (0, 1)$, and $(\pm\sqrt{3}/2, -1/2)$.

For energies less than $1/6$ the system stays near the origin and approximates a 2d simple harmonic oscillator. For larger energy the nonintegrability of the model becomes important, KAM tori break up, and chaotic behavior becomes evident. This is displayed in a series of plots in Figure 13.17.

The plots show p_y vs. y on the section $x = 0$. At low energies the Poincaré map generated by a single trajectory is quite simple and the full map must be generated from many starting points sharing the same energy. The KAM theorem states that these trajectories must be quasiperiodic if they are not chaotic. Thus each trajectory (if not chaotic) should produce lines that fill in as the system is followed in time. These plots were generated with approximately 1/2 million time steps so that the lines are nearly filled in at this resolution.

As the energy of the system is increased (lower plots) chaotic behavior becomes evident—the plot becomes dense with points and only a few trajectories are required to fill in the figure. As expected, KAM tori are dissolving and contributing to chaotic behavior. In fact, chaoticity is also present at lower energy but is much less evident since they are constrained by intact KAM tori. For example, the chaotic region in the case $E = 0.06$ is confined near the crossing points at $(p_y, y) \approx (\pm 0.25, 0.15)$.

As indicated above, one can follow a trajectory in the chaotic regime and evaluate the Lyapunov spectrum. For systems with $N = 2$ there are four exponents and the sum of these must equal zero because of Liouville's theorem. In this case two eigenvalues are zero and thus the signs of the Lyapunov eigenvalues are $(+, 0, 0, -)$ (Hilborn, 1994).

13.5.3 Billiard models

Some Hamiltonians do not exhibit chaos because of nonintegrability but because the Hamiltonian is not an analytic function of the phase space variables. A simple example is the ***billiard model*** where pointlike balls undergo elastic collisions with a fixed closed surface.

One could consider solving this problem via, say, Euler's method. The problem is that scattering off of a wall occurs at a specific time which may not be on the temporal grid. To correct for this one can

(i) step forward in time
(ii) check to see if a boundary has been passed
(iii) interpolate backward to the time the boundary is hit
(iv) reflect and continue.

Although workable, there is an onerous catch: it is quite possible that the particle has run into a corner and has made two reflections during the time dt. In D dimensions up to D reflections can be made. It is awkward to check for these!

Another approach would be to abandon the, rather pedantic, attempt to track the particle at all times. Indeed we know its motion is linear between collisions. Thus one can simply solve equations to determine the time and location of the nth collision. For a simple concave bounding surface, the location of a collision can be uniquely specified by the angle of the event with respect to the origin (placed at the center of the billiard region). Furthermore, if we are not interested in the precise temporal evolution of the system it is sufficient to label these by the collision number. Thus the goal is to find a map:

$$\theta_{n+1} = f(\theta_n, \ldots). \tag{13.30}$$

We recognize an iterative map and thus the possibility of all of the fascinating chaotic behavior that accompanies such maps.

Let us take motion in a region bounded by a circular wall of unit radius as our first example. We label the angle of reflection at the nth collision point ψ_n. Then it can be shown that any path is given by the map (Chernov and Markarian, 2006):

$$\begin{aligned} \theta_{n+1} &= \theta_n + 2\psi_n \,(\text{mod})\, 2\pi \\ \psi_{n+1} &= \psi_n \end{aligned} \tag{13.31}$$

Establish the previous equations and then show:

$$\theta_n = \theta_0 + 2n\psi_0 \ (\text{mod}) \ 2\pi,$$

hence motion in a circle is not chaotic.

Note that the angle of reflection never changes and that the distance between collision points is constant. Trajectories also never approach closer that $\cos \psi_0$ to the origin (forming a *caustic* at this radius). Finally, trajectories with ψ_0/π rational will follow periodic orbits while those with an irrational ratio will be quasiperiodic, slowly filling in the space between $r = 1$ and $r = \cos \psi_0$.

In an effort to create chaoticity, we consider placing a disk of radius $r < 1$ at the center of the circular billiard. Trajectories with $\cos \psi_0 > r$ will not collide with the disk and will remain either periodic or quasiperiodic. However those with $\cos \psi_0 < r$ will reflect from the disk and move in wildly different directions than those with $\cos \psi_0$ slightly greater than r. Although this sounds like chaotic behavior, the system is integrable because it is rotationally invariant and hence cannot be chaotic.

This issue can be simply avoided by considering a square billiard. In this case reflections from a wall reverse the component of the particle velocity that is perpendicular to the wall. All four possibilities, $(\pm v_x, \pm v_y)$, will appear infinitely often in a given trajectory. Whether the system is chaotic can be easily analyzed with a trick: instead of reflecting at a wall, we permit the trajectory to pass into an image square. The original trajectory can be recovered by mapping the image square back to the original. It is now fairly easy to establish that a trajectory with rational v_x/v_y will be periodic, while an irrational ratio of velocity components gives rise to a quasiperiodic trajectory that fills the square. Furthermore, two trajectories that differ by a small angle will simply continue through the image squares, diverging as $\delta\theta \cdot t$. Thus dynamics in a square is also not chaotic.

This time, however, if a disk is placed in the center of the square, dramatically different behavior ensues. Such a configuration, called a *Sinai billiard*[18], exhibits chaotic dynamics. It is tempting to imagine that this is due to the disk, since some trajectories will just graze it, while nearby trajectories will hit it and experience large changes. However, we have already seen that this is not sufficient to cause chaos for a disk in a circle. In fact, it appears that the necessary ingredient is a combination of *nondispersing* (i.e., flat) surfaces with *dispersing* (curved) surfaces. Thus a square in a square is non-chaotic but a circle that has been stretched into a stadium shape by inserting two straight edges is chaotic.

---

[18] Named after Yakov Grigorevich Sinai, Russian mathematician (1935 –).

Billiard models of chaotic dynamics are of interest because of their simplicity. This makes them especially useful to examine when considering **quantum chaos** (which will not be discussed here). As a final point we note that if a disk can cause chaotic motion then a moving disk must also; thus a system consisting of two mobile hard disks in a box must also be chaotic. It is therefore perhaps no surprise that many-body systems are ergodic.

13.6 Epilogue

It is valid to ask what is being learned by studying chaos. When one discovers universal behavior in physical systems it is inevitably found to be due to something fundamental in natural laws. But this cannot be the case for chaotic behavior since it applies to so many different phenomena over so many different scales. Rather, it is due to common features in abstract state space. It appears that studying chaos does nothing for the reductionist approach to science—yet one cannot avoid feeling that it does teach us about Nature.

Perhaps the reader feels a lingering unease with the concept of exponential sensitivity to initial conditions. Is it indeed possible that a butterfly in Brazil can cause a tornado in Texas? To be clear, a butterfly cannot insert enough energy into the atmosphere to cause anything of importance, but the changed initial conditions can (in principle) cause a redistribution of energy that leads to tornados. It is, of course, impossible to have complete knowledge of initial conditions; hence our ability to predict into the future must be limited by the magnitude of the relevant Lyapunov exponents. This raises several issues:

An increasingly accurate knowledge of initial conditions must eventually bring one to the quantum realm, which has its own set of fundamental limitations. How does quantum reality affect the chaotic world? Does quantum indeterminacy invalidate classical chaotic dynamics? If so, how does the classical realm emerge from the quantum? We refer the interested reader to (Rudnick, 2008).

Exponential sensitivity appears to cast doubt on many of our carefully constructed numerical techniques. For example, it is unlikely that symplectic approximations to a Hamiltonian will preserve all of the symmetries of that Hamiltonian. It is thus possible that an integrable Hamiltonian would be approximated by nonintegrable (and hence chaotic) numerical time evolution. More simply, roundoff error necessarily limits our ability to specify initial conditions. Could all of our numerical conclusions be computational artifacts? Fortunately, things are not quite as dire as this apocalyptic scenario indicates. First, we have already seen that the KAM theorem provides some succor. Second, although a computed trajectory may not be what you thought it was, it is close to *some* trajectory of the system (Sauer *et al.*, 1997). This is useful because one is often interested in quantities averaged over trajectories, and it does not matter which particular set is used. But it would be premature to assume that all computational issues are under control; for example Lai and Grebogi (1999) have argued that any nonlinear model of a physical system will not be able to reproduce a physical trajectory of the system. As usual, care is required when relying on computers to lend insight into complex problems.

13.7 Exercises

1. [T] Logistic Map Invariant Density.
   Show analytically that the invariant density is given by:
   $$\rho = \frac{1}{\pi\sqrt{x(1-x)}}$$
   in the case $\mu = 4$.

2. [T] Murphy's Eyeballs.
   Determine the phase portrait for a centrifugal governor by linearizing about the fixed points. Model this as a massless rigid rod of length ℓ with point mass m at the free end which is attached via a swivel to a vertically oriented uniformly rotating shaft. The result is called *Murphy's eyeballs*.

3. Chaotic Magnet.
   Derive the equation of motion for a magnetic needle in the presence of an oscillating magnetic field that is perpendicular to the axis of rotation of the needle. You should find:
   $$\ddot{\theta} = -\frac{\mu B}{I}\cos(\omega t)\sin(\theta)$$
   Determine the dimensionless control parameter for this equation. Verify that chaotic behavior sets in above a critical value of the control parameter. Make phase space and Poincaré map plots in the nonchaotic and chaotic regimes.

4. The Duffing Oscillator.
   Consider the dynamics of a one-dimensional damped driven oscillator in a *double well* potential: $U(x) = -\frac{1}{2}kx^2 + \frac{1}{4}bx^4$. Choose units so that $m = k = b = 1$ and introduce a damping γ. Obtain $\dot{x} = v$, $\dot{v} = x - x^3 - \gamma v + F\cos(\Omega t)$.

   a) Verify that the period for small oscillations is $\sqrt{2}\pi$.
   b) The minima of the potential correspond to attractors centered on $x = 1$ and $x = -1$. Determine the basins of attraction of the minima. Take $F = 0.25$, $\gamma = 0.25$, and $\omega = 1$. Such complexly interleaved basins of attraction are called *riddled basins*.

5. [T] Bernoulli Shift Map.
   Consider the iterative map:
   $$x_{n+1} = 2x_n \pmod{1}, \quad x_n \in (0, 1). \tag{13.32}$$
   This map is called a *shift* because if one applies it to binary numbers less than unity it shifts all digits to the left and drops the digit that appears before the decimal point.

a) Prove that the map is chaotic. Hint: what happens to trajectories starting from rational numbers, $x_0 = n/m$? What happens to trajectories that start from irrational numbers? How are the rational and irrational numbers distributed over the real axis? Notice that the sequence of 1's and 0's generated from an irrational number is indistinguishable from a random sequence generated by coin tosses. For this reason some authors consider 'deterministic chaos' to be a misnomer for *random*.

b) Prove that the Lyapunov exponent is log 2.

c) Show that the Bernoulli shift map is equivalent to the logistic map when $\mu = 4$.

6. Hénon Map Attractors.
   Map out the behavior of the attractors of the Hénon map as a function of parameters. Take $a \in (-4, 1)$ and $b \in (-2, 2)$. You should find periodic, fixed point, and chaotic behavior. Much of the parameter region will yield unbounded trajectories.

7. Lagged Logistics Map.
   Explore the properties of the logistic map with lag:

   $$x_{n+1} = \mu x_n (1 - x_{n-1}).$$

   Try making 'phase space' plots of (x_n, y_n) where $y_{n+1} = x_n$.

8. Julia Sets.
   Determine the Julia sets for the quadratic map for a variety of parameters c. Consider making a movie showing the evolution of the sets as one moves around in the parameter space.

9. Mandelbrot Set.
   Implement an *escape time* algorithm for rendering the Mandelbrot set. Hint: if the real or imaginary part of z_n exceeds 2 then that trajectory cannot be part of the Mandelbrot set. Thus one can iterate through trajectories and use $|z| > 2$ as a termination criterion. A nice step is to set the color of the point to the number of iterations required to terminate.

10. Logistic Map Lyapunov Exponent.
    Use Eq. 13.24 to plot the Lyapunov exponent for the logistic map as a function of the coupling. Try to get an decent estimate of the critical coupling.

11. Damped Driven Oscillator.
    Compute the bifurcation plot for the damped driven pendulum. Use RK4 or the Euler-Cromer method. Take $g/\ell = 1$, $b = 0.5$, $\Omega = 2/3$. Use a time step around $h = 0.04$ and evolve past the transient to about $300\,T$ ($T = 2\pi/\Omega$). Then run up to $4000\,T$ and determine the period of motion as a function of $F \in (1.35, 1.5)$. You should find $F_\star(1) \approx 1.424$ and $F_\star(2) \approx 1.459$. See if you can estimate the Feigenbaum constant.

12. **The Lorenz Attractor.**
    Solve the Lorenz equations:
    $$\dot{x} = \sigma(y - x)$$
    $$\dot{y} = x(\tau - z) - y$$
    $$\dot{z} = xy - \beta z$$

    Take $\sigma = 10$ and $\beta = 8/3$. The parameter τ is analogous to the driving force in the driven pendulum. The system is chaotic for $\tau > 470/19$. Plot the projection of the phase diagram in the xz plane for $\tau = 10$ and $\tau = 28$. The latter should look like a butterfly. Show that the attractor is simpler in the Poincaré section defined by $y = 0$ (you should obtain a V shape).

13. **Recurrence Plots.**
    A recurrence plot is a two-dimensional representation of a trajectory that is defined by the density:
    $$\rho(i,j) = H(\epsilon - ||x(i) - x(j)||)$$
    where H is the Heaviside step function, x is the trajectory of interest, and ϵ is a parameter used to define how close the trajectory points must be to register a point in the plot. Explore the utility of this concept by making plots for the logistic map in the periodic and chaotic regimes. Compare to a random walk trajectory.

14. **Feigenbaum's α.**
    There are in general many universal properties of chaotic dynamics. Define d_n as the largest distance between values of x in the logistic map at which a bifurcation occurs (thus, d_1 is determined at $\mu_\star(1)$ and is about 0.42). Determine $\alpha \equiv \lim_{n\to\infty} d_n/d_{n+1}$. You should find $\alpha \approx 2.5029$.

15. **Self-similarity.**
    Examine the bifurcation plot of the logistic map carefully to see if it is self-similar. You should find that it becomes more exactly self-similar as the number of bifurcations gets larger.

16. **[P] The FPU Problem.**
    Consider the Hamiltonian for N coupled one-dimensional nonlinear oscillators:
    $$H = \frac{1}{2m}\sum p_k^2 + \frac{K}{2}\sum_{k=0}^{N-1}(q_{k+1} - q_k)^2$$
    $$+ \alpha \sum_{k=0}^{N-1}(q_{k+1} - q_k)^3.$$

    Choose units such that $m = 1$ and $K = 1$, used fixed boundary conditions and implement the Verlet algorithm. The transformation to the mode A_ℓ is given by:

$$A_\ell = \sqrt{\frac{2}{N}} \sum_{k=1}^{N-1} q_k \sin\left(\frac{k\ell\pi}{N}\right).$$

a) Choose initial conditions so that the first mode is excited and follow the energy of the first three odd modes as a function of time to confirm the observations of FPU. Recall that:

$$E_\ell = \frac{1}{2}\dot{A}_\ell^{\,2} + \frac{1}{2}\omega_\ell^2 A_\ell^2$$

with:

$$\omega_\ell = 2\sin\left(\frac{\ell\pi}{2(N+1)}\right).$$

b) Make a Poincaré map of $\dot{A}_1$ versus A_1 on the Poincaré section defined by $A_3 = 0$. You should find a hexagonal attractor for $\alpha = 0.3$, a six-lobed spiral for $\alpha = 1.0$, and chaotic behavior for $\alpha = 3.0$.

c) Check Lyapunov exponents to determine if the system is indeed chaotic for large α.

BIBLIOGRAPHY

Barrow-Green, J. (1996). *Poincaré and the Three Body Problem*. Am. Math. Soc., New York.

Chernov, N. and R. Markarian (2006). *Chaotic Billiards*, Am. Math. Soc., New York.

Coppersmith, S. N. (1999). *A simpler derivation of Feigenbaum's renormalization group equation for the period-doubling bifurcation sequence*. Am. J. Phys. **67**, 52–4.

Cvitanovic, P., G. Gunaratne, and I. Procaccia (1988). *Topological and metric properties of Hénon-type strange attractors*. Phys. Rev. A **38**, 1503.

Duffing, G. (1918). *Erzwungene Schwingungen bei Veränderlicher Eigenfrequenz*. F. Vieweg u. Sohn, Braunschweig.

Earnshaw, J. C. and D. Haughey (1993). *Lyapunov exponents for pedestrians*. Am. J. Phys. **61**, 401.

Fatou, P. (1917). *Sur les substitutions rationnelles*. Comptes Rendus de l'Académie des Sciences de Paris, **164**, 806–808; ibid **165**, 992.

Feigenbaum, M. J. (1978). *Quantitative universality for a class of nonlinear transformations*. J. Stat. Phys. **19**, 25.

Fermi, E., J. Pasta, and S. Ulam (1955). Los Alamos preprint LA-1940.

Ford, J. (1992). *The Fermi-Pasta-Ulam Problem: paradox turns discovery*. Phys. Rep. **213**, 271–310.

Haken, H. (1983). *At least one Lyapunov exponent vanishes if the trajectory of an attractor does not contain a fixed point*. Phys. Lett. A **94**, 71–4.

Hilborn, R. C. (2004). *Sea gulls, butterflies, and grasshoppers: A brief history of the butterfly effect in nonlinear dynamics*. Am. J. Phys. **72**, 425–7.

Helleman, R. H. G. (1980). *Self-generated chaotic behavior in nonlinear mechanics*, in *Fundamental Problems in Statistical Mechanics*. vol. 5, (E.G.D. Cohen, ed.), North Holland, Amsterdam.

Helmholtz, H. L. F. (1895). *On the Sensations of Tone as a Physiological Basis for the Theory of Music.* App. XII, *Theory of Combinatorial Tones.* 411–413, translated by Ellis, A.J., 3rd edn, Longmans Green and Co. Reprinted by Dover Publications, New York.

Hénon, M. (1976). *A two-dimensional mapping with a strange attractor.* Communications in Mathematical Physics, **50**, 69.

Hénon, M. and C. Heiles (1964). *The applicability of the third integral of motion: some numerical experiments.* Astrophysical J. **69**, 73–79.

Jensen, R. V. and C. R. Myers (1977). *Images of the critical points of nonlinear maps.* Phys. Rev. A. **32**, 1222.

Julia, G. (1918). *Mémoire sur l'iteration des fonctions rationnelles.* Journal de Mathématiques Pures et Appliquées, **8**, 47.

Lai, Y.-C. and C. Grebogi (1999). *Modeling of Coupled Oscillators.* Phys. Rev. Lett. **82**, 4803–4806.

Lorenz, E. N. (1963). *Deterministic Nonperiodic Flow.* J. Atmospheric Sciences **20**, 130.

May, R. M. (1987). *Chaotic Dynamics of Biological Populations.* Nucl. Phys. B Proc. Suppl. **2**, 225–246.

Pöschel, J. (2001). *A lecture on the classical KAM theorem.* Proc. Symp. Pure Math. **69**, 707–732.

Rudnick, Z. (2008). *What is Quantum Chaos?.* Notices of the AMS, **55**, 32–34.

Sauer, T., C. Grebogi, and J. A. Yorke. (1997). *How long do numerical chaotic solutions remain valid?.* Phys. Rev. Lett. **79**, 59–62.

Sinai, Ya. G. (1970). *Dynamical systems with elastic reflections: ergodic properties of dispersing billiards.* Russ. Math. Survey **25**, 137–189.

van der Pol, B and J. van der Mark (1927). *Frequency demultiplication.* Nature **120**, 363–364.

The following books are useful to the student interested in learning more.

1. R. L. Devaney (2003). *Introduction to Chaotic Dynamical Systems.* Westview Press.
2. R. C. Hilborn (1994). *Chaos and Nonlinear Dynamics: An Introduction for Scientists and Engineers.* Oxford University Press, New York.
3. J. P. Sethna (2006). *Entropy, Order Parameters, and Complexity.* Oxford University Press, New York.
4. M. Tabor (1989). *Chaos and Integrability in Nonlinear Dynamics: An Introduction.* Wiley, New York.

14

Rotations and Lorentz transformations

14.1 Introduction	455
14.2 Rotations	456
14.2.1 Generators	457
14.2.2 Rotation matrices	458
14.3 Lorentz transformations	460
14.4 Rotations of vectors and other objects	463
14.4.1 Vectors	463
14.4.2 Spinors	464
14.4.3 Higher dimensional representations	465
14.5 Lorentz transformations of four-vectors and other objects	467
14.5.1 Four-vectors	467
14.5.2 Weyl spinors	468
14.5.3 Dirac spinors	470
14.5.4 Tensors of higher order	473
14.6 The helicity formalism	474
14.7 Exercises	479
Bibliography	482

14.1 Introduction

Among the higher mathematical objects appearing in the vocabulary of physics are vectors and tensors, which are first encountered in classical mechanics; spinors, which appear in quantum mechanics; four-vectors, which appear in relativistic classical mechanics; and Dirac or Weyl spinors, which appear in relativistic quantum theory. The dynamics of particles and fields unfolds in a four-dimensional relativistic manifold called *spacetime*. Nevertheless, when the velocities of particles are small we can ignore the synthesis of space and time and work instead in Cartesian three-space.

The reader probably regards *rotations* as a group of transformations, acting upon vectors in a three-dimensional space, having the property that the norm $\sqrt{x^2 + y^2 + z^2}$ of the vector is preserved. In fact, other objects besides vectors are also acted upon by rotations: for example Pauli spinors, the coefficients of an expansion in spherical harmonics, stress tensors, multipole tensors, and others.

Likewise, the reader probably thinks of **Lorentz transformations** as transformations on four-vectors in spacetime that preserve the invariant interval $\sqrt{(ct)^2 - x^2 - y^2 - z^2}$. But again, after a closer look, one realizes that objects like Dirac and Weyl spinors also have transformation properties under Lorentz transformations. Indeed the existence of such objects can even be inferred from the structure of the Lorentz group.

The main reason for studying rotations is that we live in a three dimensional world; we therefore need tools to study the behavior of systems like a precessing top or a precessing electron spin. On the more practical side, we frequently deal with instrumentation that has a specific orientation in space that should be properly described. Alternatively, the main reason for studying Lorentz transformations is that they are useful in the computation of fundamental processes.

There are a variety of ways to represent rotations and Lorentz transformations, which are described in a vast literature. Rotations alone can be represented as:

- An axis of rotation and a rotation angle.
- Real orthogonal matrices of dimension 3×3.
- Complex matrices of dimension 2×2.
- Unit quaternions.
- A triplet of Euler angles. There are several distinct conventions.
- Four Euler parameters.
- Cayley-Klein parameters.

Thanks to *representation theory*, we know how to represent all possible rotatable objects as a vector of complex numbers and rotation operators as complex matrices. We adopt this approach here because it is universal: a vector in Hilbert space is transformed in the same way (and with the same code) as an ordinary vector in Cartesian 3-space. As a bonus we will see that Lorentz transformations can implemented easily by forming the tensor product of two representations of the rotation group. More on that in a moment.

14.2 Rotations

A rotation can be specified by giving the axis of rotation $\hat{n}$ and an angle of rotation θ. In code we can represent these two in a single vector $\vec{n} = \theta \hat{n}$ having a length θ and pointing along the axis of rotation. We view rotations as **active**, i.e. they change the orientation of the object—in contrast to **passive** rotations, in which the coordinate axes change their orientation. The direction of a positive rotation is given by the right-hand rule. If the thumb of the right hand points along the $\hat{n}$ direction, the rotation turns the object in the direction of the fingers. Thus a positive rotation along the z axis carries x into the y direction, and y into the $-x$ direction.

Having specified the rotation, our next goal is to obtain a matrix representation. If we want to rotate a vector, this will be a 3×3 matrix; a Pauli spinor will require a 2×2 matrix. The matrices which interest us are called ***irreducible representations*** of the rotation group, SU(2). Irreducible means that they cannot be decomposed into a direct product of smaller matrices. The irreducible representations describe simple physics objects like the spin of a particle. We may also want to rotate two electrons—this is more complicated; however, composite systems like this can be built out of simpler ones, so let's start with the rotation of simple systems. These are characterized by the spin j, which is a number that can take integral or half-integral values $0, 1/2, 1, 3/2\ldots$ Corresponding to j is the dimensionality of $D = 2j + 1$ of the representation. Vectors, for example, are three dimensional objects with $j = 1$.

14.2.1 Generators

The required rotation matrices can be made from three matrix ***generators***, $\vec{J} \equiv \{\mathcal{J}_x, \mathcal{J}_y, \mathcal{J}_z\}$, which depend on the number j. To be specific we list the generators for $j = 1/2$:

$$\mathbf{J}_x^{(1/2)} \equiv \begin{pmatrix} 0 & 1/2 \\ 1/2 & 0 \end{pmatrix} \quad \mathbf{J}_y^{(1/2)} \equiv \begin{pmatrix} 0 & -i/2 \\ i/2 & 0 \end{pmatrix} \quad \mathbf{J}_z^{(1/2)} \equiv \begin{pmatrix} 1/2 & 0 \\ 0 & -1/2 \end{pmatrix},$$

for $j = 1$:

$$\mathbf{J}_x^{(1)} \equiv \begin{pmatrix} 0 & \sqrt{2}/2 & 0 \\ \sqrt{2}/2 & 0 & \sqrt{2}/2 \\ 0 & \sqrt{2}/2 & 0 \end{pmatrix} \quad \mathbf{J}_y^{(1)} \equiv \begin{pmatrix} 0 & -i\sqrt{2}/2 & 0 \\ i\sqrt{2}/2 & 0 & -i\sqrt{2}/2 \\ 0 & i\sqrt{2}/2 & 0 \end{pmatrix}$$

$$\mathbf{J}_z^{(1)} \equiv \begin{pmatrix} 1 & 0 & 0 \\ 0 & 0 & 0 \\ 0 & 0 & -1 \end{pmatrix},$$

and for $j = 3/2$:

$$\mathbf{J}_x^{(3/2)} \equiv \begin{pmatrix} 0 & \sqrt{3}/2 & 0 & 0 \\ \sqrt{3}/2 & 0 & 1 & 0 \\ 0 & 1 & 0 & \sqrt{3}/2 \\ 0 & 0 & \sqrt{3}/2 & 0 \end{pmatrix} \quad \mathbf{J}_y^{(3/2)} \equiv \begin{pmatrix} 0 & -i\sqrt{3}/2 & 0 & 0 \\ i\sqrt{3}/2 & 0 & -i & 0 \\ 0 & i & 0 & -i\sqrt{3}/2 \\ 0 & 0 & i\sqrt{3}/2 & 0 \end{pmatrix}$$

$$\mathbf{J}_z^{(3/2)} \equiv \begin{pmatrix} 3/2 & 0 & 0 & 0 \\ 0 & 1/2 & 0 & 0 \\ 0 & 0 & -1/2 & 0 \\ 0 & 0 & 0 & -3/2 \end{pmatrix}.$$

For arbitrary j, the generators can be obtained from these formulae:

$$(J_x^{(j)})_{ik} = \frac{1}{2}\sqrt{k(2j-k+1)}\delta_{i,k-1} + \frac{1}{2}\sqrt{(2j-k)(k+1)}\delta_{i,k+1}$$
$$(J_y^{(j)})_{ik} = \frac{-i}{2}\sqrt{k(2j-k+1)}\delta_{i,k-1} + \frac{i}{2}\sqrt{(2j-k)(k+1)}\delta_{i,k+1}$$
$$(J_z^{(j)})_{ik} = (j-i)\delta_{i,k}. \qquad (14.1)$$

The indices i and k run from 0 to $2j = D - 1$. In the case of $j = 1/2$ the generators $\vec{J}^{(1/2)}$ can be related to the Pauli matrices.

$$\sigma_x \equiv \begin{pmatrix} 0 & 1 \\ 1 & 0 \end{pmatrix} \quad \sigma_y \equiv \begin{pmatrix} 0 & -i \\ i & 0 \end{pmatrix} \quad \sigma_z \equiv \begin{pmatrix} 1 & 0 \\ 0 & -1 \end{pmatrix},$$

through $\vec{J}^{(1/2)} = \frac{1}{2}\vec{\sigma}$; for higher spin, the generators can be viewed as a generalization of the rescaled Pauli matrices, since they obey the same commutation relations.

14.2.2 Rotation matrices

A rotation matrix is constructed from the generators through matrix exponentiation:

$$\mathbf{R} = \exp\left(-i\theta \hat{n} \cdot \vec{\mathbf{J}}\right). \qquad (14.2)$$

Computationally, the exponentiation of square matrices is a difficult problem which is still the subject of considerable research (Moler and Van Loan, 2003). An approach that works well in the present case takes advantage of the identity:

$$\exp\left(\mathbf{A} \cdot \mathbf{D} \cdot \mathbf{A}^T\right) = \mathbf{A} \cdot \exp\left(\mathbf{D}\right) \cdot \mathbf{A}^T. \qquad (14.3)$$

When $\mathbf{D}$ is a diagonal matrix, $e^{\mathbf{D}}$ is trivial to compute. If:

$$\mathbf{D} = \begin{pmatrix} \lambda_0 & 0 & 0 & \cdots \\ 0 & \lambda_1 & 0 & \cdots \\ 0 & 0 & \lambda_2 & \cdots \\ \cdot & \cdot & \cdot & \cdots \end{pmatrix}$$

then:

$$\exp \mathbf{D} = \begin{pmatrix} \exp(\lambda_0) & 0 & 0 & \cdots \\ 0 & \exp(\lambda_1) & 0 & \cdots \\ 0 & 0 & \exp(\lambda_2) & \cdots \\ \cdot & \cdot & \cdot & \cdots \end{pmatrix}.$$

This method is prone to failure in the case of degenerate or nearly degenerate eigenvalues; however this not the case here because the eigenvalues of the generators are evenly spaced: $-j, -j+1, \ldots j-1, j$.

The composition of rotations can be computed by multiplying together the matrices of any representation and then extracting the rotation axis and angle. The most efficient is the lowest dimensional non-trivial representation, in other words, $j = 1/2$ (or $D = 2$). One can show that the vector representing the rotation axis and angle is given by:

$$\vec{n} = (-4 \operatorname{Im} (\log \lambda_1)) \, a_1^\dagger \cdot \vec{\sigma} \cdot a_1$$

where λ_1 and a_1 are the maximal eigenvalue of the 2×2 rotation matrix and its corresponding eigenspinor.

Implementation

The classes and algorithms described in this chapter are implemented in a class library called `Spacetime`, which can be downloaded and installed from the companion site:

http://www.oup.co.uk/companion/acp

The class called `Rotation` implements rotations. The main challenge in implementing rotations as described in this section is efficiency; in particular we do not want to needlessly repeat calculations. The generators of SU(2) must be computed upon demand, and then held in case they are needed later. Since these are constant matrices, they can be computed once per job. We cache them in static `std::map` which associates the integral dimensionality D with the complex-valued $D \times D$ matrix. Moreover, each rotation computes a matrix in an arbitrary representation and then holds that matrix in case it is required again later. Matrix diagonalization is handled by the `Eigen` library's `ComplexEigenSolver` class. An excerpt from the header file that implements this appears below:

```
class Rotation {

  // Describes an active rotation. Has a law of composition.
     Can
  // return the irreducible representation of a rotation
     matrix in
  // D dimensions (D=2j+1; j=1/2, j=1, j=3/2, j=2...)

  public:
```

```cpp
// Constructor.  By default it will make an identity
    element.
Rotation();

// Constructor.  Takes the rotation vector as input;
explicit Rotation(const ThreeVector & nVector);

// Composition:
Rotation operator * (const Rotation & source);

// Get the dim-dimensional representation
const Eigen::MatrixXcd & rep(unsigned int dim) const;

// Get the rotation vector.
const ThreeVector getRotationVector() const;

// Gets the inverse rotation
Rotation inverse() const;
...
};
```

The rep method provides any D dimensional representation of the rotation group. The class may appear paltry because it does not rotate *anything* yet (besides other rotations), but we will get to that later. The class ThreeVector which appears in the interface to Rotation is discussed in Section 14.4.1.

14.3 Lorentz transformations

Lorentz transformations encompass both rotations and boosts. As fortune would have it, the rotation generators discussed in Section 14.2.1 can also be used to build up irreducible representations of the proper, orthochronous Lorentz group[1]. Representations of the rotation group SU(2) are characterized by a half integer j, as previously discussed. Lorentz transformations are, by contrast, characterized by two half integers which we will call j_1 and j_2. These representations are denoted by (j_1, j_2): for example we have the trivial $(0, 0)$ representation, the so-called right-handed $(1/2, 0)$ representation, the left-handed $(0, 1/2)$ representation, the $(1/2, 1/2)$ representation, and so forth. The meaning of these numbers will become clear in a moment.

The rotation part of a Lorentz transformation can be specified as before by a vector in the direction of the rotation, with a length set equal to the rotation angle. The boost vector

---

[1] "Proper" means that the determinant of the Lorentz transformation is positive. "Orthochronous" means that the 00 component of the transformation is positive. These restrictions specify a continuous subgroup of the Lorentz transformations; the other subgroups can be obtained with parity or time-reversal transformations.

$\vec{\eta}$ points in the direction of the (active) boost and has a magnitude equal to $\eta = \tanh^{-1} \beta$, where $\beta = v/c$.

Two matrices are then constructed, the first matrix $\mathbf{R}_1^{(j_1)}$ has dimension $D_1 \times D_1$ where $D_1 = 2j_1 + 1$; the second $\mathbf{R}_2^{(j_2)}$ has dimension $D_2 \times D_2$ where $D_2 = 2j_2 + 1$. These are obtained by matrix exponentiation:

$$\mathbf{R}_1^{(j_1)} = \exp\left(-i\left(\vec{n} + i\vec{\eta}\right) \cdot \vec{\mathbf{J}}^{(j_1)}\right)$$
$$\mathbf{R}_2^{(j_2)} = \exp\left(-i\left(\vec{n} - i\vec{\eta}\right) \cdot \vec{\mathbf{J}}^{(j_2)}\right). \quad (14.4)$$

We anticipate that the same instance of a Lorentz transformation could be applied to different types of objects, (e.g. four-vectors, relativistic spinors, or tensors) and therefore require that the class build the matrices upon demand, caching them in a map for later use. Clients can then obtain a representation of the Lorentz group by requesting a pair of matrices corresponding to chosen values of j_1 or j_2. In code, we prefer to search by the integers D_1 or D_2 rather than j_1 and j_2 since there is no built-in datatype for half-integers.

Composition of Lorentz transformations is accomplished in the following way. First, one obtains the $j = 1/2$ representation of $\mathbf{R}_i^{(1/2)}$ for each of the two operands; these are 2×2 complex matrices. Then, one multiplies the two matrices to obtain $\mathbf{R}$, another two-by two matrix. To recover the boost, one can invert the relation

$$\mathbf{R} = \exp\left(\frac{-i\vec{n} + \vec{\eta}}{2} \cdot \vec{\sigma}\right)$$

yielding

$$(-i\vec{n} + \vec{\eta}) \cdot \vec{\sigma} = 2 \log \mathbf{R}$$

or

$$\begin{pmatrix} -in_z + \eta_z & -in_x + \eta_x - n_y - i\eta_y \\ -in_x + \eta_x + n_y + i\eta_y & in_z - \eta_z \end{pmatrix} = 2 \log \mathbf{R}.$$

Linear combinations of the matrix elements of $2 \log \mathbf{R}$ can be taken to obtain the components of $\vec{n}$ and $\vec{\eta}$.

Implementation

The class `LorentzTransformation` in the `Spacetime` library implements Lorentz transformations. The class is initialized with a rapidity vector $\vec{\eta}$ and a rotation vector $\vec{n}$, and has a law of composition. It returns two matrices, the return values from the methods `rep1` and `rep2`, which, as we shall outline later, can be used to transform any objects; but if either of the dimensionalities D_1 or D_2 is one, then the representation is trivial and the "matrix" is just the constant 1.0. An excerpt from the header file is shown below.

```
class LorentzTransformation {

  // Describes an active Lorentz Transformation.  Has a law
     of composition.

  // Lorentz transformations act on rank-2 tensors, in
     general:
  //
  // Weyl spinors:                       1/2
  // Dirac spinors                       1/2 + 1/2
  // Four-vectors                        1/2 x 1/2.
  // AntiSymmetric 4-Tensors             1   +   1
  //
  // It up to these objects to transcribe themselves into the
     proper
  // representation and apply the Lorentz Transformation,
     which
  // furnishes two matrices acting on the each piece of the
     rank 2 tensor.

 public:

  // Constructor.  By default it will make an identity
     element.
  LorentzTransformation();

  // Constructor.  Takes the rapidity vector and the rotation
     vector:
  LorentzTransformation(const ThreeVector & rapVector,
                        const ThreeVector & rotVector=
                            ThreeVector(0,0,0));

  // Copy Constructor:
  LorentzTransformation(const LorentzTransformation &
     source);

  // Composition:
  LorentzTransformation operator *
      (const LorentzTransformation & source);

  // Get the dim-dimensional representation, matrix 1 and
     matrix 2
  const Eigen::MatrixXcd & rep1(unsigned int dim) const;
```

```
   const Eigen::MatrixXcd & rep2(unsigned int dim) const;

   // Get the rotation vector and the rapidity vector:
   const ThreeVector getRotationVector() const;
   const ThreeVector getRapidityVector() const;

   // Gets the inverse rotation
   LorentzTransformation inverse() const;

   // Assignment operator
   LorentzTransformation & operator=(const
       LorentzTransformation & source);
...
};
```

Because the representation of the Lorentz group consists of *two* rotation matrices, we foresee two methods `rep1` and `rep2`, each taking a single parameter which is the dimensionality of the matrix. In subsequent sections we will demonstrate some uses of this class.

14.4 Rotations of vectors and other objects

In this section we will see how to apply rotations to various objects, starting with vectors. Rotating other objects, particularly those that live in Hilbert space, will also be explored.

14.4.1 Vectors

The first and most obvious object upon which rotations must act are vectors. Our rotation class can furnish 3×3 matrices; and while these matrices are not constructed to act on the three-component vector *directly*, they are homomorphic to matrices which do. This means, in practical terms, that we reassign the components x, y, z of an ordinary vector to construct[2]

$$\begin{pmatrix} \frac{-x+iy}{\sqrt{2}} \\ z \\ \frac{x+iy}{\sqrt{2}} \end{pmatrix}.$$

We can rotate *this* three-component object using our matrices, and then take linear combinations to extract the coordinates of the rotated object, x', y', z'. The class `ThreeVector` implements vectors in Cartesian 3-space; these possess all of the familiar

---

[2] If this looks peculiar consult Appendix C.V of (Messiah, 2014).

operations (access to the coordinates, dot product, cross product, addition, subtraction, scaling, unary minus, plus compound operations +=, -=, *=, and /=). Rotation of vectors is implemented in the following function:

```
ThreeVector operator * (const Rotation & R, const ThreeVector
     & v) {
  typedef std::complex<double> Complex;
  static const double sqrt2=sqrt(2.0);
  // Fill a complex vector:
  Eigen::Vector3cd
    X(Complex(-v(0), +v(1))/sqrt2,
      Complex(+v(2)),
      Complex(+v(0), +v(1))/sqrt2);

  // Rotate the complex vector:
  Eigen::Vector3cd XPrime=R.rep(3)*X;

  // Re-extract x,y,z values from the rotated complex vector.
  return ThreeVector((XPrime(2)-XPrime(0)).real()/sqrt2,
                     (XPrime(0)+XPrime(2)).imag()/sqrt2,
                     XPrime(1).real());
}
```

Note that we do not define the operation of matrices upon ThreeVectors, which are considered to be geometric rather than algebraic objects—in contrast to spinors which are discussed next.

14.4.2 Spinors

Spinors are algebraic objects upon which rotations can act. They are complex two-component column vectors representing the state of spin-1/2 particles in quantum mechanics. The two basic states are:

$$|\uparrow\rangle = \begin{pmatrix} 1 \\ 0 \end{pmatrix} \quad |\downarrow\rangle = \begin{pmatrix} 0 \\ 1 \end{pmatrix}.$$

Other orientations are possible, too: they are linear combinations $\alpha|\uparrow\rangle + \beta|\downarrow\rangle$, where α and β are complex coefficients such that $|\alpha|^2 + |\beta|^2 = 1$.

One possible implementation of spinors is to make them a subclass of Eigen::Vector2cd. To rotate the spinor, one could directly apply the matrices of the $j = 1/2$ ($D = 2$) representation to the complex two-component vector:

```
Spinor operator * (const Rotation & R, const Spinor & X) {
  return R.rep(2)*X;
}
```

Note now that the generators of Section 14.2.1 do double duty. In addition to generating the rotation, they serve as operators which act upon the spinors and which determine the spin vector $\vec{v}_s$ representing the magnitude and direction of the spin. Specifically, for spin 1/2, this is obtained from the spinor in the following way:

$$\vec{v}_s = \langle s|\vec{\mathbf{J}}^{(1/2)}|s\rangle. \tag{14.5}$$

This is implemented as follows:

```
ThreeVector spin(const Spinor & s) {
  return ThreeVector((s.adjoint()*SU2Generator::JX(2)*s)(0,0)
                       .real(),
                     (s.adjoint()*SU2Generator::JY(2)*s)(0,0)
                       .real(),
                     (s.adjoint()*SU2Generator::JZ(2)*s)(0,0)
                       .real());
}
```

Before describing our implementation, we consider yet another category of objects that are subject to rotations, namely generalizations of two-component spinors to objects of higher dimension.

14.4.3 Higher dimensional representations

Spinors describe the quantum state of spin 1/2 particles. To describe states of angular momentum beyond $j = 1/2$, we must employ higher dimensional representations of the angular momentum group SU(2). The state of a spin-j system can be represented as a vector in a $2j + 1$ dimensional complex vector space. Physics has not bestowed a great name to this object—some call it (improperly) a "representation of SU(2)", others overload the name "spinor" to refer to these higher dimensional complex vectors. We will follow the latter tradition and refer to them as spinors. Matrices of the j representation, of dimension $D = 2j + 1$, can be used to affect a rotation of these objects.

Let us require a class for all spinors, general enough to support the "normal" spin 1/2 spinors of dimension two as well as higher dimensional spinor representations. We can do this by implementing Spinor as a template class, taking the dimensionality of the representation, D, as a template parameter with a default value of 2, as follows:

```
template <unsigned int D=2>
class Spinor : public  Eigen::Matrix<std::complex<double>,
    D,1>
```

One can then instantiate a Spinor<2>, a Spinor<3>, and so forth. Our Spinor class, declared in the file Spacetime/Spinor.h, has a default constructor and an initializer-list constructor which one can use like this:

```
Spinor<> s {1/sqrt(2),1/sqrt(2)};
```

The angle brackets are necessary and if no value is enclosed then the default value $D = 2$ is taken. Spinors inherit all the basic functionality of `Eigen::VectorXcd`, i.e., of complex column vectors.

The implementation of rotations for the general D-dimensional spinor is straightforward:

```
template<unsigned int D>
inline Spinor<D> operator * (const Rotation & R, const Spinor
    <D> & X) {
  return  R.rep(D)*X;
}
```

as is the implementation of the spin function:

```
template<unsigned int D>
inline ThreeVector spin(const Spinor<D> & s) {
  return ThreeVector((s.adjoint()*SU2Generator::JX(D)*s)(0,0)
                       .real(),
                     (s.adjoint()*SU2Generator::JY(D)*s)(0,0)
                       .real(),
                     (s.adjoint()*SU2Generator::JZ(D)*s)(0,0)
                       .real());
}
```

Related objects: wavefunctions and spherical tensors

When j is an integer, the `Spinor` class can be used to represent other objects and the effect of rotations upon them. In the central force problem, particular solutions to the time-independent Schrödinger equation take the form of Eq. 11.26. The full, time-dependent solution is in general a superposition of these solutions (eigenstates), together with their time dependence, i.e.:

$$\Psi(\vec{r},t) = \sum_n \sum_l \sum_{m=-l}^{l} a_{nlm} \frac{1}{r} U_{nl}(r) Y_l^m(\theta,\phi) e^{-iE_{nlm}t/\hbar}. \tag{14.6}$$

The shape of the wavefunction is encoded in the complex coefficients a_{nlm}. To rotate the wavefunction we can act upon the coefficients: we arrange, for every value of n and l, the $2l + 1$ coefficients $a_{nl(m)}$ into a `Spinor` of dimension $D = 2l + 1$. This can then be rotated as any other spinor.

In Section 5.1.5 we discussed the multipole moments q_{lm} of a charge distribution. This is an example of a spherical tensor; for $l = 2$ it is the quadrupole tensor, for $l = 3$ the sextupole tensor, etc. Note that in general the moments are complex-valued. The components of a spherical tensor for a fixed value of l can be rotated by organizing the coefficients q_{lm} into a $D = 2l + 1$ spinor. Rotations are made as with any other spinor.

14.5 Lorentz transformations of four-vectors and other objects

In this section we investigate objects that transform covariantly under Lorentz transformations. Some of these, like four-vectors and Weyl spinors correspond to irreducible representations of the proper orthochronous Lorentz group, and others like Dirac spinors correspond to reducible representations. A lot of mathematical jargon can be thrown at the reader at this point, but we assume that you have had your fill, and focus on one point: all irreducible Lorentz covariant objects can be cast in the form of a $(2j_1 + 1) \times (2j_2 + 1)$ dimensional matrix. Lorentz transformations of these objects can be affected by simultaneously left-multiplying and right-multiplying by the matrices $\mathbf{R}_1^{(j_1)}$ and $\mathbf{R}_2^{(j_2)T}$ from Eq. 14.4. Reducible Lorentz-covariant objects can be represented as direct sums of irreducible objects.

As a quick guide to the following sections: four-vectors can be transformed as 2×2 complex matrices; Weyl spinors can be transformed exactly as they are normally represented, i.e. as two-component spinors; and Dirac spinors can also be transformed as they are represented—in this case as a sum of two Weyl spinors, or if you prefer, a four-component spinor.

14.5.1 Four-vectors

A four-vector is a basic geometric entity arising in special relativity. The archetypical four-vector is the spacetime four-vector with components (ct, x, y, z) or (x^0, x^1, x^2, x^3). A common four-vector represents energy and momentum as $(E, \vec{p})$ (in units where $c = 1$). E is the relativistic energy $E = m\gamma$ of a particle and $\vec{p}$ is the relativistic momentum $\vec{p} = m\vec{\beta}\gamma$. A decent four-vector class should include the basic operations of addition, subtraction, and scaling, the compound operations +=, -=, *=, and /=, and the dot product between two four-vectors $a \cdot b = a_\mu b^\mu$. As is likely familiar, a_μ denotes the "covariant" components $(a_0, a_1, a_2, a_3) = (a^0, -a^1, -a^2, -a^3)$ and b^μ denotes the contravariant components (b^0, b^1, b^2, b^3). Summation over repeated indices is implied. This dot product is invariant under simultaneous Lorentz transformations of the two four-vector operands.

Usually four-vectors are real-valued objects, but in one particular situation complex-valued four-vectors are needed. The polarization of a massive spin-1 particle propagating along the z direction is represented by the polarization vectors describing longitudinal (0) and transverse (x,y) polarization:

$$\epsilon_0 = (\beta\gamma, \gamma\hat{\beta})$$
$$\epsilon_x = (0, 1, 0, 0)$$
$$\epsilon_y = (0, 0, 1, 0)$$

however, often one takes longitudinal, right- and left- handed polarization vectors,

$$\epsilon_0 = (\beta\gamma, \gamma\hat{\beta})$$
$$\epsilon_R = (0, -1, -i, 0)/\sqrt{2}$$
$$\epsilon_L = (0, 1, -i, 0)/\sqrt{2}$$

as basis states; for massless spin-1 particles the longitudinal state is absent.

We implement two classes, `FourVector` and `ComplexFourVector`, by instantiating our template class `BasicFourVector`, using `double` as the underlying datatype for the former and `std::complex<double>` for the latter. These classes have the expected operations (addition, subtraction, scaling, dot product, norm, abs, plus compound operations `+=,-=,*=`). In addition they may be acted upon by Lorentz transformations (with `operator *`).

The way this works is as follows. First, one forms a complex 2 × 2 matrix from the components of x^μ:

$$\mathbf{X} = \begin{pmatrix} -x^1 + ix^2 & x^3 + x^0 \\ x^3 - x^0 & x^1 + ix^2 \end{pmatrix}$$

Then, after obtaining the rotation matrices $\mathbf{R}_1^{(1/2)}$ and $\mathbf{R}_2^{(1/2)}$ from the Lorentz transformation, the product:

$$\mathbf{X}' = \mathbf{R}_1^{(1/2)} \mathbf{X} \cdot (\mathbf{R}_2^{(1/2)})^T$$

is formed. The elements of $\mathbf{X}'$ are:

$$\mathbf{X}' = \begin{pmatrix} -x^{1'} + ix^{2'} & x^{3'} + x^{0'} \\ x^{3'} - x^{0'} & x^{1'} + ix^{2'} \end{pmatrix}.$$

Components of the transformed four-vector, $x' = (x^{0'}, x^{1'}, x^{2'}, x^{3'})$ are obtained by taking linear combinations of the elements of $\mathbf{X}'$. These are stuffed back into a new four-vector (real or complex) and returned.

14.5.2 Weyl spinors

Weyl spinors are complex-valued two-component column vectors. They are used to represent massless spin 1/2 particles. It was long thought that neutrinos provided a concrete example—but no longer! Neutrinos are now known to have a mass, leaving the Weyl spinors completely out of the game until further notice. Nonetheless, they are still an excellent approximation given that neutrino masses are exceedingly small. In addition, the Dirac spinors, considered next, are simply a direct sum of two Weyl spinors.

Weyl spinors come in two-varieties, left-handed and right-handed. In our library, these are called `WeylSpinor::Left` and `WeylSpinor::Right`. Unlike ordinary spinors,

Weyl spinors are normalized to the energy of the particle that they represent, i.e. such that $\chi^\dagger \cdot \chi = E$.

In a Lorentz transformation, the right-handed Weyl spinor is acted upon by the matrix $\mathbf{R}_1^{(1/2)}$, while the left handed Weyl spinor is acted upon by $\mathbf{R}_2^{(1/2)}$ (see Eq. 14.4). That's all there is to it. The implementations are:

```
WeylSpinor::Right operator*(const LorentzTransformation & L,
                            const WeylSpinor::Right & s) {
  return L.rep1(2)*s;
}

WeylSpinor::Left operator * (const LorentzTransformation & L,
                             const WeylSpinor::Left & s) {
  return L.rep2(2)*s;
}
```

As with ordinary spinors, it is possible to retrieve the spin orientation from the Weyl spinor; the only difference is that the alternative normalization must be accounted for, so instead of Eq. 14.5, one has:

$$\vec{v}_s = \frac{\langle s|\vec{\mathbf{J}}^{(1/2)}|s\rangle}{\langle s|s\rangle}. \tag{14.7}$$

This is implemented as:

```
ThreeVector spin(const WeylSpinor  &s)
  double E= (s.adjoint()*Eigen::Matrix2cd::Identity()*s)(0,0)
                .real();
  return ThreeVector((s.adjoint()*SU2Generator::JX(2)*s)(0,0)
                .real()/E,
                (s.adjoint()*SU2Generator::JY(2)*s)(0,0)
                .real()/E,
                (s.adjoint()*SU2Generator::JZ(2)*s)(0,0)
                .real()/E);
```

We can also obtain the four-momentum from the Weyl spinors! Conceptually, the momentum of a right-handed particle is equal to its energy, since the particle is massless. The energy therefore determines the magnitude of the momentum. The direction of momentum coincides with the direction of the spin. This can be written compactly as:

$$(E, \vec{p}) = \langle s|(\mathbf{I}, \vec{\sigma})|s\rangle. \tag{14.8}$$

In code:

```
FourVector fourMomentum(const WeylSpinor::Right &s) {
  using namespace SpecialOperators;
  return FourVector(
                    (s.adjoint()*sigma0*s)(0,0).real(),
                    (s.adjoint()*sigma1*s)(0,0).real(),
                    (s.adjoint()*sigma2*s)(0,0).real(),
                    (s.adjoint()*sigma3*s)(0,0).real());
}
```

For left-handed Weyl spinors, the four-momentum is obtained from:

$$(E, \vec{p}) = \langle s|(\mathbf{I}, -\vec{\sigma})|s\rangle. \tag{14.9}$$

The Pauli matrices `sigma0`, `sigma1`, `sigma1`, `sigma1` and other useful operators are defined in the header file `Spacetime/SpecialOperators.h` and are scoped within the namespace `SpecialOperators`. The reader can use the `Spacetime` library to verify that the two four-vectors

```
FourVector f1=L*fourMomentum(s);
FourVector f2=fourMomentum(L*s);
```

are identical as they ought to be. Here L is a Lorentz transformation and s is a Weyl spinor:

```
LorentzTransformation L(ThreeVector(0,0,1));   // e.g.
WeylSpinor::Left s(1,0);                        // e.g.
```

14.5.3 Dirac spinors

Dirac spinors represent spin 1/2 particles in relativistic quantum mechanics. They exist in two varieties, denoted as u (representing particles) and v (antiparticles). The spinors u and v represent distinct objects, though they both have four complex components. The following expressions represent spin 1/2 particles and antiparticles at rest, in the "Weyl representation", which is particularly useful for expressing the effect of rotations and Lorentz transformations:

$$u^+ = \sqrt{m}\begin{pmatrix}1\\0\\1\\0\end{pmatrix} \quad u^- = \sqrt{m}\begin{pmatrix}0\\1\\0\\1\end{pmatrix} \quad v^+ = \sqrt{m}\begin{pmatrix}-1\\0\\1\\0\end{pmatrix} \quad v^- = \sqrt{m}\begin{pmatrix}0\\1\\0\\-1\end{pmatrix}$$

$$\tag{14.10}$$

These spinors are normalized such that $u^\dagger \cdot u = v^\dagger \cdot v = 2m$, for particles at rest. The spinors can be rotated, Lorentz-boosted, or a combination of both. The generic Lorentz transformation acts on the Dirac spinor by acting on the upper two-components with the spin-1/2 with $\mathbf{R}_1^{(1/2)}$ and on the lower two components with $\mathbf{R}_2^{(1/2)}$.

Dirac spinors are represented by the following class:

```
class DiracSpinor : public  Eigen::Vector4cd
{
 public:

  // Default Constructor constructs null spinor
  DiracSpinor();

  // Constructor
  DiracSpinor(const std::complex<double> & s0,
              const std::complex<double> & s1,
              const std::complex<double> & s2,
              const std::complex<double> & s3
             );

  ...
};
```

which inherits from an `Eigen` four-component column vector. A generic Lorentz transformation can be affected using `operator *`; the implementation of this function is again simple:

```
DiracSpinor    operator * ( const LorentzTransformation & T,
   const DiracSpinor & s) {
  //
  // This computation is carried out in the Weyl
     representation. Top part
  // according to the left handed representation, bottom
     according to the
  // right handed representation:
  DiracSpinor result;
  Eigen::Vector2cd sL=T.rep2(2)*s.head(2);
  Eigen::Vector2cd sR=T.rep1(2)*s.tail(2);
  result.head(2)=sL;
  result.tail(2)=sR;
  return result;
}
```

Because the Dirac spinor represents a spin 1/2 particle, we will be interested in recovering properties of the particle from the spinor. The spin of the particle is obtained by forming the product:

$$s^\dagger \cdot \vec{S} \cdot s$$

where s is a Dirac spinor and $\vec{S}$ is a vector of matrices:

$$\vec{S} = \begin{pmatrix} \vec{\sigma} & 0 \\ 0 & \vec{\sigma} \end{pmatrix}. \tag{14.11}$$

We introduce the **Dirac matrices**, γ^μ, to extract the four-momentum of the particle:

$$(\gamma^0, \gamma^i) = \left(\begin{pmatrix} 0 & I \\ I & 0 \end{pmatrix}, \begin{pmatrix} 0 & \sigma_i \\ -\sigma_i & 0 \end{pmatrix} \right) \tag{14.12}$$

The **Dirac adjoint**, defined by:

$$\bar{s} = s^\dagger \gamma^0$$

is also useful.

The gamma matrices are constant matrices, like the Pauli spin matrices, and are declared in the header file `SpecialOperators.h`. The Dirac adjoint operation is defined in the `DiracSpinor.h` header:

```
DiracSpinorBar bar(const DiracSpinor & s);
```

A `DiracSpinorBar`, also defined in `DiracSpinor.h`, is a complex four-component row vector.

With these definitions, the four-momentum associated with a Dirac spinor is:

$$p^\mu = \frac{1}{2} \bar{s} \cdot \gamma^\mu \cdot s, \tag{14.13}$$

the factor of two being due to the normalization. The spin and four-momentum functions have been implemented in the routine:

```
// Return the Four-momentum of this Dirac Spinor
FourVector fourMomentum(const DiracSpinor &s);

// Return the Spin (magnitude and direction)
ThreeVector spin(const DiracSpinor &s);
```

Therefore, as with Weyl spinors, the operations

```
FourVector f1=L*fourMomentum(s);
FourVector f2=fourMomentum(L*s);
```

give the same thing.

14.5.4 Tensors of higher order

Finally, we consider the action of Lorentz transformations on higher order tensors. In particular, antisymmetric rank two tensors frequently occur in nature; the prime example is the electromagnetic field tensor, $F^{\mu\nu}$, $\mu, \nu = 0, 1, 2, 3$, defined by:

$$F^{\mu\nu} = \begin{pmatrix} 0 & -E_x & -E_y & -E_z \\ E_x & 0 & -B_z & B_y \\ E_y & B_z & 0 & -B_x \\ E_z & -B_y & B_x & 0 \end{pmatrix}. \tag{14.14}$$

Antisymmetric rank two tensors have six independent elements, out of which we can make two three-component complex vectors:

$$\vec{S}_1 = \begin{pmatrix} -T^{1,0} - T^{1,3} - iT^{3,2} + iT^{2,0} \\ T^{3,0} + iT^{2,1} \\ T^{1,0} - T^{1,3} + iT^{3,2} + iT^{2,0} \end{pmatrix}, \quad \vec{S}_2 = \begin{pmatrix} -T^{1,0} + T^{1,3} + iT^{3,2} + iT^{2,0} \\ T^{3,0} - iT^{2,1} \\ T^{1,0} + T^{1,3} - iT^{3,2} + iT^{2,0} \end{pmatrix}.$$

$$\tag{14.15}$$

For $T^{\mu\nu} = F^{\mu\nu}$, these vectors are just $\vec{S}_1 = \vec{E} + i\vec{B}$ and $\vec{S}_2 = \vec{E} - i\vec{B}$. A Lorentz transformation acts upon these vectors in the following way:

$$\vec{S}'_1 = \mathbf{R}_1^{(1)} \cdot \vec{S}_1 \tag{14.16}$$

$$\vec{S}'_2 = \mathbf{R}_2^{(1)} \cdot \vec{S}_2 \tag{14.17}$$

To implement Lorentz transformations on these objects, one unpacks the antisymmetric tensor into the vector, applies the above transformations to the two vectors, and then takes linear combinations of their elements to construct the transformed rank two tensor. Here is our implementation:

```
AntisymmetricFourTensor operator *(const
    LorentzTransformation & L, const AntisymmetricFourTensor
    & v) {
  typedef std::complex<double> Complex;
```

```
Complex I(0,1);
const double sqrt2=sqrt(2.0), sqrt8=sqrt(8);

Eigen::Vector3cd
  S1((-v(1,0)-I*v(3,2) + I*v(2,0)-v(1,3))/sqrt2,
     v(3,0)+I*v(2,1),
     (v(1,0)+I*v(3,2) + I*v(2,0)-v(1,3))/sqrt2);

Eigen::Vector3cd
  S2((-v(1,0)+I*v(3,2) + I*v(2,0)+v(1,3))/sqrt2,
     v(3,0)-I*v(2,1),
     (v(1,0)-I*v(3,2) + I*v(2,0)+v(1,3))/sqrt2);

Eigen::Vector3cd S1Prime=L.rep1(3)*S1, S2Prime=L.rep2(3)*
  S2;

AntisymmetricFourTensor T;
T(1,0) = real(S1Prime(2)-S1Prime(0) + S2Prime(2)-
  S2Prime(0))/sqrt8;
T(2,0) = real (-I*(S1Prime(2)+S1Prime(0)+S2Prime(2)+
  S2Prime(0)))/sqrt8;
T(3,0) = real(S1Prime(1)+S2Prime(1))/2.0;

T(3,2) = imag(S1Prime(2)-S1Prime(0) - S2Prime(2) +
  S2Prime(0))/sqrt8;
T(1,3) = imag (-I*(S1Prime(2)+S1Prime(0)-S2Prime(2)-
  S2Prime(0)))/sqrt8;
T(2,1) = imag(S1Prime(1)-S2Prime(1))/2.0;
return T;
}
```

We could continue with tensors of higher order, but at this point we have illustrated the general idea—a universal computational approach to both rotations and Lorentz transformations based on representations of SU(2)—with objects that are perhaps the most useful in physics.

14.6 The helicity formalism

One of the applications of the ideas in this chapter is the computation of angular distributions in the scattering and decay of elementary particles. The topic goes by the name of the helicity formalism and is developed in a series of papers originating with that of Jacob and Wick (1959). An accessible treatment is also available in (Richman, 1984). In this section we show how the helicity formalism can be used to predict angular distributions in two-body decays, and sequential two-body decays. The helicity formalism does not allow us to compute the distributions from first principles — for that, the Feynman calculus, described in the next chapter, is needed. However it does allow us to express angular distributions in terms of a few unknown amplitudes that depend upon the spin of mother and daughter particles, as well as their polarizations.

Consider the case of a mother particle with spin $\mathcal{J}$ decaying into two particles, daughter 1 and daughter 2, with spins j_1 and j_2. Throughout this section, we work in the center-of-mass of the mother particle, so that the daughter particles are back-to-back. The z axis of a Cartesian coordinate system will be taken as the quantization axis. Let the particle be polarized such that its spin projection along the z axis is M, and label this initial state with the quantum numbers $\mathcal{J}$ and M:

$$|i\rangle \equiv |\mathcal{J}, M\rangle.$$

The polar and azimuthal angles of daughter 1 will be denoted θ, ϕ. Each daughter particle is characterized by its **helicity**, or the projection of its angular momentum onto its direction of motion. We call these helicities λ_1 and λ_2 and note that they can take values from $-j_1, -j_1 + 1, \ldots j_i - 1, j_i$ and $-j_2, -j_2 + 1, \ldots j_2 - 1, j_2$, respectively. The final state is labelled as follows:

$$|f\rangle \equiv |\theta, \phi; \lambda_1, \lambda_2\rangle$$

The amplitude for the process $i \to f$ is the S-matrix element:

$$A_{i \to f}(\theta, \phi) \equiv \langle f|U|i\rangle \equiv \langle \theta, \phi; \lambda_1, \lambda_2|U|\mathcal{J}, M\rangle \tag{14.18}$$

where U is the time development operator. The associated probability density is:

$$\rho_{\lambda_1 \lambda_1}(\theta, \phi) = |\langle \theta, \phi; \lambda_1, \lambda_2|U|\mathcal{J}, M\rangle|^2. \tag{14.19}$$

We insert a complete set of daughter-particle states:

$$A_{i \to f}(\theta, \phi) = \sum_{j,m} \sum_{\lambda_1', \lambda_2'} \langle \theta, \phi; \lambda_1, \lambda_2 | j, m; \lambda_1', \lambda_2'\rangle \langle j, m; \lambda_1', \lambda_2'|U|\mathcal{J}, M\rangle \tag{14.20}$$

Since the interaction conserves angular momentum:

$$\langle j, m, \lambda_1', \lambda_2'|U|\mathcal{J}, M\rangle = \delta_{j\mathcal{J}} \delta_{mM} A^{\mathcal{J}}_{\lambda_1', \lambda_2'} \tag{14.21}$$

where $A^{\mathcal{J}}_{\lambda_1', \lambda_2'}$ is a complex constant. So

$$A_{i \to f}(\theta, \phi) = \sum_{\lambda_1', \lambda_2'} \langle \theta, \phi; \lambda_1, \lambda_2|\mathcal{J}M; \lambda_1', \lambda_2'\rangle A^{\mathcal{J}}_{\lambda_1', \lambda_2'}. \tag{14.22}$$

The state $\langle\theta,\phi;\lambda_1,\lambda_2|$ is related via a rotation operator to the same state with particle 1 along $\hat{z}$:

$$\langle\theta,\phi;\lambda_1,\lambda_2| = \langle 0,0,\lambda_1,\lambda_2|\mathbf{R}^\dagger \tag{14.23}$$

where $\mathbf{R}$ is a rotation by the angle θ about the axis $(-\sin\phi,\cos\phi,0)$; this rotation brings the z axis into line with the direction $\hat{n} = (\sin\theta\cos\phi, \sin\theta\sin\phi, \cos\theta)$, i.e., the direction of daughter 1. Note that $\langle 0,0,\lambda_1,\lambda_2|$ has a spin projection along the z axis of $\lambda = \lambda_1 - \lambda_2$; it is thus a pure angular momentum state with quantum numbers $j = \mathcal{J}$ and $m = \lambda$. Including an extra factor for proper normalization, we have:

$$\langle 0,0,\lambda_1,\lambda_2| = \sqrt{\frac{2\mathcal{J}+1}{4\pi}} \langle \mathcal{J},\lambda;\lambda_1,\lambda_2| \tag{14.24}$$

and therefore:

$$A_{i\to f}(\theta,\phi) = \sqrt{\frac{2\mathcal{J}+1}{4\pi}} \sum_{\lambda_1',\lambda_2'} \langle \mathcal{J}\lambda;\lambda_1,\lambda_2|\mathbf{R}^\dagger|\mathcal{J}M;\lambda_1',\lambda_2'\rangle A^{\mathcal{J}}_{\lambda_1',\lambda_2'}. \tag{14.25}$$

This simplifies because the matrix element of the rotation operator does not depend on the helicity labels, thus one can simply write:

$$A_{i\to f}(\theta,\phi) = \sqrt{\frac{2\mathcal{J}+1}{4\pi}} \langle \mathcal{J}\lambda|\mathbf{R}^\dagger|\mathcal{J}M\rangle A^{\mathcal{J}}_{\lambda_1,\lambda_2}. \tag{14.26}$$

The factor $\sqrt{(2\mathcal{J}+1)/(4\pi)}$ assures that:

$$\int |A_{i\to f}(\theta,\phi)|^2 \, d\Omega = |A^{\mathcal{J}}_{\lambda_1,\lambda_2}|^2. \tag{14.27}$$

This formalism yields a recipe for computing the angular distribution of a decay amplitude of this type (one particle to two):

- Represent the initial state as a $2\mathcal{J}+1$ dimensional spinor χ_i, configured to represent the spin orientation of the initial state. For example: for a spin-1 particle, the spin-up configuration is represented by $\chi_i = (1,0,0)^T$, spin down by $\chi_i = (0,0,1)^T$, and the "longitudinally polarized" state with $m = 0$ is represented by $\chi_i = (0,1,0)^T$. Of course any arbitrary polarization state can also be represented by the appropriate spinor.
- The (unrotated) final state is likewise represented by a spinor χ_f in the $2\mathcal{J}+1$ dimensional Hilbert space, whose magnetic quantum number is $m = \lambda$. For example, considering a final state of two spin 1/2 particles, when $\lambda_1 = +1/2$

and $\lambda_2 = +1/2$, then $\chi_f = (0,1,0)^T$; when $\lambda_1 = +1/2$ and $\lambda_2 = -1/2$, then $\chi_f = (1,0,0)^T$, and when $\lambda_1 = -1/2$ and $\lambda_2 = -1/2$, then $\chi_f = (1,0,0)^T$.

- Construct a rotation about the axis $(-\sin\phi, \cos\phi, 0)$ by the angle θ; obtain the matrix of the $\mathcal{J}$-representation from the rotation (call it **R**).
- Compute:

$$A^{i \to f}(\theta, \phi) = \sqrt{\frac{2\mathcal{J}+1}{4\pi}} \chi_f^T \cdot \mathbf{R}^\dagger \chi_i \qquad (14.28)$$

The square magnitude of this expression gives the properly normalized angular distribution.

A simple example is the decay of a W^- boson, a spin-one particle, into an electron plus an antineutrino. The electron helicity can be $+1/2$ or $-1/2$; in the former case the electron is said to **right-handed** while it the latter case it is **left-handed**. Nature is not equitable, it turns out, with regard to right-and left-handed electrons. The only possible transition from a W^- boson is to left-handed electrons and right-handed antineutrinos. So if we call the electron "particle 1" and the neutrino "particle 2", we have $\lambda_1 = -1/2$, $\lambda_2 = +1/2$ and $\lambda = \lambda_1 - \lambda_2 = -1$; therefore the final state spinor is $\chi_f = (0,0,1)^T$. The initial state depends on the polarization of the W^-; if the spin is oriented along the z axis, then $m = 1$ and the spinor for the W-boson is $\chi_i = (1,0,0)^T$. Code that computes the related squared amplitude is:

```
double ASQ (double cosTheta)   {
   Spinor<3> chiInitial={1,0,0};
   Spinor<3> chiFinal  ={0,0,1};
   Rotation R(ThreeVector(0,-acos(cosTheta),0));
   return 3.0/4.0/M_PI norm(((R*chiFinal).adjoint()*
      chiInitial)(0) );
};
```

The full program can be found in EXAMPLES/CH14/WDECAY. This example is simple but unusual because only one helicity state is present in the final state. A more complicated decay is $\mathcal{J}/\psi \to \mu^+\mu^-$; as in the previous example this is (spin 1) $\to$ (spin 1/2) + (spin 1/2), but now three configurations may exist in the final state. For a $\mathcal{J}/\psi$ polarized along z, the full angular distribution can be written as:

$$\rho(\theta, \phi) = \frac{3}{4\pi} |A_{\frac{1}{2},\frac{1}{2}}|^2 \left| \begin{pmatrix} 0 & 1 & 0 \end{pmatrix} \cdot \mathbf{R}^\dagger \cdot \begin{pmatrix} 1 \\ 0 \\ 0 \end{pmatrix} \right|^2$$

$$+ \frac{3}{4\pi} |A_{-\frac{1}{2},-\frac{1}{2}}|^2 \left| \begin{pmatrix} 0 & 1 & 0 \end{pmatrix} \cdot \mathbf{R}^\dagger \cdot \begin{pmatrix} 1 \\ 0 \\ 0 \end{pmatrix} \right|^2$$

$$+ \frac{3}{4\pi}|A_{\frac{1}{2},-\frac{1}{2}}|^2 \left| \begin{pmatrix} 1 & 0 & 0 \end{pmatrix} \cdot \mathbf{R}^\dagger \cdot \begin{pmatrix} 1 \\ 0 \\ 0 \end{pmatrix} \right|^2$$

$$+ \frac{3}{4\pi}|A_{-\frac{1}{2},\frac{1}{2}}|^2 \left| \begin{pmatrix} 0 & 0 & 1 \end{pmatrix} \cdot \mathbf{R}^\dagger \cdot \begin{pmatrix} 1 \\ 0 \\ 0 \end{pmatrix} \right|^2 \quad (14.29)$$

assuming that one does not distinguish the helicities of the final state particles.

Certain simplifications can be made from conservation laws. We have seen already that only left-handed fermions and right-handed antifermions are produced in W^- decay (more generally in charged weak decays). In the case of the J/ψ decay, which is mediated by the electromagnetic interaction, helicity conservation excludes the final states in which both muons are right-handed or both are left-handed (in the limit of massless fermions); therefore $A_{\frac{1}{2},\frac{1}{2}} = A_{-\frac{1}{2},-\frac{1}{2}} = 0$. Moreover, CP conservation requires that $A_{-\frac{1}{2},\frac{1}{2}} = A_{\frac{1}{2},-\frac{1}{2}}$. So only two terms in Eq. 14.29 are nonzero. With these considerations taken into account, the decay can be computed as:

```
double ASQ (double cosTheta)  {
  Spinor<3> chiInitial={1,0,0};
  Spinor<3> chiFinal1  ={0,0,1};
  Spinor<3> chiFinal2  ={1,0,0};
  Rotation R(ThreeVector(0,-acos(cosTheta),0));
  return 3.0/4.0/M_PI*(norm(((R*chiFinal1).adjoint()*
                      chiInitial)(0)) +
                      norm(((R*chiFinal2).adjoint()*
                      chiInitial)(0)));
};
```

The full example can be found in EXAMPLES/CH14/JSPIDECAY.

Things get even more interesting in sequential (or cascade) decays. Consider the decay $B^0 \to J/\psi K^{0*}$, which is followed by the decays $J/\psi \to \mu^+\mu^-$ and $K^{*0} \to K^+\pi^-$. The J/ψ decay has been discussed above; while the K^{0*} (a spin-1 particle), decays into a K^+ (spin 0) and a π^- (also spin 0). The final state is $\mu^+\mu^-K^0\pi^+$; however this is in fact *four* final states (two in the limit of massless muons) that are distinguished by the helicities of the muons. We take them one at a time, computing the angular distributions for each final state separately before summing them — unless we are planning to spin-analyze the decay products and distinguish the four final states, which is a practically impossible task. Each final state can be reached through any of three intermediate states S, which have the angular momentum quantum numbers of the parent B^0 meson:

- Right-handed J/ψ, right-handed K^{0*}, called $S = RR$
- Left-handed J/ψ, left-handed K^{0*}, called $S = LL$
- Longitudinal J/ψ, longitudinal K^{0*}, called $S = 00$

where "longitudinal" implies a spin projection of zero. The angular distribution is a joint probability density in six variables: θ_B and ϕ_B are the polar and azimuthal angles of the decay $B^0 \to J/\psi K^{0*}$ in the rest frame of the B^0; θ_J and ϕ_J are the polar and azimuthal angles of the decay $J/\psi \to \mu^+\mu^-$ in the rest frame of the J/ψ; and θ_K and ϕ_K are the polar and azimuthal angles of the decay $K^{-*} \to K^+\pi^-$ in the rest frame of the K^{0*}. For the decays of the daughter particles J/ψ and K^{0*}, the calculation is performed using the direction of motion of the particle as the quantization axis.

The angular distribution for each of the four final states is:

$$\rho(\theta_B, \phi_B, \theta_J, \phi_J, \theta_K, \phi_K | \lambda_{\mu^+}, \lambda_{\mu^-})$$
$$= \left| \sum_{\mathcal{S} \in \{LL, RR, 00\}} A_{\mathcal{S}}^{B^0 \to J/\psi K^{0*}}(\theta_B, \phi_B) A_{\lambda_{\mu^+}, \lambda_{\mu^-}}^{J/\psi \to \mu^+ \mu^-}(\theta_J, \phi_J | \mathcal{S}) A_{0,0}^{K^{0*} \to K^+ \pi^-}(\theta_K, \phi_K | \mathcal{S}) \right|^2$$

where the amplitudes $A_{\mathcal{S}}^{B^0 \to J/\psi K^{0*}}(\theta_B, \phi_B)$, $A_{\lambda_{\mu^+}, \lambda_{\mu^-}}^{J/\psi \to \mu^+ \mu^-}(\theta_J, \phi_J | \mathcal{S})$ and $A_{0,0}^{K^{0*} \to K^+ \pi^-}$ $(\theta_K, \phi_K | \mathcal{S})$ are calculated according to Eq. 14.28. For the amplitudes in J/ψ and K^{0*} decays, the dependence on the intermediate state $\mathcal{S}$ is through the polarization of the intermediate particle. Then, the final probability density is:

$$\rho(\theta_B, \phi_B, \theta_J, \phi_J, \theta_K, \phi_K) = \sum_{\lambda_{\mu^+}, \lambda_{\mu^-}} \rho(\theta_B, \phi_B, \theta_J, \phi_J, \theta_K, \phi_K | \lambda_{\mu^+}, \lambda_{\mu^-})$$

Most frequently one finds analytic expressions for simple decays such as these in the physics literature. They are possible because analytic expressions for elements of the rotation matrix exist and are well known. They are called the Wigner D-functions, and they are written in terms of Euler angles. Every educated person needs to know about these in order to follow developments in the quantum physics over the last half century or so. However, we have skillfully avoided any mention of them until now; which indicates that they are not essential for computational work if one is prepared to handle the rotation group abstractly as we have done in this chapter. As a learning strategem, the approach takes one's attention from the minutia of complicated formulae for the D-functions and focuses it instead on the essentials of rotations and the abstract objects upon which they act.

As a final point, we note that the rotation which carries the z axis into the unit vector $\hat{n} = (\sin\theta\cos\phi, \sin\theta\sin\phi, \cos\theta)$ is not unique. We have made the simplest choice. Other choices correspond to different overall phases in the transition amplitudes. Our rotation corresponds to the Jackson (1965) phase convention. It is less commonly used in the physics literature than the Jacob-Wick (Jacob and Wick, 1959) phase convention, but is simpler to construct.

14.7 Exercises

1. The following is a recipe for constructing the spherical harmonic $Y_l^m(\theta, \phi)$. Take $\hat{r} \equiv (\sin\theta\cos\phi, \sin\theta\sin\phi, \cos\theta)$. Take the direction of rotation to lie along $\hat{r} \times \hat{z}$

and the angle of rotation to be $\cos^{-1}\hat{r}\cdot\hat{z}$. From this axis and angle construct the rotation matrix $\mathbf{R}^{(l)}$. The middle column of this matrix multiplied by $\sqrt{(2l+1)/4\pi}$ contains the spherical harmonics $Y_l^l(\theta,\phi)$, $Y_l^{l-1}(\theta,\phi)...Y_l^{-l}(\theta,\phi)$. Try this and plot the results. Perform a comparison with a brute force computation of the spherical harmonics. Note, the basis for the rotation matrix $\mathbf{R}^{(l)}$ has $m = l$ in the first row, $m = l - 1$ in the second, $m = -l$ in the last.

2. Write a visualization program to render radiation patterns and use it to visualize dipole, quadrupole, sextupole, octupole, etc radiation patterns. Start by setting the axis of symmetry equal to the z-axis. Then, use the rotation matrices in the $2L + 1$ representation of the rotation group to rotate the radiation pattern and check with your visualization program that you obtain the expected result.

3. An eigenstate of angular momentum has $j = 3/2$ and $m = 3/2$. The state is rotated about the x-axis. On a single set of axes, plot the probability for the rotated state to have $m = 3/2$, $m = 1/2$, $m = -1/2$, $m = -3/2$, as a function of the rotation angle θ over the range $[0, 4\pi]$. Then, plot the real and imaginary parts of the amplitude. Do the same for the state $j = 2$, $m = 2$, plotting the probability and the amplitude for the rotated state to have $m = 2$, $m = 1$, $m = 0$, $m = -1$, and $m = -2$.

4. A conventional way to transform four vectors under a general proper Lorentz transformation is to act upon the four vector with a 4×4 real matrix. The antisymmetric field tensor is transformed by applying the matrix twice: $\mathbf{F}' = \mathbf{\Lambda}\mathbf{\Lambda}\mathbf{F}$. Devise a scheme for constructing this matrix from the LorentzTransformation class. Use it in a few test cases to apply

   - a pure rotation
   - a pure boost

   to

   - a purely spacelike four-vector
   - a purely timelike four-vector
   - a purely electric antisymmetric field tensor
   - a purely magnetic antisymmetric field tensor

   and check that the result agrees with the transformations as implemented in the Spacetime library.

5. A massive spin 1/2 particle has momentum along z and is also spin-polarized along z. Using Dirac spinors, verify numerically that the helicity is $\hat{p}\cdot\vec{s}$ where $\vec{p}$ is the momentum and $\vec{s}$ is the spin is invariant to rotations. Do this for both spin orientations.

6. Write a routine to compute the decay amplitude for a massive particle of mass M into two daughters of mass m_1 and m_2. Your routine should take the four-vector p_0^μ of the massive particle and return the four-vectors p_1^μ and p_2^μ of the two daughters. Let

the decay of the massive particle be isotropic in the rest frame of the massive particle. Plot the average opening angle of the two daughter particles (in the laboratory frame) as a function of the energy divided by the mass E/M of the massive particle. Check that the invariant mass-squared of the final state equals that of the initial state, i.e. $(p_1^\mu + p_2^\mu)^2 = (p_0^\mu)^2$.

7. Re-using or modifying your code from the previous problem, generate sequential decays

$$B^0(5280 \text{ MeV}) \to J/\psi(3097 \text{ MeV})K^{0*}(892 \text{ MeV})$$

followed by:

$$K^{0*}(892 \text{ MeV}) \to K^+(494 \text{ MeV})\pi^-(140 \text{ MeV})$$

and:

$$J/\psi(3097 \text{ MeV}) \to \mu^+(105 \text{ MeV})\mu^-(105 \text{ MeV})$$

under the (false) assumption that all decays are isotropic. Check the invariant masses of the muon pair, the $K^+\pi^-$ pair, and the $\mu^+\mu^-K^+K^-$ system.

8. Write a visualization program to view the four-vectors generated in Exercise 7. Color the lines so that you can recognize each particle.

9. Write a routine to compute the joint probability density $\rho(\theta_B, \phi_B, \theta_J, \phi_J, \theta_K, \phi_K)$ for the decay $B^0 \to J/\psi K^{0*}$. Build in all the constraints on the decay amplitudes in $J/\psi \to \mu^+\mu^-$ that have been discussed in the text. The routine should be parameterized on the amplitudes A_{00}, A_{RR}, and A_{LL}. Plot the following quantities:

    - the angle between the decay plane of the K^{0*} and the J/ψ, in the B^0 rest frame.
    - the angle between the direction of the μ^+ and the B^0, in the J/ψ rest frame.
    - the angle between the direction of the K^+ and the B^0, in the K^{0*} rest frame.

    Do this for various combinations of A_{00}, A_{RR}, and A_{LL}.

10. Revisit Exercise 7, except generate the decay angles according to the joint probability distribution of Exercise 9. These depend upon the complex amplitudes. Remake the plots of Exercise 9, but in two ways: first using the angles that you generate, and second, using angles that you reconstruct from the four-vectors. Compare the plots. Work out all of the bugs until they agree.

11. [P] Write a general purpose package to compute the joint probability distribution of kinematic angles for an arbitrary sequence of two-body decays. Some suggestions for writing this library follow. First, write a class to describe relevant (to this project) information about each particle: namely the spin information. Arrange for objects

of this class to hold pointers to child objects so that you can configure an arbitrary decay tree. At the top of the tree, specify the polarization of the first mother. Your code should determine the possible spins of all the daughters–and create parameters for the magnitude and phase (alternately, real and imaginary part) of each allowed transition. Thus far you have only built the control structures for your joint PDF. Now, use that to produce a probability density given the value of all the angles that describe the decay. Then build the test program that allows you to visualize the PDF.

12. [P] Rotations and Visualization.

    Computations with rotations can be confusing and visual checks are always helpful. Write an algorithm to translate the `Rotation` class of this chapter to the `SoRotations` of the Coin library. Then write a program to visualize the decay of the $H \to W^+ W^-$ followed by $W^+ \to e^+ \nu$ and $W^- \to e^- \bar{\nu}$. Display the particles as arrows and render the decay angles as they would appear in the center-of-mass frame of the decaying particle.

BIBLIOGRAPHY

Edmonds, A. R. (1957). *Angular momentum in quantum mechanics*. Princeton University Press.

Jackson, John D. (1965). in C. Dewitt and M. Jacob, *Les Houches Lectures in High Energy Physics*. Gordon and Breach.

Jackson, John D. (1998). *Classical Electrodynamics*. John Wiley and Sons.

Jacob, M. and G. C. Wick (1959). *On the general theory of collisions for particles with spin*. Ann. Phys. 7, 404.

Messiah, A. (2014). *Quantum Mechanics, vol II*, Dover, New York.

Moler, C. and C. Van Loan (2003). *Nineteen dubious ways to compute the exponential of a matrix, twenty-five years later*, SIAM Review **45**, 1, 3.

J. Richman (1984). *An experimenter's guide to the helicity formalism*. Caltech preprint CAL-68-1148.

Sakurai, J. J. (1994). *Modern quantum mechanics*. Addison-Wesley.

15
Simulation

15.1 Stochastic systems	483
15.2 Large scale simulation	484
15.3 A first example	486
15.4 Interactions of photons with matter	488
15.5 Electromagnetic processes	491
15.5.1 Bremsstrahlung	491
15.5.2 Electromagnetic showers	493
15.5.3 The need for simulation toolkits	493
15.6 Fundamental processes: Compton scattering	494
15.7 A simple experiment: double Compton scattering	501
15.8 Heavier charged particles	503
15.9 Conclusion	505
15.10 Exercises	506
Bibliography	510

15.1 Stochastic systems

The mechanical systems that we first encounter in physics have a certain appeal because of their predictable and deterministic nature. These systems are often modeled by ordinary or partial differential equations whose solutions can be regarded as a temporal simulation.

Many systems, however, are governed by stochastic or random behavior. Examples are the radioactive decay of nuclear isotopes or subatomic particles, the production of secondary particles in a shower of cosmic rays, or the scattering of alpha particles from gold nuclei. Quantum mechanics is the ultimate source of randomness in these processes. In some fully deterministic systems our ignorance of the exact state of the system (e.g., molecules in a gas) constitutes an additional source of randomness.

Physical systems are not completely random. Instead of the Newtonian certainty about the values of dynamic or thermodynamic variables, we console ourselves with the twentieth century concept of probabilities and probability densities. Instead of following a single dynamical system as it evolves according to a Hamiltonian, one is obliged to throw virtual dice as a system evolves; therefore an understanding of the system comes from studying ensembles of virtual experiments called pseudoexperiments.

15.2 Large scale simulation

The ATLAS collaboration is an international team of scientists who built and operate one of two detectors at the world's highest energy particle collider, the LHC. Prior to LHC operations, in 2008, the ATLAS collaboration at the LHC embarked on a remarkable exercise. A complete dataset of the anticipated proton-proton collisions was synthesized and distributed to hundreds of physicists to be analyzed as if it were real collider data. The result was published in a three-volume, 1800 page document (ATLAS Collaboration, 2008) called *Expected performance of the ATLAS detector*. Figure 15.1 appeared on the cover of volume three.

The figure shows the production and the decay of the Higgs boson, whose discovery was announced on July 4, 2012 (The ATLAS Collaboration, 2012, and The CMS Collaboration, 2012). This means that the image appeared a little less than four years before the discovery. The event in Figure 15.1 was born in a digital computer—the display shows the simulated digital signals left behind by simulated particles from a simulated collision, as they passed through a simulated detector.

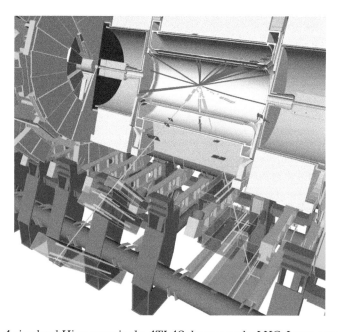

Figure 15.1 *A simulated Higgs event in the ATLAS detector at the LHC. Image courtesy of CERN.*

The simulation starts with the collision of two protons (henceforth we drop the word "simulated" from every sentence). Within the protons, quarks and gluons are moving with nearly random relativistic momenta. A pair of quarks and/or gluons are selected at random; their color and flavor are also chosen randomly, and their momenta are drawn from a probability distribution called a *parton density function*.

A fundamental process leading to the production of a Higgs is then simulated. This process is quantum mechanical, thus the production angle of the Higgs boson, the number of additional particles that are spewed out along with the Higgs boson, and their angles are all not predictable; however the *distributions* of these quantities can (more or less) be predicted by the theory of the strong, weak, and electromagnetic interactions. Thus the simulation draws these quantities from the relevant probability distributions.

Figure 15.1 displays a particular decay mode of the Higgs; $H^0 \to Z^0 Z^0$. The Z^0 is the intermediate vector boson that transmits the neutral weak force. For one of the Z^0's, the decay $Z^0 \to e^+ e^-$ was simulated, for the other $Z^0 \to \mu^+ \mu^-$. The muons are indicated by yellow lines in the figure.

The angular distribution of the decay is again not predictable, but is instead drawn at random from the applicable distribution. These particles are then propagated in a deterministic way through the magnetic field. Muons, being unstable over large distance scales, are given an opportunity to decay in flight after a random time. In traversing the detector, however, material is encountered, such as silicon sensors. When the electrons or muons enter this material, they interact in several ways. Some processes, like ionization energy loss, occur continuously along the path of the particle. Others, like bremsstrahlung of hard photons, nuclear interactions or δ–ray production, occur catastrophically at precise locations.

The energy left behind in these processes is the basis for the simulated detector response—the green dots along the yellow lines, for example, indicate the response of a straw tracker. The tracks eventually reach more dense material called a *calorimeter*, which the muons are able to penetrate. On the other hand, electrons tend to brems hard photons; these photons then either Compton scatter or pair produce electron-positron pairs. These, in turn, can brems more photons. The process continues until the entire energy of the incident electron, some tens of billions of electron volts, is degraded into a shower of low-energy particles. It is this shower that produces a visible signal in the detector, shown as red clumps of energy in the figure.

Along with this picture, the document (ATLAS, 2008) presented detailed studies of many other processes that were based on simulated data. Some of the processes have now actually been produced in *real* collisions, while others are hypothetical processes, some of which have since been proven not to have taken place at the LHC. Hundreds of Ph.D theses were produced during this time, supported by synthetic data produced on a huge and widely distributed computing grid. The synthetic data were presented and studied in much the same way as real data. They were crucial to many studies conducted during that period, and continue to be important (even with real data in hand) to predict the response of the detector to various signals.

The simulation program that produced all of this was a major endeavor on the part of a large team and a represented a substantial capital investment. We are not about to train you to start assembling an infrastructure this large, but we can show you what building blocks are required in a way that allows you to simulate simple systems and begin to understand larger ones.

15.3 A first example

The type of simulation described in the last section is called "Monte Carlo" and the main underlying computational technique behind all of this simulation is distribution sampling, such as we have already discussed in Chapter 7. The Higgs boson is a spin-0 particle which is unpolarizable and therefore the emission of any decay product occurs in a completely isotropic fashion. A simulation program, having first defined a Cartesian coordinate system, generates the cosine of the polar angle $\cos\theta$ according to a flat distribution over the interval [-1,1], and an azimuthal angle ϕ according to a flat distribution on $[0,2\pi]$. In a two body decay, either one of the daughter particles is emitted along the direction of $\hat{n} = (\sin\theta\cos\phi, \sin\theta\sin\phi, \cos\theta)$, while the other daughter particle is emitted along $-\hat{n}$.

The Higgs boson decays to multiple final states. Branching ratios are given in Table 15.1. The actual final state of a single Higgs boson is the outcome of a random process, which can be simulated most easily these days using `std::discrete_distribution`. Here is a short program which prints out a random decay mode each time it is run (and can be found in EXAMPLES/CH15/BRANCH); a typical way of testing a program like this is to histogram the frequency of decay into each of the final states (Exercise 1).

```
#include <iostream>
#include <random>
using namespace std;

enum DecayMode {BBBAR, TAUPAIR, MUONPAIR, CHARMPAIR,
    GLUONPAIR, GAMMAPAIR, ZGAMMA, WPAIR, ZPAIR, OTHER};

int main() {

  // Initialize random number generators:
  std::random_device dev;
  std::mt19937 engine(dev());

  // Initialize the generator:
  std::discrete_distribution<int> BR = { 5.809E-1,
                                         6.256E-2,
                                         2.171E-4,
                                         2.884E-2,
                                         8.180E-2,
```

```
                                   2.270E-3,
                                   1.541E-3,
                                   2.152E-1,
                                   2.641E-2};

  // Draw a random decay mode and print it out
  DecayMode mode= (DecayMode) BR(engine);
  std::cout << mode << std::endl;

  // Done
  return 0;
}
```

Now if we wish to take our simulation further, we will have to follow the daughter particles as they evolve in their surroundings, which in this case probably involves the complex instrumentation put in place to observe particles like the Higgs boson. These daughter particles do not interact with matter in a generic way; each one has specific processes. Some (like γ rays) are absolutely stable, others like ($\mu^{\pm}$) decay but—usually—far beyond the detector system; the quarks (b and c) **hadronize**, in other words, they dress themselves with light quarks to form bound states called **hadrons**. The b and c quarks and the τ lepton travel distances of a few hundred microns to a few millimeters, which are distances that can be observed in modern detector systems. The W and Z bosons with lifetimes on the order of 10^{-25} seconds decay, for all practical purposes, at the point where they are produced. Those particles which travel through significant amounts of

Table 15.1 *Higgs boson branching ratios, calculated for a Higgs mass of 125.09 GeV. Source: Heinemeyer (2014).*

Final state	Terminology	Branching ratio
$b\bar{b}$	bottom quark pairs	5.809×10^{-1}
$\tau^+\tau^-$	tau lepton pairs	6.256×10^{-2}
$\mu^+\mu^-$	muon pairs	2.171×10^{-4}
$c\bar{c}$	charm quark pairs	2.884×10^{-2}
gg	gluon pairs	8.180×10^{-2}
$\gamma\gamma$	gamma ray pairs	2.270×10^{-3}
$Z^0\gamma$	Z boson plus gamma ray	1.541×10^{-3}
W^+W^-	W boson pairs	2.152×10^{-1}
Z^0Z^0	Z boson pairs	2.641×10^{-2}

matter interact in various ways. In the next section we see how to simulate the fate of a gamma ray (γ).

15.4 Interactions of photons with matter

Photons interact with matter through several distinct processes, which we shall refer to as scattering processes. Rayleigh and Compton processes scatter the incident photon into a different angle, in the latter case with a reduced energy. Photons are absorbed through the photoelectric effect. Finally, photons can be converted into electron-positron pairs. The conservation of energy and momentum requires that this process occur in the static field of a charged particle, and we distinguish pair production in the field of the atomic nucleus from pair production in the field of electrons. The relative probability of these processes depends strongly upon the energy of the photon, as well as on the type of matter that has been struck. These probabilities can be obtained from the **cross sections**, σ, that describe the relevant processes.

When the cross section is multiplied by the *flux* of incoming particles (which is the number of particles per unit area per unit time in a beam) one obtains the number of scattered particles per unit time. Now, we are interested in the probability of a single photon interacting with a layer of matter, and this is not the most handy formulation for our purposes.

Imagine a collision between a wave packet describing a single particle (e.g. a photon), uniformly filling a square box, travelling towards a layer of material (like aluminum). Each side of the box has a length of L the thickness of the layer is dx. The interaction cross section of a single aluminum atom is σ. We are interested in the probability of an interaction in the material. One has first to take account of the fact that the "box of particle" encounters numerous scattering centers as it passes the through the material. Their number N can be calculated as the product of the number density n of scattering centers, times the volume of material $L^2 dx$ through which the box of particle passes.

$$N = nL^2 dx \qquad (15.1)$$

The number density is the mass density ρ divided by the mass m of the scattering centers:

$$n = \frac{\rho}{m} = \rho \frac{N_A}{A} \qquad (15.2)$$

where N_A is Avogadro's number and A is the atomic mass of a pure material. So that the total cross section presented by the full layer of material is:

$$\sigma_{layer} = N\sigma = \rho \frac{N_A}{A} L^2 \sigma \, dx \qquad (15.3)$$

The probability of an interaction is then the ratio of this layer cross section σ_{layer} to the cross sectional area of the "box of particle", L^2, in other words:

$$dP = \frac{\rho \frac{N_A}{A} L^2 \sigma \, dx}{L^2} = \rho \frac{N_A}{A} \sigma \, dx \tag{15.4}$$

A beam of N_γ photons evolves according to:

$$\begin{aligned} dN_\gamma &= -N_\gamma \, dP \\ &= -N_\gamma \rho \frac{N_A}{A} \sigma \, dx \end{aligned} \tag{15.5}$$

which we can integrate to obtain:

$$N_\gamma = N_\gamma(0) e^{-\rho(N_A/A)\sigma x} \tag{15.6}$$

so, if we start with one photon ($N_\gamma = 1$), the quantity $e^{-\rho(N_A/A)\sigma x}$ is the probability to be transmitted throught the material, while:

$$1 - e^{-\rho(N_A/A)\sigma x} \tag{15.7}$$

is the probability to interact within it.

The quantity $\alpha \equiv \rho(N_A/A)\sigma$ is called the ***attenuation coefficient***, and $\mu \equiv \alpha/\rho = (N_A/A)\sigma = \sigma/m$ is the called the ***mass attenuation coefficient***. Both can readily found in tables, though it is more common to find mass attenuation coefficients because they are independent of the density of the material. For aluminum, the mass attenuation coefficients for various processes are shown in Figure 15.2. These cross sections were obtained from the website

```
http://physics.nist.gov.
```

Let's consider the Higgs boson after it has decayed to two hard photons. The energy of the photon (making the not-so-accurate assumption of a Higgs decaying at rest) is half the Higgs mass, or 62.5 GeV. This photon passes through an inner detector region designed to be as transparent as possible so that the photon can be accurately measured in the calorimeter. But wait... this photon first encounters a support structure made of aluminum, with a thickness of 1/2 cm. From the data in Figure 15.2, one obtains a mass

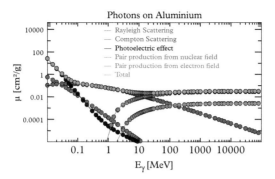

Figure 15.2 *Cross sections for the interaction of photons with aluminium. The curves are cubic spline interpolations of data points which have been downloaded from* http://physics.nist.gov, *where similar data for other elements and compounds can be found.*

attenuation coefficient of 3.19×10^{-2} cm^2/g. Using the density of aluminum, 2.70 g/cm^3, we obtain the probability of pair production in the aluminum as:

$$P = 1 - \exp\left(3.19 \times 10^{-2} \text{ cm}^2/\text{g}\right) \times 2.70 \text{ g/cm}^3 \times 0.5 \text{ cm} = 4.3\% \quad (15.8)$$

In large-scale simulations one must propagate the photon through a complicated geometry, determine the materials which are encountered, and calculate these probabilities at every step and for every process.

We return now to our photon: it either passes undisturbed through the aluminum layer (95.7% probability) or it produces an electron-positron pair somewhere within the aluminium (4.3% probability). As with our first example (Section 15.3) we throw a virtual dice. For a 62.5 GeV photon the production of an electron-positron pair is the highly favored outcome. In that case, we are finished with our photon, but now we have two daughter particles to track.

You might think that the electron+positron pair carry the energy and momentum of the original photon, but this is impossible since the magnitude of the photon's energy-momentum four-vector is zero, while that of an electron positron pair is at minimum $4m_e^2$. In fact, the process is forbidden in a vacuum, but takes place in the field the constituents of the aluminum atom (either the nucleus or the electrons). The electron and positron momenta are collinear with that of the original photon, and the energy is divided between the components, a fraction x going to the electron and $1-x$ going to the positron, where x is a random variable distributed according to

$$\rho(x) = \frac{9}{7}\left(1 - \frac{4}{3}x(1-x)\right). \quad (15.9)$$

The expression is an approximation (Tsai, 1974), valid for photons above 1 GeV. The distribution can be sampled using techniques discussed in Chapter 7, say the transformation method or von Neumann rejection.

15.5 Electromagnetic processes

The production of electron-positron pairs in matter is closely related to another process, namely bremsstrahlung of photons from electrons (or positrons). The calculation of their rates involves the complexity of the atomic and nuclear coulomb fields as well as atomic and nuclear excitations. These calculations are the work of many people; for a review see Patrignani (2016) or Tsai (1974).

From the plot in Figure 15.2, one can see that the cross section (and mass attenuation coefficient) reach a plateau at energies above a few GeV. The absorption length in this region, $\lambda = 1/\alpha$, has a fixed value, and defines another quantity of interest called the *radiation length* X_0:

$$\frac{X_0}{\rho} = \frac{7}{9}\lambda = \frac{7}{9}\frac{M}{N_A \sigma}. \tag{15.10}$$

The radiation length (like the mass absorption coefficient) is defined so that it depends only on the type of material and not on its density. Divide by the mass density ρ to obtain a characteristic length scale X_0/ρ for high-energy photon interactions, and also for another process: bremsstrahlung. On average, a high-energy electron loses a fraction $1-e$ of its energy through bremsstrahlung over a distance X_0/ρ. However, bremsstrahlung of hard photons, like pair production, is a discrete process, and we should like to know the probability of generating a photon as a function of its kinematics.

15.5.1 Bremsstrahlung

An electron (or positron) $e^{\pm}$ with energy E strikes an atomic nucleus; because the nucleus is far heavier than the $e^{\pm}$ it remains largely at rest while the $e^{\pm}$ is highly deflected; it's acceleration leads to the emission of photons. The kinematics of the process is fully specified by the energy k and polar and azimuthal angles θ, ϕ of the photon with respect to the original electron or positron direction.

The process is governed by the triple differential cross section:

$$\frac{d\sigma}{d\Omega dk} \equiv \frac{d\sigma}{d\cos\theta d\phi dk}, \tag{15.11}$$

which is the cross section per unit solid angle per unit energy (of the photon) for the process. A formula appearing in Tsai (1974) for the triple differential cross section is exceedingly difficult to work with owing to the complications arising from the structure of the atom. Instead we shall use a simpler treatment in which the correlations between the energy of the scattered photon and the scattering angle are ignored. The distribution of photon energies can be obtained from the approximate formula (Tsai, 1974):

$$\frac{d\sigma}{dk} = \frac{4\alpha r_e^2}{3k}\left[\{y^2 + 2[1+(1-y)^2]\}[Z^2(L_{rad} - f((\alpha Z)^2)) + ZL'_{rad}] + (1-y)\frac{Z^2+Z}{3}\right]. \tag{15.12}$$

In this formula, $y = k/E$, E is the energy of the incoming electron or positron, k is the energy of the scattered photon, $r_e = 2.82 \times 10^{-13}$ cm is the classical electron radius, $\alpha \approx 1/137$ is the fine structure constant, Z is the atomic number of the target nucleus, the Coulomb correction function is:

$$f(z) \approx 1.202z - 1.0369z^2 + 1.008z^3/(1+z), \tag{15.13}$$

and the elastic and inelastic form factors describing the complex structure of the atom are approximated (Tsai, 1974) by:

$$L_{rad} = \ln\left(\frac{184.15}{Z^{1/3}}\right) \quad L'_{rad} = \ln\left(\frac{1194.}{Z^{2/3}}\right) \tag{15.14}$$

for elements with $Z > 4$, while tabulated values for lighter elements can be found in (Patrignani, 2016).

The total cross section is required to simulate bremsstrahlung in material with finite thickness (Eq. 15.7), thus one must integrate Eq. 15.12. Alas, the result is infinity; however, the total *energy* lost by an $e^{\pm}$ in traversing a layer of material is finite. The result is[1] $1 - \exp \rho x/X_0$.

A practical scheme for simulating bremsstrahlung is to treat the process as discrete above some cutoff energy, and continuous below that. One possible choice is the MeV scale, since below that scale radiated photons cease to produce electron-positron pairs and will only interact through Compton scattering or the photoelectric effect.

Once the energy k has been obtained we can turn our attention to the generation of the azimuthal and polar scattering angles. The azimuthal angle ϕ is of course simple: draw from a uniform distribution in the range $[0, 2\pi]$. The polar angle θ can be generated by defining $u \equiv E\theta/m$ (where m is the mass of the electron) and generating u according to the function:

$$f(u) = \begin{cases} C\left(ue^{-au} + de^{-3au}\right), & 0 \leq u \leq \frac{E\pi}{m} \\ 0, & u < 0 \quad \text{or} \quad u > \frac{E\pi}{m}, \end{cases}$$

where:

$$C = \frac{9a^2}{9+d} \quad a = 0.625 \quad d = 27 \tag{15.15}$$

as described in The GEANT4 Collaboration (2015).

---

[1] To make this connection it is necessary to drop the term $(1-y)\frac{Z^2+Z}{3}$ from Eq. 15.12.

15.5.2 Electromagnetic showers

A high energy electron, positron, or photon develops an *electromagnetic shower* by alternating the processes of pair production and bremsstrahlung. Electromagnetic showers take place in detector systems, which concerns high-energy physicists. They also occur in the earth's atmosphere, which concerns astronomers and cosmic ray physicists, and in human tissue, which concerns medical physicists. The descriptions we have given constitute rather crude approximations to very complicated physics, and they can at best be used for crude simulations that serve as the basis for simplistic pedagogical exercises but not for serious work. So, now we turn for a moment to a brief survey of how the professionals do it.

15.5.3 The need for simulation toolkits

Simulation of radiation transport is a huge task, involving not only fundamental processes between electrons, positrons, and photons, but also other particles and interactions. The particles with which one is concerned can be categorized in several ways.

First, particles may be stable or unstable. Even if the particle is not absolutely stable, it may nonetheless be stable on the timescale of the simulation. Neutrons, for example, have a mean lifetime of 882 seconds, so for most purposes they may be considered stable. The charged pion $\pi^{\mp}$ commonly found in high energy physics experiments has a lifetime of 26 ns, so most, but not all, of them interact or escape the detector volume before they decay. Large databases containing particle properties such as mass, decay rates, branching ratios, and angular distributions are required in typical radiation transport programs.

One is seldom interested in the transport of radiation through simple idealized geometric shapes such as thin layers or ideal cylinders. Realistic applications often involve complex geometries or, in some applications, soft tissues. Particles therefore must be tracked through these geometries, often in the presence of magnetic fields, homogeneous or inhomogeneous according to the application. In such cases the equations of motion must be integrated and collisions with the geometry detected.

A second way in which particles may be categorized is by their interactions in materials. All charged particles are deflected by electric and magnetic fields; neutral particles are not. All charged particles lose energy as they pass through materials by ionizing the surrounding materials, chiefly by interacting with the electrons. In addition, their direction is continuously deflected through interactions with nuclei, an effect which is known as multiple Coulomb scattering. These effects are discussed in Section 15.8. Bremsstrahlung is predominant for light electrons, but only becomes important for heavier charged particles (the lightest of which is the muon) at extremely high energies. Thus the showers described above generally do not occur for muons below several hundred GeV.

Particles composed of quarks are called *hadrons*. Hadronic interactions in matter are the most complicated of all. In addition to the continuous processes of ionization energy loss and multiple scattering (if they are charged), hadrons can collide with nuclei, breaking them up and spewing large numbers of ground state and excited nucleons, pions, and other particles. These interactions are poorly understood and often constitute a fundamental limitation of high energy experiments. They are modeled with various degrees of success when no exact formulae can be found.

These problems are common across many domains, and therefore several toolkits have been designed to address them. A complete history of simulation toolkits would mention EGS (Ford, 1978), FLUKA (Aarnio, 1986 and 1987), PENELOPE (Sempau, 1997), and GEANT (Agostenelli et al., 2003). Today GEANT, now in its fourth version, is written in C++ and is the most commonly used toolkit for simulation. The GEANT toolkit has facilities for describing materials, particles, geometry, sensitive elements (like scintillator panels), track propagation, nearest-neighbor detection, and a complete library of physics processes, both electromagnetic and hadronic. It is widely used in HEP experiments as well as in medical physics and space physics.

This brief survey of the simulation landscape hopefully puts our efforts to simulate an electromagnetic shower into perspective: in today's complex computing environment, our shower simulation is but a small, crudely implemented illustration of a component of a massive computational task. Ultimately the criterion upon which all simulation is judged is, Does it adequately describe reality? For this purpose, comparison with real data is the "acid test".

15.6 Fundamental processes: Compton scattering

We return now from our brief interlude on simulation toolkits to our narrative of the lifecyle of the Higgs boson and its decay products. Our Higgs happens to have decayed to two photons, the photons have converted to electron-positron pairs which are losing energy through bremsstrahlung, their photons converting again to more electrons and positrons and bremsing more photons. Eventually the photons will reach an energy where Compton scattering is the dominant process, while the electrons will predominantly lose energy by ionizing the surrounding material.

Compton scattering, finally, is a fundamental process that is simple enough for us to go through the calculation from quantum electrodynamics. The cross section (Heitler, 1955) is given by the Klein-Nishina formula (Klein, 1929):

$$\frac{d\sigma}{d\Omega} = \frac{r_e^2}{2}\left(\frac{k_s}{k_0}\right)^2 \left(\frac{k_s}{k_0} + \frac{k_0}{k_s} - 2\sin^2\theta\cos^2\phi\right). \qquad (15.16)$$

Here, $k_0 = 2\pi/\lambda_0$ is the wavenumber of the incident radiation, $k_s = 2\pi/\lambda_s$ is the wavenumber of the scattered radiation, and ϕ is measured from the polarization vector (the electric field) of the incoming photon. Note that the variables θ, k_0 and k_s are related through the Compton scattering formula,

$$\lambda_s - \lambda_0 = \lambda_c(1 - \cos\theta).$$

where $\lambda_c = r_e/\alpha$ is called the Compton wavelength of the electron. Thus the scattered wavenumber k_s depends implicitly on the scattering angle. Note also that in natural units ($\hbar = c = 1$), *which we shall use from now on in this section*, the wavenumber of a photon is measured in energy units and is equal to the photon's energy.

The Klein-Nishina formula is simple enough to sample in a simulation program. Often, however, simple formulae are not available and it proves necessary to compute the cross section from quantum mechanical matrix elements. For example, Compton scattering is a 2→2 elastic collision: two particles, a photon and an electron, collide, and two particles (photon+electron) emerge from the collision. The general expression (see, e.g., Schwartz, 2104, Chapter 5) for the differential scattering cross section in the center of mass frame is:

$$\frac{d\sigma}{d\Omega_{cm}} = \frac{1}{64\pi^2 s} \frac{|\vec{p}_f|}{|\vec{p}_i|} \left| \sum_i \mathcal{M}_i \right|^2, \quad (15.17)$$

where $|\vec{p}_{i,f}|$ are the magnitudes of the incoming and outgoing momenta of the colliding particles *in the center-of-mass frame* of the collision and $\mathcal{M}_i$ is a **Lorentz invariant scattering subamplitude** for the process. These contain the dynamics and depend on the particular process under consideration. The Mandelstam variable s is $(p_A + p_B)^2$ where p_A and p_B are the four-momenta of the incoming particles. In the case of Compton scattering we shall denote the four-vector of the incoming photon as k, the four-vector of the electron as p, and the four-vector of the scattered photon as k'. The collision is elastic so we can ignore the factor p_f/p_i, and for this process $s = 2m_e(k_0 + m_e)$, where m_e is the electron mass.

Simulation programs can operate in any convenient reference frame and there is no harm in using the above formula directly. For the sake of comparison with the Klein-Nishina formula we transform Eq. 15.17 to the laboratory frame,

$$\frac{d\sigma}{d\Omega_{lab}} = \frac{d\sigma}{d\Omega_{cm}} \frac{d\Omega_{cm}}{d\Omega_{lab}}$$

$$= \frac{1}{64\pi^2 m_e^2} \left(\frac{k_s}{k_0}\right)^2 \left| \sum_i \mathcal{M}_i \right|^2,$$

using some relativistic kinematics and the chain rule.

The dynamics of Compton scattering and other processes in quantum electrodynamics are described by Feynman diagrams. The two lowest-order (tree-level) Feynman diagrams are shown in Figure 7.2. The Feynman rules appropriate for quantum electrodynamics determine how these diagrams can be used to derive matrix elements. The upper diagram in Figure 15.3 is called the *s*-channel diagram; the lower diagram is called the *t*-channel diagram.

The rules yield the following expressions for the matrix elements corresponding to each of the two diagrams:

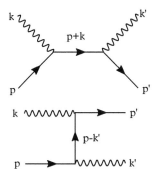

Figure 15.3 *Lowest order Feynman diagrams for Compton scattering.*

$$-i\mathcal{M}_s = \bar{u}' \left[\epsilon'^*_\nu (ie\gamma^\nu) \frac{i(\slashed{p} + \slashed{k} + m_e)}{(p+k)^2 - m_e^2} (ie\gamma^\mu) \epsilon_\mu \right] u$$

$$-i\mathcal{M}_t = \bar{u}' \left[\epsilon_\mu (ie\gamma^\mu) \frac{i(\slashed{p} - \slashed{k}' + m_e)}{(p-k')^2 - m_e^2} (ie\gamma^\nu) \epsilon'^*_\nu \right] u. \quad (15.18)$$

The objects in these expressions were discussed in Chapter 14. They include: the spinor of the incoming electron $u(p)$, the adjoint spinor of the outgoing electron $\bar{u}'(p')$, the polarization four-vector of the incoming electron ϵ, the (complex conjugate) polarization four-vector of the outging electron ϵ'^*, and the gamma matrices γ^μ. Summation over repeated indices is implied. We also use the "Feynman slash notation": $\slashed{a} \equiv \gamma^\mu a_\mu$. All of the above objects and operations can be found in the `Spacetime` library introduced in the previous chapter. The gamma matrices and a function called `slash`, taking either a `FourVector` or a `ComplexFourVector` are defined in the header file `Spacetime/SpecialOperators`.

We now show how these diagrams can be computed with the aid of rotations and Lorentz transformations. For consistency, all quantities appearing in Eq. 15.18 should be computed in the same reference frame. The center-of-mass frame is the easiest in which to work, since in that frame incoming and outgoing vectors differ by a rotation. We will boost all incoming vectors and spinors into the center-of-mass frame, then rotate to obtain the outgoing vectors and spinors. The scattering angle in the center-of-mass frame θ_{cm} can be related to the scattering angle in the lab frame θ through the equation:

$$\cos\theta_{cm} = \frac{\cos\theta - \beta}{1 - \beta\cos\theta} \quad (15.19)$$

where $\beta = k_0/(k_0 + m_e)$ is the velocity of the center-of-mass with respect to the laboratory.

We introduce a class called Compton with the public interface:

```
class Compton {

  public:

    // Constructor:
    Compton(double kLab, bool isPolarized=false);

    // Compute the cross section in units of the classical
        electron radius squared ( r_e^2)
    // where   r_e = 1/m Mev^-1 and  m=0.511 MeV.

    double sigma(double cosThetaLab, double phiLab=0.0);
    ...
};
```

This class is initialized with the energy (in MeV) of the incoming photon and an optional flag to turn on polarization. It is intended to provide the differential cross section as a function of θ, the scattering angle in the laboratory frame, and ϕ, the azimuthal angle. The incoming photon is taken to move along the positive z direction. If it is polarized, then the polarization is assumed to be along x; if not, then the scattering does not depend upon the azimuthal angle ϕ, which may be omitted in the function call.

Certain constants are declared static in the class declaration to avoid initializing them each time the Compton class is instantiated; these include constants of nature, unit vectors, constant polarization four-vectors $\epsilon^{x,y}$, spin-up and spin-down spinors, and a constant matrix $\mathbf{M} = \text{diag}(m_e, m_e, m_e, m_e)$. In the .cpp file they are iniitialzed as:

```
const double Compton::alpha=1/137.0;
const double Compton::m=0.511;
const ThreeVector Compton::nullVector(0,0,0);
const ThreeVector Compton::zHat(0,0,1);
const FourVector Compton::eps[2] = {FourVector(0,1,0,0),
    FourVector(0,0,1,0)};
const Spinor <> Compton::up={1,0};
const Spinor<> Compton::down={0,1};
const Eigen::Matrix4cd Compton::M=Compton::m*Eigen::Matrix4cd
    ::Identity();
```

These data members belong to the Compton class:

```
  // The initial state:
  const double                      k0;              // photon energy,
                                                        lab frame.
```

```
  const double                      beta;     // velocity of cm
                                              frame.
  const LorentzTransformation tCM;            // to center-of-
                                              mass frame
  const FourVector                  k;        // photon four-
                                              vector in C.M.
  const FourVector                  p;        // electron four-
                                              vector in C.M.
  const DiracSpinor::U              u[2];     // Dirac spinors
```

They represent mathematical objects appearing in Eq. 15.18, all expressed consistently in the center-of-mass frame:

$$p \quad \text{in code} \quad \text{tCM} * \text{FourVector}(m,0,0,0), \qquad (15.20)$$

the four-momentum of the incoming electron;

$$k \quad \text{in code} \quad \text{tCM} * \text{FourVector}(k0,0,0,k0), \qquad (15.21)$$

the four-momentum of the incoming photon;

$$u[0] \quad \text{in code} \quad \text{tCM} * \text{DiracSpinor} :: U(\text{up}, m), \qquad (15.22)$$

the Dirac spinor for the incoming spin-up electron,

$$u[1] \quad \text{in code} \quad \text{tCM} * \text{DiracSpinor} :: U(\text{down}, m), \qquad (15.23)$$

the Dirac spinor for the incoming spin-down electron. With regard to the spinor, we will compute the squared sum of matrix elements, and average that over both polarization states of the electron.

These objects are initialized in the constructor, shown here:

```
Compton::Compton(double k0, bool isPolarized) :
  k0(k0) ,
  beta(1.0/(1.0+m/k0)),
  tCM(ThreeVector(0,0,-atanh(beta))),
  k(tCM*FourVector(k0,0,0,k0)),
  p(tCM*FourVector(m,0,0,0)),
  u{tCM*DiracSpinor::U(up,m), tCM*DiracSpinor::U(down,m)},
  isPolarized(isPolarized)
{}
```

The outgoing four-vectors, spinors, and polarization vectors can be computed once the scattering angle is known. A Lorentz transformation consisting of a pure rotation can then be constructed. The rotation axis points in the direction of $\hat{z} \times \hat{n}$ and the rotation angle is just θ_{cm}:

```
// We calculate the scattering angle in the C.M. frame:
double cosThetaCM=(cosThetaLab-beta)/(1-beta*cosThetaLab);
double sinThetaCM = sqrt((1+cosThetaCM)*(1-cosThetaCM));
double phiCM=phiLab;

ThreeVector nHat(sinThetaCM*cos(phiCM), sinThetaCM*sin
   (phiCM), cosThetaCM);
LorentzTransformation rot(nullVector, zHat.cross(nHat).
   normalized()*acos(cosThetaCM));
```

Then we can compute the outgoing quantities:

$$k' \quad \text{in code} \quad \text{FourVectorkPrime} = \text{rot} * \text{k};, \qquad (15.24)$$

the four-momentum of the incoming photon;

$$u'[0] \quad \text{in code} \quad \text{DiracSpinor} :: \text{UuPrime} = \text{rot} * \text{u}[0];, \qquad (15.25)$$

the Dirac spinor for the outgoing positive-helicity electron,

$$u'[1] \quad \text{in code} \quad \text{DiracSpinor} :: \text{UuPrime} = \text{rot} * \text{u}[1];, \qquad (15.26)$$

the Dirac spinor for the outgoing negative-helicity electron,

$$\epsilon'[0] \quad \text{in code} \quad \text{FourVectorepsPrime} = \text{rot} * \text{eps}[0];, \qquad (15.27)$$

the polarization vector for:

$$\epsilon'[1] \quad \text{in code} \quad \text{FourVectorepsPrime} = \text{rot} * \text{eps}[1];, \qquad (15.28)$$

the Dirac spinor for the outgoing negative-helicity electron. We shall compute the squared sum of matrix elements for both spin states of the outgoing electron and both polarization states of the outgoing photon, and sum them all for the cross-section computation.

Finally, we come to the computation of the matrix elements themselves. For the s-channel, Eq. 15.18 can be realized like this:

```
Complex MS = e*e*((bar(uPrime)*(slash(epsPrime.conjugate())*
   (slash(p+k)+M)/((p+k).squaredNorm()-m*m)*slash(epsIn))*
   uIn)(0));
```

500 *Fundamental processes: Compton scattering*

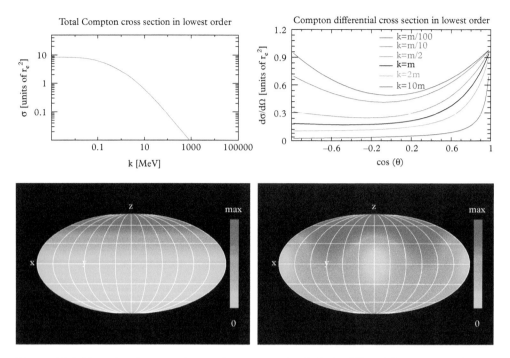

Figure 15.4 *Plots of total cross section vs. photon energy (top left), the differential cross section (top right), and angular distribution in elliptical projection for unpolarized (bottom left) and polarized (bottom right) photons at energy $k = m_e$. For the elliptical projections, the direction of the incident photon is along z.*

and for the *t*-channel, like this:

```
Complex MT = e*e*((bar(uPrime)*(slash(epsIn)*(slash(p-kPrime)
    +M)/((p-kPrime).squaredNorm()-m*m)*slash(epsPrime.
    conjugate()))*uIn)(0));
```

These two matrix elements are summed and the sum is squared. The prefactors in Eq. 15.18 are included:

```
double ks=1./(1./k0+1./m*(1.-cosThetaLab));
double re=alpha/m;
double phaseSpaceFactor=pow(ks/(8.0*M_PI*k0*m),2);
```

Finally, the cross section is for a particular polarization state of both incoming and outgoing photons and the spin state of both incoming and outgoing electrons. For this application we can't see how one would ever know or care to know the spin state of the incoming electron, the outgoing electron, or the outgoing photon. We should therefore average over the incoming spins and sum over the outgoing polarizations and spins. The polarization of the incoming photon is sometimes of interest. Averaging over the polarization state of the incoming photon is only called for if the incoming photon "beam" is unpolarized.

The class `Compton` can be found in `EXAMPLES/CH15/COMPTONXS`, together with a test program that computes the plots in Figure 15.4.

15.7 A simple experiment: double Compton scattering

In 1948, R.C. Hanna of Rutherford Laboratory in England reported the results of an important experiment in quantum mechanics (Hanna, 1948). A sketch of his apparatus from his Nature article is shown in Figure 15.5. A similar experiment was carried out two years later, more conclusively, by Wu and Shaknov of Columbia University.

In the experiment, a positron is produced from a radioactive isotope. The positron forms an "atom" of positronium, e^+e^-, which annihilates from a $^1S_0$ state (called parapositronium). In order to conserve energy and momentum the decay is required to produce into two gamma (γ) rays that are back to back and have energy $k = m$. Each γ may scatter from aluminum targets placed on either side of the source.

The theory of double Compton scattering was first correctly described by Pryce and Ward (Pryce, 1947) and almost at the same time by Snyder, Pasternack, and Hornbostel (Snyder, 1948). The $^1S_0$ positronium state has angular momentum $L = 0$ and spin $S = 0$. For positronium, a fermion-antifermion bound state, the C (charge conjugation) eigenvalue is $(-1)^{L+S} = 1$. The P (parity) eigenvalue is $(-1)^{L+1} = -1$.

Because the initial state has total angular momentum of zero, the spins of the two photons must be antiparallel. We take the spin quantization axis of each photon as its direction of travel; a right-handed photon will be designated as $|R\rangle$ and has a spin of $+1$;

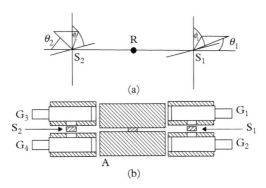

Figure 15.5 *Schematic diagram and sketch of the apparatus used by R.C.Hanna in 1948 to observe correlations in the polarization of annihilation radiation.* ©1948 Nature. Reprinted with permission.

while a left handed photon will be designated as $|L\rangle$ and has a spin of -1. The two-photon state is even under charge conjugation, so the decay is allowed, provided that parity of the final state is odd. We arrange that as follows.

The final state of two photons is a linear superposition of $|R_1\rangle|R_2\rangle \equiv |R_1R_2\rangle$ and $|L_1\rangle|L_2\rangle \equiv |L_1L_2\rangle$. The parity operation transforms a left handed photon into a right-handed photon. The initial state:

$$|\psi\rangle = \frac{|R_1R_2\rangle - |L_1L_2\rangle}{\sqrt{2}} \tag{15.29}$$

clearly has $\mathbf{P}|\psi\rangle = -|\psi\rangle$. The state can also be written as:

$$|\psi\rangle = \frac{|R_1R_2\rangle - |L_1L_2\rangle}{\sqrt{2}}.$$
$$= \frac{1}{\sqrt{2}}\left[\left(\frac{|X_1\rangle + i|Y_1\rangle}{\sqrt{2}}\right) \cdot \left(\frac{|X_2\rangle + i|Y_2\rangle}{\sqrt{2}}\right) - \left(\frac{|X_1\rangle - i|Y_1\rangle}{\sqrt{2}}\right) \cdot \left(\frac{|X_2\rangle - i|Y_2\rangle}{\sqrt{2}}\right)\right]$$
$$= i\frac{|X_1Y_2\rangle + |Y_1X_2\rangle}{\sqrt{2}}.$$

This quantum state is entangled: the state of each photon cannot be expressed independently of the other.

The annihilation radiation is unpolarized, but there is a correlation between the direction of the electric field of the two gamma rays, which are required to be perpendicular to each other. This polarization was analyzed in the Hanna experiment by Compton scattering from two heavy targets. Let's call the polar and azimuthal angles of the two scattered photons θ_1, ϕ_1, θ_2, and ϕ_2. Because of the correlation of the polarization of the two photons there will be a correlation between the two azimuthal angles ϕ_1 and ϕ_2. The experiment was designed to measure this correlation carefully.

The experiment was important because it distinguished two radically different views of quantum mechanical processes—putting different ingredients into the calculation results in different predictions for the correlation. By confronting the simulation with data, one can distinguish two scenarios:

- Scenario A: The two photons have a random (but predetermined) polarization as they move away from the source. This predetermined polarization is partially analyzed on both ends by the Compton scattering process.
- Scenario B: The two photons exist in a state $|\psi\rangle = \frac{i|X_1Y_2\rangle + |Y_1X_2\rangle}{\sqrt{2}}$ which makes a transition to a final state in which the first photon is detected with polar and azimuthal angles θ_1 and ϕ_1 while the second photon is detected at polar and azimuthal angles θ_2 and ϕ_2.

These two scenarios are rather different because in the first scenario, the annihilation radiation is *either* $|XY\rangle$ *or* $|YX\rangle$ while in the latter it is *both* $|XY\rangle$ *and* $|YX\rangle$. The two scenarios can be distinguished from an analysis of the angular distributions; an interference term in Scenario B is absent in Scenario A.

Scenario B is more unsettling than the former, because it inescapably involves nonlocal effects. If the first gamma is polarization-analyzed and determined to be in an $|X\rangle$ state, the second will be forced at the same time into a $|Y\rangle$ state—across a spacelike interval. The result of the second observer depends on the measurement of the first observer. This type of paradox was discussed in a famous paper by Einstein, Podolsky, and Rosen (EPR) (Einstein, 1935). These nonlocal effects are, in fact, actually observed in experiment, and as strange as they seem, are an established feature of quantum mechanics.

To simulate this process we consider the scattering of two gamma rays. We let $\hat{U} = \hat{U}_1\hat{U}_2$ be the time evolution operator which takes the initial state to the final state, $\hat{U}_{1,2}$ being the time evolution operators for each photon interacting with an electron from the target. We shall label the final state as $|\theta_1\phi_1\theta_2\phi_2\rangle \equiv |\theta_1\phi_1\rangle|\theta_2\phi_2\rangle$. The amplitude for this process is:

$$\langle\theta_1\phi_1\theta_2\phi_2|\hat{U}|\psi\rangle = \frac{i}{\sqrt{2}}\left[\langle\theta_1\phi_1|\hat{U}_1|X_1\rangle\langle\theta_2\phi_2|\hat{U}_2|Y_2\rangle + \langle\theta_1\phi_1|\hat{U}_1|Y_1\rangle\langle\theta_2\phi_2|\hat{U}_2|X_2\rangle\right]. \tag{15.30}$$

The amplitudes $\langle\theta_i\phi_i|\hat{U}_i|X_i\rangle$ and $\langle\theta_i\phi_i|\hat{U}_i|Y_i\rangle$ are just the amplitudes discussed in the previous section, the sum of the two Lorentz invariant scattering amplitudes in Figure 15.3 and Eq. 15.18. To compute the double-scattering fourfold differential cross section $d\sigma/d\Omega_1 d\Omega_2$, we require two phase space factors:

$$\frac{d\sigma}{d\Omega_1 d\Omega_2} = \left[\frac{1}{64\pi^2 m_e^2}\left(\frac{k_1}{k_{01}}\right)^2\right]\left[\frac{1}{64\pi^2 m_e^2}\left(\frac{k_2}{k_{02}}\right)^2\right]|\langle\theta_1\phi_1\theta_2\phi_2|\hat{U}|\psi\rangle|^2$$

$$= \frac{k_1^2 k_2^2}{(8\pi m_e^2)^4}|\langle\theta_1\phi_1\theta_2\phi_2|\hat{U}|\psi\rangle|^2 \tag{15.31}$$

where $k_{01} = k_{02} = m_e$ are the energies of the incoming photons (which are equal to the electron rest energy since we are dealing with annihilation radiation).

The reduction of this formula to an analytic expression involving the angles is a difficult task. However, simulating the process is not so difficult if one retains the amplitudes in the computation of Compton scattering processes. The simulation of this process is left as an exercise that we are certain the student will enjoy.

15.8 Heavier charged particles

The discussion of radiation transport with which we started this chapter would not be complete without a discussion of two processes that apply universally to stable charged particles traversing materials: ionization energy loss and multiple Coulomb scattering. Ionization energy loss is due to the interaction of a charged particle with the electrons in the material; while multiple Coulomb scattering is due to its interaction with nuclei.

Approximate formulae are known for both of these processes. Mean ionization energy loss over a range of $0.1 < p/m = \beta\gamma < 1000$ is described by the Bethe-Bloch equation (Patrignani, 2016):

$$\left\langle -\frac{dE}{dx} \right\rangle = Kz^2 \frac{Z}{A} \left[\frac{1}{2} \log \frac{2m_e c^2 \beta^2 \gamma^2 W_{max}}{I^2} - \beta^2 - \delta(\beta\gamma) \right] \tag{15.32}$$

where the constant K is $4\pi r_e^2 N_A m_e c^2 = 0.307075$ MeV mol^{-1}, z is the charge of the incident particle, I is the excitation energy of the material, and β and γ are the speed and Lorentz factor of the incident particle.[2] The quantity:

$$W_{max} = \frac{2m_e c^2 \beta^2 \gamma^2}{1 + 2\gamma m_e/M + (m_e/M)^2} \tag{15.33}$$

with M being the mass of the incident particle, is the maximum kinetic energy that can be transfered to an electron in a single collision. The formula calculates the mean ionization energy loss divided by the density of the material; multiply by the density of the material to obtain the energy loss per unit length. This is done to (mostly) remove the dependence of energy loss upon the density of the material; but a residual dependence persists through the term $\delta(\beta\gamma)$. This term represents the so-called "density effect", whereby the incident particle produces a polarization in the surrounding material which reduced the electric field from the incident particle. The effect is usually introduced through a parameterization due to Steinhammer (1952).

A plot of the Bethe-Bloch function is shown in Figure 15.6, for the case of muons ($z=1$, $M=105$ MeV) on aluminium, neglecting the aforementioned density effect which enters for highly relativistic particles but does not change the picture quantitatively. Several features are apparent from the plot: slow particles lose energy at a higher rate than fast particles. The energy loss curve falls to a broad minimum around $\beta\gamma \sim 3$, where (for aluminum) it takes the value 1.6 MeV/g/cm^2; note also that over the range $1.3 < \beta\gamma < 18$ the the energy loss stays in the range 1.6 MeV/g/cm$^2 < dE/dx < 2.0$ MeV/g/cm^2. A *minimum ionizing particle* (MIP) with $\beta\gamma \sim 3$, but usually particles over the broad minimum which is apparent in Figure 15.6 are taken to be MIPs. Above the minima, the density effect contributes corrections that can reach order (10%). This correction is larger for heavier materials and smaller for lighter materials and gases. Above $\beta\gamma = 1000$, heavy charged particles begin to radiate and the approximation is no longer valid.

The second effect, multiple Coulomb scattering, causes deviations in the trajectory of a particle of charge Z as it traverses a layer of material of thickness x. The angular

---

[2] Ionization energies of common materials can be found at http://physics.nist.gov.

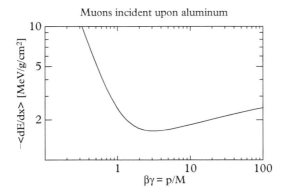

Figure 15.6 *Ionization energy loss of a muon in aluminum according to the Bethe-Bloch equation, neglecting the density effect. Eq. 15.32.*

deviation projected in any plane perpendicular to the particle's velocity is approximated (Patrignani, 2016) by a Gaussian distribution with width:

$$\theta_0 = \frac{13.6\text{MeV}}{\beta cp} Z\sqrt{x/X_0} \left(1 + 0.038\ln(x/X_0)\right). \tag{15.34}$$

Angular deviations projected into any *two* planes are independent if the two planes are mutually perpendicular. It may also be of interest to know the lateral displacement along any direction perpendicular to the original trajectory. It too may be described as a Gaussian with width $x\theta_0/\sqrt{3}$.

15.9 Conclusion

In this chapter we have given an overview of how to simulate stochastic processes, drawing examples from the decays of unstable particles and the evolution of their daughter particles through matter. The randomness in all of these examples comes ultimately from quantum mechanics, which is nature's own random number generator. We remind the reader again that every formula in this chapter is an approximation with limited accuracy and range of applicability. Simulation toolkits, discussed in Subsection 15.5.3, often start out life with simple formulae but have grown increasingly complex as the demand for higher accuracy has grown. Accurate discussion of physics processes can be found throughout the literature, but nowadays the best detailed overview of such processes can be found in the *Review of Particle Physics* (Patrignani, 2016); the best implementation is the GEANT Toolkit, and the best practical description of the physics processes is the GEANT 4 Physics reference manual (The GEANT4 Collaboration, 2016). These references are indispensable for the serious developer of radiation transport simulation code. In the exercises we invite the student to develop his or her skills using approximations which are for the most part crude.

15.10 Exercises

1. Higgs Decays.
   Generate 10K simulated decays of the Higgs boson, keeping track only of the final state, and make a histogram showing the frequency of decay of each mode.

2. [P] Gamma Ray Astronomy.
   Gamma ray astronomy relies on Cherenkov light from electromagnetic showers initiated high in the earth's atmosphere. Simulate the development of a shower initiated by a 100 GeV gamma ray striking the earth's atmosphere. Assume the density of the atmosphere falls of with altitude z as $\rho(z) = \rho_0 \exp(-z/(8\mathrm{km}))$, where ρ_0 is the density of the atmosphere at sea level. Make a graphical display of the shower. It is best to do this in two steps: simulating and saving the data in the first step, then displaying it in the second.

3. [P] Electron-Muon Scattering.
   Simulate electron-muon ($e^-\mu^-$) scattering at a center of mass energy of 10 GeV. The process involves only one t-channel diagram with the matrix element:
   $$-e^2 \frac{(\bar{u}'_1 \gamma^\mu u_1)(\bar{u}'_2 \gamma_\mu u_2)}{(p'_1 - p_1)^2},$$
   where u_1, u'_1, p_1, p'_1 are the incoming and outgoing electron spinors and four-momenta, u_2, u'_2, p_2, p'_2 are the incoming and outgoing muon spinors and four-momenta. Products such as $e^2 \bar{u}'_1 \gamma^\mu u_1$ occurring in these expressions are four-vector *currents*; they can be computed by hand or by using the class Current4 from the Spacetime library. An example of the latter is:

   ```
   #include "Spacetime/Current4.h"
   #include "Spacetime/SpecialOperators.h"
   using namespace SpecialOperators;
   ComplexFourVector jElectron=bar(u1Prime)*Gamma()*u1;
   ComplexFourVector jMuon    =bar(u2Prime)*Gamma()*u2;
   std::complex<double> jeDotjm=jElectron.dot(jMuon);
   ```

   Plot the cosine of the scattering angle and the azimuthal angle in the center of mass assuming unpolarized beams. Note that the cross section is infinite at low values of the scattering angle so impose a cut of $\cos\theta < 0.95$ when generating events.

4. [P] Polarized Electron-Muon Scattering.
   Repeat the previous exercise, but this time assume that:

   a) The electron beam is transversely polarized and the muon beam is unpolarized.
   b) The electron beam is transversely polarized and the muon beam is transversely polarized in the same direction.

c) The electron beam is transversely polarized and the muon beam is transversely polarized in an orthogonal direction.

d) The electron beam is longitudinally polarized and the muon beam unpolarized.

e) The electron beam and the muon beam are longitudinally polarized in the same direction.

f) The electron beam and the muon beam are longitudinally polarized in the opposite directions.

5. [P] Double Compton Scattering.
   Simulate the double Compton experiment of Section 15.7. Plot histograms of:

   a) the distribution of $\cos\theta_1$ and $\cos\theta_2$.
   b) the distribution of ϕ_1 and ϕ_2.
   c) the $\Delta\cos\theta = \cos\theta_1 - \cos\theta_2$ distribution.
   d) the $\Delta\phi = \phi_1 - \phi_2$ distribution.

   of the scattered radiation. Use parallel processing since event generation will be CPU intensive. Then plot the $\Delta\phi$ distribution again when both photons scatter through an angle of 90°.

6. [P] Electron-Positron Annihilation to Muons.
   In the past, several electron-positron colliders have operated at accelerator laboratories around the world. One of the processes that one observes at these colliders is $e^+e^- \to \mu^+\mu^-$. The electron has a mass of 0.511 MeV, while the muon has a mass of 105 MeV. The positively charged e^+ and μ^+ are antiparticles of the negatively charged e^- and μ^- and have the same mass as their particles. The annihilation process involves only one s-channel diagram from which the matrix element:

   $$-e^2 \frac{(\bar{v}\gamma^\nu u)(\bar{u}'\gamma_\nu v')}{(p_{e^+} + p_{e^-})^2}$$

   can be deduced. Here u, v, p_{e^+}, p_{e^-} are the incoming electron and positron spinors and four-momenta, while u', v' are the outgoing muon and antimuon spinors and four-momenta. Simulate the process and compute the total cross section σ and the differential cross section $d\sigma/d\Omega$ vs. the scattering angle θ in the center of mass frame. Average over the initial state helicity and sum over final state helicity. In so doing, note which combinations of initial state electron and positron helicities contribute to the cross section. Do the same for combinations of final state helicities. Does this suggest how you can speed up the computation?

7. [P] High Energy Electron-Positron Scattering.
   If electrons and positrons are collided at much higher energy, the neutral weak interaction becomes important. At a center of mass energy of 91.2 GeV, it completely

dominates. This interaction is mediated by the neutral Z^0 boson and thus violates parity symmetry, or invariance under spatial inversion. The amplitude for the process is:

$$-\frac{ig^2}{\cos^2\theta_w} \frac{-i(j_{in} \cdot j_{out} - (j_{in} \cdot p_z)(j_{out} \cdot p_z)/M_z^2)}{p_z^2 - M_Z^2 + iM_Z\Gamma_Z}$$

where:

$$j_{in}^\nu = \bar{v}\gamma^\nu \frac{1}{2}(c_V - c_A\gamma^5)u \qquad j_{out}^\nu = \bar{u}'\gamma^\nu \frac{1}{2}(c_V - c_A\gamma^5)v'),$$

$p_z = p_{e^+} + p_{e^-}$, u, v, p_{e^+}, p_{e^-}, u' and v' are defined in the previous problem; $M_Z = 91.2$ GeV is the mass of the Z boson; and its natural width is $\Gamma_z = 2.50$ GeV. The coupling constants are $c_v = -1/2$ and $c_a = -1/2 + 2\sin^2\theta_w$ where θ_w is the Weinberg angle, $\sin^2\theta_w = 0.231$; while the constant $g = e/\sin\theta_W$. The operator $\gamma^5 = i\gamma^0\gamma^1\gamma^2\gamma^3$, for which an implementation can be found in the `SpecialOperators.h` of the `Spacetime` library, is:

$$\gamma^5 = \begin{pmatrix} -I & 0 \\ 0 & I \end{pmatrix}.$$

a) Simulate the process $e^+e^- \to \mu^+\mu^-$ at a center of mass energy of 91.2 GeV.
b) Make a plot of the cosine of the scattering angle.
c) The matrix element for the production of bottom quark pairs, $e^+e^- \to b\bar{b}$, is the same, except that the constant $c_a = -1/2 + 2/3\sin^2\theta_w$. Show that for this process, at a center of mass energy of $E_{cm} = M_Z$,

- Polarized bottom quarks are produced from unpolarized electron beams.
- The production of bottom quarks is asymmetric in the scattering angle θ.

8. [P] Interference in Electron-Positron Annihilation to Muons.
At energies well below M_Z, both the electromagnetic and the weak interaction contribute to the process $e^+e^- \to \mu^+\mu^-$. Since the final state is identical, the two contributions interfere. Include *both* production mechanisms in your Monte Carlo simulation. Use it to compute the total cross section, and the forward-backward asymmetry $A_{FB} \equiv (N_+ - N_-)/(N_+ + N_-)$, where $N_\pm$ are the numbers of positive muons produced in the "forward" hemisphere (along the direction of the e^+) and in the backward hemisphere. Do this for various center-of-mass energies E_{cm} and plot the cross section and A_{FB} as a function of E_{cm}.

9. [P] Bhabha Scattering.
Bhabha scattering is the elastic scattering of electrons and positrons. At most e^+e^- colliders, a focused electron beam collides head-on with a focused positron beam.

Each beam has equal energy, so the laboratory frame coincides with the center-of-mass frame. Two Feynman diagrams contribute to the process. The *s*-channel diagram translates into the expression:

$$-e^2 \frac{(\bar{v}\gamma^\nu u)(\bar{u}'\gamma_\nu v')}{(p_{e^+} + p_{e^-})^2}$$

while the *t*-channel diagram translates into the expression:

$$-e^2 \frac{(\bar{u}'\gamma^\mu u)(\bar{v}\gamma_\mu v')}{(p'_{e^-} - p_{e^-})^2}$$

In this expression, p_{e^+}, p_{e^-}, p'_{e^+}, p'_{e^-} are the four-vector momenta of the incoming positron, incoming electron, outgoing positron, and outgoing electron, respectively. The spinors, $\bar{v}$, u, v, and $\bar{u}$ are the Dirac spinors representing the incoming positron, incoming electron, outgoing positron, and outgoing electron, respectively. Simulate the process; compute and plot the differential cross section $d\sigma/d\Omega$ as a function of the cosine of the scattering angle in the center-of-mass. Note that the cross section diverges in the forward direction; therefore introduce a cutoff value in $\cos\theta$ in order to generate efficiently, and in the interesting region of phase space. Take the beam energy to be 5000 MeV.

10. [P] Bragg Peak.
    Consider a monoenergetic beam of charged particles incident upon water, for which the mean ionization energy of $I = 81$ eV may be taken. Notice from Figure 15.6 that slower particles lose energy at a high rate, such that the greatest deposit of energy occurs at the end of the particle's range. Using the Bethe-Bloch equation (and ignoring the density effect), plot dE/dx (in MeV/cm) as a function of distance travelled in water. You should observe a sharp peak at the endpoint of the trajectory, known as the ***Bragg peak***. Do this for protons (mass 938 MeV) and for charged pions (mass 140 MeV). For which species of particle is the Bragg peak more pronounced?

11. [P] Proton Therapy.
    In Chapter 10, Exercise 10 you made an accurate computer description of your own head. Somewhere in there lies a pea-shaped organ called the pituitary gland, which is located behind the eyes, approximately halfway between the front and back of the skull. ***Proton therapy*** is sometimes used to treat tumours of the pituitary gland. In proton therapy the energy of a beam of protons is chosen so that the Bragg peak (see previous exercise) coincides with the tumour. To minimize damage to surrounding tissue, the orientation of the beam relative to the cranium is varied to reduce the exposure to any given point *except* the tumor. Simulate a treatment session in which the patient is slowly rotated around his/her "axis" while the energy of the beam is adjusted so that the Bragg peak always targets the pituitary gland. Use the computer description of your head for a model. Make a 3d map of the energy deposit within your cranium.

BIBLIOGRAPHY

Agostenelli S., J. Allison, K. Amako, *et al.* (2003). *Geant4 – a simulation toolkit.* Nucl. Instrum. Methods Phys. Res. A **506**, 250.

Aarnio, P. A. *et al.* (1986). *FLUKA86 User's Guide.* CERN Divisional Report TIS/RP-168.

Aarnio, P. A. *et al.* (1987). *Enhancements to the FLUKA86 Program (FLUKA87).* CERN Divisional Report TIS/RP-190.

The ATLAS Collaboration (2008). *Expected performance of the ATLAS detector.* CERN-OPEN-2008-020, arXiv:0901.0512 [hep-ex].

The ATLAS Collaboration (2012). *Observation of a new particle in the search for the standard model Higgs boson with the ATLAS detector at the LHC.* Physics Letters B **716**, 1.

The CMS Collaboration (2012). *Observation of a new boson at a mass of 125 GeV with the CMS experiment at the LHC.* Physics Letters B **716**, 30.

Einstein, A., B. Podolsky, and N. Rosen. (1935). *Can the quantum mechanical description of physical reality be considered complete?* Phys. Rev. **47**, 777–780.

Ford, R. L. and W. R. Nelson (1978). *The EGS code system: computer programs for the Monte Carlo simulation of electromagnetic cascade showers.* SLAC Report SLAC-210 UC-32.

The GEANT4 Collaboration (2015). *Physics Reference Manual.* geant4.web.cern.ch.

Hanna, R. C. (1948). *Polarization of annihilation radiation.* Nature **162**, 332.

Heinemeyer, S. (2013). *Handbook of LHC Higgs Cross Sections: 3. Higgs Properties: Report of the LHC Higgs Cross Section Working Group.* CERN Report CERN-2013-004. A spreadsheet of updated numbers is available from the website https://twiki.cern.ch/twiki/bin/view/LHCPhysics/LHCHXSWG#CERN_Reports_Handbook_of_LHC_Hig

Klein, O and Y. Nishina (1929). *Über die Streuung von Strahlung durch freie Elektronen nach der neuen relativistischen Quantendynamik von Dirac.* Z. Phys. 52 (11–12), 853 and 869.

Patrignani, C., K. Agashe, G. Aielli *et al.* (Particle Data Group) (2016). *Review of Particle Physics.* Chin. Phys. C, **40**, 100001.

Sempau J., Sempau, J., E. Acosta, J. Baro, J.M. Fernandez-Varea, and F. Salvat (1997). *An algorithm for Monte Carlo simulation of coupled electron-photon transport.* Nucl. Instr. and Meth. B. **132**, 377–90.

Pryce, M. H. L., and J. Ward (1947). *Angular Correlation of Annihilation Radiation.* Nature **160**, 435.

Schwartz, M. D. (2014). *Quantum field theory and the standard model.* Cambridge University Press.

Snyder, H. S., S. Pasternack, and J. Hornbostel (1948). *Angular Correlation of Annihilation Radiation.* Phys. Rev. **73**, 440.

Steinhammer, R. M. (1952). *The Density Effect for the Ionization Loss in Various Materials.* Phys. Rev. **88**, 851.

Tsai, Y. S. (1974). *Pair production and bremsstrahlung of charged leptons.* Rev. Mod. Phys. **46**, 815.

Wu, C. S., and I. Shaknov (1950). *The angular correlation of scattered annihilation radiation.* Phys. Rev. **77**, 136.

16
Data modeling

16.1 Tabular data	512
16.2 Linear least squares (or χ^2) fit	514
16.3 Function minimization in data modeling	516
16.3.1 The quality of a χ^2 fit	522
16.3.2 A mechanical analogy	523
16.4 Fitting distributions	523
16.4.1 χ^2 fit to a distribution	524
16.4.2 Binned maximum likelihood fit to a distribution	527
16.5 The unbinned maximum likelihood fit	529
16.5.1 Implementation	530
16.5.2 Construction of normalized PDFs	533
16.6 Orthogonal series density estimation	534
16.7 Bayesian inference	538
16.8 Combining data	540
16.9 The Kalman filter	543
16.9.1 Example: fitting a polynomial curve	545
16.9.2 Complete equations	546
16.10 Exercises	550
Bibliography	554

In this chapter we describe how to treat a problem which is in many respects the *inverse* of a simulation problem. In simulation we generate a realization of a probability density function of one, several, or even an enormous number of variables. In data modeling (or *fitting*, more colloquially) the problem is the extraction of meaningful physical parameters from data. Procedures for modeling data can be applied to either synthetic data from pseudoexperiments or real data from actual experiments; generally synthetic data is used to develop, test, and characterize fits applied to real data. This step is practically obligatory nowadays in many branches of experimental science.

Data modeling is a bit of an art, and refers to several related procedures. These have much in common—many of them are built around the idea of maximizing the probability of the data over parameters of the model. They differ by the exact measure of the probability, in other words the objective function that is to be extremized; but in each case the objective function is exactly or approximately related to a particular function of the unknown parameters $\vec{\alpha}$, which is known as the likelihood function $\mathcal{L}(\vec{\alpha})$, and which is constructed in some way from the dataset. The likelihood function itself is interesting and will have an interpretation of its own within the school of Bayesian statistics, where it determines the probability density for unknown parameters. Within the frequentist school of thought the output of our procedures, called the *maximum likelihood estimator* (MLE), is important because it possesses the desirable statistical properties of consistency, efficiency, and asymptotic normality:

- Consistency: the estimated parameter values tend to their true values in the limit of infinitely large datasets; one also says that they are *unbiased* in this limit.
- Asymptotic normality: in an ensemble of equivalent experiments, the estimated values are distributed about the true values according to a multivariate normal distribution.
- Efficiency: the variance of the estimated values from an ensemble of equivalent experiments is as low as it possibly can be. This is referred to as the Cramér-Rao bound.

This chapter deals exclusively with the large-dataset limit, and therefore in the following we regard the likelihood function as the probability density for the unknown parameters $\vec{\alpha}$. This justifies some of the steps we take in the following sections. Frequentists may take comfort in the fact that the estimators we derive from this line of reasoning are unbiased, normal, minimum-variance estimators of the unknown quantity, in precisely the same limit.

Instead of maximizing the likelihood $\mathcal{L}(\vec{\alpha})$, one instead minimizes $-2\ln\mathcal{L}(\vec{\alpha})$. For the determination of the optimal parameters $\vec{\alpha}_0$ (sometimes called the central value of the fit) the two procedures are equivalent. The uncertainties on $\vec{\alpha}$ can be determined directly from the Hessian matrix:

$$(\mathbf{H})_{ij} = \frac{\partial^2(-2\ln\mathcal{L})}{\partial\alpha_i\partial\alpha_j}$$

of the latter quantity determined numerically. These uncertainties are properties of the multivariate normal distribution, which arises in the frequentist interpretation as the asymptotic form of the likelihood function.

16.1 Tabular data

One of the simplest ways to collect data is in tabular form, and we introduce here a new class in `QatDataAnalysis` called `Table`. Each row in the table will be called a *Tuple*, and consists of several columns, each having a particular built-in datatype: `int`, `unsigned int`, `float`, or `double`.

The program in EXAMPLES/CH16/FITTING/GRAVITYTABLE is an illustration of the use of the Table class. It is excerpted here:

```
unsigned int seed=0;
std::mt19937 engine(seed);
std::normal_distribution<double> n;
double      g=9.8;
double      v0=-32.1;
double      x0=10.0;
double      sigmaV=9.0;
double      sigmaX=4.0;

Table table("Gravity");
unsigned int i=0;
for (double t=0;t<10;t+=1.0) {
  double v = v0+g*t;
  double x = x0 + v0*t + 1/2.*g*t*t;
  v += sigmaV*n(engine);
  x += sigmaX*n(engine);
  table.add("I", i++);
  table.add("Time", t);
  table.add("Velocity", v);
  table.add("Position", x);
  table.add("SigmaX", sigmaX);
  table.add("SigmaV", sigmaV);
  table.capture();
}
table.print();
```

This program is a simulation of an experiment to measure gravity, similar to the first laboratory experiment typically performed by students in elementary physics. In the simulated experiment a mass is released from rest from a great height and a clock is started after the mass passes through a gate. Thereafter, the position and velocity are measured at regular intervals by means of a radio transmitter. Both position and velocity are measured with intrinsic uncertainties (σ_x and σ_v), which are also recorded in the table. In our example we will take these uncertainties to be fixed. The deviation of the recorded values of x and v are drawn from a normal distribution with standard deviation σ_x and σ_v. The final line in this example prints the entire table as shown below:

I	Position	SigmaV	SigmaX	Time	Velocity
0	11.2112	9	4	0	-21.9948
1	-16.9078	9	4	1	-21.6623
2	-28.5197	9	4	2	-25.3009
3	-42.7324	9	4	3	-5.32255

4	-47.0466	9	4	4	5.54237
5	-22.5325	9	4	5	16.1109
6	-7.63598	9	4	6	36.8278
7	20.042	9	4	7	47.4855
8	66.3061	9	4	8	50.1554
9	117.504	9	4	9	68.8294

Next, we describe the linear least squares technique for extracting a value of the gravitational constant g, as well as the initial velocity v_0, from data such as that presented in this table.

16.2 Linear least squares (or χ^2) fit

The method of linear least squares fitting is one of the oldest, simplest and best-known techniques used to model data. We shall illustrate it by extracting the gravitational constant from the simulated acceleration experiment.

We use a linear relationship between the time and the velocity as a model for the data. The goal is the determination of the slope g and the intercept v_0 of the line. In addition, we want to determine also the uncertainty on these parameters, their correlations, and the "quality of the fit", from which one can evaluate quantitatively whether the model is a good description, or not. This is an example of a more general situation in which:

- A dataset $\mathcal{D}$ consisting of N measurements $\{y_0, y_1, \ldots y_{N-1}\}$ taken at N points $\{x_0, x_1, \ldots x_{N-1}\}$ is available; at each point the measured values of y_i deviate from the true values according to a normal distribution with standard deviation σ_i.
- The relationship between the true values of x and y is described by the data model (or fitting function) $y(x; \vec{\alpha})$, where $\vec{\alpha}$ is a vector of M unknown parameters. For a linear data model the vector $\vec{\alpha}$ could consist of the slope and intercept. However, any parameterized data model will do.

We proceed by writing down the probability density for obtaining the N measured values, which is:

$$\rho(\mathcal{D}; \vec{\alpha}) = \prod_{i=0}^{N-1} \frac{1}{\sqrt{2\pi}\sigma_i} e^{-(y_i - y(x_i; \alpha_i))^2 / 2\sigma_i^2}. \tag{16.1}$$

We then take the bold step of regarding this normalized probability density for the dataset $\mathcal{D}$ as an unnormalized probability density for the M unknown parameters $\vec{\alpha}$. Since an experiment furnishes the values of $\{y_i\}$ and their errors $\{\sigma_i\}$ at the points $\{x_i\}$, this is only a function of the M parameters $\vec{\alpha}$. Viewed in this way, the probability density is called the *likelihood function*, $\mathcal{L}(\vec{\alpha})$:

$$\mathcal{L}(\vec{\alpha}) = \rho(\mathcal{D}; \vec{\alpha}). \tag{16.2}$$

Most of the time we are interested in the part of parameter space in which $\mathcal{L}(\vec{\alpha})$ is near its peak, or equivalently, when the positive-definite quantity $-2\ln\mathcal{L}(\vec{\alpha})$ is small. We will call the most probable point in the parameter space $\vec{\alpha}_0$, which is where the likelihood is maximized and $-2\ln\mathcal{L}(\vec{\alpha})$ is minimized. Moreover, in the neighborhood of that point the likelihood function behaves as a multivariate normal distribution (in the limit of N large and σ small, a limit that we shall refer to as the *Gaussian limit*).

Expanding $-2\ln\mathcal{L}(\vec{\alpha})$ around $\vec{\alpha}_0$ yields:

$$-2\ln\mathcal{L}(\vec{\alpha}) = -2\ln\mathcal{L}(\vec{\alpha}_0) + \frac{1}{2}(\vec{\alpha} - \vec{\alpha}_0)^T \mathbf{H}(\vec{\alpha} - \vec{\alpha}_0) + \ldots$$

where $\mathbf{H}$ is the **Hessian matrix**, which is symmetric and positive-definite.

$$(\mathbf{H})_{ij} = \frac{\partial^2(-2\ln\mathcal{L})}{\partial\alpha_i \partial\alpha_j}.$$

The first term is unimportant since it merely sets an overall constant in the likelihood function and does not depend on $\vec{\alpha}$. We can write our likelihood as:

$$\mathcal{L}(\vec{\alpha}) \approx \mathcal{L}(\vec{\alpha}_0) \cdot e^{-(\vec{\alpha}-\vec{\alpha}_0)^T \mathbf{H}(\vec{\alpha}-\vec{\alpha}_0)/4}.$$
$$\approx \mathcal{L}(\vec{\alpha}_0) \cdot e^{-(\vec{\alpha}-\vec{\alpha}_0)^T \mathbf{C}^{-1}(\vec{\alpha}-\vec{\alpha}_0)/2}. \quad (16.3)$$

where:

$$\mathbf{C} \equiv \left(\frac{\mathbf{H}}{2}\right)^{-1}$$

is called the **covariance matrix**.

By comparison with Eq. 7.3, we see that the the likelihood function is approximately a multivariate normal distribution, up to an overall normalization factor which would be trivial to compute were it ever needed. In the Gaussian approximation, the important task is to determine the point $\vec{\alpha}_0$ in the parameter space where the likelihood attains its maximum value, and the covariance matrix $\mathbf{C}$. The former quantity is the *central value* of the estimate, while the latter determines the uncertainties in the parameters. Specifically, the diagonal elements $(\mathbf{C})_{ii}$ give the variance of the parameter α_i. Correlations between any two parameters α_i and α_j are quantified by the correlation coefficients:

$$\rho_{ij} \equiv C_{ij}/\sqrt{C_{ii}C_{jj}}. \quad (16.4)$$

We have just developed a rather general discussion of likelihood techniques that will be very useful later, but for the moment we can return to the case of the linear least squares fit that we started with. From Eq. 16.2 identify the quantity:

$$-2\ln \mathcal{L} = 2N\ln\left(\sqrt{2\pi}\right) + 2\sum_{i=0}^{N-1}\ln\sigma_i + \sum_{i=0}^{N-1}\frac{(y_i - y(x_i;\vec{\alpha}))^2}{\sigma_i^2}.$$

This is to be minimized over the M parameters $\vec{\alpha}$ to obtain $\vec{\alpha}_0$. We notice that the first two terms do not involve $\vec{\alpha}$ at all so they can be ignored. Instead one can minimize the third term, which we call χ^2:

$$\chi^2 = \sum_{i=0}^{N}\frac{(y_i - y(x_i;\vec{\alpha}))^2}{\sigma_i^2} \tag{16.5}$$

and determine its Hessian matrix.

In some circumstances, including data models that are polynomials of finite order, one can carry this out analytically. In the case of a linear model one computes:

$$\chi^2 = \sum_{i=0}^{N}\frac{(y_i - (mx + b))^2}{\sigma_i^2} \tag{16.6}$$

and takes partial derivatives with respect to m and b. The solution to the resulting coupled linear equations yields the central values, (m_0, b_0). The expression for the Hessian matrix:

$$\begin{pmatrix} \frac{\partial^2 \chi^2}{\partial m^2} & \frac{\partial^2 \chi^2}{\partial m \partial b} \\ \frac{\partial^2 \chi^2}{\partial m \partial b} & \frac{\partial^2 \chi^2}{\partial b^2} \end{pmatrix}.$$

is obtained and then evaluated at the minimum. Further development can be found in Press (2007) or Gibbs (2006). We do not develop the analytical method further because one rarely deals with data models that lend themselves to an analytic solution; rather a numerical approach must be taken.

We conclude this section with the observation that the function that is to be minimized is $\chi^2 \approx -2\log(\mathcal{L})$. In general other likelihood functions can be formulated that are built from other probability distributions where appropriate. In Chapter 8 we have already obtained likelihood functions built from the binomial distribution (Section 8.1.3) and from the Poisson distribution (Exercise 6). These functions can be plotted, sampled (since they appear as unnormalized PDF's), and maximized numerically, thereby extracting the parameters of the approximating multivariate Gaussian (namely the mean and covariance).

16.3 Function minimization in data modeling

To go further in our program we require a piece of functionality that can minimize a function of several variables and determine the relevant Hessian matrix. It should be able to cope with a few dozen parameters in order to be useful in typical data modeling problems. The branch of computational science devoted to this problem is called *optimization*, and there is a rather vast literature on the topic. Press (2007) devotes a substantial amount of space to various optimization algorithms.

One of the best known packages for function minimization appeared at CERN in the 1970s. Called MINUIT and written by James (1994) in Fortran, it has more recently been ported to C++ and renamed `Minuit2`. Documentation is at the website:

```
http://seal.web.cern.ch/seal/snapshot/work-packages/mathlibs/
   minuit/
```

MINUIT provides a number of minimization options along with extra functionality to determine the Hessian matrices and to extract from them statistically important information such as covariance matrices.

The examples in this book, which use `QatGenericFunctions`, use a wrapper to MINUIT found in the library `QatDataModeling`. The library defines an important class called `MinuitMinimizer`.

Experienced users (or those developing outside of the QAT framework) may wish to bypass the interface and use MINUIT2 directly.

We will not discuss the algorithms used to minimize functions in this book since these are fairly standard, well-known, and are implemented in large software projects that are publicly available. Instead we will treat function minimization as if it were carried out by a "black box" called a *minimizer*. We will regard this black box as being inhabited by an individual known as "Minimizer Man" (Figure 16.1), who turns knobs with the purpose of finding the minimum value of an objective function $f(\vec{\alpha})$[1]. For a more serious discussion one can refer to (James, 1994) or the websites above.

In data modeling, the connection between the parameters $\vec{\alpha}$ and the function to be minimized (such as χ^2 of the previous section) is as follows:

- The parameters $\vec{\alpha}$, like little knobs, control the shape of a function $y(\vec{x}; \vec{\alpha})$. We write $\vec{x}$ as a vector because in general y will be a function of one or more variables, controlled by the parameters $\vec{\alpha}$.

- Data from an experiment are used to construct a functional, meaning a mapping from the function $y(\vec{x}; \vec{\alpha})$ to real numbers. We will write this map as $f[y(\vec{x}; \vec{\alpha})]$.
  One example is the χ^2 functional. We will encounter other examples in a moment.

---

[1] "Minimizer Man" is inspired by "Transistor Man," the hero of Horowitz and Hill's classic *The Art of Electronics*. (Horowitz, 1989). Transistor Man, in contrast to Minimizer Man, has a straightforward job; namely to keep a transistor's collector current in strict proportion to its base current. Minimizer Man will sometimes fail at his job for a variety of reasons we discuss.

518 *Function minimization in data modeling*

Figure 16.1 *Minimizer man diligently adjusts a series of knobs (parameters) in order to obtain the lowest value of an objective function $f(\vec{\alpha})$. Figure by M.R. Quinn.*

In the directory EXAMPLES/CH16/GRAVITY we include an illustration of a simple χ^2 fit to the gravity simulation. A class called MinuitMinimizer carries out the minimization (using Minuit2). The objective function (χ^2) is encapsulated in a class called ChiSq which collects points and then maps a fitting function to a real-valued number. In the following excerpt, we generate a velocity smeared with measurement error, and add it to the collection of points used for the χ^2 fit:

```
PlotProfile pf;
ChiSq chi2;
// Make measurements
for (int t=0;t<10;t++) {
  double v=9.8*t+v0+gauss(engine);
  pf.addPoint(t,v, sigma);  // add to the profile plot
  chi2.addPoint(t,v,sigma); // add to the chisq objective
      function
}
```

A parameterized fitting function is constructed in the following code; the parameters are the "knobs" that MINUIT will adjust to change the shape of the function.

```
using namespace Genfun;
Variable T;
Parameter pv0("V0",0.0,   -100,    100);
Parameter pg("g",   10.0, 5.0, 15.0);
GENFUNCTION F=pv0+pg*T;
```

Then finally we construct a `MinuitMinimizer` and declare its objective function and the parameters:

```
const bool verbose=true;
MinuitMinimizer minimizer(verbose);
minimizer.addParameter(&pv0);
minimizer.addParameter(&pg);
minimizer.addStatistic(&chi2,&F);
minimizer.minimize();
```

When the `minimize` method returns results can be harvested. These include:

- a status code that can be used to determine whether the minimization was successful (`double getStatus()`). Possible return values are:
  - −1 failure
  - 0 Covariance matrix not positive-definite.
  - 1 Covariance only approximate
  - 2 Covariance matrix forced positive-definite
  - 3 Full accurate matrix.
- the value of the function at the minimum (`double getFunctionValue()`).
- the central value of each of the parameters (`Eigen::VectorXd getValues()`).
- the covariance matrix for the set of parameters.
  (`Eigen::MatrixXd getErrorMatrix()`).

The example program simulates the experiment and fits the generated data to a straight line, thereby extracting values of the parameters v_0 and g. The measurement is visualized in another plot which shows the measurement as an ellipse, whose contour is set at $\chi^2 = 2.30$, a value chosen such that the error ellipse should contain the true value 68.3% of the time, corresponding to one Gaussian standard deviation. Each run of the program starts from a different random seed. Results from a typical run are shown in Figure 16.2

The class `ChiSquare` is a functional, inheriting from `Genfun::AbsFunctional`. Its essential methods are

```
void addPoint (double x, double y, double dy);
```

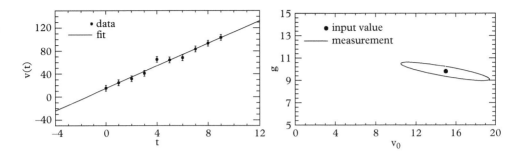

Figure 16.2 *Output from a typical run of the program in* EXAMPLES/CH16/GRAVITY. *On the left is shown the fit to the input data, on the right is shown the measurement in the space of the parameters g and v_0 (gravitational acceleration and initial velocity). The blue dot represents the input (or true) values of these parameters, i.e., those used in the simulation.*

which is amply illustrated in the above example, and

```
virtual double operator () (Genfun::GENFUNCTION f) const;
```

which maps the fit function into a real number, here the χ^2. The method is called by `MinuitMinimizer`.

Inside the `ChiSquare` class, the following code computes the value of χ^2:

```
// Function Call Operator
virtual double operator () (const Genfun::AbsFunction & f)
   const {
  double c2=0.0;
  for (unsigned int i=0;i<_points.size();i++) {
    point p = _points[i];
    double y = f(p.x);
    c2 += (p.y-y)*(p.y-y)/p.dy/p.dy;
  }
  return c2;
}
```

The χ^2 fit in this section is only one of a family of methods for fitting data that we discuss in this chapter. Most of the others are based upon a likelihood function, which, however it is constructed, is to be minimized and the Hessian matrix computed. The main difference between the methods are the objective function to be minimized. The `QatDataModeling` library implements a family of objective functions inheriting from `Genfun:AbsFunctional`, each one adapted to a different kind of fit. A class tree

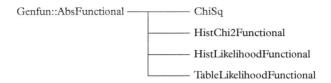

Figure 16.3 *A class tree diagram for* `AbsFunctional` *and subclasses. The subclasses are part of* `QatDataModeling` *and their header files can be found there.*

diagram is shown in Figure 16.3. Users may also implement their own objective function by extending this set through inheritance.

After the fit runs, users should check the status code; however, whether the minimizer is truly successful is not always clear; there is no guarantee that the minimization of functions converges and there is no criterion to check either. When it does converge it may not converge to the true global minimum but rather to a local minimum. In addition, the Hessian matrix may not be properly computed. The minimizer can provide rather detailed information by setting the verbose flag in the constructor call:

```
MinuitMinimizer (bool verbose=false);
```

to "true" (as in the example on page 519).

Certain circumstances can exacerbate convergence problems in MINUIT:

1. A very large number of parameters.
2. Large correlations between the parameters.
3. The existence of certain directions in the parameter space in which the function value does not change. This will certainly occur in cases where the fit function itself possesses continuous symmetries in its parameters, exact or approximate.
4. The existence of many local minima in the objective function. These are sometimes caused by discrete symmetries, exact or approximate.

The simple example which we have been discussing causes no problems for MINUIT; however complicated problems can be a source of frustration and have even been known to cause irritability, sleeplessness, headaches, nausea, and fits of vomiting.

The result of the measurement can be visualized as iso-density contours in the parameter space. These contours are elliptical in the case of Gaussian likelihood functions, and are therefore called "error ellipses". Typically one is interested in the contour that encloses 68.3% of the likelihood, corresponding to one Gaussian deviation or "one sigma", or 95.4% corresponding to two Gaussian standard deviations or "two sigma". The class `PlotErrorEllipse` in the `QatPlotting` library inherits from `Plottable`, and is therefore an object which can be added to a graph (`PlotView`). This is shown together with the fit in Figure 16.2.

16.3.1 The quality of a χ^2 fit

While a χ^2 fit is a little pedestrian for most serious data modeling applications, it has a feature that other techniques lack—the value of χ^2 at the minimum can be used to evaluate the quality of the fit. Given N true values $y(x_i)$ for the state of a system at various values of x and also N measurements y_i described by independent Gaussian errors σ_i, the deviations $y(x_i) - y_i$ for the full set of measurements follow a multivariate Gaussian distribution with a mean of $(0, 0, 0...0)$ and a covariance matrix

$$\mathbf{C} = \begin{pmatrix} \sigma_0^2 & 0 & 0 & \cdots & 0 \\ 0 & \sigma_1^2 & 0 & \cdots & 0 \\ 0 & 0 & \sigma_2^2 & \cdots & 0 \\ . & . & . & . & 0 \\ 0 & 0 & 0 & 0 & \sigma_{N-1}^2 \end{pmatrix}$$

The probability for a fluctuation less than that observed in the dataset is the cumulative χ^2 distribution $P(\chi^2|N)$ described in Section 7.3. The complement $Q(\chi^2|N) = 1 - P(\chi^2|N)$ is the probability to observe an even greater fluctuation. This is also colloquially (but quite commonly) called the *Probability* χ^2. The logic behind this figure of merit is the following: while large fluctuations can occur, fluctuations whose probability χ^2 is 10% or less should occur only 10% of the time, those whose probability χ^2 is 1% or less should occur only 1% of the time, etc., if the model describes

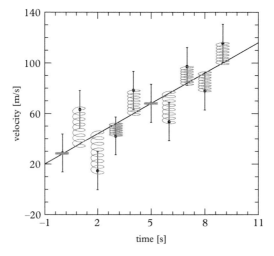

Figure 16.4 *In a mechanical analogy to a χ^2, the data model is connected to the data points with springs whose stiffness depends on the error bars, and the model adjusts itself in order to minimize the energy stored in the springs.*

the data. Thus very low values of probably χ^2 should be considered suspicious. When parameters of the model have been adjusted during the fit, the number of degrees of freedom $N - P$ where P is the number of adjustable parameters should be used in calculating the probability χ^2, i.e. $Q(\chi^2|N - P)$ should be taken. A useful rule of thumb is that the "reduced χ^2", or χ^2/dof where $dof = N - P$, should be approximately one unit for a reasonably good fit.

16.3.2 A mechanical analogy

The behavior of a χ^2 fit can be understood intuitively with a simple mechanical analogy. We imagine that the data model is connected to the data points via small springs, as in Figure 16.4. The springs are stiff if the error bar is small and loose if the error bar is large: the spring constant should be related to the error via the relationship $k_i = 2/\sigma_i^2$, so that:

$$\chi^2 = \sum_{i=0}^{N} \frac{(y_i - y(x_i|\vec{\alpha}))^2}{\sigma_i^2} \rightarrow \sum_{i=0}^{N} \frac{1}{2} k_i (y_i - y(x_i|\vec{\alpha}))^2$$

and the χ^2 is then simply the fictional energy stored in the fictional springs. The fitter will minimize this energy. This analogy can give some intuitive feeling for how the fit will act in response to a particular set of data points.

16.4 Fitting distributions

Distributions of random variables can be very sensitive to underlying physics—one has only to think of the spectral distribution of blackbody radiation, which triggered a huge paradigm shift in the physical sciences, to find a dramatic example. Depending upon the circumstances, we could be dealing with univariate data or multivariate data. There are two common ways of fitting the data; first, we can collect the data into the bins of a one, two, or multidimensional histogram and fit the bin contents, and second, we can fit the data without putting it into bins at all. The first method becomes difficult when the number of variables becomes large, since the number of bins grows exponentially with the number of dimensions, while the amount of data is usually fixed. However for univariate data it is still a useful technique.

When looking at a histogram of binned data, we need to keep two probability distributions in mind: first, the distribution according to which the data is distributed, and second, the distribution that the contents of a particular bin would follow if the experiment could be repeated. The latter is called the Poisson distribution and is given by:

$$P(N; \lambda) = e^{-\lambda} \frac{\lambda^N}{N!}.$$

Fitting distributions

The Poisson distribution is a discrete probability distribution; i.e. the outcome can only be an integer, $N = 0, 1, 2...$ It gives the probability that N random events are observed in some interval (of time, space, frequency, or whatever) when the expected number is λ. If one were to record a stable source of natural background radiation with a Geiger counter, for example, the number of counts in a five second interval would follow a Poisson distribution. Both the mean and the variance of the Poisson distribution are λ. When N events are actually observed in some interval or bin, we will therefore take the uncertainty in the number of observed events to be $\sqrt{N}$. Thus, error bars for a bin in a histogram are often set to be the square root of the number of events in that bin. This is a good choice except that it is clearly wrong for an empty bin, where $N = 0$. Finally, for large values of λ, the Poisson distribution approaches a normal distribution.

We discuss in the following sections a series of useful techniques for fitting distributions. We take, as an example, data which is distributed according to a gamma distribution, which is a function of one variable whose shape is controlled by two parameters, called α and β. The probability density for the gamma distribution is:

$$\rho(x; \alpha, \beta) = \frac{x^{\alpha-1} e^{-x/\beta}}{\Gamma(\alpha) \beta^\alpha}. \tag{16.7}$$

To generate according to this distribution one can use `std::gamma_distribution`, which takes parameters α and β in its constructor call.

16.4.1 χ^2 fit to a distribution

If we have binned data (i.e. a histogram) then according to the previous section the histogram itself can provide a series of points and error bars needed to compute the χ^2 for a particular data model. Eq. 16.5 can be written as:

$$\chi^2 = \sum_{i=0}^{N} \frac{(M_i - n \cdot \Delta x \cdot \rho(x_i; \vec{\alpha}))^2}{M_i} \tag{16.8}$$

where i is the bin number and M_i is the bin contents, and x_i is the coordinate of the bin, n is the total number of events in the sample, Δx is the width of a bin in the histogram, and $\rho(x; \vec{\alpha})$ is the probability density for the variable x. The class `HistChiSquareFunctional` of the `QatDataModeling` library provides an implementation.

An example program, gammaChiSq, can be found in EXAMPLES/CH16/GAMMA/ CHISQ and is excerpted below. The program uses `std::gamma_distribution` to fill a histogram (`hist` in the excerpt below), and then fits the histogram to a gamma distribution, determining the values of the parameters α and β. This is repeated in multiple runs, and in the end the program makes summary plots relevant to determining the quality and accuracy of the fits.

```
#include "QatDataModeling/HistChi2Functional.h"
  //
  Hist1D hist;              // histogram of binned data
  .
  .
  using namespace Genfun;

  // PREPARE THE FIT.  THIS IS A TWO-PARAMETER FIT.
  Parameter pAlpha("Alpha", 0.5, 0.1, 4.0);
  Parameter pBeta ("Beta",  0.5, 0.1, 4.0);


  // MAKE A FIT FUNCTION AND CONNECT IT TO ITS PARAMETERS:
  Variable X;
  Log log;
  Gamma gamma;
  Exp   exp;
  GENFUNCTION gammaDistribution=
  exp(-pAlpha*log(pBeta))/gamma(pAlpha)*exp(-X/pBeta+(pAlpha
      -1)*log(X));

  // MAKE A FUNCTIONAL,
  HistChi2Functional objectiveFunction(&hist);
  GENFUNCTION f = hist.sum()*gammaDistribution;

  // MAKE AN ENGINE FOR MINIMIZING THE LIKELIHOOD BY VARYING
       THE PARAMETERS.
  bool verbose=true;
  MinuitMinimizer minimizer(verbose);
  minimizer.addParameter(&pAlpha);
  minimizer.addParameter(&pBeta);
  minimizer.addStatistic(&objectiveFunction,&f);
  minimizer.minimize();
```

Results from the fit are shown in Figure 16.5. On the left is the fit itself, and on the right is the fit result, shown as an error ellipse, together with the true value of the parameters α and β used in the simulation. Two kinds of χ^2 can be obtained from the information which is displayed. First, from the information in the left-hand plot one derives the χ^2 for the fit, which can be used to judge whether the model used for the data is a good one. We will call this the "fit χ^2". Second, from the information on the right-hand plot, one can obtain the χ^2 for the measured value of the parameters to match the input values, we will call this the "matching χ^2" which is used to assess the accuracy of the fit. Obviously, this second can only be determined for pseudoexperiments, for

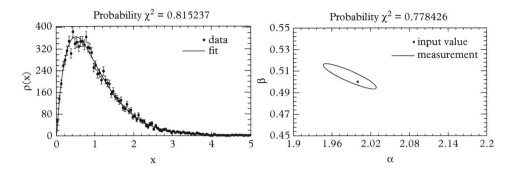

Figure 16.5 *Results of a χ^2 fit to binned data following a γ distribution. On the left, the fit is shown together with the probability χ^2 for the fit; on the right is the $\Delta\chi^2 = 1$ error ellipse, shown together with the probability χ^2 for the result to match the input value.*

which the true input values are known. From either value of χ^2, one can compute the "*probability χ^2*" (the language is somewhat colloquial but quite common). This is the the complementary cumulative χ^2 distribution function for N degrees of freedom. For the fit χ^2, N must be taken as the number of bins in the histogram, minus the number of parameters in the fit. For the matching χ^2, N is simply the number of parameters in the fit.

Is the fit good? This is a statistical question, and statistical questions can only be answered with "probably" or "probably not". *If the model used to fit the data is correct*, the distribution of the probability of the fit χ^2 over many pseudoexperiments should be flat over the entire range, i.e. [0,1]. On the other hand, bad data models will usually lie far from the data points, produce large values for the fit χ^2, and small values for the probability of the fit χ^2.

The probability of the matching χ^2 gives the probability of a deviation of the measured values from the input values larger than the one which is observed. This is also a random variable between zero and one which is flat if deviations from the measured central values are well described by their estimated errors.

Both probabilities can be used as indicators of other sources of trouble in the fits. One possible issue is simple errors in normalization or the numerical procedures. Another source of trouble is the approximation of the Poisson distribution by a normal distribution in the χ^2 fitting procedure. Low statistics in the experiments may result in a failure of the likelihood function to reach its asymptotically limiting form as a multivariate normal distribution. By performing large numbers of pseudoexperiments and plotting the distribution of fit probabilities and matching probabilities one can frequently spot this kind of problem. An example of this type of exercise is shown in Figure 16.6 for low-statistics (10K events) and high-statistics (500K events) χ^2 fits.

In addition to plots of probability χ^2 distributions for the fitted parameters in the full multidimensional parameter space, one can check the accuracy one parameter at a time. Usually this is done by normalizing the deviation $\Delta\alpha_i = \alpha_i - \alpha_{i0}$ by the estimated error $\sigma(\alpha_i)$ returned by the fit. This normalized deviation $\Delta\alpha_i/\sigma(\alpha_i)$ is called the *pull*. Pull distributions for an ensemble of pseudoexperiments should be normally distributed in the limit of large-statistics pseudoexperiments. Any shift in the mean is usually indicative of a bias in the fit, and departures from unit variance is a sign that errors are over- or underestimated.

16.4.2 Binned maximum likelihood fit to a distribution

The minimization of χ^2 is based on the minimization of a likelihood function, Eq. 16.2, which is a product of normal distributions. The use of this approximation affects the accuracy of the fit, and is largely responsible for the accuracy problems in Figure 16.6. In the case of fitting binned data, the true likelihood function is a product of Poisson distributions. In this case the likelihood is:

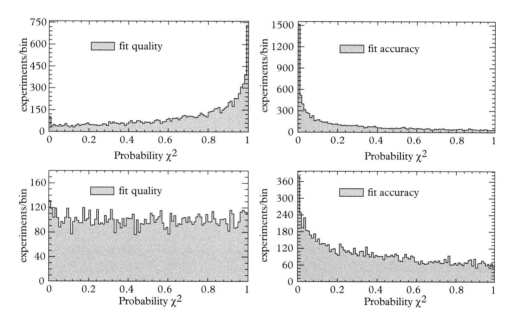

Figure 16.6 *Probability χ^2 plots for fit quality and fit accuracy. Top: 10K pseudoexperiments with 10K events per experiment. Bottom: 10K pseudoexperiments with 500K events per experiment. Ideally, all of these plots should be flat as departures from flatness indicate defects in the χ^2 fit. Top: with few events per experiment, the content of each bin is Poisson-distributed with too few entries to be adequately described by a normal distribution. This causes distortion of the fit quality plot and the fit accuracy plot. Bottom: with higher statistics the first problem is eliminated and the second is greatly reduced.*

$$\mathcal{L}(\vec{\alpha}) = \prod_{i=0}^{N-1} e^{-n \cdot \Delta x \cdot \rho(x_i; \vec{\alpha})} \frac{(n \cdot \Delta x \cdot \rho(x_i; \vec{\alpha}))^{M_i}}{M_i!},$$

where M_i is the bin count in the ith bin. The log likelihood is then:

$$-2 \ln \mathcal{L}(\vec{\alpha}) = 2 \sum_{i=0}^{N-1} \left[n \cdot \Delta x \cdot \rho(x_i; \vec{\alpha}) - M_i \ln (n \cdot \Delta x \cdot \rho(x_i; \vec{\alpha})) + \ln(M_i!) \right].$$

The last term in the expression is a constant and can be neglected.

The parameters $\vec{\alpha}_0$ that minimize this functional are the central values of the estimate. As before, the covariance matrix of the estimate is $\mathbf{C} = (\mathbf{H}/2)^{-1}$, where $\mathbf{H}$ is the Hessian matrix of the likelihood function calculated at the minimum. The binned likelihood fit does not provide an estimate of goodness-of-fit; instead, the value of $-2\ln \mathcal{L}(\vec{\alpha}_0)$ after minimization may be compared to the distribution of values from a large ensemble of pseudoexperiments.

In order to perform a binned likelihood fit, a different functional is chosen to map the fitting function into a objective function; specifically:

```
HistChi2Functional objectiveFunction(&hist);
```

is changed to

```
HistLikelihoodFunctional objectiveFunction(&hist);
```

The accuracy of the fit is much improved, as can be seen in Figure 16.7.

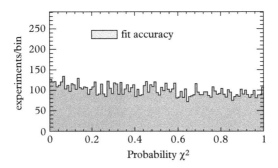

Figure 16.7 *Probability χ^2 distribution for 10K binned likelihood fits to a gamma distribution, with 10K events per experiment. The fit quality cannot be assessed in a binned likelihood fit. The use of the Poisson distribution in describing the bin counts has solved the fit accuracy trouble of the χ^2 fit (compare to Figure 16.6 top right), as can be seen from the flatness of the distribution.*

In comparing binned likelihood and χ^2 fitting algorithms, several observations can be made. The χ^2 fit is not expected to be very accurate when the histogram contains empty bins or bins containing less than about ten counts because the Gaussian approximation to the Poisson distribution is inaccurate in such cases. Indeed, bins with zero counts cannot be used at all since the corresponding term in the Eq. 16.8 is not defined (and is therefore usually omitted). Clearly; however, the population of every bin carries information, even those with no counts. The binned likelihood fit, on the other hand, correctly accounts for empty bins.

A binned likelihood fit can be very accurate if the correct data model is used, but it can also incur large systematic errors if the data shows minor discrepancies with the model. This is because bins with low statistics are taken into account by the binned likelihood fit and have a important effect on the overall fit. In real data, the presence of a few events in the tail of the distribution that are not described by the model can grossly distort the fit. This introduces large discrepancies in the major modes of the probability density function in order to obtain a more satisfactory agreement with the tails. The χ^2 fit on the other hand is less heavily influenced by the tails of the distribution. In any case, the statistical properties of the fit should be checked by running tests involving large numbers of pseudoexperiments.

16.5 The unbinned maximum likelihood fit

The *unbinned maximum likelihood* fit is the most powerful of all data modeling techniques used in the business. This method is used by the pros because it is simple, easily accomodates multivariate data, and does not require the data to be binned. Thus the rapid increase in the number of bins with dimension, and the corresponding decrease in bin counts, is avoided.

Imagine that multivariate data, which is assumed to come from a normalized parent probability density $\rho(\vec{x}; \vec{\alpha})$, is collected. As usual, $\vec{x}$ is a set of measured data points and $\vec{\alpha}$ is a vector of unknown parameters. The likelihood for the entire dataset, $\mathcal{D} = \{\vec{x}_0, \vec{x}_1 ... \vec{x}_{N-1}\}$, is given by:

$$\mathcal{L}(\vec{\alpha}) = \rho(\mathcal{D}; \vec{\alpha}) = \prod_{i=0}^{N-1} \rho(\vec{x}_i; \vec{\alpha}), \tag{16.9}$$

which is considered as an unnormalized probability density for the parameters $\vec{\alpha}$. In the Gaussian limit, near to the maximum, this quantity behaves (except in pathological cases) like a multivariate probability density. Central values and errors can be extracted from the log likelihood function:

$$-2 \ln \mathcal{L}(\vec{\alpha}) = -2 \sum_{i=0}^{N-1} \ln \rho(\vec{x}_i; \vec{\alpha}). \tag{16.10}$$

530 *The unbinned maximum likelihood fit*

As usual, the central value occurs at $\vec{\alpha}_0$, the point in parameter space for which $-2\ln\mathcal{L}(\vec{\alpha})$ is minimized, and the covariance matrix for the parameter $\vec{\alpha}$ is $\mathbf{C} = (\mathbf{H}/2)^{-1}$, where $\mathbf{H}$ is the inverse of the Hessian matrix calculated at the minimum.

16.5.1 Implementation

The input to the likelihood function is a set of multivariate data together with a model for the data, called a fitting function. The two are combined into an objective function which is minimized and the Hessian matrix is determined. QatDataModeling provides functionality to carry out the fit using a Table and a GENFUNCTION. The multivariate data is stored in the Table, while the fitting function is a GENFUNCTION whose shape is controlled by parameters. The ith row of the table contains the variables x_i in Eq. 16.10. The fitting function can be constructed from the variables through the use of the method Table::symbol as follows:

```
Table table;
Genfun::Variable  X=table.symbol("X");
```

This creates a variable associated with the column labeled "X" in the Table that can be used in symbolic expressions. For example, to create a gamma distribution controlled by two parameters in the variable X, the following lines of code can be used:

```
Genfun::Parameter pAlpha("Alpha", 0.5, 0.1, 3.5);
Genfun::Parameter pBeta ("Beta",  0.5, 0.1, 3.5);

Genfun::Variable  X=table.symbol("X");
// MAKE A FIT FUNCTION AND CONNECT IT TO ITS PARAMETERS:
Genfun::Log log;
Genfun::Gamma gamma;
Genfun::Exp   exp;
Genfun::GENFUNCTION gammaDistribution=exp(-pAlpha*log(pBeta))
        /gamma(pAlpha)*exp(-X/pBeta+(pAlpha-1)*log(X));
```

In case a probability density that is a function of more than one variable is desired, two variables can be extracted from the table and used together in symbol expressions. The following example sets up a function of two variables (call them X and Y) that is controlled by four parameters:

```
Genfun::Parameter pAlpha0("Alpha0", 0.5, 0.1, 3.5);
Genfun::Parameter pBeta0 ("Beta0",  0.5, 0.1, 3.5);
Genfun::Parameter pAlpha1("Alpha1", 0.5, 0.1, 3.5);
Genfun::Parameter pBeta1 ("Beta1",  0.5, 0.1, 3.5);
Genfun::Variable  X=table.symbol("X");
```

```
Genfun::Variable  Y=table.symbol("Y");

Genfun::Log log;
Genfun::Gamma gamma;
Genfun::Exp   exp;

// Function of X
Genfun::Variable _x; // Dummy variable
Genfun::GENFUNCTION gammaDistribution0=
exp(-pAlpha0*log(pBeta0))/gamma(pAlpha0)*exp(-_x/pBeta0+
   (pAlpha0-1)*log(_x));

// Function of Y
Genfun::GENFUNCTION gammaDistribution1=
exp(-pAlpha1*log(pBeta1))/gamma(pAlpha1)*exp(-_x/pBeta1+
   (pAlpha1-1)*log(_x));

// Make a multivariate PDF
Genfun::GENFUNCTION f = gammaDistribution0(X)*
   gammaDistribution1(Y);
```

The objective function is constructed using the class `TableLikelihoodFunctional`; the constructor for this class takes a table containing the input data:

```
Table *table;
 ...
TableLikelihoodFunctional objectiveFunction(table);
```

As usual the fitting function (a normalized probability density) and the functional are declared to the minimizer so that it may construct and minimize an objective function:

```
bool verbose=true;
MinuitMinimizer minimizer(verbose);
minimizer.addParameter(&pAlpha1);
minimizer.addParameter(&pBeta1);
...
minimizer.addStatistic(&objectiveFunction,&f);
minimizer.minimize();
```

Several examples can be found in EXAMPLES/CH16/GAMMA:

- EXAMPLES/CH16/GAMMA/UNBINNEDLIKELI/ contains code which fits a single random variate. This is an unbinned version of the fits we have discussed in previous sections. The variate is constructed using `std::gamma_distribution`.

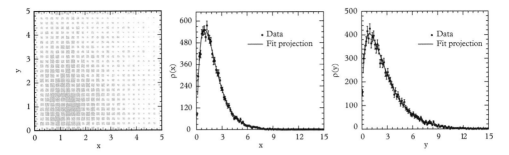

Figure 16.8 *Example of an unbinned maximum likelihood in two variables. Left: the data in an x vs. y scatterplot. Center and Right: histograms of the x and y variables alone, compared to the fitted functions. The example is available in EXAMPLES/CH16/GAMMA/MULTIDIM.*

- EXAMPLES/GAMMA/CH16/MULTIDIM contains code to generate bivariate data according to:

$$\rho(x, y) = \Gamma(x; \alpha_x, \beta_x) \cdot \Gamma(y; \alpha_y, \beta_y)$$

and then to fit the distribution, determining the unknown parameters α_x, β_x, α_y, and β_y. Since it is difficult to compare the fitted distribution to the data in two dimensions, we check the fit by projecting both the generated data and the fitted PDF onto one-dimensional subspaces. This is always easy for the data, since it just consists of making a histogram of the x or y variables separately. It is also easy for the PDF used in this example because it is a direct product of two one-dimensional PDFs. This so-called *likelihood projection* may be more demanding in the general case. The joint PDF $\rho(x, y)$ and the two projections are shown in Figure 16.8

- EXAMPLES/CH16/UNBINNED_EXP is a fit to an exponential distribution that can be written:

$$\rho(t) = \Theta(t) \frac{1}{\tau} e^{-t/\tau} \qquad (16.11)$$

where $\Theta(x)$ is the Heaviside (or unit step) function. The exponential distribution describes, for example, the number of radioactive decays versus time produced in a radioactive substance. This example allows you to experiment with the size of the dataset by giving the command line parameter NPER=value. With this example you can see that the unbinned maximum likelihood fit works with very few events, and even with only *one* event! However careful consideration must always be given to measurements performed with very small statistical samples.

16.5.2 Construction of normalized PDFs

The PDFs used in fitting functions to an unbinned maximum likelihood fit must be normalized. While constants like $1, 2, \pi$, represent only constant offsets to the $-2\ln \mathcal{L}(\vec{\alpha})$ objective function, normalization constants like $1/\tau$ in Eq. 16.11 depend upon one or more of the parameters $\vec{\alpha}$ and so vary during the minimization—they cannot be ignored.

Normalized probability densities can be obtained in several ways. The probability densities can be born normalized, or normalized through analytic or numerical integration. Numerical integration in more than one dimension is especially tricky; Monte Carlo integration should be avoided unless it can be carried out to machine precision, since any statistical fluctuation of the objective function during the minimization step will confuse the minimizer.

Two normalized probability distributions $G_1(x)$ and $G_2(x)$ may be used to construct a summed normalized probability density, $\rho(x) = f_1 G_1(x) + (1-f_1) G_2(x)$. The fraction f_1 is parameter that usually floats in the fit, and ranges between 0 and 1. If a third normalized component $G_3(x)$ is added to the mix, a PDF defined in the following way:

$$\rho(x) = f_1 G_1(x) + f_2 G_2(x) + (1 - f_1 - f_2) G_3(x)$$

could easily fail as the minimizer explores parameter space. This is because both f_1 and f_2 must vary between 0 and 1, which means that negative values of $(1 - f_1 - f_2)$ can occur. If this happens the PDF need not be positive definite in this region of parameter space. The practical consequence is usually the occurrence of an `inf` or a `NaN` in the calculation that evaluates $-2\ln \mathcal{L}(\vec{\alpha})$. A better parameterization is:

$$\rho(x) = f_1 G_1(x) + (1 - f_1)(f_2 G_2(x) + (1 - f_2) G_3(x)),$$

which does not suffer from the same problem because the entire rectangular region of parameter space with $0 < f_1 < 1$ and $0 < f_2 < 1$ is physically meaningful. Figure 16.9 shows an example fit; code for which is available in EXAMPLES/CH16/GAMMA/UNBINNED_GAUSS.

Given two normalized probability densities $f(x)$ and $g(x)$, the convolution $(f \circ g)(x)$ is normalized and represents (as we have seen in Chapter 7) the distribution of the sum of two random variables, one distributed according to $f(x)$ one according to $g(x)$.

Orthogonal polynomials in any number of dimensions can be used to construct normalized PDFs through a procedure which we illustrate in one dimension. Let $\{\phi_i(x)\}$ be a set of orthogonal polynomials such that $\int_a^b \phi_i(x)\phi_j(x)dx = \delta_{ij}$. If $\{a_i\}$ are complex coefficients satisfying:

$$\sum |a_i|^2 = 1$$

then the function $\psi(x) = \sum a_i \phi_i(x)$ satisfies the condition $\int \rho(x) = \int_a^b |\psi(x)|^2\, dx = 1$. By choosing a parameterization for a set of coefficients a_i satisfying Eq. 16.5.2 one can

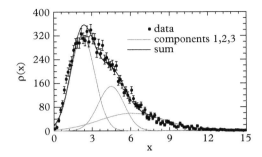

Figure 16.9 *This plot shows a fit of three Gaussians to a gamma distribution. The sum of the Gaussians data model is constrained to integrate to unity by the parameterization which has been adopted. The three fitted components are shown in grey, and their sum is shown in black. This example is accessible in EXAMPLES/CH16/GAMMA/UNBINNED_GAUSS.*

construct a normalized $\rho(x)$. Do not, however, be tempted to enforce the normalization condition using a Lagrange multipliers to modify the objective function. This technique produces saddle-point extrema rather than minima.

For functions of more than one variable, normalized multivariate PDFs can be constructed as the direct product of normalized univariate PDFs.

Finally these tricks can be used in combination. If, for example, we have two normalized PDFS $F_1(x)$ and $F_2(x)$ for the variable x, and two normalized PDFs $G_1(y)$ and $G_2(y)$ for the variable y then the direct products $F_1(x)G_1(y)$ and $F_2(x)G_2(y)$ are normalized joint PDFs, and so is $f_1 F_1(x)G_1(y) + (1 - f_1)F_2(x)G_2(y)$, where f_1 is a parameter that can be adjusted between 0 and 1. Tricks like this are of great practical use in devising complicated data models.

It's remarkable how easy it is to get the hang of this, and with a little practice simple data modeling exercises can be refined with more and more detail with little additional effort. The main general difficulties are coaxing the minimizer into convergence and satisfying yourself that the minimization has found the correct minimum. Practice makes perfect, and experienced numerical programmers can succeed in obtaining good convergence, even when thousands of parameters are involved.

16.6 Orthogonal series density estimation

When fit parameters happen to be the coefficients of orthogonal functions, as above, another approach can be used to determine their values. *Orthogonal series density estimation* (Watson, 1969) is a simple approach related to Fourier analysis. The fitting function is expressed in a basis expansion with undetermined coefficients. For example, a two-dimensional PDF can be modeled with a Fourier expansion in a rectangular region (of size $[0, L_x] \times [0, L_y]$) as follows:

$$\rho(x,y) = \frac{1}{\sqrt{L_x L_y}} \sum_n a_{m,n} \exp(im\pi x/L_x) \exp(in\pi y/L_y). \tag{16.12}$$

The coefficients of the expansion are given by:

$$a_{m,n} = \frac{1}{4} \frac{1}{\sqrt{L_x L_y}} \int_0^{L_x} \int_0^{L_y} \rho(x,y) \exp(-im\pi x/L_x) \exp(-in\pi y/L_y) \, dx dy. \tag{16.13}$$

By determining the coefficients $a_{m,n}$ we fully determine the PDF in Eq. 16.12. Those coefficients are obtained by recognizing the integral in Eq. 16.13 as the average of the quantities

$$A_{m,n} = \frac{1}{4} \frac{1}{\sqrt{L_x L_y}} \exp(-im\pi x/L_x) \exp(-in\pi y/L_y) \tag{16.14}$$

over the dataset. This average is quite simple to compute.

An illustration of this process is given in Figure 16.10. The left panel shows a photograph of Abraham Lincoln, but we will imagine that it defines a two-dimensional distribution that we will whimsically call the "Abraham Lincoln distribution". We imagine that this PDF is to be reproduced from data. The dataset consists of a number of points distributed according to the unknown PDF, as shown in the center panel. This data is used to compute $\langle A_{m,n} \rangle_{\mathcal{D}}$ up to $m = 40$ and $n = 40$, and the resulting joint PDF is shown in the right panel of the figure. Some details of the image have been lost, both because of the limited number of points in the dataset and because of the limited number of Fourier components used to reconstruct the image. In physics, many distributions lack

Figure 16.10 *Orthogonal series density estimation. Left: an original photograph of Abraham Lincoln (public domain) which defines the Abraham Lincoln distribution, and two dimensional joint probability density. Center: data generated from this distribution using Monte Carlo techniques. Right: an approximation to the original density function is recovered using orthogonal series density estimation.*

the sharp features of this photograph and the smoothing that can be observed here is actually a desirable feature.

Some data of interest is directional in nature and can be mapped onto a spherical surface; these include celestial data on one hand, and data from scattering experiments on the other. This data is characterized by a polar angle θ and an azimuthal angle ϕ and the appropriate basis functions are the spherical harmonics. In this case the PDF expansion is:

$$\rho(\theta, \phi) = \sum_{l=0}^{\infty} \sum_{m=-1}^{l} a_{lm} Y_l^m(\theta, \phi) \tag{16.15}$$

where:

$$a_{lm} = \int \rho(\theta, \phi) Y_l^{m*}(\theta, \phi) \, d\Omega, \tag{16.16}$$

which can again be estimated by taking averages.

Formally, all of these series are infinite; in practice they have been terminated, either by truncation or by "rolling off" the coefficients smoothly to zero after some cutoff. It is of interest to examine the infinite series too, even if it cannot be calculated on a computer. Our estimate of the coefficients is the average:

$$a_{lm} = \frac{1}{N} \sum_{i=0}^{N-1} Y_l^{m*}(\theta_i, \phi_i) \tag{16.17}$$

and therefore the PDF we would (conceptually) construct with an infinite series is:

$$\begin{aligned}
\tilde{\rho}(\theta, \phi) &= \sum_{l=0}^{\infty} \sum_{m=-1}^{l} a_{lm} Y_l^m(\theta, \phi) \\
&= \sum_{l=0}^{\infty} \sum_{m=-1}^{l} \left(\frac{1}{N} \sum_{i=0}^{N-1} Y_l^{m*}(\theta_i, \phi_i) \right) Y_l^m(\theta, \phi) \\
&= \frac{1}{N} \sum_{i=0}^{N-1} \delta(\phi - \phi_i) \delta(\theta - \theta_i).
\end{aligned}$$

Thus if the sum is not truncated, one would obtain a superposition of δ-functions, each one with equal weight and centered directly upon one of the points in the realization. This is not useful. The truncation, on the other hand, broadens these contributions so that a "blob" is laid down at each point rather than a δ-function.

Note that the reconstructed PDF can be thought of as a scalar product of abstract vectors. For example, with a spherical harmonic basis, we express the probability density for a single trial point (after having determined it from an independent, "training" dataset):

$$\rho(\theta,\phi) = \sum_{l=0}^{l_{max}} \sum_{m=-l}^{l} \langle Y_l^{m*}(\theta,\phi) \rangle_{\mathcal{D}} \, Y_l^m(\theta,\phi). \tag{16.18}$$

In this case the scalar product is between "vectors" composed of a finite number of spherical harmonics evaluated for the point θ, ϕ, in the case of the trial point, and averaged over θ_i, ϕ_i in the case of the training sample, $\mathcal{D}$. This result can be useful in techniques requiring this probability, such as data modeling, or in techniques designed to classify data, such as character recognition.

Are two datasets different?

We consider the N real coefficients for a dataset determined through orthogonal series density estimation; in case the N coefficients are complex, as in both of our the previous examples, we have $2N$ real numbers, the real and imaginary parts of the coefficients. These are obtained by taking averaging the N basis functions $e_i(\vec{x})$ (for complex basis functions: the $2N$ real and imaginary parts) over the dataset.

While computing the averages $\langle e_i(\vec{x}) \rangle$ in the previous equation, one can also compute the average of their squares:

$$\langle e_i^2(\vec{x}) \rangle \tag{16.19}$$

as well as the products.

$$\langle e_i(\vec{x}) e_j(\vec{x}) \rangle \tag{16.20}$$

and form a matrix $\mathbf{C}_s$ called the ***sample covariance matrix***:

$$(\mathbf{C}_s)_{i,j} \equiv \langle e_i(\vec{x}) e_j(\vec{x}) \rangle - \langle e_i(\vec{x}) \rangle \langle e_j(\vec{x}) \rangle.$$

Dividing this by the number of events N in the dataset yields the covariance of the mean:

$$\mathbf{C} = \mathbf{C}_s / N. \tag{16.21}$$

Given two data sets for which we have determined two coefficient sets $\vec{a}^1$ and $\vec{a}^2$, with covariance $\mathbf{C}^1$ and $\mathbf{C}^2$, we can compute the difference $\vec{\Delta} = \vec{a}^1 - \vec{a}^2$, and its covariance $\mathbf{C}^\Delta$ with elements:

$$C_{ij}^\Delta = \frac{\partial \Delta_i}{\partial a_k^1} \frac{\partial \Delta_j}{\partial a_l^1} C_{kl}^1 + \frac{\partial \Delta_i}{\partial a_k^2} \frac{\partial \Delta_j}{\partial a_l^2} C_{kl}^2 = C_{ij}^1 + C_{ij}^2 \tag{16.22}$$

(summation implied over repeated indices) such that $\mathbf{C}^\Delta = \mathbf{C}^1 + \mathbf{C}^2$. Finally, we can calculate the χ^2 for the two sets to be consistent, which is the standard distance between the two vectors $\vec{a}^1$ and $\vec{a}^2$:

$$\chi^2 = \vec{\Delta}^T \cdot \mathbf{C}^\Delta \cdot \vec{\Delta} = (\vec{a}_1 - \vec{a}_2)^T \cdot (\mathbf{C}_1 + \mathbf{C}_2)^{-1} \cdot (\vec{a}_1 - \vec{a}_2). \qquad (16.23)$$

The probability of this χ^2 can now be calculated for N degrees of freedom and used as a test of the consistency of the two datasets. It is also possible to fit a *dataset* to pseudodata (rather than a function) by minimizing this χ^2 over the parameters used to generate the pseudodata.

This recipe is vague on several points. Which functions should be used as basis functions? This answer is usually determined by the space on which the analysis is carried out, e.g. for functions on a one-dimensional interval, either harmonic functions or Legendre polynomials could be used, whereas spherical harmonics are more appropriate for functions defined on a sphere. Then, how many coefficients should one use? Generally a smooth function without oscillations that serves only to follow fluctuations in the data is desired, so this depends on how much smoothing is required and how much statistical fluctuation is present in the dataset, the latter depending mostly on the amount of data and the number of dimensions.

A common approach to comparing two datasets is to fill two histograms with the same quantity and compare the histograms by computing the χ^2 (the ***template method***). See for example (Press, 2007, Section 14.3). We note that this is in fact a special case of the method we have just described, provided that the basis functions are defined to be:

$$e_i(x) = \begin{cases} \frac{1}{\sqrt{x_{i+1} - x_i}}, & x_i < x \leq x_{i+1} \\ 0, & \text{otherwise} \end{cases}. \qquad (16.24)$$

Thus what probably strikes the reader as a completely different technique turns out to be simply a particular choice of basis functions. Which choice is best? Depending on the distributions, some descriptions of the data may be more economical than others. This is particularly true when the distribution is known to contain only a small number of harmonics, and also with smooth multidimensional distributions. This case arises, for example, in the decay and scattering of particles.

16.7 Bayesian inference

In previous sections we have treated the likelihood as a probability distribution for unknown parameters. This view, which is based in Bayes' theorem, is justifiable with caveats. Consider the joint probability distribution for the dataset $\mathcal{D}$ and the parameters $\vec{\alpha}$. This is a purely abstract notion that can be expressed in terms of conditional probabilities in two ways:

$$\rho(\mathcal{D}, \vec{\alpha}) = \rho(\vec{\alpha}|\mathcal{D})\rho(\mathcal{D}) = \rho(\mathcal{D}|\vec{\alpha})\rho(\vec{\alpha}), \tag{16.25}$$

from which we may obtain:

$$\rho(\vec{\alpha}|\mathcal{D}) = \frac{\rho(\mathcal{D}|\vec{\alpha})\rho(\vec{\alpha})}{\rho(\mathcal{D})} \tag{16.26}$$

which is Bayes' theorem. We identify the conditional probability $\rho(\mathcal{D}|\vec{\alpha})$ with the likelihood $\mathcal{L}(\vec{\alpha})$. As for the denominator in Eq. 16.26, it can expressed as:

$$\rho(\mathcal{D}) = \int \rho(\mathcal{D}|\vec{\alpha})\, d\vec{\alpha} = \int \mathcal{L}\, d\vec{\alpha}.$$

The density $\pi(\vec{\alpha}) \equiv \rho(\vec{\alpha})$ is the notorious **prior density**, while $\rho(\vec{\alpha}|\mathcal{D})$ is called the **posterior density**, which may be written as:

$$\rho(\vec{\alpha}|\mathcal{D}) = \left[\frac{\mathcal{L}(\vec{\alpha})}{\int \mathcal{L}(\vec{\alpha}) d\vec{\alpha}} \right] \pi(\vec{\alpha}). \tag{16.27}$$

In brackets, the factor in the denominator is a normalizing factor for the likelihood, so that the entire expression within the brackets is a normalized probability density. The prior density $\pi(\vec{\alpha})$ represents one's prior knowledge about the parameters $\vec{\alpha}$. This prior density may be taken as flat, but note that probability densities are not invariant under a simple change of variables $\vec{\alpha} \to \vec{\alpha}'$, which introduces additional terms due to the Jacobian of the transformation.

In previous sections we have ignored the prior density $\pi(\alpha)$. The basis for this approximation is as follows. Usually we will assume flat priors with respect to a certain choice of parameters $\vec{\alpha}$ to describe the likelihood function. Other parameterizations will not be flat, but they will be smooth. The higher the sample size, the more sharply peaked the likelihood function. As it becomes more sharply peaked, the variation of the prior over the range in which the likelihood function is non-negligible becomes smaller. Eventually the variation can be neglected, at which point one says that the "data overwhelms the prior".

If the data does not overwhelm the prior, then the prior should be specified in Bayesian analyses. For example, information from previous measurement may be incorporated in the prior. In this case Bayes' theorem describes a *refinement* of previous knowledge with new information. The Kalman filter, described in the next section, is an example of this philosophical point of view. In the case of small sample sizes, Bayesian analysis involves a certain degree of subjectivity because results will depend on the prior.

The likelihood function has many uses. It may be used, as we have illustrated, in Bayesian inference, interpreted usually as the probability density for $\vec{\alpha}$, or it may be combined with other likelihood functions derived from other datasets. Additionally, it may be used as the basis for a test statistic in frequentist analysis, see for example (Feldman, 1997).

Communicating likelihood functions

Often it may be desired to exchange the posterior density with the scientific community. In these cases the question is how? The likelihood function, being defined by data, has no analytic form. One can exchange a numerical procedure together with necessary configuration files. Alternatively, one can find an analytic expression that closely approximates the likelihood, through interpolation or some other procedure. Lastly, one can capture the shape of the likelihood function in a plot, or characterize it by determining intervals containing a specified fraction of the integrated probability density.

The problem of communicating the likelihood function is simplest when one is mainly concerned with only one of the parameters at a time. In that case Monte Carlo integration can be performed to integrate out the unobserved parameters. The function:

$$\rho(\alpha_j) = \int \rho(\vec{\alpha})\, d\alpha_0 d\alpha_i \ldots d\alpha_{j-1} \ldots d\alpha_{j+1} \ldots d\alpha_{N-1} \tag{16.28}$$

represents the probability density for the single parameter α_j independently of the value of all of the other $N-1$ parameters. A simple way to perform the integration is using Markov chain Monte Carlo, discussed in Section 7.5. Markov chain Monte Carlo is first used to generate a realization of the full N dimensional posterior density $\rho(\vec{\alpha})$. Then, a histogram is made of the variable α_j. As usual, care must be taken when generating the Markov chain to ensure the elimination of burn-in and spurious results from insufficiently long runs. If the Markov chains are stored in a file, they can be reweighted later with other priors, to determine the sensitivity of a result to assumptions about the prior. This is referred to as a sensitivity analysis and is a standard part of Bayesian analysis.

16.8 Combining data

When two or more datasets which constrain the same set of parameters are available, one can obtain the constraints from the combined datasets by multiplying together the two likelihoods, or, equivalently, adding their logarithms. Defining $\mathcal{L}_1(\vec{\alpha})$ and $\mathcal{L}_2(\vec{\alpha})$ as the likelihood functions for datasets $\mathcal{D}_1$ and $\mathcal{D}_2$ and $\mathcal{L}_{12}(\vec{\alpha})$ as the likelihood function for the combined dataset, one has:

$$-2\ln\mathcal{L}_{12}(\vec{\alpha}) = -2\ln\mathcal{L}_1(\vec{\alpha}) + -2\ln\mathcal{L}_2(\vec{\alpha}). \tag{16.29}$$

An example is given in Figure 16.11; here measurements from two segments of a single straight line are fit for the slope m and the intercept b using a χ^2 fit. This type of problem can arise in reconstructing a charged particle track in a segmented tracking chamber, for example. The measurements are then combined according to the above Eq. 16.29.

This amounts to the equation:

$$\chi^2_{12}(m,b) = \chi^2_1(m,b) + \chi^2_2(m,b).$$

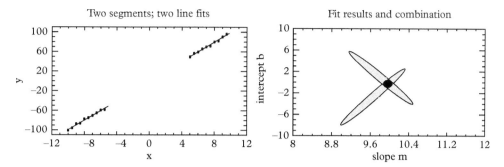

Figure 16.11 *Combining data. Left: two fits to two datasets. Right: error ellipses in the parameters m (slope) and b (intercept) from the two fits and a contour at the intersection representing the combination of the two datasets.*

A substantial reduction in the error on both parameters is observed when the measurements are combined; intuitively this is because, as one can see from Figure 16.11, one requires measurements on both ends of the line to accurately determine the slope and the intercept. Alternately, looking at the error ellipses in Figure 16.11, one sees that the two measurements constrain the parameters in different ways and are thus complementary. This is due to choice of data points used in each fit. Example code lives in EXAMPLES/CH16/COMBO.

In the general case, Eq. 16.29 is written as:

$$\chi^2_{12}(\vec{\alpha}) = \chi^2_1(\vec{\alpha}) + \chi^2_2(\vec{\alpha})$$

where $\chi^2(\vec{\alpha})$ now denotes the Taylor series expansion of $-2\ln \mathcal{L}_{12}(\vec{\alpha})$ about the point $\vec{\alpha}_0$, up to terms of up to second order in $\vec{\alpha}$. In this approximation, the combination can be carried out analytically

$$\chi^2_{12}(\vec{\alpha}) = (\vec{\alpha} - \vec{\alpha}_1)^T \mathbf{C}_1^{-1}(\vec{\alpha} - \vec{\alpha}_1) + (\vec{\alpha} - \vec{\alpha}_2)^T \mathbf{C}_2^{-1}(\vec{\alpha} - \vec{\alpha}_2) + \text{constant}. \quad (16.30)$$

In this expression $\vec{\alpha}_1$ and $\mathbf{C}_1$ represent the central value and covariance matrix of the first fit and $\vec{\alpha}_2$ and $\mathbf{C}_2$ represent the central value and covariance matrix of the second fit. They are outputs from a fit as well as inputs to a combined fit, so it is also proper to call them measurements. Measurements go into the combined fit and more refined measurements come out. The constant in Eq. 16.30 is of no consequence.

We seek a similar approximation for $\chi^2_{12}(\vec{\alpha})$, in other words:

$$\chi^2_{12}(\vec{\alpha}) = (\vec{\alpha} - \vec{\alpha}_{12})^T \mathbf{C}_{12}^{-1}(\vec{\alpha} - \vec{\alpha}_{12}),$$

where $\vec{\alpha}_{12}$ will be identified with the central value of the combined measurement and $\mathbf{C}_{12}$ with its covariance matrix. The covariance matrix is the inverse of the Hessian of $\chi^2_{12}(\vec{\alpha})$,

542 *Combining data*

$$\mathbf{C^{-1}}_{12} = \frac{1}{2}\mathbf{H}_{12} = \frac{1}{2}(\mathbf{H}_1 + \mathbf{H}_2) = \mathbf{C}_1^{-1} + \mathbf{C}_2^{-1}$$

or:

$$\mathbf{C}_{12} = \left(\mathbf{C}_1^{-1} + \mathbf{C}_2^{-1}\right)^{-1}.$$

To compute the central value of the combined measurement, $\vec{\alpha}_{12}$, we need to minimize the χ^2, Eq. 16.30, over the parameters $\vec{\alpha}$. The derivatives with respect to the parameters $\vec{\alpha}$ can be taken analytically:

$$\frac{\partial \chi_{12}^2(\vec{\alpha})}{\partial \vec{\alpha}} = \frac{\partial}{\partial \vec{\alpha}}(\vec{\alpha} - \vec{\alpha}_1)^T \mathbf{C}_1^{-1}(\vec{\alpha} - \vec{\alpha}_1) + \frac{\partial}{\partial \vec{\alpha}}(\vec{\alpha} - \vec{\alpha}_2)^T \mathbf{C}_2^{-1}(\vec{\alpha} - \vec{\alpha}_2)$$
$$= 2\mathbf{C}_1^{-1}(\vec{\alpha} - \vec{\alpha}_2) + 2\mathbf{C}_2^{-1}(\vec{\alpha} - \vec{\alpha}_2)$$

and set to zero:

$$\mathbf{C}_1^{-1}(\vec{\alpha} - \vec{\alpha}_2) + \mathbf{C}_2^{-1}(\vec{\alpha} - \vec{\alpha}_2) = 0. \qquad (16.31)$$

Thus:

$$\vec{\alpha} = \left(\mathbf{C}_1^{-1} + \mathbf{C}_1^{-1}\right)^{-1} \cdot \left(\mathbf{C}_1^{-1}\vec{\alpha}_2 + \mathbf{C}_2^{-1}\vec{\alpha}_2\right). \qquad (16.32)$$

We summarize the procedure in two equations:

$$\mathbf{C}_{12} = \left(\mathbf{C}_1^{-1} + \mathbf{C}_2^{-1}\right)^{-1}$$
$$\vec{\alpha}_{12} = \mathbf{C}_{12} \cdot \left(\mathbf{C}_1^{-1}\vec{\alpha}_2 + \mathbf{C}_2^{-1}\vec{\alpha}_2\right) \qquad (16.33)$$

These two equations become intuitive in the case of one parameter, which we could call y. Assume we have two measurements, $y_1 \pm \sigma_1$ and $y_2 \pm \sigma_2$, and that we denote the combined measurement $\bar{y}$ and its error σ. Our two equations read:

$$\sigma^2 = \left(1/\sigma_1^2 + 1/\sigma_2^2\right)^{-1}$$
$$\bar{y} = \frac{y_1/\sigma_1^2 + y_2/\sigma_2^2}{1/\sigma_1^2 + 1/\sigma_2^2},$$

which is a common formula for the weighted average of two measurements.

16.9 The Kalman filter

In Section 16.8 and Figure 16.11 we fit a straight line in two steps, first by fitting the segments and then by combining the two fits. In this section we will go the extreme case in which the entire dataset is fit by combining individual data points $(x_i, y_i \pm \sigma_i)$, one-by-one. The reward is the development of a technique that can be used to update a fit with new data points as they are made available. This permits tracking an evolving system in the presence of both measurement error and Gaussian perturbations on the parameters. The technique, called the **Kalman filter**, was invented in the early 1960s at NASA where it was used to land the lunar module. It has also been applied to estimate the trajectories of charged particles in tracking detectors in the presence of both measurement error and multiple Coulomb scattering. There is a vast literature on the Kalman filter and there are many extensions to the basic method, to which we limit ourselves in this text.

We will derive the Kalman filter method in analogy to the combination of measurements made above. The method combines information about the measured parameters from the points indexed $0, 1, 2 \ldots N-1$ with additional information from the Nth point. To be clear: using a Kalman filter to fit a straight line is like killing a fly with an elephant gun. However the example is designed to soften our introduction to a complicated subject.

There are two main differences between this type of combination and that of the previous section with which we will have to cope. The first is that the individual data points, $(x_i, y_i \pm \sigma_i)$, constrain a higher-dimensional parameter space and thus do not have a full 2×2 dimensional covariance matrix in the space of the parameters m and b (slope and intercept). So the formulae of Eq. 16.33 cannot be used.

The second issue is how one initiates the procedure. In the case of a straight line the parameters are not known until two points have been incorporated, for second order polynomials three are required, etc. The first of these two issues will be clarified in the derivation of the Kalman filter equations, while the second will be sidestepped for the moment by assuming that after the first n points, central values and covariance matrices for the fit are already available. When the form of the Kalman filter equations is available we will return to the issue of how to start the fit.

Before we start the derivation we generalize a bit by assuming that more than one variable is measured at each point x_i, such that the variable y_i is itself a vector, designated $\vec{y}_i$. It matters little if the measured quantities are correlated or uncorrelated, because in any case we will express the contribution of the full set of independent measurements to χ^2 as

$$\left(\vec{y}_i - \vec{f}(x_i; \vec{\alpha})\right)^T \mathbf{R}_i^{-1} \left(\vec{y} - \vec{f}(x_i; \vec{\alpha})\right)$$

where $\mathbf{R}_i$ is the covariance matrix for the measurements $\vec{y}_i$. The equation for generating the prediction for the value of $\vec{y}_i$ is denoted $\vec{f}(x_i; \vec{\alpha})$ and is called the *measurement equation*. For example, for the straight line with parameters m and b, the measurements y_i consist of only one value and the covariance matrix $\mathbf{R}_i = (\sigma_i^2)$ has dimension 1×1. We define $\vec{\alpha} = (b, m)$ and $f(x; \vec{\alpha}) = mx + b$.

The Kalman filter

We choose now some particular point $i = n$ and assume that the previous n points $0, 1, \ldots n-1$ have already yielded central values $\vec{\alpha}_{n-1}$ for the parameters and a covariance matrix C_{n-1}. Combining the measurement $\vec{y}_n$ with the covariance matrix $\mathbf{R}_n$ gives a new χ^2 which we write as:

$$\chi_n^2(\vec{\alpha}) \approx (\vec{\alpha} - \vec{\alpha}_{n-1})^T \mathbf{C}_{n-1}^{-1} (\vec{\alpha} - \vec{\alpha}_{n-1}) + (\vec{y}_n - \vec{f}(x_n; \vec{\alpha}))^T \mathbf{R}_n^{-1} (\vec{y}_n - \vec{f}(x_n; \vec{\alpha})).$$

We seek an approximate solution in the form:

$$\chi_n^2(\vec{\alpha}) = (\vec{\alpha} - \vec{\alpha}_n)^T \mathbf{C}_n^{-1} (\vec{\alpha} - \vec{\alpha}_n).$$

This is facilitated by expanding the measurement equation to first order in the parameters:

$$\vec{f}(x; \vec{\alpha}) = \vec{f}(x; \vec{\alpha}_{n-1}) + \mathbf{A}_{n-1} \cdot (\vec{\alpha} - \vec{\alpha}_{n-1}) + \ldots$$

where $\mathbf{A}_{n-1}$ is a matrix of derivatives,

$$(\mathbf{A}_{n-1})_{ij} = \left. \frac{\partial f_i(x; \vec{\alpha})}{\partial \alpha_j} \right|_{x=x_n; \vec{\alpha}=\vec{\alpha}_{n-1}}.$$

With this we can write:

$$\begin{aligned}
\chi_n^2 &\equiv (\vec{\alpha} - \vec{\alpha}_n)^T \cdot \mathbf{C}_n^{-1} \cdot (\vec{\alpha} - \vec{\alpha}_n) \\
&= (\vec{\alpha} - \vec{\alpha}_{n-1})^T \cdot \mathbf{C}_{n-1}^{-1} \cdot (\vec{\alpha} - \vec{\alpha}_{n-1}) \\
&\quad + \left(\vec{y}_n - \vec{f}(x_n; \vec{\alpha}_{n-1}) - \mathbf{A}_{n-1} \cdot (\vec{\alpha} - \vec{\alpha}_{n-1})^T \right) \cdot \mathbf{R}_n^{-1} \\
&\quad \cdot \left(\vec{y}_n - \vec{f}(x_n; \vec{\alpha}_{n-1}) - \mathbf{A}_{n-1} \cdot (\vec{\alpha} - \vec{\alpha}_{n-1}) \right).
\end{aligned}$$

The covariance matrix $\mathbf{C}_n$ can be obtained immediately from the inverse of the Hessian matrix of the right-hand side, which is:

$$\mathbf{C}_n^{-1} = \mathbf{C}_{n-1}^{-1} + \mathbf{A}^T \mathbf{R}_n^{-1} \mathbf{A}_{n-1}$$

or:

$$\mathbf{C}_n = \left(\mathbf{C}_{n-1}^{-1} + \mathbf{A}^T \mathbf{R}_n^{-1} \mathbf{A}_{n-1} \right)^{-1}.$$

The central value $\vec{\alpha}_n$ can be determined by minimizing the right hand side over the parameters $\vec{\alpha}_n$. The relevant derivatives are:

$$\frac{\partial \chi_n^2(\vec{\alpha})}{\partial \vec{\alpha}} = 2\mathbf{C}_{n-1}^{-1} \cdot (\vec{\alpha} - \vec{\alpha}_{n-1}) - 2\mathbf{A}_{n-1}^T \mathbf{R}_n^{-1} \cdot \left(\vec{y}_n - \vec{f}(x_n; \vec{\alpha}_{n-1}) - \mathbf{A}_{n-1} \cdot (\vec{\alpha} - \vec{\alpha}_{n-1})\right),$$
(16.34)

which we set to zero to obtain:

$$\mathbf{C}_{n-1}^{-1} \cdot \vec{\alpha} + \mathbf{A}_{n-1}^T \mathbf{R}_n^{-1} \mathbf{A}_{n-1} \cdot \vec{\alpha} = \mathbf{C}_{n-1}^{-1} \cdot \vec{\alpha}_{n-1} + \mathbf{A}_{n-1}^T \mathbf{R}_n^{-1}$$
$$\cdot \left(\vec{y}_n - \vec{f}(x_n; \vec{\alpha}_{n-1}) + \mathbf{A}_{n-1} \cdot \vec{\alpha}_{n-1}\right)$$

or:

$$\vec{\alpha} = \left(\mathbf{C}_{n-1}^{-1} + \mathbf{A}^T \mathbf{R}_n^{-1} \mathbf{A}_{n-1}\right)^{-1} \cdot \left(\mathbf{C}_{n-1}^{-1} \vec{\alpha}_{n-1} + \mathbf{A}_{n-1}^T \mathbf{R}_n^{-1} \cdot (\Delta \vec{y}_n + \mathbf{A}_{n-1} \cdot \vec{\alpha}_{n-1})\right)$$
(16.35)

where $\Delta \vec{y}_n = \vec{y}_n - \vec{f}(x_n; \vec{\alpha}_{n-1})$.

In summary, the Kalman filter equations are:

$$\mathbf{C}_n = \left(\mathbf{C}_{n-1}^{-1} + \mathbf{A}^T \mathbf{R}_n^{-1} \mathbf{A}_{n-1}\right)^{-1},$$
$$\vec{\alpha}_n = \mathbf{C}_n \cdot \left(\mathbf{C}_{n-1}^{-1} \cdot \vec{\alpha}_{n-1} + \mathbf{A}_{n-1}^T \mathbf{R}_n^{-1} \cdot (\Delta \vec{y}_n + \mathbf{A}_{n-1} \cdot \vec{\alpha}_{n-1})\right).$$
(16.36)

These are the Kalman filtering equations in their preliminary form, a set of equations for combining a fit to the first n points with information from the $(n+1)^{th}$ point. The equations are used at each step to refine an existing fit by updating it with new information. This is, in fact, an excellent illustration of Bayes' theorem at work.

The philosophical problem we encountered in the discussion of Bayesian inference, namely, the dependence upon the prior density, becomes a technical issue in the context of Kalman filtering. In particular, the Kalman filter equations only specify how to *update* a fit. Prior information is needed to start the fit. This prior information comes in the form of a starting estimate, as well as a starting covariance. At the first few steps in the fit, the parameter values are not well estimated since insufficient information has been provided to the fit. However, with more steps in the fit the data should eventually overwhelm the initial assumptions. Sensitivity to these assumptions should be examined, and tests should be carried out with simulated data before applying the procedure to real data.

If the starting covariance is too small or the starting estimate is too far from the mark, it can pull the fitted values of the parameters; usually this will result in bad pull distributions for one or more parameter. If the covariance is too large on the other hand, the Kalman filtering equations may fail due to roundoff error. These considerations will become clear in the example of the following section.

16.9.1 Example: fitting a polynomial curve

The Kalman filter equations, Eq. 16.36, are still incomplete but we take a break from the development to see a simple application. The application seeks to fit data generated according to a polynomial with the addition of Gaussian noise. The measurement equation will be taken to be a polynomial of form:

$$f(x) = \alpha_0 + \alpha_1 x + \alpha_2 x^2 \ldots \alpha_{n-1} x^{n-1},$$

which has a $1 \times n$ derivative matrix:

$$\mathbf{A} = \begin{pmatrix} 1, & x, & x^2, & x^3, \ldots, & x^{n-1} \end{pmatrix}.$$

The quantity $\Delta \vec{y}_n + \mathbf{A} \cdot \vec{\alpha}_{n-1}$ appearing in Eq. 16.36 reduces to just $\vec{y}_n$ in this case. The matrix $\mathbf{R}_i$ is one-dimensional since only one variable, y, is measured at each point x along the curve. Thus $\mathbf{R}_i = \delta_i^2$, where δ_i is the measurement error on the ith point. Large diagonal covariance matrices are taken to start the fit, together with an arbitrary guess for the initial values of the parameters.

Three directories in the EXAMPLES/KALMAN area, called KALMAN-FLAT, KALMAN-LINEAR, and KALMAN-THIRD, fit zeroth-order, first-order, and third-order polynomials to the synthetic data. For the third order polynomials, the progression of the fitted curve as more points are added is shown in Figure 16.12.

The zeroth-order polynomial fit is worth examining. In this case the measurement equation is $y = a_0$; the matrix $\mathbf{A}$ is a 1×1 matrix whose value is unity, and Eqs. 16.36 are:

$$\sigma_n^2 = \left(\frac{1}{\sigma_{n-1}^2} + \frac{1}{\delta_n^2} \right)^{-1}$$

$$a_{0,n} = \sigma_n^2 \cdot \left(\frac{a_{0,n-1}}{\sigma_{n-1}^2} + \frac{y_n}{\delta_n^2} \right). \quad (16.37)$$

This represents a procedure for averaging a set of measured values $y_i \pm \delta_i$ by taking, at each step, the weighted average of the new measurement with the weighted average of all the previous measurements.

A second example uses the sigmoid (S-shaped) function:

$$f(x; w, \theta) = \tanh(wx + \theta).$$

The matrix $\mathbf{A}$ is $\left(\frac{\partial f}{\partial w}, \frac{\partial f}{\partial \theta} \right) = (x \cdot \text{sech}^2(wx + \theta), \text{sech}^2(wx + \theta))$. Again $\mathbf{R}_i = \sigma_i^2$, where σ_i is the measurement error on the ith point. The progression of the fit is similar to that of Figure 16.12. The code for this example is available in the directory EXAMPLES/CH16/KALMAN/KALMAN-SIGMOID.

16.9.2 Complete equations

The equations in Eq. 16.36 are expressed in a form that is called the ***weighted means formalism***. This form is well adapted for recognizing the relationship between a Kalman

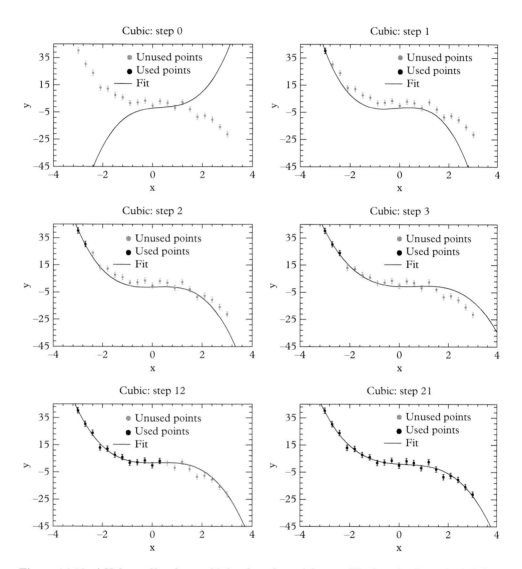

Figure 16.12 *A Kalman-filter fit to a third-order polynomial curve. The first plot shows the fit before any points have been added—this corresponds to "prior knowledge" on the input parameters. The second plot shows the fit after a single point is added. Then two and three points are added (third and fourth plots), and finally 12 points (fifth plot) and 21 points (sixth plot). At each stage, points which are used are shown in black and unused points are shown in grey.*

filter and a weighted mean. A calculationally superior formalism known as the **Kalman gain matrix formalism** can be found by giving a slightly different treatment to Eq. 16.35. We have:

$$\mathbf{C}_{n-1}^{-1} \cdot \vec{\alpha} + \mathbf{A}_{n-1}^T \mathbf{R}_n^{-1} \mathbf{A}_{n-1} \cdot \vec{\alpha} = \mathbf{C}_{n-1}^{-1} \cdot \vec{\alpha}_{n-1} + \mathbf{A}_{n-1}^T \mathbf{R}_n^{-1}$$
$$\cdot \left(\vec{y}_n - \vec{f}(x_n; \vec{\alpha}_{n-1}) + \mathbf{A}_{n-1} \cdot \vec{\alpha}_{n-1} \right)$$

$$\left(\mathbf{C}_{n-1}^{-1} + \mathbf{A}_{n-1}^T \mathbf{R}_n^{-1} \mathbf{A}_{n-1} \right) \cdot \vec{\alpha} = \left(\mathbf{C}_{n-1}^{-1} + \mathbf{A}_{n-1}^T \mathbf{R}_n^{-1} \mathbf{A}_{n-1} \right) \cdot \vec{\alpha}_{n-1}$$
$$+ \mathbf{A}_{n-1}^T \mathbf{R}_n^{-1} \cdot \left(\vec{y}_n - \vec{f}(x_n; \vec{\alpha}_{n-1}) \right)$$

$$\vec{\alpha} = \vec{\alpha}_{n-1} + \left(\mathbf{C}_{n-1}^{-1} + \mathbf{A}_{n-1}^T \mathbf{R}_n^{-1} \mathbf{A}_{n-1} \right)^{-1}$$
$$\mathbf{A}_{n-1}^T \mathbf{R}_n^{-1} (\vec{y}_n - \vec{f}(x_n; \vec{\alpha}_{n-1}))$$

$$\vec{\alpha} = \vec{\alpha}_{n-1} + \mathbf{K}_n \cdot (\vec{y}_n - \vec{f}(x_n; \vec{\alpha}_{n-1}))$$

where $\mathbf{K}_n$ is the **Kalman gain matrix**; there are several equivalent ways of writing it:

$$\mathbf{K}_n = \left(\mathbf{C}_{n-1}^{-1} + \mathbf{A}_{n-1}^T \mathbf{R}_n^{-1} \mathbf{A}_{n-1} \right)^{-1} \mathbf{A}_{n-1}^T \mathbf{R}_n^{-1}$$
$$= \mathbf{C}_n \mathbf{A}_{n-1}^T \mathbf{R}_n^{-1}$$
$$= \mathbf{C}_{n-1} \mathbf{A}_{n-1}^T \left(\mathbf{A}_{n-1} \mathbf{C}_{n-1} \mathbf{A}_{n-1}^T + \mathbf{R}_n \right)^{-1}$$

The Kalman gain matrix involves a single matrix inversion; having obtained that, we can now update the covariance and the parameters without further inversions:

$$\mathbf{C}_n = \left(\mathbf{C}_{n-1}^{-1} + \mathbf{A}^T \mathbf{R}_n^{-1} \mathbf{A}_{n-1} \right)^{-1}$$
$$= (\mathbf{I} - \mathbf{K}_n \mathbf{A}_{n-1}) \cdot \mathbf{C}_{n-1}$$
$$\vec{\alpha}_n = \vec{\alpha}_{n-1} + \mathbf{K}_n \cdot (\vec{y}_n - \vec{f}(x_n; \vec{\alpha}_{n-1})).$$

The χ^2 of the fit may also be of interest. The change in χ^2 is:

$$\Delta \chi^2 = \chi_n^2 - \chi_{n-1}^2$$
$$= (\vec{\alpha}_n - \vec{\alpha}_{n-1})^T \cdot \mathbf{C}_{n-1}^{-1} \cdot (\vec{\alpha}_n - \vec{\alpha}_{n-1})$$
$$+ \left(\vec{y}_n - \vec{f}(x_n; \vec{\alpha}_{n-1}) - \mathbf{A}_{n-1} \cdot (\vec{\alpha}_n - \vec{\alpha}_{n-1}) \right)^T$$
$$\cdot \mathbf{R}_n^{-1} \cdot \left(\vec{y}_n - \vec{f}(x_n; \vec{\alpha}_{n-1}) - \mathbf{A}_{n-1}(\vec{\alpha}_n - \vec{\alpha}_{n-1}) \right).$$

The system we have described until now can be considered a deterministic dynamical system which evolves in a trivial way with random measurement noise. This can be represented as:

$$\vec{y}_i = \vec{f}(x_i; \vec{\alpha}) + \vec{v}_i$$

where $\vec{v}_i$ is Gaussian distributed measurement noise with covariance matrix $\mathbf{R}_i$. Using these equations we *track* the development of the system. With a small modification, the equations can track a less trivial system, where the system itself is subject to random perturbation, obtaining what are usually called the **Kalman filter equations**, first introduced by R.E. Kalman in 1960 during the Apollo program (Kalman, 1960). The mathematical foundations of the Kalman filtering algorithms can be found in Chui (1991). Applications of the algorithm include real-time tracking of the motion of an object (e.g., a satellite) in the presence of deterministic and/or random changes to the trajectory, or tracking charged particles through a detector system. In this case the system is described by a different set of equations:

$$\vec{\alpha}_i = \vec{g}(\vec{\alpha}_{i-i}) + \vec{w}_i$$
$$\vec{y}_i = \vec{f}(x_i; \vec{\alpha}) + \vec{v}_i$$

The measurement noise $\vec{v}_i$ is still Gaussian distributed with covariance matrix $\mathbf{R}_i$. New ingredients include, the determinisic development of the system at each step, governed by the function $\vec{g}(\vec{\alpha})$, and the *system noise* $\vec{w}_i$, which will be taken to be Gaussian distributed with covariance matrix $\mathbf{Q}_i$. Such a system is called a deterministic-stochastic system. In this case the predicted parameter values may evolve at the beginning of each step, and in addition the covariance matrix needs to be incremented by adding the system noise. We change our notation to accommodate applications with these effects, recognizing that there are two phases to the procedure, a prediction phase and a correction phase. The prediction phase takes estimates of parameter values and their covariance at the output of one fitting stage and predicts their values at the input of another. We designate the predicted values of $\vec{\alpha}_n$ and $\mathbf{C}_n$ after point $n-1$ has been added as $\vec{\alpha}_{n|n-1}$ and $\mathbf{C}_{n|n-1}$. The Jacobian of the function $\vec{g}(\vec{\alpha})$, is designated as $\mathbf{P}$:

$$(\mathbf{P}_k)_{ij} \equiv \left. \frac{\partial g_i}{\partial \alpha_j} \right|_{\vec{\alpha} = \vec{\alpha}_k}. \tag{16.38}$$

The full set of Kalman filter equations now becomes:

$$\mathbf{C}_{n|n-1} = \mathbf{P}_{n-1} \mathbf{C}_{n-1|n-1} \mathbf{P}_{n-1}^T + \mathbf{Q}_n$$
$$\vec{\alpha}_{n|n-1} = \vec{g}(\vec{\alpha}_{n-1|n-1})$$
$$\mathbf{K}_n = \mathbf{C}_{n|n-1} \mathbf{A}_{n-1}^T \left(\mathbf{A}_{n-1} \mathbf{C}_{n|n-1} \mathbf{A}_{n-1}^T + \mathbf{R}_n \right)^{-1}$$

$$\mathbf{C}_{n|n} = (\mathbf{I} - \mathbf{K}_n \mathbf{A}_{n-1}) \mathbf{C}_{n|n-1}$$

$$\vec{\alpha}_{n|n} = \vec{\alpha}_{n|n-1} + \mathbf{K}_n \cdot \left(\vec{y}_n - \vec{f}(x_n; \vec{\alpha}_{n|n-1})\right)$$

$$\Delta \chi^2 = \left(\vec{\alpha}_{n|n} - \vec{\alpha}_{n|n-1}\right)^T \cdot \mathbf{C}_{n-1|n-1}^{-1} \cdot \left(\vec{\alpha}_{n|n} - \vec{\alpha}_{n|n-1}\right)$$
$$+ \left(\vec{y}_n - \vec{f}(x_n; \vec{\alpha}_{n|n-1}) - \mathbf{A}_{n-1} \cdot (\vec{\alpha}_{n|n} - \vec{\alpha}_{n|n-1})\right)^T$$
$$\cdot \mathbf{R}_n^{-1}$$
$$\cdot \left(\vec{y}_n - \vec{f}(x_n; \vec{\alpha}_{n|n-1}) - \mathbf{A}_{n-1} \cdot (\vec{\alpha}_{n|n} - \vec{\alpha}_{n|n-1})\right)$$

This formulation is not the most general one; more general formulations, which include, for example, the incorporation of control input used to correct the trajectory of, say, an airborne projectile or a robotic arm, can be found in the references. Fruehwirth (1987) describes the application of Kalman filtering to track and vertex reconstruction in charged particle detectors; Luchsinger (1993) shows how the method can be used to carry out a fit to a whole cascade of particles decaying in such a detector, and Avery (1991a; 1991b; 1991c; 1992a; 1992b), in a series of on-line lecture notes, gives an excellent overview of fitting theory with applications to track reconstruction. In most applications it is the computation of the derivative matrix $\mathbf{A}_{n-1}$ that is the most challenging and time-consuming part of the task. For the charged particle tracking applications, these computations are carried out in Fruehwirth (1987), Luchsinger (1993), and Avery (1991a; 1991b; 1991c; 1992a; 1992b).

16.10 Exercises

1. Iron Decay.
   The iron isotope Fe^{55} decays by electron capture to Mn^{55}. After the electron capture the atom has an empty K-shell orbital. As the excited atom fills this vacancy, several forms of radiation are emitted: Auger electrons (60%), and X-rays at 5.89875 keV (16.2%), 5.88765 keV (8.2%), and 6.49045 keV (2.85%). New devices can measure the energy of gamma rays with a resolution in the range of a few eV. In the directory

   DATA/CH16/EX1

   you will find a text file Fe55-1KCounts.txt containing a set of simulated values for measured energies of gamma rays in the energy range of the first two decays. Plot these data and note the presence of two spectral lines. Determine from these data,

   a) The energy of each spectral line and its error.
   b) The energy resolution of the device, and its error.

c) The natural width of the spectral line.

d) Use the uncertainty principle to obtain the lifetime of the atomic state giving rise to the X-ray.

2. Mossbauer effect.
   The Mossbauer effect is the recoiless emission and absorption of gamma radiation from a nucleus. In the file

   DATA/CH16/EX2/Fe57.dat

   is raw Mossbauer data recorded at the National Bureau of Standards in 1967 (Spijkerman, 1967) for the 14.4 KeV line of Fe^{57}. The file consists of a list of photon counts for each bin of a multichannel analyzer, arranged sequentially by bin number. The emission line is split by 6.22×10^{-8} eV due to the electric quadrupole field of the compound, sodium nitroprusside, at the Fe^{57} nucleus. Fit the absorption lines to a Voigt distribution and determine the natural line width and lifetime of Fe^{57}.

3. Smeared Spectral Lines.
   The following exercises refer to data files that can be found in the directory:

   DATA/CH16/EX3.

   The files contain unbinned data from a series of spectral lines that have been measured using a mass spectrometer with a finite resolution. We would like to ask the question: does the width of the spectral lines come from the natural width of the resonance or is it due to measurement uncertainty? The answer is provided by the shape of the line. A pure spectral line without energy resolution effects is described by a Lorentzian or Breit-Wigner lineshape whose width is the natural width of the transition. On the other hand, resolution effects add measurement noise which we assume to be Gaussian.

   a) The dataset `data00.dat` contains a set of measured photon energies (in eV). From this data: decide whether the data follows a Lorentzian distribution or a Gaussian distribution, and, determine the energy of the spectral line (and its error). If the lineshape is Lorentzian, determine the lifetime of the state; if it is Gaussian, determine the experimental resolution.

   b) The dataset `data01.dat` is more complicated since it contains two partially overlapping spectral lines that are not very well resolved. Your job is to determine the relative fraction of light in each of the two spectral lines, and the energies (in eV) of either line.

   c) The dataset `data02.dat` is the same as before, except that these spectral lines have a contribution from experimental resolution and a natural line width, with neither effect completely dominating or completely negligible. Your job is to determine the same parameters as in the previous problem, but, additionally to determine the amount of smearing due to experimental resolution.

d) The dataset `data03.dat` contains hypothetical data describing the thermal radiation from a black body. Determine the temperature from this data. Your answer should be in electron volts. It should include an error estimate.

e) Finally, the data in the `file data04.dat` contains two spectral lines with a background from thermal radiation. Determine the energy and width of the two spectral lines in addition to the temperature of the blackbody radiation. Is it possible to determine the energy resolution of the spectrometer?

4. Synthetic Poisson.

   a) Generate data according to a Poisson distribution. Make a histogram of the generated data and observe the shape of the function for various values of the mean λ of the distribution. Notice how the shape of the distribution becomes similar to a Gaussian for $\lambda \gg 1$.

   b) Pretend you did not know the input value of λ and determine it from a fit to the data. Do this for several values of λ between 0.1 and 10.0, and plot the extracted value versus the input value. Include error bars on the plot.

   c) Fit the plot to a straight line. Plot the results as an error ellipse showing the 1σ (68%CL) contour.

5. Multivariate Gaussian
   Generate five independent random numbers $\{x_1, x_2 \ldots x_5\}$ from a normal distribution, and from these generate the correlated variables:

$$\begin{aligned} v_1 &= 9x_1 + 3x_2 + 4x_3 - 2x_4 + 5x_5 \\ v_2 &= 3x_1 + 2x_2 + 5x_3 - 2x_4 + 2x_5 \\ v_3 &= 4x_1 + x_2 + 5x_3 - 2x_4 + 3x_5 \\ v_4 &= -2x_1 - 2x_2 - 2x_3 5x_4 + 0x_5 \\ v_5 &= 5x_1 + 2x_2 + 3x_3 + 0x_4 + 7x_5 \end{aligned} \quad (16.39)$$

Then create a data set of approximately 1000 points and determine from a single unbinned maximum likelihood fit the covariance matrix of the five variables $\{v_1, v_2 \ldots v_5\}$. Check your fit by plotting a projection of the multivariate Gaussian distribution onto each of these variables.

6. Particle Decay.
   In Exercise 6 of Chapter 7, you generated the angular distribution of decaying particles according to a series in Legendre polynomials. Apply orthogonal series density estimation to extract the function:

$$\rho(\cos\theta) = \sum_{l=0}^{14} a_l P_l(\cos\theta).$$

Determine and tabulate the coefficients a_l with their errors, and overlay the function you determine onto the data you have generated.

7. Mars.
The text file `Mars01.dat` in the directory `DATA/CH16/EX5` contains data on the celestial coordinates (the right ascension and the declination) of the planet Mars taken over a two year period starting Jan 1, 2015.

This type of data can be obtained from the website

`http://ssd.jpl.nasa.gov/horizons.cgi`

which is maintained by the Jet Propulsion Laboratory. From these data, determine

a) The eccentricity of the earth's orbit.
b) The eccentricity of Mars's orbit.
c) The semi-major axis of Mars, in astronomical units.
d) The orbital period of Mars.
e) The inclination of Mars's orbit relative to that of the earth.
f) The days in the two-year period during which the earth was at perhelion.
g) The days in the two-year period during which Mars was at perhelion.
h) Graph both the raw data and your fitted curve for both of the celestial coordinates.

See (Goldstein, 2001) for a discussion of the Kepler problem.

8. [P] Superball.
A superball is a familiar kid's toy and also is frequently used in physics demonstrations. The rebound of the superball is characterized by two numbers, α and β, called *coefficients of restitution*, which relate the linear and angular velocity after a rebound, v'_x, v'_y, and ω', to their values v_x, v_y, and ω before the bounce. The radius of the superball is a, x is the horizontal direction and y is the vertical direction. In the model put forth by Garwin (1969), the relation is:

$$\begin{pmatrix} v'_x \\ v'_y \\ \omega' \end{pmatrix} = \begin{pmatrix} \frac{1}{7}(5 - 2\alpha) & 0 & -\frac{2a}{7}(1+\alpha) \\ 0 & -\beta & 0 \\ -\frac{5}{7a}(1+\alpha) & 0 & \frac{1}{7}(2 - 5\alpha) \end{pmatrix} \cdot \begin{pmatrix} v_x \\ v_y \\ \omega \end{pmatrix} \qquad (16.40)$$

Values of the restitution coefficients obtained for a superball of radius 3.0 cm by Cross (2002) were $\alpha = 0.76$ and $\beta = 0.86$. In this exercise, we assume these values.

- An air cannon is used to accurately fire a superball onto the hardwood floor of a large gymnasium. The cannon fires superballs such that they strike the floor with an average speed of 7.3 m/s and a standard deviation of 1.5 m/s. The average angle of incidence is 45° with a standard deviation of 4°. There is no forespin/backspin on the initial impact, on average, but the angular velocity varies about zero with a standard deviation of 5 radians/sec. Write a Monte Carlo program to simulate the location and time of 25 successive impact points and plot these on a graph. Record this information on an output file or in a database.

- The time between impacts is measured with an acoustic device that records the rebounds and achieves a precision of 100 ms. In addition, the horizontal component of the ball's velocity is measured with a camera and strobe light with precision of 1.0 m/s. Simulate these measurements and add them to the output file.

- Now reconstruct, using the Kalman filter, the optimal estimate of v_x, v_y, and ω immediately before each bounce, using all of the available information up to that point. Also compute the covariance of these three quantities. Check your estimates by producing pull distributions, computed for a few thousand simulated shots, verifying that unit Gaussian distributions centered at zero are obtained. Then, plot the expected precision on each of these quantities as a function of bounce number.

- Suppose that, in addition to the above mentioned effects, flaws in the surface cause the x-component of the velocity to vary by 3% on each rebound, and the y-component varies by 1.2%. Add these effects into the simulation, then treat them properly in the Kalman filter fit. Repeat the checks of the previous section.

BIBLIOGRAPHY

Avery, P. (1991a). *Applied fitting theory I General least squares theory*. Preprint CBX 91-72, CLEO Experiment.

Avery, P. (1991b). *Applied fitting theory II Determining systematic effects by fitting*. Preprint CBX 91-73, CLEO Experiment.

Avery, P. (1991c). *Applied fitting theory III Non-optimal least squares fitting and multiple scattering*. Preprint CBX 91-74, CLEO Experiment.

Avery, P. (1992a). *Applied fitting theory IV Formulas for track fitting*. Preprint CBX 92-45, CLEO Experiment.

Avery, P. (1992b). *Applied fitting theory V Track fitting using the Kalman filter*. Preprint CBX 92-39, CLEO Experiment.

Chui, C. K. and G. Chen (1991). *Kalman filtering with real-time applications*. Springer-Verlag.

Cross, R. (2002). *Measurements of the horizontal coefficient of restitution for a superball and a tennis ball*. Am. J. Phys. **70** 5, 482.

Feldman, G. J. and R. D. Cousins, *A unified approach to the classical statistical analysis of small signals*. Phys. Rev. D **57**, 3873.

Fruehwirth, R. (1987). *Application of Kalman filtering to track and vertex fitting*. Nucl. Instrum. Meth. Phys. Rev. A **262** 2–3, 444.

Garwin, R. (1969). *Kinematics of an ultraelastic rough ball*. Am. J. Phys. **37** 88–92, 1969.

Gibbs, W. R., (2006). *Computation in Modern Physics*. 3rd ed. World Scientific.

Goldstein, H., C. P. Poole, and J. Safko (2001). *Classical Mechanics*. 3rd ed. Addison-Wesley.

Kalman, R. E. (1960). *A new approach to linear filtering and prediction problems*. Journal of Basic Engineering **82**, 35.

Horowitz, P., and W. Hill (1989). *The Art of Electronics.* 3nd ed. Cambridge University Press.

James, Fred (1994). *MINUIT Function Minimization and Error Analysis Reference Manual Version 94.1.* CERN Report CERN-D-506.

Luchsinger, R. and C. Grab (1993). *Vertex reconstruction by means of the method of Kalman filtering.* Comp. Phys. Comm. **76** 3, 263.

Press, W., S. Teukolsky, W. Vetterling, and B. Flannery (2007). *Numerical Recipes, the Art of Scientific Computing.* 3rd ed. Cambridge University Press.

Spijkerman, J. J., D. K. Snediker, F. C. Ruegg, and J. R. DeVoe. (1967). *Mossbauer Spectroscopy Standard for the Chemical Shift of Iron Compounds.* National Bureau of Standards Miscellaneous Publication 260–13.

Watson, G. S. (1969). *Density Estimation by Orthogonal Series.* Ann. Math. Statist. **40** 4, 1496.

17

Templates, the standard C++ library, and modern C++

17.1 Generic type parameters	557
17.2 Function templates	558
17.3 Class templates	560
17.3.1 Class template specialization	561
17.4 Default template arguments	563
17.5 Non-type template parameters	563
17.6 The standard C++ library	565
17.6.1 Containers and iterators	566
17.6.2 Algorithms	575
17.7 Modern C++	577
17.7.1 Variadic templates	578
17.7.2 Auto	580
17.7.3 Smart pointers	581
17.7.4 Range-based for loop	582
17.7.5 Nullptr	583
17.7.6 Iterators: nonmember begin and end	583
17.7.7 Lambda functions and algorithms	584
17.7.8 Initializer lists	586
17.8 Exercises	590
Bibliography	593

In Chapter 12 we discussed polymorphism, a key ingredient to a style of programming called object-oriented programming. In this chapter we will explore another style of programming called *generic programming*. In generic programming, functions or classes that work with different data types are created by leveraging the templating mechanism provided by the C++ programming language. We also introduce the Standard C++ Library, which provides a set of templated classes and utility functions useful in all types of programming, including scientific programming. Finally, we discuss some new features and functionality introduced in the most recent C++ standards.

17.1 Generic type parameters

Functions and classes are the two main building blocks of any program. These are built upon constants and variables having specific datatypes. It happens frequently that some of the classes and functions we write repeatedly are nearly identical, differing only by the datatypes on which they are built, or upon which they act. For this reason C++ introduces two mechanisms, *function templates* and *class templates* that are built from `generic type parameters`. A function template is not really a function, but a recipe for generating a function once the generic type parameter(s) are specified. Likewise, a class template is a recipe for generating a C++ class, given the generic type parameter.

Class templates are already familiar. The class `std::vector` is our beloved collection class that allows us to put almost anything into a vector. The type of object to be collected, a generic type parameter, is specified in one of two ways, e.g.

```
#include <vector>
  ...
  std::vector<std::string> vectorOfStrings;
      // declare an object
```

or

```
#include <vector>
  typedef std::vector<std::string> VectorOfStrings;
      // declare a data type
  VectorOfStrings vectorOfStrings;
      // declare an object
```

`std::vector` is a class template, while the entity `std::vector<std::string>` is called a *parameterized class*, whose generic type parameter is `std::string`. The `std::string` class itself is a parameterized type; it is defined as

```
typedef basic_string<char> string
```

The `string` class stores alphanumeric characters as `char` data.

Function templates are less familiar. An example is the function `std::min`, which one may invoke like this:

```
#include <algorithm>
...
  double x=std::min<double>(2.5, 3.8);
```

As might be guessed, this returns the value 2.5. A different result is returned if we write

```
#include <algorithm>
...
  double x=std::min<int>(3.5, 3.8);
```

Can you guess what it is? If not, give it a try. Also, note that we may omit mention of the generic type parameter in the function call:

```
#include <algorithm>
...
  double x=std::min(3.5, 3.8);
```

The compiler is able to guess that the generic type parameter is double. When the generic type parameter cannot be determined, don't worry—you can count on the compiler to complain about it.

17.2 Function templates

Function templates allow us to build functions that expect different kinds of arguments. Below, for example, is a function that can be used for ordering many things, including float, int, double, but also std::string; in the latter case the comparison is based upon lexicographical order of the two strings.

```
template <typename T> bool comparesGreater(T a, T b) {
  if (a > b)
    return true;
  return false;
}
```

We have defined the datatype of the input variables in an abstract way, as *generic types* using the keywords typename T[1]. The compiler will generate a parameterized function according to the datatypes appearing in the function call.

```
  std::string name1="Philemon", name2="Leontine";
  std::cout << comparesGreater(3.8,    3.4) << std::endl;
     // --> true
  std::cout << comparesGreater(name1,name2) << std::endl;
     // --> true
  std::cout << comparesGreater('c',    'f') << std::endl;
     // --> false
```

---

[1] <class T> appears in older code and you are likely to encounter it. It is still legal but typename is preferred in modern C++.

When compiling the above code, the compiler first searches for a non-templated function matching the signature of the function call; if it fails, it attempts to deduce the template parameters from the arguments to the function.

Not all datatypes can be used with this function. The greater-than operator (>) required by our function is not defined for `std::complex<double>`, so ***template unification*** fails for the following lines

```
std::complex<double> c1 = -1; // -1 + 0i
std::complex<double> c2(1,2); // 1 + 2i
std::cout << "a > b: " << comparesGreater( c1, c2 )
   << std::endl; // --> compilation error!!
```

and the compiler generates an error. One can fix the error by defining the missing operation, but in some cases, including this one, it may be better to decide if the comparison operation really makes sense.

In the previous example, template parameters were used only as input arguments; but a function may also have a generic return type. In the example below, we define a function in which the two input parameters and the return value share the same generic type T:

```
template <typename T> T getMin(T a, T b) {
  if (a < b)
    return a;
  return b;
}
std::cout << "getMin(a,b): " << getMin( 2.8, 1.5 )
   << std::endl; // --> 1.5
std::cout << "getMin(a,b): " << getMin( 2, 7 )
   << std::endl;      // --> 2
```

The parameters a and b, plus the return value, will also be of the same type in the parameterized function generated by the compiler. The following code:

```
std::cout << "getMin(a,b): " << getMin( 2, 7.2 )
   << std::endl; // --> ERROR!
```

generates a compilation error, because argument deduction fails for a signature of type (int, float). Automatic type conversion is not applied for template functions. On the other hand, more than one template parameter may be used if needed:

```
template <typename Ta, typename Tb> Ta getMin(Ta a, Tb b) {
  if (a < b)
    return a;
  return b;
```

```
}
std::cout << "getMin(a,b): " << getMin( 2, 7.2 )
    << std::endl; // --> 2
```

In this case, argument deduction succeeds and the code compiles successfully.

Function overloading, wherein two functions with the same name are allowed and considered to be distinct by the compiler, is also allowed for function templates. More than one function may match the signature of the function call, in which case the compiler does its best to find the best match by applying *overload resolution rules*, generating an error in case it does not succeed. See

http://en.cppreference.com/w/cpp/language/overload_resolution

for more details.

17.3 Class templates

The second type of template is the class template, which allows the programmer to parameterize a class on a generic type. The templates `std::vector` and `std::complex` are two familiar examples. The class `std::ostream`, moreover, is an example of a parameterized class, of which `std::cout` is an instance. In this section we demonstrate how to write a class template. Our class template, which we call `Pair`, will handle pairs of values, exposing methods to calculate the minimum and the maximum, plus methods to insert and pop values.

To declare a class template, as with a function template, we use a template parameter which acts as a placeholder for the template argument. The `typename` keyword defines the generic type parameter `T`, which is used as any other type in the declaration of member data and member functions, as well as any local variables required in the implementation of the latter. The declaration of the class template looks like this:

```
template <typename T> class Pair {
  private:
    T items[2];      // array of two items

  public:
    Pair();                // constructor
    setA( const T& );  // set first element
    setB( const T& );  // set second element
    T getA();              // get first element
    T getB();              // get second element
    T getMin();            // get minimum
    T getMax();            // get maximum
};
```

We have declared two methods that accept one argument to set the two values of the pair. The definition of the member functions has a somewhat special syntax:

```
template <typename T> T Pair<T>::getMin() {
  if (items[0] < items[1])
    return items[0];
  return items[1];
}
```

The type of our class is `Pair<T>`, which has to be used in defining the scope of member functions that are template functions. The other member functions are defined in a similar way.

The class we defined above can now be used to instantiate a pair of values of any type, provided they have the same type. However, when an object of a parameterized class is declared the compiler cannot deduce the template argument, which therefore has to be specified explicitly:

```
Pair<int> iPair;              // pair of int
Pair<double> dPair;           // pair of double
Pair<std::string> sPair;      // pair of std::string
```

The parameterized class (e.g. `Pair<double>`) can now be used like any other class.

It may be convenient to employ `typedef` to avoid writing long lists of template parameters. The keyword `typedef` defines an alias for the actual type as shown in the following example.

```
// define the alias 'IntPair' as a 'Pair<int>'
typedef Pair<int> IntPair;

// a Pair<int>
IntPair ip;

// alternative definitions of functions accepting two pairs
   of int values
void foo( IntPair a, IntPair b );
void bar( Pair<int> a, Pair<int> b );
```

An example which is likely familiar comes from the `Eigen` library. The datatype `Vector3d` (in the namespace `Eigen`) is a typedef, defined as `Matrix<double, 3, 1>`. Several other typedefs are defined by the library for convenience, and others can be added by programmers to simplify a particular project or use case.

17.3.1 Class template specialization

Class templates can be specialized to optimize the implementation for certain datatypes or to address issues. Imagine, for example a class template designed to represent a square matrix, with a method to return the Hermitian conjugate. If the matrix is parameterized on `doubles` rather than `std::complex` datatypes, a special implementation could be desirable to dispense with complex conjugation and simply perform the transpose. Whereas a parameterized class is derived from a class template with a standard recipe, a *class template specialization* provides an exception from the recipe.

To specialize a templated class, one has to specialize all its templated members. The keyword `template<>` is used to declare a class template specialization. The empty parameter list tells the compiler that we are specializing the class for a certain type. For example, if we want to specialize our class for `std::string`, we can add the class below to our program. The compiler will choose the right candidate based on the type used in the call, with priority given to any existing specialized class.

```
template<> class Pair<> {
  private:
    std::string items[2];        // array of two items

  public:
    Pair();                      // constructor
    setA( const std::string& );  // set first element
    setB( const std::string& );  // set second element
    std::string getA();          // get first element
    std::string getB();          // get second element
    std::string getMin();        // get minimum
    std::string getMax();        // get maximum
};
```

In case the class template takes multiple parameters, as in this variant:

```
template <typename Ta, typename Tb> class Pair {
  private:
    Ta valueA;
    Tb valueB;

  public:
    Pair();                 // constructor
    set( const Ta&, const Tb& )
      ...
};
```

the template can be partially specialized, such that only a given subset of the arguments are specialized, while the others are left as template parameters. In the code below we partially specialize our modified class for different types: first, for a generic type plus an `int`, and second, for pointers to two generic types:

```
template <typename T> class Pair<T, int> {
  private:
    T valueA;
    int valueB;
    ...
};
template <typename Ta, typename Tb> class Pair<Ta*, Tb*> {
  private:
    Ta* valueA;
    Tb* valueB;
    ...
};
```

The compiler will choose the right candidate at compile-time based on the signature of the function call. If there is no clear best candidate, the compiler will generate an *ambiguous declaration* error. The programmer then has to resolve the ambiguity.

17.4 Default template arguments

In some cases, it can be useful to define *default template arguments*:

```
template <typename Ta, typename Tb, typename RETURN = Ta>
   class Pair {
  private:
    Ta valueA;
    int valueB;
  public:
    RETURN getMin();
    ...
};
```

Here we have defined a class in which the return type of the member function `getMin` is assigned to a third template parameter; and, if not specified, its default argument is the first parameter Ta. Here we use as default value another template parameter; but we could have used another value as well, such as `int`, `double`, `std::complex<double>`.

17.5 Non-type template parameters

It is possible to incorporate *non-type parameters* (generally variables of integer type) in class definitions.

For example, let us extend our Pair class into a more generic Sequence class that can handle an array of items and that offers methods to get the minimum and maximum values.

```
template <typename T, int LENGTH> class Sequence {
  private:
    T items[ LENGTH ]; // array of LENGTH items

  public:
    Sequence();        // constructor
    T getMin();        // get minimum
    T getMax();        // get maximum
};
```

Here we have used the non-type template argument int LENGTH to set the length of a variable array. The value of the parameter is specified in the object instantiation, as below:

```
Sequence<int, 10> iTenSeq;              // a Sequence of 10 int
    values
Sequence<std::string, 5> sFiveSeq; // a Sequence of 5 strings
```

The class can also be slightly modified to define a default Sequence:

```
template <typename T = int, int LENGTH = 10> class Sequence {
    ...
};
```

Thus if no LENGTH argument is specified the code will default to instantiating Sequence with an array of 10 integers:

```
Sequence defSeq;                        // a Sequence of 10 int values
    - DEFAULT
Sequence<double, 3> dThreeSeq; // a Sequence of 3 double
```

Non-type parameters may also be used for function templates as well, as in the following brief function implementing a Legendre polynomial whose order is specified at run time:

```
template <typename T, int ORDER>  T Legendre(T x) {
  T p[3]={1,0,0};
  for (int i=0;i<ORDER;i++) {
```

```
      p[2]=p[1];
      p[1]=p[0];
      p[0]=((2.0*(i+1.0)-1.0)*x*p[1]-((i+1.0)-1.0)*p[2])/
           (i+1.0);
   }
   return p[0];
}
```

The function can be defined with defaults and called with default or specified parameters:

```
template <typename T=double, int ORDER=0>  T Legendre(T x) {
   T p[3]={1,0,0};
   for (int i=0;i<ORDER;i++) {
      p[2]=p[1];
      p[1]=p[0];
      p[0]=((2.0*(i+1)-1)*x*p[1]-((i+1)-1.0)*p[2])/(i+1);
   }
   return p[0];
}
...
double p0= Legendre(x);                    //  gives P_0(x)
double p4= Legendre <double,4>(10.0);      //  gives P_4(10.0);
```

As of this writing, the C++ standard allows only integer datatypes to be used as non-type template parameters.

17.6 The standard C++ library

The standard C++ library is a powerful collection of classes and functions that provides a convenient and efficient way of accomplishing many routine programming tasks. These range from simple tasks like taking the minimum of two values, to more complicated procedures such as memory management, list ordering, and many others. It offers a range of high-performance, optimized classes including *data containers* to store and handle data objects, *iterators* to navigate through the containers, and *algorithms* to execute a wide range of operations on objects stored in containers. We discuss the standard C++ library here because much of it is based upon templates, a feature that makes the library easy to apply in many contexts. Originally, the C++ standard library was referred to as the Standard Template Library (Stepanov, 1994), or STL, a fact that underscores the importance of the templating mechanism to the library. Some people continue to confuse the names. Here, we will call it either the "standard C++ library" or the "standard library".

We have already used some of the most common objects provided by today's standard library: `std::string` and `std::vector`. The purpose of this section is to give the reader an overview of the rest of the library. The best documentation of the C++ standard library that we have found is online, in the reference section of the website

    www.cplusplus.com.

Other references can be found in the bibliography.

The design of the standard template library was the work of Alexander Stepanov of Silicon Graphics and Ming Lee of Hewlett-Packard. Their original paper (Stepanov, 1994) describes the motivation for the library in the following terms:

> The following description helps clarify the structure of the library. If software components are tabulated as a three-dimensional array, where one dimension represents different data types (e.g. int, double), the second dimension represents different containers (e.g. vector, linked-list, file), and the third dimension represents different algorithms on the containers (e.g. searching, sorting, rotation), if i,j, and k are the size of the dimensions, then i*j*k different versions of code have to be designed. By using template functions that are parameterized by a data type, we need only j*k versions. Further, by making our algorithms work on different containers, we need merely j+k versions.

Moreover, *you* do not need to write most of these, since many are already provided in the standard library. The standard library is not part of C++ itself, but is installed along with the compiler in a standard location.

17.6.1 Containers and iterators

Perhaps the easiest part of the standard library to use are the container classes, (actually class templates), into which one can insert almost any type of object and from which one can later retrieve a specified object. The containers grow and shrink as needed while handling all the required memory allocation and deallocation operations. The standard containers consist of *sequence containers* such as `std::vector` and *associative containers*, typified by the `std::map` class.

Sequence containers

A vector is a collection of items stored in a contiguous block of memory. The contiguity makes random access to the elements very efficient. `std::vector` provides methods to insert elements, to access a given element, and to get the length of the container. Any type of element can be stored in a vector, from base types to instances of custom classes. In fact, like all other containers in the standard library, `vector` is a class template that defines a container of elements of some generic type T, that is, a `vector<T>`.

The basic usage of `std::vector` was discussed in Section 3.7. Empty vectors are constructed with the default constructor. The `push_back` method adds a new element at the end of the vector and increases the size of the vector. Since the new standard

C++11, a useful initialization mechanism named "list initialization" has been introduced: a curly-braces initializer can be used to directly initialize containers with starting elements at the same time as declaring them, as below:

```
vector<double> dVec = {1.5, 2.1, 2.7, 9}; // a vector of 4
    doubles
```

An empty `std::vector` has a size of zero. If we insert an element in it with `push_back()`, its size is increased by one; when an element is removed by using the method `pop_back()`, its size is decreased by one. If a `push_back` operation cannot allocate adjacent memory the entire contents of the vector are automatically moved to newly allocated memory where all the elements can be stored contiguously. This can be costly. If performance is important one can reserve the right amount of contiguous memory slots. The method `reserve()` can be used to make or change the allocation.

```
std::vector<int> iVec;
iVec.reserve(1000);                           // memory
                                                 allocation

std::cout << iVec.size() << std::endl;       //prints 0
std::cout << iVec.capacity() << std::endl;   //prints 1000

iVec.push_back( 5 );
iVec.push_back( 50 );

std::cout << iVec.size() << std::endl;       //prints 2
std::cout << iVec.capacity() << std::endl;   //prints 1000
```

The `size()` method returns the number of elements in the `vector` while the `capacity()` method returns the number of elements that will fit into the vector before a memory reallocation becomes necessary.

Two removal operations are defined on a `vector`: `pop_back` and `erase`. The former removes only one element from the back of the `vector` while the latter erases one or more elements at an arbitrary position. This forces the container to relocate all the elements after the deleted segment to fill the gap. This is a costly operation. If frequent insert or erase operations are necessary, it may be better to use a different container such as `std::list`.

The container `std::list` implements a doubly-linked list of elements. Each element contains a payload and pointers to the next and previous elements. Insertions into, and deletions from a `list` are efficient because only pointers need to be updated. This has to be weighed against the lack of random access: the only way to access an element in the `list` is iteratively, starting either at the beginning or at the end of the `list`.

Alternatively, the complexity of an `erase` call on a vector is $O(n)$, (Josuttis, 2012); while the same call on a list has a constant complexity of $O(1)$, so that the speed of the erasing operation does not depend on the size of the container.

While vectors only allow efficient insertion and deletion from the end of the container, the standard library provides the primitive class deque, whose name stands for "double-ended queue". Beside fast insertion and deletion at *both* ends of the container, deque also offers automatic dynamical resizing of the container storage space[2]. On the contrary, deque does not offer contiguous memory storage, its typical implementation being a sequence of individually-allocated arrays.

Since the release of C++11, two new sequence containers have been introduced: array and forward_list. The former is a static contiguous array (in contrast to vector which is a dynamic contiguous array), which offers the performance and accessibility of a C-style array. The latter offers fast insertion and removal of elements from any position within the container when fast random access (vector) and bidirectional iteration (list) are not needed.

Iterators

Here is a simple piece of code that instantiates an array of five integers and prints them out:

```
int anArray[]={1, 4, 9, 16, 25};
for (int i=0;i<5;i++) cout << anArray[i] << end;
```

The loop is written in perhaps the most natural way, but not necessarily the best way. An alternative is:

```
int anArray[]={1, 4, 9, 16, 25};
for (int *a=anArray;a<anArray+5;a++) cout << *a << end;
```

This accomplishes the same thing. We can carry out the loop in a function, passing in only two pointers which one would write like this:

```
void printArray (int *begin, int *end) {
    for (int *a=begin;a<end;a++) cout << *a << end;
}
```

invoked like this:

```
int anArray[]={1, 4, 9, 16, 25};
printArray(anArray, anArray+5);
```

---

[2] For comparison, when using the erase method on elements of a vector, the elements are deleted, but the memory slots are not released, and the memory footprint of the container does not change; one has to invoke the methods resize or shrink_to_fit to resize the vector and free the unused memory. deque handles this automatically and its expansion is cheaper, because it does not involve relocating the elements of the container.

Now we make a simple modification with very important consequences. We rewrite `printArray` as a function template:

```
template <typename T> void printArray (T begin, T end) {
  for (int *a=begin;a<end;a++) cout << *a << end;
}
```

This requires no change to the client code, but because it is a template function, it is much more general. It is able to print out `int`, `double`, `std::complex<double>`, and user-defined types from an array of these datatypes. For template unification to succeed, the operations `operator *`, and `operator++` must be defined for type `T`. Also, in this example, `cout « (*T)` must be defined.

A *smart pointer* is a C++ object that responds to the dereferencing operation, i.e. `operator *`, which return a reference to an actual object, but may do more, according to the objectives of the programmer. It might, for example, manage the lifetime of the object it points to. Or it might retrieve the object from a sophisticated memory management system or a database.

An *iterator* is a smart pointer that, in addition, responds to `operator ++()`. *Iterators exist for the sequential containers defined above.* The containers all have a `begin` method returning an iterator (think: pointer) to the first element of the array, and an end method returning an iterator to the end of the array. Conceptually, this is one element past the block in which the data is stored; we have remarked already, however, that the data is not always actually stored in a block of contiguous data.

The way in which our `printArray` method is invoked upon an `std::vector` is as follows:

```
  std::vector<double> vectorOfDouble={1.0, 4.0, 9.0, 16.0,
    25.0};
  printArray(vectorOfDouble.begin(), vectorOfDouble.end());
```

The return type of the `begin` and `end` methods is the same: `std::vector<double>::iterator`. It gives read/write access to the elements of the container. If the vector were declared `const` these overloaded methods return a `std::vector<double>::const_iterator`, giving only read access to the elements.

Programmers should be aware that certain operations on containers may invalidate iterators. For example, inserting an element into a vector may trigger a reallocation of memory and a relocation of its elements to the newly allocated memory. This invalidates any existing iterators previously returned from the vector. This behavior varies from container to container so check the documentation if you are in doubt.

One can now imagine many algorithms (k), written as template functions so that the (j) containers of (i) different datatypes can be written, realizing the design goals of Stepanov and Lee, as quoted in Section 17.6. Some of the many convenient routines for sorting, searching, partitioning, shuffling, etc, which make up the algorithm section of the standard library, are described in Section 17.6.2.

Associative containers

In a bygone era, a *telephone book* was used to obtain the telephone number of a person residing in a specific metropolitan area. This ingenious piece of technology consisted of hundreds of pages of bound paper with alphabetically listed names and their associated phone numbers.

The alphabetical order was important; it implied that searching for a name in the phone book was an operation of complexity $O(\log(n))$, a significant improvement over an unsorted list whose complexity is $O(n)$. One only has to imagine the amount of time required to find a name in an unsorted list containing phone numbers for all the households in a major city.

The telephone book has a number of close equivalents in the standard library called *associative containers*. The `std::map` is the closest parallel; it associates a key to a value; the key can be any datatype which is ordered by the less than (<) operator. An `std::string`, for example, is a valid key; the less-than operation orders strings lexographically. The `std::set` is just a sorted set of keys, like a list of names without phone numbers. One can only add one instance of a key to either the `map` or the `set`; any other additions are ignored. The standard library provides the `multiset` and `multimap` as well, where the notion of key uniqueness is dropped.

In the following code we use an `std::map`, to record the mass of a few particles, which can then be retrieved by name:

```
#include <map>
#include <string>
std::map<std::string,double> mass;

mass["electron"]=0.511;
mass["proton"]=938.0;
mass["neutron"]=940.0;
mass["W"]=80.4E3;
mass["Z"]=91.2E3;
mass["H"]=125.E3;

std::cout << "size: " << mass.size() << std::endl;
    //prints 6
std::cout << "mass of proton: " << mass["proton"]
    << std::endl;   //prints 938
std::cout << "mass of neutron: " << mass["neutron"]
    << std::endl;   //prints 940
```

To illustrate the list of the `std::set`, we create two empty sets, one for *leptons*, and one for *hadrons*; we fill these two sets with the names of a few of the particles that belonging to each category, and then access the particles in each set:

```
#include <set>
  std::set<std::string> leptons, hadrons;
  leptons.insert("electron");
  leptons.insert("electron neutrino");
  leptons.insert("muon");
  leptons.insert("muon neutrino");

  hadrons.insert("pi plus");
  hadrons.insert("pi minus");
  hadrons.insert("pi zero");
  hadrons.insert("proton");
  hadrons.insert("neutron");
  hadrons.insert("antiproton");
  hadrons.insert("antineutron");

  std::set<std::string>::iterator it;
  for (it=leptons.begin();it!=leptons.end();it++) {
    std::cout << *it << " is a lepton" << std::endl;
  }

  for (it=hadrons.begin();it!=hadrons.end();it++) {
    std::cout << *it << " is a hadron" << std::endl;
  }
```

Container adaptors

The standard library also provides so-called *container adaptors*, which implement various commonly used interfaces for sequential containers: queue, stack and priority_queue. Primitive data structures are not part of these adaptors, which merely provide different interfaces to the previously discussed sequence containers.

The class templates for those adaptors have two template parameters: one represents the type of the variables stored in the container; the second one is the primitive container that underlies the implementation.

The queue container implements the logic of a FIFO ("first in, first out") list. Values are inserted into the container from one end and retrieved from the opposite end. A FIFO queue is a very common data structure, used in many patterns. The generation of classical orthogonal polynomials using three-term recurrence relations, for example, can be efficiently implemented with a FIFO. By default, the queue is implemented using the deque as an underlying container, though a different container type may be specified as a template parameter.

The stack interface, instead, implements LIFO ("last in, first out") logic, which is the logic of a stack: elements are removed from the top of the stack in the reverse order they have been inserted, as in a stack of books. Like the queue, the stack is implemented using the deque as the underlying container.

The `priority_queue` is a container providing access to sorted elements. The elements may be inserted in any order, and then retrieved the "largest" (by default) of these values at any time. `priority_queue` provides constant time lookup of the largest element, at the expense of logarithmic insertion and extraction. Priority queues work like a heap, which in turn is basically a reversed array. They are implemented, by default, in terms of the `vector` class template. In the following example we use `priority_queue` to put a set of integers in decreasing, then in increasing, order:

```cpp
#include <queue>

std::priority_queue< int > q1;
for(int n : {2,9,6,5,4,3,0,8,7,1})
    q1.push(n);

std::priority_queue< int, std::vector<int>,
    std::greater<int> > q2;
for(int n : {1,5,8,9,3,0,2,6,7,4})
    q2.push(n);

while ( ! q1.empty() ) {
    std::cout << q1.top() << " ";   // get the top element
    q1.pop();                        // delete the top element
}
std::cout << std::endl;

while ( ! q2.empty() ) {
    std::cout << q2.top() << " ";   // get the top element
    q2.pop();                        // delete the top element
}
std::cout << std::endl;
```

In the above code, we declare two `priority_queues` to store a collection of `ints`. For the first one we only declare the type of the collection items `int`; while for the second we also define the two optional template parameters, the first of which declares the container type to be `std::vector<int>`, the second of which declares the class which is used to set the ordering of `ints`. The method `top` retrieves the top element, without modifying the priority queue. The method `pop` removes the top element from the queue. This will make the container push another element to the top: the largest value for the first queue and the smallest for the second one. The output will look like this:

```
9 8 7 6 5 4 3 2 1 0
0 1 2 3 4 5 6 7 8 9
```

The class declaration illustrates how the template mechanism is used in a real example from the standard library:

```
template<
    class T,
    class Container = std::vector<T>,
    class Compare = std::less<typename Container::value_type>
> class priority_queue;
```

As we can see, the class template is declared with three template parameters[3]: T, which specifies the type of the items stored in the container; Container, which specifies the container type. The third template parameter is a comparison class used to order the elements in the priority queue; while constructing the second queue, we have set this parameter to the class std::greater<int>. The default behaviour of the priority queue, obtained with the default comparison class std::less<int>, is to sort from the largest to the smallest.

There is another container-like templated class provided by the standard library, which is the commonly-used std::string. Since many examples can be found in other parts of the book, we will not repeat them here.

More on iterators

All of the standard containers support iterators and have common methods like begin and end that return iterators. Moreover, the algorithms implemented in the standard library, which will be discussed in the next section, use iterators as input arguments (like the example printArray function discussed in Section 17.6.1).

Iterators are grouped in five categories according to the functionalities they offer: Input and Output iterators are the most limited in that they only perform uni-directional sequential input or output operations on the container; the Forward iterator, which has both input and output functionalities but can incremented only in one direction; the Bidirectional iterator, which acts forward and backward; and the RandomAccess iterator, which, in addition, can access elements non-sequentially by applying an offset to the current pointer, like an ordinary pointer.

Not all containers offer all the types of iterators. For example, vector offers a RandomAccess iterator, which is not provided by map.

All types of iterators can be "dereferenced" with the dereference operator * to access the element to which they currently point[4], just as with a ordinary pointer. They can also be incremented to point to the next element using the increment operator ++. All iterators can be compared using the comparison operators == and !=. Other types of operations, like decrementing the iterator with the operator--, can be applied only to the Bidirectional and RandomAccess iterator types.

The definitions of iterators are contained within the header file of the container with which they are associated. Let us now turn our attention to the vector iterator.

---

[3] We remind the reader that the usage of the keyword class T is equivalent to the keyword typename T, even if the latter is preferred nowadays because it is less ambiguous.

[4] The Input iterator can be dereferenced only as *r-value* (that is, a *right value*), to be used in reading operations, like in var = (*iter); while the Output iterator can be only dereferenced as *l-value* (*left value*), to be used in assignment (writing) operations, like in (*iter) = var.

```cpp
#include <vector>                              // the iterator is
    defined here
std::vector<int> vec;

vec.push_back(1);
vec.push_back(2);
vec.push_back(3);

std::vector<int>::iterator vecIt;              // a RandomAccess
    iterator

// forward loop through '++' increment
for ( vecIt = vec.begin(); vecIt != vec.end(); vecIt++) {
  std::cout << (*vecIt) << std::endl; // read the element
      value (r-value)
}

std::vector<int>::reverse_iterator rVecIt; // a reversed
    RandomAccess iterator

// backward loop through 'reversed' iterator
for ( rVecIt = vec.rbegin(); rVecIt != vec.rend();
    rVecIt++) {
std::cout << (*rVecIt) << std::endl; // read the element
    value (r-value)
}

// write to vector
for ( vecIt = vec.begin(); vecIt != vec.end(); vecIt++) {
  (*vecIt) = 10;  // write to the vector element (l-value)
    std::cout << (*vecIt) << std::endl;
}
```

In the code above, a vector is declared and populated, and then a vector iterator is obtained. By default a vector returns a `RandomAccess` iterator[5], which can be incremented, decremented, read, and assigned.

The first loop iterates over all elements from `begin` to `end`, which are functions that return pointers to the initial and final plus one elements. The iterator is dereferenced in the loop body to read the relevant element.

---

[5] You can always check the default type of iterator returned by the `iterator` method by looking at the code of the container header in the on-line documentation of the C++ classes.

The vector class offers a `reverse_iterator` as well, which is designed to loop over the container in reverse order, as in the second loop in the code. This loop makes use of the `rbegin` and `rend` functions, which return the reversed start and end points.

Finally, the third loop uses the *l-value* dereferencing offered by the `RandomAccess iterator` to assign a value to the vector's elements.

By default the map container uses a mutable `BidirectionalIterator`, as shown in the following example.

```
#include <map>
std::map< std::string, int> iMap;
std::map< std::string, int >::const_iterator mapIt;

iMap["A"] = 5;
iMap["B"] = 50;

for ( mapIt = iMap.begin(); mapIt != iMap.end(); ++mapIt ) {
  std::cout << "key: " << (*mapIt).first << " - value: "
       << (*mapIt).second << std::endl;
}
```

As we have seen, the map contained stores pairs of values: the first item is usually called the *key* and the second one the *value*. In the code above we first declare a map which handles a string key and an integer value and we populate it with a couple of example values. In this case we do not plan to use the iterator to modify the elements stored in the container, so a `const_iterator` is requested. This is an example of applying the ***principle of least privilege*** to avoid unintended misbehavior.

While the `for` loop for the map looks the same as that of the vector example, access to the elements is slightly different, namely, because a map stores pairs of values, the map iterator has two different handles to access them: `first` and `second`.

17.6.2 Algorithms

The standard C++ library offers a set of algorithms ready to be used for different kinds of operations on containers. They are not part of the containers themselves but, leveraging the generality introduced with the template mechanism, they can be used on containers through their iterators. Many of them can be even used on different containers. Here we give a short overview only, referring the reader to the bibliography for more information.

Some of the most useful algorithms are those that sort elements and those that search for a specific item in the container. Let us take a look:

```
#include <algorithm>

std::vector<int> intVec;
std::vector<int>::iterator it;
```

```
intVec.push_back(3);
intVec.push_back(2);
intVec.push_back(5);
intVec.push_back(4);
intVec.push_back(1);

// sort the elements
std::sort( intVec.begin(), intVec.end() );

// print all container's elements
for ( std::vector<int>::iterator vecIt = intVec.begin();
    vecIt != intVec.end(); ++vecIt )
  std::cout << (*vecIt) << std::endl;

// find the minimum value
it = std::min_element( intVec.begin(), intVec.end() );
std::cout << "min: " << *it << std::endl;

// find the maximum value
it = std::max_element( intVec.begin(), intVec.end() );
std::cout << "max: " << *it << std::endl;

// find the element whose value is 2
it = std::find( intVec.begin(), intVec.end(), 2 );
std::cout << "item: " << *it << std::endl;

// declare a new empty vector
std::vector<int> intVec_copy;
// resize its capacity to be able to store the element from
   the first vector
intVec_copy.resize( intVec.size() );

std::copy( intVec.begin(), intVec.end(),
    intVec_copy.begin() );
// print all container's elements
for ( std::vector<int>::iterator vecIt = intVec_copy.begin();
     vecIt != intVec_copy.end(); ++vecIt )
  std::cout << (*vecIt) << std::endl;
```

In the code above we make use of some of the standard algorithms. Their usage is self-explanatory, but there are a few points to which we can draw attention. The header <algorithms> contains all the standard algorithms. Most of the algorithms return iterators, thus we have declared a vector iterator at the beginning. The algorithm min_element takes two iterators and looks for the minimum among the elements contained between them. The analogue is provided by `max_element`. To search for a specific element within a container we can use the algorithm find, which accepts three arguments: two iterators to set the input range and the item to search for; in this case the algorithm searches for the integer equal to two. The `copy` algorithm is used to copy all the elements from one container to another; before copying the elements, we make use of the vector's `resize` function to increase its capacity. Then, three arguments are passed to `copy`: again, two iterators that define the set of items to be copied, and a third iterator that sets the position of the first element of the destination vector which accepts the copied elements.

The standard library offers many other algorithms and utility functions. For example, a collection of numerical algorithms are defined in the `numeric` header. Among these are `accumulate`, which can be used for performing iterator-based summation, as follows:

```
#include <numeric>
...
std::cout << "sum: " << std::accumulate( intVec_copy.
   begin(), intVec_copy.end(), 0 ) << std::endl;
   // prints '15'
std::cout << "sum(-10): " << std::accumulate( intVec_copy.
   begin(), intVec_copy.end(), -10 ) << std::endl;
   // prints '5'
```

As can be seen, the `accumulate` function used above accepts three arguments: two iterators to define the set of input elements, and a third argument which sets the base value used in the computation of the sum.

One final remark on algorithms: those defined as general algorithms are sometimes less efficient than those that are container-specific. For example, if a search operation is made on a `set` using the generic `std::find`, it will be slower than the same search made with the container-specific `set::find` method. In fact, `std::find` performs a linear search, with an execution time of $O(N)$, while the execution time of `set::find` goes as $O(\log N)$. We suggest that the user check the standard library documentation when choosing a container or an algorithm, particularly in time-critical operations.

17.7 Modern C++

So far, we have mostly used older C++ syntax because it is universally accepted and because it is often encountered in legacy code. In this section we will present some of the new features introduced by the most recent revisions of the C++ standard: C++11 and C++14[6]. According to Bjarne Stroustrup, the creator of C++, "*C++11 feels like a new language*". Presenting all these features would require an entire book—you will find

---

[6] At the time of writing, proposals have been presented for C++17.

some of these in the bibliography. Here we will discuss the features you are most likely to encounter.

17.7.1 Variadic templates

C++11 introduced functions that accept a variable number of template parameters. These are resolved by the compiler at compile time to assure their type-safety. The following excerpt shows a basic example of a function template that uses the new mechanism:

```
// base version, with single argument
template <typename T> T add(T a) {
  return a;
}
// The recursive version, with multiple arguments
template <typename T, typename... ARGS> T add(T a, ARGS...
    args) {
  return a + add(args...);
}
```

Two function templates are defined in the code above: one that handles a single template parameter, and a second that handles a variable number of parameters. In fact, variadic templates are written like recursive functions: one writes a base version, which will be used by the recursive general version of the same template. The parameter list contains the construct "typename... ARGS", which is called a *template parameter pack*, while the construct "ARGS... args" in the function signature is called a *function parameter pack*. As for the parameter T, any name can be chosen for the variadic parameter ARGS. The use of the variadic templated function is demonstrated in the next example:

```
// sum of 5 integers
int iSum = add(1, 2, 3, 4, 5);

// sum of 3 doubles
double dSum = add(1.5, 2.3, 3.5);

// sum of 2 std::string
std::string a = "c++";
std::string b = "11";
std::string c = add(a, b);

// output
std::cout << "out: " << iSum << ", " << dSum << ", " << c <<
    std::endl;
```

with output:

```
out: 15, 7.3, c++11
```

A call to the variadic function template with N arguments triggers the recursive version of the template definition. In this version, $N-1$ arguments are used to build a further recursive call. At each level of recursion the argument list is reduced until it finally consists of a single element, at which point the base version is called. The summation occurs one step at a time as the recursion unwinds.

The mechanism by which all this happens can be exposed by using the __PRETTY_FUNCTION__ variable provided by gcc on Linux/Mac platforms, which holds the name and the signature of the function in which these variables appear. We modify our functions to print out this information:

```
// base version, with single argument
template <typename T> T add(T a) {
  std::cout << __PRETTY_FUNCTION__ << "\n";
  return a;
}
// The recursive version, with multiple arguments
template <typename T, typename... ARGS> T add(T a, ARGS...
    args) {
  std::cout << __PRETTY_FUNCTION__ << "\n";
  return a + add(args...);
}
```

This allows us to observe all the function calls that have been created recusively by the compiler for the example above:

```
T add(T, ARGS...) [T = int, ARGS = <int, int, int, int>]
T add(T, ARGS...) [T = int, ARGS = <int, int, int>]
T add(T, ARGS...) [T = int, ARGS = <int, int>]
T add(T, ARGS...) [T = int, ARGS = <int>]
T add(T) [T = int]

T add(T, ARGS...) [T = double, ARGS = <double, double>]
T add(T, ARGS...) [T = double, ARGS = <double>]
T add(T) [T = double]

T add(T, ARGS...) [T = std::__1::basic_string<char>, ARGS =
    <std::__1::basic_string<char>>]
T add(T) [T = std::__1::basic_string<char>]
out: 15, 7.3, c++11
```

17.7.2 Auto

C++ is a statically-typed language, i.e., the type of each variable must be defined statically; and the compiler assumes the type of a variable never changes throughout the variable lifetime[7]. Thus, if we require an integer and a string variable, their types must be specified at declaration:

```
int iVar = 1;
std::string sVar = get_string();
```

With the introduction of the auto keyword in C++11 it is possible to let the compiler deduce type, as with template parameters. Of course, one cannot declare a variable with auto without any initialization because the compiler would not be able to deduce its type:

```
auto iVar = 1;              // ok
auto sVar = get_string();   // ok
auto kVar;                  // compilation error!
```

With C++11 the keyword auto could not be used to define the return type of functions. However, the new C++14 standard allows the use of auto as a return type (for non-void functions). This is shown in the example below. The point, of course, is that the compiler is able determine the appropriate return type based on information in the function body.

```
auto get_string() {
    ...
    std::string myStr;
    return myStr;
}
```

The auto keyword was introduced to permit cleaner and less error-prone code. It is particularly useful when handling iterators, which can be complicated objects. Consider this example:

```
std::map<std::string, std::vector<double>> stringToVDouble;
for (std::map<std::string, std::vector<double>>::
        const_iterator i=
            stringToVDouble.begin(); i<stringToVDouble.end();
                i++) {
```

---

[7] This is not true in dynamically-typed languages like Python, where the type of some variables can be changed during their lifetime.

```
      ...
    }
```

Using `auto` this rather obtuse construct can be written as:

```
    for (auto i=stringToVDouble.begin(); i<stringToVDouble.
         end();i++) {
      ...
    }
```

which is more economical and readable, and less likely to be mistyped or cause compilation errors.

17.7.3 Smart pointers

We have used the operator `new` to create a new object in the heap and to return a pointer to the newborn object. We have also seen how to delete allocated memory with the operator `delete`. And we have stressed that it is the programmer's responsibility to allocate and free memory in such a way that memory leaks do not occur.

With C++11 a new set of *smart pointers* have been introduced which relieve the programmer of some of this responsibility. These pointers also feature different ways of exposing the ownership of the allocated memory. Examples in old and new versions of C++ follow:

```
#include <memory>   // for smart pointers

// C++98
void useRawPointer() {
  Widget* widget = new Widget();
  // ... make something with the pointer
  delete widget; // do not forget to delete! if not, a memory
      leak occurs!
}

// C++11
void useSmartPointer() {
  std::unique_ptr<Widget> widget( new Widget() );
  // ... make something with the pointer
} // the pointer is automatically deleted here!

// C++11, with 'auto'
void useSmartPointerAuto() {
```

```
    auto widget = std::make_shared<Widget>(); // 'auto'
        declaration
} // automatically deleted
```

In the code above we have used the declaration `std::unique_ptr` to define a pointer that owns and manages a single object; when the pointer goes out-of-scope (as at the end of our function), the object is automatically destroyed and its memory is freed. There is another smart pointer, `std::shared_ptr`, which retains shared ownership of the relevant object. Different `shared_ptrs` can manage the same object. In this case the object is destroyed and its memory deallocated when the last `shared_ptr` pointing to it is destroyed or assigned to another object.

In the final example we have used the `auto` keyword, together with the `make_shared` templated function to create a shared object efficiently. There are other types of pointers that offer different levels of ownership and that can be used in different use-cases. These can be found in the C++ documentation on-line or in the references suggested at the end of the chapter.

17.7.4 Range-based for loop

The new C++ standard has introduced a *range-based for loop*. This kind of syntax is called a `foreach` loop in other languages, like Python or Java. Until C++11 the `foreach` loop was only available through external C++ libraries, like Boost or Qt.

The new loop declaration can be used, for example, to iterate over all items in a container or in an array, as below:

```
std::vector<int> iVec;
iVec.push_back(5);
iVec.push_back(11);

// C++98
for( vector<int>::iterator it = iVec.begin(); it
    != iVec.end(); ++it ) {
    std::cout << *it << std::endl;
}

// C++11
for( auto item : iVec ) {
    cout << item << endl;
}

// C++11 on arrays
int iArr[] = {1,-3,2,4,10};
```

```
for (int i : iArr)
{
  cout << i << endl;
}
```

It is worth noticing that the variable in the new `foreach` loop does not have to be dereferenced to access the underlying item, as it is for the iterator in the first example. The new syntax leads to code that is more concise, readable, and easier to understand.

17.7.5 Nullptr

The `nullptr` keyword is now the preferred way to instantiate a null pointer.

```
// C++98
int* iPointer1 = 0;
int* iPointer2 = NULL;

// C++11
int* iPointer3 = nullptr;
```

17.7.6 Iterators: nonmember begin and end

The new C++ standards have introduced non-member `begin(v)` and `end(v)` iterators. They are now preferred compared to the member methods offered by a container, `v.begin()` and `v.end()`, because they can be used with different containers or with classes other than those in the standard library, like arrays, or with user-defined classes, provided they offer a `begin` and an `end` method.

```
std::vector<int> iVec;
int iArr[10];

// C++98
sort( iVec.begin(), iVec.end() ); // container member methods
sort( &iArr[0], &iArr[0] + sizeof(iArr)/sizeof(iArr[0]) );
    // array does not offer 'begin' and 'end' methods

// C++11
sort( begin(iVec), end(iVec) ); // container non-member
    methods
sort( begin(iArr), end(iArr) ); // the new methods handle
    arrays as well
```

17.7.7 Lambda functions and algorithms

We have seen how to define functions to perform actions and to encapsulate functionalities. Functions must be declared with a function name and its body is normally defined in a separate piece of code. Also, a function cannot be declared within another function or within the `main()` function of the program.

Certain algorithms in the Standard Library require a function to modify their default behavior, like `sort`. Unfortunately, defining this additional functionality can lead to code cluttered with a plethora of tiny functions.

To deal with this problem, the C++11 standard introduces *anonymous functions*, often called called *lambda expressions* or, simply, *lambdas*. Lambda functions are a convenient way to encapsulate small functions and to define them exactly at the place where they are used. This is especially useful when they have to be used by other functions or algorithms. An example with the old and new function paradigms follows.

```
// C++98

// defining a custom sorting function outside 'main()'
bool absReverse(int a, int b) {
  return (std::abs(a) > std::abs(b));
}

int main() {
  ...
  sort( begin(iVec), end(iVec), absReverse );
  ...
}
```

```
// C++11
int main() {
  ...
  // defining the custom sorting function at the place where
     it is used
  sort( begin(iVec), end(iVec), [](int a, int b) { return (
     std::abs(a) > std::abs(b)); } );
  ...
}
```

In the code above we have defined a custom sorting function for the `sort` algorithm. In C++98 we are forced to define a whole function outside the main function of the program and then use the function name in the sort call. While this is fine if we use the absReverse function in more than one place, it is redundant if the function is only needed for one sorting operation.

C++11 lets one define the sorting function as a lambda expression exactly at the place where it is used, within the sort call. The symbol [] is called the *lambda-introducer*, or *capture clause*: it introduces the definition of the lambda expression and it accepts capture options as we will see later. After that, the parameter list, also called the *lambda declarator*, specifies the input arguments of the functions. The body of the function follows.

The general form of a lambda expression is:

```
[] () mutable throw() -> return-type { ... }
```

Only the lambda-introducer [] and the body are needed, all the others are optional.

Lambdas are useful when combined with Standard Library utility functions. In the example below we generate vector elements with the help of the std::generate function and a lambda expression:

```
std::vector<int> iV(3); // --> 0 0 0
int value = 0;
std::generate(iV.begin(), iV.end(), [=] () mutable { return
    value++; }); // --> 0 1 2
```

The capture clause [] also defines the variables the lambda function can *capture* from the enclosing scope: an empty [] means that no variables from the enclosing scope are accessible within the lambda. Two symbols are accepted by the capture clause: = and &: [=] which means that all variables from the enclosing scope are accessible by value, while [&] means that they are accessible by reference. A list of specific parameters can be passed to the capture clause, too: for example [=, &sum] means that all variables can be accessed by value except sum which can be accessed by reference. The mutable optional keyword lets the programmer modify the variables captured by value within the lambda function. The optional throw() statement may be used to provide a list of exceptions that the function can throw, which may be left empty or, equivalently, omitted. The last optional parameter -> return-type sets the return type of the lambda function, which is by default deduced by the compiler when the return statement is used in the lambda body.

In the example below we make use of the std::for_each algorithm to compute a sum of the elements in a vector augmented by a quantity grain. The variable grain is captured by value with the usage of the global = and the mutable keyword let us increment it before adding it to the vector element in the lambda body; the variable sum is explicitly captured by reference using the & and therefore its value in the enclosing scope can be modified by the lambda:

```
int grain = 0;
int sum = 0;
for_each(iV.begin(), iV.end(), [=, &sum] (int& element)
   mutable throw() { element += ++grain;
   sum += element; } );
```

Since C++14, new variables can be declared within the lambda functions as well.

17.7.8 Initializer lists

With C++98 there was no direct way to initialize a vector or a map to a given set of values. C++11 introduced the keyword *initializer lists*, which permits the use of braces to define an initializing list.

```
// C++98
int iArr[] = { 1, 2, 3, 4 };    // arrays work

std::vector<int> iVec;
for( int i = 1; i <= 3; ++i )    // must explicitly initialize
    vector elements
  v.push_back(i);

std::map<int, std::string> sMap;
sMap[0] = "A";                    // and map elements
sMap[1] = "B";
sMap[2] = "C";

// C++11
vector<int> iVec { 1, 2, 3 };    // everything works
int iArr[] { 1, 2, 3, 4 };
std::map<int,std::string> sMap { {0,"A"}, {1,"B"}, {2,"C"} };
```

Custom classes can make use of the new initializer lists as well, provided they accept an object of the `std::initializer_list<T>` template class in their constructor.

```
#include <initializer_list>

template <class T> class MyClass {
  public:
    MyClass(std::initializer_list<T> l) : vec(l) {
        std::cout << "MyClass object constructed with a
            " << l.size() << "-element list\n";
    }
```

```
      typedef typename std::vector<T>::iterator myIterator;
      myIterator begin() {return vec.begin();}
      myIterator end() {return vec.end();}
   private:
      std::vector<T> vec;
};

   ...

MyClass<int> myC{ 10, 15, 20};
for( auto item : myC ) {
   cout << item << " ";
}
cout << endl;
```

Uniform initialization

C++11 also introduced the concept of **uniform initialization**, which expands on the braces of the initializer lists to provide a uniform type initialization that works with any object. For example, let us consider a simple structure and a simple class that holds three coordinates:

```
struct PosStruct {
   int x, y, z;
};

class PosClass {
   public:
      PosClass(int x, int y, int z): m_x(x), m_y(y), m_z(z) {};
   private:
      int m_x;
      int m_y;
      int m_z;
};
```

The struct is an **aggregate class** because it has no user-provided constructor, no private or protected members, and no virtual functions. `PosClass`, on the other hand, is an ordinary (i.e. **non-aggregate**) class. Consider the following code that creates a `PosStruct` in correct and incorrect ways:

```
// C++98
PosStruct pos1;                  // invokes default constructor
cout << pos1.x << endl;
```

```
PosStruct pos2();          // a function declaration (not
   a constructor)
cout << pos2.x << endl;    // error!
PosStruct pos4 = {2,5,11}; // invokes default constructor
```

With the old C++98 standard, an attempt to construct a PosStruct object with the standard parentheses (as in pos2()) actually declares a new function[8].

A new C++11 {} syntax has been introduced to help avoid the ambiguity. Now the default constructor can be called with curly braces, as shown in the code snippet below.

```
// C++11
PosStruct pos5;
PosStruct pos6 {};
PosStruct pos7 {2,5,11};
PosStruct pos8 = {2,5,11};
```

Also, the curly braces can be safely used to declare all types of variables in a uniform way. Thus one should compare the old standard, as shown here:

```
// --- C++98
PosStruct ps1;
PosStruct ps2 = {2,5,11};

PosClass pc1(2,5,11);
PosClass pc2 = {2,5,11};

std::vector< PosClass > vecPosClass;
vecPosClass.push_back( PosClass(1,2,3) );
vecPosClass.push_back( PosClass(4,5,6) );
vecPosClass.push_back( PosClass(7,8,9) );

// PosStruct is an aggregate class, we can initialize an
   array this way
PosStruct arrOfPosStruct[] = { {1,2,3}, {4,5,6}, {7,8,9} };

// PosClass is an ordinary non-aggregate class, we have to
   write a call to the constructor three times
PosClass arrOfPosClass[] = { PosClass(1,2,3),
   PosClass(4,5,6), PosClass(7,8,9) };
```

---

[8] Depending on the compiler and on the compilation settings, you can get a compilation warning when using empty parentheses, something like warning: empty parentheses interpreted as a function declaration. Just after the warning, some compilers print a suggestion like: note: replace parentheses with an initializer {} to declare a variable.

```
// ordinary types
int ii = 3;
int jj = 1;
std::string str("ABC");
```

to the new standard:

```
// --- C++11
PosStruct ps1{};
PosStruct ps2{2,5,11};
PosStruct ps3 = {2,5,11};

PosClass pc1{2,5,11};
PosClass pc2 = {2,5,11};

std::vector< PosClass > vecPosClass = { {1,2,3}, {4,5,6},
    {7,8,9} };

// PosStruct is an aggregate class, we can initialize an
    array this way
PosStruct arrOfPosStruct[] = { {1,2,3}, {4,5,6}, {7,8,9} };

// PosClass is an ordinary non-aggregate class, now we can
    initialize the same way as the array
PosClass arrOfPosClass[] = { {1,2,3}, {4,5,6}, {7,8,9} };

// we can use it with ordinary types as well
int ii {3};
int jj {1};
std::string str{"ABC"};
```

In particular, notice that it is possible to initialize non-aggregate classes with an iterated curly brace construct.

In-class initialization of data members

Finally, we mention that C++11 now supports in-class initialization of data members, as the following example illustrates:

```
class MyClass
{
    int ii = 7; //C++11 only
    ...
  public:
    MyClass();
};
```

There are many more features of modern C++ that we have not discussed here. The intrepid reader is invited to explore these in the references given below. New capabilities and constructs are regularly added to the C++ language—in general these are either useful or permit more elegant and error-proof code to be written. Either way, it is advisable to keep up as these features are introduced.

17.8 Exercises

1. Write a class that represents a floating point number to an arbitrary precision. Overload `operator` « so that your floating point number can be printed to an `std::ostream`. Using this class, compute the value of π using the techniques of Chapter 2. Check your answer by comparing to the built-in datatypes `float` and `double`.

2. Make a dataset of 1MB size containing the digits of a decimal representation of π. You will have one million decimal characters in this file. Then, count the number of times that each digit appears. Follow up on this by counting the frequency of occurrence of every two-digit sequence, every three-digit sequence, and every four-digit sequence. Compare these to determine whether there is any statistically significant difference in the frequency of occurrence of the possible sequences.

3. Write a container class template representing a `tree`, i.e., a data structure of nodes; each node may have children, and every node has a parent except for the "root node". The tree container should provide a way of organizing objects of a given type in a hierarchical manner. Provide a means of navigating the tree (an iterator, for example). Then test this by filling it with objects. These could include strings identifying the names of officers in your organization, particle four-momenta in a decay chain, or subroutine call graphs.

4. Indexing.
   Use variadic templates to write an indexing function that maps D integers of ranges $[0, N_i]$ to a single integer. This is useful for dealing with D-dimensional discrete spaces.

5. [P] Young Tableaux.
   The special unitary group in N dimensions, SU(N), underlies the structure of fundamental interactions and is useful in classifying hadronic matter. Irreducible representations of these groups (henceforth *irreps*) can be studied with the aid of *Young tableaux*. These are graphical representations of irreps that correspond to tensors. Tableaux (the singular is tableau) are drawn as connected boxes. For example, the fundamental rep is drawn as □. A rank n symmetric tensor is written as a row of n boxes: $S^{ijk} = \boxed{i\,j\,k}$, whereas a rank n antisymmetric tensor is a column of n boxes, eg, $A^{ij} = \boxed{\genfrac{}{}{0pt}{}{i}{j}}$. A general tableau will be of mixed symmetry. A tableau can consist of any number of boxes provided that no more than N boxes occur in any column (for

SU(N)), that the rows start at the left, and that no row is longer than a row above it. Thus the following are valid tableaux:

whereas the following are not:

The dimension of an SU(N) irrep is given by the formula:

$$D = \prod_{\text{boxes}} \frac{(N + d_{\text{box}})}{h_{\text{box}}}$$

where d_{box} is assigned as shown here:

0	1	2	3
-1	0	1	2
-2	-1	0	1

etc. The h_{box} are called the *hook lengths* and are given by one plus the number of boxes to the right of the box plus the number of boxes below the box. For example, hook lengths for  are given by:

4	2	1
1		

Thus, in SU(N) we have the following dimensions (for example):

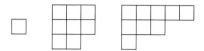

$$\square = N \qquad = \frac{1}{3}N(N+1)(N-1) \qquad = \frac{1}{80}N^2(N^2-1)(N+2)(N+3)$$

Finally, tableaux can be combined to determine irreps of products of irreps, say $T_1 \otimes T_2$. To do this:

(i) assign distinct labels to boxes in each row of T_1 as follows, eg,.

(ii) attach boxes labelled by a to T_2 in all possible ways such that no two as appear in the same column and the result is still a Young tableau.

(iii) repeat with bs, cs, etc.:

(iv) after all the boxes of T_1 have been added to T_2 form the *sequence* for each tableau by reading each row from right to left while going from top to bottom.

Eliminate tableaux with inadmissable sequences, namely, when reading the sequence (from left to right) there must be at least as many *a*s as *b*s, *b*s as *c*s, etc. at any point in the sequence. Any resulting tableaux with the same structure and sequence are copies of each other and only one tableau should be considered.

Some examples:

$$N \otimes \frac{1}{2}N(N+1) = \square \otimes \square\square = \boxed{a} \otimes \square\square = \square\square\boxed{a} \oplus \begin{array}{c}\square\square\\\boxed{a}\end{array}$$

$$= \frac{1}{6}N(N+1)(N+2) + \frac{1}{3}N(N^2-1).$$

For SU(3):

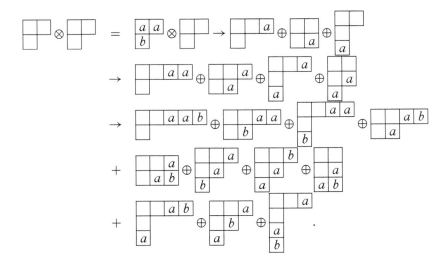

The sequences for these are *baa, aab, aab, baa, aba, aab, baa, aba, baa, aba, aab*. All are admissable except *baa* so we remove tableaux 1, 4, 7, and 9. The remaining have the following dimensions in SU(3): 27, 10, 10 (a different 10!), 8 (different from ⊞), 1, 8 (notice another way to write the singlet), null. Thus

$$8 \otimes 8 = 1 \oplus 8' \oplus 8' \oplus 10 \oplus 10' \oplus 27.$$

The two 8′ irreps have a different topology than the 8 and are distinguished by their unique sequences, *aab* and *aba*.

Example irreps are the fundamental, $F = \square$, the conjugate of the fundamental, $\bar{F}$, which is represented as a column of $N-1$ boxes, the adjoint, which is a column of $N-1$ boxes with an additional box at the top of the second column, and the singlet, which is a column of N boxes.

Write a program to compute the dimension of an irrep of SU(N), and to determine the irreps of products of irreps. Use it to:

a) Compute $F \otimes F$, $F \otimes A$, and $A \otimes A$ for SU(2). Confirm that your results agree with the spin algebra familiar from undergrad.
b) Compute $3 \otimes \bar{3}$, $3 \otimes 3$, $3 \otimes 3 \otimes 3$, $3 \otimes 6$, and $6 \otimes 6$ in SU(3).
c) Compute $F \otimes F$, $\overline{20} \otimes \bar{4}$ in SU(4). Use $\overline{20} = \square\!\square\!\square$ and $\bar{4} = \square\!\square\!\square$.

•••

BIBLIOGRAPHY

ISO/IEC 9899:TC3, *INTERNATIONAL STANDARD - Programing languages - C*. WG14/N1256, http://www.open-std.org/JTC1/SC22/WG14/www/docs/n1256.pdf.
Josuttis, N. M. (2012). *The C++ standard library: a tutorial and reference*. 2nd ed. Addison-Wesley.
Meyers, S. (2001). *Effective STL: 50 Specific Ways to Improve Your Use of the Standard Template Library*. Addison-Wesley.
Meyers, S. (2015). *Effective modern C++: 42 specific ways to improve your use of C++11 and C++14*. O'Reilly.
Stroustrup, B. (2013). *The C++ programming language*. 4th ed. Addison-Wesley.
Vandevoorde, D. and N. M. Josuttis (2003). *C++ Templates: The Complete Guide*. Addison-Wesley.

18
Many body dynamics

18.1 Introduction	594
18.2 Relationship to classical statistical mechanics	595
18.3 Noble gases	597
18.3.1 The Verlet method	598
18.3.2 Temperature selection	603
18.3.3 Observables	605
18.4 Multiscale systems	608
18.4.1 Constrained dynamics	611
18.4.2 Multiple time scales	614
18.4.3 Solvents	616
18.5 Gravitational systems	616
18.5.1 N-Body simulations of galactic structure	618
18.5.2 The Barnes-Hut algorithm	620
18.5.3 Particle-mesh methods	625
18.6 Exercises	627
Bibliography	640

18.1 Introduction

Chemists, astrophysicists, cosmologists, and plasma physicists are interested in the bulk properties of large numbers of particles. This is often called the ***N-body problem*** or ***molecular dynamics*** and is primarily treated as a problem in classical dynamics. The general approach will be direct integration of the equations of motion for all N bodies, which is referred to as a "simulation" of the system. Because the system is followed in time, molecular dynamic simulations are also useful for investigating non-equilibrium problems, which Monte Carlo computations cannot address.

The first attempt at directly computing the N-body problem was an ingenious simulation of 37 particles using an analogue computer built from lightbulbs and galvanometers (Holmberg, 1941). Computational work started in the 1960s with simulations of a few hundred particles. The effective treatment of the N-body problem is a relatively recent development in science—it was only in the 1980s that computers became powerful enough to simulate systems of useful size. At the same time efficient algorithms were developed that converted a rather onerous $O(N^2)$ problem to a more manageable $O(N \log N)$ problem.

Contemporary effort takes advantage of specialized hardware and parallel computing to simulate up to 10^6 bodies for planetary dynamics or an astounding 10^{10} bodies for cosmological computations (Springel, 2005). Alternatively, only approximately 10^4 bodies can be accommodated in computations of the evolution of globular clusters, since these require very long time scales.

In principle, the methods of Chapter 11 can be used to implement a simulation; however, the existence of many degrees of freedom necessitates employing more specialized techniques. In this chapter these techniques will be presented as a means to describe (i) properties of simple substances such as noble gases, (ii) gravitating systems, (iii) systems with complex multi-scale dynamics. Most practical large-scale problems fall into the latter category (climate modelling is a good example), which makes a general discussion difficult because useful techniques depend on the details of the system. Additional techniques discussed include the stochastic Langevin equation, particle-mesh methods, and hierarchical models.

18.2 Relationship to classical statistical mechanics

The central goal is to follow the temporal evolution of many interacting bodies by integrating the equations of motion. This is called performing a *simulation*, as we have mentioned. A good computational method will (approximately) conserve energy during this evolution and thus generates elements of a *microcanonical ensemble*. The ergodic theorem implies that averages of quantities over their (real space) trajectories are equivalent to expectation values of those quantities in microcanonical ensembles at a temperature determined by the system energy. If the temperature is kept constant by coupling the system to a heat bath then the time average is a weighted average of configurations with fixed volume and particle number with weight given by the Boltzmann factor, which yields the *canonical ensemble* average (Callaway and Rahman, 1982). Thus one has:

$$\lim_{\tau_f \to \infty} \frac{1}{\tau_f} \int_0^{\tau_f} d\tau \, \mathcal{O}(\tau) \underset{ergodic}{\to} \langle \mathcal{O} \rangle (E = E(T))_{micro} \underset{thermo}{\to} \langle \mathcal{O} \rangle (T)_{can}. \tag{18.1}$$

This equivalence establishes a connection to statistical mechanics which can inform the numerical work to follow. For example, averages over trajectories will equilibrate as the simulation proceeds, which means that the distribution of particle speeds will approach the Maxwell-Boltzmann density:

$$\rho_{MB}(v) \, dv = \left(\frac{m}{2\pi k_B T}\right)^{3/2} 4\pi v^2 \exp\left(-\frac{mv^2}{2k_B T}\right) dv. \tag{18.2}$$

Evidently this expression can be used to obtain the system temperature. We recall a few other quantities of interest that will be useful later.

The specific heat is related to fluctuations in the internal energy, which do not exist in microcanonical simulations:

$$C_V = \left(\frac{\partial E}{\partial T}\right)_{N,V} = \frac{1}{k_B T^2} \frac{\partial^2 \log Z}{\partial \beta^2} = \frac{1}{k_B T^2}\left(\langle E^2 \rangle_{NVT} - \langle E \rangle_{NVT}^2\right).$$

In this case one can obtain the specific heat from trajectory-averages of the fluctuations in the total kinetic energy via Lebowitz (1967):

$$\langle \delta K^2 \rangle = k_B T \left(1 - \frac{3N}{2C_V}\right) \langle K \rangle.$$

Recall that the equipartition theorem relates the total (nonrelativistic) kinetic energy and the temperature via (Pathria, 1984):

$$\langle K \rangle = (dN - N_c)\frac{1}{2}k_B T$$

where we consider N particles in d dimensions with N_c constraints. For example, if the system is at rest (so its center of mass velocity is zero) then $N_c = d$. Alternatively $\langle \frac{1}{2}m_i v_i^2 \rangle = \frac{d}{2}k_B T$.

The *pair correlation function* is defined by a thermodynamic average over all degrees of freedom (momentum has been integrated) except two positions (Pathria, 1984):

$$g(r, r') = \frac{V^2}{N! \, h^{3N} Z} \int_V d^3 r_3 \ldots d^3 r_N \exp[-\beta V(r, r', r_3, \ldots, r_N)].$$

In an isotropic homogeneous system this is a function of $|r - r'|$. This is a useful quantity because integrals over the correlation function give thermodynamic observables. For example, the total energy is given by Pathria (1984):

$$E = \frac{3}{2}Nk_B T \left(1 + \frac{4\pi\rho}{3k_B T} \int V(r)g(r)r^2 dr\right) \tag{18.3}$$

Similarly,

$$PV = Nk_B T\left(1 - \frac{2\pi\rho}{3k_B T}\int \frac{\partial V}{\partial r} g(r) r^3 dr\right).$$

The *virial theorem* will prove useful many times in the following discussion. The theorem applies to averages of kinetic and potential energies for systems that are in equilibrium or are stable and bound. For classical and nonrelativistic systems with two-body velocity-independent interactions one has:

$$2\langle T\rangle + \left\langle \sum_i \vec{F}_i \cdot \vec{r}_i \right\rangle = 2\langle T\rangle - \left\langle \sum_{i<j} r_{ij} \cdot \nabla V(r_{ij}) \right\rangle = 0. \tag{18.4}$$

over sufficiently long times. For a power law potential $V(r) \propto r^n$ the latter relationship gives the familiar result:

$$2\langle T\rangle = n\langle U\rangle$$

where $U = \sum_{i<j} V(r_{ij})$. Lastly, the virial theorem yields an expression for the pressure in terms of the virial:

$$PV = Nk_B T - \frac{1}{6}\left\langle \sum_{ij} \vec{r}_{ij} \cdot \nabla V(r_{ij}) \right\rangle.$$

18.3 Noble gases

Our goal will not be to construct world-class N-body simulation code as this would require highly specialized algorithmic techniques and hardware. We will instead focus on relatively simple methodologies that address most of the issues that arise in the N-body problem. Our starting point will be in the context of the simplest physical N-body system, namely the noble gases. The quantum mechanical stability of these elements permits treating the system as a collection of identical classical particles that interact via a short-ranged potential. Indeed, the equipartition theorem implies that the average kinetic energy of any molecule at room temperature is around 0.03 eV. This is much lower than typical electronic excitation energies which are of order 10 eV[1].

The interaction between noble atoms is traditionally taken to be the **Lennard-Jones potential**:

$$V(r) = 4\epsilon \left[\left(\frac{\sigma}{r}\right)^{12} - \left(\frac{\sigma}{r}\right)^6 \right]. \tag{18.5}$$

The second term represents a nonrelativistic induced dipole-dipole interaction, while the first is a convenient way to incorporate a short range repulsion. Fitting to the properties of argon yields $\sigma = 3.405\text{Å}$ and $\epsilon/k_B = 119.8$ K.

As usual, it is useful to work in appropriate units, in this case given by:

$$r_0 = \sigma = 3.4 \text{ Å}$$

and:

$$t_0 = \sigma\sqrt{m/\epsilon} = 1.9 \text{ ps}.$$

---

[1] This energy is sufficient to excite rotational modes of molecules, but, of course, we do not need to worry about that for a noble gas.

Finally, although we refer to Noble gases, we shall also have opportunity to consider phase transitions to the solid and liquid states.

18.3.1 The Verlet method

Our method, and the workhorse in the field, will be to discretize the equations of motion:

$$m_i \frac{d^2 \vec{r}_i}{dt^2} = -\nabla \sum_{j \neq i} V(r_{ij}).$$

where the index i refers to the ith particle, $r_{ij} = |\vec{r}_i - \vec{r}_j|$, and the sum is over all other particles in the system. It is important that the discretization be accurate and easy to compute since time evolution must typically be carried out for many time steps. Fortunately, Verlet (1967) constructed an algorithm that meets these criteria. The method steps forward in time according to:

$$\vec{r}(t+h) = 2\vec{r}(t) - \vec{r}(t-h) + h^2 \vec{F}(\vec{r}(t))/m + O(h^4). \tag{18.6}$$

The error is at worst $O(h^2)$ when iterated over many time steps. This simple algorithm meets our criteria because it is both reversible and symplectic (refer to Section 11.7 if you have forgotten what this means) and therefore exactly conserves a good approximation to the total energy of the system.

Notice that the initial step requires knowledge of $\vec{r}(-h)$ which is not known. Thus the algorithm really starts at $t + 2h$; we recommend setting $\vec{r}(h)$ according to:

$$\vec{r}(h) = \vec{r}(0) + h\vec{v}(0) + \frac{h^2}{2m} \vec{F}(r(0)) + O(h^3). \tag{18.7}$$

If the velocity is required during the simulation (and it almost always is) use:

$$\vec{v}(t) = \frac{\vec{r}(t+h) - \vec{r}(t-h)}{2h} + O(h^2). \tag{18.8}$$

The desirable qualities of the Verlet scheme exist in the fantasy world of perfect arithmetical computations where no round off error occurs. Variants of the Verlet scheme that are slightly more stable with respect to round-off errors are the **Verlet leap frog** algorithm (Hockney, 1970):

$$\begin{aligned} \vec{v}(t+h/2) &= \vec{v}(t-h/2) + h\vec{F}(r(t))/m \\ \vec{r}(t+h) &= \vec{r}(t) + h\vec{v}(t+h/2), \end{aligned} \tag{18.9}$$

or the *velocity Verlet* algorithm, defined by Swope (1982):

$$\vec{v}\left(t + \frac{1}{2}h\right) = \vec{v}(t) + h\vec{F}(r(t))/2m$$

$$\vec{r}(t+h) = \vec{r}(t) + h\vec{v}\left(t + \frac{1}{2}h\right)$$

$$\vec{v}(t+h) = \vec{v}'(t) + h\vec{F}(r(t+h))/2m. \qquad (18.10)$$

Both versions are useful because the velocity is manifestly available; however, the leap frog algorithm does not evaluate the position and velocity at the same time, and therefore is slightly less accurate when tracking total energy.

Pseudocode for the velocity Verlet algorithm is:

```
step
   v = v + h a/2
   r = r + h v
   a = a(r)
   v = v + h a/2
end
```

It will not have escaped the reader that the first and fourth statements in the loop do precisely the same thing and can be safely combined. The only effect of splitting these statements is in the first and last passes through the loop. The last pass is not important so this algorithm is equivalent to

```
v = v + h a/2
step
   r = r + h v
   a = a(r)
   v = v + h a
end
```

Furthermore, the velocity Verlet and standard Verlet algorithms produce the exact same sequence of positions if $\vec{r}(h)$ is chosen according to Eq. 18.7. This is also true for the leap frog algorithm if one sets $v(-h/2) = v(0) - ha(0)/2$. If the initial velocity statement is extracted from the leap frog loop this algorithm becomes equivalent to that above, and therefore to velocity Verlet. There is therefore no substantive difference at all between these variants of the Verlet method.

Coding considerations

It is clear that molecular dynamics of any type must deal with many three-vectors (or, more generally, D-vectors): position, momentum, acceleration, angular momentum, and forces all play major roles in the method. For this reason it is desirable to create a class that

Noble gases

permits simple and encapsulated manipulation of three vectors. One could for example employ `Vector3d` from the `Eigen` library or the class `ThreeVector` mentioned in Section 14.4.1.

Initial conditions

We return to the task of implementing a Verlet solution to the molecular dynamics problem.

Some care should be exercised when setting the initial configuration. The low temperature spatial configuration is a face-centered-cubic lattice, and it can be useful to chose this option[2]. Alternatively N^3 particles can be placed on a uniform grid in a box of size L^3. An initial period of thermalization will then seek out the preferred spatial configuration. In this case it is useful to offset the initial coordinates by random small amounts to help thermalize. Initial velocities can be set by drawing from a Maxwell-Boltzmann distribution. An alternative is to set the velocity of each particle to:

$$\vec{v} = \sqrt{\frac{Dk_B T}{m}}\, \hat{v} \qquad (18.11)$$

where D is the number of spatial dimensions and $\hat{v}$ is a random unit vector. Subsequent thermalization will be bring the system into a Maxwell-Boltzmann distribution.

Boundary conditions

It is traditional to implement many-body simulations with periodic boundary conditions. This is conceptually easy in systems with short ranged interactions. In this case; however, we have N_p particles interacting via, say, the Coulomb force on a torus. If one considers two particles separated by a distance r along the x axis, the closed topology implies that they will interact infinitely many times, once at a distance r, again at a distance $r+L$, and so on. In view of this, it is common to adopt the ***minimum image convention*** where one evaluates interactions between particle pairs using the minimum distance between the particles:

$$r_{ij}^{(min)} = \min_{\vec{n}} |\vec{r}_i - \vec{r}_j + L\vec{n}|.$$

An alternative is to place a system in a physical box, which requires implementing collisions with walls. Issues involved in taking this approach are explored in Ex. 4. It is sometimes convenient to implement periodic boundary conditions that correspond to other geometrical forms. See Ex. 10 for more on this.

---

[2] Although the FCC configuration is preferred in nature, a Lennard-Jones potential yields the closely related hexagonal-close-packed ground state.

Parameter selection

It is easy to break many-body code with poor parameter selection. To start, choosing the number of particles to simulate is something between you, your code, and your computer. The box size L is fixed by the desired particle density. This leaves the time step h and the total run times to fix. A prime cause of trouble is when the step size is chosen too large: a particle can suddenly find itself very near another, and the subsequent large acceleration can send the particle streaking out of the torus. It is therefore a good idea to monitor particle acceleration, especially in the code development stage. In this regard it helps to know the mean free path which is given in terms of the particle density $n = N_p/V$ and the collision cross section in D dimensions $\sigma \sim a^{D-1}$:

$$\ell = \frac{v_{rel}}{n v_{rel} \sigma} \sim \frac{V}{N_p} \frac{1}{a^{D-1}}.$$

One typically does not want a particle moving beyond the mean interaction length in a time step, so issuing a warning when $h^2 a_i > \ell$ is a good idea. In the case of argon it is traditional to choose $h = 0.005$ for "typical" run conditions (not too dense or hot).

As the time step is increased, the fluctuations in total energy become larger, leading to a drift in the energy (Hansen and McDonald, 2006). Notice that *any* finite time step leads to error in the temporal evolution of the system. These are analogous to uncertainty in the initial conditions and therefore give rise to chaotic behavior (Chapter 13). In principle this does not cause a problem as long as the dynamics is ergodic and one only asks statistical questions (i.e., one makes measurements with ensembles or long time sequences).

Thermalization should proceed for long enough that a particle can traverse the torus, thus $t_{therm} > L/v$ where v is a typical particle speed (which is set by the temperature). It is important to monitor quantities of interest to detect long term drift in observables (recall the Monte Carlo maxim: make a few long runs). This is because of the *quasi-ergodic problem* where smaller systems can become trapped in a localized region of phase space.

Finally, measurements should include enough time steps that thermodynamic averages can be evaluated to the desired level of precision. Using $t_{meas} \approx t_{therm}$ is not a bad choice. One should be aware, however, that periodic boundary conditions imply that temporal correlation functions will have spurious effects over time intervals that are larger than the "recurrence" time for phenomena of interest. Thus, for example, a disturbance that propagates through the system with a characteristic speed c will reappear at a time of order L/c and make corresponding contributions to an autocorrelation function.

Speeding things up

The chief expense in solving a molecular dynamics problem lies in computing the $N^2/2$ pair-wise forces that contribute to the acceleration of the particles, which swamps the $O(N)$ computations required in the Verlet steps. The problem becomes even worse if three-body forces are present as these represent an additional $N^3/6$ computations per time step[3]. We trust that the reader will have noted that the force calculation loops are an ideal opportunity to apply the omp parallel construct of openMP (see Section 9.3.2), or any of a host of other parallelization strategies.

If one is willing to make approximations it is possible to ameliorate the situation. The simplest step would be to neglect the interaction beyond a certain cutoff, r_c. One would still need to loop over all $N^2/2$ force pairs, but at least the computation of most of the pairwise forces would be avoided. Since the Lennard-Jones potential is an approximation to a complicated inter-atomic interaction, one could simply regard the truncated potential as a different model and declare victory. Alternatively, if one insisted on computing with the full Lennard-Jones potential then one could use relationships like Eq. 18.3 to correct for the error induced by the cutoff. Thus, for example, the internal energy can be computed as:

$$\langle U \rangle = \langle U \rangle_c + 2\pi \frac{N(N-1)}{V} \int_{r_c}^{\infty} V(r) g(r) r^2 dr$$

with the additional approximation that g is 1 for large r. Lastly, all this fancy stuff can be avoided by evaluating observables for a few values of r_c and either extrapolating or confirming that errors are within desired limits. We note that this trick comes with some cost as it introduces an infinity in the force at $r = r_c$. This can be avoided by shifting the potential so that the derivative at r_c is smooth (although, of course, higher derivatives will not be smooth).

A more elaborate way to speed up the algorithm is to maintain a table of particle pairs with separation less than some cutoff $r_n > r_c$ (Verlet, 1967). The table can be updated occasionally (clearly the update frequency will depend on the system density and the temporal step size) and has the great advantage that the force calculation is now $O(M^2)$ where M is the typical number of particles in a sphere of radius r_n. A clean way to form the neighbors list is to use the push_back member function of the vector <int> container class, which has the great advantage of not requiring one to manage indices. The following code can be used to create a neighbors list:

```
// obtain the neighbors list
vector <vector <int>> Nlist (Np);
for (int i=0;i<Np;i++) NList[i].clear();
for (int i=0;i<Np;i++) {
  for (int j=i+1;j<Np;j++) {
    if (rsqMin(i,j) < rcut*rcut) {
      NList[i].push_back(j);
      NList[j].push_back(i);
    }
  }
}
```

---

[3] Pair-wise interactions tend to be inadequate for transition metals and semiconductors (Leach, 2001).

Obtaining the forces acting on particle *i* amounts to iterating over elements of the *i*th neighbors list, as follows.

```
// obtain F
for (int i=0;i<Np;i++) {
  f[IJ(i,i)].setV(0.0,0.0,0.0);
  for (int it=0; it<NList[i].size();it++) {
    int j = NList[i][it];
    // obtain force(r(i,j)) and rhat
    f[IJ(i,j)] = force*rhat;
    f[IJ(j,i)] = (-1.0)*f[IJ(i,j)];
  }
}
```

Finally, the force f is an array of ThreeVectors that can be accessed via a superindex, or managed as the coder wishes.

More elaborate schemes for speeding up molecular dynamics computations are presented in Sections 18.5.2 and 18.5.3.

18.3.2 Temperature selection

Although we have recommended setting the initial particle speeds according to the temperature (via either Eq. 18.11 or Eq. 18.2) there will be some exchange of kinetic and potential energy as the system equilibrates; thus the temperature will drift from its initial value. This is no surprise: we are implementing a microcanonical procedure, not a canonical one. Here we discuss three techniques of varying degrees of rigor that are used to set a desired temperature.

Speed rescaling

The simplest approach is to adjust the average particle speed towards the desired temperature as the system equilibrates. Thus one executes code such as

```
double lambda = sqrt(temp0/temp());
for (int i=0;i<Np;i++) v[i] = lambda*v[i];
```

Here temp0 is the desired temperature and temp() returns the temperature of the system as determined by the equipartition theorem. This method works for the leap-frog or velocity variants of the Verlet method. For the canonical method one can reset position vectors so that the desired speed is achieved.

While simple and effective, speed rescaling induces deviations from the canonical Boltzmann distribution that are of order $1/\sqrt{N_p}$ (Nosé, 1984).

Nosé-Hoover thermometer

A method that drives the system to the desired temperature while maintaining canonical dynamics has been derived by Nosé (1984) and Hoover (1985). The idea to add a frictional force that injects energy when the temperature is below the desired point and absorbs energy when the temperature is high. Thus the dynamics is specified by:

$$m\ddot{\vec{x}} = F(x) - \xi\dot{\vec{x}}$$

where ξ is independent of particle and specifies the frictional force. In the Nosé-Hoover approach the parameter is determined by:

$$\dot{\xi} = (K - K_0)/Q$$

where K is the total kinetic energy, $K_0 = 3Nk_BT/2$ is the desired kinetic energy, and Q is a parameter that controls the rate at which the frictional force changes. Unfortunately, the Nosé-Hoover method appears to induce strange dynamics in certain conditions. For example, the temperature can start to oscillate with a magnitude that is much larger than expected (Holian, 1995).

Langevin dynamics

A physical solution to the temperature selection problem is to couple the system in question to a heat bath. If the heat bath degrees of freedom have a much shorter time scale than the ones in which we are interested, the details of the heat bath do not matter and they can be "integrated out" (see Section 8.4 for the same concept in a different context).

Paul Langevin (French physicist and anti-facist campaigner, 1872—1946) first applied these ideas to Brownian motion (Langevin, 1908). He postulated that Brownian particles experience their characteristic motion because they are suspended in a solvent and are subject to perturbations induced by short time scale molecular impacts. Langevin separated this force into an aggregate viscous drag, represented by a velocity-dependent frictional force, and a rapidly fluctuating stochastic, or noise, force. Thus the equation of motion for a particular Brownian particle of mass m is:

$$m\ddot{\vec{x}} = \vec{F}(x) - \frac{m}{\tau}\dot{\vec{x}} + \vec{\eta}(t). \tag{18.12}$$

The parameter τ is a measure of the strength of the frictional force and is the characteristic time associated with the diffusion induced in the system.

In the simplest models the stochastic force obeys $\langle \eta_i(t) \rangle = 0$ and

$$\langle \eta_i(t)\,\eta_j(t') \rangle = 2\frac{m}{\tau}k_BT\,\delta_{ij}\,\delta(t-t'). \tag{18.13}$$

This equation relates the frictional dissipative force with stochastic fluctuations, and hence is called the ***fluctuation-dissipation theorem*** (Pathria, 1984). For the industrious, a more general account of integrating out degrees of freedom can be found in Hansen and McDonald (2006).

Our goal is to use the Langevin equation to couple an argon gas to an effective heat bath, thereby maintaining the system at a desired temperature. To start, we note that the molecular dynamics algorithm steps in temporal units h, which implies that one should consider the average noise over a time interval $(t, t+h)$:

$$\bar{\eta}_i = \frac{1}{h} \int_{t_n}^{t_n+h} \eta_i(t) dt. \tag{18.14}$$

This implies:

$$\langle \bar{\eta}_i^2 \rangle = \frac{2 m_i k_B T}{\tau h}, \tag{18.15}$$

which indicates that average noise forces can be chosen from a Gaussian distribution:

$$\rho(\bar{\eta}) = \frac{1}{\sqrt{2\pi \langle \bar{\eta}^2 \rangle}} \cdot e^{-\bar{\eta}^2/(2\langle \bar{\eta}^2 \rangle)}. \tag{18.16}$$

Some minor changes need to be made to the Verlet formalism when frictional forces are present (recall that the force in Eq. 18.6 must be velocity independent). Dynamics described by Eq. 18.12 yields the formula (see Ex. 6 for the derivation):

$$\left(1 + \frac{h}{2\tau}\right)\vec{r}(t+h) = 2\vec{r}(t) - \left(1 - \frac{h}{2\tau}\right)\vec{r}(t-h) + h^2\bar{\eta} + h^2\vec{F}(r(t)) + O(h^4). \tag{18.17}$$

The leap-frog version of this is:

$$\left(1 + \frac{h}{2\tau}\right)\vec{p}(t+h/2) = \left(1 - \frac{h}{2\tau}\right)\vec{p}(t-h/2) + h\bar{\eta} + h\vec{F}(r(t))$$

$$\vec{r}(t+h) = \vec{r}(t) + \frac{h}{m}\vec{p}(t+h/2)$$

Figure 18.1 shows thermalization for a system of 1000 argon atoms in a box of size 10^3 at a nominal temperature of 0.6. The particles are rather tightly packed and thus both methods must shed excess energy; the rescaling run does so by lowering the kinetic energy every 20 time steps; whereas the Langevin method dissipates the energy according to the fluctuation-dissipation theorem.

18.3.3 Observables

In the development stage all simulations should track the temperature and the total energy to verify the code and confirm that nothing untoward, such as quasi-ergodic motion, is occurring. In this regard, it is useful to make a histogram of the speed or v_x distribution to confirm that it follows the dictates of Maxwell and Boltzmann (Eq. 18.2).

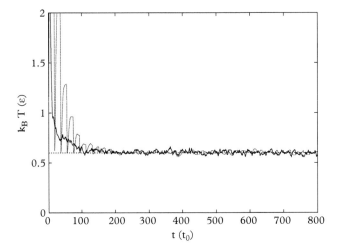

Figure 18.1 *Thermalization of an argon gas. Solid line: Langevin. Dotted line: rescaling.*

Figure 18.2 *2d argon.*

As usual, error estimation is important. Do not make the mistake of using measurements of observables at different times to compute errors as these are highly correlated. In fact, Eq. 18.1 indicates that observables need to be evaluated over long temporal spans—combining many of these measurements is a safe way to proceed.

It is amusing to plot (or make movies) of the particle positions. For example, Figure 18.2 shows the configuration of argon atoms in a two-dimensional lattice at relatively high density and low temperature. A hexagonal close-packed structure is evident although several dislocations mar the landscape. The physics behind this is explored in Exercise 11.

The pair correlation function is a central element of the description of simple substances and is directly accessible experimentally. This can be obtained from the number density as a function of distance via

$$g(r) = \frac{2V}{N_p(N_p - 1)} \frac{n(r)}{4\pi r^2 \Delta r}$$

where Δr is the binning radius used to form $n(r)$. Don't forget to divide by the number of measurements you make as well.

Useful physical parameters for the three-dimensional system are: liquid ($T = 1$, $\rho = 0.8$), solid ($T = 0.5$, $\rho = 1.2$), and gas ($T = 3.0$, $\rho = 0.3$). Simulation results for these parameters for 2744 particles with $h = 0.005$, 2000 thermalization steps, and 1000 measurements are shown in Figure 18.3. Notice that the solid line has clear peaks that correspond to distances on a hexagonal close-packed lattice and these peaks are substantially softened in the liquid phase. In the gas phase all peaks disappear except an enhancement at r somewhat greater than 1.0 that appears due to the strong repulsion for $r < 1$. All correlation functions tail off from the expected large distance behavior due to the finite size of the confining torus. This is explored in more detail in Exercise 12.

Information about the phase structure can also be obtained from the average displacement of a particle from its equilibrium position. At early times particle motion is ballistic so that $\langle |\vec{r} - \vec{r}_0| \rangle \sim t$, whereas for larger times particles should diffuse, hence $\langle |\vec{r} - \vec{r}_0| \rangle \sim \sqrt{t}$. This behavior is evident in Figure 18.4. A clear distinction between the phases is evident. For particles in a solid there is no dispersion at all past approximately 1 ps. Alternatively, particles in a liquid execute random walks with a small diffusion coefficient, while those in a gas have a much larger diffusion coefficient.

Another possible observable is the Fourier transform of the particle positions, $\sum_i \cos \vec{k} \cdot \vec{r}_i$. If the system is a solid the Fourier transform will peak strongly at a reciprocal lattice vector that reflects the lattice structure. The peak will be of order the number particles in strength. If the system is in the liquid phase the Fourier transform will

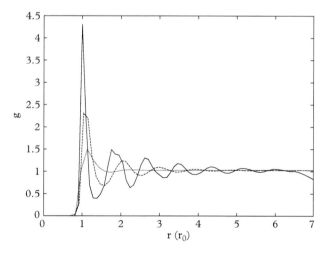

Figure 18.3 *Pair correlation function, $g(r)$, for solid (solid line), liquid (dashed line), and gas (dotted line) phases.*

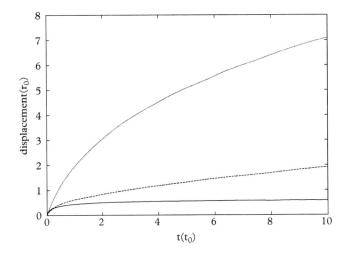

Figure 18.4 Average particle displacement. Top to bottom: gas, liquid, solid.

oscillate (in time) about zero with an amplitude of order the square root of the number of particles (Verlet, 1967).

Finally, measuring time-dependent observables is both possible and of interest. A common example is the velocity auto-correlation, $\langle \vec{v}(0) \cdot \vec{v}(t) \rangle$, which is sensitive to *memory effects*. This refers to the possibility that the environment of a particle depends on its past behavior; for example a moving particle will create a zone of high density in front of it and leave one of low density in its wake. Memory effects can be incorporated in the Langevin theory by generalizing Eq. 18.13 to read (Kubo, 1966):

$$\langle \eta_i(t)\eta_j(t') \rangle = M_{ij}(t - t'),$$

where the function M encapsulates the physical effects discussed. Exercise 13 explores memory effects in a three-dimensional argon liquid (where it does indeed exist).

18.4 Multiscale systems

While we all love argon, it is not especially representative of the array of substances that interest the scientific community. In particular, most substances are made of constituents (we are thinking of molecules) that have important internal degrees of freedom. Simple examples are diatomic molecules such as N_2 and O_2 that have rotational and vibrational excitation modes. More complicated molecules also have torsional excitation modes. Simulating enormous biological molecules is of obvious interest to the medical and pharmacological communities.

Things can get complicated quickly: even a relatively simple molecule like propane (C_3H_8) is typically modelled with ten bond-stretching, eighteen angle, eighteen torsional, and twenty-seven non-bonded interactions. The first three refer to possible motions of three-dimensional collections of objects, while the latter are weak interactions and are often modelled as an electrostatic Coulomb interaction or as a van der Waals interaction. These interactions are typically written in terms of very simple functions such as the Lennard-Jones potential for the van der Waals force, quadratic interactions for angular motion, and functions of torsional angles such as:

$$V(\omega) = \sum_n \frac{v_n}{2} \left[1 + \cos(n\omega - \gamma)\right]. \tag{18.18}$$

Vibrational modes are often modeled with the ***Morse potential*** (named after American physicist and operations research pioneer, Philip Morse, 1903—1985):

$$V(r) = D_e \left(1 - \exp(-a(r - r_c))\right)^2. \tag{18.19}$$

This model accounts for anharmonicity observed in physical systems and permits bond breaking. Some molecular forces are illustrated in Figure 18.5. A wide variety of force models specialized in biomolecules, hydrocarbons, organics, etc., exist (Cramer, 2004).

In principle, interactions in large molecules need not be simply related to those in smaller systems; it is therefore convenient that parameters determined in small molecules approximate those in larger—a property called ***transferability*** (Leach, 2001). Thus, for example, the Lennard-Jones potential that describes the vibrational spectrum of N_2 can also be used to describe the pair-wise interaction of nitrogen atoms in large biomolecules. For a discussion of the state of the art, see Boas (2007).

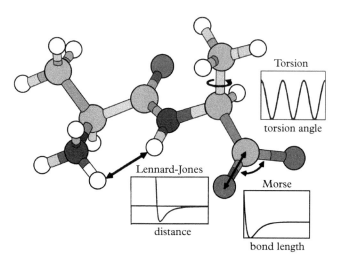

Figure 18.5 *Some molecular forces. Figure by R. Williams.*

Naively, simulating large molecules (or systems of large molecules) is a matter of applying the Verlet (or other) techniques in elaborate environments. However, the differing scales associated with the different forces can cause serious problems. For example, kinetic energies are 0.03 eV at room temperature while typical electronic excitation energies are 1–10 eV, rotational energies are 0.001–0.01 eV, and vibrational energies are 0.1–1 eV. These differing excitation energies correspond to different natural time scales that must be accounted for in the simulation. Practical suggestions for time steps are listed in Table 18.1. We see that integrating a system with vibrational degrees of freedom can take ten to one hundred times longer than a comparable simple atomic system (such as argon gas). This problem can become dramatically worse for systems with widely varying scales. For example, climate models must deal with phenomena ranging from energy transfer in atmospheric gases (angstroms) to hurricanes, spanning an astonishing fifteen orders of magnitude.

Figure 18.6 shows the xy positions of 36 of the 216 N_2 molecules in a leap frog Verlet simulation. The system was simulated for 0.4 ps with a step size of 0.0002 ps and N_2 bond vibration was modelled with a Morse potential fit to the experimentally measured ground state vibrational spectrum (see Ex. 5 for details). As can be seen, the nitrogen atoms tend to pair up with some deviation permitted in their inter-atomic distance (this is enhanced in the figure since it presents a projection of the three-dimensional

Table 18.1 *Time scales associated with different atomic forces (Leach, 2001).*

System	Motion	Time step (10^{-14} s)
atoms	translation	1
rigid molecule	translation and rotation	0.5
molecules, fixed bonds	translation, rotation, torsion	0.2
molecules, extensible bonds	translation, rotation, torsion, vibration	0.1 – 0.01

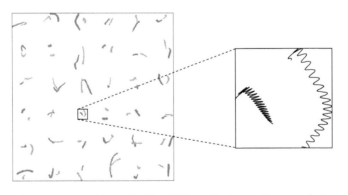

Figure 18.6 *Positions of a slice of N_2 gas in the xy plane over 0.4 ps.*

atomic coordinates). The enlargement shows the path of each atom in the 15th molecule, revealing the near harmonic motion of the nitrogen atoms.

18.4.1 Constrained dynamics

The behavior shown in Figure 18.6 leads one to suspect that, at least in some cases, internal motion in a molecule is not important to bulk properties of a gas composed of molecules. This observation motivates the strategy of freezing degrees of freedom that are associated with very short time scales relative to those scales of interest. In the case of the N_2 molecule this implies fixing the interatomic distance to the equilibrium value of 1.08 Å. One option at this stage is to switch to internal (i.e., intra-molecular) and center of mass (inter-molecular) coordinates and simulate a gas of rigid diatomic molecules by integrating the equations of motion for the center of mass motion and the torque. However, this is already somewhat awkward, even for this simple case; and it rapidly becomes unmanageable for more complicated objects. Even defining the internal coordinates can be difficult for systems with cyclic constraints. Nevertheless, using "internal coordinates" is efficient and should be considered as an option. More information can be found in Jain (1993) or Fincham (1984).

A simpler approach is to implement (holonomic) constraints with Lagrange multipliers. In this case constraints such as fixed bond lengths are imposed with subsidiary equations:

$$\sigma_k(\{r\}) = 0 \qquad (18.20)$$

where the index k labels the constraint. In the case of diatomic molecules there would be one constraint per molecule. Constraints are incorporated in the system dynamics via an additional term in the Lagrangian:

$$-\sum_k \int \lambda_k(t)\sigma_k \, dt$$

and yield an additional fictitious force in the equations of motion:

$$m_i\ddot{\vec{r}}_i = \vec{f}_i - \nabla_i \sum_k \lambda_k \sigma_k. \qquad (18.21)$$

In this case the form of the equations of motion remain the same, but constraint equations must be solved at each time step.

To illustrate we consider two particles A and B to be held a fixed distance d apart. The constraint is:

$$\sigma = (\vec{r}_A - \vec{r}_B)^2 - d^2 \qquad (18.22)$$

and the equations of motion are:

$$m_A \ddot{\vec{r}}_A = \vec{f}_A - 2\lambda(\vec{r}_A - \vec{r}_B)$$
$$m_B \ddot{\vec{r}}_B = \vec{f}_B + 2\lambda(\vec{r}_A - \vec{r}_B).$$

The standard Verlet procedure reads:

$$\vec{r}_A(t+h) = 2\vec{r}_A(t) - \vec{r}_A(t-h) + h^2 \vec{a}_A(t) - 2h^2 \frac{\lambda}{m_A}(\vec{r}_A(t) - \vec{r}_B(t))$$

$$\vec{r}_B(t+h) = 2\vec{r}_B(t) - \vec{r}_B(t-h) + h^2 \vec{a}_B(t) + 2h^2 \frac{\lambda}{m_B}(\vec{r}_A(t) - \vec{r}_B(t)).$$

The final step is to impose the constraint of Eq. 18.22 by substituting the coordinates at time $t + h$ and solving the resulting quadratic equation for λ. This value is then used to advance the coordinates to their new values.

While plausible, this approach becomes impractical for constraints that are even slightly more complicated. In such cases one can consider solving the (coupled and nonlinear) equations for the Lagrange multiplier with numerical procedures such as the Newton-Raphson method. However this requires evaluating the inverse of a matrix in the middle of a production loop, which is to be avoided if possible.

An elegant procedure for avoiding this headache was proposed by Ryckaert and colleagues (Ryckaert, 1985). The idea is to linearize the constraint equations and solve them by iteration. The algorithm starts with unconstrained coordinates defined by:

$$\tilde{\vec{r}}_i(t+h) = 2\vec{r}_i(t) - \vec{r}_i(t-h) + h^2 \vec{a}_i(t). \tag{18.23}$$

One proceeds by setting $r_i^{old} = \tilde{r}_i$ and then iterates:

$$\vec{r}_i^{new} = \vec{r}_i^{old} - h^2 \lambda_k \nabla_i \sigma_k(\{r(t)\}) \tag{18.24}$$

where:

$$h^2 \lambda_k = \frac{\sigma_k(\{r(t)\})}{\sum_i \frac{1}{m_i} \nabla_i \sigma_k(\{r^{old}\}) \cdot \nabla_i \sigma_k(\{r(t)\})}. \tag{18.25}$$

The latter equation is obtained by expanding $\sigma_k(r_i^{new})$ to first order in h^2 and solving for λ_k. Once the iteration has converged to the desired tolerance the new particle coordinates are simply $\vec{r}_i^{new}$.

We illustrate the general procedure for the simple case of a fixed diatomic molecule bond length. The constraint is given by Eq. 18.22 and the Ryckaert iteration loop is given by:

$$\vec{r}_A^{new} = \vec{r}_A^{old} - 2\frac{h^2}{m_A}\lambda(\vec{r}_A - \vec{r}_B)$$

$$\vec{r}_B^{new} = \vec{r}_B^{old} + 2\frac{h^2}{m_B}\lambda(\vec{r}_A - \vec{r}_B) \tag{18.26}$$

with:

$$h^2\lambda = \frac{\left(\vec{r}_A^{old} - \vec{r}_B^{old}\right)^2 - d^2}{\frac{4}{\mu_{AB}}(\vec{r}_A^{old} - \vec{r}_B^{old}) \cdot (\vec{r}_A - \vec{r}_B)}. \tag{18.27}$$

Notice that the gradient ∇_i selects only those particles that are involved in the kth constraint; thus the sum over constraints simplifies dramatically and the totality of constraints can be resolved sequentially.

Implementation code is:

```
// impose Lagrange constraints via Ryckaert iteration
for (int m=0;m<Nmol;m++) {
  ThreeVector Dr;
  Dr.setV(r[IM(m,0)]-r[IM(m,1)]);
  ThreeVector r0 = rnew[IM(m,0)];
  ThreeVector r1 = rnew[IM(m,1)];
  ThreeVector Drold=r0-r1;  // starting value for Ryckaert
      iteration
  double lambda = 1000.0;
  int it=0;
  while (abs(lambda) > tol) {
    lambda = (Drold.magSq() - d*d)*mu/(dt*dt*4.0*Drold*Dr);
    r0 = r0 - 2.0*dt*dt*lambda/mass*Dr;
    r1 = r1 + 2.0*dt*dt*lambda/mass*Dr;
    Drold = r0-r1;
    it++;
    if (it > 100) {
      cout << " iteration trouble " << lambda << " " << m
        << " " << t << endl;
      abort();
    }
  }
  rnew[IM(m,0)] = r0;   // rA(m)
  rnew[IM(m,1)] = r1;   // rB(m)
```

We see that some bookkeeping effort is required to keep track of (i) the ith particle, (ii) the mth molecule, and (iii) the ath atom of a molecule. The function IM maps the indices

$(m, a) \to i$. Finally, a way to bail out if convergence appears to be failing has been added to the iteration loop.

During development it is a good idea to track the distance between atoms A and B to ensure that convergence is being achieved and that no bugs have crept into the code. We have found that the method converges very quickly (only a few iterations are needed in most cases). However, there is no guarantee that the iteration will converge. A crucial point is that the starting coordinates, $\tilde{r}_i$, are close to the final constrained coordinates. Since the coordinates obey the constraints at earlier times, this is assured if the time step is sufficiently small (in fact, the panacea for all molecular dynamics difficulties appears to be "reduce your step size").

A collection of variants of the Ryckaert algorithm has developed over the years. These have names like RATTLE, WIGGLE, and MSHAKE and address other time evolution schemes (such as velocity Verlet), more complicated constraints, or improved iteration procedures. See Leach (2001) for more details.

While it is easy to feel satisfied with this resolution to the problem of constrained dynamics, difficulties lurk at the boundaries. For example, imposing constraints in some systems leads to inconsistencies or singular matrices and further steps must be taken when implementing numerical procedures. See Chapter 3 of Allen and Tildesley (1987) for more details.

The reader should also bear in mind that a constrained system often does *not* yield the same thermodynamic quantities as the analogous system with constraints replaced by stiff springs. This is referred to as the **metric tensor problem**; it arises because changing to generalized coordinates to resolve constraints can lead to coordinate-dependent momentum tensors:

$$\sum_i \frac{1}{m_i} \dot{r}_i \dot{r}_i \to \sum_{\alpha\beta} \dot{q}_\alpha G_{\alpha\beta}(q) \dot{q}_\beta = \sum_{\alpha\beta} p_\alpha (G^{-1})_{\alpha\beta}(q) p_\beta.$$

This tensor is different (in coordinate-dependence and in size) depending on whether constraints have been resolved or implemented as stiff springs, which gives rise to different values of thermodynamic observables (Allen and Tildesley, 1987, Section 15.1.1). If the metric tensor is not coordinate-dependent, such as when all molecular bonds are constrained, there is no difference. In the case of the commonly used bond-length constraints, differences appear to be small in practice.

18.4.2 Multiple time scales

The metric tensor problem raises the original issue of integrating equations of motion with differing time scales. We wager that the reader's first thought upon running into the multiple-time-scale problem was "Let's try integrating the fast and slow degrees of freedom with different time steps". A possible leap frog implementation of this idea is:

```
loop in t
    compute a_long
    v = v + (nh)a_long
    loop in i=1,n
        compute a_short
        v = v + ha_short
        r = r + hv
    end loop
end loop
```

Here we imagine that the total force can be split into forces with fast and slow components. For example, in the case of the N_2 molecule, the Lennard-Jones interactions comprise the slow forces while the Morse potentials comprise the fast. Note that "slow" and "fast" have been rebranded as "long" and "short" respectively. This will be explained shortly.

In fact a very similar algorithm has been suggested by Tuckerman *et al.* (1992):

```
compute a_short, a_long
loop in t
    v = v + ½(nh)a_long
    loop in i=1,n
        v = v + ½ha_short
        r = r + hv
        compute a_short
        v = v + ½ha_short
    end loop
    compute a_long
    v = v + ½(nh)a_long
end loop
```

Recall; however, that these statements can be rearranged to reproduce completely the standard or leap frog Verlet algorithms. Thus our naive guess corresponds to the algorithm of Tuckerman.

Some care must be taken when implementing this algorithm. Consider first an N_2 molecule with nitrogen atoms that are bound by a strong and short range Morse potential. This molecule can interact with other molecules via weak Lennard-Jones potentials that are assumed to act only between non-bonded atoms. One might be tempted to associate f_{fast} with the Morse potential; however, rapid motion in the intramolecular coordinate will cause rapid fluctuations in the Lennard-Jones potential that cannot be ignored in the inner loop. This is the reason that the forces have been split into long range and short range components. The inner loop integrates over the short range Morse potential *and* the short range component of all Lennard-Jones potentials. Because of this, there is no savings in computational effort, except by leveraging the short range nature of the "inner" force with an interaction cutoff. For perspective see (Schlick, 1999).

18.4.3 Solvents

Medical and pharmaceutical researchers are rarely interested in the properties of large molecules *in vacuo* because the environment can play an important role in determining the molecular *conformation*. This refers to the particular spatial (spin, charge, ...) arrangement that atoms take in a given molecule. The conformation of proteins is central to their biological function and is a function of how they *fold* as they are synthesized. The subject of protein folding is an ongoing and challenging subtopic in the field of molecular dynamics.

Of course, the "environment" mentioned above is often water (since humans are, to a good approximation, bags of salty water that produce waste heat). The most direct way to account for water as a solvent is to include a number (typically several hundred) of water molecules in the simulation. The exact meaning of "water molecule" can change depending on the simulation requirements. At the simplest level H_2O is modelled as a hard sphere. More complicated models assume three atoms with fixed bond angles, or four or five "objects" to account for nuclear and some electronic properties of the molecule.

An alternative to the direct approach is to incorporate solvent features analytically. For example, the permittivity of space can be increased to account for electrostatic screening induced by the H_2O charge distribution. More elaborate schemes employ forces that reproduce the average behavior of a water molecule. We have already seen one such model in our discussion of Langevin dynamics. In fact, the model was built to address Brownian motion, which describes a large molecule in a solvent. In applications to the problem at hand the relaxation time τ can be set according to Stoke's law for a spherical particle of mass m and radius a moving through a fluid with viscosity η: $\tau = m/(6\pi a \eta)$.

18.5 Gravitational systems

The observable universe is a gravitational system with an estimated 10^{23} stars arranged in superclusters of galaxy clusters of galaxies. Larger scale structures are suspected, while smaller scale ones such as solar systems are important to ephemeral carbon-based life forms. The total number of galaxies depends on the minimum size one takes in their definition, with estimates ranging from 10^{11} to 10^{12} (Conselice, 2016). A particularly famous spiral galaxy and a nearby dwarf elliptical galaxy are shown in Figure 18.7.

This section is concerned with simulating point particles that interact via Newton's gravitational force. If general relativity can be ignored then this is a classical problem and the methods of the previous sections apply (with some drastic changes of scale). Remarkably, this remains true for cosmological simulations, where homogeneity and isotropy of large scale structure implies that classical methods can be used in the appropriate coordinate system.

Figure 18.7 *Messier 51 and NGC 5195 as imaged by the Hubble Space Telescope. Credit: HST/STScI/AURA/NASA/ESA.*

When dealing with sub-cosmological scale problems the equations of motion will be taken to be:

$$m_i \ddot{\vec{r}}_i = -G \sum_j \frac{m_i m_j}{(r_{ij}^2 + \epsilon^2)^{3/2}} \vec{r}_{ij}. \qquad (18.28)$$

Typically the index i refers to a star, an aggregate of stars, or a galaxy. Notice that the equations of motion represent a drastic simplification of the dynamics; interstellar gas, radiative energy transfer due to electromagnetic processes, stellar structure and evolution, and a host of other processes are neglected. The parameter ϵ is a rough accommodation of some of this reality. Namely, if one is modeling galaxy clusters then binary interactions between galaxies should be softened at short distances (Hockney and Eastwood, 1988; Binney & Tremaine, 1987). In this case ϵ is roughly the radius of a galaxy. See Ex. 20 for more on this.

Units are removed by introducing a distance scale r_0 and a mass scale m_0. In this case Eq. 18.28 becomes:

$$\ddot{\vec{r}}_i = -\sum_j \frac{1}{(r_{ij}^2 + \epsilon^2)^{3/2}} \vec{r}_{ij}$$

where now all quantities are dimensionless and a time scale:

$$t_0 = \sqrt{\frac{r_0^3}{Gm_0}} \qquad (18.29)$$

has been introduced. Notice that equal masses have been assumed, the conditions under which this is true will be explained shortly. If m_0 is the total mass of a system of linear size r_0 then t_0 is the **virial time** needed for a system to evolve to a state that obeys the virial relationship. Notice that this is consistent with our earlier estimate (page 601) $t_{therm} \sim L/\langle v \rangle$ if one takes $L = r_0$ and $\langle v \rangle \sim \sqrt{\langle U \rangle / m_0} \sim \sqrt{Gm_0/r_0}$.

The Milky Way contains about 10^{11} stars and another 10^{10} $m_\odot$ in gas. Most of the stars move in nearly circular orbits in a disk of radius 10 kpc (1 kpc is $3 \cdot 10^{19}$ m). The average speed for a star is 200 km/s so that the orbital time is around 300 million years. Thus the Milky Way has had plenty of time to equilibrate during its evolution.

The rate at which our galaxy's stars interact is approximately given by the mean free path divided by the random component of the average star velocity, which is 40 km/s (the free streaming velocity is not relevant to scattering). The mean free path is $\lambda \sim 1/(n\sigma)$; with a mean density of $n \sim 0.3/\text{pc}^3$ and a cross section of $4\pi r_\odot^2$ we obtain a mean collision time of 10^{19} years. The conclusion is that stars do not collide in the Milky Way. Such systems are called **collisionless**. The motion of stars in the Milky Way (and in a typical galaxy) is set by the mean local gravitational field, and is not influenced by the local mass distribution. Alternatively, open clusters typically have hundreds of stars and one must account for collisions. Globular clusters have 10^4 to 10^6 stars and hence are intermediate in their collisionality.

Optimal numerical methods depend on the collisionality of the system in question. The smooth response of collisionless systems and the great number of bodies present implies that methods based on the gravitational potential are preferred to those based on force summation. These will be discussed briefly below and in more detail in Chapter 19. Alternatively, highly collisional systems require the machinery of N-body simulations to account for the frequent interactions present.

18.5.1 N-Body simulations of galactic structure

Although we have stated that potential methods are well-suited for collisionless systems like galaxies, direct simulations of galaxies were historically important in the development of the *cosmological concordance model*. In particular, spiral galaxies are common and stable at cosmic time scales[4]. For many decades it was thought that the spiral arms were magnetic in origin and not due to gravitational dynamics. Further confusion existed

---

[4] They comprise about 80% of all galaxies outside of the dense regions at the core of clusters (Binney & Tremaine, 1987).

because, counter to evidence, spiral arms should "wind up" as the galaxy spins. It is now believed that spiral arms are *density waves*, thereby resolving both issues.

A third issue concerned the stability of spiral galaxies. In particular, early numerical investigations indicated that disk-like distributions of stars are unstable to the formation of bars. This was most famously investigated by Ostriker and Peebles (1973), who simulated several hundred "stars", and noted that the bar instability forms when the ratio of rotational kinetic energy to potential is larger than 0.14. More importantly, they observed that adding a spherical "halo" to the initial mass distribution stabilized the disk. These days the halo is regarded as being comprised of *cold dark matter* in a sphere that is substantially larger than the visible disk.

We seek to reproduce the Ostriker-Peebles results with the Verlet formalism. Code development is simplified with respect to molecular physics because boundaries need not be considered. On the demerit side, the attractive nature of the gravitational interaction implies that large accelerations can be experienced for masses in close contact. One way to deal with this problem is to introduce variable temporal step sizes associated with each particle. Of course the cutoff introduced in Eq. 18.28 helps as well. We remark that if the simulation particles were stars the cutoff would not be relevant, since different dynamics affects close stellar encounters. However, in practice a simulation particle represents aggregates of millions or billions of stars, so the cutoff remains useful and valid. Because of this it is also sensible to consider all particles to be of the same mass.

Before launching into the simulation it is prudent to confirm that two bodies interact sensibly. This can be achieved by starting two particles at rest and following their subsequent motion. If the step size is too large the particles will approach each other, artificial energy will be injected into the system, and the particles will recoil to infinity (rather than execute a closed orbit). Further details can be found in Ex. 19.

Our simulation is made with 1000 bodies in units $G = 1$ and $m_i = 1$. Following Ostriker and Peebles, we set $\epsilon = 0.05$, assume no halo, and choose a step size $h = 0.001$. The initial configuration was generated by placing particles randomly in a disk of radius 1 and width 0.1 with a mass density proportional to $1/r$. Ostriker and Peebles implemented a rather elaborate initial velocity distribution; we simply set the initial velocity for each particle to be $20\hat{\phi}$ where the unit vector indicates purely rotational motion. This gives an initial ratio of rotational kinetic energy to potential energy of 0.2, above the critical value of 0.14. Figure 18.8 shows the initial mass configuration on the left and the configuration at time 0.256. It is apparent that a bar instability has developed.

It is useful (and pleasing) to visualize the temporal evolution of simulated gravitational systems. Developing custom visualization programs from a graphics toolkit, such as we discussed in Chapter 10, particularly Section 10.5 and Exercise 5, gives the most flexibility. Alternately, the *VMD* (Visual Molecular Dynamics) package, which is free software for animating and analyzing large biomolecular systems developed at the University of Illinois, can be installed. While designed primarily for sophisticated manipulation of complex molecules, the code also functions well for displaying dynamical many-body systems such as discussed in this chapter. A plugin can be used to create movies in a variety of formats.

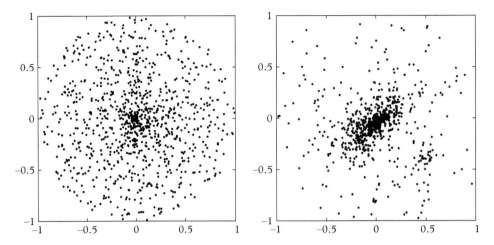

Figure 18.8 *Evolution of a disk to a bar morphology. Top view; 1000 particles.*

As stated before, the addition of a halo increases the stability of a disk galaxy. This is, however, rather indirect evidence for dark matter. More convincing evidence comes from measurements of ***rotation curves***, namely the average orbital speed of stars as a function of distance from the disk center. In collionless sytems rotation curves are a function of the mass density (see Ex. 18), and therefore provide a readily accessible measure of the distribution of visible matter in a galaxy. A long history of astronomical observations, most famously associated with Fritz Zwicky and Vera Rubin, have demonstrated that the rotation curve for spiral galaxies tends to be flat, even beyond the bulk of the galaxies. This implies the presence of gravitating matter with a mass density, $\rho \sim 1/r^2$, out to very large distances from the disk center. We stress, however, that the dynamical imperative for such a mass distribution is not established.

Finally, the reader will have noticed that we have not produced any spiral galaxies. This is a rather difficult problem that relies on the coupling of gravitational dynamics at several length scales. For example, the presence of supermassive black holes is thought to play a role, as are galaxy-galaxy collisions. In fact, compared to other fields covered in this book, numerical gravity is relatively nascent, with many unanswered questions for enquiring numericists.

18.5.2 The Barnes-Hut algorithm

There are a lot of stars out there so it would be nice to avoid employing the $O(N^2)$ direct summation method for evaluating forces. If one is stuck with this method, parallelization can be easily achieved because terms in the force sum (the most expensive part of the computation) are independent. Thus, as mentioned above, this code block is an ideal candidate for parallelization with the openMP directive `#pragma omp parallel for` or other techniques, such as multithreading.

A variety of other methods that are $O(N \log N)$ are available if one is willing to accept approximations. Unfortunately, the simplest approximation of employing a force cutoff is unacceptably inaccurate for long range forces such as gravity.

It may have occurred to the reader that computing the force due to a distant clump of matter on a given particle by summing over all the individual force pairs is rather wasteful. It would certainly be more efficient to consider the clump as an aggregate particle, possibly with multipole moments. This idea forms the basis of the **Barnes-Hut algorithm** (Barnes & Hut, 1986), whose novel feature is the aggregation of particles in a recursive tree structure. Recursion is central to the algorithm, so the reader may wish to consult a reference such as Weiss (2014) before pressing on.

The Barnes-Hut algorithm uses a tree structure containing nodes that specify the total mass, center of mass, and spatial size of a group of particles. The nodes also contain pointers to subgroups within the group. This information is contained in a `struct` as follows:

```
struct BHnode {         // two dimensions
   double mass;         // total mass in the cell
   double dx;           // linear size of the cell
   TwoVector r;         // center of mass coordinate
   BHnode *node0;       // pointers to subcells in order
   BHnode *node1;       //  3 2
   BHnode *node2;       //  0 1
   BHnode *node3;
};
```

In two dimensions each node is associated with a (square) region of the space. The four subnodes are associated with the four quadrants of the parent square. This procedure is repeated recursively until each square contains zero or one particles. In this way a given node will contain the center of mass and total mass of all of the subnodes within it.

The procedure is illustrated with an especially simple case of five particles in two dimensions in Figure 18.9. If each particle has mass 1, the root node will have mass 5; its four subnodes will have masses of 1, 1, 1, and 2 respectively. The fourth subcell only has one subcell with particles, and that has two subcells with one particle each. Thus the entire tree has eight nodes arranged as shown in Figure 18.10. Center of mass information has been suppressed in this figure. The left number refers to the total mass in a cell, while the right is the linear size of the cell (with the entire system set to be of size 1).

Some care must be exercised when designing an algorithm to create Barnes-Hut trees. For example, if one simply scans all particles and recursively places them into a tree the resulting algorithm is $O(N^2)$. Rather, particles should be added sequentially as is traditional with binary search trees. Thus one defines a confining area and adds the first particle, thereby filling in the coordinate and mass fields of the root node. If subsequent particles fall within an occupied cell, that cell is subdivided and the original and additional

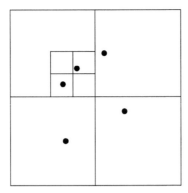

Figure 18.9 *A Barnes-Hut recursive decomposition of a simple mass distribution.*

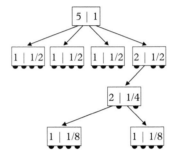

Figure 18.10 *The Barnes-Hut tree corresponding to Figure 18.9.*

particles must then be assigned to the appropriate subcells. If both land in the same subcell the procedure must be repeated. Notice that this procedure creates empty nodes. A useful feature of the BH tree is that all leaves must represent a single particle, which we take to be of mass 1.

A generic implementation of the algorithm relies on recursion and looks something like

```
BHinsert(r,n)
  // insert particle r_i at node n
  if n is empty
     add r to n
  else if n has multiple particles
     determine the subcell ,n', in which r belongs
     BHinsert(r,n')
  else (n has exactly one particle)
     create subnodes
     move the original particle to its quadrant
     determine the subcell , n', in which r belongs
     BHinsert(r,n')
  end
end
```

Many body dynamics

It is likely that this is a little obscure so we provide an explicit implementation in C++. First, a class representing two-vectors is useful. We call this `TwoVector`; one possible implementation is simply

```
typedef Eigen::Vector2d TwoVector
```

or the user can roll his own. C++ code to implement the pseudocode is

```
void BHinsert (BHnode * rt, TwoVector r, TwoVector Rnode,
   double dx) {
  /*
     [two dimensions]
     rt = pointer to relevant BHnode
     r = particle position
     Rnode defines the cell via its center of mass
     dx = cell radius ie    R-dx < x < R+dx
  */
  if (rt->mass == 0.0) { // insert at current cell
     rt->mass = 1.0;
     rt->r = r;
     rt->dx = dx;
  } else if (rt->mass > 1.0) {    // more than one particle,
        so an internal node
     dx /= 2.0;
     if (in(r,Rnode+dx*x0,dx)) {
        BHinsert(rt->node0,r,Rnode+dx*x0,dx);
     } else if (in(r,Rnode+dx*x1,dx)) {
        BHinsert(rt->node1,r,Rnode+dx*x1,dx);
     } else if (in(r,Rnode+dx*x2,dx)) {
        BHinsert(rt->node2,r,Rnode+dx*x2,dx);
     } else {
        BHinsert(rt->node3,r,Rnode+dx*x3,dx);
     }
     rt->r = (rt->mass*rt->r + r)/(rt->mass + 1.0);
        // update CM and mass
     rt->mass = rt->mass + 1.0;
  } else { // so m=1 therefore a leaf
     dx /= 2.0;
     rt->node0 = new BHnode;                   // split
     rt->node1 = new BHnode;
     rt->node2 = new BHnode;
     rt->node3 = new BHnode;
     if (in(rt->r,Rnode+dx*x0,dx)) {     // move original
           mass
        rt->node0->mass = 1.0;
```

```
            rt->node0->dx = dx;
            rt->node0->r = rt->r;
        } else if (in(rt->r,Rnode+dx*x1,dx)) {
            rt->node1->mass = 1.0;
            rt->node1->dx = dx;
            rt->node1->r = rt->r;
        } else if (in(rt->r,Rnode+dx*x2,dx)) {
            rt->node2->mass = 1.0;
            rt->node2->dx = dx;
            rt->node2->r = rt->r;
        } else {
            rt->node3->mass = 1.0;
            rt->node3->dx = dx;
            rt->node3->r = rt->r;
        }
        if (in(r,Rnode+dx*x0,dx)) {
            BHinsert(rt->node0,r,Rnode+dx*x0,dx);
        } else if (in(r,Rnode+dx*x1,dx)) {
            BHinsert(rt->node1,r,Rnode+dx*x1,dx);
        } else if (in(r,Rnode+dx*x2,dx)) {
            BHinsert(rt->node2,r,Rnode+dx*x2,dx);
        } else {
            BHinsert(rt->node3,r,Rnode+dx*x3,dx);
        }
        rt->r = (rt->mass*rt->r + r)/(rt->mass + 1.0);
            // update CM and mass
        rt->mass = rt->mass + 1.0;
    }
} // end of BHinsert
```

Notice that the center of mass of the cell associated with a node is tracked so that locating subquadrants can be achieved. The TwoVectors x0, x1, x2, x3 point to the relevant subquadrants. Thus, for example, $x0 = (-1, -1)$. The variable dx keeps track of the size of the cell for use in locating subquadrants and in force evaluation. As promised, whenever cells are shared four subcells are created. In principle only filled subcell nodes could be created, but this complicates the subsequent insertion of the new particle, so we accept the mild inefficiency. Finally, it is useful to set $dx = 0$ in leaf nodes because that eliminates an awkward conditional in the force evaluation, to which we turn now.

The force on a particle at position R is evaluated by traversing the Barnes-Hut tree from the root. If the size of the mass distribution at a given point in the tree divided by $|R - r_{CM}|$ is less than an apparent angle θ then the force is evaluated using that distribution. If this is not the case, the mass distribution is close enough to R that it should be decomposed at the next level. Recursion makes for a compact implementation of this method:

```
TwoVector BHforce(BHnode * rt, TwoVector R) {
  TwoVector f;
  if ((rt != NULL) && (rt->mass != 0.0)) {
    TwoVector r0;
    r0 = R - rt->r;
    double rsq = r0.magSq();
    if (rsq == 0.0) return r0;
    if ((rt->mass == 1.0) || (rt->dx/sqrt(rsq) < theta)) {
      f = (-1.0)* rt->mass*r0/(sqrt(rsq)*rsq);
    } else { // go deeper
      f = f + BHforce(rt->node0,R);
      f = f + BHforce(rt->node1,R);
      f = f + BHforce(rt->node2,R);
      f = f + BHforce(rt->node3,R);
    }
  }
  return f;
}
```

The original implementation of force evaluation by Barnes and Hut required 100 lines of code—thus this code snippet illustrates the efficiency of modern object oriented languages. We wish to stress, however, that line counts are a particularly meaningless metric for code quality. Barnes and Hut applied the algorithm to a simulation of two galaxies with 4096 particles. They noted that a large opening angle of $\theta = 1$ was sufficient to gain 1% accuracy in the energy. The code itself ran in about one tenth the time of the analogous direct force evaluation code.

We test the algorithm by making one force evaluation for all N unit mass particles in a disk or radius 1. The direct summation method scales as N^2, as expected. This is shown as a line in Figure 18.11. When the Barnes-Hut algorithm is run with a small opening angle all masses are resolved and the algorithm devolves to the direct summation method (with additional overhead). Results for $\theta = 0.001$ are shown as open squares in the figure and confirm these expectations. As the opening angle is increased the algorithm becomes dramatically faster since fewer force evaluations are required (tree creation is surprisingly fast, even for trees with millions of nodes). Fits to the curves reveal that they follow the expected $N \log N$ behavior very closely for $N \gtrsim 5000$.

18.5.3 Particle-mesh methods

Another class of methods for speeding up N-body simulations employs the field equations to evaluate forces on particles. We do not have space to describe the

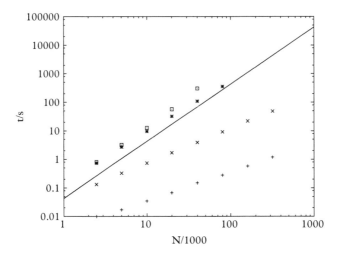

Figure 18.11 *Time for complete force evaluation. Solid line: direct summation. Points: Barnes-Hut results for $\theta = 0.001, \theta = 0.01, \theta = 0.1, \theta = 1$ (top to bottom).*

techniques in detail; rather we will sketch the ideas and point the interested reader to the literature.

Field-based methods are useful because, if the field is known, the $O(N)$ operation of computing the force on a particle at position r is replaced by the $O(1)$ operation of evaluating:

$$\vec{f}(\vec{r}) = -\nabla \phi(\vec{r}). \tag{18.30}$$

The problem, of course, is that one needs the potential. This can be obtained by the formula:

$$\phi(\vec{r}) = -G \sum_i \frac{m_i}{|\vec{r} - \vec{r}_i|},$$

but this reduces the formalism to the $O(N^2)$ direct particle method and we have achieved nothing.

But now a way forward becomes apparent: if one were to obtain the gravitational potential via Poisson's equation

$$\nabla^2 \phi(x) = 4\pi G \rho(x).$$

by solving on a grid that is relatively small (of linear size N_g), then Eq. 18.30 can be used with an ensuing savings in computational effort. In fact efficient methods of $O(N_g \log N_g)$ exist for solving Poisson's equation, so the whole thing is $O(NN_g \log N_g)$. This approach is called the ***particle mesh method***. The cost is a, possibly drastic, misrepresentation of short range forces. However, collisionless systems tend to have "smooth" matter distributions which imply smooth potentials and hence reasonable accuracy of the method.

The source for Poisson's equation is normally written

$$\rho(\vec{r}) = \sum_i m_i \delta(\vec{r} - \vec{r}_i);$$

however, this tends to lead to large fluctuations in the force, so in practice one "smears" particles over several (say eight) nearby grid cells.

If collisions are important in a given system it is natural to extend the method by employing direct summation for nearby particles. This technique is called the *particle-particle particle-mesh method* (Hockney and Eastwood, 1988).

18.6 Exercises

Molecular Dynamics

1. [T] Potential Cutoff.
   Many sources say that cutting off the potential "creates problems, especially in molecular dynamics simulations where energy conservation is required". What is really going on?

2. [T] Boundary Conditions.
   Consider one liter of water in a cubic container.

   a) How many water molecules are present (assume STP)?
   b) If interactions with the wall extend to 10 molecular diameters (2.8Å), how many water molecules "feel the boundary conditions"?
   c) Would it be reasonable to simulate water in a box with, say, 1000 molecules?

3. [T] Initial Conditions.
   Discuss whether *any* initial conditions can be used in a molecular dynamics simulation. Can you come up with any that do not lead to ergodic motion?

4. [T] System in a Box.
   Consider placing a system of N particles in a confining box. One possibility for implementing this would be to model the box walls as a collection of static particles interacting with the gas via some short ranged force. However, this is clearly impractical since an N-body problem would be transformed into an $N + M$-body problem where $M \gg N$. A cleaner approach would be to model the box walls as static surfaces interacting with the gas particles via a hard core (i.e., particles bounce when they hit the wall).

a) Come up with a computational strategy to implement hard walls.

b) Are there rare situations where your strategy runs into trouble?

5. [T] N_2 Morse Potential.
   Parameters for the Morse potential can be fit to the vibrational spectrum for a given molecular configuration. Parameters for the ground state of N_2 ($X\,^1\Sigma_g^+$) are $D_e = 9.904$ eV, $a_e = 2.85/\text{Å}$, and $r_e = 1.080$ Å (Konowalow, 1961).

   a) Convert these to natural units for argon.
   b) Determine the frequency of oscillation in the harmonic approximation.
   c) What time step (in argon units) would be appropriate for this problem?

6. [T] Friction in the Verlet Formalism.
   Derive Eq. 18.12 by expanding the equation of motion to obtain expressions for $\vec{r}(h)$ and $\vec{r}(-h)$ in terms of $\vec{r}(0)$ and the right hand side. Add the results and make a sensible substitution for $\dot{\vec{r}}(0)$ to obtain Eq. 18.17.

7. Energy Fluctuations.
   Code up argon in three dimensions with one of the Verlet methods. Plot the fluctuation in total energy as a function of the step size. What is the general behavior?

8. Particle Diffusion.
   Compute the diffusion of argon in three dimensions as in Section 18.3.3. Is it possible to discern the location of the phase transitions with this observable? Try to make a connection to **Lindemann's law**, which states that melting occurs when the root mean square deviation from equilibrium exceeds a critical fraction of the crystal interatomic distance (Hansen and McDonald, 2006, pg 173). What is this fraction?

9. Reversibility.
   Code up argon in three dimensions and confirm that the Verlet method is time reversible. You can do this by running for N steps, reversing the velocities of all particles, and running for an additional N steps. Try plotting the trajectories of individual particles, or determine $\langle \vec{r}(0) - \vec{r}(2N) \rangle$, or similar.

10. Octahedral Boundary Conditions.
    Implement **truncated octahedron boundary conditions** for argon in three dimensions. These are more appropriate than cubic boundary conditions for a problem with spherical symmetry such as argon. Use the following algorithm (Leach, 2001):

```
r_i = r_i - 2L⌊r_i/L⌋
if (∑_i |r_i| ≥ 3/2L) then
    r_i = r_i - sgn(L, r_i)
end if
```

where r_i is the ith component of the position vector, $\lfloor x \rfloor$ is the integer part of x, $\mathrm{sgn}(x,y)$ returns $|x|$ if $y > 0$ and $-|x|$ if $y < 0$, and the octahedron is contained in a cube of side $2L$. Other geometries are discussed in Leach (2001).

11. Ordering in Two-Dimensional Argon.
    Peierls (1935) has argued that conventional long range order cannot exist in two-dimensions because phonons cause the mean square deviation of a particle from its equilibrium position to diverge logarithmically in the system size. Explore this claim numerically. The subject does not end here: Kosterlitz and Thouless (1973) have famously argued that long range *topological order* can exist in two-dimensional systems. The numerical work can get fiddly; see Toxvaerd (1978) for help.

12. Particle Correlation Function.
    Explore the particle correlation function for argon in two dimensions. Compare your results to the analytic expression for a uniform particle distribution. This will be a good approximation for large radii and will show how g drops off for $r > L/2$. Also plot the expected location of peaks assuming a triangular crystal structure.

13. Velocity Autocorrelation.
    Measure the velocity autocorrelation for a three dimensional argon gas:

    $$A_v(t) = \frac{\langle \vec{v}(0) \cdot \vec{v}(t) \rangle}{\langle v^2 \rangle}.$$

    Try the densities $\rho = 0.86$ and $\rho = 1.4$. Compare your result to the **Langevin velocity autocorrelation**:

    $$\langle v(0)v(t) \rangle = \langle v(0)^2 \rangle \exp(-t/\tau).$$

    Comment on what you see. What is going on for $\rho = 1.4$? See if you can determine the large time behavior of the autocorrelation. Does it follow the Langevin result?

14. [P] DNA Under Duress.
    The **PBD model** of DNA is written in terms of the deviation of the ith base pair length from equilibrium, $y_i/\sqrt{2}$ as:

    $$m_i \ddot{y}_i = -U'_i(y_i) - W'(y_{i+1}, y_i) - W'(y_i, , y_{i-1}).$$

    The Morse potential:

    $$U_i(y) = D_i[\exp(-a_i y) - 1]^2$$

    describes the hydrogen bonding between base pairs. Inter-pair interactions are described by the **stacking potential**:

    $$W(x,y) = \frac{1}{2} k (x-y)^2 \{1 + \rho \exp[-\beta(x+y)]\}.$$

Consider coupling DNA to an external electric field and placing it in a dissipative medium. Model these effects with the additional term:

$$-m_i \gamma \dot{y}_i + A \cos \Omega t.$$

Use the following parameters; $m = 300$ amu, $D_{AT} = 0.05$ eV, $a = 4.2$ 1/Å, $k = 0.025$ eV/Å^2, $\beta = 0.35$ 1/Å, and $\rho = 2.0$. A relaxation time typical for water is $\gamma = 1.0$/ps.

a) Assume 64 AT base pairs and a driving force of $A = 144$ pN and examine the behavior of the system in the neighborhood of $\Omega = 1$ THz. Inject energy into the system by perturbing it at $t = 0$. For sufficiently large perturbations you should find long-lived 'breather' resonance modes at frequencies of 1.0 and 1.5 THz (amongst others) (Swanson, 2011).

b) Consider a more realistic model with random AT/GC base pairs. Show that the breather modes disappear. Take $D_{AT} = 0.05$ eV and $D_{GC} = 0.075$ eV.

c) Couple the system to a Langevin heat bath and explore the DNA melting transition.

15. [P] Two-dimensional Electron Film.
Consider a two-dimensional film of electrons interacting via the Coulomb potential in the presence of a uniform positive background that maintains electric neutrality of the system. Compute the internal energy at a number density of $10^{14}/m^2$ in the temperature range from 1 to 10 K. You should find a phase transition. Refine your result by computing the specific heat C_V, which should exhibit a sharp "lambda transition". Determine the temperature of the phase transition. Choose a time step of order 10^{-12} s. It might also be useful to truncate the Coulomb interaction at short distances (i.e., set $V(r) = V(r_c)$ for $r < r_c$). Try $r_c = 6 \cdot 10^{-8} m$. A handy way to explore temperature-dependence is to equilibrate the system at low temperature, gather measurements, increment the temperature using rescaling, and repeat.

16. [P] X-ray Scattering from Argon.
When light scatters from liquids the radiation scattered by each molecule tends to cancel. Thus it is the *fluctuations* of the particles that drive the scattering intensity. Fluctuations, on the other hand, are related to the pair correlation function in a beautiful way (see Pathria, 1972, Section 13.2). We explore this numerically.

a) Evaluate the pair correlation function for argon at 94 K and a density of 1.474 gm/cm^3.

b) Evaluate the scattering intensity obtained from the Fourier transform:

$$I(q) = \frac{N}{V} \int_0^\infty \frac{\sin qr}{qr} [g(r) - 1] \, 4\pi r^2 dr.$$

c) Compare the peaks in intensity to those measured in (Eisenstein & Gingrich, 1942): $q\sigma = 6.8, 12.3, 18.4, 24.4$.

d) Measure the mean square particle displacement and determine the diffusion coefficient using $\langle r^2 \rangle = 6Dt$ for large t. Compare your result to the experimental value of $2.43 \cdot 10^{-5}$ cm^2/s.

Gravitational Dynamics

17. [T] Galaxies, Galaxy Clusters, and Scale.
    Nondimensionalizing gravitational interactions implies that N galaxies in a galaxy cluster should experience the same dynamics as N stars in a galaxy. But there are no spiral galaxy cluster. Speculate on reasons for this.

18. [T] Mass and Velocity Profiles.
    Consider matter with a mass density $\rho(r)$ and a velocity profile $v(r)$. Derive the following relationship

    $$\rho(r) = \frac{v^2(r)}{4\pi G r^2}\left(1 + 2\frac{r}{v}\frac{dv}{dr}\right).$$

    What mass profile gives rise to a linear rotation curve?

19. Step Size.
    Check the minimal step size required for the sensible interaction of two particles of unit mass interacting gravitationally. Set $G = 1$ and use the Verlet algorithm. Start the particles at rest and follow their trajectories. Plot the total energy and comment on its conservation. You should find $h_{min} \sim \epsilon^\nu$. Determine the power ν (this will not be a precision measurement).

20. Galactic Force.
    Compute the force between two spherically symmetric mass distributions and compare your result to Newton's law and its softened version. You should get results like those in Figure 18.12.

    a) Determine ϵ and the scales r_0 for both fits.

    b) Evidently our fit models are better representations of the force between galaxies. Why would they not be used in practice?

21. [P] Tumbling of Hyperion.
    The only satellite in the solar system that is not synchronized with its host planet is Hyperion, a satellite of Saturn. This is likely because Hyperion has a highly elliptical orbit and is shaped like a potato, which combine to make its motion chaotic.

    We model this behavior by assuming that Hyperion consists of two particles connected by a rigid rod. Follow the angle the mass dipole makes as a function of time by either

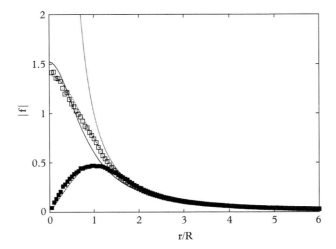

Figure 18.12 *Force between spherical galaxies of radius R. Open squares: $1/r^2$ mass density; solid squares: uniform mass density; dotted line: Newton's law; solid line: softened Newton law; short dash line: fit $(1 - \exp(-r^2/r_0^2))/r^2$; dashed line: fit $(1 - \exp(-r^3/r_0^3))/r^2$.*

(i) implementing the constrained dynamics of Section 18.4.1

(ii) solving for the torque and angular velocity directly.

In the latter case, if one calls the coordinate of the center of mass of Hyperion $\vec{r}$ and the relative coordinate between Hyperion's masses $\vec{d}$, then the expression for the rate of change of the angular velocity is:

$$\frac{d\vec{\omega}}{dt} = -\frac{3GM_{Sat}}{r^3}(\hat{r} \times \hat{d})(\hat{r} \cdot \hat{d}) + O(d^2).$$

You should derive this and use it to simplify the numerical code.

Use the orbital radius of Hyperion to set the length scale and its orbital period to set the time scale. In these units $h = 0.0001$ should suffice for accurate calculations. Make plots of θ ($\hat{d} \equiv (\cos(\theta), \sin(\theta), 0)$) out to 10 Hyperion-years for

a) a circular orbit (what initial speed must be taken for this?)

b) and for an elliptical orbit (take $v_0 = 5$).

c) The elliptical orbit will display chaotic behavior. Use the methods of Section 13.4.1 to estimate the Lyapunov exponent.

22. **Disk Galaxies.**

Write Verlet code to simulate a disk galaxy. Set the disk radius to unity, the disk height to 0.1, the initial velocity distribution to $v_0\hat{\phi}$, and the initial mass profile to $1/r$.

Determine the following properties of the disk:

a) $t = T_{mean}/U$; where U is the total gravitational potential energy and T_{mean} is the bulk rotational kinetic energy. Ostriker and Peebles estimated this as:

$$T_{mean} = \frac{1}{2}\sum_\alpha n_\alpha v_\alpha^2,$$

where α refers to an annular ring, n is the number of particles in the ring and v is the average angular speed in the ring;
b) the pair correlation function;
c) the rotation curve;
d) the mass density profile.

23. Realistic Stars.
Examine the effect of employing a realistic stellar mass distribution on disk galaxies. A possible mass distribution is given by $\rho(m) \propto m^{-\alpha}$ with

$$\alpha = \begin{cases} 0.3, & m < 0.08 M_\odot \\ 1.3, & 0.08 M_\odot < m < 0.5 M_\odot. \\ 2.3, & m > 0.5 M_\odot \end{cases}$$

(Kroupa, 2002). Simulate a disk galaxy with the parameters of Ex. 22 with uniform and Kroupa mass distributions. Notice that this implies simulating very small galaxies. Does the Kroupa distribution change stability with respect to bar formation?

24. Cuspy Halos.
Observations seem to indicate an approximately constant dark matter density in the inner parts of galaxies, while simulations indicate a matter density that scales as $\rho \sim 1/r$. This difference has become known as the "core-cusp problem". Simulate a spherical matter distribution with mass density $\rho \sim 1/r^2$ and track the mass density profile as the system equilibrates. Confirm that $\rho \sim r^{-\alpha}$ for small r with $\alpha \sim 1$. Consult De Blok (2009) for more information.

25. [P] Galaxy Clustering in an Expanding Universe.
We examine structure formation in an FLRW (Friedmann-Lemaître-Robertson-Walker) universe. Consult (Carroll, 2004, chapter 8), or better, (Weinberg, 2008) if you need to remind yourself about the basics of cosmology. If one assumes an isotropic and homogeneous cosmological background, the spacetime metric can only be a function of a single time-dependent variable. This is termed the **FLRW scale factor**, $a(t)$. The scale factor is given by reducing the Einstein gravitational field equations to the **Friedmann equation**:

$$\dot{a} = aH_0\left[\Omega_\Lambda + \Omega_k\frac{a_0^2}{a^2} + \Omega_M\frac{a_0^3}{a^3} + \Omega_R\frac{a_0^4}{a^4}\right]^{1/2}.$$

The Hubble constant is defined as $H = \dot{a}/a$ and present values of quantities are denoted with "0" subscripts. The constants under the square root are parameters that specify the relative density of differing materials in the universe. These are vacuum (Ω_Λ), matter (Ω_M), and radiation (Ω_R). The curvature parameter, Ω_k, is introduced as a convenience. A good approximation to the current universe is $\Omega_\Lambda = 0.73$, $\Omega_M = 0.23$, and $\Omega_R = 0$. Assume a flat universe ($\Omega_k = 0$) in what follows.

The equations of motion for a point particle in *co-moving coordinates* are (Davis and Peebles, 1977; Efstathiou and Eastwood, 1981):

$$\ddot{\vec{x}}_i + 2\frac{\dot{a}}{a}\dot{\vec{x}}_i = -\frac{Gm}{a^3}\sum_{j \neq i}\frac{\vec{x}_{ij}}{(x_{ij}^2 + \epsilon^2/a^2)^{3/2}}.$$

where $\vec{x}_{ij} = \vec{x}_i - \vec{x}_j$. Co-moving coordinates are defined by $x = r/a$, which is convenient for numerical simulation. We have taken the liberty of adding a smoothing parameter, ϵ, to the force law.

Solve the equations of motion using the friction Verlet algorithm of Eq. 18.18. Of course, you should also integrate the Friedmann equation as you go. Assume a uniform initial distribution of galaxies with zero peculiar velocities. Evolve to scale factors of at least $a = 10$ and compute the pair correlation function $g(x) - 1$.

a) Why is ϵ/a rather than ϵ used in the force law?
b) How does the pair correlation function evolve with a?
c) Compare your correlation function to the theoretical expectations of Davis and Peebles (1977), who find:

$$g(x) - 1 = \begin{cases} x^{-\frac{9+3n}{5+n}}, & \text{small } x \\ x^{-(3+n)}, & \text{large } x \end{cases}.$$

Experimentally, $n \approx 0$.

d) Repeat your analysis for a matter dominated universe.
e) Test sensitivity of your results to ϵ.

Further information can be found in Hockney and Eastwood (1988), Section 11.4

Algorithms

26. [T] Verlet Methods.
    Verify that the variant Verlet methods yield the same sequence of position vectors.

27. [T] Symplectic Verlet.

   a) Show that the Verlet algorithm can be written as:
   $$\vec{p}(t+h/2) = \vec{p}(t-h/2) + h\vec{F}(r(t))$$
   $$\vec{r}(t+h) = \vec{r}(t) + h\vec{p}(t-h/2) + h^2 \vec{F}(r(t)).$$

   b) Find the Jacobean matrix (see Eq. 11.60 for a reminder) and hence show that the Verlet algorithm is symplectic.

28. [T] Leap-Frog Angular Momentum.
    Conservation of angular momentum can be problematic. For example, it is ruined by periodic images.

    A convenient representation for angular momentum is:
    $$\vec{L}(t+h/2) = m \sum_i \vec{r}_i(t) \times \vec{v}_i(t+h/2).$$

    Show that:
    $$\frac{d\vec{L}}{dt} \equiv \frac{\vec{L}(t+h/2) - \vec{L}(t-h/2)}{h}$$

    gives the desired torque, $\sum_i \vec{r}_i(t) \times \vec{f}_i(t)$, under Verlet leap frog evolution.

29. Barnes-Hut Tree Size.
    Determine the functional relationship between the number of particles and the number of nontrivial nodes in a Barnes-Hut tree in two dimensions.

30. [P] Binary Expression Trees.
    A **binary expression tree** is a data structure that facilitates the analysis and evaluation of arithmetic expressions. An example tree is given in Figure 18.13. Notice that numerical values are stored in leaves while operators reside in internal nodes.

    Evaluating the expression tree is reasonably simple, as the following code illustrates.

```
double getValue(ExpNode *node) {
   if ((node->left == NULL) && (node->right == NULL)) {
      return node->number;
   } else {                                        // internal node
                                                   // = operator
      double leftVal = getValue(node->left);
      double rightVal = getValue(node->right);
      switch (node->op) {
```

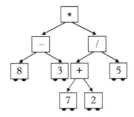

Figure 18.13 *Binary expression tree for* $(8 - 3) * ((7 + 2)/5)$.

```
            case '+':   return leftVal + rightVal;
            case '-':   return leftVal - rightVal;
            case '*':   return leftVal * rightVal;
            case '/':   return leftVal / rightVal;
            case '^':   return pow(leftVal,rightVal);
          }
        }
      }
```

a) Creating an expression tree is tougher. Consult Wikipedia or `cplusplus.com` to get help on the algorithm and code it up.

b) Extend the method to handle trigonometric functions.

31. [P] Boids.
This chapter has focused on collective behavior of identical (or similar) particles that obey Newtonian dynamics. It is, however, interesting to examine systems of particles that do not obey physical laws as models of collective behavior of living things. In particular, we seek to study flocking in everything from bacteria to birds to humans.

The original model concerned schooling in fish (Aoki, 1982) and postulated three "laws of motion": (i) individuals are similar (ii) individuals move at nearly constant speeds and attempt to match velocities (iii) individuals interact within a characteristic length scale.

The model was subsequently popularized by C. Reynolds, who coined the word *boids* to describe the individuals. Unlike physical collisions, boid "collisions" seek to maintain velocities; thus these systems do not equilibrate and the velocities do not approach a Maxwell-Boltzmann distribution. Nevertheless, they display interesting collective behavior, including phase transitions, which we explore in this exercise and the next.

We construct a boids algorithm with three rules for "collisions": movement towards the center of mass, (ii) collision avoidance, and (iii) velocity matching. A specific implementation is:

```
for (int t=0;t<T;t++) {
  TwoVector R;
  R = RCM();
  for (int b=0;b<N;b++) {
    v[b] = v[b] + (R - r[b])/movefactor + avoid(b) +
       match(b);
    r[b] = r[b] + v[b];
    getBC(b);           // enforce periodic boundary
                           conditions
  }
} // end of the time loop
```

with routines

```
TwoVector RCM () {
  TwoVector R;
  for (int b=0;b<N;b++) {
    R = R + r[b];
  }
  return R/N;
}
```

```
TwoVector match(int B) {
  TwoVector vavg;
  for (int b=0;b<N;b++) {
     if (b != B) vavg = vavg + v[b];
  }
  vavg = vavg/(N-1);
  return (vavg-v[B])/matchfactor;
}
```

```
TwoVector avoid (int B) {
  TwoVector R;
  for (int b=0;b<N;b++) {
    if (b != B) {
      if ((r[b]-r[B]).mag() < avoiddistance)
         R = R - r[b] + r[B];
    }
  }
  return R/avoidfactor;
}
```

Figure 18.14 *Twenty-five two-dimensional boids on an android phone.*

Useful parameters are movefactor = 50, avoiddistance = 6, avoidfactor = 1, and matchfactor = 6. The box size can be taken to be several hundred.

Write your own version that displays N boids in real time in a periodic two-dimensional box. Consider adding flourishes like: (i) display the boids direction (see Figure 18.14), (ii) add a color attribute that is a function of the boid speed, (iii) a maximum boid speed, (iv) perching (attraction to a given point), (v) obstacle avoidance. Plot the center of mass location as a function of time. Note the random behavior that emerges from a completely deterministic system!

32. [P] Vicsek Model.
A simple variant of the boids model, called the **Vicsek model**, combines the three laws above and adds a stochastic component. The algorithm can be summarized by (Vicsek and Zafeiris, 2012):

$$
\begin{Vmatrix}
\text{for all boids} \\
\quad \vec{v}_i(t) = v_0 \frac{\langle \vec{v}_j(t-h) \rangle_R}{|\langle \vec{v}_j(t-h) \rangle_R|} + \hat{\eta} \\
\quad \vec{r}_i(t+h) = \vec{r}_i(t) + h\vec{v}_i(t) \\
\text{end}
\end{Vmatrix}
$$

The angle brackets refer to averaging over velocities within a radius R of boid i. The stochastic element is denoted by $\hat{\eta}$. In two dimensions it can be chosen as a uniform random angle in the range $(-\eta/2, \eta/2)$, which is then added to $\theta = \arctan(v_x/v_y)$.

Remarkably, this system is rich enough to experience phase transitions with the parameter η acting as temperature. A useful order parameter is the average normalized velocity:

$$\varphi = \frac{1}{Nv_0} \left| \sum_{i=1}^{N} \vec{v}_i \right|$$

where v_0 is the boid speed.

The approach to the bulk limit is very slow. Runs with $N \sim 10^4$ will give a good idea of what is going on, but you really want $N \sim 10^5$ and very long runs of order 10^5 steps. On my laptop a single-process run with $N = 80000$, at $\rho = 2$, for 10k steps takes about 45 hours. Do what you can. As a final point, it is useful to define the product of two unit vectors in your two-vector class:

```
TwoVector TwoVector::operator*(const TwoVector & other){
  // multiply two unit vectors
  return TwoVector(x*other.x - y*other.y,x*other.y
      + y*other.x);
}
```

a) Code up!
b) Set $R = 1$ and $h = 1$, work in a 100×100 box, take order 10^5 boids, $v_0 = 0.1$ in these units. Fix the boid density to $\rho = N/L^2 = 2$ and evaluate φ over very long runs for η between 0 and 6. You should be able to locate a phase transition near $\eta = 3$.
c) Extrapolate in $1/\sqrt{L}$ to determine η_c to two digits accuracy.
d) Determine the phase diagram in the $\eta - \rho$ plane.
e) For amusement, plot the boid positions and velocity vectors (determined from, say, $\vec{r}(T) - \vec{r}(T-40)$ so that you can see what is going on) for $N = 4000$, $L = 40$, $v_0 = 0.01$, $\eta = 0$, at time $T = 50$. You should get something vortexy like Figure 18.15; this pattern will disappear after several hundred time steps.

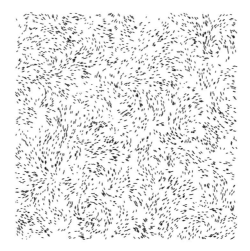

Figure 18.15 *4000 boids shortly after initialization.*

BIBLIOGRAPHY

Allen, M. P. and D. J. Tildesley (1987). *Computer Simulation of Liquids.* Clarendon Press.

Aoki, I. (1982). *A simulation study on the schooling mechanism in fish.* Bulletin of the Japanese Society of Scientific Fisheries 8, 1081.

Boas, F. E. and P. B. Harbury (2007). *Potential Energy Functions for Protein Design.* Current Opinion in Structural Biology 17, 199.

Barnes, J. and P. Hut (1986). *A hierarchical O(N log N) force-calculation algorithm*, Nature 324, 446.

Binney, J. and S. Tremaine (1987). *Galactic Dynamics.* Princeton Series in Astrophysics, Princeton NJ.

Callaway, D. J. E. and A. Rahman (1982). *Microcanonical Ensemble Formulation of Lattice Gauge Theory*, Phys. Rev. Lett. 49, 613.

Carroll, S. M. (2004). *An Introduction to General Relativity, Spacetime, and Geometry.* Addison Wesley.

Conselice, C. J., A. Wilkinson, K. Duncan, and A. Mortlock (2016). *The Evolution of Galaxy Number Density at $z < 8$ and its Implications.* Ap. J. 830, 83.

Cramer, C. (2004). *Computational Chemistry: Theories and Models.* Wiley and Sons.

Davis, M. and P. J. E. Peebles (1977). *On the Integration of the BBGKY Equations for the Development of Strongly Nonlinear Clustering in an Expanding Universe.* Ap. J. Suppl. 34, 425.

De Blok, W. J. G. (2009). *The Core-cusp Problem*, arXiv:0910.3538.

Efstathiou, G. and J. W. Eastwood (1981). *On the Clustering of Particles in and Expanding Universe.* Mon. Not. Royal Astro. Soc. 194, 503.

Eisenstein, A. and N. S. Gingrich (1942). The Diffraction of X-rays by Argon in the Liquid, Vapor, and Critical Regions. Phys. Rev. 62, 261.

Fincham, D. (1984). *More on Rotational Motion of Linear Molecules.* CCP5 Quaterly, 12, 47.

Hansen, J.-P. and I. R. McDonald (2006). *Theory of Simple Fluids.* Academic Press.

Hockney, R. W. (1970). *The Potential Calculation and Some Applications.* Meth. Comp. Phys. 9, 136.

Hockney, R. W. and J. W. Eastwood (1988). *Computer Simulation Using Particles.* Adam Hilger, Bristol.

Holian, B. L., A. F. Voter, and R. Ravelo (1995). *Thermostatted Molecular Dynamics: how to avoid the Toda Demon hidden in Nosé-Hoover Dynamics.* Phys. Rev. E25, 2338.

Hoover, W. G. (1985). *Canonical Dynamics: Equilibrium Phase Space Distributions.* Phys. Rev. A31, 1695.

Jain, A., N. Vaidehi, and G. Rodriguez (1993). *A Fast Recursive Algorithm for Molecular Dynamics Simulation.* J. Comp. Phys. 106, 258.

Konowalow, D. D. and J. O. Hirschfelder (1961). *Morse Potential Parameters for O-O, N-N, and N-O Interactions.* Phys. Fluids. 4, 637.

Kosterlitz, J. M. and D. J. Thouless (1973). *Ordering, Metastability, and Phase Transitions in Two-dimensional Systems.* J. Phys. C 6, 1181.

Kroupa, P. (2002). *The initial mass function of stars: evidence for uniformity in variable systems*, Science 295, 82.

Kubo, R. (1966). *The Fluctuation-Dissipation Theorem.* Rep. Prog. Phys. 29, 255.

Langevin, P. (1908). *Sur la théorie du mouvement brownien.* C. R. Acad. Sci. (Paris) 146, 530.

Leach, A. R. (2001). *Molecular Modelling, Principles and Applications.* Prentice-Hall.

Lebowitz, J. L., J. K. Percus, and L. Verlet (1967). *Ensemble Dependence of Fluctuations with Application to Machine Computations.* Phys. Rev. 153, 250.

Lennard-Jones, J. E. (1924). *On the Determination of Molecular Fields*. Proc. R. Soc. London A, **106**, 463.

Meyers, S. (2008). *Effective C++*. Addison-Wesley Professional Computing Series.

Nosé, S. (1984). *A Unified Formulation of Constant Temperature Molecular-Dynamics Methods*. J. Chem. Phys. **81**, 511.

Ostriker, J. P. and P. J. E. Peebles (1973). *A Numerical Study of the Stability of Flattened Galaxies: or, can cold galaxies survive?*. Ap. J. **186**, 467.

Pathria, R. K. (1984). *Statistical Mechanics*, Pergamon Press.

Peierls, R. E. (1935). *Quelques propriétés typiques des corps solides*. Ann. Inst. Henri Poincaré, **5**, 177.

Ryckaert, J. P. (1985). *Special geometrical constraints in the molecular dynamics of chain molecules*, Mol. Phys. **55**, 549.

Schlick, T., R. D. Skeel, A. T. Brunger, L. V. Kale, J. A. Board, J. Hermans, and K. Schulten (1999). *Algorithmic challenges in computational molecular biophysics*. J. Comp. Phys., **151**, 9.

Springel, V. *et al.* (2005). *Simulations of the formation, evolution, and clustering of galaxies and quasars.* Nature **435**, 629.

Swanson, E. S. (2011). *Modeling DNA Response to Terahertz Radiation*. Phys. Rev. E **83**, 040901R.

Swope, W. C *et al.* (1982). *A Computer Simulation Method for the Calculation of Equilibrium Constants for the Formaion of Physical Clusters of Molecules: Application to Small Water Clusters*. J. Chem. Phys. **76**, 637.

Toxvaerd, S. (1978). *Melting in a Two-dimensional Lennard-Jones System*. J. Chem. Phys. **69**, 4750.

Tuckerman, M. B. J. Berne, and G. J. Martyna (1992). *Reversible Multiscale Molecular Dynamics*. J. Chem. Phys. **97**, 1990.

Verlet, L. (1967). *Computer Experiments on Classical Fluids. I. Thermodynamical Properties of Lennard-Jones Molecules*. Phys. Rev. **159**, 98.

Vicsek, T. and A. Zafeiris (2012). *Collection Motion*. Phys. Rep. **517**, 71.

Weinberg, S. (2008). *Cosmology*. Oxford University Press.

Weiss, Mark A. (2014). *Data Structures and Problem Solving using C++*. 4th ed., Prentice-Hall

Zwicky, F. (1933). *Die Rotverschiebung von Extragalaktischen Nebeln*. Helvetica Physica **6**, 110.

19
Continuum dynamics

19.1 Introduction	643
19.2 Initial value problems	643
19.2.1 Differencing	644
19.2.2 Continuity equations	645
19.2.3 Second order temporal methods	648
19.2.4 The Crank-Nicolson method	648
19.2.5 Second order equations	650
19.2.6 Realistic partial differential equations	651
19.2.7 Operator splitting	652
19.3 The Schrödinger equation	655
19.4 Boundary value problems	658
19.4.1 The Jacobi method	659
19.4.2 Successive over-relaxation	662
19.5 Multigrid methods	664
19.6 Fourier techniques	667
19.6.1 The fast Fourier transform	669
19.6.2 The sine transform	670
19.6.3 An application	672
19.7 Finite element methods	673
19.7.1 The variational method in one dimension	674
19.7.2 Two-dimensional finite elements	675
19.7.3 Mesh generation	677
19.8 Conclusions	684
19.9 Exercises	685
Bibliography	699

*Big whirls have little whirls that feed on their velocity,
and little whirls have lesser whirls and so on to viscosity.*

—Lewis Fry Richardson.

19.1 Introduction

Most physical phenomena are described by fields, and it is partial differential equations that specify how those fields behave. It is therefore no exaggeration to claim that the ability to obtain numerical solutions to partial differential equations underpins much of the technological revolution that has swept the world in the past seventy years. Thus, for example, the ability to obtain accurate solutions to Maxwell's equations in complex environments has transformed the electronics industry. Similarly, the *finite element method* for solving the partial differential equations of structural engineering has revolutionized that field. The *Navier-Stokes* equations occupy a central place in the science of fluid dynamics and find application in automotive and aeronautical development, modeling weather, combustion research, heat flow, and a host of other fields. Applications in chemistry, biology, geology, and social sciences such as economics abound.

This chapter will discuss the main workhorses of the field, namely the finite difference and finite element methods, in sufficient detail that the reader will be able to write code from scratch that is capable of solving an enormous range of problems. We will also learn how to analyze the stability of numerical solution techniques, methods for speeding up convergence and for working in multiple dimensions, efficient ways to apply the Fourier transform, and a powerful approach to dealing with multi-scale problems.

In the following, the generic, physics, and mathematics notations:

$$\frac{\partial u}{\partial x} = \partial_x u = u_x$$

will be used interchangeably with little attempt at uniformity or even rationality.

19.2 Initial value problems

We shall be primarily concerned with solving linear partial differential equations of the sort:

$$R(x,y)f_{xx} + 2S(x,y)f_{xy} + T(x,y)f_{yy} = F(x,y;f,f_x,f_y).$$

The equation can be generalized to higher dimensions and made nonlinear as desired. Our partial differential equation is often categorized according to the sign of the determinant of:

$$\begin{pmatrix} R & S \\ S & T \end{pmatrix},$$

with positive, negative, and zero determinants giving *elliptic*, *hyperbolic*, and *parabolic* equations respectively (generalization to higher dimensions is possible).

This classification refers to the way in which information is propagated. In particular, hyperbolic and parabolic equations permit *causal* propagation and therefore are suitable for describing *initial condition problems*. The boundary conditions relevant to initial value problems specify the unknown function on three surfaces (i.e., $f(t = t_0, x)$, $f(t, x = x_L)$, and $f(t, x = x_R)$) and permit computing the solution on the fourth. In contrast, elliptic equations do not permit evolving the solution in one coordinate and require that all boundaries are fixed. Equations like this are referred to as *boundary value problems*. Thus from the numerical point of view, it is boundary conditions that are relevant and that determine solution strategies. In this regard, we remind the reader that *Dirichlet conditions* specify the field on the boundary while *Neumann conditions* specify normal gradients at the boundary.

19.2.1 Differencing

For the moment we choose to emulate the techniques of quadrature (Chapter 5) and ordinary differential equations (Chapter 11) by *differencing*. Thus we discretize spacetime with a temporal grid spacing of h and a spatial spacing Δ. The notation for a function of x and t becomes:

$$\rho(x, t) \to \rho(\Delta \ell, hn) \to \rho_\ell^n.$$

With this notation, a temporal partial derivative can be written in terms of a *forward difference*:

$$\partial_t \rho(x, t) = \frac{\rho_\ell^{n+1} - \rho_\ell^n}{h} + O(h),$$

a *backward difference*:

$$\partial_t \rho(x, t) = \frac{\rho_\ell^n - \rho_\ell^{n-1}}{h} + O(h),$$

or a *central difference*:

$$\partial_t \rho(x, t) = \frac{\rho_\ell^{n+1} - \rho_\ell^{n-1}}{2h} + O(h^2).$$

Notice that the central difference is more accurate but requires knowledge of the relevant function at time slices that are separated by two units.

Second derivatives can be obtained by iterating the expressions above. Thus, for example,

$$\partial_t^2 \rho(x,t) = \frac{\rho_\ell^{n+1} - 2\rho_\ell^n + \rho_\ell^{n-1}}{h^2} + O(h^2). \tag{19.1}$$

Higher order expressions can be obtained by expanding the range of indices considered (ρ_ℓ^n, $\rho_\ell^{n\pm 1}$, $\rho_\ell^{n\pm 2}$, etc). Whether this is worth the additional computational effort is a strong function of the specific problem (in our experience it is rarely useful to go beyond Eq. 19.1). Even greater flexibility in constructing finite difference approximations to derivatives exists in multiple dimensions. Some of these are discussed in the following sections and in the exercises.

19.2.2 Continuity equations

A fundamental class of differential equations deals with the temporal and spatial evolution of conserved quantities. Mass, energy, momentum, and charge are typical examples of these quantities. At an intuitive level, the rate of change of a conserved quantity in a region must balance the rate at which the quantity is leaving the region. If the quantity is denoted by a scalar field, $\rho(\vec{x}, t)$, and the flux is specified by a vector field $\vec{j}(\vec{x}, t)$, then the *continuity equation* is:

$$\frac{\partial \rho}{\partial t} + \nabla \cdot \vec{j} = 0. \tag{19.2}$$

Consider first the motion of a fluid. In this case ρ is the fluid mass density and the flux is given by $\vec{j} = \rho \vec{v}$, where $\vec{v}$ is the fluid velocity field. For uniform motion in one dimension Eq. 19.2 reads:

$$\frac{\partial \rho}{\partial t} = -v \frac{\partial \rho}{\partial x}. \tag{19.3}$$

This is called an *advection equation* since the solution is of the form $\rho(x - vt)$.

The continuity equation is typically an initial value problem and we seek the temporal evolution of the density. It is therefore convenient to choose the forward difference for the time derivative, while central differencing can be used for the spatial derivative. Thus Eq. 19.3 is:

$$\rho_\ell^{n+1} - \rho_\ell^n = -v \frac{h}{2\Delta} \left(\rho_{\ell+1}^n - \rho_{\ell-1}^n \right) + O(h) + O(\Delta^2). \tag{19.4}$$

This is called the *forward time centered space* (FTCS) method. The usefulness of forward differencing in time is now apparent since this is an equation for ρ_ℓ^{n+1} in terms of quantities known at time hn.

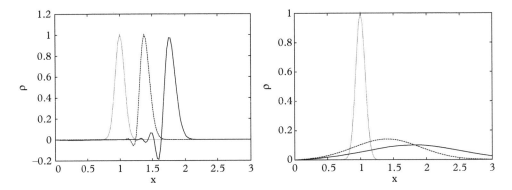

Figure 19.1 *Left: FTCS solution to $\partial_t \rho = -\partial_x \rho$. Dotted line: $t = 0$, dashed line: $t = 0.4$, solid line: $t = 0.8$. Notice the appearance of a spurious mode by $t = 0.4$. Right: Lax solution to $\partial_t \rho = -\partial_x \rho$. The lines are as in the left panel. Notice the rapid decrease in the solution amplitude.*

The left panel of Figure 19.1 shows the time evolution of a pulse that is initially set to be $\rho(x, 0) = \exp(-100(x - 1)^2)$. We set $v = 1$ and algorithmic parameters were $\Delta = 0.025$ and $h = 0.001$. As expected, the pulse moves to the right at speed $v = 1$. However, problems are apparent: the magnitude of the pulse appears to be slowly decaying and, more alarmingly, a spurious mode is evident at $t = 0.4$ and is quite large by $t = 0.8$. Some exploration in parameter space reveals that this anomalous behavior is strongly dependent on parameter values. For example, using $\Delta = 0.008$ resolves (to a large extent) the spurious mode problem, but enhances the magnitude problem.

von Neumann stability analysis

It appears that we need to slow down and consider the consequences of differencing more carefully. A central tool in this effort is *von Neumann stability analysis*, named after the famed Hungarian polymath.

If the coefficients of a partial differential equation are slowly varying one can solve the equation with a spatial Fourier transform. This also applies to the discrete equation, which implies that solutions are of the form:

$$\rho_\ell^n = \xi^n(k) \exp(-ik\ell\Delta) \tag{19.5}$$

where k is a wave number and $\xi(k)$ is a complex function. The form of this equation makes it clear that modes with $|\xi(k)|>1$ will grow exponentially in time, leading to instabilities. The wiggles that appear in the left panel of Figure 19.1 indicate that a high frequency mode has become unstable in precisely this manner.

Substituting the Ansatz of Eq. 19.5 into Eq. 19.4 yields:

$$\xi = 1 - i\frac{vh}{\Delta}\sin(k\Delta)$$

which has a magnitude greater than unity for all wave numbers. The FTCS scheme is therefore *unconditionally unstable*. Evidently, we must rethink our approach to numerically solving the advection equation.

The Lax method

The **Lax method** resolves this issue by replacing ρ_ℓ^n on the right hand side of the time evolution equation with:

$$\rho_\ell^n \to \frac{1}{2}(\rho_{\ell+1}^n + \rho_{\ell-1}^n).$$

This has the effect of replacing the "1" in the previous equation for ξ with $\cos(k\Delta)$:

$$\xi(k) = \cos(k\Delta) - i\frac{vh}{\Delta}\sin(k\Delta) \tag{19.6}$$

which implies stability if:

$$\frac{h}{\Delta} \leq \frac{1}{|v|}. \tag{19.7}$$

This is known as the **CFL condition** and is generally true for all **explicit differencing** schemes[1]. Notice that the CFL condition is a statement concerning numerical causality wherein the speed at which information propagates on the grid must be greater than the physical speed of advection.

The Lax method solution to the advection equation is displayed in the right panel of Figure 19.1. It is apparent that instabilities no longer bedevil the solution. However, the initial pulse (dotted line) suffers from severe dispersion with increasing time. Since the advection equation admits no such effect, we appear to have broken something in the numerical procedure. In fact, the Lax difference equation is equivalent to the FTCS difference equation for:

$$\frac{\partial \rho}{\partial t} = -v\frac{\partial \rho}{\partial x} + \frac{\Delta^2}{2h}\frac{\partial^2 \rho}{\partial x^2},$$

which contains a dispersion term that is a numerical artefact. Thus enforcing stability with the Lax method has the unfortunate corollary of introducing dispersion.

Dispersion is closely related to the damping that is evident in the figure. Indeed, the amplitude of Eq. 19.6 is less than unity when the CFL condition is upheld, and therefore solutions die out. Setting $v = \Delta/h$ resolves the problem.

---

[1] The distinction in differencing schemes will be explained shortly.

19.2.3 Second order temporal methods

So far we have focussed on forward temporal differencing. Are other differencing methods worth considering?

Backwards differencing the advection equation gives:

$$\rho_\ell^n - \rho_\ell^{n-1} = -\frac{hv}{2\Delta}(\rho_{\ell+1}^n - \rho_{\ell-1}^n) + O(h) + O(\Delta^2).$$

In this case one must solve for all $\rho_{\{\ell\}}^n$ in terms of ρ_ℓ^{n-1} at a stroke. For this reason the backward difference equation is called an *implicit method*. Although the equation can be solved by inverting a tridiagonal matrix, which is simple and fast, it is still substantially more onerous than forward differencing. Nevertheless, the method is *unconditionally stable* and therefore can be run for large grid spacings, which is good reason to take on the additional overhead.

The other option is to employ central temporal differencing, which has the advantage of yielding an algorithm that is second order accurate in time:

$$\rho_\ell^{n+1} - \rho_\ell^{n-1} = -\frac{hv}{\Delta}(\rho_{\ell+1}^n - \rho_{\ell-1}^n) + O(h^2) + O(\Delta^2).$$

This is called the *leap frog method*; notice that it does not require inverting matrices but does require storing an additional array of densities. Von Neumann analysis reveals that the CFL condition is again required for stability. An important bonus is that the method does not suffer from numerical dispersion; however, it does suffer from instabilities that are similar to those of the FTCS method because the even and odd mesh points are uncoupled at neighboring time steps (Press, 2007).

19.2.4 The Crank-Nicolson method

A clever way to combine the stability of implicit methods with the accuracy of second order methods was suggested long ago by John Crank and Phyllis Nicolson (1947). The idea is to shift the backward method by one unit in time ($n \to n+1$) and average the result with the FTCS scheme. The *Crank-Nicolson scheme* then reads:

$$\rho_\ell^{n+1} = \rho_\ell^n - \frac{vh}{4\Delta}\left(\rho_{\ell+1}^n - \rho_{\ell-1}^n + \rho_{\ell+1}^{n+1} - \rho_{\ell-1}^{n+1}\right) + O(h^2) + O(\Delta^2). \quad (19.8)$$

Both sides of this equation are centered at $nh + h/2$ (once ρ_ℓ^n has been shifted to the left hand side) so that the method is indeed second order in the temporal spacing. The Crank-Nicolson method is unconditionally stable and exhibits much better control over dispersion than the Lax method. Furthermore, $|\xi| = 1$ so the amplitude of any mode does not decay with time, which implies that numerical dispersion is minimal.

Tridiagonal Matrix Inversion

Implementing the Crank-Nicolson algorithm requires solving the tridiagonal equation $T\vec{v} = \vec{r}$ for $\vec{v}$ many times. Fortunately, an algorithm due to Thomas (of precession fame) provides an efficient way to achieve this. Since the algorithm is not implemented in `Eigen` we provide code here.

```
void triDiag (const std::vector <double> &d,
              const std::vector <double> &u,
              const std::vector <double> &l,
              const std::vector <double> &r,
              std::vector <double> &v) {
  /*
   Thomas algorithm for solving T v = r for v. T is
       tridiagonal:                            d u 0
   with the last element of u                  l d u
   and the first element of l unused           0 l d
  */
  std::vector <double> gamma (v);
  double beta = d[0];
  v[0] = r[0]/beta;
  for (int i=1;i<v.size();i++) {
    gamma[i] = u[i-1]/beta;
    beta = d[i] - l[i]*gamma[i];
    if (beta == 0.0) exit(0);
    v[i] = (r[i]-l[i]*v[i-1])/beta;
  }
  for (int i=v.size()-2;i>=0;i--) v[i] -= gamma[i+1]*v[i+1];
  return;
}
```

Evidently the Thomas algorithm will fail if it should happen that `beta` is zero. In practice this is an unusual occurence, and as Press *et al.* remark, "The tridiagonal algorithm is the rare case of an algorithm that, in practice, is more robust than it appears to be in theory."

Application to one-dimensional advection

We test the Crank-Nicolson method on our advection problem with $\Delta = 0.01$ and $h = 0.01$. The result is shown in the left panel of Figure 19.2, where a beautifully stable advecting density profile is evident.

In spite of this triumph, the Crank-Nicolson scheme is not immune to problems of another sort. The right panel shows the temporal evolution of a square wave pulse under the same conditions as the left panel. The rapid emergence of spurious modes is clear and indicates that the method has difficulty dealing with discontinuities. A resolution to the

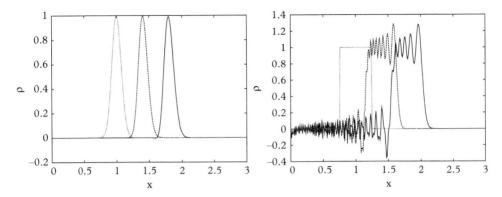

Figure 19.2 *Crank-Nicolson solutions to $\partial_t \rho = -\partial_x \rho$. Left: Gaussian initial distribution. Right: square wave initial distribution. Dotted line: $t = 0$; dashed line $t = 0.4$; solid line $t = 0.8$.*

problem is to employ **upwind differencing**, which is designed to propagate information from grid site ℓ to $\ell + 1$ (i.e., in the direction of the advection). In this case, the difference equation is

$$\rho_\ell^{n+1} - \rho_\ell^n = -v \frac{h}{\Delta} \begin{cases} \left(\rho_\ell^n - \rho_{\ell-1}^n \right) + O(h) + O(\Delta), & v > 0 \\ \left(\rho_{\ell+1}^n - \rho_\ell^n \right) + O(h) + O(\Delta), & v < 0 \end{cases}. \tag{19.9}$$

Notice that the equation is lower order in the spatial spacing, which impacts accuracy; however, this differencing enhances the *fidelity* (the physical reasonableness of the solution) of the solution. Consult Press (2007) for more information and references on this technique.

19.2.5 Second order equations

Many partial differential equations involve second derivatives; in particular the Laplacian operator is ubiquitous because it is used to construct the local representation of net flux through a surface. Typical equations of interest include the heat equation:

$$\frac{\partial T}{\partial t} = \alpha \nabla^2 T$$

where α is the thermal diffusivity, or the advection-diffusion equation,

$$\frac{\partial \rho}{\partial t} = \nabla (D \nabla \rho) - \nabla (\vec{v} \rho).$$

To start the discussion let us consider the diffusion equation in one dimension with constant diffusivity:

$$\frac{\partial \rho}{\partial t} = D\nabla^2 \rho.$$

Implementing FTCS differencing yields the equation:

$$\rho_\ell^{n+1} - \rho_\ell^n = \frac{Dh}{\Delta^2}(\rho_{\ell+1}^n - 2\rho_\ell^n + \rho_{\ell-1}^n) + O(h) + O(\Delta^2). \tag{19.10}$$

Stability analysis for this equation leads to the condition:

$$\frac{2Dh}{\Delta^2} \leq 1, \tag{19.11}$$

which says that the "numerical diffusion" should be greater than the physical diffusion, which is a reasonable analogue of the CFL condition. This result is problematic because one generally wishes to allow the system to diffuse over large spatial scales, L. This requires a time proportional to L^2/D, and Eq. 19.11 then implies that L^2/Δ^2 time steps are required. Clearly it is desirable to make larger time steps, which is only possible with stable (and therefore implicit) methods.

As with the advection equation, the way forward is to average forward and backward differencing, yielding the Crank-Nicolson scheme:

$$\rho_\ell^{n+1} - \rho_\ell^n = \frac{Dh}{2\Delta^2}(\rho_{\ell+1}^{n+1} - 2\rho_\ell^{n+1} + \rho_{\ell-1}^{n+1} + \rho_{\ell+1}^n - 2\rho_\ell^n + \rho_{\ell-1}^n) + O(h^2) + O(\Delta^2) \tag{19.12}$$

This method is unconditionally stable and is our recommended technique for solving second order parabolic or hyperbolic partial differential equations with simple boundaries.

19.2.6 Realistic partial differential equations

So far the partial differential equations we have considered have been rather restricted in form. Strategies for dealing with more realistic cases (that require numerical solution!) are discussed here. Multidimensional partial differential equations are discussed in the next subsection.

If the velocity field in the advection equation is a function of space or time the natural generalization of the FTCS equation (Eq. 19.4) is:

$$\rho_\ell^{n+1} - \rho_\ell^n = -\frac{h}{2\Delta}\left(v_{\ell+1}^n \rho_{\ell+1}^n - v_{\ell-1}^n \rho_{\ell-1}^n\right) + O(h) + O(\Delta^2). \tag{19.13}$$

If v is slowly varying, this gives the stability condition:

$$\frac{h}{\Delta} < \frac{1}{|v_\ell^n|},$$

which is a reasonable extension of the CFL condition. When imposing this equation one should ensure that the left hand side is smaller than the right hand side over all t and x.

Similar considerations apply for a position-dependent diffusion coefficient. In this case it is convenient to discretize $\partial_x D$ at half-time steps. Doing so gives the FTCS equation:

$$\rho_\ell^{n+1} - \rho_\ell^n = \frac{h}{\Delta^2}\left[D_{\ell+1/2}^n(\rho_{\ell+1}^n - \rho_\ell^n) - D_{\ell-1/2}^n(\rho_\ell^n - \rho_{\ell-1}^n)\right]. \tag{19.14}$$

See Ex. 5 for alternatives. Precisely the same expression can be used in any of the other algorithms discussed here. As expected, the generalized stability condition for this equation is:

$$\frac{h}{\Delta^2} < \frac{1}{2|D_\ell^n|},$$

which, of course, needs to hold for all n and ℓ.

Nonlinear diffusion in which $D = D(\rho)$ is not uncommon. Since $\rho_{\ell\pm1/2}$ is not available, one can map:

$$D_{\ell+1/2} \to \frac{1}{2}\left[D\left(\rho_{\ell+1}^n\right) + D\left(\rho_\ell^n\right)\right]$$

with a similar expression for $D_{\ell-1/2}$. This approach presents no problem for explicit schemes. However, for implicit schemes one must solve nonlinear equations of the sort

$$\rho_\ell^{n+1} = \rho_\ell^n + f(D(\rho_{\{\ell\}}^{n+1}); D(\rho_{\{\ell\}}^n)),$$

which is not recommended in code that must be repeated a great many times. A simple expedient is to linearize this equation, which should suffice for sufficiently small grid spacings.

19.2.7 Operator splitting

Until now the discussion has focussed on problems with one spatial dimension. This is for good reason, as multiple dimensions introduce a set of new difficulties that must be addressed. At the simplest level, in going from one to D dimensions storage requirements of order $1/\Delta$ become order $1/\Delta^D$. It is therefore expedient to use methods that permit large spatial grid spacings. For this reason we focus on the implicit Crank-Nicolson scheme in the following.

First, we introduce the differencing operator $\mathcal{D}_\mu^2$ to ease notational burden. This is defined by:

$$\mathcal{D}_\mu^2 \rho_{\vec{\ell}}^n = \rho_{\vec{\ell}+\hat{\mu}}^n - 2\rho_{\vec{\ell}}^n + \rho_{\vec{\ell}-\hat{\mu}}^n.$$

Notice that the spatial index $\vec{\ell}$ is now a vector and that differencing can happen along a certain direction, $\hat{\mu}$, in space.

Applying this to the two-dimensional Laplacian gives:

$$\frac{1}{\Delta^2}\mathcal{D}^2 f_{\ell,m} = \frac{1}{\Delta^2}(f_{\ell+1,m} + f_{\ell-1,m} + f_{\ell,m+1} + f_{\ell,m-1} - 4f_{\ell,m}) + O(\Delta^2),$$

where we have set $\vec{\ell} = (\ell, m)$ to make the two-dimensional nature of the problem explicit. The difference operator on the right hand side is called a *five-point stencil* and is often written as:

$$\mathcal{D}^2 = \begin{pmatrix} 0 & 1 & 0 \\ 1 & -4 & 1 \\ 0 & 1 & 0 \end{pmatrix} \quad (19.15)$$

in what we hope is a clear notation. The analogue in three dimensions is of course a seven-point stencil. As in one dimension, differencing is not unique and other possibilities can sometimes be useful for various reasons. For example the nine-point stencil:

$$\mathcal{D}_9^2 = \frac{1}{6}\begin{pmatrix} 1 & 4 & 1 \\ 4 & -20 & 4 \\ 1 & 4 & 1 \end{pmatrix} \quad (19.16)$$

better reflects the spherical symmetry of the Laplacian operator.

With this notation, the Crank-Nicolson version of the two-dimensional diffusion equation:

$$\frac{\partial \rho}{\partial t} = D\nabla^2 \rho \quad (19.17)$$

is (we consider a constant diffusivity and equal x and y grid spacing for simplicity):

$$\begin{aligned} \rho_{\ell,m}^{n+1} - \rho_{\ell,m}^n &= \frac{Dh}{2\Delta^2}\left(\mathcal{D}_x^2 \rho_{\ell,m}^{n+1} + \mathcal{D}_x^2 \rho_{\ell,m}^n + \mathcal{D}_y^2 \rho_{\ell,m}^{n+1} + \mathcal{D}_y^2 \rho_{\ell,m}^n\right) \\ &= \frac{Dh}{2\Delta^2}\left(\mathcal{D}^2 \rho_{\ell,m}^{n+1} + \mathcal{D}^2 \rho_{\ell,m}^n\right). \end{aligned} \quad (19.18)$$

654 Initial value problems

The problem with this implementation is that the formerly manageable implicit tridiagonal matrix inversion has been replaced with a more onerous matrix inversion (although the matrix remains sparse).

An elegant solution to this issue is to update each dimension separately. We develop the formalism with the aid of the ***displacement operator*** defined by

$$\rho(t+h) = e^{h\partial_t}\rho(t), \tag{19.19}$$

which is likely familiar to the reader as the translation operator from quantum mechanics. The relationship can be established by Taylor expanding both sides of the equation. The starting point will be the simple equation:

$$\rho^{n+1} = e^{h\partial_t}\rho^n$$

or, adopting the Crank-Nicolson trick:

$$e^{-\frac{h}{2}\partial_t}\rho^{n+1} = e^{\frac{h}{2}\partial_t}\rho^n.$$

The diffusion Eq. 19.17 implies $\partial_t = D\partial_x^2 + D\partial_y^2$ which is approximated by

$$\partial_t \approx \frac{D}{\Delta^2}\mathcal{D}_x^2 + \frac{D}{\Delta^2}\mathcal{D}_y^2.$$

Let us refer to the right hand side of this expression generically as $\mathcal{L}_1 + \mathcal{L}_2 + \ldots$. Substituting into the Crank-Nicolson expression gives

$$e^{-\frac{h}{2}\mathcal{L}_1 - \frac{h}{2}\mathcal{L}_2 + \ldots}\rho^{n+1} = e^{\frac{h}{2}\mathcal{L}_1 + \frac{h}{2}\mathcal{L}_2 + \ldots}\rho^n.$$

Expanding both sides to order h^2 then gives the Crank-Nicolson Eq. 19.18.

Operator splitting is implemented by factoring the displacement operators with the aid of the Baker-Campbell-Hausdorff formula:

$$\begin{aligned} e^{\frac{h}{2}\mathcal{L}_1 + \frac{h}{2}\mathcal{L}_2 + \ldots} &= e^{\frac{h}{2}\mathcal{L}_1} e^{\frac{h}{2}\mathcal{L}_2} \ldots e^{\frac{h^2}{8}[\mathcal{L}_1, \mathcal{L}_2]} \ldots + O(h^3) \\ &= e^{\frac{h}{2}\mathcal{L}_1} e^{\frac{h}{2}\mathcal{L}_2} \ldots + O(h^2). \end{aligned} \tag{19.20}$$

Crucially, this expression simplified drastically because central differencing in time is order h^2 and we therefore do not need to worry if some of the operators do not commute. We now rewrite the Crank-Nicolson difference equation with the aid of the definition:

$$\mathcal{U}_i = e^{\frac{h}{2}\mathcal{L}_i};$$

with K operators we obtain:

$$\mathcal{U}_1^\dagger \mathcal{U}_2^\dagger \ldots \mathcal{U}_K^\dagger \cdot \rho^{n+1} = \mathcal{U}_1 \mathcal{U}_2 \ldots \mathcal{U}_K \cdot \rho^n. \tag{19.21}$$

The *operator splitting* procedure consists of replacing this equation with the equivalent set of equations:

$$\begin{aligned}
\mathcal{U}_{K-1}^\dagger \cdot \rho^{n+1/K} &= \mathcal{U}_K \cdot \rho^n \\
\mathcal{U}_{K-3}^\dagger \cdot \rho^{n+2/K} &= \mathcal{U}_{K-2} \cdot \rho^{n+1/K} \\
&\ldots \\
\mathcal{U}_{K-2}^\dagger \cdot \rho^{n+(K-1)/K} &= \mathcal{U}_{K-3} \cdot \rho^{n+(K-2)/K} \\
\mathcal{U}_K^\dagger \cdot \rho^{n+1} &= \mathcal{U}_{K-1} \cdot \rho^{n+(K-1)/K}
\end{aligned} \tag{19.22}$$

Applying this scheme to Eq. 19.18 and rearranging slightly gives the equations:

$$\rho_{\ell,m}^{n+1/2} = \rho_{\ell,m}^n + \frac{Dh}{2\Delta^2}\left(\mathcal{D}_x^2 \rho_{\ell,m}^{n+1/2} + \mathcal{D}_y^2 \rho_{\ell,m}^n\right) \tag{19.23}$$

$$\rho_{\ell,m}^{n+1} = \rho_{\ell,m}^{n+1/2} + \frac{Dh}{2\Delta^2}\left(\mathcal{D}_y^2 \rho_{\ell,m}^{n+1} + \mathcal{D}_x^2 \rho_{\ell,m}^{n+1/2}\right).$$

This trick is called the *alternating direction implicit* (ADI) method. Notice that each half time step requires the inversion of a tridiagonal matrix, as desired.

It is not necessary to split operators based on coordinates. For example an application to the advection-diffusion equation could use an implicit update scheme for the diffusion term while a faster explicit method could be used for the advection portion of the equation.

An alternative splitting scheme updates fully according to $\mathcal{U} = \exp(\frac{h}{2}\mathcal{L}_1 + \frac{h}{2}\mathcal{L}_2 + \ldots)$ but employs a different update method for each suboperator $\mathcal{U}_i$. For example, $\mathcal{U}_i$ may be stable with respect to the ith term in the difference operator. In this case updating is done according to Eq. 19.22 with the replacement $\mathcal{U}_i(h) \to \mathcal{U}_i(h/K)$.

Finally, although Eq. 19.22 is equivalent to Eq. 19.21, it is not necessary to follow this precise sequence. Alternate orderings will induce terms like $[\mathcal{U}_1, \mathcal{U}_2]$ that behave as $\exp(\frac{h^2}{8}[\mathcal{L}_1, \mathcal{L}_2])$, which are $1 + O(h^2)$ and therefore do not affect the difference equations. An example of this will be considered in the next section.

19.3 The Schrödinger equation

The Schrödinger equation is a parabolic equation that describes the spatio-temporal evolution of a complex function. While it is common to separate the temporal evolution via the relation $\psi(x,t) = \exp(-iEt)\phi(x)$ sometimes it is convenient to solve the time-dependent Schrödinger equation. This is of interest, for example, to chemists who seek to understand reaction dynamics.

While stability and accuracy remain desirable goals when differencing the Schrödinger equation, retaining *unitarity* under evolution is also of paramount importance. Thus we require that $\int |\psi|^2 = 1$ at all times so that "probability" is not lost from the system. Conveniently, the Crank-Nicolson scheme meets this additional constraint! If the differenced version of the Hamiltonian is written as:

$$\mathcal{H} = -\frac{1}{2\mu\Delta^2}\mathcal{D}^2 + V_{\vec{\ell}}$$

then the Crank-Nicolson differenced Schrödinger equation can be written as:

$$\psi_{\vec{\ell}}^{n+1} = \psi_{\vec{\ell}}^n - i\frac{h}{2}\mathcal{H}\psi_{\vec{\ell}}^{n+1} - i\frac{h}{2}\mathcal{H}\psi_{\vec{\ell}}^n.$$

Solving for $\psi_{\vec{\ell}}^{n+1}$ gives:

$$\psi_{\vec{\ell}}^{n+1} = \frac{1 - i\frac{1}{2}\mathcal{H}h}{1 + i\frac{1}{2}\mathcal{H}h}\psi_{\vec{\ell}}^n \approx \exp(-i\mathcal{H}h)\,\psi_{\vec{\ell}}^n \approx \exp(-iHh)\psi(\vec{x},t),$$

wherein we recognize the exact quantum mechanical evolution operator.

We shall use the motion of a free Gaussian wavepacket in one dimension for code verification. In this case the initial condition can be taken to be:

$$\psi(x,0) = \frac{1}{\sqrt{a\pi^{1/4}}}\exp(ik_0 x)\exp(-(x-x_0)^2/(2\alpha^2)).$$

The free time-dependent wavefunction is given by:

$$\psi(x,t) = \int_{-\infty}^{\infty} f(k)\exp(i(kx - \frac{k^2}{2m}t))\,dk$$

where f is the Fourier transform of the initial wavepacket.

A comparison of the exact result to the numerical one is shown in Figure 19.3. This was evaluated with $\alpha = \sqrt{2}$, $m = 1$, $k_0 = 1$, at a time of $t = 4$. Algorithm parameters were $\Delta = 0.4$ and $h = 0.01$ and the wavefunction was set to zero at the boundaries. As can be seen, the agreement is excellent.

Let us move on to the more difficult problem of scattering off of a cylindrical barrier of radius a in two dimensions. As we have just discussed, a variety of options for implementing ADI updating are possible. In this case it is convenient to split the

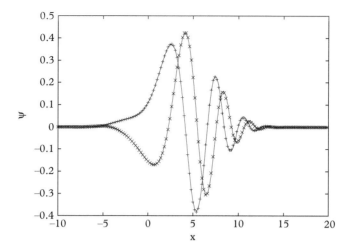

Figure 19.3 *Crank-Nicolson solution of the time-dependent Schrödinger equation for a free one-dimensional wavepacket. Plusses: real part. Crosses: imaginary part. Lines are the exact solution.*

Hamiltonian into three pieces, $-\partial_x^2/(2m)$, $-\partial_y^2/(2m)$, and V. We also choose a slightly different update order than that of Eq. 19.22. Namely:

$$\left(1 + i\frac{h}{2\Delta^2}\mathcal{D}_x^2\right)\psi^{(n+1/3)} = \left(1 - i\frac{h}{2\Delta^2}\mathcal{D}_x^2\right)\psi^n$$
$$\left(1 + i\frac{h}{2\Delta^2}\mathcal{D}_y^2\right)\psi^{(n+2/3)} = \left(1 - i\frac{h}{2\Delta^2}\mathcal{D}_y^2\right)\psi^{(n+1/3)}$$
$$\left(1 + i\frac{h}{2}V\right)\psi^{(n+1)} = \left(1 - i\frac{h}{2}V\right)\psi^{(n+2/3)}. \qquad (19.24)$$

This order is convenient because the two dimensional indexing is simplified when the same operator appears on the left and right hand sides of the equations. Notice that the final equation is diagonal and is thus trivial to implement. It is, of course, possible to include the potential in the first two steps. This has the minor disadvantage of requiring a vector of diagonal matrix elements in the tridiagonal inversion rather than a single number.

Figure 19.4 shows the magnitude of the wavefunction after scattering from a cylindrical barrier of radius 0.5 and height 10 that is placed at the origin. Other parameters were $m = 1$ and $\vec{k}_0 = (1, 2)$. The packet was centered at $(-4, -4)$ originally, thus the collision was rather oblique. Algorithmic parameters were $h = 0.01$ and $\Delta = 0.4$. After the collision the resulting probability density shows a complicated structure with a depletion to the right (where the wave packet has impacted the barrier) and reflected waves moving towards the lower left.

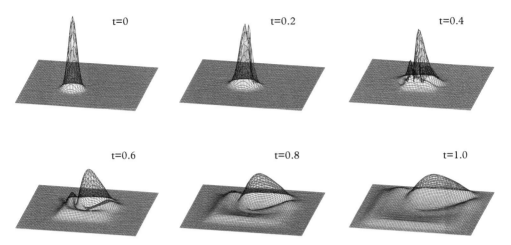

Figure 19.4 *Two dimensional scattering from a cylindrical barrier.*

It is useful to remember that many solution strategies that trade one advantage for another exist. Thus, for example, the ***Visscher method*** is unitary, second order accurate in time, and has explicit updating, which makes it simple to code and fast to run. This method is explored in Ex. 23.

Finally, unlike classical field problems that require discretization on a three-dimensional grid, quantum problems with N particles must be solved (in general) on a $3N$-dimensional grid. This is a severe limitation that renders even the two- or three-particle problem beyond reach. Thus when solving practical quantum problems more sophisticated methods with varying degrees of approximation must be employed. Some of these are discussed in Chapter 23.

19.4 Boundary value problems

Static field configurations are given by solutions to ***boundary value problems***. The canonical exemplar is the elliptic Poisson equation:

$$\nabla^2 u = \rho \qquad (19.25)$$

where u is a scalar field to be determined and ρ is a known source term. A solution is selected via boundary conditions—as mentioned above, these are often Dirichlet conditions that give the value of u on the boundary of the region of interest, or Neumann conditions that specify the normal gradient u at the boundary. Equations of this sort are ubiquitous in the sciences; they appear in the description of heat distributions, electrostatics, gravitational fields, structural mechanics, fluid mechanics, and a variety of other applications.

Boundary value problems differ from initial value problems in that one cannot "integrate inwards" and expect to obtain a consistent solution to the partial differential equation. The problem is, in fact, a *global* one in the sense that the function u at one location must be consistent with values at all neighboring locations.

19.4.1 The Jacobi method

We attempt a solution to the two-dimensional version of Eq. 19.25 by discretizing on a square grid with lattice spacing Δ:

$$u_{\ell+1,m} + u_{\ell-1,m} + u_{\ell,m+1} + u_{\ell,m-1} - 4u_{\ell,m} = \Delta^2 \rho_{\ell,m}. \tag{19.26}$$

Evidently we seek the solution to the problem $A\vec{u} = \vec{\rho}$ where A is a large and sparse matrix. Carl Jacobi (1804 – 1851) invented an iterative scheme to solve matrix problems such as this called the *Jacobi method*. The idea is to split the matrix A into its diagonal, upper, and lower components:

$$A = D + U + L.$$

One then rewrites the matrix problem as:

$$\vec{u} = D^{-1}[\vec{\rho} - (U+L)\vec{u}]$$

and solves it iteratively via the equation:

$$\vec{u}^{(k+1)} = D^{-1}[\vec{\rho} - (U+L)\vec{u}^{(k)}].$$

The method works best for diagonally dominant matrices (which means that A_{ii} is greater than the sum of all the other elements in the ith row). More generally, Jacobi iteration converges if the spectral radius[2] of the *iterator*, $-D^{-1}(U+L)$, is less than unity. This is important because the rate of convergence of the method depends on how close the spectral radius is to unity, and generally this radius approaches unity as the grid size grows. For a d-dimensional grid the spectral radius is $1 - C_d/N^2$ where C_d is a constant and N is the number of grid points along one direction. Thus the algorithm requires $O(N^{d+2})$ steps to converge in d dimensions. We shall see shortly that it is possible to do much better.

The iteration takes on an especially simple form for the two-dimensional Poisson equation. In this case all the elements of D are $-4/\Delta^2$ and

$$(U+L)\vec{u} \to (U+L)u_{\ell,m} = (u_{\ell+1,m} + u_{\ell-1,m} + u_{\ell,m+1} + u_{\ell,m-1})/\Delta^2.$$

---

[2] This is defined as the largest value of the magnitude of all iterator eigenvalues.

Thus Jacobi iteration becomes:

$$u^{(k+1)}_{\ell,m} = \frac{1}{4}\left[u^{(k)}_{\ell+1,m} + u^{(k)}_{\ell-1,m} + u^{(k)}_{\ell,m+1} + u^{(k)}_{\ell,m-1}\right] - \frac{\Delta^2}{4}\rho_{\ell,m}. \qquad (19.27)$$

This is an intuitively pleasing result: new values of the field are obtained from averages over neighboring field values.

Perhaps it has occurred to the reader that iterating as is done in the Jacobi method looks a lot like temporal evolution (to join notation, just let $k \to n$ in Eq. 19.27). This is correct: one can replace an equation such as:

$$\mathcal{L}u = \rho,$$

where $\mathcal{L}$ is some differential operator, with

$$\frac{\partial u}{\partial t} = \mathcal{L}u - \rho. \qquad (19.28)$$

With an additional step (see Ex. 1), FTCS differencing this equation yields the Jacobi iteration equation.

This observation raises a subtle question. In general there are no uniqueness theorems for boundary value problems. In contrast, initial value problems *do* have unique solutions under quite general conditions. Thus Jacobi iteration converges to some solution, but it is not necessarily unique. In practice, this seems to not be much of a concern; however, the careful numericist might want to check the stability of her solution under a variety of different initial function choices.

Despite the inefficiency of the Jacobi method, it still can provide useful results. As an example, we compute the electric potential, V, in and around a parallel plate capacitor. The capacitor is surrounded by a conducting box, hence V is zero there. The box and the plates extend to infinity in the z direction, so this is an effective two-dimensional problem. The capacitor and the potential derived from the Jacobi method are shown in Figure 19.5.

It is convenient to track the global error

$$\Delta V = \frac{1}{N^2} \sum_{\ell,m} |V^{(k)}_{\ell,m} - V^{(k-1)}_{\ell,m}|$$

and terminate iteration once this falls below some tolerance. Values for this quantity are plotted versus iterations for three grid sizes in Figure 19.6. With this one can confirm that achieving a given error scales roughly as $1/N^2$, and hence the entire scheme scales as N^4 (in two dimensions).

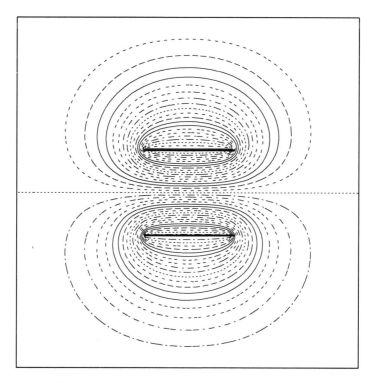

Figure 19.5 *Parallel plate capacitor with electric potential.*

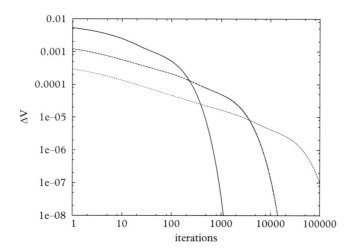

Figure 19.6 *Error in jacobi iteration for the two-dimensional capacitor. Solid line: $N = 40$; dashed line: $N = 160$; dotted line: $N = 640$.*

19.4.2 Successive over-relaxation

A simple improvement on the Jacobi algorithm, called ***Gauss-Seidel iteration***, amounts to overwriting values of the new field $\vec{u}^{(k+1)}$ as they are calculated. It transpires that this can be accommodated in the iterator formalism by splitting the matrix A according to the following (Press, 2007):

$$(L+D)\vec{u}^{(k+1)} = -U\vec{u}^{(k)} + \vec{\rho}. \tag{19.29}$$

In this case the spectral radius of the iterator (in two dimensions with typical boundary conditions) behaves as $1 - C_d/(2N^2)$ and thus a factor of two in efficiency is gained over Jacobi iteration.

This is not much of an improvement; however, Gauss-Seidel iteration is important because it is an element of an algorithm, called ***successive over-relaxation***, that is much more efficient than the Jacobi method. Successive over-relaxation was introduced by Lewis Fry Richardson (of extrapolation fame)[3] in 1911 (Reddy, 2006). The idea is to "over correct" by weighting the proposed update with a factor, ω. The old value of the iterand is also included with a factor of $1 - \omega$. Richardson's original method applied over-relaxation to the Jacobi method, but this was subsequently deprecated because it is unstable. The technique is, however, stable and powerful when combined with Gauss-Seidel iteration. In this case Eq. 19.29 becomes

$$\vec{u}^{(k+1)} = (1-\omega)\,\vec{u}^{(k)} - \omega\,(L+D)^{-1}\,(U\,\vec{u}^{(k)} - \vec{\rho}). \tag{19.30}$$

This version of successive over-relaxation converges for any ω between zero and two. It is important to choose the optimal value of ω, but if one has it, the method converges to a given accuracy in $O(N)$ iterations. This is an enormous advantage over Gauss-Seidel or Jacobi iteration since the saved factor of N is typically much larger than one hundred.

For a square two-dimensional problem with homogeneous Neumann or Dirichlet boundary conditions, the optimal value of the over-relaxation parameter is given by (Press, 2007)

$$\omega_\star = \frac{2}{1 + \sqrt{1 - \cos^2 \frac{\pi}{N}}}.$$

In general the optimal value is not known and one is advised to search for it numerically. An example with Dirichlet boundary conditions and $N = 200$ is shown in Figure 19.7 (left). Over-relaxation exhibits a massive improvement over the Gauss-Seidel ($\omega = 1$) algorithm. The theoretical optimum parameter is 1.969 in this case, which is verified

---

[3] British mathematician, physicist, and pacifist (1881 – 1953). He paid for his pacifism by being denied academic positions for most of his life.

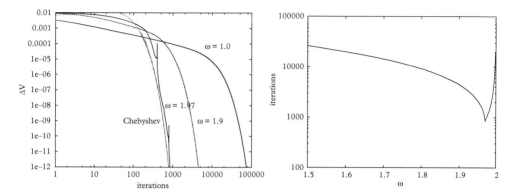

Figure 19.7 *Left: Error for a two-dimensional field problem by iterations. Right: Iterations required to reach an error of* 10^{-12} *vs. the over-relaxation parameter.*

numerically. Notice that the curve for $\omega = 1.97$ exhibits a few "glitches". These glitches grow as ω approaches 2.0; at $\omega = 2.0$ the algorithm fails and the error remains above unity. The numericist should also be aware that the number of iterations required to reach a given tolerance is a sensitive function of the over-relaxation parameter. Figure 19.7 (right) illustrates this by showing the number of iterations required to reach a tolerance of 10^{-12} for a two-dimensional system with $N = 200$.

Although successive over-relaxation is a superior method, it is not the last word in the subject. It is possible, for example, to resurrect Jacobi iteration by choosing an update sequence that mixes over-relaxation and under-relaxation ($0 < \omega < 1$) (Yang and Mittal, 2014). It is also possible to improve efficiency by varying the over-relaxation parameter as the iteration occurs. This is called ***Chebyshef acceleration***. The algorithm requires the user to ***checkerboard update***, which means (in two dimensions) that grid locations are divided into even and odd squares and only one set is updated at each iteration. The sequence of over-relaxation parameters obey the following:

$$\omega^{(0)} = 1$$
$$\omega^{(1)} = \frac{1}{1 - \rho^2/2}$$
$$\omega^{(n+1)} = \frac{1}{1 - \rho^2 \omega^{(n)}/4} \qquad (19.31)$$

with $\rho = \cos(\pi/N)$ on a uniform grid.

Implementing the checkerboard update is simple with the aid of the modulus operator (%). An example follows.

```
for (int i=1;i<N;i++) {        // leave boundaries untouched
  for (int j=1;j<N;j++) {
```

```
  if ((i+j) % 2 == it % 2) { // checkerboard
    double xi = -V[i][j] + ( V[i+1][j] + V[i-1][j] + V[i][j
       +1] + V[i][j-1] )/4.0;
    V[i][j] += omega*xi;
  }
}}
```

Figure 19.7 (left) compares the efficacy of Chebyshev acceleration to successive over-relaxation. Evidently it does somewhat better than over-relaxation in all cases (one must remember to compare sweeps when making this comparison).

19.5 Multigrid methods

Let us look once more at Eq. 19.27: the averaging over nearest neighbors that is apparent in this equation can be interpreted as *smoothing* the solution over short wavelengths. In fact iterative methods are generally effective at reducing high frequency errors. Alternatively, errors at longer wavelengths tend to be retained over very many sweeps (naively, L sweeps would be required to affect wavelengths of order L). If the grid spacing were doubled, the minimum wavelength supported by the grid would also double, and iteration would reduce errors at twice the wavelength. Leveraging this simple observation leads to the ***multigrid method***, which is especially useful for multiscale problems (Wesseling, 1992).

We shall develop the multigrid method by considering a problem:

$$\mathcal{L}u = \rho \tag{19.32}$$

on two grids, one with spacing Δ and one with spacing $\Delta' = 2\Delta$. Here $\mathcal{L}$ is some elliptic differential operator and ρ is a fixed source term. Let $\mathcal{L}_\Delta$ be some appropriate discretization of the differential operator so that the discretized version of Eq. 19.32 is:

$$\mathcal{L}_\Delta u_\Delta = \rho_\Delta. \tag{19.33}$$

One can solve this equation with a large number of smoothing operations (in practice, checkerboard Gauss-Seidel iterations), but the idea is to only perform a few of these to save computational effort. Call the resulting approximate solution to Eq. 19.33, $\tilde{u}_\Delta$ and introduce the *error* by:

$$\epsilon_\Delta \equiv u_\Delta - \tilde{u}_\Delta \tag{19.34}$$

and the *defect* via:

$$\delta_\Delta \equiv \mathcal{L}_\Delta \tilde{u}_\Delta - \rho_\Delta. \tag{19.35}$$

Notice that:

$$\mathcal{L}_\Delta \epsilon_\Delta = -\delta_\Delta. \tag{19.36}$$

Now consider a coarse grid with, say, $\Delta' = 2\Delta$. We shall define all the quantities above on the new grid; thus, for example,

$$\mathcal{L}_{\Delta'} \epsilon_{\Delta'} = -\delta_{\Delta'}. \tag{19.37}$$

The multigrid strategy will be to compute a correction to $\tilde{u}_\Delta$, namely ϵ_Δ, that brings the numerical solution closer to the exact result. The error at scale Δ is to be obtained after several smoothing steps on the fine grid and several more on the coarse grid, which requires being able to map information between the grids. In particular, we will require the defect on the coarse grid after having obtained it on the fine grid. This can be accomplished via an operator that *downsamples* the defect:

$$\delta_{\Delta'} = \mathcal{R}_{\Delta',\Delta} \delta_\Delta, \tag{19.38}$$

known as the *restriction operator*.

We shall also need the error function on the fine grid as obtained from the coarse grid. Evidently this requires interpolation, hence we define an interpolation operator via

$$\epsilon_\Delta = \mathcal{P}_{\Delta,\Delta'} \tilde{\epsilon}_{\Delta'}. \tag{19.39}$$

The notation is because the interpolation operator is often called the *prolongation operator*.

With the notation in place we are ready to specify a simple multigrid algorithm. In pseudocode:

```
iterate
    approximately solve ℒ_Δ u_Δ = ρ_Δ for ũ_Δ
    compute δ_Δ = ℒ_Δ ũ_Δ − ρ_Δ
    compute δ_Δ' = ℛ_{Δ',Δ} δ_Δ
    approximately solve ℒ_Δ' ε_Δ' = −δ_Δ' for ε_Δ'
    compute ε_Δ = 𝒫_{Δ,Δ'} ε_Δ'
    set ũ_Δ = ũ_Δ + ε_Δ
end
```

Again, the recommended method for smoothing is checkerboard Gauss-Seidel iteration. Successive over-relaxation is *not* recommended because it does not smooth high frequency modes.

This algorithm makes many passes through a pair of grids to arrive at a solution. Of course, it is possible to extend the idea and introduce a series of ever coarser grids, each of which is visited according to some protocol (notice the conceptual similarity to the

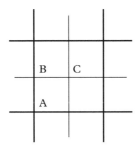

Figure 19.8 *Two-dimensional prolongation.*

spatial renormalization group discussed in Section 8.4). An additional option is to start the algorithm on a very coarse grid and progress to finer grids after several iterations. Thus there are many multigrid methods; which to choose is up to the user and his or her specific problem.

The final task is to specify the restriction and prolongation operators. A simple choice for prolongation is to employ bilinear interpolation (in two dimensions). Referring to Figure 19.8 we see that the field at the point A will map to itself, thus, calling this point (x, y),

$$u_\Delta(x, y) = u_{\Delta'}(x, y).$$

For point B we obtain:

$$u_\Delta(x, y + \Delta) = \frac{1}{2}\left(u_{\Delta'}(x, y) + u_{\Delta'}(x, y + \Delta')\right).$$

And finally, the expression at point C is:

$$u_\Delta(x+\Delta, y+\Delta) = \frac{1}{4}\left(u_{\Delta'}(x, y) + u_{\Delta'}(x + \Delta', y) + u_{\Delta'}(x, y + \Delta') + u_{\Delta'}(x + \Delta', y + \Delta')\right).$$

These equations are often represented in terms of a ***prolongation stencil***

$$\mathcal{P}_{\Delta,\Delta'} = \frac{1}{4}\begin{bmatrix} 1 & 2 & 1 \\ 2 & 4 & 4 \\ 1 & 2 & 1 \end{bmatrix}, \qquad (19.40)$$

in a notation that we hope is clear.

Restriction is even simpler: one can set:

$$u_{\Delta'}(x, y) = u_\Delta(x, y).$$

However, it turns out that this choice is not ideal, and a better option is to set:

$$\mathcal{R} = \frac{1}{4}\mathcal{P}^\mathsf{T}.$$

This works in well in cases where $\Delta' = 2\Delta$ and can be generalized to other cases (Press, 2007).

The multigrid method is readily applicable to nonlinear problems if a nonlinear relaxation method is available for smoothing errors. Updating is done via:

$$\tilde{u}'_\Delta = \tilde{u}_\Delta + \mathcal{P}_{\Delta,\Delta'}(\tilde{u}_{\Delta'} - \mathcal{R}_{\Delta',\Delta}\tilde{u}_\Delta)$$

where $\tilde{u}_{\Delta'}$ is obtained by solving a nonlinear equation on the coarse grid (Press, 2007; Brandt, 1977).

Finally, we comment that parallelization of the multigrid method as presented is not simple because Gauss-Seidel iteration overwrites values in place. However, one can pay the factor of two price in speed and employ Jacobi iteration, which permits trivial parallelization of the updating loops.

19.6 Fourier techniques

Differencing is not the only path to the solution of partial differential equations. It is possible, for example, to expand the field in a basis and determine the coefficients of the expansion via the relevant partial differential equation. An expansion in Fourier modes is especially useful for differential equations with constant coefficients and boundaries that coincide with the coordinate system. While it is possible to set, say, $u(x) = \sum_i a_i \sin(i\pi x)$ we shall develop the Fourier expansion for difference equations.

One may be tempted to expand in plane waves; however, it is important to respect the problem boundary conditions. A sine expansion forces the solution to zero at the boundaries. This is convenient for Dirichlet boundary conditions because nonzero boundary values can be incorporated with a trick, as we shall see. Alternatively, Neumann conditions can be accommodated with a cosine expansion. The technique for solving partial differential equations with Dirichlet conditions will be developed here. Neumann conditions follow a similar scheme; details can be found in Press (2007).

The method follows the traditional approach of undergraduate mathematical physics by using a remarkable discrete analogue of the orthogonality relationship:

$$2\int_0^1 \sin(i\pi x)\sin(j\pi x)\,dx = \delta_{i,j};$$

namely,

$$\frac{2}{N+1} \sum_{n=1}^{N} \sin\left(\frac{in\pi}{N+1}\right) \sin\left(\frac{jn\pi}{N+1}\right) = \delta_{i,j}. \qquad (19.41)$$

Let us consider the problem:

$$\nabla^2 u = \rho \qquad (19.42)$$

in the two dimensional region $[0, 1] \times [0, 1]$ with Dirichlet boundary conditions. We take these to be $u(0, y) = u(1, y) = u(x, 0) = u(x, 1) = 0$ for the moment. Discretize in the usual way

$$\frac{u_{i+1,j} - 2u_{i,j} + u_{i-1,j}}{\Delta_x^2} + \frac{u_{i,j+1} - 2u_{i,j} + u_{i,j-1}}{\Delta_y^2} = \rho_{i,j}.$$

Expand the field and the source in a sine series:

$$u_{i,j} = \sum_{n=1}^{N} \sum_{m=1}^{N} \hat{u}_{n,m} \sin\left(\frac{ni\pi}{N+1}\right) \sin\left(\frac{mj\pi}{N+1}\right) \qquad (19.43)$$

with a similar equation for $\rho_{i,j}$. Substituting into Eq. 19.42 and performing some simple trigonometric manipulations yields the solution in Fourier space:

$$\hat{u}_{m,n} = \frac{\hat{\rho}_{m,n}}{\frac{2}{\Delta_x^2}\left(\cos(\frac{n\pi}{N+1}) - 1\right) + \frac{2}{\Delta_y^2}\left(\cos(\frac{m\pi}{N+1}) - 1\right)}. \qquad (19.44)$$

Substitution into Eq. 19.43 then gives the desired solution.

Dirichlet boundary conditions

So far we have obtained a solution to the Poisson equation that evaluates to zero on the boundary. To incorporate boundary conditions we take the grid to run from $i = 0$ to $i = N + 1$ with the former corresponding to $x = 0$ and the latter to $x = 1$. Consider the case $u(0, y) = w(y)$ so that $u_{0,j} = w_j$. Evaluating the difference equation at $i = 1$ gives

$$\frac{u_{2,j} - 2u_{1,j} + w_j}{\Delta_x^2} + \frac{u_{1,j+1} - 2u_{1,j} + u_{1,j-1}}{\Delta_y^2} = \rho_{1,j}.$$

Dirichlet conditions correspond to setting $u_{0,j} = 0$ so this equation can be brought to the previous form by translating w_j to the right hand side. Thus the condition on the bottom surface can be incorporated into the formalism by making the replacement

$$\rho_{1,j} \to \rho_{1,j} - \frac{w_j}{\Delta_x^2}.$$

Of course a similar transformation can be made for the other three surfaces.

19.6.1 The fast Fourier transform

The method provides strikingly accurate solutions in cases where the source and boundary functions are smooth. However, it is rather slow, requiring $O(N^{2d})$ operations in d dimensions. Performance can be improved considerably by employing the famed *fast Fourier transform* (FFT) method, which performs one dimensional discrete Fourier transforms in $N \log N$ time. FFT algorithms are traditionally designed for signal processing; we therefore briefly review some of the conventions in that field.

FFT input is described by a function $f(t)$ in the time domain that is sampled at uniform discrete points, $t_j = \Delta j$. The *sampling frequency* is defined to be $1/\Delta$. A sawtooth pattern represents the shortest wavelength possible on the temporal grid and therefore the highest frequency that can be supported is $1/(4\Delta)$. The *Nyquist frequency* is the minimum rate at which a signal can be sampled without introducing errors and is twice the highest frequency, $f_N = 1/(2\Delta)$. Thus, if one wishes to reproduce the spectrum of a signal with highest frequency $1/(4\Delta)$, the signal must be sampled with at least twice this frequency.

Consider N signal points f_j with $j \in [0, N-1]$. These correspond to N points in the frequency domain given by

$$\hat{f}_n = \sum_{j=0}^{N-1} f_j \exp\left(-\frac{2\pi i j n}{N}\right) \tag{19.45}$$

with $n \in [0, N-1]$.

It is useful to keep the definition of the FFT in mind; for example, one might be tempted to test FFT code by implementing the Fourier transform:

$$\hat{f}(\omega) = \int_{-\infty}^{\infty} e^{-i\omega t} \exp(-t^2)\, dt = \sqrt{\pi}\, \exp(-\omega^2/4).$$

To obtain this one can set $t = \Delta(j - N/2)$ and sum over j as in Eq. 19.45. The result is:

$$\hat{f}(\omega) = \Delta(-1)^n \hat{f}_n^{FFT}$$

with $\omega = 2\pi n/(\Delta N)$. Note also that this is a periodic function in the domain $\omega \in (0, 2\pi/\Delta)$, which, when mapped to $(-\pi/\Delta, \pi/\Delta)$, does indeed look like the expected result.

The FFT method has a long history; however, it was popularized by James Cooley and John Tukey in the 1960s at about the same time that digital signal processing became an industrial activity (Elliot and Rao, 1982). The **Cooley-Tukey algorithm** is based on a series of clever observations that was made twenty years before by Gordon Danielson and Cornelius Lanczos. First, note that the discrete Fourier transform is a linear transformation:

$$\vec{\hat{f}} = \mathbf{Z}\vec{f}$$

where the matrix $\mathbf{Z}$ has a special structure, namely:

$$(\mathbf{Z})_{nj} = z^{nj}$$

with $z = \exp(2\pi i/N)$.

Next we observe that the Fourier sum can be broken into even and odd subsums:

$$\hat{f}_n = \sum_{j=0}^{N/2-1} f_{2j} \exp\left(-\frac{2\pi i j n}{(N/2)}\right) + z^n \sum_{j=0}^{N/2-1} f_{2j+1} \exp\left(-\frac{2\pi i j n}{(N/2)}\right)$$
$$\equiv \hat{f}_n^{(e)} + z^n \hat{f}_n^{(o)}.$$

A student who has just read the section on multigrid methods or Chapter 8 will not fail to see the possibility of recursing this expression down to sums over a single element, for which the FFT transform is simply the signal value itself, f_j. Thus for every j there is a pattern of even and odd decompositions that maps to it.

The requisite pattern is given by **bit reversal**, namely the sequence of even and odd sums is reversed, even sums are assigned 0 and odd sums are assigned 1, and the resulting pattern is j in binary notation.

Given its ubiquity and power, the Cooley-Tukey algorithm has justifiably been called one of the most important algorithms of all time. We cannot stress enough that the FFT method is *fast* compared to brute force summing. A simple test on a mid 2014 MacBook Pro reveals that naively summing to obtain the sine transform of a series with one million points takes approximately 4.2 hours, whereas the FFT algorithm takes 0.18 seconds. The numericist should never use brute force summing except in very simple cases.

19.6.2 The sine transform

The discussion makes it clear that it is useful to implement the sine transform using FFT methods. Ideally, this will be provided by a software package; however, there appears to be a paucity of quality C++ FFT software. We therefore use unsupported code[4] from `Eigen` that partially implements the `KissFFT` code suite of Mark Borgerding (2013).

---

[4] `Eigen` contains contributed modules that are provided "as is". Documentation and support for these can be sparse.

This code executes fast Fourier transforms and inverse transforms of one-dimensional data. In skeletal form the protocol is as follows:

```
#include <unsupported/Eigen/FFT>
...
Eigen::FFT<double> fft;
std::vector<double> f(N,0.0);
std::vector<std::complex<double> > fhat(N,0.0);
...
fft.fwd(fhat,f);    // Fourier transform f -> fhat
...
fft.inv(f,fhat);    // inverse Fourier transform fhat -> f
```

`Eigen::FFT` scales the inverse transform such that the inverse of a transform yields the initial function. Other datatypes supported are float and `long double`.

The sine transform of a series f_j can be obtained from the FFT of a series that has been doubled in length and is antisymmetric about the midpoint. Working through this carefully (see Ex. 25) yields the following equality:

$$\hat{f}_m \equiv \sum_{j=0}^{2N-1} f_j \exp\left(-\frac{2\pi i j m}{2N}\right) = -\frac{1}{2i}\hat{f}_m^{\text{sine}}$$

when one extends the series according to

$$f_0 = 0, \quad f_N = 0, \quad \text{and } f_{2N-j} = -f_j \text{ for } j \in [1, N-1].$$

An implementation using `std::vector` is:

```
void sineTransform(vector<double> & fk, const vector<double>
    & fx) {
  int N = fx.size();
  Eigen::FFT<double> fft;
  std::vector<double> f(2*N,0.0);
  std::vector<std::complex<double> > fhat(2*N,0.0);
  for (int i=1;i<N;i++) {
    f[i] = fx[i];
    f[2*N-i] = -fx[i];
  }
  f[0] = 0.0;
  f[N] = 0.0;
  fft.fwd(fhat,f);
  for (int i=0;i<N;i++) fk[i] = -1.0*imag(fhat[i])/2.0;
  return;
}
```

The inverse of the sine transform is also a sine transform, so additional code is not necessary. However, the user should divide a sine transform that is acting as an inverse by $N/2$ to maintain normalization (see Eq. 19.41 for the explanation).

We seek to solve a two-dimensional Poisson equation and thus require the two-dimensional sine transform. This can be achieved by transforming the rows of the input matrix and then the columns of the resulting matrix (or vice versa).

19.6.3 An application

We will illustrate the Fourier transform method by solving the problem of viscous fluid flow in a channel with a uniform rectangular cross section. The channel will be oriented in the z direction and the fluid will be under pressure p. In this case the speed of the fluid in the z direction is governed by the partial differential equation:

$$\nabla^2 v(x,y) = \frac{1}{\mu}\frac{\partial p}{\partial z}.$$

A constant viscosity μ has been assumed, and we will also assume that the rate of change of pressure in the z direction is constant in x and y. The fluid speed at the channel walls will be taken to be zero, while the speed at the top of the channel will be set to v_z. Finally, if the channel width is W and its height is H, the flow rate is given $R = \int_0^W \int_0^H dx\, dy\, v(x,y)$.

Following the procedure leading to Eq. 19.44 gives the following code. The boundaries have been labelled by i or $j = 0$ or $N+1$, thus the degrees of freedom have indices running from $i/j = 1$ to $i/j = N$.

```
Eigen::MatrixXd rho(N+1,N+1);
Eigen::MatrixXd rhoHat(N+1,N+1);
Eigen::MatrixXd uHat(N+1,N+1);
Eigen::MatrixXd u(N+1,N+1);

// obtain rho
for (int i=1;i<N+1;i++) {
for (int j=1;j<N+1;j++) {
  rho(i,j) = Rho;
  if (j == N) rho(i,j) = Rho - vz/(dely*dely); // implement
                                               top BC
}}
// obtain rhoHat
sineTransform2D(rhoHat,rho);
// obtain uHat
for (int n=1;n<N+1;n++) {
for (int m=1;m<N+1;m++) {
```

```
      uHat(n,m) = rhoHat(n,m)/(2.0*(cos(n*M_PI/(N+1))-1.0)/
                                   (delx*delx) +
                              2.0*(cos(m*M_PI/(N+1))-1.0)/
                                   (dely*dely)) ;
   }}
   // invert to obtain u
   sineTransform2D(u,uHat);
   // multiply by 4.0/((N+1)*(N+1)) to normalize
```

The velocity field produced by this code for a modest 80 × 80 grid is shown in Figure 19.9. Physics parameters were $\mu = 1$, $v_z = 0.1$, $H = 0.5$, $W = 1$, and $p_z = -6$. We stress that the code is extremely fast and stable; running with millions of grid points is entirely feasible on a laptop.

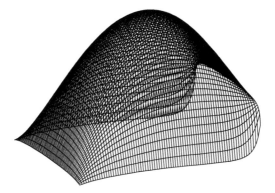

Figure 19.9 *Velocity field for a rectangular channel.*

19.7 Finite element methods

Finite differencing is an effective and efficient method for solving partial differential equations; however, its usefulness is severely hampered in applications with non-rectilinear integration regions. This can exclude entire fields, such as engineering, where one is typically interested in properties of objects of unlimited variety in shape and composition. In this case the standard tool is the ***finite element method***, which is applied to stresses in materials, heat flow, fluid dynamics, and many other topics in science and engineering.

At its simplest, the idea of the finite element method is to expand the unknown field in a basis and choose the parameters of that basis variationally (in a way to be defined shortly). In practice the basis is chosen to be a simple function with support over a small region. Regions can be arbitrary but are often taken to be ***simplexes***, which are shapes that contain volume with the smallest possible number of sides. Simplexes are particularly useful in approximating complex boundaries, which is desirable in engineering applications. In one dimension a simplex is a line segment, in two dimensions it is a triangle, and in three it is a tetrahedron. All of these are called *elements* when applied to the finite element method.

19.7.1 The variational method in one dimension

We introduce the finite element method by considering the solution to a one dimensional differential equation such as:

$$\frac{d^2 u}{dx^2} = f \qquad (19.46)$$

with Dirichlet boundary conditions $u(0) = u_L$ and $u(1) = u_R$. The function f is taken to be a constant for the time being.

The first step is to convert the differential equation into a variational problem. One can verify that the functional derivative of

$$\mathcal{J}[u] = \int_0^1 dx \left[\frac{1}{2}\left(\frac{du}{dx}\right)^2 + fu\right] \qquad (19.47)$$

does indeed yield Eq. 19.46. For the moment we do not worry about boundary conditions.

Finite elements are introduced by considering a basis of linear functions over line segments. We take these elements to have length Δ for simplicity. Element j is defined to extend from $x_j = x_0 + j\Delta$ to x_{j+1}, while the last element runs from x_N to x_{N+1}. In our case $x_0 = 0$ and $x_{N+1} = 1$. Similarly, the values of the function at the node points are denoted $u_j = u(x_j)$.

The next step is to introduce basis functions that are defined over each element. There is benefit in keeping these as simple as possible, and we therefore consider linear functions. In general, an nth order partial differential equation should employ basis functions that ensure continuity across interfaces of derivatives up to order $n-1$ (Fenner, 1996). Setting notation, we have:

$$\phi_j(x) = \begin{cases} a_j + b_j x, & x \in (x_j, x_{j+1}) \\ 0, & \text{elsewhere} \end{cases} \qquad (19.48)$$

and

$$u(x) = \sum_{j=0}^{N} \phi_j(x) \equiv \sum_{j=0}^{N} \tilde{\phi}_j(x)\,\theta_j, \qquad (19.49)$$

where θ_j is a step function on the interval (x_j, x_{j+1}) so that $\tilde{\phi}_j$ is ϕ_j without the constraint to the jth interval.

Boundary conditions are imposed by requiring $\phi_0(x_0) = u_L$ and $\phi_N(x_{N+1}) = u_R$. Furthermore, we require that $u(x)$ is continuous so that $u_j(x_{j+1}) = u_{j+1}(x_{j+1})$. The net result is a piecewise continuous approximation to $u(x)$ that is determined by the values u_j for j between 1 and N. In detail, the basis functions are:

$$\tilde{\phi}_j(x) = u_j + \frac{u_{j+1} - u_j}{\Delta}(x - x_j), \tag{19.50}$$

with $u_0 = u_L$, and $u_{N+1} = u_R$.

Inserting Eqs. 19.49 and 19.50 into Eq. 19.47 gives:

$$\mathcal{J}(\{u_j\}) = \frac{1}{2}\Delta \sum_j \left(\frac{u_{j+1} - u_j}{\Delta}\right)^2 + f\frac{\Delta}{2}\sum_j (u_{j+1} + u_j). \tag{19.51}$$

The variational principle is now implemented by minimizing with respect to the unknowns, $u_1 \to u_N$. Taking the derivative with respect to u_ℓ gives:

$$\frac{1}{\Delta}(u_{\ell+1} - 2u_\ell + u_{\ell-1}) = \Delta f. \tag{19.52}$$

This is a remarkable result: minimizing the functional $\mathcal{J}$ over a space of piecewise linear functions is equivalent to finite differencing! Because of this equivalence we do not pursue the one-dimensional case any further, except to note that generalizations such as varying element sizes and higher order basis functions can be easily (in principle) incorporated in the method. It is also possible to forget about elements entirely and use an expansion such as $u = \sum_j c_j \phi_j(x)$ where the basis functions are now defined over the entire region. This is called the ***Ritz method***.

19.7.2 Two-dimensional finite elements

It is in higher dimensions, where boundaries can be complicated, that the power of the finite element method becomes apparent. We will develop the formalism using the simple Poisson equation

$$\nabla^2 u = \rho \tag{19.53}$$

as a reference point. The solution is required over a region Ω subject to boundary conditions of some type.

The solution to Eq. 19.53 is an extremum of the functional:

$$\mathcal{J}[u] = \int_\Omega dx\, dy \left[\frac{1}{2}(\nabla u)^2 + \rho u\right].$$

We ignore boundary conditions once more. As in one dimension, a solution will be produced by dividing the region of integration into simplexes—triangles—and expanding $u(x,y)$ in linear functions on these regions. If an element is labelled e then the basis function can written as:

$$\phi_e(x,y) = \begin{cases} a_e + b_e x + c_e y, & (x,y) \in e\text{th element} \\ 0, & \text{elsewhere.} \end{cases} \quad (19.54)$$

Thus

$$u(x,y) = \sum_e \phi_e(x,y).$$

The three constants in the basis function map to the values of u at the three nodes of the element, which we will label u_{e0}, u_{e1}, and u_{e2}. Specifying the basis function in this way serves the important role of guaranteeing continuity over the element boundaries because shared edges will also share values of u_{ei}.

For a given element, the contribution to the equivalent functional evaluates to:

$$\mathcal{J}[u_e] = \frac{A_e}{2}(b_e^2 + c_e^2) + \rho A_e \bar{u}_e$$

where A_e is the area of the eth element and:

$$\bar{u}_e = \frac{1}{3}(u_{e0} + u_{e1} + u_{e2}).$$

If ρ is a function of location this will be replaced with

$$\rho A_e \bar{u}_e \to a_e \langle \rho \rangle_e + b_e \langle x\rho \rangle_e + c_e \langle y\rho \rangle_e$$

where we have introduced some new notation that is hopefully clear.

Since a_e, b_e, and c_e depend linearly on u_{e0}, u_{e1}, and u_{e2} the derivatives of $\mathcal{J}[u_e]$ are easy to obtain:

$$\frac{\partial \mathcal{J}[u_e]}{\partial u_{ei}} = \sum_{j=0}^{2} K_{ij}^{(e)} u_{ej} + f^{(e)} u_{ei}$$

where $K^{(e)}$ is the **local stiffness matrix** and $f^{(e)}$ is the **local force**. Expressions for these are:

$$K_{ij}^{(e)} = \frac{1}{4A_e}(\beta_i^{(e)} \beta_j^{(e)} + \gamma_i^{(e)} \gamma_j^{(e)}) \quad (19.55)$$

and:

$$f^{(e)} = \rho \frac{A_e}{3}. \qquad (19.56)$$

The subsidiary variables are geometrical factors associated with the simplex e and are given by

$$\beta_0^{(e)} = y_{e1} - y_{e2}, \quad \beta_1^{(e)} = y_{e2} - y_{e0}, \quad \beta_2^{(e)} = y_{e0} - y_{e1},$$
$$\gamma_0^{(e)} = x_{e2} - x_{e1}, \quad \gamma_1^{(e)} = x_{e0} - x_{e2}, \quad \gamma_2^{(e)} = x_{e1} - x_{e0}. \qquad (19.57)$$

19.7.3 Mesh generation

The next step in the finite element method consists of gathering the local stiffness matrices into a global stiffness matrix. This depends crucially on the layout of the elements, which brings us to the topic of **mesh generation**. A mesh is defined by the coordinates of its nodes and their triangulation and this information must be carefully tracked to construct the global stiffness matrix.

A typical mesh for a two-dimensional magnetic field problem with a ring of shielding material (at the lower left) is shown in Figure 19.10. Several things should be noted about this figure: (i) all internal nodes must have at least three edges emerging from them, and can have arbitrarily many (although more than six is rare), (ii) triangles that are overly elongated are not preferred as they do not yield reasonable distributions of nodes,

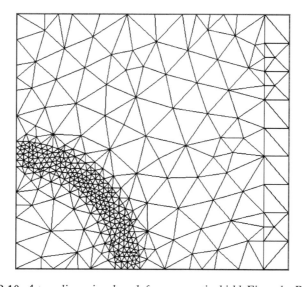

Figure 19.10 *A two-dimensional mesh for a magnetic shield. Figure by R. Williams.*

678 *Finite element methods*

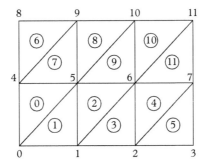

Figure 19.11 *Element and node labels for a rectangular region.*

(iii) most importantly, every element has exactly three nodes associated it. Elements that obey the last condition are called ***conforming***; see Exercise 32 for why this property is important.

Generating complex meshes such as in Figure 19.10 is not an easy task. We therefore start with meshes for simple geometries, such as a rectangle. In this case one can create a mesh in a uniform pattern by placing evenly spaced nodes throughout the region, as shown in Figure 19.11. Here we have chosen to place $N_x = 4$ nodes along the x direction and $N_y = 3$ along the y axis. There are $N_x N_y$ nodes and $2(N_x - 1)(N_y - 1)$ elements in total. Nodes are numbered according to the order in which they were created and their coordinates are stored. Lastly, elements should be numbered and, most importantly, the nodes associated with each element need to retained. One way to do this is to assign three node numbers to each element. These can be stored in a custom Eigen array as follow:

```
typedef Matrix<int,Dynamic,3> eMat;   // custom Eigen matrix
                                      //    of ints with
                                      // dynamically
                                      //    determined row size
eMat nL;                              // track nodes
                                      //    associated with
                                      //    element
                                      // 0,1,2 label nodes in
                                      //    the CCW direction
```

For example nL(0,0)=0, nL(0,1)=5, and nL(0,2)=4.

An alternative method is to associate elements with a node. For example, node 5 is related to elements as follows: $5 \to \{0, 1, 2, 9, 8, 7\}$. We will employ the first method in the following since it is slightly less awkward. Assigning node numbers is a matter of indexing for regular meshes.

Once node coordinates and node labels have been assigned it is simple to assemble the global force vector:

```
// assemble force vector   [constant rho]
for (element e=0;e<Nelements;e++) {
   double ff = rho*area(e)/3.0;
   for (int i=0;i<3;i++) f(nL(e,i)) += ff;   // sum force
      contribution each node
}
```

It is also relatively simple to evaluate the full stiffness matrix:

```
// assemble FEM stiffness matrix, K
for (element e=0;e<Nelements;e++) {
   double beta[3],gamma[3];
   for (int i=0;i<3;i++) {
      int j = (i+1)%3;
      int k = (i+2)%3;
      beta[i]  = rNode[nL(e,j)].y - rNode[nL(e,k)].y;
      gamma[i] = rNode[nL(e,k)].x - rNode[nL(e,j)].x;
   }
   for (int i=0;i<3;i++) {
   for (int j=0;j<3;j++) {
      K(nL(e,i),nL(e,j)) += (beta[i]*beta[j] + gamma[i]*
         gamma[j])/(4.0*area(e));
   }}
}
```

Here rNode is a vector of TwoVectors that is indexed by node number.

The next step is to apply boundary conditions. In the case of Dirichlet conditions one way to achieve this is to set boundary nodes to their appropriate values and eliminate them from the stiffness matrix. If n labels a boundary node a simpler approach is to set K(n,n) equal to a large number and set f(n) = K(n,n) * α(n), where α is the boundary function. This has the effect of forcing the solution vector to take on the value α at node n, which is the desired result.

Finally, the solution is obtained by matrix inversion:

```
u = K.inverse()*f;
```

We have found this to be effective for matrices with linear dimension as large as several tens of thousands. In general, the stiffness matrix is sparse (since only adjacent elements contribute) and therefore employing sparse matrix methods can be profitable, or even necessary (millions of nodes are not uncommon in industrial applications). Of course it is also possible to use iterative solution methods such as Gauss-Seidel or over-relaxation. Other techniques such as the *conjugate gradient method* are also commonly employed. The latter is a powerful iterative method for solving large sparse matrix problems of the type $K\vec{u} = \vec{f}$ that leverages the fact that $\vec{u}$ minimizes $\frac{1}{2}\vec{u}^T K\vec{u} - \vec{u}^T \vec{f}$. The library Eigen contains a class that implements the technique. A skeleton showing its application follows.

```
int n=100000;
VectorXd u(n), f(n);
SparseMatrix<double> K(n,n);
// fill K and f
ConjugateGradient<SparseMatrix<double>, Lower|Upper> cg;
cg.compute(K);
u = cg.solve(f);
```

Futher information on the class can be found in the Eigen documentation while more information on the conjugate gradient method can be found in (Saad, 2003).

An application

We will illustrate the finite element method by considering the fluid flow problem of Section 19.6.3. Thus we seek to solve:

$$\nabla^2 v(x, y) = \frac{1}{\mu} \frac{\partial p}{\partial z} \tag{19.58}$$

where the viscosity and pressure are taken to be constant. Boundary conditions are $v = 0$ along the bottom and side walls and $v = v_z$ on the top surface. Finally, the flow rate through the channel is given by $R = \int_0^W \int_0^H dx\,dy\,v(x,y)$, where H and W are the channel height and width.

We solve the problem with the triangular mesh of Figure 19.11 with $N_x N_y$ nodes. Simple rescaling yields a rectangle of size $H \times W$. The velocity profile and the mesh for the case $H/W = 1/2$, $\mu = 1$, $p' = -6$, $v_z = 0.1$, and $N_x = N_y = 20$ are shown in Figure 19.12. We remark that the velocity is only known on the nodes; the contours are interpolated by the plotting code.

Figure 19.13 shows a variety of algorithmic and solution properties relevant to this problem as a function of $N_x = N_y$. The number of elements in the stiffness matrix trivially scales as N_x^4, while the number of elements (simplexes) and the number of nonzero elements in K scale as N_x^2. The clock time to obtain a solution scales as $N_x^{5.5}$, which is somewhat worse than the expected N_x^4.

Lastly, the curve labeled "1/relative error" refers to the (inverse) relative difference between the estimated flow rate and the exact result (Fenner, 1996):

$$R = \frac{1}{2} WHV_z F_D - \frac{WH^3 p'}{2\mu} F_P$$

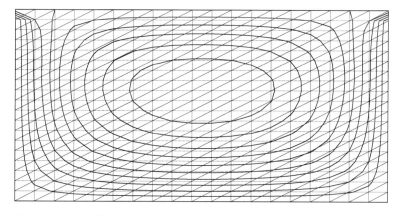

Figure 19.12 *Velocity profile in a rectangular channel.*

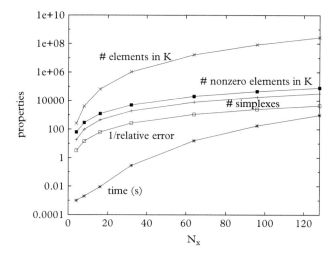

Figure 19.13 *Various properties of the channel flow problem as a function of $N_x = N_y$.*

with

$$F_D = 16\frac{W}{H}\sum_{n=0}^{\infty}\frac{\tanh((2n+1)\pi H/(2W))}{\pi^3(2n+1)^3}$$

and

$$F_P = 1 - 192\frac{H}{W}\sum_{n=0}^{\infty}\frac{\tanh((2n+1)\pi W/(2H))}{\pi^5(2n+1)^5}.$$

Finite element methods

The relative error scales as $1/N_x^2$ which is as expected (since the discretization error follows this relationship).

Nonuniform meshes

Our application, while useful and illustrative, does not concern variegated shapes where the finite element method excels. Creating meshes for general surfaces is a complex problem, hence the discussion will be restricted to some simple observations and pointers to references. One of the key aspects of mesh generation is the ability to refine meshes by creating more elements at crucial areas in the surface. This will be discussed in the next subsection.

Perhaps the simplest way to tile a general shape is to fill in a bounding rectangle with a uniform mesh, as in Figure 19.11. Nodes outside of the shape can be removed, and then the nodes closest to the boundary can be translated to the boundary. This method will work well for shapes with smooth boundaries.

Another, more robust, algorithm starts by enclosing the region in a bounding simplex. Then a point within the region is added to the mesh and connected to each vertex. A second point will lie within one of the simplexes, which should then be divided. Once as many points as desired have been added, the enclosing triangle and connecting edges can be deleted. The result will be a mesh with conforming elements, as desired.

This is fine; however, it is possible that some of the elements will be malformed in that one angle is highly acute. This is an undesirable situation since the basis function must interpolate over a large distance along two edges of the element. The solution is to adjust the nodes to form elements that are as equilateral as possible. Formally, this is called a *Delauney triangulation*, which is a mesh in which the minimum angle in each triangle has been maximized.

Some thought, which we need not replicate, leads to the following algorithm for forming a Delauney triangulation of an arbitrary mesh:

```
repeat until converged
  select an edge
    if the sum of the edge angles exceeds 180 degrees
      flip the edge
```

The meaning of "edge angles" and "flip" is clarified in Figure 19.14.

This algorithm has complexity $O(N^2)$, which can be improved. At this stage it is probably best to rely on professionally constructed code as provided by, for example, the Computational Geometry Algorithms Library. This library, located at http://www.cgal.org, is the product of a global collaboration of academic and business developers who seek to create quality open source C++ code dealing with geometrical algorithms.

An attractive alternative is provided by code developed by Persson and Strang (2004). The code relies on a *signed distance function*, which gives the distance of any point to the nearest boundary and is negative inside the region of interest. For example, for a circle one can use $d(x, y) = \sqrt{(x^2 + y^2)} - R$. The signed distance function is used

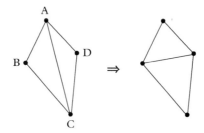

Figure 19.14 *Delauney triangulation of an edge. The edge AC has associated angles (ABC) and (ADC) that sum to an angle greater than 180 degrees. Thus edge AC is flipped to BD.*

to define the region and to move nodes to the boundary surface. The algorithm starts with a uniform grid that covers the entire region of interest. External nodes are then truncated and an iterative procedure that places nodes and fixes the mesh is initiated. Node placement is achieved by connecting nodes with an artificial force and stepping the many-body difference equation forward in artificial time. Persson and Strang have found that an asymmetric repulsive spring law works well for this purpose:

$$f(\ell) = \begin{cases} k(\ell_0 - \ell) & \text{if } \ell < \ell_0 \\ 0 & \text{if } \ell > \ell_0 \end{cases}. \tag{19.59}$$

Here ℓ_0 is an equilibrium edge length (or node separation) that can be set by the user to obtain the desired element density. Once the nodes have been placed, a Delauney triangulation is made to establish the mesh geometry. The entire process is iterated until convergence.

Mesh refinement and error estimation

Take a look at Figure 19.12 again: notice that the velocity contour lines bend sharply near the top left and right corners. This is because the speed is constrained to be zero on the vertical surfaces and v_z on the top surface. Thus all contours with $v(x, t)$ between zero and v_z must merge at the top left or top right corners. It therefore appears prudent to place more nodes near these locations.

In general, one would like to be able to increase the density of nodes in places where the field is expected to vary rapidly. With some intuition, these areas can be anticipated and the density of nodes can be adjusted accordingly. For example, in the Persson-Strang method one can make the equilibrium node lengths a function of position, $\ell_0 \to \ell_0(x, y)$.

Absent intuition, *error estimation* is useful, both for assessing the reliability of a numerical result and for deciding where additional elements are needed. A simple way to estimate errors on a uniform grid is to halve the grid scale, thus for example, one can use $2N_x \times 2N_y$ nodes in Figure 19.11. Since the original nodes will correspond to one half of the nodes on the refined grid, one can compare solutions and determine where the solution has most changed.

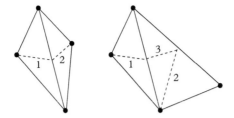

Figure 19.15 *Left: Bisecting an element with shared longest edge. Edges 1 and then 2 are added to the mesh. Right: Bisecting an element without a shared longest edge. Edge 1 and 2 are added, at some later time edge 3 is added, making the visible elements conforming. Additional bisections must occur in the remainder of the grid to ensure that the elements along the top right edge are also conforming.*

For general meshes it is expeditious to use an algorithm that bisects triangles along their longest edges. This will make a nonconforming element in the neighboring triangle so if that edge is also the neighbor's longest, it must be bisected as well (see Figure 19.15 left). If it is not the longest, then the neighboring element should be bisected along its longest edge, and the process should be repeated until convergence (Figure 19.15 right). All of this can be achieved with a recursive algorithm, such as that in the following pseudocode.

```
refine(E) {
  let L = longest edge of E; N = neighbor element with the
      shared edge
  if L != longest neighbor edge {
     refine(N)
  }
  bisect L: {
    add new node
    add new element
    reset node labels for the old element
    set node labels for the new element
  }
}
```

For more detailed information, the reader is directed to Mitchell (1989), who summarizes the properties of a variety of error estimation and adaptive mesh generation methods.

19.8 Conclusions

The breadth of physical problems that are defined in terms of fields is nearly matched by the range of relevant numerical techniques. The method to be used will depend on the way in which information is propagated (initial or boundary values), the dimensionality of the problem, the complexity of the region of interest, coordinate-dependence in the partial differential equation, and whether the problem is nonlinear. We recommend starting with Fourier transform techniques because they are fast and robust. If boundary conditions are uncooperative then finite differencing, especially the Crank-Nicolson scheme, can be considered. Finally, complex problems can be solved with finite element methods, although this will require some investment in grid generation and refinement technology. It is clear that choosing wisely will require some experience on the part of the numericist — hopefully this chapter will have started the reader on his way to gaining this experience.

19.9 Exercises

Finite Difference Methods

1. [T] The Jacobi Method and FTCS Discretization.
   We pursue the analogy made in the lead up to Eq. 19.28, wherein it was claimed that iterative methods are related to time-dependent partial differential equations.
   Perform an FTCS differencing of Eq. 19.28 with $\mathcal{L} = \nabla^2$. Impose the two-dimensional version of the CFL condition and show that the Jacobi scheme results.

2. [T] Nine-point stencil.
   Perform a Taylor series analysis of the nine-point stencil (Eq. 19.16) and show that it is of order Δ^2. Show that the coefficient of Δ^2 is proportional to $\nabla^2(\nabla^2 \rho)$ and comment on the importance of this when solving Laplace's equation.

3. [T] Alternate Nine-point Stencil.
   Consider a nine-point stencil that is defined on the points $(x, y), (x \pm \Delta, y), (x \pm 2\Delta, y), (x, y \pm \Delta)$, and $(x, y \pm 2\Delta)$. Obtain the coefficients at these points such that the Laplacian operator is estimated to an accuracy of $O(\Delta^4)$.

4. [T] CFL condition in d dimensions.
   Perform stability analysis for a typical explicit scheme in d dimensions.

   a) Show that the CFL condition becomes
   $$\frac{h}{\Delta} \leq \frac{1}{\sqrt{d}|v|}$$
   for the advection problem. Use $\rho_\ell^n = \xi^n(\vec{k}) \exp(i\vec{k} \cdot \vec{\ell}\Delta)$.

   b) What is the analogous expression for diffusion?

5. [T] Position-dependent Diffusion.
   Consider discretization of the equation $\partial_t \rho = \partial_x(D(x)\partial_x \rho)$.

a) Confirm that Eq. 19.14 is a viable discretization.
b) Find the analogous form of the FTCS difference equation if one were to use $\partial_x D = (D_{\ell+1} - D_{\ell-1})/2\Delta$.

6. [T] Stability of the Wave Equation.

   a) Write the one-dimensional wave equation
   $$\frac{\partial^2 u}{\partial t^2} = c^2 \frac{\partial^2 u}{\partial x^2}$$
   as coupled first order equations of the advection type.
   b) Perform von Neumann stability analysis of the coupled equation and determine the stability condition.

7. [T] Efficiency in ADI.
   Consider the ADI method applied to the problem
   $$\partial_i D_{ij} \partial_j u - \vec{v} \cdot \nabla u = \partial_t u$$
   with a constant diffusivity tensor in two dimensions.

   a) Obtain the stability condition from von Neumann analysis.
   b) Consider the x step portion of the ADI method. How would you obtain $u_{ij}^{n+1/2}$? Specifically, is it better to map the matrix to a vector, $u_{ij} \to u_{I(i,j)}$, or to solve the tridiagonal problem many times in the index j?

8. Complex Tridiagonalization.
   Write a version of the tridiagonalization routine that operates on tridiagonal matrices with constant complex diagonal entries.

9. Circular Symmetry.
   Use finite differencing to solve for the electric field of a point particle in a large box. Box boundary conditions ruin the circular symmetry of the problem. Compare your solutions using five-point and nine-point stencils and comment on the symmetry.

10. FTCS Instability.
    Solve the one-dimensional advection equation $\partial_t \rho = -\partial_x \rho$ in the region $x \in (0, 1)$ with initial condition $\rho(x, 0) = \sin(\pi x)$.

    a) What boundary conditions should be used for ρ_0 and ρ_N?
    b) Plot the solution at $t = 0, 5$, and 10. Try $h = 0.01$ and $\Delta = 0.01$. Note the appearance of unstable modes as time increases.

11. **Jacobi Successive Over-relaxation.**
    Implement successive over-relaxation with Jacobi iteration and try it on the capacitor problem of Section 19.4.1. You should find that the method fails. Examine the von Neumann amplification factor and comment.

12. **Successive Over-relaxation Efficiency.**
    Code up the two-dimensional capacitor of Section 19.4.1 with SOR and Gauss-Seidel iteration.

    a) Plot the error vs. iteration and confirm that Gauss-Seidel iteration is $O(N^2)$ while SOR (with optimal over-relaxation parameter) is $O(N)$.
    b) How does the SOR algorithm scale if ω is fixed to, say, 1.95?

13. **The Heat Equation.**
    Consider the heat equation in one dimension
    $$u_t = \frac{1}{\pi^2} u_{xx}$$
    in the region $x \in [0, 1]$ with boundary condition $u(x, 0) = \sin(\pi x)$ (and therefore $u(0, t) = u(1, t) = 0$).

    a) Find the analytic solution.
    b) Solve the equation using the Crank-Nicolson scheme and compare to the exact result.

14. **The Black-Scholes Equation.**
    The Black-Scholes equation is a financial model that yields the price of a stock option (Black and Scholes, 1973). The equation reads
    $$\partial_t V + \frac{1}{2}\sigma^2 S^2 \frac{\partial^2 V}{\partial S^2} + rS\frac{\partial V}{\partial S} - rV = 0,$$
    where $V(S, t)$ is the price of the stock option, S is the price of the underlying stock, σ is the stock volatility, and r is the annualized interest rate.

    The boundary conditions are $V(S, t) \to S$ as $S \to \infty$, $u(0, t) = 0$, and $V(S, T) = \max(S - K, 0)$ where K is the *strike price* of the option and T is the strike time. The first condition need not be applied (although it is a useful check). The constraint on the option price at time T is somewhat awkward at first sight, but can be incorporated as a final condition, with the finite difference equation going backwards in time.

    a) Implement a backwards-time finite difference algorithm and obtain the solution to the Black-Scholes equation. Take $K = 1$, $T = 1$, $\sigma = 0.2$, and $r = 0.1$. Plot $V(S, t)$ for various times between zero and T.

b) An alternative is to nondimensionalize by setting $t = T - \frac{\tau}{\sigma^2/2}$, $S = K\exp(x)$, and $V = Kv(x, \tau)$. Show that the resulting equation is:

$$\partial_t v = \partial_x^2 v + (k-1)\partial_x v - kv.$$

Identify k. Obtain the initial conditions and solve.

Lastly, the Black-Scholes equation can be reduced to the heat equation, but we are not interested in that here.

15. **SIR Disease Dynamics Model.**
A simple model of population dynamics was introduced in Ex. 15 of Chapter 11. Here we generalize this to account for differing disease dynamics among different age groups. In the limit in which age is considered a continuous variable, a, one obtains coupled *integro-differential equations* (Rock et al., 2014):

$$\partial_t S(a, t) = b\delta(a) - S \int_0^\infty \beta(a, a')I(a', t)\, da' - dS - \partial_a S$$

$$\partial_t I(a, t) = S \int_0^\infty \beta(a, a')I(a', t)\, da' - \gamma I - dI - \partial_a I$$

$$\partial_t R(a, t) = \gamma I - dR - \partial_a R$$

a) FTCS difference these equations, dealing with the integrals in the natural way. Find a way to handle the delta function.

b) Solve the SIR equations assuming $\beta = \beta_0 \exp(-(a-a')^2)$ with $\beta_0 = 0.2$, $d = 10^{-4}$, $\gamma = 0.1$, $b = 10^{-4}$. Take the initial condition to be $I_0 = 10^{-6}$, $R_0 = 0$, and $S_0 = 1 - I_0$.

16. **[P] Stochastic Advection-Diffusion.**
Diffusion is simply particles executing a random walk. Advection is particles in a current. Thus it appears natural to forego the partial differential description of advection-diffusion and directly simulate the process by following many particles as they walk randomly in a moving background. Implement this idea in one dimension and compare to the analytic solution for a simple source term. See Sections 21.3.1 and 22.4 for more information on random walks in a different context.

17. **[P] A Nonlinear Partial Differential Equation.**
This chapter has almost exclusively dealt with linear equations; however, differencing can be usefully applied to nonlinear partial differential equations. As an illustration consider:

$$\nabla^2 u + \frac{1}{\lambda^2} u^2 = \rho.$$

Work in the unit square with $u = 0$ on the boundary. Set $\rho = -1$ and $\lambda = 0.1$.

Discretize to obtain:

$$D^2 u_{\ell,m} + \frac{\Delta^2}{\lambda^2} u_{\ell,m}^2 - \Delta^2 \rho_{\ell,m} = 0.$$

We solve this equation by linearizing to implement the Newton-Raphson method and by isolating the diagonal operators on the left hand side to implement Jacobi iteration. To facilitate this let $\mathcal{D}^2 = (D + U + L)$ as before, and:

$$\frac{\Delta^2}{\lambda^2} u_{\ell,m}^2 \equiv N(u)\vec{u}.$$

a) Linearize and obtain

$$\vec{u}' = (D + 2N(u))^{-1} \left(\Delta^2 \rho + N(u)\vec{u} - (U+L)\vec{u} \right).$$

b) Implement this scheme and confirm that the code converges to a solution.
c) Think about the Gauss-Seidel and SOR algorithms. Can you implement these methods for this problem?
d) You might be tempted to forego linearization. Show that the resulting Jacobi equation is now

$$\vec{u}' = (D + N(u))^{-1} \left(\Delta^2 \rho - (U+L)\vec{u} \right).$$

Does this lead to a viable solution? How does the code differ in behavior from part (b)?

18. [P] Burger's Equation.
Burger's equation is a simple model of nonlinear effects in fluid flow that is based on the Navier-Stokes equations for a homogeneous incompressible fluid. The one-dimensional equation reads:

$$u_t + u u_x = D u_{xx}.$$

The equation is capable of describing shocks because the nonlinearity in the ∂_x term causes the top of a wavepacket to move faster than the base of the packet. Eventually the top catches the base and a cusp is formed. These tend to occur at high **Reynolds number**, $R \equiv 1/D$. The existence of possible discontinuities is a concern since differencing is predicated on the smoothness of the fields! Nevertheless, let us press ahead with hope in our hearts.

a) Forward-time-upwind-difference Burger's equation. Remove the nonlinearity by mapping

$$u\partial_x u \to u_\ell^{n-1} \frac{u_\ell^n - u_{\ell-1}^n}{\Delta}.$$

Thus a linear problem is recovered the expense of an additional vector. What are we assuming about u in this scheme?

b) Set $u(x, 0) = \exp(-(x-3)^2)$ and $D = 0.01$ and explore the numerical properties of the solution. Does the expected cusp develop? Make a simple estimate for when this happens. What happens if you run for longer?

c) Intuitively the CFL condition becomes $h/\Delta < 1/u$. Check this by performing a von Neumann stability analysis. Simplify by setting $D = 0$ if you wish.

d) The **Lax-Wendroff scheme** provides a more stable difference equation. Consider Burger's equation in the *inviscid limit*:

$$u_t + \partial_x F(u) = 0,$$

where $F = u^2/2$. The associated Lax-Wendroff difference equation is:

$$u_\ell^{n+1} = u_\ell^n - \frac{h}{2\Delta}(F_{\ell+1}^n - F_{\ell-1}^n) + \frac{h^2}{2\Delta^2}\Big[F'(u_{\ell+1/2}^n)(F_{\ell+1}^n - F_\ell^n) - F'(u_{\ell-1/2}^n)(F_\ell^n - F_{\ell-1}^n)\Big].$$

Code this up and compare to the naive approach of part (a).

e) The stability condition for the Lax-Wendroff scheme is:

$$|F'(w_\ell^n)| < \frac{\Delta}{h}$$

for all n and ℓ. Use this to implement a variable time-step algorithm and compare to (d).

19. [P] The FitzHugh-Nagumo Model for Cardiac Dynamics.
    Heart muscle can be thought of as an **excitable medium**, meaning that small external stimuli can induce large responses. In the case of the heart, the *sinoatrial node* produces regular electrical impulses that are amplified and travel through the heart causing a contraction.

    The FitzHugh-Nagumo equations provide a simple model of this phenomenon via the nonlinear partial differential equations (Winfree, 1991):

    $$\partial_t V = \nabla^2 V + \frac{1}{\epsilon}\left(V - \frac{1}{3}V^3 - W\right)$$
    $$\partial_t W = \epsilon(V - \gamma W + \beta)$$

where V is the *transmembrane potential*, W is the *recovery variable*, and the Greek letters are parameters with values $\epsilon = 0.2$, $\gamma = 0.8$, and $\beta = 0.7$.

We will explore this nonlinear system in a series of steps.

a) Neglect the Laplacian term in the model and plot the lines along which $\partial_t V = \partial_t W = 0$ in the $V - W$ plane. The point where these lines cross, $(V_\star, W_\star)$, is the resting state of the heart (and the fixed point of the dynamics).

b) Stimulate the heart by starting a simulation at the point $(V_\star + \delta V, W_\star)$. Plot V and W as a function of time. How large must δV be to obtain a pulse?

c) Communication between heart cells is emulated with the Laplacian term. Model the heart as a two-dimensional box. Set $V = W = 0$ at the boundary. Solve the full equations with the initial condition $(V_\star, W_\star)$ except in a small region at the origin, where you should use $(V_\star + 3, W_\star)$. Describe the temporal evolution of the system.

d) Mimic the sinoatrial node by periodically stimulating the heart in the region at the origin. Make the period long enough that pulses finish before new ones start.

e) Stimulate random small regions of the heart with $\delta V = 3$ while pulses are active. You should be able to generate **spiral waves**, which are interpreted as heart arrhythmias.

f) Ponder extensions to the model: dead tissue, spatially-dependent parameters, pacemakers, etc.

g) Consider making a realistic finite element model of the heart.

20. [T] Separable Two-particle Scattering.
Show that a two-particle wavepacket that is separable in coordinates

$$\psi(\vec{r}_1, \vec{r}_2) \propto \exp(-(\vec{r}_1 - \vec{a})^2 + i\vec{k}_1 \cdot \vec{r}_1) \cdot \exp(-(\vec{r}_2 - \vec{b})^2 + i\vec{k}_2 \cdot \vec{r}_2)$$

is also separable in the relative coordinates. You should obtain an expression like:

$$\psi(\vec{R}, \vec{r}) \propto \exp(-\alpha(\vec{R} - \vec{A})^2 + i\vec{K} \cdot \vec{R}) \cdot \exp(-\beta(\vec{r} - \vec{B})^2 + i\vec{k} \cdot \vec{r}).$$

Specify the constants $\alpha, \beta, \vec{A}$, and $\vec{B}$ in this formula.

This result implies that simulating a two-particle scattering process is no more difficult than a one-particle process.

21. One-dimensional Scattering.
Solve the time-dependent Schrödinger equation for the scattering of a Gaussian wavepacket from a square potential barrier. Set the barrier height equal to the average energy. Study the behavior of the system for sufficient time that the transmitted and reflected waves have left the region of the potential. You should observe residual probability density within the potential barrier. Follow this density over time and describe its behavior.

Exercises

22. **[P] The Double Slit.**
Use the Crank-Nicolson method to solve for the time evolution of a wavepacket interacting with a double slit. This might need some industrial strength computing since the problem becomes quite large in three dimensions.

23. **[P] Visscher Method for Quantum Scattering.**
It is useful to remember that many solution strategies that trade one advantage for another exist. Thus, while the Crank-Nicolson scheme is second order accurate, stable, and unitary, other schemes that are not unconditionally stable, but are fast, exist. We describe one such method, due to Visscher (1991), that has explicit updating, is unitary, and is second order accurate in time.

The idea is to evolve the real and imaginary portions of the wavefunction in alternating steps. Thus one sets $\psi = R + iI$ and rewrites the Schrödinger equation as:

$$\frac{\partial R}{\partial t} = \mathcal{H} I$$
$$\frac{\partial I}{\partial t} = -\mathcal{H} R$$

The equations are discretized as follows:

$$I_\ell^{n+1} = I_\ell^n - h\mathcal{H} R_\ell^n$$
$$R_\ell^{n+1} = R_\ell^n + h\mathcal{H} I_\ell^{n+1}$$

with boundary conditions:

$$R_\ell^0 = \text{Re}\,\psi(x, 0)$$

and:

$$I_\ell^0 = \text{Im}\,\psi\left(x, -\frac{h}{2}\right).$$

Thus the algorithm actually makes half steps. The wavefunction norm defined by

$$|\psi|^2 = R_\ell^{n+1} R_\ell^n + I_\ell^{n+1} I_\ell^{n+1}$$

is conserved under the discrete time evolution of the method. Finally, the stability condition is:

$$-\frac{2}{h} < V_\ell < \frac{2}{h} - \frac{2}{\mu \Delta^2}$$

($\hbar$ has been set to unity).

Code up the Visscher algorithm for one-dimensional scattering from a square well and compare the results to those of the Crank-Nicolson scheme. What values of h are required with respect to the Crank-Nicolson scheme to achieve comparable results? What happens if Δ is made "too" small? How does the speed compare to the Crank-Nicolson case?

Fourier Methods

24. Gaussian Fourier Transform.
    Use the `Eigen` FFT package to compute the Fourier transform of $\exp(-t^2)$ (recall the discussion after Eq. 19.45). Explore the error as a function of N and Δ.

25. [T] The Sine Transformation.
    Show that the relationship of the sine transformation to the FFT developed in Section 19.6.2 is correct. Hint: start with the definition of the FFT, double N, split the sum into two pieces, relabel the index in the second piece according to $j' = 2N - j$, notice that the limits don't quite agree so that a sine function is not readily computable, fix this by separating certain summands.

26. [T] Discrete Green Functions.
    The solution for the field $\hat{u}_{m,n}$ given in Eq. 19.44 contains the discrete Green function for the Poisson equation.

    a) Confirm that the two-dimensional Green function for the discrete Laplacian is as given in Eq. 19.44.
    b) Obtain the solution to the equation $\partial_x u(x) = \rho(x)$ with the discrete sine transform method.
    c) Derive the Green function for the nine-point stencil of Eq. 19.16 with the discrete sine transform.

27. A Simple Equation.
    Consider the boundary value problem:
    $$u'' = f$$
    where f is a constant and $u(0) = u(1) = 0$.

    a) Derive the exact solution.
    b) Make a sine expansion $u = \sum_n a_n \sin(n\pi x)$ and obtain the expression for the coefficients.
    c) Graph the first few Fourier estimates of u along with the exact result.
    d) Discretize the equation and solve it using the discrete sine transformation method. Compare to the exact answer.

e) Generalize the formalism to handle the boundary conditions $u(0) = u_L$ and $u(1) = u_R$ and verify that it works.

28. **Two-dimensional Sine Transform.**
    Create a routine to compute the two-dimensional sine transform,

    $$\hat{f}_{mn} = \sum_{ij=1}^{N-1} f_{ij} \sin(im\pi/N) \sin(jn\pi/N),$$

    that leverages `Eigen::FFT`.

29. **Two-Dimensional Poisson Equation.**
    Solve the equation $\nabla^2 u = \exp(-(x^2 + y^2)/\beta)$ on the unit square using the two-dimensional sine transform.

    a) Set $\beta = 0.1$ and $u = 0$ on the boundaries.
    b) Set $\beta = 0.1$ and $u(x, 0) = \alpha \sin(\pi x)$ with $\alpha = 0.004$. Make u zero on the other boundaries.

30. **[P] Image Processing.**
    Image data created by a camera can be characterized by a response function:

    $$r(x, y) = \frac{1}{T} \int_0^T \int dt dx' dy'\, d(x, y; x', y', t)\, s(x', y', t) + \eta(x', y')$$

    where T is the exposure time, η is random noise, and d is a distortion function. Under ideal conditions there is no noise and no distortion so that:

    $$d(x, y; x', y', t) = \delta(x - x')\delta(y - y').$$

    Furthermore, if the subject is not moving then s is independent of time and one recovers the signal, $r = s$. Examples of distortion are atmospheric turbulence that can be modeled with:

    $$d(x, y; x', y', t) = \exp(-\alpha(x'^2 + y'^2))$$

    or blurring due to lack of camera focus with

    $$d(x, y; x', y', t) = \begin{cases} 1, & x'^2 + y'^2 < r^2 \\ 0, & \text{elsewhere} \end{cases}.$$

Here we consider blurring due to uniform motion along the x axis. In this case the response function is given by:

$$r(x,y) = \frac{1}{T}\int_0^T s(x - vt, y)\, dt = \int_{-\infty}^{\infty} d(x')s(x - x', y)\, dx'$$

where:

$$d(x) = \begin{cases} \frac{1}{vT} & 0 < x < vT \\ 0 & \text{otherwise} \end{cases}.$$

We seek to recover an unblurred image s from the camera data r. This is a difficult process in space, but is simple in the frequency domain. In particular:

$$\hat{r}(f_x, f_y) = \int \exp(-2\pi i f_x x)\, \exp(-2\pi i f_y y)\, r(x,y)\, dx\, dy$$

is given by:

$$\hat{r} = \hat{s} \cdot \hat{d}$$

by the convolution theorem. Thus the desired signal is given by the inverse Fourier transform of $\hat{r}/\hat{d}$.

Find an appropriate blurred image, discretize it, perform the indicated Fourier analysis, and obtain the unblurred image. You will need to adjust v to maximize the image sharpness.

Finite Element Method

31. [T] Two-element Stiffness Matrix.
    Consider a rectangular region with two elements and four nodes ($N_x = N_y = 2$ in Figure 19.10).

    a) Obtain the analytic expressions for the local and global stiffness matrices.
    b) What happens for larger N_x and N_y?
    c) Can you make sense of your findings?

32. [T] Bad Elements.
    Explain why a mesh such as shown in Figure 19.16 is unacceptable in the finite element method. Such a configuration is called **nonconforming**.

33. [T] One-dimensional Finite Elements i.
    Generalize Eqs. 19.51 and 19.52 to the case $f = f(x)$. How does your result compare to the analogous finite difference equation?

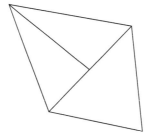

Figure 19.16 A poorly constructed mesh.

34. [T] One-dimensional Finite Elements ii.
    Generalize Eqs. 19.51 and 19.52 to the case:

    $$\frac{d^2u}{dx^2} + gu = f.$$

    How do your results compare to the analogous difference equations?

35. [T] One-dimensional Tent Functions.
    Repeat the analysis of the one-dimensional finite element method with a *tent function* basis. The tent function is defined to be:

    $$\phi_j(x) = \begin{cases} (x - x_{j-1})/(x_j - x_{j-1}), & x_{j-1} < x < x_j \\ (x_{j+1} - x)/(x_{j+1} - x_j), & x_j < x < x_{j+1} \\ 0, & \text{elsewhere} \end{cases}.$$

    Let $u = \sum_j c_j \phi_j$.

    a) Use the finite element variational approach to determine the equation for the $\{c_j\}$ for the equation $u'' = f$, where f is constant. How does the result compare to the finite difference equation?
    b) How do things change if f is a function of x?

36. A Circular Mesh.
    Write code to generate the nodes and node labels for a mesh in a circular region. Node locations are easy to get; node labels are a bit of a bear. Consult Fenner (1996) if you need help.

37. Laplace's Equation i.
    Solve Laplace's equation in a unit square with $u(0, y) = 0$, $u(1, y) = 0$, $u(x, 0) = 0$, and $u(x, 1) = \sin(\pi x)$ using finite elements. Compare your answer to the exact solution:

$$u(x,y) = \frac{\sin(\pi x)\,\sinh(\pi y)}{\sinh(\pi)}.$$

38. **Laplace's Equation ii.**
    Solve Laplace's equation in a unit square with $u(0,y)=0$, $u(1,y)=0$, $u(x,0)=0$, and $u(x,1)=1$ using finite elements. Compare your answer to the exact solution:

    $$u(x,y) = \frac{4}{\pi}\sum_{n=0}^{\infty}\frac{\sin[(2n+1)\pi x]\,\sinh[(2n+1)\pi y]}{(2n+1)\,\sinh[(2n+1)\pi]}.$$

39. **Beam under Torsion.**
    St. Venant's theory of torsion yields the partial differential equation

    $$\nabla^2 u(x,y) = 2Gk$$

    for the torsional stress in a beam. Here k is the rate of change of twist $k = d\theta/dz$ along the beam (defined in the z direction), θ is the twist angle, u is the stress function (Prandtl's function) defined on the beam cross section, and G is the shear modulus of the beam. With these definitions, the total torsion is given by $T = -2G\int u(x,y)dxdy$.
    Use finite elements to find the stress function for a beam with a uniform triangular cross section. Set $u=0$ on the beam boundary.

40. **[P] Fluid Flow Past a Disk.**
    Consider fluid flow through a uniform channel with square cross section, so that Eq. 19.58 applies. Place a disk of radius one half the square dimension in the center of the channel and use the finite element method to solve for the fluid velocity field assuming that $v=0$ on the surface of the disk and at the channel boundaries.

41. **[P] Laplacian Pac-Man.**
    We seek the solution to Laplace's equation in a circular region of radius unity centered at the origin. The region has its lower right quadrant removed and has Dirichlet boundary conditions of $u(r,\theta=0)=0$, $u(r,\theta=3\pi/2)=0$, and $u(1,\theta)=\sin(2\theta/3)$ for $\theta\in(0,3\pi/2)$.

    a) Show that the analytical solution is $u = r^{2/3}\sin(2\theta/3)$. You will want to consider the Laplacian in circular coordinates.

    b) The rapid change in u at small r indicates that a nonuniform mesh is desirable. Construct such a mesh using a reasonable node density in the radial direction.

    c) Code, solve, and compare to the analytic result.

42. Convex Hulls.

A *convex hull* is the subset of points that contains all other points in a set. The convex hull is an important component of many algorithms that deal with geometrical computation, with applications to finite element analysis, imaging, graphics, and robotics. The simplest approach to obtaining a convex hull is called **gift wrapping** or the *Jarvis march*. In two dimensions the idea is to start with the left-most point of a set of nodes and sweep a line from this point until another node is encountered. The process is repeated with the next node and terminates when the original node is encountered. If there are N nodes in total and M nodes in the hull, it is evident that the algorithm is $O(NM)$. A (possibly) more efficient algorithm, called the **Graham scan**, can achieve $O(N \log N)$ complexity via an analogue of sorting. **Chan's algorithm** achieves an optimal $O(N \log M)$ complexity by combining the Jarvis march with the Graham scan. For more information see (Corman *et al.*, 2001).

The latter algorithms are fairly complex, and the numericist interested in professional grade code is advised to consult online packages such as **Qhull** (available at www.qhull.org) or the packages at www.cgal.org. Alternatively, the Jarvis march is relatively simple to implement in two dimensions. C++ code to collect node labels (in the vector hullPoints) for a set of nodes is given below. This code requires finding the left-most node and storing its node label in hullStart. This is easy to do during node creation. The algorithm given below does not handle collinear points; the required extension is not difficult.

The function left returns true if the node m is left of the line joining nodes hullPoints[n] and endPoint. Implement this function and the Jarvis march and test your code on sets of random points in the plane. Output should look like Figure 19.17.

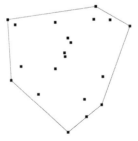

Figure 19.17 *The convex hull for 20 random points.*

```
  // convex hull via Jarvis march
  node endPoint;
  node hullPt = hullStart;
  node n=0;
  do {
   hullPoints.push_back (hullPt);
   endPoint = 0;
   for (node m=1;m<N;m++) {
      if ( (endPoint == hullPt) || left(hullPoints[n],
         endPoint,m) ) {
         endPoint = m;
      }
   }
   n++;
   hullPt = endPoint;
  } while (endPoint != hullStart);
  cout << n << "  hull  points  found  " << endl;
```

⋯⋯⋯⋯⋯⋯⋯⋯⋯⋯⋯⋯⋯⋯⋯⋯⋯⋯⋯⋯⋯⋯⋯⋯⋯⋯⋯⋯⋯⋯⋯⋯⋯⋯⋯⋯

BIBLIOGRAPHY

Black, F. and M. Scholes (1973). *The Pricing of Options and Corporate Liabilities*. Journal of Political Economy, **81**, 637.

Borgerding, M. (2013). *Kiss FFT*. https://sourceforge.net/projects/kissfft/.

Brandt, A. (1977). *Multilevel Adaptive Solutions to Boundary Value Problems*. Math. of Comp. **31**, 333.

Chorin, A. J. (1968). *Numerical solution of the Navier-Stokes equations*. Math. Comp. **22**, 745.

Cormen, T. H., C. E. Leiserson, R. L. Rivest, and C. Stein (2001). *Introduction to Algorithms*. MIT Press.

Crank, J. and P. Nicolson (1947). *A Practical Method for Numerical Evaluation of Solutions of Partial Differential Equations of the Heat-Conduction Type*. Proc. Camb. Phil. Soc. **43**, 50.

Elliott, D. F. and K. R. Rao (1982). *Fast Transforms: Algorithms, Analysis, and Applications*. Academic Press.

Fenner, R. T. (1996). *Finite Element Methods for Engineers*. Imperial College Press.

Mitchell, W. F. (1989). *A Comparison of Adaptive Refinement Techniques for Elliptic Problems*. ACM Trans. Math. Software, **15**, 326.

Persson, P. -O. and G. Strang (2004). *A Simple Mesh Generator in MATLAB*. SIAM Rev. **46**, 329.

Press, W. H., B. P. Flannery, S. A. Teukolsky, and W. T. Vetterling (2007). *Numerical Recipes: third edition*. Cambridge University Press.

Reddy, J. N. (2006). *An Introduction to the Finite Element Method*. McGraw-Hill.

Richardson, L. F. (1911). *The approximate arithmetical solution by finite differences of physical problems involving differential equations, with an application to the stresses in a masonry dam*. Phil. Trans. R. Soc. Lond., Ser. A, **210**, 307.

Rock, K., S. Brand, J. Moir, and M. J. Keeling (2014). *Dynamics of Infections Diseases*. Rep. Prog. Phys. 77, 026602.

Saad, Y. (2003). *Iterative methods for sparse linear systems*. SIAM.
Visscher, P. B. (1991). *A fast explicit algorithm for the time-dependent Schrödinger equation*. Computers in Physics **5**, 596.
Wesseling, P. (1992). *An Introduction to Multigrid Methods*. Wiley.
Winfree, A. T. (1991). *Varieties of Spiral Wave Behavior: an Experimentalist's Approach to the Theory of Excitable Media*. Chaos, **1**, 303.
Yang, X. I. J. and R. Mittal (2014). *Acceleration of the Jacobi iterative method by factors exceeding 100 using scheduled relaxation*. J. Comp. Phys. **274**, 695.

20
Classical spin systems

20.1 Introduction	701
20.2 The Ising model	702
20.2.1 Definitions	703
20.2.2 Critical exponents and finite size scaling	704
20.2.3 The heat bath algorithm and the induced magnetization	706
20.2.4 Reweighting	709
20.2.5 Autocorrelation and critical slowing down	710
20.2.6 Cluster algorithms	713
20.3 The Potts model and first order phase transitions	717
20.4 The planar XY model and infinite order phase transitions	720
20.5 Applications and extensions	724
20.5.1 Spin glasses	724
20.5.2 Hopfield model	725
20.6 Exercises	726
Bibliography	730

20.1 Introduction

This chapter studies the behavior of collections of localized classical degrees of freedom, which we generically call spin systems. Some, like the Ising model, are almost trivial to describe, yet embody an amazing diversity of emergent phenomena. This is reminiscent of the percolation models of Chapter 8, and indeed all of the general techniques introduced there find application in the spin systems to be discussed here. In terms of physics, the chief difference is that spin systems can be thermalized and often provide accurate models of physical materials or more theoretical constructs. The original application was, of course, to magnetic materials. Classical spin models also serve as simplified exemplars of complex dynamical problems such as voter dynamics, disease propagation, binary alloys, neural networks, cracking, avalanches, and the liquid-gas phase transition. As discussed in Chapter 8, it is *universality* that underpins this breadth of applicability.

The Monte Carlo method is ideal for analyzing the properties of spin systems and we shall focus exclusively on its application to a variety of problems in this chapter. Among the topics addressed will be determining the order of a phase transition, finite size scaling, and critical exponents. We will also explore a method to simultaneously compute an observable at many parameter values called "reweighting"; examine the phenomenon of "critical slowing down", where algorithmic efficiency is impaired near critical points; take a close look at error estimation; and introduce specialized Monte Carlo techniques where blocks of spins are updated at the same time. These issues will be examined in the context of various models known as the two-dimensional nearest neighbor Ising model, the q-state Potts model, and the XY model, but are very general in applicability. Analytic approximations and applications to a variety of problems are presented in the exercises.

20.2 The Ising model

In many ways the Ising model is the simplest possible implementation of a classical many body system.[1] In spite of this, its thermodynamic properties display an astonishing richness. The Ising model can be considered as a simplified classical limit of Heisenberg's model of quantum ferromagnetism (see Chapter 22). In this case the quantum mechanical spin operator is replaced with a classical "spin" that can take on two possible values:

$$\vec{S} = \frac{\hbar \vec{\sigma}}{2} \to \frac{\hbar \sigma_3}{2} \to \frac{\sigma}{2}. \qquad (20.1)$$

Since σ_3 commutes with itself, it can be considered a classical variable and quantum mechanics no longer applies. The model is then:

$$H = -\sum_{ij} \mathcal{J}_{ij} \sigma_i \sigma_j - \sum_i h_i \sigma_i \qquad (20.2)$$

where $\sigma_i = \pm 1$ is a spin variable at site i. An external magnetic field is denoted h_i in this expression. Often the model is simplified by assuming a uniform external magnetic field and by choosing coupling constants to be uniform and zero except when the sites i and j are nearest neighbors. In this way the system becomes translationally invariant, simplifying the analysis substantially. Standard notation for this "nearest neighbor" problem is:

---

[1] Ernst Ising (1900–1998), (pronounced "ee-zing"), German physicist. The eponymous Ising model was suggested as a PhD thesis topic by his supervisor, Wilhelm Lenz. His paper on the model was published in 1925 and the model was subsequently developed by such prominent physicists as Rudolph Peierls, Hendrik Kramers, Gregory Wannier, and Lars Onsager. Ising spent the years before WWII as a teacher before the Nazi pogroms forced him to flee for Luxembourg, where he worked for some time as a shepherd. He settled in Peoria, Illinois after the war and it was only at this time that he discovered that his model had become famous. He never published again.

$$H = -\mathcal{J}\sum_{(ij)} \sigma_i\sigma_j - h\sum_i \sigma_i, \tag{20.3}$$

where the symbol (ij) means sum over distinct nearest neighbor pairs.

When the coupling $\mathcal{J}$ is positive, spins tend to align at low temperature and the system is called *ferromagnetic*. Alternatively, if the coupling is negative spins anti-align and the system is called an *antiferromagnet*.

20.2.1 Definitions

We seek to evaluate the thermodynamic partition function

$$Z = \sum_{\{\sigma\}} e^{-\beta H} \equiv e^{-\beta F} \tag{20.4}$$

where $\beta = 1/k_B T$, the sum is over all 2^N possible spin configurations (N is the number of spins), and F is the free energy. The reader is reminded that derivatives of the free energy yield all thermodynamic quantities of interest. Henceforth we work in units $\mathcal{J} = 1$ and often suppress writing the Boltzmann constant, k_B.

The one dimensional Ising model can be solved exactly (see Ex. 2) and does not exhibit a phase transition. Alternatively, the model defined on a two dimensional square lattice has a second order phase transition at the critical temperature:

$$\frac{k_B T_c}{\mathcal{J}} = \frac{2}{\log(1+\sqrt{2})} \approx 2.269185 \tag{20.5}$$

with an induced magnetic field of strength:

$$M = \left(1 - \left[\sinh\left(\frac{2\mathcal{J}}{k_B T}\right)\right]^{-4}\right)^{1/8}. \tag{20.6}$$

Recall that the induced magnetic field is defined as:

$$M = -\frac{\partial F}{\partial h} = \frac{1}{Z}\sum_{\{\sigma\}}\sum_i \sigma_i e^{-\beta H} \equiv \langle S \rangle. \tag{20.7}$$

We have introduced the thermal average of a quantity as:

$$\langle \mathcal{O} \rangle = \frac{1}{Z}\sum_{\{\sigma\}} \mathcal{O}(\{\sigma\}) e^{-\beta H}. \tag{20.8}$$

The magnetic field is an ***order parameter*** for the Ising model. Specifically, at large temperature the spins flip independently and randomly, hence $M \to 0$, while for a ferromagnet at low temperature the spins tend to align (the state with all spins up ($\sigma_i = +$) or down ($\sigma_i = -1$) minimizes the energy), and hence $M = \pm 1$. For finite lattices one expects a smooth transition between $M = \pm 1$ and $M = 0$ as the temperature is increased. As the lattice size goes to infinity this transition sharpens and a singularity develops at $T = T_c$. When the system has zero magnetization (i.e., when $T > T_c$), it is said to be in the ***symmetric phase***, which means that the rotational symmetry of the Hamiltonian is manifest in expectation values of observables. Alternatively, when the system is magnetized (i.e., $T < T_c$), full rotational symmetry is no longer evident because a special direction (of magnetization) has been selected. In this case the system is said to be in the ***broken phase*** and the symmetry is ***spontaneously broken***.

Other quantities of interest are the ***entropy***:

$$S = -\frac{\partial F}{\partial T} = -\frac{1}{T^2}\langle H \rangle, \qquad (20.9)$$

while from $F = U - TS$ we derive the ***internal energy***:

$$U = -T^2 \frac{\partial (F/T)}{\partial T}. \qquad (20.10)$$

The ***isothermal magnetic susceptibility*** is defined by:

$$k_B T \chi = k_B T \frac{\partial M}{\partial h} = \langle S^2 \rangle - \langle S \rangle^2, \qquad (20.11)$$

and the ***specific heat*** is:

$$k_B T^2 C_V = k_B T^2 \frac{\partial U}{\partial T} = \langle H^2 \rangle - \langle H \rangle^2. \qquad (20.12)$$

20.2.2 Critical exponents and finite size scaling

At low temperature all the spins tend to align, and thus the correlation between widely separated spins is close to unity. At high temperature β is small and there is little to distinguish random spin configurations, thus the correlation between spins approaches zero. In between there is a temperature at which the qualitative behavior of the system changes from correlated to uncorrelated, which is the critical temperature, T_c, discussed previously.

In general, correlations are characterized by a ***correlation length*** ξ, which is a function of the size of the system and the temperature (we consider the case of zero external magnetization). The completely correlated low temperature phase has a correlation length of infinity (or L if the system is of linear size L). Alternatively, the

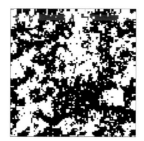

Figure 20.1 *An ising spin configuration at $k_B T = 2.5$.*

correlation length is small at high temperatures. As with percolation (see Chapter 8), the system fluctuates at all scales at the critical temperature, and hence ξ diverges at T_c. This is illustrated in Figure 20.1, which shows a spin configuration near T_c that is dominated by fluctuations that appear to occur over a wide range of length scales.

Systems whose dynamics are dominated by a single scale obey *scaling relations* that specify simplified behavior for various quantities. This concept was introduced in Section 8.3, which the reader may want to peruse. In the case of spin (and other) systems, this occurs near phase transitions. The functional form of the behavior is often written as simple power laws. Thus, for example, the way in which the correlation length diverges is parameterized as:

$$\xi \to \epsilon^{-\nu} \quad \epsilon \to 0, \tag{20.13}$$

with:

$$\epsilon = \frac{|T - T_c|}{T_c}. \tag{20.14}$$

A *critical exponent* ν has been introduced to express the rapidity of the divergence of the correlation length as the critical temperature is approached.

At criticality one expects $\xi \sim L$, where L is the linear size of the system. Thus the combination $L\epsilon^\nu$ is invariant. A similar argument for the external magnetic field reveals that the free energy must scale as (Fisher, 1971):

$$F(T, h; L) = L^{-(2-\alpha)/\nu} \hat{F}(\epsilon L^{1/\nu}, h L^{(\gamma+\beta)/\nu}) \tag{20.15}$$

Differentiation with respect to h (and setting $h = 0$) then gives the finite scaling laws:

$$M = L^{-\beta/\nu} \hat{M}(\epsilon L^{1/\nu}), \tag{20.16}$$

$$\chi = L^{\gamma/\nu} \hat{\chi}(\epsilon L^{1/\nu}), \tag{20.17}$$

$$C_V = L^{\alpha/\nu} \hat{C}_V(\epsilon L^{1/\nu}). \tag{20.18}$$

The powers in these expressions tend to depend solely on global properties of a system (like its dimensionality) and thus their values label the ***universality class*** to which the

Table 20.1 *Ising model critical exponents (Pelisetto and Vicari, 2002).*

Definition	Exponent	$d=2$	$d=3$	MFT
$C \to |T-T_c|^{-\alpha}$	α	0	0.110(1)	0
$M \to |T-T_c|^{\beta}$	β	1/8	0.3265(3)	1/2
$\chi \to |T-T_c|^{-\gamma}$	γ	7/4	1.2372(5)	1
$G \to \frac{e^{-r/\xi}}{r^{d-2+\eta}}$	η	1/4	0.0364(5)	0
$\xi \to |T-T_c|^{-\nu}$	ν	1	0.6301(4)	1/2

system belongs. Thus determining critical exponents is an important part of analyzing the structure of a given system.

The critical exponent η makes its appearance in spatial correlation functions as:

$$\langle \rho(0)\rho(r) \rangle \sim \frac{1}{r^{d-2+\eta}} \qquad r \to \infty, \qquad T \to T_c \qquad (20.19)$$

where it is dubbed the ***anomalous dimension***. Here ρ is typically a density or, in this case, the spin at position r. Away from the critical temperature one writes:

$$\langle \rho(0)\rho(r) \rangle \sim \frac{e^{-r/\xi}}{r^{(d-1)/2}} \qquad r \to \infty, \qquad T > T_c. \qquad (20.20)$$

Part of the business of Monte Carlo computations is determining the critical exponents of a model, and hence the universality class to which it belongs. A summary of critical exponents for the Ising model is presented in Table 20.1. The column labelled "MFT" gives *mean field theory* expressions for critical exponents. These can be thought of as critical exponents in the limit of a large number of dimensions and are typically obtained by considering the behavior of a single spin in the average environment of its neighboring spins.

20.2.3 The heat bath algorithm and the induced magnetization

Our immediate goal is to evaluate the induced magnetization with the Monte Carlo method in an attempt to determine the critical temperature. We will use the *heat bath algorithm* of Chapter 7; although it should be noted that the Ising model is a rare instance of a case where the *Metropolis algorithm* is more efficient.

Recall that the heat bath algorithm makes transitions to new configurations with probability:

$$p(\mathcal{S} \to \mathcal{S}') \propto \exp[-\beta H(\mathcal{S}')] \qquad (20.21)$$

We implement this with the Gibb's sampling method by updating a single spin at a time. Thus the heat bath probabilities are:

$$p(\sigma_I) = \frac{e^{\beta \mathcal{J} \sigma_I (h_I + \sum_{i \in I} \sigma_i)}}{\sum_{\sigma_I} e^{\beta \mathcal{J} \sigma_I (h_I + \sum_{i \in I} \sigma_i)}} \qquad (20.22)$$

where the sum in the exponential extends over nearest neighbors of spin I. In two dimensions there are only five possible such spin sums and the heat bath probabilities can be precomputed and stored in a matrix, saving much computational time.

As with the percolation problem, finite size effects are best minimized by adopting *periodic boundary conditions*, wherein the system is placed on a torus. This is not the only choice, and others may be more appropriate to the problem at hand. A short list of possibilities includes:

periodic boundary conditions Map neighbors on the edges of the volumes.

free boundary conditions Spins at the edge of the volume have no neighbors.

fixed boundaries Set boundary spins to fixed values, which may depend on their location. This is useful for creating interior interfaces for studying wetting and similar effects.

twisted boundary conditions Map spins as with periodic boundary conditions but invert the coupling for these spins. This is a translationally invariant boundary condition that sets up a kink in the interior of the lattice.

We shall only consider periodic boundary conditions in the following.

Common choices for the spin configuration updating protocol are:

deterministic sweeps Scan through the lattice in a fixed order.

stochastic updates Choose a spin at random.

checkerboard update Split the lattice into even and odd sublattices and update one sublattice at a time. This is very useful in parallel computing environments.

And lastly, the initial state of the simulation must be set. Two obvious choices are

cold start All spins are set to $+1$ or -1.

hot start Spins are assigned random states.

As a first exercise, we compute the magnetization M at $k_B T = 2.0$ and $k_B T = 2.5$ with hot and cold starts. The results are displayed in Figure 20.2 as a function of *sweep* (one sweep is one heat bath update per spin for the entire lattice). The two curves for $k_B T = 2.0$ start at $M = 0$ (hot start) and $M = 1$ (cold start), as expected. By around 200 sweeps they have converged and thereafter fluctuate around $M \approx 0.9$. This is an example

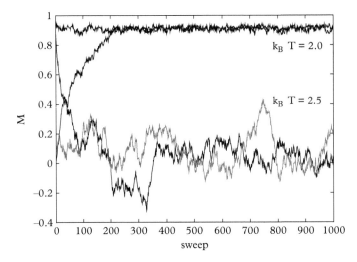

Figure 20.2 *Magnetization measured at $k_B T = 2.0$ and $k_B T = 2.5$. Sequences with hot and cold initial states are shown.*

of **thermalization**. All Monte Carlo computations require an initial period during which the configuration loses memory of its starting state, called the **thermalization time**, τ_{th}. Clearly it is undesirable to average quantities during this time. It is a good idea to estimate τ_{th} and run the heat bath algorithm for at least this many sweeps before collecting data. Notice that the thermalization time depends on the algorithm, the lattice size, the temperature, and the operator being measured.

The two curves labelled $k_B T = 2.5$ behave qualitatively differently from the lower temperature data. In this case the thermalization time is roughly $\tau_{\text{th}} = 100$, and the equilibrium magnetization fluctuates around zero. Thus it appears the $k_B T = 2.0$ belongs to the broken phase of the model, while $k_B T = 2.5$ lies in the symmetric phase, in agreement with Eq. 20.5. Finally, notice that the fluctuations present in the curves at $k_B T = 2.0$ are much smaller than those at $k_B T = 2.5$. This is because $k_B T = 2.5$ lies just above the critical temperature, where the system undergoes large-scale fluctuations.

Armed with this preliminary result, we now compute the induced magnetization by thermalizing, sweeping many times through the lattice, and averaging the total spin to obtain $M = \langle S \rangle$. Doing this for many temperatures yields the result shown in Figure 20.3. It is clear that the algorithm is working as intended, and that we can estimate that the critical temperature is between $T_c = 2.2$ and 2.4. More importantly, it appears that $L = 64$ is a good approximation to $L = \infty$! A close look at the figure reveals that M approaches zero smoothly as T gets large. This is as expected for a finite system. Obtaining a more accurate estimate of T_c will require performing computations on larger lattices and carefully tracking the rapid transition to $M = 0$.

Because the Ising Hamiltonian is symmetric under spin reflection, it is possible for the system to make excursions from a positive value of $S = \sum_i \sigma_i$ to a negative value. In fact very long runs of the type shown in Figure 20.2 would show tunneling transitions

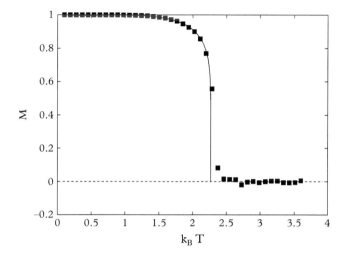

Figure 20.3 *Magnetization vs. T with the Onsager exact result. These results were obtained on a lattice of size* 64 × 64.

between $\pm S$. Thus $M = 0$ must hold (this was originally used as an argument against the possibility of long range order in the Ising model). We avoid this awkward situation in the following by considering the observable $|M| = \langle |\sum_i \sigma_i| \rangle$. That short Monte Carlo runs fail to obtain zero for the magnetization can be regarded as a flaw of the algorithm, but it reflects how symmetry breaking is manifested in nature. Namely, all physical systems are finite in extent and therefore are analytic in their parameters. Discontinuous symmetry breaking transitions only *appear* to happen due to inevitable experimental imprecision and extraordinarily long metastability times.

It is evident that this method for determining the critical temperature will not yield very precise results. We thus turn to the magnetic susceptibility (Eq. 20.11), which has a peak at $T_c(L)$. Plotting the susceptibility with respect to T for various lattice sizes then permits extrapolation to the bulk limit via

$$T_{\max}(L) = T_c - aL^{-1/\nu}. \tag{20.23}$$

Justify this relationship using the discussion of Section 20.2.2.

20.2.4 Reweighting

An accurate determination of T_c now requires an accurate determination of the peak of the susceptibility, which requires measurements at many temperatures. This is very expensive so it is fortunate that another trick, called **reweighting**, is available.

The idea of reweighting is to use a single Monte Carlo run to evaluate expectation values at many parameter values. In this example the parameter is temperature and reweighting is achieved by shifting some of the Boltzmann measure into the observable. Thus,

$$\langle \mathcal{O} \rangle_{\beta'} \equiv \frac{\text{tr}\, \mathcal{O} e^{-\beta' H}}{\text{tr}\, e^{-\beta' H}} = \frac{\text{tr}\, \mathcal{O} e^{-(\beta'-\beta)H} e^{-\beta H}}{\text{tr}\, e^{-(\beta'-\beta)H} e^{-\beta H}} = \frac{\langle \mathcal{O} e^{-(\beta'-\beta)H} \rangle_\beta}{\langle e^{-(\beta'-\beta)H} \rangle_\beta}. \tag{20.24}$$

The new notation, "tr", represents the sum over all possible spin configurations, $\sum_{\{\sigma\}}$.

If one starts in the symmetric phase the only way in which sensible results can be obtained in the broken phase is if the system size is relatively small. Thus the range over which one can expect to use reweighting reliably must decrease with system size. Furthermore, $\beta' - \beta$ cannot be too large or one swamps the Boltzmann measure and the Monte Carlo method fails.

The method is easily extended to other parameters, for example one can measure quantities at many external magnetizations by including $\exp(\beta \mathcal{J}(h' - h) \sum \sigma_i)$ in the observable.

With this refinement we are able to compute decent estimates of $T_{\max}(L)$. Since reweighting must be done over a restricted temperature range, the strategy will be to compute χ on a small lattice of size L, obtain $T_{\max}$ from this lattice, and center a new computation of the susceptibility at $T_{\max}(L)$ on a lattice of size $2L$. In this way the computation is kept in the region of the best estimate of the critical temperature.

20.2.5 Autocorrelation and critical slowing down

Unfortunately our quest to obtain the critical temperature now runs into a new problem, namely it becomes exceedingly difficult to obtain good estimates for χ once the size of the system grows beyond $L \approx 256$. The problem arises due to *critical slowing down*, which is an increase in the autocorrelation time due to proximity to a second order phase transition.

As a quick reminder, the autocorrelation function for an observable $\mathcal{O}$ is:

$$C(t) = \frac{\langle \mathcal{O}_i \mathcal{O}_{i+t} \rangle - \langle \mathcal{O}_i \rangle^2}{\langle \mathcal{O}_i^2 \rangle - \langle \mathcal{O}_i \rangle^2} \tag{20.25}$$

and the integrated autocorrelation time is:

$$\tau_{\text{int}} = \sum_t C(t).$$

More information is contained in Chapter 7.

Generally one computes autocorrelation times as measurements are made to ensure that statistical inefficiency is not invalidating one's efforts. A simple way to do this is to store all measurements $\mathcal{O}_t$ and then perform the sums indicated to obtain the autocorrelation function. This can be quite wasteful for large runs (we often perform 10^6

sweeps in this chapter), especially when correlations tend to die off quickly. Thus it is preferable to maintain a list of the past nXvalues measurements and compute the correlation function as new values of the observable (called X below) are obtained. An elegant way to do this is provided by the C++ list container class:

```
#include <list>;

double autoCorr[nXvalues+1];
list<double> Xvalues;                    // could also use a deque

void getAutoCorrs(double X) {
  if (Xvalues.size() == nXvalues) {// we've filled the list
    autoCorr[0] += X*X;
    list<double>::const_iterator  it = Xvalues.begin();
    for (int i=1; i<=nXvalues; i++) {
      autoCorr[i] +=   *it++ * X;
    }
    Xvalues.pop_back();                  // pop the oldest X
  }
  Xvalues.push_front(X);                 // push the new X
}
```

A computation of the autocorrelation function for $L = 64$ lattices at various temperatures is shown in Figure 20.4. One observes an exponential decay (which can

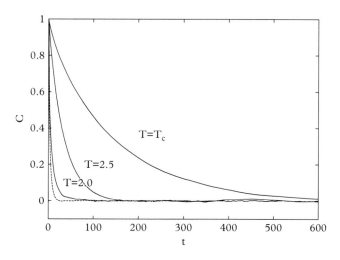

Figure 20.4 *Autocorrelation functions for the heat bath algorithm at different temperatures. Autocorrelation for the Wolff algorithm (Section 20.2.6) at $T = T_c$ is shown as a dashed line.*

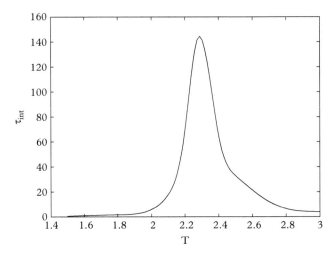

Figure 20.5 *Integrated autocorrelation time for the heat bath algorithm.* ($L = 32$, $\mathcal{O} = |M|$, 10^6 *measurements*).

also be used to characterize the autocorrelation time[2]), with a much slower decay at the critical temperature. We stress that autocorrelation times depend on lattice volume, the algorithm employed, the temperature, and the observable.

Figure 20.5 displays the autocorrelation time as a function of temperature. The peak at the critical temperature is due to large scale fluctuations that occur in the vicinity of T_c. Because the error in the magnetization grows as $\sqrt{2\tau_{\text{int}}/N}$ one needs to increase the number of measurements by a factor of approximately 100 to achieve the same precision as obtained at $T \lesssim 2.0$. The situation rapidly becomes worse for larger lattices and temperatures near T_c. Unfortunately, it is precisely large lattices near T_c in which we are interested!

The rate at which the autocorrelation time increases with system volume is of interest. This is specified with a ***dynamical exponent*** as:

$$\tau_{\text{int}} \propto L^z \tag{20.26}$$

(other definitions of the autocorrelation time can be used as well.) Overall, the computational effort scales as L^{2+z}. For local algorithms like heat bath and Metropolis $z \approx 2$, thus if N_{32} sweeps are required to achieve a given precision on an $L = 32$ lattice, $2^{12} N_{32}$ sweeps are required on a 256×256 lattice! Factors like this can rapidly bring progress to a halt. A computation of the dynamical exponent for the Metropolis algorithm as measured with $|M|$ is shown in Figure 20.6. A fit gives $z \approx 1.95$.

---

[2] See (Landau and Binder, 2000) for a detailed discussion of various correlation times.

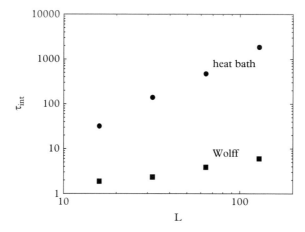

Figure 20.6 *Integrated autocorrelation time vs. lattice size at $T = T_c$. (10^6 measurements, $\mathcal{O} = |M|$). Rough fits give $z(HB) = 1.95$ and $z(W) = 0.5$.*

20.2.6 Cluster algorithms

Critical slowing down is clearly a substantial impediment in our attempt to compute T_c. Continuing progress relies on accommodating the large scale fluctuations seen near T_c. In the case of the Ising model, this can be accomplished with algorithms that flip large groups of spins at one time, called *cluster algorithms*.

The breakthrough was initiated by Fortuin and Kasteleyn (1972) who realized that the Ising model[3] could be mapped to a percolation model with percolating probability $p_{FK} = 1 - \exp(-2\beta J)$ by considering the bonds formed between like-spin states. The importance of this observation is that percolation clusters are uncorrelated (because percolation sites are uncorrelated) and hence percolation simulations suffer no critical slowing down.

Swendsen-Wang algorithm

The first practical algorithm to employ these ideas was due to Swendsen and Wang (1987). The idea is to sweep through a lattice placing bonds between pairs of spins with probability p_{FK}. Clusters of connected spins are then flipped with probability 1/2. Clusters can be identified using the percolation algorithm presented in Chapter 8. The crucial point is that many spins can be flipped at one time, and critical slowing down is largely evaded.

Wolff algorithm

The Swendsen-Wang algorithm considers all clusters, including single-site ones. An alternative, first suggested by Ulli Wolff (1989), concentrates effort on identifying and

---

[3] More generally, the q-state Potts model, which is discussed in the next section.

flipping large clusters and is generally more efficient than the Swendsen-Wang algorithm. The procedure is as follows:

```
// Wolff Algorithm

choose a seed spin at random, flip it, mark it as a cluster
    spin

grow the cluster:
  visit neighbors
    if neighbor is not a cluster spin and if it is parallel
       to the original seed
       add it to the cluster with probability p_FK = 1 - e^{-2βJ}
       flip the spin
       visit this spin's neighbors
```

As with the percolation cluster algorithm, clusters are labelled with elements of a matrix, `cluster`. Spins in the cluster are tracked with a stack:

```
int spin[L][L];
int cluster[L][L];
int Istack[L*L];
int Jstack[L*L];

void growC (int i, int j, int seed) {
  if ((cluster[i][j] == 0) && (spin[i][j] == seed))
// not in the cluster and parallel to seed
    if (ran() < p_FK) {
      spin[i][j] = -spin[i][j];          // flip spin
      cluster[i][j] = 1;                 // place in cluster
      stackp++;                          // and in the stack
      Istack[stackp] = i;
      Jstack[stackp] = j;
    }
}
```

The full update process is

```
void wolffUpdate (int N) {
  int i, j, seed;
  for (int nn=0;nn<N;nn++) {
    // zero cluster spin markers
    for (i = 0; i < L; i++) {
```

```
      for (j = 0; j < L; j++) {
        cluster[i][j] = 0;
  }}
  // initialize stack
  stackp = -1;
  // choose a random spin
  i = (int) (L*ran());
  j = (int) (L*ran());
  // grow cluster
  seed = spin[i][j];      // save the seed spin
  spin[i][j] = -spin[i][j];    // flip the seed spin
  cluster[i][j] = 1;
  stackp++;                      // push the seed spin onto the
                                 //    stack
  Istack[stackp] = i;
  Jstack[stackp] = j;
  while (stackp > -1) {
    i = Istack[stackp];
    j = Jstack[stackp];
    stackp--;
    // try to add each neighbor to cluster
    growC(pl[i], j, seed);
    growC(mn[i], j, seed);
    growC(i, pl[j], seed);
    growC(i, mn[j], seed);
  }
 }
}
```

We still need to prove that the Wolff algorithm is a valid Monte Carlo procedure. It is clearly Markovian and ergodic. In order to demonstrate detailed balance consider a cluster, $\mathcal{C}$, that has as its seed a spin up at site r. Assume that the cluster has $n^\uparrow$ ($n^\downarrow$) neighbors of spin up (down). The change in energy due to flipping the cluster is $\Delta E(\mathcal{C} \to \mathcal{C}') = 2(n^\uparrow - n^\downarrow)\mathcal{J}$. Let P_r be the probability that the cluster in question grew from site r. The transition rates for the cluster flip $\mathcal{C} \to \mathcal{C}'$ is then

$$W_{\mathcal{C} \to \mathcal{C}'} = \sum_r P_r (1 - p_{FK})^{n^\uparrow} \qquad (20.27)$$

and

$$W_{\mathcal{C}' \to \mathcal{C}} = \sum_r P_r (1 - p_{FK})^{n^\downarrow}. \qquad (20.28)$$

The second factors in these expressions count the number of times the cluster did not grow when starting from the up seed spin (for $\mathcal{C} \to \mathcal{C}'$) or from the down seed (for $\mathcal{C}' \to \mathcal{C}$). We then confirm that:

$$\frac{W_{\mathcal{C} \to \mathcal{C}'}}{W_{\mathcal{C}' \to \mathcal{C}}} = e^{-\beta \Delta E} = \frac{e^{-2\beta \mathcal{J} n^\uparrow}}{e^{-2\beta \mathcal{J} n_\downarrow}} \tag{20.29}$$

holds when $p_{FK} = 1 - \exp(-2\beta \mathcal{J})$ and hence have established that the Wolff algorithm correctly thermalizes the Ising model.

We return to the computation of the susceptibility for $|M|$. Using reweighting and the Wolff algorithm we are now able to generate estimates of $T_{\max}$ with relatively small effort up to lattices of size 512×512. Results are shown in Figure 20.7. The final step is extrapolating to the bulk limit and is shown in Figure 20.8. The result is $T_c \approx 2.26927(21)$ which is in nice agreement with the exact expression. This was obtained using the Ansatz $T_{\max} = T_c - a/L$. Using $T_{\max} = T_c - a/L^{1/\nu}$ (see Eq. 20.23) gives the same result for T_c with a slightly larger error and $1/\nu = 1.00(7)$, in agreement with the theoretical value for this critical exponent.

As expected, the Wolff algorithm is much more efficient than the heat bath algorithm, especially near the critical temperature. This is illustrated in Figure 20.4 where the autocorrelation function for the Wolff algorithm is displayed as a dashed line. The rapid decrease in autocorrelation is evident. Scaling of the integrated autocorrelation time versus lattice size is shown in Figure 20.6. Again, the autocorrelation times are always

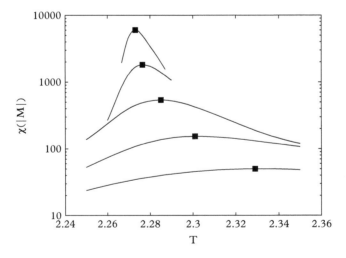

Figure 20.7 *Susceptibility vs. T. From bottom to top $L = 32, 64, 128, 256,$ and 512. The squares indicate the location of maxima.*

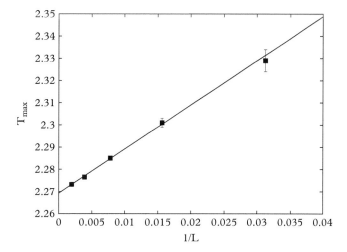

Figure 20.8 *Extrapolation of the peak susceptibilities (see Figure 20.7) to infinite volume. The result is $T_c \approx 2.26927(21)$ (exact result: 2.269185).*

much smaller than those for the heat bath algorithm, and they grow more slowly (with a dynamical exponent of approximately 0.5) with lattice size.

20.3 The Potts model and first order phase transitions

So far our attention has been focused on systems exhibiting second order phase transitions. These are characterized by self-similarity, fractal dynamics, and infinite correlation length at the critical point. First order phase transitions are quite different: correlation lengths remain finite, scaling tends to be with the system volume rather than with critical exponents, phase coexistence is common (think of boiling water), and latent heat is associated with the transition between phases.

The two-dimensional Ising model exhibits a first order phase transition in the magnetization as the external field is brought from positive to negative values. Recall that the system is symmetric under inversion of the spins, and in fact, will make excursions from $+M$ to $-M$ over large algorithmic times. Setting $h > 0$ breaks this symmetry and forces the magnetization to be positive, regardless of the magnitude of h. Once h passes through zero, the $-M$ state is preferred and the system makes a first order transition to the $-M$ phase. Things need not work this way, for example the induced magnetization could smoothly run to zero as the external field is sent to zero. In fact this does happen for $T > T_c$, and there is no phase transition above the critical temperature. This is illustrated in Figure 20.9, which shows magnetization curves for various temperatures. Evidently $M(h)$ shows no dramatic change in behavior near $h = 0$ when $T = 4$, well above the

The Potts model and first order phase transitions

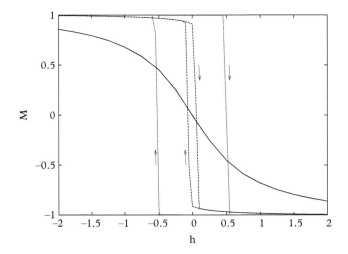

Figure 20.9 *Magnetization in the Ising model while sweeping in h for $k_B T = 4$ (solid line), 2 (dashed line), and 1 (dotted line). $L = 128$; 1000 sweeps per point.*

critical temperature. However, a sudden jump from $+M$ to $-M$ is apparent for the cases $T = 2$ and $T = 1$.

The **Potts model** is a well-known spin system that exhibits first and second order phase transitions[4]. Like the Ising model, degrees of freedom are restricted to take on discrete values and to reside on a fixed lattice. The Hamiltonian is

$$H = -\mathcal{J} \sum_{(i,j)} \delta_{\sigma_i, \sigma_j} \tag{20.30}$$

where the sum is over nearest lattice neighbors and there are q degrees of freedom at each site: $\sigma_i = 1, 2, \ldots, q$ (Potts, 1952). Many experimental realizations of the Potts model exist. An interesting one represents the $q = 3$ model as a mixture of ethylene glycol, water, lauryl alcohol, nitromethane, and nitroethane (Wu, 1982). Notice that the Potts model reduces to the Ising model in the case $q = 2$ and thus exhibits a second order phase transition in the magnetization as the temperature is increased. As with the Ising model, many properties are known analytically, especially for low dimensional systems. For example, the critical temperature for 2d square lattices is (Wu, 1982):

$$\frac{k_B T_c}{\mathcal{J}} = \frac{2}{\log(1 + \sqrt{q})}. \tag{20.31}$$

---

[4] Named after R.B. Potts, the model was suggested by his supervisor Cyril Domb as a D.Phil. research project.

It is also conjectured that the Potts model exhibits a first order phase transition for $q > 4$ in two dimensions (and $q > 2$ in three dimensions).

The order of the phase transition at $q = 4$ can be difficult to determine numerically because the latent heat is very small. It is therefore fortunate that a simple extension of Wolff's Ising algorithm permits efficient computation with the Potts model: the algorithm should form clusters of a given q-value and should flip to a new cluster with a uniformly chosen q-value.

Although it is traditional to assign site states to the integers $1 \ldots q$, the connection to the standard definition of magnetization is better made by labeling the states according to elements of Z_q (the group formed by the qth roots of -1):

$$\sigma_j = \exp\left(i\frac{2\pi j}{q}\right), \qquad j = 0 \ldots q - 1. \qquad (20.32)$$

Using this definition leads to ready confirmation of Eq. 20.31, with all magnetization curves rapidly running to zero near $k_B T_c$. The rapidity with which this happens increases with larger q, however, this observation is not sufficient to determine the order of the phase transition. A more telling observable is the internal energy, shown in Figure 20.10. A smooth transition is evident for $q = 4$, while a definite latent energy of around 1.1 (in units of the coupling) is visible for $q = 20$, with something more ambiguous at $q = 8$.

Phase separation and latent heat are associated with metastability in the system; namely, the system can exhibit memory due to long transition times between local minima of the free energy. This behavior can be exposed in stochastic computations by adjusting an external variable while collecting measurements. The resulting curves demonstrate **hysteresis**, wherein the system remains in a 'false' phase once passing

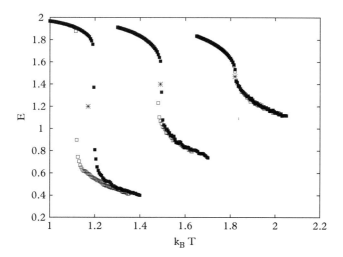

Figure 20.10 *Internal energy for the q-state Potts model, showing hysteresis. From left to right $q = 20$, 8, 4. Critical temperatures are indicated with bursts.*

through a transition point. This behavior is made more evident by passing through the phase transition in both directions, making a hysteresis loop. Figure 20.9 shows hysteresis loops for the Ising model obtained by making a relatively small number of measurements (in this case 1000) for each value of the external field, h, without rethermalizing. As expected, the curve obtained at $k_B T = 4$ shows no evidence of hysteresis; however those at $T < T_c$ show an increasing memory effect as the temperature is dropped, which agrees with the statement that the phase transition at $h = 0$ for $T < T_c$ is first order. Similar curves can be obtained with the Potts model, except that it is the temperature, not the field, that is changed. Examples are shown in Figure 20.10 where one sees little or no evidence for a first order phase transition for $q = 4$ and definite evidence for one at $q = 20$ and $q = 8$.

We stress that hysteresis is a nonequilibrium effect and thus is present in any system in which the rate at which the system thermalizes is slower than the rate at which the external parameter is changed. The physical effect is observed for systems that exhibit macroscopically long times tunneling between local minima. Real magnets, for example, contain many domains that are magnetized in different directions. A slowly varying external field flips one domain after another as it increases in strength. The resulting induced magnetization exhibits small discrete jumps in its value – an effect known as **Barkhausen noise**. Kuntz et al. (1999) have developed a method to simulate this with a Hamiltonian:

$$H = -\mathcal{J} \sum_{\langle ij \rangle} \sigma_i \sigma_j - \sum_i (H(t) + h_i) \sigma_i. \tag{20.33}$$

The idea is that each spin represents a coarse-grained domain that is coupled to a random external field, h_i. The other external field H is increased in magnitude, flipping domains as it goes. Since the model is meant to simulate nonequilibrium physics, the authors set the temperature to zero and deterministically flip spins as H increases when this decreases the energy:

$$\Delta E = \mathcal{J} \sum_{j \in i} \sigma_j + h_i + H(t) > 0. \tag{20.34}$$

Notice that this can happen when the increasing magnitude of H or a spin flip causes the local field to change sign. Thus a single spin flip can lead to an **avalanche** of spin flips and Barkhausen noise (Sethna, 2006).

20.4 The planar XY model and infinite order phase transitions

The Ising and related models possess a discrete inversion symmetry $\sigma_i \to -\sigma_i$. This can be generalized to a continuous symmetry by considering the interaction of classical vectors:

$$H = -\mathcal{J} \sum_{(ij)} \vec{\sigma}_i \cdot \vec{\sigma}_j - \sum_i \vec{h}_i \cdot \vec{\sigma}_i. \tag{20.35}$$

We specialize to the case of two-dimensional vectors in the following. Thus this model is often called the "planar XY" or "2d XY" model. If the vectors are taken to be of unit length and the external field is aligned along the $\hat{z}$ axis then the Hamiltonian can be rewritten as:

$$H = -\mathcal{J} \sum_{(ij)} \cos(\theta_i - \theta_j) - \sum_i h_i \cos(\theta_i). \tag{20.36}$$

When $\vec{h} = \vec{0}$, the Hamiltonian possesses a $U(1)$ (2d) or $SO(3)$ (3d) symmetry under rotations $\sigma_\nu \to R_{\nu\mu} \sigma_\mu$. Magnetizing the system (either with an external field or spontaneously) permits fluctuations of spins called ***spin waves***. The quantized versions of these are called ***magnons***, and are generically called ***Goldstone bosons***. In two-dimensions these fluctuations are sufficient to completely disorder the system, thus preventing a second order phase transition from occurring. This result goes by the name of the ***Mermin-Wagner theorem***.

Nevertheless the 2d XY model[5] exhibits a low temperature ordered phase with a power law spin-spin correlation function and a high temperature disordered phase with an exponentially decaying correlation function. This phase transition was discovered by British physicists J. Michael Kosterlitz and David Thouless (1973) and is *topological* in nature. Specifically, the XY model generates vortex spin configurations in which spin fields rotate through 2π as they circle a given point. Vortices tend to pair in configurations with net zero vorticity below the critical temperature, while above it the vortices are free. This novel type of phase transition is called an ***infinite order phase transition***.

A simple Metropolis code run with sufficient care does indeed demonstrate that the system does not magnetize. This makes it difficult to determine the critical temperature; however, a plot of the internal energy, Figure 20.11, shows a suspicious change in curvature near $k_B T = 1.0$. This transition can be clarified with the aid of other observables. For example, the spin-spin correlation function

$$G(r) = \langle \vec{s}(r_0) \cdot \vec{s}(r_0 + r) \rangle \tag{20.37}$$

is expected to follow a power law below T_c:

$$G(r) \sim r^{-\eta}, \quad T < T_c \tag{20.38}$$

---

[5] The XY model was introduced by Matsubara and Matsuda as a paradigm for 2d quantum lattice gases in 1956. (Matsubara, 1956).

The planar XY model and infinite order phase transitions

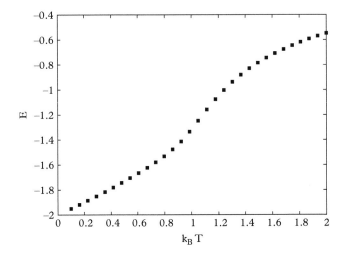

Figure 20.11 *Internal energy for the XY model.* $L = 128$, 10000 *sweeps per point.*

and an exponential law with diverging correlation length for high temperature:

$$G(r) \sim e^{-r/\xi(T)}, \qquad T > T_c \tag{20.39}$$

with

$$\xi(T) \sim e^{b(T/T_c-1)^{-\nu}}. \tag{20.40}$$

One can implement the XY model with n discrete values for the angles (ie., the Z_n spin model). This would, for example, be useful if one wished to employ the heat bath algorithm. Under which conditions will this be a good approximation to the XY model? Where should it fail?

Figure 20.12 shows correlation functions computed for $k_B T = 2.0, 1.5, 1.2, 1.1, 1.0$, and 0.5. On a log-log plot the correlation functions for $k_B T = 1.0$ and 0.5 are very nearly linear indicating that $T_c > 1.0$. Correlation functions for $k_B T \geq 1.1$ show curvature and require an exponential fit. As expected, the correlation lengths for these curves grow as $k_B T = 1.1$ is approached. Thus it appears that there is indeed a phase transition in the 2d XY model and that it occurs at $T_c \approx 1.05 J$.

Nevertheless, the student is reminded of the *prime directives for Monte Carlo work*:

1. evaluate errors
2. check autocorrelation

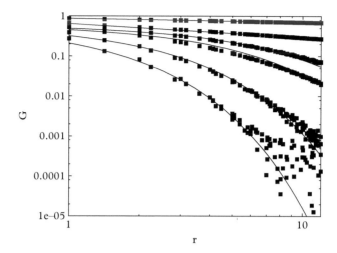

Figure 20.12 *Spin-spin correlation function.* $L = 64$, 10000 *sweeps per curve. From bottom to top* $k_B T = 2.0, 1.5, 1.2, 1.1, 1.0, 0.5$. *Fits give* $\xi(2.0) = 0.95, \xi(1.5) = 1.6, \xi(1.2) = 3.5, \xi(1.1) = 5.0$, $\eta(1.0) = 0.35$, *and* $\eta(0.5) = 0.095$.

3. do some long runs
4. check finite size effects.

In this case a few runs at $k_B T = 1.0$ with $L = 128, 256$, and 512 indicate that curvature does creep into the correlation functions and hence $k_B T_c < 1.0$. And in fact, Tobochnik and Chester (1979) find $T_c \approx 0.89$, $\nu \approx 0.7$, and $b \approx 1.5$.

Finally, as pointed out by Wolff in his original paper, it is possible to employ a cluster algorithm for XY models with spins of dimension d as follows:

```
// Wolff Algorithm for SO(d) spins

choose a random vector V in SO(d)
choose a seed spin (i) at random, flip it(*), mark it as a
    cluster spinA

grow the cluster:
  visit neighbors (j)
    if neighbor is not a cluster spin and if it is parallel
        to the original seed
      add it to the cluster with probability p = 1 − e^{−2βJ_{ij}}
      where J_{ij} = (V · σ_i)(V · σ_j)
      flip the spin (*): σ_i → σ_i − 2(V · σ_i)
      visit this spin's neighbors
```

20.5 Applications and extensions

The simplicity and rich phenomena associated with classical spin models have led to their widespread adoption in many fields. We discuss two additional topics here, namely spin glasses and neural networks. Research is ongoing in both areas, with an enormous and growing literature. We must be brief, but trust that the following gives a flavor of the fields.

20.5.1 Spin glasses

The discussion of avalanches and hysteresis led us to consider a model with spins coupled to an external magnetic field that fluctuates randomly in space. This is called a *random field Ising model* (introduced by Imry and Ma (1975)). Physical realizations of this model typically involve materials with impurities, but the model also describes binary liquids in porous media, mixed Jahn–Teller systems, diluted frustrated antiferromagnets, hydrogen in metals, and mixed crystals undergoing structural or ferroelectric transitions (Nattermann, 1997).

Spin glasses form another class of random spin models in which the couplings are random. They model magnetic systems with impurities whose effect is to essentially randomize the sign and strength of the interactions between spins. At high temperatures spin glasses behave much like 'regular' spin systems; at low temperatures the spins attempt to align according to their couplings, which gives rise to a disordered, but quenched, state. The time it takes to seek out the local minima in this case can be very long (or infinite), hence spin glasses are always out of equilibrium. This effect is called *broken ergodicity*. It is this analogy between low temperature magnetic disorder and the positional disorder in conventional glass that gives rise to the term 'spin glass'.

The underlying cause of broken ergodicity is *frustration*. Frustration occurs when a single low temperature spin configuration does not satisfy all the spin couplings. A simple example of this effect is four spins at the corners of a square with three antiferromagnetic and one ferromagnetic coupling. It is possible to make three antiferromagnetic bonds in this case, but the remaining bond must then also be antiferromagnetic, ruining the attempt to satisfy all of the bond constraints. Frustration leads to many peculiar effects such as broken ergodicity and the failure of mean field theory. In fact, it appears that mean field theory fails for all dimensions, which in turn implies that the infinite dimension limit is singular (Stein, 2003). Remarkably it is not definitively known if phase transitions occur for low dimensions. It is even possible that spin glasses have chaotic temperature dependence, where small temperature changes can induce large and unpredictable changes in correlation functions (see Chapter 13 for much more on chaotic dynamics).

A simple paradigm for spin glasses is provided by the *Edwards-Anderson model* (1975). The model assumes Ising spins interacting with random nearest neighbor couplings chosen from a Gaussian distribution (typically with zero mean):

$$H = -\sum_{(ij)} \mathcal{J}_{ij}\sigma_i\sigma_j \quad (20.41)$$

$$p(\mathcal{J}_{ij}) = \frac{1}{\sqrt{2\pi}\mathcal{J}} \exp\left(-\frac{1}{2\mathcal{J}^2}(\mathcal{J}_{ij} - \mathcal{J}_{\text{avg}})^2\right) \quad (20.42)$$

The randomness of the couplings precludes an ordered ground state, so it is impossible to use the average magnetization as an order parameter for this model. One could attempt to avoid this by considering the average local magnetization, $m_i = \langle\sigma_i\rangle$. However the Edwards-Anderson model is symmetric under spin reflection, so this quantity is zero over long equilibration times. Thus Edwards and Anderson proposed using the following observable as an order parameter:

$$q_{EA} = \frac{1}{N}\sum_{i=1}^{N} \overline{\langle\sigma_i\rangle^2} \quad (20.43)$$

Here the angle brackets refer to thermodynamic averages while the bar represents an average over coupling realizations. This two step method of forming observables is necessary because the disorder in the system is **quenched**, meaning it is a property of the sample in question. Of course one is not interested in properties of a specific magnetic sample, so one averages over ensembles of samples to obtain observables.

A mean field theory estimate of the order parameter can be made (see Ex. 18); the result is called the **Sherrington-Kirkpatrick equation**, which must be solved for q.

$$q = \int_{-\infty}^{\infty} \frac{dx}{\sqrt{2\pi}} e^{-x^2/2} \tanh^2(\beta\mathcal{J}\sqrt{q}x). \quad (20.44)$$

The solution indicates that the (mean field estimate of the) critical temperature of the Edwards-Anderson model is at $T_c = \mathcal{J}$. Although this is a mean field result, which is not regarded as reliable for spin glasses, more sophisticated computations yield the same estimate. Monte Carlo calculations (see Ex. 18) agree well with this result.

20.5.2 Hopfield model

A simple mathematical model of a nerve cell or neuron was introduced by McCulloch and Pitts (1943). The idea is to subsume the complex of states that a neuron can take into a binary choice: firing or quiescent. The state of the ith neuron at time t is then written $\sigma_i(t) = 1$ or 0. It is further assumed that the system evolves synchronously at regular unit time intervals. The state of the ith neuron in a network is determined at the next time step by:

$$\sigma_i(t+1) = \theta\left(\sum_j \mathcal{J}_{ij}\sigma_j(t) - f_i\right) \quad (20.45)$$

where θ is a step function, f_i is the firing threshold for neuron i, and $\mathcal{J}_{ij}$ is the synaptic strength between the axon of neuron j and the dendrite of neuron i. Note that the couplings can be positive or negative, depending on whether the neural link is excitatory or inhibitory.

John Hopfield (1982) built on this idea to introduce an *associative neural network*, called the *Hopfield model*. The model postulates the same dynamical relationship between neural "spins" as in the McCulloch-Pitts model subject to an appropriate update protocol (e.g., synchronous update, random, or sequential). Hopfield demonstrated that the evolution of the system tends towards the global minimum of the function:

$$E = -\frac{1}{2} \sum_{ij} \mathcal{J}_{ij} \sigma_i \sigma_j + \sum_i f_i \sigma_i \tag{20.46}$$

and hence the model is equivalent to an Ising spin system with long range (possibly random and asymmetric) interactions.

Hopfield realized that because the energy function contains many local minima in its configuration space it can serve as model of *memory*. In particular, the system can be trained (or memories can be implanted) by setting its couplings according to *Hebb's rule*:

$$\mathcal{J}_{ij} = \sum_{m=1}^{M} (1 - 2\sigma_i^{(m)})(1 - 2\sigma_j^{(m)}) \tag{20.47}$$

where M is the number of patterns to be stored and $\sigma_i^{(m)}$ is the value of the ith bit of the information to be stored. Hebb's rule builds M local minima in the energy landscape at the bit patterns to be recovered. Information is recalled when the system evolves to one of these minima. This process will be more robust if the minima are widely separated, or equivalently, have large basins of attraction. One expects that the maximum number of patterns that can be reliably recalled is order N^2/N; in practice it is around $0.13N$.

20.6 Exercises

1. **Another Critical Exponent.**
   For the Ising model one has $M \propto h^{1/\delta}$. In 2d $\delta = 15$. See if you can verify this numerically. Hint: δ is quite large, so you should consider small values of h.

2. **Trivial Renormalization Group.**

   a) Solve the 1 d Ising model using the transfer matrix method (see Ex. 13 of Chapter 8 for more on this).

b) Obtain the partition function by first summing over the even sublattice. Notice that this yields a simple iterative procedure. Naively doing the sum over configurations requires 2^N additions, how many are required with your iterative method?

c) Code up and test!

3. **Swendsen-Wang Algorithm.**
   Code up the SW cluster algorithm using the cluster identification routine of Chapter 8 for the 2d Ising model. Compare its efficiency to the Wolff algorithm for large lattices.

4. **Finite Size Scaling.**
   Confirm finite size scaling for the 2d Ising model by showing that the magnetization curves collapse onto a universal curve once plotted in terms of the scaling arguments (see Figure 8.8 for a percolation example).

5. **Reweighting and χ_{max}.**
   Use reweighting and the Wolff algorithm to obtain the peak values of the susceptibility as a function of lattice size for the 2d Ising model. Fit your results to $\chi_{max} = aL^b$ and hence determine the ratio of critical exponents γ/ν. How do your results compare to the exact ratio?

6. **Algorithms.**
   Compare Metropolis and heat bath algorithm run times using both random and stochastic sweeps for the 2d Ising model. Use the time per L^2 updates multiplied by the integrated autocorrelation time as your figure of merit. Which algorithm performs better?

7. **Cluster Algorithms.**

   a) Implement the Wolff algorithm but only consider flipping the cluster rooted at (0,0). Is this a valid method?

   b) Identify clusters, flip the largest cluster. Is this a valid method?

8. **Spin-spin Correlation Function.**
   The *spin-spin correlation function* is defined by

   $$G(r) = \langle \sigma_i \sigma_{i+r} \rangle - \langle \sigma_i \rangle^2. \tag{20.48}$$

   The second term is what we have called previously called M^2. Rather than binning in the distance r it is simpler to store values of $G[r_x][r_y]$ and then plot these versus $\sqrt{r_x^2 + r_y^2}$. If rotational invariance is achieved (for sufficiently large lattices) the points should fall on a universal curve. Obtain the spin-spin correlation function for the 2d Ising model for temperatures that straddle T_c. Interpret your results.

9. **[T] Potts Model on a Cayley Tree.**

a) Show that the free energy for the q-state Potts model on a Cayley tree is given by $F = \log(\exp(2\beta\mathcal{J}) + q - 1)$. A Cayley tree is defined in Exercise 4 of Chapter 8.

b) Obtain the spin-spin correlation function:

$$G(r) = \frac{q-1}{q^2}\left(\frac{e^{2\beta\mathcal{J}}-1}{e^{2\beta\mathcal{J}}+q-1}\right)^r. \tag{20.49}$$

10. **The q-state Potts Model.**
As $q \to 4$ from above, the latent heat and correlation length behave as

$$\mathcal{L} \sim 3\pi/\sqrt{x} \tag{20.50}$$

and

$$\xi \sim \frac{x}{8\sqrt{2}} \tag{20.51}$$

where

$$x = \exp(\pi^2/2\cosh^{-1}(\sqrt{q}/2)). \tag{20.52}$$

Confirm these expressions numerically. Warning: x gets enormous very quickly. Work with $q > 6$ to keep correlation lengths below 100.

11. **[T] One-dimensional XY Model.**
Assume free boundary conditions and $h = 0$ for the 1d L-chain XY model and obtain the exact result $Z = (2\pi)^L K(\beta\mathcal{J})^{(L-1)}$. Here K is the Bessel function, $\int_{-\pi}^{\pi} d\theta \exp(\beta\mathcal{J}\cos(\theta))$.

12. **XY Model and the Kosterlitz-Thouless Phase Transition.**
Evaluate the susceptibility $\chi = \left\langle \left(\frac{1}{L^2}\sum_i \vec{\sigma}_i\right)^2 \right\rangle$ in the 2d XY model. It has been suggested that this scales as $\chi \sim \xi^{2-\eta}$, which means that the peak in the susceptibility locates the Kosterlitz-Thouless phase transition. Confirm this. Try to estimate the coefficient η.

13. **Baxter-Wu Model.**
Consider the Hamiltonian

$$H = -\mathcal{J}\sum_{(i,j,k)} \sigma_i\sigma_j\sigma_k \tag{20.53}$$

where the system is defined in a 2d triangular grid of Ising spins and the sum is over nearest neighbors. See (Novotny and Landau, 1981) for help.

a) Show that the ground state is four-fold degenerate.

b) Simulate the system for various MC times and visualize the states. Note that domains of different structure can dominate at different times, illustrating that multiple fluctuations with differing time scales contribute to the partition function.

14. Neural Nets.
Set $\mathcal{J}_{ij} = \text{sign}[\mathcal{J}_{ij}(Hebbs)]$ in the Hopfield model. How well does the system perform at recall? How many patterns can be stored with this algorithm?

15. [T] Hopfield Memory.
How does the Hopfield model behave as memories are added to it? Is there a phase transition to a memoryless state or are memories gradually lost?

16. Random Field Ising Model.
A theorem says that the RFIM has no long range order for $d \leq 2$. Confirm this numerically by measuring the magnetization in the 2d RFIM. Note that you will need to perform a thermodynamic average, as usual, and an average over random field configurations (denoted thus $\overline{}$). Any random field distribution will do, but ensure that $\overline{h_i} = 0$ and $\overline{h_i h_j} = H^2 \delta_{ij}$.

17. [T] Glauber Updates.
Show that the Glauber update $W_{n \to m} = 1 + \sigma_i \tanh(\beta E_i)$ is a valid Monte Carlo thermalization algorithm.

18. [P] Edwards-Anderson Model.

    a) Verify the Sherrington-Kirkpatrick equation as follows.

        i. Use mean field arguments to obtain $m_i = \tanh \beta h_i$ with $h_i = \sum_j \mathcal{J}_{ij} m_j$.
        ii. Argue that $\overline{h_i^2} = \mathcal{J}^2 q^2$.
        iii. Apply (ii) to (i), change variables, and obtain Eq. 20.44. The neglect of fluctuations necessary to obtain this result is a bad approximation. However, more sophisticated methods (the replica method) yield the same expression.

    b) Run a suitable Monte Carlo algorithm, averaging over many random fields, work in units of $\mathcal{J}$, plot $q(t)$ vs. T for $T \in (0, 1.5)$. Here t is the number of Monte Carlo sweeps. Make your plots at several large values of t and compare to the SK result.

19. [P] Avalanches á là Sethna.
Code up the avalanche problem of Eq. 20.33. A way to do this is to search for the unflipped spin that is due to flip next, push H to the required value, and continue propagating the avalanche. An algorithm for this is:

```
//brute force avalanche propagation

while there are spins to flip
```

```
    find the next triggering spin (*)

    increment H to minus f_i and push the
    spin onto a FIFO queue

    while there are spins in the queue
      pop the top spin
      if the spin has not been flipped
        flip it
        push all unflipped neighbors with positive
        local fields onto the queue
      end if
    end while

end while
    (*) the spin with the largest value of the
        local field f_i ≡ J ∑_{j∈i} σ_j + h_i
```

Generate random fields h_i with a Gaussian distribution of mean zero and standard deviation R. Run your code on a 300 × 300 system for $R = 1.4$, 0.9, and 0.7. Try plotting the number of flips per time step. You might want to measure other observables such as $M(H)$, the average avalanche size, the avalanche size distribution, or avalanche correlation functions. If you run into trouble consult Sethna (2006), Chapter 8.

The algorithm above is rather inefficient since it must search the entire lattice for every avalanche. This can be improved by keeping a sorted list of the spins in order of their local field strengths. See Sethna for more details. Lastly, enormous systems can be considered by using bit arithmetic and other tricks. See (Kuntz *et al.*, 1999).

···

BIBLIOGRAPHY

Edwards, S. F. and P. W. Anderson (1975). *Theory of spin glasses*. J. Phys. **F5**, 965.
Essam J. W., D. S. Gaunt, and A. J. Guttmann (1978). *Percolation theory at the critical dimension*. J. Phys. A **11**, 1983.
Fisher, M. E. (1964). *Correlation Functions and the Critical Region of Simple Fluids*. J. Math. Phys. **5**, 944.
Fisher, M. E. (1971). in *Critical Phenomena*. ed. M.S. Green. Academic Press, London.
Hopfield, J. J. (1982). *Neural networks and physical systems with emergent collective computational abilities*. Proceedings of the National Academy of Sciences of the USA **79**, 2554–8.
Imry, Y. and S. K. Ma (1975). *Random-Field Instability of the Ordered State of Continuous Symmetry*. Phys. Rev. Lett. **35**, 1399.
Fortuin, C. M. and P. W. Kasteleyn (1972). *On the random-cluster model: I. Introduction and relation to other models*. Physica **57**, 536.

Kosterlitz, J. M. and D. J. Thouless (1973). *Ordering, metastability and phase transitions in two-dimensional systems.* J. Phys. **C6**, 1181.

Kuntz, M. C., O. Perkovic, K. A. Dahmen, B. W. Roberts, and J. P. Sethna (1999). *Hysteresis, Avalanches, and Noise: Numerical Methods.* Comp. Sci. Eng. **1**, 73–81.

Landau, D. P. and K. Binder (2000). *A Guide to Monte Carlo Simulations in Statistical Physics.* Cambridge University Press.

Matsubara, T. and H. Matsuda (1956). *A Lattice Model of Liquid Helium, I.* Prog. Theor. Phys. Japan **16**, 416.

McCulloch, W. and W. Pitts (1943). *A logical calculus of the ideas immanent in nervous activity.* Bulletin of Mathematical Biophysics 7, 115.

Nattermann, T. (1997). "Theory of the random field Ising model", *Spin Glasses and Random Fields.* ed. P. Young, World Scientific.

Novotny, M. and D. Landau (1981). *Critical behavior of the Baxter-Wu model with quenched impurities.* Phys. Rev. **B24**, 1468.

Onsager, L. (1944). *Crystal statistics. I. A two-dimensional model with an order-disorder transition.* Phys. Rev. **65**, 117–49.

Pelissetto, A. and E. Vicari (2002). *Critical Phenomena and Renormalization-Group Theory.* Phys. Rept. **368**, 549–727.

Potts, R. B. (1952). *Some generalized order-disorder transformations.* Proc. Camb. Phil. Soc. **48**, 106.

Sethna, J. P. (2006). *Statistical Mechanics.* Oxford University Press.

Stein, D. L. (2003). *Spin glasses: still complex after all these years?.* SFI 2003-01-001.

Swendsen, R. and J.-S. Wang (1987). *Nonuniversal critical dynamics in Monte Carlo simulations.* Phys. Rev. Lett. **58**, 86.

Tobochnik, J. and G. V. Chester (1979). *Monte Carlo study of the planar spin model.* Phys. Rev. **B20**, 3761.

Wolff, U. (1989). *Collective Monte Carlo Updating for Spin Systems.* Phys. Rev. Lett. **62**, 361.

Wu, F. Y. (1982). *The Potts Model.* Rev. Mod. Phys. **54**, 235.

21
Quantum mechanics I–few body systems

21.1 Introduction	732
21.2 Simple bound states	733
21.2.1 Shooting methods	734
21.2.2 Diagonalization	735
21.2.3 Discretized eigenproblems	735
21.2.4 Momentum space methods	737
21.2.5 Relativistic kinematics	739
21.3 Quantum Monte Carlo	740
21.3.1 Guided random walks	740
21.3.2 Matrix elements	745
21.4 Scattering and the T-matrix	750
21.4.1 Scattering via the Schrödinger equation	750
21.4.2 The T-matrix	752
21.4.3 Coupled channels	757
21.5 Appendix: Three-dimensional simple harmonic oscillator	761
21.6 Appendix: scattering formulae	763
21.7 Exercises	763
Bibliography	771

21.1 Introduction

Although we live in a quantum universe, the realm in which nonrelativistic quantum mechanics is pertinent is relatively small. Quantum mechanics applies to systems that are small or at low temperatures and hence its usefulness is restricted to atomic, molecular, and nuclear physics. Two types of quantum mechanical problems are of interest, depending on whether the boundary conditions permit Hermitian or nonHermitian representations of the Hamiltonian. These are of course the bound-state problem and scattering.

Quantum mechanical problems are famously difficult and numerical methods are required to solve all but the very simplest problems. A large variety of methods for dealing with quantum mechanical problems of various types have been developed. Because many of these are well documented in other places, we will focus on techniques that are somewhat more unusual.

This chapter deals with few-particle (one or two, for the time being) interactions. The first half will introduce two novel (or at least, less well-known) ways to solve the bound state problem: naive discretization and quantum Monte Carlo. These methods are either simple or powerful, and sometimes both. The second half of the chapter covers few-body scattering. This is traditionally done by integrating the Schrödinger equation with a variety of trial initial conditions. We will, of course, eschew the traditional approach and instead study the far more efficient momentum space T-matrix formalism for scattering.

Quantum mechanical problems with more degrees of freedom are discussed in different chapters scattered throughout the text. Techniques designed to address the difficulties associated with atoms and molecules are presented in Chapter 23. The many-body problem of quantum magnetism is the topic of Chapter 22. That chapter introduces the powerful Lanczos algorithm and extends the quantum Monte Carlo method discussed below to systems with discrete degrees of freedom. Solving the time-dependent Schrödinger equation that is so important to chemists is addressed in Chapter 18. Finally, methods tuned for quantum field theory are presented in Chapter 24.

21.2 Simple bound states

We start with the central problem of undergraduate quantum mechanics: the bound state of two nonrelativistic particles. A time-independent interaction is assumed; we furthermore assume that the system has spatial symmetry and thus can be represented by a Hamiltonian of reduced dimensionality. A familiar example is a particle in a central potential in which the full Schrödinger equation:

$$\left[-\frac{\hbar^2}{2m}\nabla^2 + V(|\vec{r}|) \right] \psi(\vec{r}, t) = i\hbar \frac{\partial}{\partial t} \psi(\vec{r}, t) \tag{21.1}$$

is replaced by the one-dimensional radial equation

$$-\frac{\hbar^2}{2m}\left[\frac{d^2}{dr^2} - \frac{\ell(\ell+1)}{r^2} \right] u(r) + V(r)u(r) = Eu(r). \tag{21.2}$$

The radial wavefunction u is related to the full wavefunction via

$$\psi(\vec{r}, t) = \frac{u(r)}{r} Y_{\ell m}(\hat{r}) \exp(-iEt/\hbar) \tag{21.3}$$

where $Y_{\ell m}$ is a spherical harmonic wavefunction and u obeys the boundary condition

$$\lim_{r \to 0} r^{-l} u(r) = 0. \tag{21.4}$$

In addition, if the system forms a bound state then $\lim_{r \to \infty} u = 0$.

21.2.1 Shooting methods

It is natural to look at Eq. 21.2 as a problem in differential equations but with the additional complication that it must be solved as an eigenvalue problem. In other words eigenvalues are specified by requiring solutions that respect the boundary conditions at the origin and infinity. This is quite different from the usual differential equation where two initial conditions, say $u(0)$ and $u'(0)$, are given. Nevertheless, it is possible to make progress by setting $u(0)$ and guessing a value of $u'(0)$; one then integrates out to some large value of $r = R$ and checks if:

$$|u(R)| < \epsilon \tag{21.5}$$

where ϵ is some tolerance deemed acceptable and R is sufficiently large to render it a useful approximation to infinity (in practice this means $R \gg \max(\langle r \rangle, r_0)$ where r_0 is a definition of the range of the interaction and $\langle r \rangle$ is the mean size of the wavefunction of interest).

This strategy is called the **shooting method**. It can be implemented with an ordinary differential equation solver such as those discussed in Chapter 11. Alternately, it may be performed with the aid of **Numerov's method**, which is a technique for solving second order differential equations in which the first derivative does not appear. The method can be applied on a uniform grid and yields results that are good to sixth order in the grid size h. Specifically, if:

$$u'' + f(x)u(x) = 0 \tag{21.6}$$

then:

$$\left(1 + \frac{h^2}{12} f_{n+1}\right) u_{n+1} = \left(2 - \frac{5h^2}{6} f_n\right) u_n - \left(1 + \frac{h^2}{12} f_{n-1}\right) u_{n-1} + O(h^6). \tag{21.7}$$

To start the integration, take $u_0 = 0$ as required by boundary conditions. The next value u_1 is arbitrary, as it only determines the normalization of the solution. Subsequent grid points are computed by the recurrence relation of Eq. 21.7. In a bound-state problem, the energy is adjusted until the radial wavefunction $u(r)$ is seen to satisfy condition 21.5.

Although this method is efficient, the necessity to solve the differential equation many times, the interpolation required to obtain an accurate estimate of the eigenvalue, and the dependence on R and the tolerance make this method slightly awkward for many problems (if one wishes to press ahead nonetheless, consult Stoer and Bulirsch (1980), Section 7.3 or Press *et al.* (2007), Section 18.1 for more information).

21.2.2 Diagonalization

While the shooting method is admirable (and all code is "simple" once it has been developed and debugged), a cleaner method is to simply use one of the many high quality diagonalization routines such as those available in the package `Eigen`. In other words we eschew Schrödinger's wave mechanics for Heisenberg's matrix mechanics. Thus Eq 21.2 is replaced with

$$\langle n|H|m\rangle = E\langle n|m\rangle \tag{21.8}$$

where $\{|n\rangle\}$ is an appropriately chosen set of N basis states. Diagonalizing immediately yields N eigenvectors and eigenvalues E_k, $k = (1 \ldots N)$ for the system. If the basis is a complete set, the solution is exact, otherwise it is an approximation whose extrapolation to the exact solution is controlled by the single parameter N.

A theorem (MacDonald, 1933) asserts that the eigenvalues of the approximate solution are always greater than or equal to the true eigenvalues. This can be exploited by combining diagonalization with minimization over parametric dependence in the basis. With a parameterized set of basis functions, the elements $\langle n|H|m\rangle$ of the matrix H and therefore its eigenvalues depend upon the parameters. Minimizing an eigenvalue over these parameters always achieves a better approximation. This technique is a form of the Rayleigh-Ritz variational method, whose simplest and most familiar application is the determination of ground-state energies. The minimization can be carried out using the techniques of Chapter 16.

Some care must be taken when implementing the diagonalization-variational method. For example if the basis is chosen to depend on a single parameter, β, then minimizing $E_0(\beta)$ determines an optimal value $\beta = \beta_0$. This same value must be used when examining other eigenvalues if one wishes to maintain orthogonality of the eigenvectors. Alternatively, more accurate estimates of higher eigenvalues can be obtained by minimizing those eigenvalues independently. Finally, this approach often necessitates the use of nonorthogonal bases and one must recall that in this case the completeness relation is expressed as:

$$\mathbb{1} = \sum_{nm} |n\rangle (\mathcal{N}^{-1})_{nm} \langle m| \tag{21.9}$$

where $\mathcal{N}_{nm} \equiv \langle n|m\rangle$.

21.2.3 Discretized eigenproblems

While the direct diagonalization method is simple and efficient it suffers from at least two serious issues:

- the matrix element must be evaluated and this can be extremely difficult;
- the form of the basis function can dominate the approximate eigenfunction at some scales.

To expand the second point, if one were interested in the large distance behavior of an eigenfunction that was expanded in a Gaussian or simple harmonic oscillator (SHO) basis one would inevitably obtain a—not necessarily correct—behavior that approximates $\exp(-r^2)$, even when the eigenvalue is accurately computed.

Fortunately there is a simple solution to both problems. If one chooses the basis

$$\langle r|n\rangle = \delta(r - r_n) \tag{21.10}$$

then the short and large distance behavior of eigenfunctions can be accurately approximated and matrix elements are trivial to evaluate. Of course, this basis is equivalent to simply discretizing the Hamiltonian on an appropriate grid if a discretization of the Laplacian is used. The method is extremely simple to code and, somewhat surprisingly, it is also very effective. Its flaw is that it is only useful for a limited set of problems. Specifically, the boundary conditions imposed by the method are $u(0) = 0$ and $u(L) = 0$. See Exercise 7 if you are curious about this. This implies that the method is only applicable for more than one spatial dimension (in fact, it only works for three or more dimensions—see Exercise 8).

If one discretizes on a uniform grid, the radial eigenproblem reduces to diagonalizing H where:

$$H_{ij} = -\frac{\hbar^2}{2m\Delta^2}\left(-2\delta_{ij} + \delta_{i-1,j} + \delta_{i+1,j}\right) + \frac{\hbar^2 l(l+1)}{2\mu r_i^2}\delta_{ij} + V_i \delta_{ij}. \tag{21.11}$$

Here $r_i = i\Delta$ and Δ is the grid spacing. V_i denotes $V(r_i)$. End points are handled by truncating the matrix, which is equivalent to setting $u_{-1} = 0$ and $u_{N+1} = 0$ (indexing runs from zero to N; see Ex. 7 for more on this point). One can also employ higher-point approximations to the second derivative, although we have found that the gain is not worth the effort. See Chapter 19 for much more on discretization of derivative operators.

One might be tempted to avoid dealing with an infinite region by mapping the half plane to a finite region. But then additional analytic work must be done and a nonlinear grid should be employed to probe the appropriate distance scales. We therefore recommend that one simply uses a uniform grid in a box of size L. The box size should be chosen to be several times larger than the radius of the largest wavefunction in which one is interested.

Figure 21.1 displays the efficacy of the method when applied to the three dimensional simple harmonic oscillator problem. Notice that using a small box size improves the accuracy of the ground energy for a given matrix size (or grid spacing). This is because

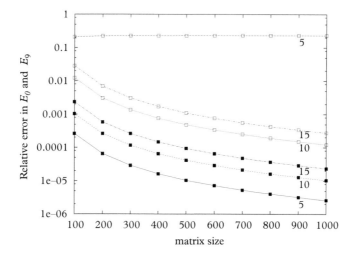

Figure 21.1 *Relative errors in E_0 (squares) and E_9 (open squares) for the 3d SHO vs. matrix size. The numbers to the right refer to the box size in natural units.*

the ground state "fits" in a box of size 5. This remains true for the tenth state except in this case a box size of at least 10 is required.

21.2.4 Momentum space methods

It is often useful, or even necessary, to solve the Schrödinger equation in momentum space. For example, a particular interaction may be simple to write in momentum space or relativistic kinematics may be required. Fourier transforming the time-independent Schrödinger equation yields:

$$\frac{p^2}{2m}\psi(\vec{p}) + \int \frac{d^3q}{(2\pi)^3} V(\vec{p}-\vec{q})\psi(\vec{q}) = E\psi(\vec{p}) \qquad (21.12)$$

where:

$$V(\vec{p}) = \int d^3r\, V(\vec{r})e^{-i\vec{p}\cdot\vec{r}}. \qquad (21.13)$$

Project onto partial waves by setting $\psi(\vec{p}) = R_{nl}(p)Y_{lm}(\hat{p})$ to obtain:

$$\frac{p^2}{2m}R_{nl}(p) + \int q^2 dq\, V^{(l)}(p,q)R_{nl}(q) = ER_{nl}(p). \qquad (21.14)$$

This expression introduces the partial wave projected potential given by (note that we again assume a central potential):

$$V^{(l)}(p,q) = \frac{1}{4\pi^2} \int_{-1}^{1} dx\, V(\sqrt{p^2 + q^2 - 2pqx})\, P_l(x). \quad (21.15)$$

Our method will be to diagonalize the Hamiltonian on a momentum grid. An elegant way to do this is to map the half-plane onto [0 : 1] via the transformation $p = \eta/(1-\eta)$ and then generate integration points and weights with the Gauss-Legendre method (see Chapter 5). Eq. 21.14 becomes:

$$\frac{p_i^2}{2m} R_i + \sum_j w_j V_{ij}^{(l)} R_j = E R_i \quad (21.16)$$

The weight factors are given by:

$$w_j = \frac{p_j^2}{(1-\eta_j)^2} \cdot GL(j) \quad (21.17)$$

where $GL(j)$ refers to the jth Gauss-Legendre weight, η_j is the jth Gauss-Legendre abscissa, and $p_j = \eta_j/(1-\eta_j)$. The indices nl have been suppressed in Eq. 21.16; also R_i refers to $R(p_i)$ and $V_{ij}^{(\ell)}$ is short for $V^{(\ell)}(p_i, p_j)$.

The asymmetric form of the Hamiltonian is inconvenient so we symmetrize by multiplying through by w_i yielding the final form of the Schrödinger equation:

$$H_{ij} R_j = E \mathcal{N}_{ij} R_j \quad (21.18)$$

where:

$$H_{ij} = \frac{1}{2m} p_i^2 w_i \delta_{ij} + w_i V_{ij}^{(l)} w_j \quad (21.19)$$

and the normalization kernel is given by (no summation):

$$\mathcal{N}_{ij} = w_i \delta_{ij}. \quad (21.20)$$

In the case of Coulombic interactions one has $V(\vec{r}) = -e^2/(4\pi\epsilon_0 r)$ with Fourier transform $V(\vec{p}) = -e^2/(\epsilon_0 p^2)$. Diagonalizing this interaction yields the following results (Table 21.1) for S-wave energies on grids with 100 or 400 points (angular integrals were also evaluated numerically with 100 or 400 points). Evidently one hundred momentum space grid points are sufficient to obtain quite accurate energies up to the third or fourth level. After this the method rapidly becomes useless and larger grids must be employed. If very large quantum numbers must be explored it is best to design a momentum grid

Table 21.1 *Percent deviation for S-wave Coulomb energies.*

n	N=100	N=400
1	0.25	0.065
2	0.37	0.09
3	1.7	0.42
4	3.8	0.94
5	6.6	1.6
6	10.1	2.5
7	14.5	3.5
8	19.6	4.7
9	25.5	6.1
10	32.2	7.7

that adequately probes the required momentum scale. An example calculation can be found in EXAMPLES/CH21/PSPACE.

21.2.5 Relativistic kinematics

It is sometimes desirable to implement a 'semi-relativistic' approach to a quantum mechanical problem in which the kinetic energy operator is replaced by its relativistic version:

$$\hat{T} \to \sqrt{m^2 + k^2}, \qquad (21.21)$$

where k is the relative momentum in the system. A typical application is in nuclear physics where relativistic pions can play an important role.

While it is easy to implement a relativistic kinetic energy in momentum space, it is often desirable to avoid the Fourier transform of the potential and to work in configuration space. In this case we recommend diagonalizing the Laplacian operator $-\nabla^2$ in the preferred basis. If we call the resulting eigenvalues $\{\lambda_n\}$ and the eigenvectors $\{|n\rangle\}$ then

$$\langle n|H|m\rangle = \sqrt{m^2 + \lambda_n^2}\,\delta_{nm} + \sum_i \langle n|i\rangle\, V_i\, \langle i|m\rangle. \qquad (21.22)$$

Notice that the sum over the potential is tantamount to performing a numerical Fourier transform in a generalized basis.

21.3 Quantum Monte Carlo

The relationship of the Schrödinger equation to diffusion has been mentioned in Section 7.6.2. As a brief reminder, if one replaces time with 'Euclidean time' via $t \to -i\tau$ then one obtains an advection-diffusion equation with diffusion coefficient $D = \hbar/(2\mu)$:

$$\frac{\partial}{\partial \tau}\psi = \frac{\hbar}{2\mu}\nabla^2\psi - \frac{1}{\hbar}V\psi. \tag{21.23}$$

This observation permits obtaining quantum mechanical energies and matrix elements by directly simulating a process in which particles perform random walks with an additional "drift" term (the potential). Many variants of quantum mechanical random walk algorithms exist; here we present a particularly simple algorithm called **guided random walks**.

21.3.1 Guided random walks

Let us reverse the direction of the previous discussion and start with a simple diffusion problem wherein random steps of size Δx are taken every $\Delta \tau$. In the continuum limit this process is described by the differential equation:

$$\frac{\partial \rho_0}{\partial \tau} = \frac{1}{2}\frac{\Delta x^2}{\Delta \tau}\frac{\partial^2 \rho_0}{\partial x^2}. \tag{21.24}$$

Here ρ_0 is the particle density at position x and time τ. Of course, this equation is the free Schrödinger equation if $t = -i\tau$ and:

$$\frac{\hbar}{m} = \frac{\Delta x^2}{\Delta \tau}. \tag{21.25}$$

Now consider a process where a diffusing particle can be absorbed with probability $a(x)$. In this case the diffusion equation reads

$$\rho(x, \tau + \Delta\tau) = (1 - a(x))\left[\frac{1}{2}\rho(x - \Delta x, \tau) + \frac{1}{2}\rho(x - \Delta x, \tau)\right], \tag{21.26}$$

with a continuum limit of

$$\frac{\partial \rho}{\partial \tau} = \frac{1}{2}\frac{\Delta x^2}{\Delta \tau}\frac{\partial^2 \rho}{\partial x^2} - a(x)\rho(x, \tau). \tag{21.27}$$

Observe that this equation maps to the interacting Schrödinger equation with the aid of Eq. 21.25 and:

$$V(x) = \hbar a(x). \tag{21.28}$$

This simple result means that it is possible to simulate the Schrödinger equation by executing random walks while allowing for the possibility that walks can be eliminated at site x with probability $a(x)$.

However, while feasible, the proposed algorithm is cumbersome to code and wasteful because random walks can end early. A more elegant way to implement a potential in diffusion is to associate a weight with each walk and then to construct the average of ρ by binning the weights as a function of x and τ. The correct weight is given by (Barnes, 1986):

$$w(x(\tau)) = e^{-\int_0^\tau a(x(\tau')) \, d\tau'}, \tag{21.29}$$

and thus

$$\rho(x,\tau) = w(x,\tau) \rho_0(x,\tau). \tag{21.30}$$

Recall that ρ_0 is the solution to the free diffusion equation.

To summarize, one may simulate the Euclidean time Schrödinger equation by performing a large number of random walks; the histogram of average weights associated with these walks is then proportional to the time-dependent wavefunction of the theory.

While the algorithm is perfectly capable of simulating systems with a few degrees of freedom, it must incorporate importance sampling to permit its application to systems with many (say, greater than 5) degrees of freedom. This is a common attribute of all Monte Carlo algorithms, as we have discussed in Chapter 7.

The problem with the random walk algorithm presented so far is that the walks uniformly explore a vast Hilbert space and thus the weights, w, have a huge variance. Most walks end up far in the classically disallowed region of configuration space and hence do not contribute to the weight histogram.

The solution is to bias the walks towards the classically allowed region. This may be achieved by introducing a **guidance function**, $g(x)$, and by preferentially walking towards minima of g. In this approach one therefore sets the random walk stepping probabilities to be:

$$\begin{aligned} &\frac{1}{2} - \Delta x \frac{dg}{dx}, \quad \text{stepping right} \\ &\frac{1}{2} + \Delta x \frac{dg}{dx}, \quad \text{stepping left}. \end{aligned} \tag{21.31}$$

The difference equation for $\rho(x, \tau)$ is now:

$$\rho(x,\tau+\Delta\tau) = \left(\frac{1}{2} + \Delta x \frac{dg(x+\Delta x)}{dx}\right)\rho(x+\Delta x,\tau) + \left(\frac{1}{2} - \Delta x \frac{dg(x-\Delta x)}{dx}\right)\rho(x-\Delta x,\tau). \tag{21.32}$$

In differential form this is:

$$\frac{\partial \rho}{\partial \tau} = \frac{\Delta x^2}{\Delta \tau}\left(\frac{1}{2}\frac{\partial^2 \rho}{\partial x^2} + 2\frac{\partial g}{\partial x}\frac{\partial \rho}{\partial x} + 2\frac{\partial^2 g}{\partial x^2}\rho\right). \tag{21.33}$$

We now add the process of evaluating the weight

$$w = e^{-\int_0^\tau a(x(\tau'))\,d\tau'}$$

down a guided random walk and compute

$$\rho_g = \langle w \rangle \, \rho.$$

The net result is the solution to the continuum equation:

$$\frac{\partial \rho_g}{\partial \tau} = \frac{\Delta x^2}{\Delta \tau}\left(\frac{1}{2}\frac{\partial^2 \rho_g}{\partial x^2} + 2\frac{\partial g}{\partial x}\frac{\partial \rho_g}{\partial x} + 2\frac{\partial^2 g}{\partial x^2}\rho_g\right) - a(x)\,\rho_g(x). \tag{21.34}$$

To establish contact with the Schrödinger equation let:

$$\rho_g = e^{-2g}\psi,$$
$$\frac{\Delta x^2}{\Delta \tau} = \frac{\hbar}{m},$$
$$a(x) = \frac{1}{\hbar}V(x) + \frac{\Delta x^2}{\Delta \tau}\left(\frac{\partial^2 g}{\partial x^2} - 2\left(\frac{\partial g}{\partial x}\right)^2\right). \tag{21.35}$$

The resulting equation is the Euclidean time Schrödinger equation as desired:

$$-\hbar \frac{\partial \psi}{\partial \tau} = -\frac{\hbar^2}{2m}\frac{\partial^2 \psi}{\partial x^2} + V(x)\psi(x). \tag{21.36}$$

We have reached a satisfying point: the solution to the Euclidean time Schrödinger equation can be obtained by performing guided random walks and evaluating:

$$\psi(x,\tau) = \eta\, e^{2g(x)} \left\langle \exp\left(-\int_0^\tau [V(x(\tau'))/\hbar + \frac{\hbar}{m}(g'' - 2(g')^2)]\,d\tau'\right)\right\rangle \rho(x,\tau). \tag{21.37}$$

(A normalization factor, η, has been restored to the equation.)

The ground state energy

It is now a simple matter to evaluate the ground state energy of the system. The eigenmode expansion of ψ is:

$$\psi(x, \tau) = \sum_n \alpha_n \phi_n(x) e^{-E_n \tau} \tag{21.38}$$

where ϕ_n is an eigenstate of the system and α_n is the overlap of ϕ_n with ψ. Thus the ground state energy may be projected out of the ratio of the wavefunctions as the Euclidean time becomes large:

$$E_0 = \lim_{\tau_1, \tau_2 \to \infty} \frac{1}{\tau_2 - \tau_1} \ln \frac{\psi(x, \tau_1)}{\psi(x, \tau_2)}. \tag{21.39}$$

A subtle problem arises at this stage: the value of x must be the same in the numerator and denominator for this equation to hold. But when performing random walks the functions obtained will be $\psi(x(\tau), \tau)$ and hence Eq. 21.39 will not be useful. The resolution to this problem is to integrate over x to obtain (N_{RW} is the number of random walks):

$$\begin{aligned} E_0 &= \lim_{\tau_1, \tau_2 \to \infty} \frac{1}{\tau_2 - \tau_1} \ln \frac{\int \psi(x, \tau_1) \, dx}{\int \psi(x, \tau_2) \, dx}, \\ &= \lim_{\substack{\tau_1, \tau_2 \to \infty \\ N_{RW} \to \infty \\ \Delta x, \Delta \tau \to 0}} \frac{1}{\tau_2 - \tau_1} \ln \frac{\sum_{\text{walks}} e^{2g(x)} e^{-\int_0^{\tau_1} a(x(\tau')) \, d\tau'}}{\sum_{\text{walks}} e^{2g(x)} e^{-\int_0^{\tau_2} a(x(\tau')) \, d\tau'}}. \end{aligned} \tag{21.40}$$

It is important that the sum over random walks approximates the spatial integral well since this is used to cancel the x-dependence in the estimate of the energy.

Notice that the spatial dependence in $\exp(2g(x))$ is unimportant in extracting E_0 because the integral over $\exp(-2g(x))\psi(x, \tau)$ and $\psi(x, \tau)$ produces the same temporal dependence. Thus our final expression for the ground state energy is:

$$E_0 = \lim_{\substack{\tau_1, \tau_2 \to \infty \\ N_{RW} \to \infty \\ \Delta x, \Delta \tau \to 0}} \frac{1}{\tau_2 - \tau_1} \ln \frac{\sum_{\text{walks}} e^{-\int_0^{\tau_1} a(x(\tau')) \, d\tau'}}{\sum_{\text{walks}} e^{-\int_0^{\tau_2} a(x(\tau')) \, d\tau'}}. \tag{21.41}$$

This simplification is useful as it considerably reduces the statistical fluctuations in estimates of E_0.

Extrapolation and guidance parameters

The limits in the expression for E_0 must be taken by extrapolating results for finite step size, Euclidean time, and statistics. The behavior of this extrapolation can be anticipated; for example, the central limit theorem leads one to expect that corrections due to a finite number of random walks behave as $1/\sqrt{N_{RW}}$. Also, the corrections to the diffusion equation (Eq. 21.24) imply that step size corrections are linear in Δx; this is mostly true although Δx^2 corrections are also seen. Finally, the eigenmode expansion of ψ implies that:

$$E_0(\tau_1, \tau_2) \sim E_0 + \frac{1}{\tau_2 - \tau_1} \frac{\alpha_1 \int \phi_1\, dx}{\alpha_0 \int \phi_0\, dx}(e^{-(E_1-E_0)\tau_1} - e^{-(E_1-E_0)\tau_2}), \quad (21.42)$$

so that the expression 21.41 approaches the ground state energy exponentially in time with a scale set by the energy gap, $E_1 - E_0$. Thus accurate measurements require one to run for a time, $\tau \gg 1/(E_1 - E_0)$. We note, however, that in practice one cannot run for too long in τ because fluctuations tend to build up. This point is evident in Figure 21.2, which shows the ground state energy of the one-dimensional simple harmonic oscillator as a function of τ_1 (τ_2 has been set to $\tau_1 + 1$). The approach to the exact result is clear, as is the need to run for sufficiently long Euclidean time to project out the excited state contamination in the wavefunction.

The remaining chore for the numericist is choosing the guidance function. Ideally one would like to reduce the scatter in the weights to zero. This may be achieved if $\int a(x) d\tau = const$, which implies that the optimal guidance is a solution to

$$\frac{V(x)}{\hbar} + \frac{\hbar}{\mu}\left(g'' - 2(g')^2\right) = const.$$

Not surprisingly, this equation is equivalent to the Schrödinger equation, with solution

$$g(x) = -\frac{1}{2}\log \phi_0.$$

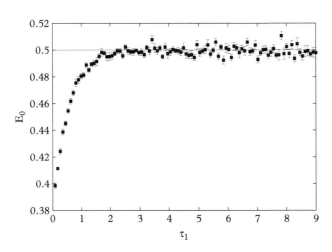

Figure 21.2 *Ground state energy for the 1d SHO as a function of τ_1.*

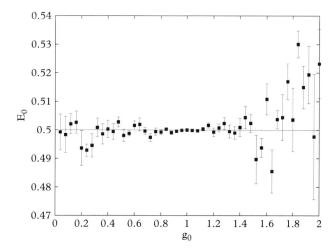

Figure 21.3 *Ground state energy for the 1d SHO as a function of the guidance parameter.*

Although we have not got very far, this equation indicates that guidance functions can be constructed from perspicacious variational Ansätze. Variance in energy levels (or matrix elements) can then be tuned to a minimum by adjusting the Ansatz parameters. This is illustrated in Figure 21.3, which shows the ground state energy for our test problem as evaluated with the guidance function $g = g_0 x^2/4$. Notice that this function is exact when $g_0 = 1$. This is reflected in the figure by the tiny error bars near optimal guidance. In fact, running a guided random walk code with exact guidance will produce the exact energy with no error for any number of time steps, which is useful for code verification.

21.3.2 Matrix elements

The guided random walk algorithm would not be very satisfying if one could not evaluate matrix elements of observables. This cannot be done by simply averaging the appropriate operator over random walks because the probability of arriving at x is proportional to $\phi_0(x)$ and not $|\phi_0(x)|^2$. Fortunately it is possible to circumvent this problem; two methods for evaluating ground state matrix elements are presented in this section. Both methods are only applicable to matrix elements of the form $\int |\phi_0(x)|^2 f(x)$.

Matrix elements with the double walk method

A way to generate a weight proportional to $|\phi_0|^2$ is to exploit the convolution property of quantum mechanical evolution kernels. These are typically used to evolve wavefunctions in space and time (Feynman and Hibbs, 2005):

$$\psi(x_f, \tau_f) = \int dx_i K(x_i, \tau_i; x_f, \tau_f) \psi(x_i, \tau_i). \tag{21.43}$$

Evolution from $0 \to x_f$ can be split into evolution from $0 \to x$ and $x \to x_f$ as follows:

$$K(0,0;x_f,\tau_f) = \int dx\, K(0,0;x,\tau_1)\, K(x,\tau_1;x_f,\tau_f). \tag{21.44}$$

The latter relationship is useful because the (Euclidean time) eigenmode expansion of the kernel,

$$K(x_i,\tau_i;x_f,\tau_f) = \sum_n \phi_n^*(x_i)\phi_n(x_f)e^{-E_n(\tau_2-\tau_1)}$$

$$\to \phi_0^*(x_i)\phi_0(x_f)e^{-E_0(\tau_2-\tau_1)} \qquad \text{large } \tau_2-\tau_1 \tag{21.45}$$

implies that $|\phi_0(x)|^2$ will appear under the integral of Eq. 21.44.

The next task is to relate the quantum mechanical kernel to guided random path weights:

$$\langle w(x_i,\tau_i;x_f,\tau_f) \rangle = e^{2g(x_i)} K(x_i,\tau_i;x_f,\tau_f) e^{-2g(x_f)}. \tag{21.46}$$

We will not go into details deriving this, but note that it can be done by comparing the integral equation version of the diffusion problem to quantum mechanical wavefunction evolution given by Eq. 21.43.

Imagine walking for a time, τ_1, evaluating the operator of interest $\mathcal{O}(x(\tau_1))$, and then continuing the walk to τ_2. The amplitude for this process is:

$$\mathcal{A} \equiv \langle w(0,0;x_f,\tau_2)\mathcal{O}(x(\tau_1)) \rangle. \tag{21.47}$$

Now use Eqs. 21.44 and 21.46 to obtain:

$$\mathcal{A} = \left\langle e^{2g(0)} \int dx\, K(0,0;x,\tau_1)\, e^{-2g(x)}\, \mathcal{O}(x(\tau_1))\, e^{2g(x)}\, K(x,\tau_1;x_f,\tau_2)\, e^{-2g(x_f)} \right\rangle. \tag{21.48}$$

Finally, use Eq. 21.45 to extract the behavior of $\mathcal{A}$ at large times. One obtains:

$$\langle \mathcal{O} \rangle = \lim_{\tau_1,\tau_2 \to \infty} \frac{\langle w(0,0;x_f,\tau_2)\mathcal{O}(x(\tau_1)) \rangle}{\langle w(0,0;x_f,\tau_2) \rangle}. \tag{21.49}$$

Notice that, as with the energy measurement, the x_f dependence in the end point arrival weights is redundant and can be dropped.

While this method is certainly feasible, it has much larger errors than energy measurements do. In fact, it is possible and desirable to extract matrix elements using energy measurements. This is described in the next section.

Matrix elements with the Feynman-Hellmann method

The starting point for this approach is the Feynman-Hellmann theorem, which relates matrix elements of operators to derivatives of energies:

$$\langle \phi_0 | \mathcal{O} | \phi_0 \rangle = \frac{\partial}{\partial \lambda} E_0(\lambda)|_{\lambda=0}, \qquad (21.50)$$

where

$$E_0(\lambda) | \phi_0(\lambda) \rangle = (H + \lambda \mathcal{O}) | \phi_0(\lambda) \rangle. \qquad (21.51)$$

If $\mathcal{O}$ is diagonal (i.e., in our application, if $\mathcal{O}$ is only a function of position) this expression provides a simple way to evaluate its expectation value via the expression:

$$E_0 = \lim_{\tau_1, \tau_2 \to \infty} \frac{1}{\tau_2 - \tau_1} \ln \frac{\langle w_\mathcal{O}(\tau_1) \rangle}{\langle w_\mathcal{O}(\tau_2) \rangle} \qquad (21.52)$$

where

$$w_\mathcal{O}(\tau) = \exp\left(-\int_0^\tau \left[V/\hbar + \lambda \mathcal{O}/\hbar + \frac{\Delta x^2}{\Delta \tau}(g'' - 2(g')^2)\right] d\tau \right). \qquad (21.53)$$

Substituting and taking the derivative in Eq. 21.50 gives the final expression:

$$\langle \phi_0 | \mathcal{O} | \phi_0 \rangle = \lim_{\tau_1, \tau_2 \to \infty} \frac{1}{\tau_2 - \tau_1} \left[\frac{\langle w \int_0^\tau \mathcal{O} d\tau' \rangle}{\langle w \rangle} \bigg|_{\tau = \tau_2} - \frac{\langle w \int_0^\tau \mathcal{O} d\tau' \rangle}{\langle w \rangle} \bigg|_{\tau = \tau_1} \right]. \qquad (21.54)$$

The algorithm

Enough formalism—let's see the algorithm! The following code obtains ground state energies and matrix values for the one-dimensional simple harmonic oscillator with a guidance function given by $g(x) = g_0 x^2/4$. Recall that a large number of random walks are required to reliably extract the ground energy, thus statistical errors are computed by running the main algorithm `Nstats` times.

```
...
using namespace std;

double V(double x) {
  // the 1-d SHO, m=1, m omega^2=1
  return x*x/2.0;
}

double gp(double g, double x) {
  // g'(x) for g(x) = g*x*x/4
```

```cpp
    return g*x/2.0;
}

double gpp(double g, double x) {
  // g''(x)
  return g/2.0;
}

int main() {
  double tau1,tau2,g,dtau;
  int seed,NRW, Nstats;
  cout << " enter NRW, dtau, tau1, tau2, g, Nstats " << endl;
  cin >> NRW >> dtau >> tau1 >> tau2 >> g >> Nstats;
  cout << " enter a random seed " << endl;
  cin >> seed;
  mt19937 e(seed);                          // initiate a
      Mersenne twistor engine
  uniform_real_distribution<double> uDist(0,1);
  double dx = sqrt(dtau/1.0);
      // ie dx^2/dt = 1/m
  cout << " dx = " << dx << endl;
  int T1 = tau1/dtau;
  int T2 = tau2/dtau;
  double E0avg=0.0;
  double E0sq=0.0;
  double Oavg = 0.0;
  double Osq = 0.0;
  for (int ns=0; ns < Nstats; ns++) {
   double w1=0.0,w2=0.0,Osum=0.0;
   for (int  nr=0; nr < NRW; nr++) {
     double x = 0.0;
     double a = 0.0;
     for (int t=0; t<T1; t++) {
      if ( uDist(e) < 0.5 - dx*gp(g,x) )  {
        x += dx;
      } else {
        x -= dx;
      }
      a += V(x) + gpp(g,x) - 2.0*gp(g,x)*gp(g,x);
     }  // end of tau1 loop
     double O = x*x*x*x;                     // evaluate
                                                O(tau1)
     w1 += exp(-dtau*a);
     for (int t=T1; t<T2; t++) {
      if ( uDist(e) < 0.5 - dx*gp(g,x) )   {
```

```
          x += dx;
       } else {
          x -= dx;
       }
       a += V(x) + gpp(g,x) - 2.0*gp(g,x)*gp(g,x);
    } // end of tau2 loop
    w2 += exp(-dtau*a);
    Osum += exp(-dtau*a)*O;                      // <exp(-a)
                                                  O(tau1)>
  } // end of NRW loop
  double E0 = log(w1/w2)/(tau2-tau1);
  E0avg += E0;
  E0sq += E0*E0;
  Oavg += Osum;
  Osq += Osum*Osum;
} // end of Nstats loop
E0avg /= Nstats;
E0sq /= Nstats;
E0sq = sqrt(abs(E0sq - E0avg*E0avg))/sqrt(Nstats-1);
cout << "E0 = " << E0avg << " +/- " << E0sq << endl;
Osum /= Nstats;
Osq /= Nstats;
Osq = sqrt(abs(Osq - Oavg*Oavg))/sqrt(Nstats-1);
cout << "<x^4> = " << Oavg << " +/- " << Osq << endl;
return 0;
}
```

Of course, it is easy to parallelize this code using the methods of Chapter 9. It is also an easy exercise to generalize the code to any number of dimensions. See Ex. 13 for the generalization required when dealing with problems with multiple particle masses.

SHO matrix elements

Figure 21.4 shows the matrix elements $\langle\phi_0|x^n|\phi_0\rangle$ for $n = 2, 4, 6,$ and 8 vs. τ_1. These results were computed with $d\tau = 0.001$, $g_0 = 0.5$, and 10 groups of 10000 random walks. The Feynman-Hellmann code was approximately 7 times slower than the double walk code, although errors are somewhat smaller. Notice that the convergence in τ_1 to the exact values (shown as horizontal lines) is slower for higher powers of x. It is also evident that the errors (and corresponding fluctuations) become larger with the power. Finally, these data were generated with $\tau_2 = \tau_1 + 3$ because it was found that using smaller values of τ_2 led to consistent over-estimates of the matrix elements for higher n. As usual, the numericist is strongly cautioned to check all parametric dependence of results before assuming that her work is done!

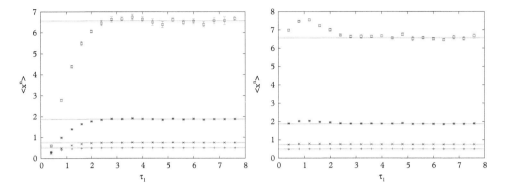

Figure 21.4 *Ground state matrix elements of x^n. Left: double walk method; right: Feynman-Hellmann method. From bottom to top: $n = 2, 4, 6, 8$.*

21.4 Scattering and the T-matrix

The computation of scattering amplitudes forms the other major numerical task in our program. As with the bound state problem, it is tempting to address this issue by numerically integrating the Schrödinger equation. This approach will be addressed in the following subsection. We will then advocate a more efficient method based on the Lippmann-Schwinger equation for the transition (or T) matrix. Extensions to more complicated scattering processes are also discussed.

21.4.1 Scattering via the Schrödinger equation

Let us recall some of the terminology and main results of nonrelativistic scattering theory. Scattering wavefunctions obey boundary conditions with an outgoing spherical wave:

$$\psi_k^{(+)}(r) \to A \left[e^{i\vec{k}\cdot\vec{r}} + f_E(\theta) \frac{e^{ikr}}{r} \right]. \tag{21.55}$$

The center of mass scattering angle is denoted as θ in this formula and A is a normalization constant.

A differential cross section is defined in terms of the scattering amplitude by:

$$\frac{d\sigma}{d\Omega} = |f_E(\theta)|^2. \tag{21.56}$$

If one makes the partial wave expansion

$$\psi(r) = \sum_{\ell m} \frac{u_\ell(r)}{r} Y_{\ell m(\theta,\varphi)} \tag{21.57}$$

then:

$$u_\ell(k,r) \to A_\ell(k) \sin\left(kr - \frac{1}{2}\ell\pi + \delta_\ell(k)\right) \quad (21.58)$$
$$= kr A_\ell(k) \cos\delta_\ell \left[j_\ell(kr) - \tan\delta_\ell n_\ell(kr)\right]$$
$$= A_\ell \, e^{i\delta_\ell + i\ell\pi/2} \left(e^{-ikr} + (-)^{\ell+1} S_\ell \, e^{ikr}\right)$$

at large distance. These expressions define the **phase shift**, denoted as δ_ℓ. The second form is written in terms of the spherical Bessel function and the Neumann function, which are the regular and irregular solutions to the free scattering problem in the ℓth partial wave. The third form introduces the **scattering matrix**:

$$S_\ell(k) = e^{2i\delta_\ell}. \quad (21.59)$$

Admittedly, this is not much of a matrix at present. The full structure of the S-matrix will emerge when we consider coupled channel (where an initial state can scatter into distinct final states) scattering. Finally, at short distance one has the boundary condition:

$$u_\ell(r) \to r^{\ell+1}. \quad (21.60)$$

The partial wave expansion relates the phase shift and the scattering amplitude:

$$f_E(\theta) = \sum_{\ell=0}^{\infty} (2\ell + 1) \, a_\ell(k) P_\ell(\cos\theta) \quad (21.61)$$

where

$$a_\ell(k) = \frac{1}{2ik}(S_\ell(k) - 1). \quad (21.62)$$

A simple numerical method for obtaining phase shifts is to integrate beyond the range of the interaction and use Eq 21.59 to extract δ_ℓ. In general the slope at the origin must be adjusted so that the normalization of the asymptotic solution follows that of Eq 21.59. However, it is simpler to set $u'_\ell(0) = \delta_{\ell,0}$ and compare the solution at two points to extract the phase shift. Since most numerical implementations will keep track of u'_ℓ, it is even simpler to integrate to a large distance $r = R$ and use:

$$\delta_\ell(k) = \frac{1}{2}\ell\pi - kR - \tan^{-1}\left(ku_\ell(R)/u'_\ell(R)\right). \quad (21.63)$$

It would of course be prudent to test the stability of the phase shift for several values of R. This extrapolation can be combined with extrapolation in the step size to yield final estimates for the phase shift.

21.4.2 The T-matrix

Although the methods of the previous section are straightforward, they grow in complexity when coupled channels are considered (this topic will be discussed below). It is therefore convenient that a method that is analogous to the matrix formulation of the bound state problem exists for scattering. This can be established by recasting the Schrödinger equation as the *Lippmann-Schwinger equation*:

$$\psi_k^{(+)}(r) = \phi_k(r) + \int d^3r' G_E^{(+)}(r,r') V(r') \psi_k^{(+)}(r'). \tag{21.64}$$

This is obtained from Eq 21.1 by formally dividing by $H_0 - E - i\epsilon$, where H_0 is the kinetic energy operator and ϵ enforces the outgoing boundary conditions. The Green function is related to H_0 by:

$$G_E^{(+)} = \frac{1}{E - H_0 + i\epsilon}. \tag{21.65}$$

The inhomogeneous term in the Lippmann-Schwinger equation is a solution to:

$$(H_0 - E)\phi_k(r) = 0 \tag{21.66}$$

and specifies the scattering source. This source is of course a plane wave, which we shall denote as $\phi_k(r) = \langle r|k\rangle$ in the following.

The large distance limit of the Green function,

$$G_E^{(+)}(r,r') = -\frac{\mu}{2\pi} \frac{e^{ik|r-r'|}}{|r-r'|}, \tag{21.67}$$

permits the identification (see Eq. 21.55) of the scattering amplitude:

$$f_E(\theta) = -\frac{\mu}{2\pi} \int d^3r \, \phi_{k'}^*(r) V(r) \psi_k^{(+)}(r) \tag{21.68}$$

where $\phi_k(r) = \exp(\vec{k}\cdot\vec{r})$ and $\vec{k}'$ is a vector of the same magnitude as k with $\hat{k}'\cdot\hat{r} = \hat{r}'\cdot\hat{r}$.

We now define an operator whose diagonal matrix elements in plane waves, $|k\rangle$, are proportional to the scattering amplitude. This operator is the *transition matrix*:

$$\langle k'|T|k\rangle = \langle k'|V|\psi_k^{(+)}\rangle. \tag{21.69}$$

Writing the Lippmann-Schwinger equation in Dirac bracket notation and multiplying by $\langle k'|V$, yields the *T*-matrix equation:

$$\langle k'|T_E|k\rangle = \langle k'|V|k\rangle + \langle k'|VG_E^{(+)}T_E|k\rangle. \tag{21.70}$$

This expression can be written in operator form as:

$$T_E = V + V G_E T_E.$$

The subscripts serve as reminders of the parametric dependence on the scattering energy $E = k^2/(2\mu)$ in the Green function and the T-matrix.

As with the Schrödinger equation, the three-dimensional problem can be reduced to one dimension by projecting onto partial waves. To effect this let

$$V(k', k) \equiv \langle k'|V|k\rangle = \sum_\ell (2\ell + 1)\, V_\ell(k', k)\, P_\ell(\cos\theta) \quad (21.71)$$

where $\hat{k}' \cdot \hat{k} = \cos\theta$. A similar expansion for $T(k', k)$ then yields

$$T_\ell(k', k) = V_\ell(k', k) + \frac{1}{2\pi^2} \int q^2 dq\, V_\ell(k', q)\, G_E^{(+)}(q)\, T_\ell(q, k). \quad (21.72)$$

For a local potential the expression for the partial wave-projected interaction is:

$$V_\ell(k', k) = 4\pi \int_0^\infty r^2 dr\, j_\ell(k'r) j_\ell(kr) V(r). \quad (21.73)$$

We will need the relationship between the phase shift and the T-matrix. This can be obtained by comparing Eqs. 21.68 and 21.69 to obtain:

$$T_E(k, k) = -\frac{2\pi}{\mu} f_E(k). \quad (21.74)$$

Now use Eqs. 21.61, 21.59, and 21.62 to get:

$$T_\ell(k, k) = -\frac{2\pi}{\mu k} e^{i\delta_\ell(k)} \sin\delta_\ell(k). \quad (21.75)$$

Notice that the T-matrix has been evaluated **on-shell**, where $E = k^2/(2\mu) = k'^2/(2\mu)$.

In the jargon, Eq 21.72 is an "inhomogeneous linear Fredholm equation of the second kind with a singular kernel". The first step in attempting a numerical solution is to tame the singularity in the Green function via the relationship

$$\frac{1}{x + i\epsilon} = P\left(\frac{1}{x}\right) - i\pi\delta(x), \quad (21.76)$$

where P represents the ***principal value***. Eq. 21.72 then becomes:

$$T_\ell(k',k) = V_\ell(k',k) + \frac{1}{2\pi^2} P \int q^2 dq \, V_\ell(k',q) \frac{1}{E - E(q)} T_\ell(q,k) - i\frac{\mu k}{2\pi} V_\ell(k',k) \, T_\ell(k,k) \tag{21.77}$$

The next step is to implement the principal value prescription with a trick (Haftel and Tabakin, 1970):

$$P\int_0^\infty dk \, \frac{f(k)}{q^2 - k^2} = \int_0^\infty dk \, \frac{f(k) - f(q)}{q^2 - k^2}. \tag{21.78}$$

This equation follows because the last term integrates to zero but serves to regulate the singularity at $k = q$.

It is convenient to map the half real axis to a finite region with a transformation such as $\kappa = \tan^{-1}(k)$ or $\kappa = k/(1+k)$. Don't forget to reset the integration measure. We next turn the integral equation into a matrix equation by discretizing momentum space over N Gaussian quadrature points, $\{q_i\}$, with associated weights $\{w_i\}$. The right hand side of the equation involves the quantities $T_\ell(q_i, k)$ and $T_\ell(k,k)$ and thus represents an equation for $N+1$ unknowns. These can be obtained by extending the momentum grid to include the *half on-shell point*, k. We choose to implement this by setting $q_0 = k$ and label the range $[0, \ldots, N]$ with the index a. With this notation $T_a = T_\ell(k_a, k_0)$ and one has $N+1$ linear equations:

$$T_a = V_a + \frac{1}{2\pi^2} \sum_{i=1}^N w_i k_i^2 \frac{V_{ai} T_i}{(k_0^2 - k_i^2)/2\mu} - \frac{1}{2\pi^2} \sum_{i=1}^N \frac{w_i}{(k_0^2 - k_i^2)/2\mu} \cdot k_0^2 V_{a0} T_0$$
$$- i\frac{\mu k_0}{2\pi} V_{a0} T_{00}. \tag{21.79}$$

This can be written compactly as:

$$T_a - \sum_b V_{ab} D_b T_b = V_a \tag{21.80}$$

where:

$$D_a = \begin{cases} \frac{1}{2\pi^2} \frac{w_i k_i^2}{(k_0^2 - k_i^2)/2\mu}, & a = i = [1 \ldots N] \\ -\frac{1}{2\pi^2} k_0^2 \sum_{i=1}^N \frac{w_i}{(k_0^2 - k_i^2)/2\mu} - i\frac{\mu k_0}{2\pi}, & a = 0 \end{cases}. \tag{21.81}$$

Inversion of Eq 21.80 then yields $T_a = T_\ell(k_a, k_0)$ and hence the on-shell T-matrix element $T_\ell(k_0, k_0)$. It is convenient to choose the set of k_0 values to be on a linear grid since this is often desired for plotting and since it avoids the zero in the Green function denominator. Furthermore, the use of complex arithmetic can be avoided by working in terms of the *R*-matrix. Finally, it is possible to extend the formalism to the case of relativistic kinematics. The latter points are pursued in the exercises.

Example: square well scattering

We will illustrate the algorithm by obtaining the phase shift for scattering from a three-dimensional square well in S-wave. The S-wave interaction is given by Eq. 21.73 with the choice $V(r) = V_0 \theta(r - r_0)$ and evaluates to:

$$V_S(k', k) = -2\pi \frac{V_0}{kk'} \left(\frac{\sin(k+k')r_0}{k+k'} - \frac{\sin(k-k')r_0}{k-k'} \right). \tag{21.82}$$

This is implemented in C++ with the aid of std::complex<double> as follows (notice that the case $k = k'$ must be handled separately):

```
typedef std::complex<double> Complex;
double V0,r0,mu;

Complex SqWell(double k, double p, int ell) {
  // V = V0*theta(r-r0) S-wave only
  double SqReal;
  if (k != p) {
    SqReal = -2.0*M_PI*V0/(k*p)*(sin((k+p)*r0)/(k+p) -
                       sin((k-p)*r0)/(k-p));
  } else {
    SqReal = -2.0*M_PI*V0/(k*k)*(sin(2.0*k*r0)/(2.0*k)-r0);
  }
  return Complex(SqReal,0.0);
}
```

The phase shift can be obtained from Eq. 21.75 via either

$$\delta_\ell(k) = \tan^{-1}\left(\frac{\operatorname{Im}(T_\ell(k,k))}{\operatorname{Re}(T_\ell(k,k))} \right) \tag{21.83}$$

or

$$\delta_\ell(k) = \frac{1}{2}\sin^{-1}\left[-\frac{\mu k}{\pi} \operatorname{Re}(T_\ell(k,k)) \right]. \tag{21.84}$$

We use Eigen to manage matrices and to effect matrix inversion. Complete code to obtain the scattering phase shift is:

```cpp
int main () {
  double m1,m2;
  int Ngrid,ell;
  Complex D;

  cout << " enter m1, m2, ell " << endl;
  cin >> m1 >> m2 >> ell;
  cout << " enter V0 (<0 for attractive), r0 " << endl;
  cin >> V0 >> r0;
  cout << " enter Ngrid " << endl;
  cin >> Ngrid;

  mu = m1*m2/(m1+m2);
  VectorXd k(Ngrid+1),w(Ngrid);
  VectorXcd V(Ngrid+1),T(Ngrid+1);
  MatrixXcd M(Ngrid+1,Ngrid+1);

  GaussLegendreEigen(k,w,Ngrid,0.0,M_PI_2);   // create GL
      grid & weights
  for (int i=0;i<Ngrid;i++) {                 // map grid
    w(i) = w(i)/pow(cos(k(i)),2);
    k(i) = tan(k(i));
  }

  for (int ik=1;ik<50;ik++) {
      // loop in scattering momentum
    double k0 = ik*0.04;
    k(Ngrid) = k0;                            // add  k0 to the
        grid
    for (int a=0;a<=Ngrid;a++) {
      V(a) = SqWell(k(a),k0,ell);
    }
    for (int a=0;a<=Ngrid;a++) {
    for (int b=0;b<=Ngrid;b++) {
      if (b == Ngrid) {
        double Dr = 0.0;
        for (int i=0;i<Ngrid;i++)
          Dr -= w(i)*k0*k0/(k0*k0-k(i)*k(i));
          D = Complex(Dr,-M_PI*k0/2.0);
      } else {
          D = Complex(w(b)*k(b)*k(b)/(k0*k0-k(b)*k(b)),0.0);
      }
      M(a,b) = -mu/(M_PI*M_PI)*SqWell(k(a),k(b),ell)*D;
```

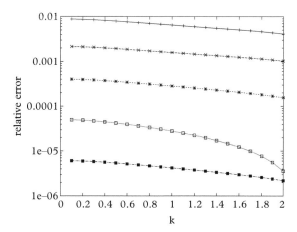

Figure 21.5 *Relative error for the attractive square well. From top to bottom $N = 10, 20, 40, 80, 160$. ($m = 1, V_0 = 1, r_0 = 1, \ell = 0$).*

```
      if (a == b) M(a,b) = Complex(1.0,0.0) + M(a,b);
   }} // end of a/b loops
   T = M.inverse()*V;
      // invert to obtain  T(k,k0)
   Complex Tonshell = T(Ngrid);
   double delta = atan(Tonshell.imag()/Tonshell.real())
       *180.0/M_PI;
   cout << k0 << " " << delta << endl;
  } // end of k0 loop
  return 0;
}
```

The results can be compared to the exact phase shift (see Appendix 2 of this chapter) and are shown in Figure 21.5. It is remarkable that even momentum grids with as few as ten values are capable of obtaining the phase shift to better than one percent, which indicates the power of this method.

21.4.3 Coupled channels

So far the discussion has focussed on elastic scattering of the type $AB \to AB$. However most physical problems are inelastic and involve many coupled channels. For example, in low energy hadronic physics one cannot consider $\pi\pi$ scattering in isolation since processes such as $\pi\pi \to \eta\eta$, $\pi\pi \to K\bar{K}$, and $\pi\pi \to \rho\rho$ can occur. Let us consider a two channel scattering problem labelled by a ***channel index***, α, where $\alpha = 1$ denotes channel AB and $\alpha = 2$ denotes channel CD. The differential equation describing this system is:

$$\left[\begin{pmatrix} T_1 & 0 \\ 0 & T_2 \end{pmatrix} + \begin{pmatrix} V_{11} & V_{12} \\ V_{21} & V_{22} \end{pmatrix} \right] \begin{pmatrix} \psi_1 \\ \psi_2 \end{pmatrix} = E \begin{pmatrix} \psi_1 \\ \psi_2 \end{pmatrix}, \qquad (21.85)$$

where, eg., $V_{12} = V_{AB:CD}$ and:

$$T_1 = m_A + m_B - \frac{\hbar^2}{2\mu_{AB}} \nabla^2 \qquad (21.86)$$

with a similar expression for T_2. It is important that the rest masses of the system be included in the kinetic energy since they set the *threshold* for a channel opening; namely, if $m_C + m_D > m_A + m_B$ scattering into the second channel cannot occur until

$$\frac{k^2}{2\mu_{AB}} + m_A + m_B > m_C + m_D. \qquad (21.87)$$

In this case channel 2 is said to be **open**. If the kinetic energy lies above $m_A + m_B$ and below $m_C + m_D$ then channel 2 is said to be **closed**.

We shall first consider a traditional approach wherein differential equations are solved and then move on to the more efficient T-matrix methodology.

From a computational standpoint, solving Eq. 21.85 as a differential equation is no more complicated than solving the single channel case—one merely works with four first order equations rather than two. Unfortunately, solving the differential equation also means implementing appropriate boundary conditions, and it is here that some additional complexity arises.

In the case where channel 1 is the incoming channel, boundary conditions are:

$$\begin{aligned} u_\ell^{(1)} &\to e^{-ikr} + (-)^{\ell+1} S_\ell^{(11)} e^{ikr} \\ u_\ell^{(2)} &\to \begin{cases} (-)^{\ell+1} S_\ell^{(12)} e^{ikr}, & \text{2 open} \\ 0, & \text{2 closed} \end{cases} \end{aligned} \qquad (21.88)$$

The S-matrix must be generalized to the two-channel case. A common parameterization is:

$$S_\ell^{(ij)} = \begin{pmatrix} \eta_\ell e^{2i\delta_\ell^{(1)}} & i\sqrt{1-\eta_\ell^2}\, e^{i(\delta_\ell^{(1)}+\delta_\ell^{(2)})} \\ i\sqrt{1-\eta_\ell^2}\, e^{i(\delta_\ell^{(1)}+\delta_\ell^{(2)})} & \eta_\ell e^{2i\delta_\ell^{(2)}} \end{pmatrix}. \qquad (21.89)$$

Here η_ℓ is a new parameter called the *inelasticity* that represents the loss of flux in an elastic channel due to the presence of other channels.

Unlike the single channel case, the boundary conditions can only be arranged if the initial slope takes on a precise, unknown, value. Fortunately, the Schrödinger equation is linear, and hence the solution to the coupled equations with two sets of initial conditions can be used to construct any solution required. One can, for example, compute two solutions, denoted A and B with boundary conditions

$$\vec{u}_A(0) \equiv (u_\ell^{(1)}(0), u_\ell^{(1)'}(0), u_\ell^{(2)}(0), u_\ell^{(2)'}(0)) = (0, 1, 0, 0) \quad (21.90)$$

and

$$\vec{u}_B(0) = (0, 0, 0, 1). \quad (21.91)$$

In practice one will want to start the computation at a small non-zero value of r and set $u_\ell^{(i)}(r) = r^{\ell+1}$. At large distance (for channel 2 closed) one must obtain:

$$\vec{u}_A \to \begin{pmatrix} X_A \sin kr + Y_A \cos kr \\ k(X_A \cos kr - Y_A \sin kr) \\ x_A e^{kr} + y_A e^{-kr} \\ k(x_A e^{kr} - y_A e^{-kr}) \end{pmatrix} \quad (21.92)$$

where the coefficients X_A, Y_A, x_A, and y_A are known and a similar parameterization holds for solution B. One seeks to match the solutions to $\vec{u} = \alpha \vec{u}_A + \beta \vec{u}_B$ such that:

$$\vec{u} \to \begin{pmatrix} e^{-ikr} + (-)^{\ell+1} S_\ell^{(11)} e^{ikr} \\ -ik(e^{-ikr} + (-)^\ell S_\ell^{(11)} e^{ikr}) \\ Ce^{-kr} \\ -kCe^{-kr} \end{pmatrix}. \quad (21.93)$$

The complex parameters α, β, C and $S_\ell^{(11)}$ must be determined by solving eight linear equations obtained by equating Eqs 21.92 and 21.93. In the case when channel 2 is open one must solve eight equations obtained by matching $\vec{u} = \alpha \vec{u}_A + \beta \vec{u}_B$ with

$$\vec{u}_A \to \begin{pmatrix} X_A \sin kr + Y_A \cos kr \\ k(X_A \cos kr - Y_A \sin kr) \\ x_A \sin kr + y_A \cos kr \\ k(x_A \cos kr - y_A \sin kr) \end{pmatrix} \quad (21.94)$$

and

$$\vec{u} \to \begin{pmatrix} e^{-ikr} + (-)^{\ell+1} S_\ell^{(11)} e^{ikr} \\ -ik(e^{-ikr} + (-)^\ell S_\ell^{(11)} e^{ikr}) \\ (-)^{\ell+1} S_\ell^{(12)} e^{ikr} \\ ik(-)^{\ell+1} S_\ell^{(12)} e^{ikr} \end{pmatrix}. \quad (21.95)$$

Again eight equations in terms of the known parameters $X_{A,B}$, etc. yield the eight desired quantities, α, β, $S^{(11)}$, and $S^{(12)}$.

In general an N-channel problem requires obtaining N independent solutions and solving a $4N \times 4N$ linear matrix problem for the desired S-matrix elements. Although this is certainly feasible, the coding is cumbersome, and we turn to the T-matrix formalism to see if it can provide a cleaner computational solution to the problem.

The generalization of the T-matrix equation to the coupled channel case can be simply made by supplying channel labels to the scattering potential and the T-matrix. Thus one writes, for example, $V \to V_{\alpha\beta}$ where α and β denote a particular channel (called 1=AB and 2=CD previously). The T-matrix equation then reads:

$$T_{\alpha\beta} = V_{\alpha\beta} + \sum_{\gamma} V_{\alpha\gamma} G^{\gamma} T_{\gamma\beta} \qquad (21.96)$$

in operator notation. The only new element here is that the Green function is now channel-dependent:

$$G^{\gamma} = \frac{1}{E - H_0^{\gamma} + i\epsilon\theta(E > E_{\gamma})}. \qquad (21.97)$$

Here H_0^{γ} is the free Hamiltonian in the γ channel (thus if $\gamma = 1$ then $H_0^{\gamma} = k^2/2\mu_{AB} + m_A + m_B$). Notice that a step function has also been introduced into the Green function. This implements scattering boundary conditions only in the case that the channel is open (thus E_{γ} is the threshold energy for the channel γ). If the channel γ is closed then the $+i\epsilon$ prescription is not used and outward spherical waves are not generated. The location of poles in the Green function is now channel-dependent and located at the solution to:

$$m_A + m_B + \frac{k_0^2}{2\mu_{AB}} = m_C + m_D + \frac{k^2}{2\mu_{CD}} \qquad (21.98)$$

in the two-channel case. We label the general pole location $k_{\gamma}^{\star}$. Notice that $k_1^{\star} = k_0$ in reference to the single channel notation. The single-channel principal value prescription must now be generalized as follows:

$$\int_0^{\infty} dq \frac{f(q)}{E - E_{\gamma}(q) + i\epsilon} = \int_0^{\infty} dq \left[\frac{f(q)}{E - E_{\gamma}(q)} - \frac{f(q_{\gamma}^{\star})\theta(E > E_{\gamma})}{(q_{\gamma}^{\star 2} - q^2)/(2\mu_{\gamma})} \right]$$
$$- i\pi \frac{\mu_{\gamma}}{q_{\gamma}^{\star}} f(q_{\gamma}^{\star}) \theta(E > E_{\gamma}) \qquad (21.99)$$

Notice that a theta function has been factored into the principle value subtraction. This is because the critical momentum $q_{\gamma}^{\star}$ does not exist if the channel is closed, and no subtraction is needed. Lastly, the generalized principle value prescription implies that $T(k_{\gamma}^{\star}, k)$ is required to construct observables, which means that the momentum grid must be expanded to $\{k_i\} \cup \{k_\gamma^\star\}$. Thus if there are N_{ch} channels and N Gauss-Legendre grid points, then there are $N_{ch}(N + N_{ch})$ unknowns. In the following the momenta $k_\gamma^\star$ will be assigned grid labels $N+1\ldots N+N_{ch}$.

The procedure for solving the coupled channel problem follows the single channel case with the exception of a larger matrix and a slightly more complicated expression for D:

$$D_a^\gamma = \begin{cases} \frac{1}{2\pi^2} \frac{w_i k_i^2}{(k_{N+\gamma}^2 - k_i^2)/2\mu_\gamma}, & a = i = [1\ldots N] \\ \left[-\frac{1}{2\pi^2} k_{N+\gamma}^2 \sum_{i=1}^N \frac{w_i}{(k_{N+\gamma}^2 - k_i^2)/2\mu_\gamma} - i\frac{\mu_\gamma}{2\pi} k_{N+\gamma} \right] \theta(E > E_\gamma), & a = N+\gamma \\ 0, & a > N, a \neq N+\gamma \end{cases} \quad (21.100)$$

In the end, the coupled channel method is a relatively simple extension of the single channel method and we recommend using a numerical implementation of the T-matrix equation whenever one is presented with quantum mechanical scattering problems. Further exploration of the coupled channel case is presented in the exercises.

21.5 Appendix: Three-dimensional simple harmonic oscillator

Consider the spherically symmetric three-dimensional simple harmonic oscillator defined by

$$-\frac{\nabla^2}{2\mu}\psi + \frac{1}{2}\mu\omega^2 r^2 \psi = E\psi. \quad (21.101)$$

The energy eigenvalues are $E_{nlm} = \frac{1}{2}\omega(4n + 2l + 3)$, where n is the number of nodes in the radial wavefunction. The eigenfunctions are

$$\psi_{nlm}(r) = N_{nlm} r^l Y_{lm}(\hat{r}) L_n^{l+1/2}(\beta^2 r^2) e^{-\frac{1}{2}\beta^2 r^2} \quad (21.102)$$

where $\beta^2 = \mu\omega$ and the normalization is given by

$$N_{nlm} = \sqrt{\frac{2 \cdot n!}{(n+l+1/2)!}} \beta^{l+3/2}. \quad (21.103)$$

The functions L_n^k are associated Laguerre polynomials (Arfken, pg 620). These obey the differential equation

$$xL_n^{k''} + (k+1-x)L_n^{k'} + nL_n^k = 0 \quad (21.104)$$

and the Rodrigues formula

$$L_n^k(x) = \frac{e^x x^{-k}}{n!} \frac{d^n}{dx^n} \left(e^{-x} x^{n+k} \right). \tag{21.105}$$

Recurrence relations for Laguerre polynomials are

$$(n+1)L_{n+1}^k(x) = (2n+k+1-x)L_n^k(x) - (n+k)L_{n-1}^k(x) \tag{21.106}$$

and

$$xL_n^{k'}(x) = nL_n^k(x) - (n+k)L_{n-1}^k(x). \tag{21.107}$$

Special values for the Laguerre polynomials are

$$L_0^k(x) = 1, \tag{21.108}$$

$$L_1^k(x) = k+1-x, \tag{21.109}$$

and

$$L_2^k(x) = \frac{x^2}{2} - (k+2)x + \frac{(k+2)(k+1)}{2}. \tag{21.110}$$

Setting $\psi_{nlm} = R_{nl}(\beta^2 r^2) Y_{lm}(\hat{r})$ gives the following radial wavefunctions:

$$R_{00}(x) = \frac{2}{\pi^{1/4}} e^{-x^2/2}, \tag{21.111}$$

$$R_{10}(x) = \frac{\sqrt{6}}{\pi^{1/4}} \left(1 - \frac{2x^2}{3} \right) e^{-x^2/2}, \tag{21.112}$$

$$R_{01}(x) = \sqrt{\frac{8}{3}} \frac{1}{\pi^{1/4}} x e^{-x^2/2}, \tag{21.113}$$

and

$$R_{11}(x) = \sqrt{\frac{5}{3}} \frac{2}{\pi^{1/4}} \left(1 - \frac{2}{5} x^2 \right) x e^{-x^2/2}. \tag{21.114}$$

Finally, the Fourier transform is given by the simple rule:

$$\tilde{\psi}_{nlm}(\beta; k) = (2\pi)^{3/2} (-i)^{2n+l} \psi_{nlm}(\beta \to 1/\beta; x \to k). \tag{21.115}$$

21.6 Appendix: scattering formulae

When writing code it is important to test against known cases. We therefore offer the following formulae:

- **Hard Sphere** (radius r_0): $\tan \delta_\ell = -\frac{j_\ell(kr_0)}{n_\ell(kr_0)}$.
- **Square Well** (radius r_0, depth V_0, $K^2 = k^2 + 2\mu|V_0|$): $\tan \delta_\ell = \frac{j'_\ell(kr_0) - j_\ell(kr_0)\gamma_\ell}{n'_\ell(kr_0) - n_\ell(kr_0)\gamma_\ell}$, $\gamma_\ell = j'_\ell(Kr_0)/j_\ell(Kr_0)$.
- **Separable Potential**: If $V(r',r) = \lambda v(r')v(r)$ then $T(k',k) = \lambda g(k')g(k)/(1 - \lambda \mathcal{J}(E))$ with $\mathcal{J}(E) = \int \frac{d^3p}{(2\pi)^3} \frac{g(p)^2}{E - E(p) + i\epsilon}$ and $g(p) = \int d^3r \, e^{ikr} v(r)$.
- **Born Approximation**: $\tan \delta_\ell \approx -2\mu k \int_0^\infty r^2 dr \, j_\ell^2(kr) V(r)$.

The first two spherical Bessel and Neumann functions are

$$j_0(x) = \frac{\sin(x)}{x}, \quad j_1(x) = \frac{\sin(x)}{x^2} - \frac{\cos(x)}{x} \quad (21.116)$$

$$n_0(x) = -\frac{\cos(x)}{x}, \quad n_1(x) = -\frac{\cos(x)}{x^2} - \frac{\sin(x)}{x}. \quad (21.117)$$

21.7 Exercises

1. [T] Radial Basis.
   Consider the basis

   $$\langle r|n\rangle = \exp(-n\beta r^2)$$

   where β is a variational parameter. Given that it can only represent even functions, is this a viable basis for radial problems?

2. 1d SHO.
   Solve the Schrödinger equation for the one-dimensional quantum mechanical simple harmonic oscillator problem. Compare Runge-Kutta integration with Numerov's method. Notice that $\psi(0) = 0$ is not a requirement of the wavefunction, however it is nonetheless obeyed by all solutions with negative parity. For positive parity solutions, a different strategy for starting Numerov's recurrence relation is required. Can you find it?

3. Charmonia (i).
   Charmonia are mesons that can be modeled as quantum mechanical bound states of a charm quark and a charm antiquark. Charm quarks have a mass of approximately 1.5 GeV and are spin-1/2 fermions. Charmonia bound states form a spectrum of particles with varying quantum numbers, $n^{(2S+1)}L_J$, where n is the radial quantum

number, S is the total spin of the quark and antiquark, L is their relative orbital angular momentum, and $\mathcal{J}$ is the state's total spin (this notation is the same as that for the hydrogen atom).

Experiments do not measure S and L as these are not observable quantities; they can determine, however, total spin, parity, and charge conjugation of a given state. This information is represented in the notation $\mathcal{J}^{PC}$. For a fermion-antifermion system $P = (-)^{L+1}$ and $C = (-)^{L+S}$. Thus, a vector state with $\mathcal{J}^{PC} = 1^{--}$ can be a linear combination of $^3S_1$ and $^3D_1$ quark-antiquark configurations.

Reasonably well-known charmonia states are tabulated in Table 21.2. These figures are obtained from the *Particle Data Group* website http://pdg.lbl.gov, which is a government-funded entity tasked with tracking experimental information in particle physics. The last column in the table are (reasonable) guesses about the internal structure of the experimental states listed.

Our goal is to reproduce this spectrum with a simple nonrelativistic quantum mechanical model. Assume that the quarks interact according to the potential

$$V(r) = -\frac{4\alpha_s}{3r} + br.$$

The first term is a strong interaction analogue of the Coulomb potential, while the second is a model of a ***confinement interaction*** that is meant to bind quarks into

Table 21.2 *Experimental charmonium spectrum.*

state	mass (GeV)	J^{PC}	$n^{(2S+1)}L_J$
η_c	2979.2	0^{-+}	$1^1 S_0$
J/ψ	3096.8	1^{--}	$1^3 S_1$
η_c'	3637.7	0^{-+}	$2^1 S_0$
ψ'	3686.0	1^{--}	$2^3 S_1$
$\psi(3770)$	3770	1^{--}	$1^3 D_1$
$\psi(4040)$	4040	1^{--}	$3^3 S_1$
$\psi(4160)$	4159	1^{--}	$2^3 D_1$
χ_{c0}	3415.3	0^{++}	$1^3 P_0$
χ_{c1}	3510.5	1^{++}	$1^3 P_1$
χ_{c2}	3556.2	2^{++}	$1^3 P_2$
χ_{c0}'	3918.4	0^{++}	$2^3 P_0$
χ_{c2}'	3927.2	2^{++}	$2^3 P_0$
$\psi(3823)$	3822.2	2^{--}	$1^3 D_2$
$\psi(4415)$	4421	1^{--}	$4^3 S_1$

hadrons. Take $m_c = 1.48$ GeV, $\alpha_s = 0.546$, and $b = 0.143$ GeV^2. Determine the masses of all the states listed in the table.

4. **Charmonia (ii).**
   If you completed the previous question, you will have noticed that the predicted spectrum only roughly agrees with the experimental data. The discrepancy is largely due to missing spin-dependent interactions. Here we address the *hyperfine interaction* contribution to the spectrum (the nomenclature comes from relativistic corrections to the spectrum of hydrogen) which is given by a spin-spin interaction:

$$V_{hyp}(r) = \frac{32\pi \alpha_s}{9 m_c^2} (\sigma/\sqrt{\pi})^3 \exp(-\sigma^2 r^2) \vec{S}_c \cdot \vec{S}_{\bar{c}}.$$

The new constant is given by $\sigma = 1.095$ GeV.

The spin-dependent portion of the interaction can be evaluated exactly using the standard trick:

$$(\vec{S}_c + \vec{S}_{\bar{c}})^2 = S_{tot}(S_{tot} + 1) = S_c(S_c + 1) + S_{\bar{c}}(S_{\bar{c}} + 1) + 2\vec{S}_c \cdot \vec{S}_{\bar{c}}.$$

Determine the masses of all the states in Table 21.2 with the new model (i.e., $V + V_{hyp}$). You should set S_{tot} and L in the code and diagonalize. Radially excited states will be automatically generated as higher states. What is the effect of the hyperfine interaction?

5. **Charmonia (iii).**
   The hyperfine interaction introduced in the last question improves the agreement with experiment; however, it is evident that additional mass splittings are still missing (such as in the χ_{cJ} states). These can be accommodated by including *fine structure interactions*; we model these as:

$$V_{fs} = \left(\frac{2\alpha_s}{r^3} - \frac{b}{2r}\right) \frac{1}{m_c^2} \vec{L} \cdot \vec{S} + \frac{4\alpha_s}{m_c^2 r^3} T$$

where T is the tensor operator:

$$T = \vec{S}_c \cdot \hat{r}\, \vec{S}_{\bar{c}} \cdot \hat{r} - \frac{1}{3} \vec{S}_c \cdot \vec{S}_{\bar{c}}.$$

Compute the masses of all the states in Table 21.2 using the model defined by $V + V_{hyp} + V_{fs}$. Use only the diagonal matrix elements of the spin-orbit and tensor interactions.

6. **[P] Charmonia (iv).**
   Use the model defined by the potential $V + V_{hyp} + V_{fs}$ to *fit* the charmonium spectrum in Table 21.2. The fitting program should allow the parameters m_c, α_s, σ, and b to

float. You should also permit mixing between states as generated by V_{fs}. Consult Barnes et al. (2005) if you need help.

7. [T] Boundary Conditions for Discretized Hamiltonians.
   The method discussed in Section 21.2.3 is very powerful but imposes constraints on acceptable boundary conditions. We choose to work in a box defined by $r \in (0, 1)$; grid points start at $r = \Delta$ and run up to $r = 1 - \Delta$.

   a) What is the value of the wavefunction at $r = 0$?
   b) Argue that there are reasons not to include $r = 0$ in numerical work.
   c) Imagine implementing the box constraints by imposing an infinite strength potential at $r = 1$. How will the wavefunction behave near $x = 1$?
   d) Do numerical solutions follow the expected behavior?

8. Discretized Hamiltonians in D Dimensions.
   Form the radial Schrödinger equation in D spatial dimensions and discretize it according to the prescription of Section 21.2.3.

   a) Analytically determine the behavior of $u(r)$ at the origin as a function of ℓ and D.
   b) Determine the boundary conditions being imposed.
   c) Establish that the method fails for $D = 1$ and $D = 2$ and works for $D = 3$ and $D = 4$ by carefully obtaining the ground state wavefunction for the D-dimensional simple harmonic oscillator.

9. Linear Potential in Momentum Space.
   We consider the problem of solving the Schrödinger equation with a linear interaction $V(r) = br$. This is an easy task in configuration space, but we shall take on the more difficult job of solving for the spectrum in momentum space. According to Section 21.2.4 this requires the interaction in momentum space. The difficulty is that the Fourier transform of a linear function does not exist. We can proceed by regulating the integral to obtain:

$$V_\epsilon(k) = \int d^3r\, r\, e^{-\epsilon r} e^{i\vec{k}\cdot\vec{r}} = \frac{-8\pi\epsilon}{(k^2 + \epsilon^2)^3}(k^2 - 3\epsilon^2).$$

   a) Argue that $\int d^3k\, V_\epsilon(k)$ should be zero. Confirm that it is by doing the integral.
   b) Thus establish that one cannot take $\epsilon \to 0$ at this stage.
   c) Code up the momentum space Schrödinger equation and confirm that one cannot obtain a sensible spectrum if one diagonalizes and then takes ϵ to zero.

   This establishes that our attempt at regulating the momentum space potential fails. Another way to regulate is to leverage the result of point (a) and subtract:

$$\int \frac{d^3q}{(2\pi)^3} V(p-q)\psi(q) = \int \frac{d^3q}{(2\pi)^3} V(p-q)[\psi(q) - \psi(p)]$$
$$\equiv \int \frac{d^3q}{(2\pi)^3} V(p-q)\psi(q) - \Sigma(p)\psi(p),$$

where Σ is a *self-energy* term defined implicitly above. Thus the idea is to solve:

$$\left(\frac{p^2}{2\mu} - \Sigma(p)\right)\psi(p) + \int \frac{d^3q}{(2\pi)^3} V(p-q)\psi(q) = E\psi(p). \qquad (21.118)$$

d) Show that the new regulator works by solving the modified Schrödinger equation and comparing to configuration space solutions.

10. [P] Resonance in a Box.
It may be surprising to learn that it is possible to study resonances with the discretized bound state technique of Section 21.2.3. We explore this here by considering two particles of mass 1 interacting via a 3d square well of depth V_0 and range r_0. Set the scale by choosing $r_0 = 1$ in the following.

a) Obtain the analytic solution to the square well problem and determine the bound state energy for the case $m_1 = m_2 = 1$ and $V_0 = -4$.

b) Implement a particle-in-a-box code and confirm the previous result.

c) Now couple this state to a decay channel. Call the first channel α and the second β. Masses for α and 1, as above; set the masses for β to be 0.5. Couple the channels with the following $V_{\alpha\alpha} = -4$, $V_{\alpha\beta} = 0.1$, and $V_{\beta\beta} = 0$. This induces a weak coupling between the bound state and the β continuum. Implement this situation in a coupled-channel particle-in-a-box code.

d) Obtain the lowest 10 or 20 eigenvalues of the coupled system. Print the RMS radius of each state along with the eigenvalues. You should obtain a spectrum of distorted plane waves with energies starting just above 1.0. However, you should also notice an unusual state "embedded" in the continuum. How do you know it is unusual? How do you interpret it?

e) Repeat with a larger box. How do the energy levels and RMS radii shift? Explain what this reveals about the eigenstates.

f) Improve your code by computing the wavefunction components of the eigenfunctions (i.e., $|\langle\alpha|\psi\rangle|^2$, where α is one of the channels). How do wavefunction components differ for the embedded state?

11. Quantum Monte Carlo and the D-dimensional SHO.
Implement the GRW algorithm of Section 21.3.1 in D-dimensions. Confirm that your code reproduces the ground state energy $E_0 = D/2$ in the units used in the text.

12. **Quantum Monte Carlo Excited States.**
    Computing the energy of the first excited state can be difficult with quantum Monte Carlo methods. One could study $E(\tau)$ carefully to extract the subleading τ-dependence. In practice this is quite noisy and unreliable (one tends to get an average of many excited states), thus direct access to E_1 is preferable. This can be achieved if the nodal surface of the excited state is known. Implement this idea in the 1d SHO by putting a node at zero in the guidance function. Test your code by confirming you recover $E_1 = 3/2$ in the appropriate units.

13. **Many-body Quantum Monte Carlo.**
    Generalize the guided random walk algorithm to the case of many particles of different masses. Be aware that Eq. 21.25 will need to be modified in this case! Test your code in the independent particle limit. Then turn on the interactions and confirm that the results behave as expected.

14. **[T] Quantum Monte Carlo Excited State Contamination.**
    Argue that the first excited state does not contribute to wavefunction contamination in the guided random walk solution to the 1d SHO (in other words the relevant energy gap is $E_2 - E_0$ not $E_1 - E_0$). Generalize your conclusions to other interactions.

15. **GRW Matrix Elements.**
    Compare the efficiency of the double walk and Feynman-Hellmann methods for computing $\langle \phi_0 | x^6 | \phi_0 \rangle$ for the 1d SHO. You should consider a figure of merit that accounts for the accuracy of the measurement and the time it takes to make the measurement.

16. **Single Channel Scattering.**

    a) Solve the Schrödinger equation scattering problem by using one of the methods in Chapter 11 and the formula for the phase shift, Eq. 21.63.
    b) Test your code on the square well.
    c) Compute the S- and P-wave phase shifts for the potential $V = V_0 \exp(-r^2/r_0^2)$.
    d) Compare your results to the Born approximation for this potential. When do they agree?

17. **[P] Coupled Channel Scattering.**

    a) Implement the coupled channel T-matrix algorithm of Section 21.4.3 and solve the two-channel square well problem of Ex. 10 for $\beta \to \beta$ scattering.
    b) Plot the magnitude of the T-matrix as a function of energy. You should see a resonance peak at the same location as in Ex. 10.
    c) Plot the phase shift as a function of energy. Compare your result to the *Breit-Wigner* resonance approximation:

    $$\tan \delta_\ell \approx \frac{\Gamma/2}{E_R - E}$$

and determine the resonance parameters, Γ and E_R. These are the called the Breit-Wigner width and resonance mass respectively.

d) Plot the imaginary part of the T-matrix vs. the real part while varying the scattering energy. (You will need a lot of points close to the resonance location. Alternatively, try increasing the size of $V_{\alpha\beta}$.) This is called an *Argand plot*. How does the trajectory vary as one passes through the resonance location?

18. **The R Matrix.**
    It is, somewhat surprisingly, possible to obtain the T-matrix with an entirely real formalism. The idea is to neglect the imaginary part of the T-matrix equation that is implicit in Eq. 21.4.2 or explicit in Eq. 21.78, which amounts to adopting standing wave boundary conditions in the scattering problem. The R-matrix (or *reaction matrix*) equation thus reads:

$$R_E(k',k) = V(k',k) + P\int \frac{d^3q}{(2\pi)^3} V(k',q) \frac{1}{E-E(q)} R_E(q,k)$$

or, in operator form:

$$R = V + VG^{(P)}R$$

where $G^{(P)}$ is the Green function with standing wave boundary conditions.

It is possible to prove that:

$$R_E = T_E + i\pi T_E \delta(E - H_0) R_E$$

by comparing the T-matrix and R-matrix equations. Thus the R-matrix can be used to reconstruct the T-matrix and the simpler equation for R can be solved in place of that for T. In particular,

$$R_\ell(k,k) = -\frac{2\pi \tan \delta_\ell}{\mu k}.$$

Implement an R-matrix code for a simple potential and confirm that it produces the same phase shift as T-matrix code.

19. **Relativistic Green Function.**
    Sometimes it is desirable to use relativistic kinematics in a scattering problem. In this case the Green function becomes

$$G(k) = \frac{1}{E - m_1 - m_2 - k^2/(2\mu) + i\epsilon} \to \frac{1}{E - \sqrt{k^2 + m_1^2} - \sqrt{k^2 + m_2^2} + i\epsilon}.$$

a) Generalize the Haftel-Tabakin trick (Eq. 21.78) to the relativistic case. If you are stuck consult Landau (1996), Chapter 18.

b) Solve for the single channel phase shift for your favorite problem. Under what conditions does relativistic kinematics become important?

20. [T] Poles in the S-matrix.
The S-matrix contains all information for a given quantum mechanical scattering problem. It is natural to ask, does it also contain information about the bound states of the system? The answer is yes—we illustrate this here in the case of S-wave scattering in a three dimensional square well of depth V_0 and range a.

a) Establish that the equation for an S-wave bound state is:

$$1 + \frac{k}{K} \tan Ka = 0,$$

where $k^2 = 2\mu E$ and $K^2 = 2\mu(|V_0| - |E|)$.

b) Show that the S-wave phase shift is given by:

$$\delta_0 = -ka + \tan^{-1}\left(\frac{k}{K} \tan Ka\right).$$

c) Consider single channel scattering and write the S-matrix as:

$$S = \frac{1 + i \tan \delta_\ell}{1 - i \tan \delta_\ell}.$$

Substitute the S-wave phase shift and show that the equation for a pole in the S-matrix corresponds to that for a bound state if one replaces k with ik. Thus poles in the S-matrix along the positive imaginary momentum axis correspond to bound states.

21. Scattering with Two Degenerate Channels.
In general it is very difficult to make analytic calculations in the coupled channel scattering problem (and thus it is difficult to verify coupled channel code). We discuss two special cases that *do* admit analytic solutions in this problem and the next.

Consider a two-channel problem that interacts via a potential that is common to all channels, with the exception of an overall strength. Thus the coupled-channel interaction can be written as $V_{\alpha\beta}(r) = A_{\alpha\beta} V(r)$, where the As are constants. We specialize further by considering the case where the particle masses in channel one are the same as those in channel two. Under these conditions it is natural to diagonalize the matrix A to obtain decoupled single-channel problems. This would be the complete solution except for one complication: one must implement scattering boundary conditions on the channel wavefunctions, and by diagonalizing one mixes up these boundary conditions.

a) Unravel the mess to obtain the complete solution to the scattering problem, including all elements of the S-matrix.
b) Test your solution against your coupled-channel code.

22. Coupled Channels with Separable Potentials.
Another important class of coupled-channels problems that permit analytic solution involves ***separable interactions***. A separable potential is maximally nonlocal and can be written as

$$V(r', r) = \langle r'|V|r\rangle = \lambda \tilde{g}(r') \tilde{g}(r).$$

Alternatively,

$$V(k', k) = \langle k'|V|k\rangle = \lambda g(k') g(k).$$

This property is convenient because it permits solution to the scattering equation $T = V + VGT$. This can be affected by forming the Ansatz $T_E(k', k) = t_E g(k') g(k)$ and substituting. Solving for t_E and substituting gives:

$$T_E(k', k) = \lambda \frac{g(k') g(k)}{1 - \lambda \mathcal{J}(E)}$$

where

$$\mathcal{J}(E) = \int \frac{d^3q}{(2\pi)^3} \frac{g(q)^2}{E - E(q) + i\epsilon}.$$

a) Confirm the above results.
b) Generalize the solution to the case of coupled channels. Take the interaction to be $V_{\alpha\beta} = \lambda_{\alpha\beta} \, g(k') g(k)$.
c) Confirm your solution by comparing to your coupled channel scattering code.

....

BIBLIOGRAPHY

Arfken, G. B., H. J. Weber, and F. E. Harris (2012). *Mathematical Methods for Physicists*. Academic Press.
Barnes, T., G. J. Daniell, and D. Storey (1986). *An Improved Guided Random Walk Algorithm for Quantum Field Theory Computations*. Nucl. Phys. B **265**, 253.

Barnes, T., S. Godfrey, and E. S. Swanson (2005). *Higher Charmonia*. Phys. Rev. D **72**, 054026.
Feynman, R. P. and A. R. Hibbs (2005). *Quantum Mechanics and Path Integrals*. Dover, New York.
Haftel, M. I. and F. Tabakin (1970). *Nuclear Saturation and the Smoothness of Nucleon-Nucleon Potentials*. Nucl. Phys. **158**, 1.
Joachain, C. J. (1975). *Quantum Collision Theory*. North-Holland.
Landau, R. L. (1996). *Quantum Mechanics II*. R. H. Landau. John Wiley and Sons.
MacDonald, J. K. L. (1933). *Successive Approximations by the Rayleigh-Ritz Variational Method*. Phys. Rev. **43**, 830.
Press, W. H., and S. A. Teukolsky, W. T. Vetterling, and B. P. Flannery (2007). *Numerical Recipes*. Cambridge University Press.
Stoer, J. and R. Bulirsch (1980). *Introduction to Numerical Analysis*. Springer-Verlag.

22

Quantum spin systems

22.1 Introduction	773
22.2 The anisotropic Heisenberg antiferromagnet	775
22.3 The Lanczos algorithm	778
22.3.1 The Lanczos miracles	779
22.3.2 Application of the Lanczos method to the Heisenberg chain	781
22.4 Quantum Monte Carlo	785
22.5 Exercises	791
Bibliography	800

22.1 Introduction

Although magnetism has been known to the world since the ancient Greeks, it has yet to be fully understood because it is a collective quantum effect. Heisenberg (1928) was the first to realize that the low energy behavior of a magnet may be characterized by the interplay of the Pauli exclusion principle and electrostatic exchange forces. This is something of a surprise, since one may expect that the interaction of magnetic dipoles would dominate the system. However, the typical energy of two magnetic dipoles is $O(10^{-4})$ eV (Ashcroft and Mermin, 1976) while the electrostatic energy of a pair of magnetic ions is typically 1 eV. Thus if the electrostatic energy of the ions can depend on the direction of their moments then Heisenberg's contention is correct. It is the exclusion principle which provides this dependence.

To illustrate, consider a spin-independent Hamiltonian describing two electrons. An eigenstate of the system will be a product of a spatial wavefunction and a singlet or triplet spin wavefunction:

$$\psi(\vec{r}_1, \vec{r}_2; s_1, s_2) = \phi(\vec{r}_1, \vec{r}_2)\chi(s_1, s_2). \qquad (22.1)$$

Generally ground states have spatially symmetric wavefunctions. Fermi statistics (the Pauli principle) requires that ψ is antisymmetric under interchange of the electrons, thus the antisymmetric spin singlet state is preferred, and we conclude that Heisenberg's contention is correct.

774 Introduction

An effective Hamiltonian that separates the spin singlet state from the spin triplet can be constructed from spin degrees of freedom in the following manner:

$$H_{eff} = \frac{1}{4}(E_s + 3E_t) - (E_s - E_t)\vec{S}_1 \cdot \vec{S}_2$$
$$= \mathcal{J}\vec{S}_1 \cdot \vec{S}_2 + \text{const}$$
$$\mathcal{J} \equiv (E_t - E_s). \tag{22.2}$$

The first line is designed to yield an eigenvalue of E_s (E_t) in the singlet (triplet) states, while the second is a simple redefinition.

> Show that H_{eff} works; i.e., show $\langle SM|\vec{S}_1 \cdot \vec{S}_2|SM\rangle = -3/4$ for $S = 0$ and $+1/4$ for $S = 1$. Hint: use the old trick $(S_1 + S_2)^2 = S(S + 1)$.

If $\mathcal{J}$ is written in terms of the appropriately symmetrized wavefunctions (and one ignores issues of normalization, etc) then it is approximately $-2\langle \phi(\vec{r}_1, \vec{r}_2)|H|\phi(\vec{r}_2, \vec{r}_1)\rangle$ which is an *exchange* energy and is purely a quantum effect. If $\mathcal{J}$ is positive then the spatially symmetric ground state is preferred so that the spins tend to anti-align; this is referred to as **antiferromagnetism**. The configuration with spins aligned in opposing directions on alternating lattice sites is called the **Néel state**. If $\mathcal{J}$ is negative the antisymmetric ground state is preferred and the spins tend to align. This is *ferromagnetism*.

It is remarkable that in many cases (Herring, 1965) the spin Hamiltonian for N bodies is simply

$$H_{HAF} = \sum_{ij} \mathcal{J}_{ij} \vec{S}_i \cdot \vec{S}_j. \tag{22.3}$$

where $\mathcal{J}_{ij}$ is the exchange energy between bodies i and j. Notice that H_{HAF} is isotropic in spin space, however if dipolar or spin orbit couplings are present the rotational symmetry may be broken. A simple way to incorporate some of these effects is by introducing anisotropy into the spin coupling. Thus we shall deal with the **anisotropic Heisenberg antiferrromagnet** (aHAF) described by the Hamiltonian

$$H_{aHAF} = \mathcal{J} \sum_{(ij)} (S_i^z S_j^z + g\vec{S}_i^\perp \cdot \vec{S}_j^\perp) \tag{22.4}$$

where (ij) represents a nearest neighbor in the lattice of interest (usually a 2d square lattice) and $\vec{S}^\perp$ is (S^x, S^y). The limit $g \to 0$ represents the Ising model and is trivially solvable at zero temperature. The $g \to \infty$ limit is the **XY model** which has achieved some fame due to the existence of a novel topological phase transition (see Chapter 20 for much more on these models).

The exact value of the ground state energy per spin (in units of $\mathcal{J}$) for the one-dimensional $S = 1/2$ isotropic spin system was found by Hulthèn in 1938 with the aid of the famous *Bethe Ansatz* (Bethe, 1931) and is given by:

$$e_0 = \frac{1}{4} - \ln 2 \approx -0.4431471. \tag{22.5}$$

The solution for the 1d $S = 1/2$ anisotropic case was obtained in 1959 and is:

$$e_0 = \frac{1}{4} - \frac{1}{2}\left(1 + \sum_{n=1}^{\infty} \frac{4}{e^{2n\lambda} + 1}\right)\tanh\lambda, \tag{22.6}$$

where $\text{sech}\lambda = g$.

The conditions under which *long range order* (i.e., magnetization of one sort or another) occur has been a longstanding problem in the field. In 1966, Mermin and Wagner proved that there exists no spontaneous (sublattice) magnetization at non-zero temperature for the 1d or 2d isotropic Heisenberg model (Mermin, 1966). In fact, it is the presence of low energy fluctuations that disrupts long range order at finite temperature and low dimensionality. The existence of these fluctuations is underpinned by *Goldstone's theorem* (Goldstone, 1961; Bludman, 1963), explored in Exercise 11, which states that a relativistic theory with a broken symmetry and corresponding degenerate vacuum must have a massless particle. Lange (1966) has shown that a version of this theorem applies to non-relativistic many-body systems which have finite-range interactions.

Goldstone's theorem is based on the usual set of field theoretic axioms, in particular the norm must be positive definite and the theory must be manifestly Lorentz covariant. One cannot have *both* of these in locally gauge invariant field theories. The resulting breakdown in Goldstone's theorem allows the Goldstone boson to be 'eaten' to form the extra degree of freedom needed for the, now massive, gauge particles. This process, originally explored in condensed matter physics, is known as the *Higgs mechanism* (Higgs, 1964; Anderson, 1958). The analogous process occurs in non-relativistic systems when long range forces are present. Indeed, for infinite range interactions spins with a finite relative rotation will produce a positive excitation energy at any distance. Hence the energy at zero wave number will be finite and the spin wave will be massive (Negele and Orland, 1988).

22.2 The anisotropic Heisenberg antiferromagnet

We seek the ground state (and perhaps a few excited states) of the 1d and 2d anisotropic $S = 1/2$ Heisenberg antiferromagnet at zero temperature. Thus a large (how large will be specified shortly) quantum mechanical problem must be diagonalized. In this section our main tool will be the Lanczos and related iterative algorithms.

The starting point is the anisotropic Heisenberg antiferromagnet of Eq. 22.4. We consider $S = 1/2$ spins, set $\vec{S}_i = \hbar\vec{\sigma}_i/2$, and absorb a factor of $\hbar^2$ into the coupling.

Nearest neighbor couplings will be considered henceforth. For the moment we consider a one-dimensional chain of spins of length L. Periodic boundary conditions will be used because they minimize surface effects. Finally, we seek the behavior of the system in the *bulk limit* (i.e., a very large number of spins).

Thus:

$$H = \frac{1}{4}\mathcal{J} \sum_{(ij)} [\sigma_i^z \sigma_j^z + g\sigma_i^\perp \cdot \sigma_j^\perp] \tag{22.7}$$

The Pauli matrices are defined in the usual way:

$$\vec{\sigma} \equiv (\sigma^x, \sigma^y, \sigma^z) = \left[\begin{pmatrix} 0 & 1 \\ 1 & 0 \end{pmatrix}, \begin{pmatrix} 0 & -i \\ i & 0 \end{pmatrix}, \begin{pmatrix} 1 & 0 \\ 0 & -1 \end{pmatrix} \right]. \tag{22.8}$$

Before launching into a full-blown investigation of the model it is worth spending some effort developing intuition with small systems. Consider, therefore, a two-spin system. A complete set of states is $\{|++\rangle, |+-\rangle, |-+\rangle, |--\rangle\}$ where, for example,

$$|+-\rangle \equiv |+\rangle_1 \otimes |-\rangle_2 \equiv \begin{pmatrix} 1 \\ 0 \end{pmatrix}_1 \otimes \begin{pmatrix} 0 \\ 1 \end{pmatrix}_2. \tag{22.9}$$

We have denoted spin locations with subscripts 1 and 2.

A straightforward way to proceed is to evaluate the matrix elements of H and diagonalize. This is best accomplished with the **step operators**:

$$\sigma^\pm = \frac{1}{2}(\sigma^x \pm i\sigma^y) \tag{22.10}$$

for which:

$$\sigma^+|+\rangle = 0 \qquad \sigma^+|-\rangle = |+\rangle \tag{22.11}$$

and:

$$\sigma^-|+\rangle = |-\rangle \qquad \sigma^-|-\rangle = 0. \tag{22.12}$$

Furthermore:

$$\sigma^z|+\rangle = |+\rangle \qquad \sigma^z|-\rangle = -|-\rangle. \tag{22.13}$$

The Hamiltonian is thus:

$$H = H_0 + V = \frac{\mathcal{J}}{4} \sum_{(ij)} \sigma_i^z \sigma_j^z + \frac{g\mathcal{J}}{2} \sum_{(ij)} [\sigma_i^+ \sigma_j^- + \sigma_i^- \sigma_j^+]. \tag{22.14}$$

The notation signifies that H_0 is diagonal in our z-diagonal basis. Thus:

$$H_0|s_1 s_2 \ldots s_L\rangle = \frac{\mathcal{J}}{4} \sum_i (s_i s_{i+1})|s_1 s_2 \ldots s_L\rangle \tag{22.15}$$

where $s_i = \pm$ denotes the spin state at site i.

Alternatively, the interaction V is a *spin flip* operator:

$$V|\pm\mp\rangle = \frac{g\mathcal{J}}{2}|\mp\pm\rangle, \tag{22.16}$$

while $V|\pm\pm\rangle = 0$.

With this formalism it is easy to evaluate the Hamiltonian for our two-spin trial problem. In the basis $|++\rangle, |+-\rangle, |-+\rangle, |--\rangle$ the result is:

$$H = \frac{\mathcal{J}}{4} \begin{pmatrix} 1 & & & \\ & -1 & 2g & \\ & 2g & -1 & \\ & & & 1 \end{pmatrix}. \tag{22.17}$$

Notice that the Hamiltonian is block diagonal, with $|++\rangle, |--\rangle$, and $\{|+-\rangle, |-+\rangle\}$ falling into separate sectors. This occurs because the total z-component of spin

$$\sigma^z_{\text{tot}} = \sum_i \sigma^z_i \tag{22.18}$$

is conserved (see Ex. 4). Furthermore the total spin is also conserved when $g = 1$

$$\left[\sum_i \vec{\sigma}_i, H\right] = 0. \tag{22.19}$$

The eigenvalues for the two-spin problem are $1/4, 1/4$, and $-(1\pm 2g)/4$. In the isotropic case we thus obtain a triplet with eigenvalue $\mathcal{J}/4$ and a singlet with eigenvalue $-3\mathcal{J}/4$, as expected. Notice that the ground state of the antiferromagnet is in the $\sigma^z_{\text{tot}} = 0$ sector. This will turn out to be always true. Alternatively the ferromagnet always has as a ground state the degenerate pair $|S_{\max}, \pm M_{\max}\rangle = |\pm\pm\ldots\pm\rangle$.

With periodic boundary conditions the ground state energy per spin for $L = 2$ is:

$$e_0 = \frac{E_0}{\mathcal{J}L} = -3/2. \tag{22.20}$$

Repeating this exercise for $L = 4$ and $g = 1$ (Ex. 1) confirms that the ground state lies in the singlet sector and gives $e_0(4) = -1/2$. Evidently we are making progress towards the expected bulk limit result of $e_0 \approx -0.4431$.

It is clear that our strategy of forming and diagonalizing the Hamiltonian cannot be carried out much further. In general one must deal with a $2^L \times 2^L$ matrix which is impractical for even relatively small L. If one restricts attention to the $\sigma^z_{\text{tot}} = 0$ sector, the problem is reduced to diagonalizing a

$$\binom{L}{\frac{L}{2}} \times \binom{L}{\frac{L}{2}}$$

matrix, which is still a daunting task. For example, for $L = 10$ this is a manageable 252×252 matrix, but for $L = 20$ one must diagonalize a $184{,}756 \times 184{,}756$ matrix, which will require about 68 gigabytes of memory to store. For $L = 30$ the matrix is of size $(1.6 \cdot 10^8)^2$ and will require 48 petabytes of memory.

22.3 The Lanczos algorithm

The *Lanczos algorithm*[1] is *the* standard method for diagonalizing large matrices. The method acts iteratively on an initial trial state to produce a sequence of new states within which the Hamiltonian is tridiagonal. Pseudocode for the algorithm follows.

```
// Lanczos Algorithm (pseudocode)

set |φ₋₁⟩ = 0;  b₀ = 0
select a trial state |φ₀⟩
generate |φₙ₊₁⟩ = H|φₙ⟩ − aₙ|φₙ⟩ − bₙ²|φₙ₋₁⟩

    with  aₙ = ⟨φₙ|H|φₙ⟩/⟨φₙ|φₙ⟩
          bₙ² = ⟨φₙ|φₙ⟩/⟨φₙ₋₁|φₙ₋₁⟩

repeat
```

The end result is that the Hamiltonian takes on the form

$$H = \begin{pmatrix} a_0 & b_1 & & & \\ b_1 & a_1 & b_2 & & \\ & b_2 & a_3 & b_3 & \\ & & \cdots & \cdots & \end{pmatrix}. \qquad (22.21)$$

The orthogonal basis that achieves this is given by

$$\left\{ |\phi_0\rangle, \frac{|\phi_1\rangle}{b_1}, \frac{|\phi_2\rangle}{b_1 b_2}, \ldots \right\}$$

---

[1] Named after Cornelius Lanczos (1893–1974) (pronounced "Lahn-ts-osh"), who was a Hungarian mathematician and physicist. He was driven from Hungary by the Nazis and from America by McCarthyism. He is also known for work on the fast Fourier transform and general relativity.

22.3.1 The Lanczos miracles

In principle one can obtain the complete spectrum of an $N \times N$ matrix with N iterations of the Lanczos algorithm. This will prove impractical for the extremely large matrices considered here. We are thus fortunate that the *first Lanczos miracle* occurs: namely one often obtains very good estimates of the low lying spectrum with a relatively small Lanczos matrix (say, 10^2 or 100^2). We know of no reason why this should be and it is easy to come up with counter examples. Nevertheless we shall see that extremely accurate estimates of the few lowest eigenvalues of the Heisenberg chain can be obtained with Lanczos matrices that are as small as $O(L) \times O(L)$ in size.

The *second Lanczos miracle* concerns numerical stability. Namely, the vast number of numerical operations quickly renders Lanczos basis states nonorthogonal (typically only a few iterations are required for round off error to accumulate to this point). In principle this should make the Lanczos algorithm useless. In practice, however, the effects of the breakdown of the algorithm are benign. Typically, redundant copies of eigenvalues and eigenvectors are obtained as the iterations proceed (Collum and Willoughby, 1985). This can be a problem if one is interested in a degenerate portion of the spectrum, but is often not an issue in the low lying spectrum. Furthermore, the generation of spurious eigenvalues can easily be followed, allowing them to be removed from the spectrum.

We illustrate this effect by diagonalizing the matrix $A_{ij} = i\delta_{ij}$ (in other words, we tridiagonalize a diagonal matrix and then rediagonalize it). As expected, one finds that orthogonality of the Lanczos states is quickly lost (in fact, single precision arithmetic is entirely useless; one must employ extended precision). Figure 22.1 shows the obtained eigenvalues as a function of the number of Lanczos iterations for a 200×200 matrix. Evidently the first three eigenvalues are being accurately recovered once 30 iterations have been made. Notice, however, that the second trajectory makes a transition to the lowest state ($e_0 = 1$) at around 80 iterations. This is followed by a transition of the third

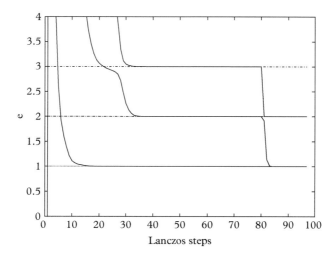

Figure 22.1 *Lanczos eigenvalues for $A_{ij} = \mathrm{diag}(1, 2, \ldots, 200)$.*

trajectory to the second eigenvalue. In fact, one might justifiably suspect that the second trajectory has hung up on the third eigenvalue around 20 iterations before moving to its correct value. These transitions occur because of accumulating round off error and generating the redundant eigenvalues discussed above.

> Are there situations where trajectory jumping can be useful?

If the eigenvector coefficients of the nth eigenvalue are z_{nm} in the Lanczos basis then the eigenvector can be reconstructed as:

$$|\psi_n\rangle = \sum_m \frac{z_{nm}}{\prod_{i=0}^m b_i} |\phi_m\rangle. \tag{22.22}$$

This can be computed while iterating, but requires storing N_{it} Lanczos vectors. The accepted way to deal with this issue is to run the Lanczos iterations a second time, performing the sum indicated in Eq. 22.22 as one proceeds. The lowest eigenvector ($v_0 = (1, 0, 0, \ldots)$) has been generated in this way for the diagonal test case and is displayed in Figure 22.2. One sees that the first element of the eigenvector is indeed unity, however the remaining components can be as large as 1% when 40 Lanczos iterations have been made. Fortunately, the situation improves rapidly with additional iteration; 80 iterations yield components with strength less than 10^{-8}. Similar behavior is obtained for other test matrices; for example $A_{ij} = \delta_{ij}/i$ behaves as above but with much slower convergence in both the eigenvalues and eigenvectors. In general the Lanczos algorithm performs better if eigenvalues are well-separated (Collum and Willoughby, 1985).

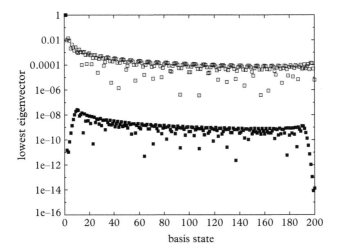

Figure 22.2 *First eigenvector for $A = \text{diag}(i)$. Top points: 40 Lanczos steps, bottom points: 80 Lanczos steps.*

22.3.2 Application of the Lanczos method to the Heisenberg chain

A typical diagonalization problem will require storing the (hopefully sparse) matrix H and performing many matrix-vector multiplies. However, a great advantage of the algorithmic definition of the Heisenberg antiferromagnet Hamiltonian we have developed is that it permits generating the new basis vectors on the fly. Thus memory requirements are reduced to that required to store a few basis vectors (two or three) of length $L!/(L/2)!^2$, which is an enormous advantage.

> Show that the Lanczos algorithm can be implemented with only two basis vectors in memory. At what cost?

The states of the Lanczos algorithm are written in terms of our $\hat{z}$-diagonal basis:

$$|\phi\rangle = \sum_n \alpha_n |n\rangle \tag{22.23}$$

where $|n\rangle$ is the nth basis state. It is convenient to leverage the binary nature of the spins to label these basis states. For example for $L = 4$ one could set:

$$|0\rangle = |----\rangle, \quad |1\rangle = |---+\rangle, \quad \ldots, \quad |15\rangle = |++++\rangle. \tag{22.24}$$

Of course the point is that each state can now be referred to by its associated integer. In the following we call this a ***statecode***. In this way a state such as:

$$\frac{1}{\sqrt{5}}|----\rangle - \frac{2}{\sqrt{5}}|--++\rangle \tag{22.25}$$

would be stored as the vector $(\frac{1}{\sqrt{5}}, 0, 0, -\frac{2}{\sqrt{5}}, 0, \ldots, 0)$. State overlaps are now easy to compute, if $|\chi\rangle = \sum \beta_n |n\rangle$ then[2]

$$\langle \chi | \phi \rangle = \sum_n \sum_m \beta_n^* \alpha_m \langle n | m \rangle = \sum_n \beta_n^* \alpha_n. \tag{22.26}$$

Reducing the size of the Hilbert space by taking advantage of the block-diagonal structure of the Hamiltonian is crucial (further reduction is possible: see Ex. 5). We implement this by scanning statecodes, determining configurations with a specific value of σ_{tot}^z, and keeping an array that maps 'relabeled statecodes' to statecodes.

---

[2] In spite of the notation, state coefficients are always real because the Clebsch-Gordan matrix is unitary.

```
// set up relabeled state codes
unsigned long int maxnum = 1 << L;    // 2^L
int rl[size];                          // size = L!/(L/2)!(L/2)!

rsc = 1;
for (unsigned long int sc=0; sc<maxnum; sc++) {
  if (Sz(sc) == Sztot) {
    rl[rsc] = sc; // push the statecode onto the "relabelled
        statecode" array
    rsc++;
  }
}
```

As an illustration, for $L = 4$ the statecodes for configurations with $\sigma_{tot}^z = 0$ are 3, 5, 6, 9, 10, and 12. The relabeling array thus is `rl(1)=3`, `rl(2)=5`, ..., `rl(6)=12`. It will be necessary to invert this relabeling to access spin configurations, thus one must define a function `inv`, such that `inv(3)=1`, `inv(5)=2`, ...`inv(12) = 6`. Notice that the range of this function runs to 2^L whereas the domain runs to $L!/(L/2)!^2$, thus it is impractical to implement `inv` as a look-up table for anything but very small L. A better approach is to compute the inverse using a binary search. This is an example of trading memory for speed that occurs often in algorithm design.

The next step is to choose an initial state. An obvious choice that is appropriate for small g is the Néel state $|\phi_0\rangle = (|+-+-\ldots\rangle + |-+-+\ldots\rangle)/\sqrt{2}$:

```
// initial Neel state (pseudocode)
zero phi
neel1 = 2^L/3
neel2 = 2*neel1
phi(inv(neel1)) = 1/sqrt(2)
phi(inv(neel2)) = 1/sqrt(2)
```

Another possibility is to assign random numbers to the elements of `phi`.

Verify that `neel1` and `neel2` as specified above actually give the Néel state.

The central element of the Lanczos algorithm is the computation of $H|\phi\rangle$. Obtaining $H_0|\phi\rangle$ may be simply done with Eq. 22.15 and with the aid of a function to extract a spin from a state code:

```
int spin(unsigned long int sc, int bt) {
  // return the bt'th bit in sc converted to +-
  int S = -1;
  int k = sc & 1 << bt;
```

```
    if (k != 0) S=1;
    return S;
}
```

Evaluating $V|\phi\rangle$ requires flipping spins according to Eq. 22.16. The first step is to determine if a spin pair is flippable (i.e., it is either $|+-\rangle$ or $|-+\rangle$). A routine that does this follows.

```
bool flippable(unsigned long int rsc, int bt) {
   // check if bit bt != bit bt+1
   bool f = false;
   int b0 = (rl[rsc] >> bt) & 1;
   int b1 = (rl[rsc] >> pl[bt]) & 1;
   if (b0 != b1) f = true;
   return f;
}
```

Once a flippable spin pair has been determined the flip operation can proceed:

```
unsigned long int flip(unsigned long int rsc, int bt) {
   // flip bits bt and bt+1; return a relabeled statecode
   unsigned long int sc = rl[rsc];
   sc ^= 1 << bt;
   sc ^= 1 << pl[bt];
   return inv(sc);
}
```

Finally, obtaining $H|\phi\rangle$ is not difficult:

```
// H0{phi}
for (rsc=1; rsc<=maxlen; rsc++) {
   Hphi[rsc] = H0(rsc)*phi[rsc];
}
// V{phi}
for (rsc=1; rsc<=maxlen; rsc++) {
   for (int bt=0;bt<L;bt++) {
      if (flippable(rsc,bt)) {
         Hphi[flip(rsc,bt)] += g/2.0*phi[rsc];
      }
   }
}
```

Running a typical code is straightforward until the number of spins approaches 30. At this point 100 Lanczos iterations can take several hours and memory requirements grow

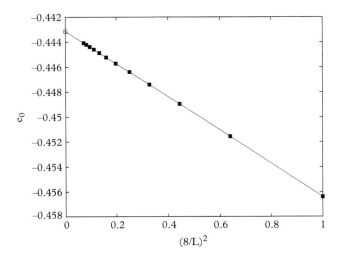

Figure 22.3 $S = 1/2$ isotropic Heisenberg chain ground state energy.

to several gigabytes. Pressing ahead without taxing our meager laptops too strenuously yields results like those of Figure 22.3, which have been obtained for $L = 8$ up to $L = 30$. One sees that the ground state energy $e_0(g = 1)$ scales very nearly as L^{-2}. A cubic fit to the Lanczos eigenvalues is also shown in the figure and yields the estimate

$$e_0(L = \infty; g = 1) = -0.4431461(3).$$

The exact energy in one dimension (to seven digits) is -0.4431471. This remarkable result illustrates the first Lanczos miracle, namely a bulk limit quantity has been computed to six digit accuracy by diagonalizing a (at most) 120×120 matrix!

Once the low lying eigenvalues have been obtained the corresponding eigenvectors can be computed using the technique discussed in Section 22.3.1. We use this method to compute the expectation of the ***staggered magnetization***:

$$N_z = \frac{1}{L} \sum_i (-)^i \sigma_i^z \qquad (22.27)$$

where the sum is over all spins. This serves as an order parameter for Néel ordering in the ground state. Reflection symmetry implies that $\langle N_z \rangle = 0$ so it is advisable to measure a related quantity such as $\langle |N_z| \rangle$. Here we choose to compute $\langle N_z^2 \rangle$. The results are shown in Figure 22.4.

The solid line in the figure is the perturbative result, $\langle N_z^2 \rangle = 1 - 2g^2$ (see Ex. 13), which demonstrates that the numerical work is functioning correctly. It is possible to obtain the finite size (i.e., L-dependent) corrections to this result so that an even more detailed check of code can be made. The circles in the plot are extrapolations in $1/L$ of

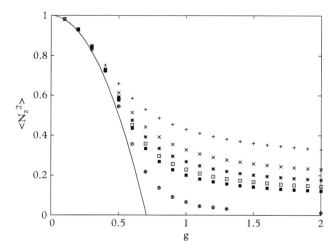

Figure 22.4 $\langle N_z^2 \rangle$ vs. coupling for various chain lengths. From top to bottom $L = 8, 12, 16, 20, 24$. The lowest points are a linear extrapolation to the bulk limit in $1/L$.

the points computed at $L = 8, 12, 16, 20$ and 24. One sees a rapid transition from large staggered magnetization to small staggered magnetization around $g = 1$. Indeed one expects that there is no long range order in one dimension for $g = 1$[3]. The rather large estimate, $\langle N_z^2 \rangle (L = \infty; g = 1) \approx 0.04$, indicates that substantial finite size corrections remain in the extrapolation. Indeed, the estimate for this quantity drops consistently as larger chains are included in the fit, providing strong evidence that there is a magnetic phase transition at $g = 1$.

22.4 Quantum Monte Carlo

Obtaining matrix elements and eigenenergies for large chains with the Lanczos algorithm is a difficult task. One could attempt to employ parallel computers by partitioning the state vector $|\phi\rangle$ among the processors. This works well for the computation of overlaps (a_n and b_n^2) and $H_0|\phi\rangle$, but runs into trouble when computing $V|\phi\rangle$ since the result for a given relabeled state code can lie in any partition. Thus communication latency severely restricts the effectiveness of a parallel Lanczos code.

An alternative is to generalize the guided random walk (GRW) quantum Monte Carlo algorithm described in Chapter 21 to problems with discrete degrees of freedom (Barnes and Daniell, 1988).

---

[3] Despite the fact that the ground state of the anisotropic chain is known (the Bethe Ansatz), an analytic expression for the long range order has not been obtained.

Consider a problem with a discrete set of basis states $\{|S\rangle\}$. We wish to solve

$$-\frac{\partial}{\partial \tau}\psi = H\psi \tag{22.28}$$

as a diffusion problem in configuration space. To start, split H into $|S\rangle$-diagonal and interaction pieces, H_0 and H_I respectively. Define the diagonal potential by:

$$H_0 = \sum_S |S\rangle V_0(S)\langle S|. \tag{22.29}$$

Let us also define the nearest neighbors of $|S\rangle$ in configuration space as the set $\{|S'\rangle\}$ such that $\langle S|H_I|S'\rangle \neq 0$. Now we say that if the walk is at $|S\rangle$ at time τ then it may step to a nearest neighbor at time $\tau + \Delta\tau$ with probability:

$$p(S, \tau \to S', \tau + \Delta\tau) = N_{SS'} r_{SS'} \Delta\tau. \tag{22.30}$$

The transition rates $r_{SS'}$ are a set of arbitrary positive numbers which constitute the guidance in this algorithm (see Section 21.3.1 for the importance of guidance in this technique); $N_{SS'}$ is the number of states that can be reached in one time step from $|S\rangle$.

A solution of the Euclidean Schrödinger equation is generated by constructing a histogram of weight factors in configuration space. In this case the weight factor associated with the mth random walk is the product of two factors, an exponential "diagonal weight" that depends on the path followed in configuration space and a "transition weight".

$$w_{total}^{(m)}(0 \to \tau) = w_{trans}^{(m)} \cdot w_{diag}^{(m)}, \tag{22.31}$$

$$w_{trans} = \prod_{\substack{S \to S' \\ \text{transitions}}} \left(-\frac{\langle S'|H_I|S\rangle}{r_{SS'}}\right), \tag{22.32}$$

$$w_{diag} = \exp\left(-\int_0^\tau \left(V_0(S(\tau')) - \sum_{S'} r_{SS'}\right) d\tau'\right). \tag{22.33}$$

Recall that one is able to optimize the continuum algorithm provided that one can find the "perfect" guiding potential $g^p(x) = -\ln(\psi_0)/2$, and that this choice allows exact energy measurements from a single walk. We enquire whether there exists an analogous "perfect" choice for the guiding matrix $r_{SS'}$. The answer is in the affirmative: it is given by:

$$r_{SS'}^p = -\langle S'|H_I|S\rangle \frac{\psi_0(S')}{\psi_0(S)}. \tag{22.34}$$

The condition

$$V_0(S) - \sum_{S'} r^p_{SS'} = \text{const}, \qquad (22.35)$$

reproduces the Schrödinger equation as in the continuum case. To obtain energy measurements with zero statistical error given Eq. 22.34 one must discard the transition weight factor in Eq. 22.31 and calculate energies using only the diagonal weight of Eq. 22.33. This simplification is justified for any $r_{SS'}$ of the general form:

$$r_{SS'} = -\langle S'|H_I|S\rangle \frac{\psi_0^g(S')}{\psi_0^g(S)} \qquad (22.36)$$

because the transition weight is a function of the arrival configuration $S(\tau)$ only and has no explicit time dependence. The function ψ_0^g may be regarded as a parametrized trial Ansatz for ψ_0, analogous to $\psi_0^g \equiv \exp(-2g)$ used for importance sampling in the continuum algorithm. Substitution into Eq. 22.32 shows that with the form Eq. 22.36 the transition weight is a function of $S(\tau)$ only; since $p(S, \tau)$ has a limiting distribution $p(S, \infty)$ at large τ, the asymptotic time dependence $e^{-E_0\tau}$ evidently arises exclusively from the diagonal weight.

An important issue arises if transition weights are required in a given calculation. As is evident from Eq. 22.32, the sign of w_{trans} can change with each transition, depending on the signs of $r_{SS'}$ and $\langle S'|H_I|S\rangle$. This leads to large cancellations in weight averages that can overwhelm the signal. This is called the *fermion sign problem* and can be very severe. The fermion sign problem is present in the case of the Heisenberg model because the guidance and matrix elements are positive. Fortunately, it is simple to avoid the issue by performing a unitary transformation that flips the sign the transition matrix element (see Ex. 6). Unfortunately such tricks are not available for more complicated problems such as the two-dimensional Hubbard model (Ex. 22).

As with the continuum version of the guided random walk algorithm, care must be taken to obtain the limits

$$\Delta\tau \to 0, \ \tau \to \infty, \ \text{stats} \to \infty. \qquad (22.37)$$

A number of approaches to obtain the ground state energy are possible. For example, since

$$\langle w(\tau)\rangle \to \alpha e^{-E_0\tau}(1 + \beta e^{-(E_1-E_0)\tau} + \ldots) \qquad (22.38)$$

where α and β are normalization constants, a fit of w to a sum of (say two) Gaussians will yield an estimate of the ground state energy. (We warn that the fit to $E_1 - E_0$ generally does *not* provide a good estimate of the excited state energy because the effects of the sum over many exponentials are absorbed into that fit parameter). The reader is additionally cautioned that the errors in a random walk are strongly correlated in time, and hence care must be taken in evaluating the mean and error (see Chapter 7 for information on dealing with correlated data).

An alternative is to extract the ground state energy with the ***effective energy***:

$$e_{\text{eff}}(\tau; \Delta) = \frac{1}{\Delta} \log \frac{\langle w(\tau) \rangle}{\langle w(\tau + \Delta) \rangle}. \tag{22.39}$$

Again, the effective energy should approach the ground state energy as τ gets large. An example is shown in Figure 22.5; the curves are computed with $\Delta = \Delta\tau = 0.01$ on a 16 spin chain. A chief concern with the method is determining the window in which the effective energy reaches equilibrium (one says it ***plateaus***). The figure shows the effective energy obtained from three starting points, a Néel state, a 'dimer' state ($|++--++--\ldots\rangle$), and an 'iceberg' state ($|++++\ldots----\rangle$). The data were obtained with 10^5 random walks for the Néel and dimer initial configuration and 10^6 for the iceberg. All three approach the correct energy and indicate plateau times of approximately 3, 4, and 6 for the dimer, Néel, and iceberg states respectively (using the symbol size as a rough accuracy criterion). Such explorations are crucial when the ground state energy of the system is not known.

The plateau time must be chosen such that the error induced by averaging over measurements at finite Euclidean time is less than the desired accuracy in the computation. This can be difficult to achieve since errors in the effective energy values grow with τ. A more sensible way to proceed is to incorporate as much information as possible, and extract the desired estimates with a fit. In this case Eqs. 22.38 and 22.39 provide a simple functional form that can be fit to $e_{\text{eff}}(\tau)$. However, the effective energies are correlated and care must again be taken. Dealing with correlations can be time consuming (a great

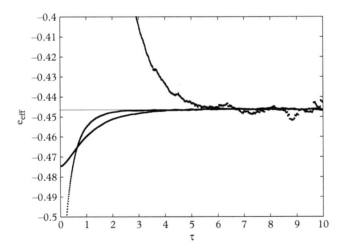

Figure 22.5 *Effective energy for dimer (at the bottom for low τ), Néel (middle), and iceberg (top) initial configurations.*

many correlation matrix elements must be evaluated), hence we choose to evaluate the effective energy at temporal separations larger than the autocorrelation time (see Chapter 7 for more on autocorrelation). In this case choosing $\Delta = 100\Delta\tau$ works quite well. The resulting data series is uncorrelated and errors can be estimated in the usual way.

The next task is to choose and optimize the guidance function. First we note that the unguided diagonal weight is given by:

$$w_{diag} = \int_0^\tau d\tau' \left(V_0(S(\tau')) - \frac{g}{s}\sum_{S'} 1 \right)$$

$$= \int_0^\tau d\tau' \left(\frac{L}{4} - \frac{1}{2}\bar{N}(\tau) - \frac{g}{2}\bar{N}(\tau) \right) \quad (22.40)$$

where $\bar{N}$ is the number of antialigned spin pairs in the given configuration.

For the walk guidance function we employ the Ansatz:

$$\psi_g(\{S\}) = \frac{g}{2}\exp(-BV_0(S)). \quad (22.41)$$

With this choice $r_{SS'}$ is a function of the change in V_0 in going from the state S to S', δE. In general a function of V_0 is preferable since it can be evaluated quickly in the walk basis.

Summing over possible states gives:

$$N_{SS'}\sum_{S'} r_{SS'} = \sum_{\delta E_{SS'}} -\langle S'|H_I|S\rangle \exp(-B\delta E) = \sum_{\delta E_{SS'}} \frac{g}{2}\exp(-B\delta E). \quad (22.42)$$

Verify that the guided weight reverts to the unguided weight as $B \to 0$.

Figure 22.6 displays a scan of the ground state energy in B for a 12-spin chain. A clear optimum near $B = 0.4$ is evident in the plot. This guidance has been used for all of the results presented in Table 22.1. The reader is reminded that the triple limit indicated in Eq. 22.37 needs to be monitored carefully. Each of these gives rise to a systematic error that must be incorporated in fits and extrapolations: typical statistical errors will be much smaller than true errors. The guided random walk results of the table were generated with a variety of number of walks, fit windows, and temporal step size. Typical values were 4 million walks, $\Delta\tau = 0.002$, $\Delta = 100\Delta\tau$, and fit windows of [2 : 14] and [4 : 14]. Algorithmic parameters were usually varied by factors of two to check stability and more walks were required for larger chains.

Recall that the Lanczos results were only able to hint that the isotropic model has no long range order. The guided random walk method provides more information because it yields useful results at larger L. A fit to guided random walk data is shown in Figure 22.7.

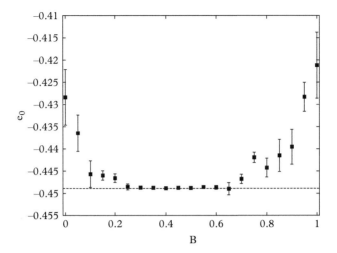

Figure 22.6 *Ground state energy vs. guidance parameter. The exact result is shown as a line.*

Table 22.1 *Ground state energy and staggered magnetization squared of the Heisenberg antiferromagnetic chain ($g = 1$).*

L	e_0(Lanczos)	e_0(GRW)	N_z^2(Lanczos)	N_z^2(GRW)
12	−0.4489492	−0.448933(15)	0.3276	0.3284(4)
16	−0.4463935	−0.446398(36)	0.2683	0.2672(6)
20	−0.4452193	−0.445267(21)	0.2289	0.2310(5)
24	−0.4445839	−0.444560(15)	0.2007	0.1980(7)
28	−0.4442017	−0.444226(15)	0.1793	0.180(1)
30	−0.4440654	−	−	−
32	−	−0.44396(2)	−	0.1682(7)
36	−	−0.44368(4)	−	0.1485(15)
40	−	−0.44375(4)	−	0.138(3)
44	−	−0.44356(3)	−	0.12(1)

This fit is cubic in $1/L$, which was required to incorporate the higher L values. The fit gives $\langle N_z^2\rangle(L \to \infty) = 0.014(33)$, indicating that there is indeed no long range order at isotropy for the 1d Heisenberg model.

A comparison of the main results of this chapter indicates that the Lanczos method is very powerful if the system approaches the bulk limit rapidly and smoothly and if several low lying eigenvalues and eigenvectors are of interest. It also has the great benefit

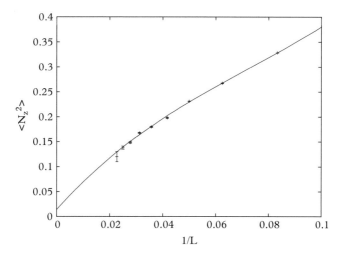

Figure 22.7 $\langle N_z^2 \rangle$ vs. $1/L$ for $g = 1$.

that very little extrapolation is required, one simply iterates until convergence is reached. Alternatively, the Monte Carlo method is preferred if larger systems need to be simulated, since memory requirements are minimal and the algorithm can be trivially parallelized. On the negative side, greater care must be taken to monitor systematic errors, only a very few low lying states are accessible, and the method fails when fermionic signs destroy the signal to noise ratio.

22.5 Exercises

1. [T] $L = 4$ Chain.
   Obtain the matrix elements of the anisotropic Heisenberg antiferromagnet for the $L = 4$ chain in the sectors σ_{tot}^z=4,2,0,-2,-4. Verify that the minimum energy in the antiferromagnetic case has $\sigma_{\text{tot}}^z = 0$.

2. Lanczos Check.
   Implement the Lanczos algorithm and apply it to the matrix $A = \text{diag}(1, 2, \ldots, 200)$.

   a) Confirm the results of Section 22.3.1.
   b) Test the efficacy of the code in single precision.
   c) Track the convergence of the second and tenth eigenvectors as a function of Lanczos iteration.

3. Alternative Lanczos.
   A simple version of the Lanczos algorithm starts with a guess $|\phi_0\rangle$ and diagonalizes the Hamiltonian in the basis $|\phi_0\rangle$ and

$$|\phi_1\rangle \propto H|\phi_0\rangle - |\phi_0\rangle \langle\phi_0|H|\phi_0\rangle.$$

The ground state of this system is taken as the new guess, $|\phi_0\rangle'$, and the process is repeated.

a) Confirm that

$$|\phi_0\rangle' = \frac{1}{\sqrt{1+\alpha^2}}\left(|\phi_0\rangle + \alpha|\phi_1\rangle\right)$$

where

$$\alpha = f - \sqrt{1+f^2}$$

and

$$f = \frac{1}{2}\frac{\langle H^3\rangle - 3\langle H\rangle\langle H^2\rangle + 2\langle H\rangle^3}{\left(\langle H^2\rangle - \langle H\rangle^2\right)^{3/2}},$$

where $\langle H\rangle = \langle\phi_0|H|\phi_0\rangle$.

b) Compare the efficacy of this method to the Lanczos algorithm in a variety of cases.

4. [T] Conservation Laws.

   a) Show that:

   $$\left[\sum_i \sigma_i^z, H\right] = 0.$$

   for Hamiltonians of the form of Eq. 22.4.

   b) Show that (for $g = 1$):

   $$\left[\sum_i \vec{\sigma}_i, H\right] = 0.$$

   Thus the total spin and the total z-component of spin are conserved.

5. Translation Invariance.
   The importance of using symmetry to reduce the size of the Hilbert space has been stressed. Another symmetry present in the periodic chain is *translation invariance*. States such as $|++--\rangle$ and $|-++-\rangle$ are related by shifting the origin and thus

contribute to eigenvectors in the same way. In this way the six $\sigma^z_{tot} = 0$ states for $L = 4$ can be reduced to 2.

a) Define a translation operator T such that $T|s_1 s_2 \ldots s_L\rangle = |s_L s_1 \ldots s_{L-1}\rangle$. Prove that $[T, H] = 0$.
Assume that the ground state has total momentum zero. Filter statecodes so that only states with $\sigma^z_{tot} = 0$ and states related by $T, T^2, \ldots T^{L-1}$ are accepted.

b) What is the size of the Hilbert space for the periodic chain of length L?

c) Implement the algorithm and compare your performance to the σ^z_{tot} filter.

6. [T] Unitary Transformation.
We consider the unitary transformation that changes the spin orientation on a checkerboard of lattice sites

$$U_\alpha = e^{-i \sum_j \pi (j_x + j_y) S^j_\alpha}$$
$$= \prod_j \left(I \cos \frac{1}{2}\pi(j_x + j_y) - i\sigma_\alpha \sin \frac{1}{2}\pi(j_x + j_y) \right),$$

where a lattice site is denoted by $\vec{j} = (j_x, j_y)$. Now consider $H' = U^\dagger_\alpha H U_\alpha$. Observe that U_α will have no effect until j in U_α and $U^\dagger_\alpha$ is equal to j in $\sum_{(ij)} H_{(ij)}$.

a) Confirm that ($n \equiv j_x + j_y$):

$$\left[I \cos \frac{\pi n}{2} + i\sigma_\alpha \sin \frac{\pi n}{2} \right] \frac{\sigma_\beta}{2} \left[I \cos \frac{\pi n}{2} - i\sigma_\alpha \sin \frac{\pi n}{2} \right]$$

$$= \begin{cases} \frac{\sigma^j_\beta}{2}, & n = 2m \\ \frac{\sigma^j_\beta}{2}, & n = 2m+1, \alpha = \beta \\ -\frac{\sigma^j_\beta}{2}, & n = 2m+1, \alpha \neq \beta. \end{cases}$$

b) Finally if

$$H = \sum_{(ij)} S^z_i S^z_j + g(S^x_i S^x_j + S^y_i S^y_j) \tag{22.43}$$

show that

$$U^\dagger_x H U_x = \sum_{(ij)} -S^z_i S^z_j + g(S^x_i S^x_j - S^y_i S^y_j)$$

$$U^\dagger_y H U_y = \sum_{(ij)} S^z_i S^z_j + g(-S^x_i S^x_j + S^y_i S^y_j)$$

$$U^\dagger_z H U_z = \sum_{(ij)} S^z_i S^z_j - g(S^x_i S^x_j + S^y_i S^y_j).$$

Notice that these results imply that the energy of the anisotropic Heisenberg model is even in g. The last form of H is useful because the transition weights in the path weight all become positive, thereby avoiding the infamous minus sign problem.

7. Alternative Guided Random Walks Algorithm.
   Two possibilities for the unguided random walks algorithm are:

$$\begin{Vmatrix} \vdots \\ \text{compute } \bar{N} \\ \text{if (rand} < \frac{1}{2}g\delta\tau\bar{N}) \, \{ \\ \quad \text{flip spins} \\ \} \\ w = w + \frac{L}{4} - \frac{1}{2}(1+g)\bar{N}_{\text{pre-flip}} \\ \text{OR} \\ w = w + \frac{L}{4} - \frac{1}{2}(1+g)\bar{N}_{\text{post-flip}} \end{Vmatrix}$$

Discuss the difference between these implementations of the random walk algorithm.

8. [T] Discrete and Continuous Guided Random Walks Algorithm.
   Establish that the continuous GRW algorithm of Chapter 21 maps to the discrete version presented here.

   a) Start with:

$$p(x, \tau \to x \pm \Delta x, \tau + \Delta \tau) \equiv r_{x, x \pm \Delta x} \Delta \tau = 1/2 \mp \Delta \frac{\partial g(x)}{\partial x},$$

and verify that the diagonal weight is given by:

$$V_0(x) - \sum_{x'} r_{xx'} = \langle x | \left(-\frac{1}{2m} \nabla^2 + V(x) \right) | x \rangle - r_{x, x - \Delta x} - r_{x, x + \Delta x} = V(x).$$

What choice for $\Delta \tau$ gives this result?
Thus the continuum weight factor,

$$w^{(m)} = \exp\left(-\int_0^\tau V(x^{(m)}(\tau')) \, d\tau' \right)$$

is recovered as a special case of the discrete algorithm.

b) Use

$$\langle S'|H_I|S\rangle \equiv \langle x'| \left(-\frac{1}{2m} \nabla^2 \right) |x\rangle \bigg|_{x' \neq x} = -\frac{1}{2m\Delta_x^2} (\delta_{x', x-\Delta x} + \delta_{x', x+\Delta x})$$

to obtain

$$w_{\text{transition}} = e^{2g(x_f)-2g(x_i)} e^{-\frac{1}{m}\int_0^\tau (g''-2(g')^2)d\tau'}.$$

Thus the product of the transition and diagonal weights reproduces the continuum algorithm.

9. **Haldane's Conjecture.**
   In 1982 Haldane suggested that chains of half-integer spins behave in a fundamentally different manner than chains of integer spins (Haldane, 1983). In particular, the half-integer spin chain is gapless (the ground state and first excited state are degenerate) while integer spin chains have a gap.

   a) Test the conjecture numerically by computing $E_1 - E_0$ for a spin 1/2 and spin 1 chain of length L. Take the first excited state to be the ground state of the spin-triplet sector (i.e., take $\sigma^z_{\text{tot}} = 2$). Carefully extrapolate to $L = \infty$.
   b) Does the conjecture generalize to two dimensions?

10. **[T] Spin Wave Theory.**
    We address the triplet-singlet gap again, this time analytically with the aid of spin-wave theory.

    a) Set

    $$S_z^2 = S(S+1) - S_x^2 - S_y^2 \qquad (22.44)$$

    so that

    $$S_z = \pm\left[\sqrt{S(S+1)} - \frac{S_x^2 + S_y^2}{2\sqrt{S(S+1)}} + O(1/S^2)\right]. \qquad (22.45)$$

    Choose opposite signs for S_z on opposite sublattices, expand H_{aHAF} to quadratic order in $S_\perp^2$, and diagonalize. Show that the ground state energy is

    $$\epsilon_0 = -dS(S+1) + \frac{2dS}{N} \sum_{\vec{q} \in \text{FBZ}'} \sqrt{1 - \frac{g^2}{d^2}\left(\Sigma_{i=1}^d \cos q_i\right)^2}$$

    for a d-dimensional square lattice of spin-S ions. The notation "$\vec{q} \in \text{FBZ}$" signifies reciprocal vectors which are in a reduced Brillouin zone consisting of $\frac{1}{2}$ of the usual Brillouin zone.

    b) Show that the triplet-singlet energy gap is given by

    $$E_1 - E_0 = 2\sqrt{1 - g^2}.$$

11. **[T] Goldstone's Theorem.**
    Consider a theory with a global symmetry, G. In quantum mechanics the ground state of the theory, $|\psi_0\rangle$, is unique and has the symmetry of G. However, in a theory with infinitely many degrees of freedom this need not be true. Say that the ground state is only invariant under a subgroup, H, of G. Thus $H|\psi_0\rangle = |\psi_0\rangle$. The action of G on $|\psi_0\rangle$ does not in general give $|\psi_0\rangle$ but some state $|\psi_0'\rangle$ which has the same energy as $|\psi_0\rangle$. Thus $G|\psi_0\rangle$ generates a manifold of degenerate vacua. Any given state in this manifold is unchanged by a group of transformations isomorphic to H. Thus the set of degenerate vacua is isomorphic to the coset space G/H.

    In general there will exist some quantity whose value distinguishes between an invariant or non-invariant vacuum. This quantity is called the **order parameter**. Transforming $|\psi_0\rangle$ in G/H corresponds to variations of the order parameter in which the effective action is kept at its minimum. Quantizing the excitations along these directions produces zero mass particles if G/H is continuous. These are the **Goldstone bosons**, one for each direction in G/H. The quantum numbers of the Goldstone bosons are those of the generators of G/H.

    Consider the isotropic three-dimensional Heisenberg antiferromagnet with the symmetry group $G = O(3)_{\text{spin}}$. Below the Néel temperature, there is spontaneous magnetization and hence a particular direction in spin space is singled out (say the $+\hat{z}$ axis). Establish the quantum numbers and number of Goldstone modes in this case.

12. **[T] Perturbative Mass Gap in the 2d anisotropic Heisenberg antiferromagnet.**
    Perform traditional perturbation theory in the anisotropy g to obtain the mass gap in the two-dimensional anisotropic Heisenberg model. You should obtain:

    $$\Delta = E_1 - E_0 = 2 - \frac{5}{3}g^2 + O(g^4). \quad (22.46)$$

    Here E_1 refers to the ground state energy in the $\sigma_{\text{tot}}^z = 2$ sector.

13. **[T] Perturbative Staggered Magnetization in the 2d anisotropic Heisenberg antiferromagnet.**
    Use standard perturbation theory to verify that:

    $$\langle N_z^2 \rangle = 1 - \frac{4}{9}g^2 + O(g^4). \quad (22.47)$$

14. **Finite Size Scaling of the 1d anisotropic Heisenberg Antiferromagnet.**
    We have attempted the extrapolation of the ground state energy to the infinite chain length limit by assuming that $e_0(L) = e_0(\infty) + a/L^2$. The Bethe Ansatz suggests a more effective fitting Ansatz, namely

    $$e_0(L) = e_0(\infty) + \frac{a}{L^2} + \frac{b}{L^2 \log(L)^3} + \frac{c}{L^2 \log(L)^4} + \cdots.$$

Compute the values of $e_0(L)$ up to $L = 30$ using the Lanczos algorithm and compare the results of both fits to the exact result. Comment on systematic errors.

15. Spin-spin correlation function.
    One expects
    $$\langle \sigma_n^z \sigma_0^z \rangle \sim (-1)^n \frac{\sqrt{\log(n)}}{n}$$
    for large n in the one-dimensional isotropic antiferromagnetic Heisenberg model (Mikeska and Kolezhuk, 2004). Test this result numerically. See Liang (1990) for help.

16. 2d Heisenberg antiferromagnet.
    Implement the Lanczos algorithm for the 2d square Heisenberg antiferromagnet. Since it is difficult to go beyond a 6×6 lattice, use the unusual configurations of Figure 22.8 to evaluate the ground state energy in periodic systems of size 4, 8, 10, 16, and 18. See if you can get to 25 and 36.

17. 2d Heisenberg antiferromagnet.
    Implement the guided random walk algorithm for the 2d square Heisenberg antiferromagnet and obtain the ground state energy in the bulk limit. Consult (Barnes and Daniell, 1988) if you need help.

18. Alternating Chain.
    The isotropic alternating 1d chain Heisenberg model is defined by:
    $$H = \sum_{i \text{ even}} \vec{\sigma}_i \cdot \vec{\sigma}_{i+1} + \lambda \sum_{i \text{ odd}} \vec{\sigma}_i \cdot \vec{\sigma}_{i+1}.$$

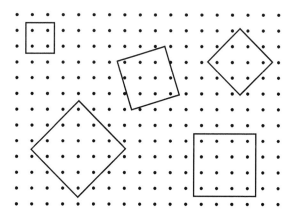

Figure 22.8 *Tiling the square lattice with tiles of size* 4, 8, 10, 16, *and* 18.

Take the system to be periodic with L spins. Notice that for $\lambda = 0$ the ground state is a collection of $L/2$ spin singlets. Thus $e_0(\lambda = 0) = -3/2$.

a) Compute the perturbative correction to e_0.
b) Solve this model with the Lanczos algorithm. Verify your perturbative result for small coupling. Verify that the Heisenberg chain is recovered as $\lambda \to 1$.

19. Code Like you Mean it.
Section 22.3 presents several code snippets that refer to statecodes and relabelled state codes. Both of these variables are represented as `unsigned long ints` in the snippets. This is inelegant: it would be better to differentiate the variables. This could, for example, be implemented with

```
typedef unsigned long int Statecode;
typedef unsigned long int RelabelledStatecode;
```

and functions could pass variables of the appropriate type. However, it would be better to leverage the power of C++ to make these variables truly different (so that, for example, a statement like `sc = 3*rsc;` throws an error).

a) Write a Lanczos code that implements this idea with classes.
b) Use `<vector>` for state vectors.
c) The code snippets perform many bit-based operations. Rewrite this functionality with the aid of the class template `<bitset>` (which you should research at cpluplus.com).
d) Replace the bisection `inv` routine with one based on the functionality found in the standard library class template `<algorithm>`.

20. [P] Heisenberg Ladder.
The antiferromagnetic Heisenberg spin ladder is a quasi-two-dimensional system consisting of L pairs of spins arranged in $2 \times L$ lattice. The Hamiltonian is:

$$H = \sum_{ij\,||} \vec{\sigma}_i \cdot \vec{\sigma}_j + g \sum_{ij\,-} \vec{\sigma}_i \cdot \vec{\sigma}_j$$

in a notation we hope is clear.

As $g \to 0$ one simply has two Heisenberg chains with ground state energy $e_0 = 1/2 - 2\ln 2$. For large coupling the ground state consists of L spin singlets with total energy $e_0 = -3/4\,g$. Excited states have rungs in the spin triplet state. For a large but finite coupling the rungs interact and hence the system looks like a spin-1 chain. Thus the ladder interpolates spin-1/2 and spin-1 chains (see Ex. 9 for why this is interesting).

a) Compute the ground state energy as a function of g for this system for as large L as your machine can handle using the Lanczos algorithm.

b) Compute the perturbative ground state energy to order g^2 and check your previous results.

c) Implement a guided random walk code to solve this problem. Verify its results against perturbation theory and the Lanczos calculation.

See (Barnes, 1994) if you need help.

21. [P] Structure Factor.
Evaluate the **structure factor** (an important ingredient in the description of scattering from magnetic materials) for the 1d anisotropic Heisenberg antiferromagnet. This is defined by:

$$S(k) = \langle \psi_1(k)|\sigma_0^+|\psi_0\rangle,$$

where ψ_0 is the ground state with $\sigma_{\text{tot}}^z = 0$ and $\vec{k} = 0$ and ψ_1 is the ground state with $\sigma_{\text{tot}}^z = 2$ and momentum k. A state of good momentum is defined by:

$$|n(k)\rangle = \frac{1}{\sqrt{\ell(n)}} \sum_{j=0}^{\ell(n)-1} e^{ikj}\, T_j\, |n\rangle$$

Here the number of independent states associated with a "root" state $|n\rangle$ is denoted $\ell(n)$; for example $\ell(|+-+-\rangle) = 2$ while $\ell(|++--\rangle) = 4$.

22. [P] Hubbard Model in One Dimension.
Hubbard has proposed a model that contains the bare minimum of features necessary to yield both bandlike and localized behavior in suitable limits (Hubbard, 1963).

The **single band Hubbard model** is written in terms of electron creation and annihilation operators as:

$$H_{\text{HUB}} = -t\sum_{\langle i,j\rangle}(c_{i\sigma}^\dagger c_{j\sigma} + \text{H.c.}) + U\sum_i n_{i+}n_{i-}.$$

The operator $c_{i\sigma}^\dagger$ creates an electron of spin $\sigma = \pm 1/2$ at ionic site i. The number operator is defined by:

$$n_{i\sigma} = c_{i\sigma}^\dagger c_{i\sigma}.$$

Thus the second term counts the number of electrons at site i and applies an energy penalty of $+U$ if the site is doubly occupied. The first term permits "hopping" between sites with an amplitude given by t.

The states of the model are given by specifying the four possible configurations on each ion: empty ($|0\rangle$), one spin up electron ($|\uparrow\rangle$), one spin down electron ($|\downarrow\rangle$), a spin up and a spin down electron ($|\uparrow\downarrow\rangle$).

Implement the Lanczos algorithm for this system and attempt to confirm the exact result for the ground state energy (Lieb and Wu, 1968):

$$e_0 = -4|t| \int_0^\infty d\omega \frac{\mathcal{J}_0(\omega)\mathcal{J}_1(\omega)}{\omega(1 + \exp(\omega U/|t|))}. \tag{22.48}$$

BIBLIOGRAPHY

Anderson, P. W. (1958). *Random-Phase Approximation in the Theory of Superconductivity*. Phys. Rev. **112**, 1900.

Ashcroft, N. and D. Mermin (1976). *Solid State Physics*. Cengage Learning.

Barnes, T., E. Dagotto, J. Riera, and E. S. Swanson (1994). *Excitation Spectrum of Heisenberg Spin Ladders*. Phys. Rev. B **47**, 3196.

Barnes, T. and G. Daniell (1988). *Numerical solution of spin systems and the S=(1/2 Heisenberg antiferromagnet using guided random walks*. Phys. Rev. B **37**, 3627.

Barnes, T. and E. S. Swanson (1988). *Two-dimensional Heisenberg antiferromagnet: a numerical study*. Phys. Rev. B **37**, 9405.

Bethe, H. A. (1931). *Theorie der Metalle. Erster Teil. Eigenwerte und Eigenfunktionen der lineären atomischen Kette*. Z. Phys. **71**, 205–226.

Bludman, S. A. and A. Klein (1963). *Broken Symmetries and Massless Particles*. Phys. Rev. **131**, 2364.

Collum, J. K. and R. A. Willoughby (1985). *Lanczos Algorithms for Large Symmetric Eigenvalue Computations*. Birkhäuser-Boston.

Goldstone, J. (1961). *Field Theories with "Superconductor" Solutions*. Nouvo Cimento, **19**, 154.

Haldane, F. D. M. (1983). "Nonlinear field theory of large spin Heisenberg antiferromagnets. Semiclassically quantized solitons of the one-dimensional easy axis Neel state". Phys. Rev. Lett. **50**, 1153.

Heisenberg, W. (1928). *Zur Theorie der Ferromagnetismus*, Z. Phys. **49**, 619.

Higgs, P. (1964). *Broken Symmetries and the Masses of Gauge Bosons*. Phys. Rev. Lett. **13**, 508.

Herring, C. (1965). *Direct Exchange Between Well Separated Atoms* in *Magnetism. 2B* ed. G.T. Rado, H. Suhl, Academic Press, New York.

Hubbard, J. (1963). *Electron Correlations in Narrow Energy Bands*. Proc. Royal Soc. Lond. **A276**, 238.

Lange, R. V. (1966). *Nonrelativistic Theorem Analogous to the Goldstone Theorem*. Phys. Rev. **146**, 301.

Liang, S. (1990). *Monte Carlo Calculations of the Correlation Functions for Heisenberg Spin Chains at T=0*. Phys. Rev. Lett. **64**, 1597.

Lieb, E. and F. Y. Wu (1968). *Absence of Mott Transition in an Exact Solution of the Short-Range, One-Band Model in One Dimension*. Phys. Rev. Lett. **20**, 1445.

Mermin, N. D. and H. Wagner (1966). *Absence of Ferromagnetism or Antiferromagnetism in One- or Two-Dimensional Isotropic Heisenberg Models.* Phys. Rev. Lett. **17**, 1133.

Mikeska, H.-J. and A. K. Kolezhuk (2004). *One-Dimensional Magnetism*, Lect. Notes Phys. **645**, 1–83.

Negele, J. W. and H. Orland (1988). *Quantum Many-Particle Systems.* Addison-Wesley, Don Mills.

Parlett, B. N. (1997). *The Symmetric Eigenvalue Problem.* SIAM, Philadelphia.

23
Quantum mechanics II–many body systems

23.1 Introduction	802
23.2 Atoms	803
23.2.1 Atomic scales	803
23.2.2 The product Ansatz	804
23.2.3 Matrix elements and atomic configurations	805
23.2.4 Small atoms	807
23.2.5 The self-consistent Hartee Fock method	809
23.3 Molecules	814
23.3.1 Adiabatic separation of scales	815
23.3.2 The electronic problem	817
23.4 Density functional theory	824
23.4.1 The Kohn-Sham procedure	827
23.4.2 DFT in practice	828
23.4.3 Further developments	830
23.5 Conclusions	831
23.6 Appendix: Beyond Hartree-Fock	831
23.7 Exercises	836
Bibliography	842

23.1 Introduction

For many students the exploration of the quantum world stops with the hydrogen atom. Although this is sufficient to illustrate the basic ideas of quantum mechanics, it does not reveal the wealth of profound phenomena that underpin familiar and unfamiliar features of nature. The quantum behavior of the humble hydrogen atom must surely pale compared to a protein that climbs along a strand of DNA to detect and repair flaws. Most of this structure arises because new features emerge when many bodies are present ("more is different" – P.W. Anderson) and because the large mass ratio between electrons and nuclei leads to a hierarchy of quantum phenomena.

Large inhomogeneous quantum systems are difficult to solve and progress has traditionally been predicated on developing methods that are specialized to the system at hand. In this chapter we will explore methods relevant to atoms and molecules. Similar techniques also apply to the structure of nuclei and nuclear matter. We will quickly learn that any aspiration to exact calculations must be abandoned since the computational demands of exact methods overwhelm present capabilities. Thus the chief topics will be methods that approximate many-body systems as effective single-body problems. These are the self-consistent Hartree-Fock method and density functional theory. We will also briefly touch on techniques that permit higher accuracy computation.

23.2 Atoms

The bewildering variety of chemical phenomena associated with the elements hints at the difficulty of computing accurate atomic properties. Nevertheless, the regularities revealed in the periodic table indicate that phenomenological rules of thumb — and hence simplifying approximations — must be present.

23.2.1 Atomic scales

The first simplification arises due to the very different energy scales associated with atomic and nuclear phenomena (ev vs. MeV). Thus at atomic scales the nucleus is essentially an inert body that serves as the source of a static Coulomb potential. In this case, nondimensionalizing the hydrogenic Schrödinger equation with central charge Z yields a distance scale of:

$$\bar{r} = \frac{\hbar}{Z\alpha m_e c} \tag{23.1}$$

and an energy scale

$$\bar{E} = Z^2 \alpha^2 m_e c^2. \tag{23.2}$$

Thus one obtains a speed scale

$$\bar{v} = Z\alpha c. \tag{23.3}$$

The fact that the *fine structure constant*, $\alpha = e^2/(4\pi\epsilon_0 \hbar c) \approx 1/137$, is small is vital to our program because it implies that nonrelativistic dynamics can be used throughout. Indeed, all of these scales are consistent with the assumed nonrelativistic dynamics *except* when $Z \sim 1/\alpha$. The nonrelativistic approximation can also break down for inner electrons. Thus relativistic dynamics is needed to understand, among other things, the color of gold and the liquid nature of mercury (Scerri, 1998). Finally, we remark that it is inconsistent to attempt to account for nuclear corrections by replacing the electron mass with the reduced mass without including relativistic corrections since these are of a similar size.

23.2.2 The product Ansatz

If one neglects nuclear and relativistic effects, an atom can be modelled as N electrons coupled to a static central charge and to each other via the Coulomb interaction. The Hamiltonian is thus:

$$H = H_0 + V \tag{23.4}$$

with:

$$H_0 = \sum_i h_i = \sum_i \left[-\frac{1}{2}\nabla_i^2 - \frac{Z}{r_i} \right] \tag{23.5}$$

and:

$$V = \sum_{i<j} \frac{1}{r_{ij}}. \tag{23.6}$$

The origin has been placed at the nucleus, thus r_i is the distance of the ith electron from the nucleus and r_{ij} is the distance between electrons i and j.

These quantities have been expressed in **atomic units** that coincide with the scales just introduced in the case $Z = 1$, $\hbar = 1$, and $c = 1$. Thus the atomic units length scale is the Bohr radius,

$$a_0 = \frac{1}{m_e \alpha} \approx 0.529 \text{Å}$$

and the energy scale is the *hartree*, defined by:

$$\bar{E} = \alpha^2 m_e \approx 27.2 \text{ eV}.$$

These units will be used throughout this chapter.

Show that the speed of light in atomic units is $1/\alpha$.

The exact eigenvalues of H_0 are given in terms of the hydrogenic *single particle* wavefunctions:

$$h_i \varphi_j(q_i) = \epsilon_j \varphi_j(q_i). \tag{23.7}$$

The coordinates are labeled q_i where $q_i = \{\vec{r}_i, s_i\}$ and s_i represents the spin of the ith electron. The index j includes all spatial and spin quantum numbers; for example $j = n\ell m_\ell m_s$ where n is the principle quantum number, ℓ is the angular momentum, m_ℓ specifies its projection, and m_s is a spin projection. In nuclear physics an *isospin* index is also required.

The eigenstates of H_0 are given by:

$$\Psi_{j_1, j_2, \ldots, j_N}(q_1, q_2, \ldots, q_N) = \frac{1}{\sqrt{N!}} \mathcal{A} \Psi_{\text{prod}}(q_1, q_2, \ldots, q_N) \tag{23.8}$$

with

$$\Psi_{\text{prod}}(q_1, q_2, \ldots, q_N) = \varphi_{j_1}(q_1) \cdot \varphi_{j_2}(q_2) \cdot \ldots \cdot \varphi_{j_N}(q_N) \tag{23.9}$$

and

$$\mathcal{A} = \sum_P (-)^P. \tag{23.10}$$

The total wavefunction has been antisymmetrized with the operator $\mathcal{A}$ which is given by a sum over permutations, denoted by P in the previous equation. We shall call this wavefunction the ***product Ansatz***; it is also often called the ***Slater determinant***. The antisymmetrization operator is Hermitian and idempotent:

$$\mathcal{A}^2 = \mathcal{A}. \tag{23.11}$$

It also commutes with H_0 and V

$$[\mathcal{A}, H_0] = [\mathcal{A}, V] = 0. \tag{23.12}$$

Finally, the eigenenergy is:

$$E_{j_1 \ldots j_N} = \sum_{i=1}^{N} \epsilon_{j_i}. \tag{23.13}$$

23.2.3 Matrix elements and atomic configurations

The simplicity of the product Ansatz makes it a popular starting point for more involved calculations. In fact, this approach is further simplified because matrix elements typically are only required for **one body** or **two body** operators. These are operators that depend on a single coordinate q_i or two coordinates, q_i, q_j, respectively.

The properties of the antisymmetrization operator imply the following:

$$\langle\Psi|H|\Psi\rangle = \langle\Psi_{\text{prod}}|\mathcal{A}H\mathcal{A}|\Psi_{\text{prod}}\rangle = \langle\Psi_{\text{prod}}|H\mathcal{A}^2|\Psi_{\text{prod}}\rangle = \langle\Psi_{\text{prod}}|H\mathcal{A}|\Psi_{\text{prod}}\rangle. \quad (23.14)$$

From this one obtains:

$$\langle\Psi'|h_a|\Psi\rangle = \langle\varphi_{j'_a}(q_a)|h_a|\varphi_{j_a}(q_a)\rangle \cdot \delta_{j'_1,j_1}\cdots\delta_{j'_N,j_N} \quad (23.15)$$

where the product of delta functions does not include the ath term. Similarly, a two-body matrix element is given by (additional delta functions are implied):

$$\langle\Psi'|O(a,b)|\Psi\rangle = \langle\varphi_{j'_a}(q_a)\varphi_{j'_b}(q_b)|O(a,b)|\varphi_{j_a}(q_a)\varphi_{j_b}(q_b)\rangle$$
$$- \langle\varphi_{j'_a}(q_a)\varphi_{j'_b}(q_b)|O(a,b)|\varphi_{j_b}(q_a)\varphi_{j_a}(q_b)\rangle. \quad (23.16)$$

Thus the matrix element of the full Hamiltonian in a product Ansatz wavefunction is:

$$\langle\Psi|H|\Psi\rangle = \sum_{j=1}^{N}\langle\varphi_j|h|\varphi_j\rangle + \sum_{i<j}\left[\langle\varphi_i\varphi_j|V|\varphi_i\varphi_j\rangle - \langle\varphi_i\varphi_j|V|\varphi_j\varphi_i\rangle\right], \quad (23.17)$$

where a somewhat more compact notation has been introduced. In particular, the indices i and j represent both a particle and the quantum numbers associated with that particle. We also refer to the single particle Hamiltonian as h since it is of the same form for each particle. Finally, this result serves as a variational upper bound to the exact energy and is also the expression for the eigenenergy in first order perturbation theory.

Applying the product Ansatz to atoms or ions requires choosing values for Z, N, and the quantum numbers of the wavefunction. The set of quantum numbers is called an **atomic configuration**. This is not known *a priori* but extensive experimental and theoretical experience indicates that atomic configurations follow **Madelung's rule**, namely a configuration can be constructed by placing electrons into hydrogenic orbitals in a specified order and in accordance with the Pauli exclusion principle. Orbitals are given in terms of $(n\ell)$ with $\ell = 0, 1, 2, 3, \ldots$ traditionally labelled s, p, d, f, etc. This order is:

$$1s, 2s, 2p, 3s, 3p, (3d, 4s), 4p, (4d, 5s), 5p, 6s, 4f, 5d, 6p, 7s, 5f, 6d, \ldots. \quad (23.18)$$

Orbitals in brackets are close in energy and can sometimes have their order permuted.

The number of electrons in an orbital is specified by a superscript. Since there are two spin states available per electron, the maximum that this number can be is $2(2\ell+1)$. Thus, for example, the lowest lying configuration describing beryllium ($Z = 4$) is written as

$$1s^2 2s^2 \quad (23.19)$$

while for sodium ($Z = 11$) it is

$$[Be]\, 2p^6 3s^1 = [Ne]\, 3s^1. \tag{23.20}$$

We remark that the simple configuration picture need not be accurate. For example, it was recently demonstrated that plutonium has a mixture of four, five, or six electrons in its 5f outer orbital, thereby explaining its metallic, but nonmagnetic, behavior (Janoshek et al., 2015).

In general the projections m_ℓ and m_s must also be specified. Guidance is provided by **Hund's rules**: choose quantum numbers to (in order)

1. Maximize S
2. Maximize L
3. Minimize J if the valence orbital is less than or equal to half filled, otherwise maximize J.

Hund's rules apply to outer orbitals because only these contribute to atomic quantum numbers. For example, carbon has $Z = 6$ thus two electrons must be placed into the $2p$ orbital. According to Hund's rules these should form an $S = 1$ state; hence,

$$|SM_S\rangle_{2p} = |11\rangle = |\tfrac{1}{2}\tfrac{1}{2}; \tfrac{1}{2}\tfrac{1}{2}\rangle. \tag{23.21}$$

The symmetry of the spin state forbids placing the $2p$ electrons into an angular momentum state with $L = 2$, thus Hund's rule requires that they have $L = 1$:

$$|LM_L\rangle_{2p} = |11\rangle = \frac{1}{\sqrt{2}}|11; 10\rangle - \frac{1}{\sqrt{2}}|10; 11\rangle. \tag{23.22}$$

Finally, the third rule implies that $J = 0$ and one has:

$$|JM_J\rangle_{2p} = |00\rangle = \frac{1}{\sqrt{3}}|11; 1-1\rangle - \frac{1}{\sqrt{3}}|10; 10\rangle + \frac{1}{\sqrt{3}}|1-1; 11\rangle. \tag{23.23}$$

23.2.4 Small atoms

As a first step, let us consider obtaining eigenvalues for a small atom by direct diagonalization with the discrete basis method that was advocated in Section 21.2.3. Because the many body interaction depends on the angles between coordinates, $2^N \cdot N_g^{3N}$ basis expansion coefficients are required to specify the Hamiltonian, where N_g is the grid size and N is the number of electrons. Since typical general matrix diagonalization routines slow down drastically beyond a basis size of order 1000, this places severe constraints on the basis size or the number of electrons. We shall not consider approaches such as this henceforth.

In spite of this setback, diagonalization in a carefully chosen basis set has been successfully applied to helium and other small atoms (Hylleraas, 1929). For example, Pekeris (1958) assumed an exact helium wavefunction of the type

$$\Psi = \Psi(r_1, r_2, r_{12}).$$

It is, however, difficult to expand this wavefunction in a set of orthogonal functions because the variables are related. Thus Pekeris switched to "parimetric coordinates":

$$u = r_2 - r_1 + r_{12}, \tag{23.24}$$
$$v = r_1 - r_2 + r_{12}, \tag{23.25}$$

and

$$w = 2(r_1 + r_2 - r_{12}), \tag{23.26}$$

and set

$$\Psi = e^{-\epsilon(u+v+w)/2} F(u, v, w) \tag{23.27}$$

where $\epsilon = \sqrt{-E}$ and E is the binding energy.

The function F was then expanded in a triple series of Laguerre polynomials. The resulting expression for the lowest eigenstate contained a 33 term recurrence relation for the expansion coefficients. Solving these equations up to approximately 200 terms and extrapolating gave results for ground state energies and matrix elements as shown in Table 23.1.

Results such as this are useful for checking methods that make additional approximations, but are not sufficiently general to warrant further attention here.

Table 23.1 *Ground state matrix elements for two-electron ions for various Z. All quantities in appropriate atomic units (Pekeris, 1958).*

Z	ϵ	$\langle p_1^4 \rangle$	$\langle \delta(r_2) \rangle$	$\langle \delta(r_{12}) \rangle$
1	0.726 464 701 2	2.462	0.164 5	0.002 74
2	1.704 031 793 6	54.088 2	1.810 403	0.106 36
3	2.698 131 462 0	310.548	6.851 97	0.533 78
4	3.695 343 849 6	1047.29	17.198 1	1.523 0
10	9.690 552 434 4	46 098.77	297.622	32.621

An approach that can be successfully applied to larger atoms leverages the reasonable accuracy of product Ansatz solutions by evaluating the Hamiltonian in a basis of atomic configurations. This is called the *configuration interaction* approach. Thus one sets:

$$|\Psi\rangle = \sum_\alpha c_\alpha |\alpha\rangle \qquad (23.28)$$

where $|\alpha\rangle$ denotes a specific atomic configuration. Matrix elements can be evaluated with the aid of Eqs. 23.15 and 23.16.

The difficulty with this method is that selecting and setting up appropriate configurations is a relatively demanding task and must be repeated for every atom or ion under consideration. Automation is very difficult under these circumstances.

23.2.5 The self-consistent Hartee Fock method

The long-established standard for computing atomic properties is the *self-consistent Hartree-Fock method*. The original formulation, due to Hartree (1928), was motivated by the observation that an electron in a large atom would interact with the nuclear charge and an effective charge due to all the other electrons[1]. The latter interaction can be approximated by summing the electric charge density at the site of the ith electron:

$$V_H(r_i) = \int d^3x \sum_{j\neq i} |\varphi_j(x)|^2 \frac{1}{|\vec{r}_i - \vec{x}|}. \qquad (23.29)$$

Hartree further simplified by considering the spherical average of the interaction V_H. This idea was subsequently extended to account for antisymmetry in the wavefunction by Fock.

An alternative point of view (first advocated by Slater) is to seek the single particle wavefunctions that minimize the atomic energy in an antisymmetrized product Ansatz trial function. Thus one seeks to solve

$$\frac{\delta}{\delta \varphi_j} \langle \Psi[\varphi]|H|\Psi[\varphi]\rangle = 0 \qquad (23.30)$$

subject to the constraint that $\langle \varphi_j|\varphi_k\rangle = \delta_{jk}$. Taking the derivative of Eq. 23.17 yields the collection of *Hartree-Fock equations*:

$$h(r)\varphi_j(r) + V_D(r)\varphi_j + \int d^3x \, V_E^{(j)}(r,x)\varphi_j(x) = \epsilon_j \varphi_j(r) \qquad (23.31)$$

---

[1] A similar "mean field" idea was also employed by Weiss in describing the Ising model; see Chapter 20.

where

$$h(r) = -\frac{1}{2}\nabla^2 - \frac{Z}{r} \qquad (23.32)$$

as before[2].

The effective local interaction is called the ***direct interaction*** and is given by:

$$V_D(\vec{r}) = \int d^3x \sum_j |\varphi_j(\vec{x})|^2 \, V(\vec{r} - \vec{x}). \qquad (23.33)$$

We temporarily generalize the two-body interaction to $V(\vec{r})$. Unlike in Hartree's original formulation, the sum does not exclude any particles and no spherical averaging is made. A spin-independent interaction has been assumed. Thus if spin degrees of freedom are removed from the index j, say $j \to (\mathcal{J}, s)$, one has:

$$V_D(\vec{r}) = 2\int d^3x \sum_{\mathcal{J}} |\varphi_{\mathcal{J}}(\vec{x})|^2 \, V(\vec{r} - \vec{x}).$$

Finally, antisymmetrization introduces a nonlocal interaction, called the ***exchange interaction***, that is given by:

$$V_E^{(s_j)}(\vec{r}, \vec{x}) = -\sum_i \delta_{m_{s_i}, m_{s_j}} \, \varphi_i^\dagger(\vec{r}) \varphi_i(\vec{x}) V(\vec{r} - \vec{x}). \qquad (23.34)$$

With spin-independent interactions a delta function in spin arises, as indicated. This induces a dependence on the spin of the state j, which explains the superscript in the notation for the exchange potential. In more detail, the form of the exchange term in the energy, $\langle \varphi_i \varphi_j | V | \varphi_j \varphi_i \rangle$, implies that the spin of the ith and jth states must be the same. Thus the expanded expression for the exchange interaction is:

$$V_E^{(s_j)}(\vec{r}, \vec{x}) = -\sum_I \varphi_{I, s_j}^*(\vec{r}) \varphi_{I, s_j}(\vec{x}) \, V(\vec{r} - \vec{x}). \qquad (23.35)$$

We remind the reader that the sums over indices refer to those single particle orbitals that appear in the product wavefunction. Of course, $V(r) = 1/r$ for atomic systems. The exchange potential tends to be small for Coulombic interactions because the wavefunction overlap restricts the range of integration. It is, however, very important for short range potentials like those that occur in nuclear physics.

---

[2] The reader not familiar with the beautiful second-quantized version of the derivation of this equation is encouraged to consult a source such as (Fetter and Walecka, 2003).

The total binding energy of the atom is given by the expression in Eq. 23.17; however, in this case φ is a Hartree-Fock wavefunction satisfying Eq. 23.31. Thus one can also write:

$$E = \sum_j \epsilon_j - \frac{1}{2} \sum_{ij} \left[\langle \varphi_i \varphi_j | V | \varphi_i \varphi_j \rangle - \langle \varphi_i \varphi_j | V | \varphi_j \varphi_i \rangle \right]. \tag{23.36}$$

It is possible that the rather general notation has caused the student to lose the thread of the development. We therefore present detailed versions of the Hartree-Fock equations for the case of lithium. We shall model lithium with the single atomic orbital $1s^2 2s^1$, and take the spin to be "up". The product wavefunction is then:

$$\Psi = \frac{1}{\sqrt{3!}} \mathcal{A} \varphi_{1s+} \varphi_{1s-} \varphi_{2s+}.$$

The Hartree-Fock equations read:

$$h\varphi_\ell(\vec{r}) + V_D(\vec{r})\varphi_\ell(\vec{r}) + \int V_E^{(\ell)}(\vec{r}, \vec{x})\varphi_\ell(\vec{x}) = \epsilon_\ell \varphi_\ell(\vec{r}) \tag{23.37}$$

where $\ell \in (1s+, 1s-, 2s+)$ and

$$V_D(\vec{r}) = \int d^3x \frac{|\varphi_{1s+}(\vec{x})|^2 + |\varphi_{1s-}(\vec{x})|^2 + |\varphi_{2s+}(\vec{x})|^2}{|\vec{r} - \vec{x}|}.$$

The exchange interactions are given by:

$$V_E^{(1)}(\vec{r}, \vec{x}) = V_E^{(3)}(\vec{r}, \vec{x}) = -\frac{\varphi_{1s+}^*(\vec{x}) \varphi_{1s+}(\vec{r}) + \varphi_{2s+}^*(\vec{x}) \varphi_{2s+}(\vec{r})}{|\vec{r} - \vec{x}|}$$

and

$$V_E^{(2)}(\vec{r}, \vec{x}) = -\frac{\varphi_{1s-}^*(\vec{x}) \varphi_{1s-}(\vec{r})}{|\vec{r} - \vec{x}|}.$$

Finally, the Hartree-Fock expression for the total binding energy of lithium is:

$$E = \sum_{\ell=1}^{3} \left[\epsilon_\ell - \frac{1}{2} \int d^3r \, |\varphi_\ell(\vec{r})|^2 V_D(\vec{r}) - \frac{1}{2} \int d^3x \, d^3y \, \varphi_\ell^*(\vec{x}) V_E^{(\ell)}(\vec{x}, \vec{y}) \varphi_\ell(\vec{y}) \right]. \tag{23.38}$$

The value of $\epsilon_1 = \epsilon_{1s+}$ is given by the ground state of the Hartree-Fock equation labelled by $\ell = 1$, while $\epsilon_3 = \epsilon_{2s+}$ is given by the first excited state. The ground state of the second Hartree-Fock equation gives $\epsilon_2 = \epsilon_{1s-}$.

The Hartree-Fock equations comprise a set of coupled nonlinear integro-differential equations. In general these can be difficult to solve. Nevertheless, they specify single particle wavefunctions and therefore drastically simplify the N-body problem (at the expense of approximation and of imposing atomic configurations). Furthermore, the equations can be robustly solved with iteration. This obviates the nonlinearity and decouples the equations, which are enormous savings at very little expense since iteration tends to converge quickly. The strategy (due to Hartree) is thus

```
// iterative solution to the Hartree-Fock equations

set initial single particle wavefunctions
//hydrogenic wavefunctions obtained by setting $V_D = V_E = 0$ are
    a good choice

while (dE > tolerance)
  compute $V_D$ and $V_E$
  solve the set of Hartree-Fock equations for $\{\varphi_j\}$
  compute the ground state energy and dE = $||E_\text{old} - E_\text{new}||$
end while
```

The Hartree-Fock method improves the single particle Ansatz in an intuitive way. It is reliable enough to account for the periodic table of the elements and the generic size and energies of atoms. It has also proven capable of explaining the spectrum of the alkalies in fair detail. Finally, Hartree-Fock single particle wavefunctions serve as the starting point for more sophisticated (and far more complicated) methods. A word of caution: the Hartree-Fock potential for an excited state of the atom is different from the ground state potential, thus excited state solutions to the Hartree-Fock equations do *not* represent excited atomic orbitals.

Because the single particle wavefunctions are fixed in the direct and exchange interactions, solving the Hartree-Fock equations is equivalent to solving a single particle Schrödinger equation and we recommend the methods of Chapter 21. In particular, it is convenient to employ a given basis and diagonalize. The result is called the **Roothaan equations**. The implementation is straightforward and we do not go into details. Specific examples are given in the exercises.

The convergence of the iterative process is illustrated in Figure 23.1. The case considered here is helium in the $1s^2$ configuration (see Ex. 6) with a point basis of size 1600—consult Section 21.2.3 for more on the point basis). The box size is taken to be 10 a.u. and the code ran in a few seconds. One of the difficulties in implementing Hartree-Fock algorithms is in evaluating two-body matrix elements. A substantial benefit of the point basis is that this problem is completely eliminated.

The left panel of Figure 23.1 shows the initial (hydrogenic) and final (Hartree-Fock) radial wavefunctions—substantial corrections are visible. The right panel displays the convergence of the direct potential in a few iterations. If the direct and exchange potentials are set to zero for the initial iteration, hydrogenic wavefunctions should be

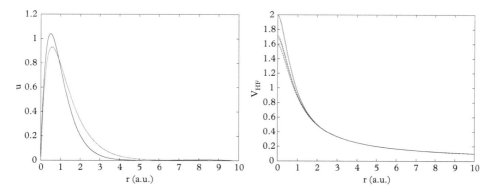

Figure 23.1 *Hartree-Fock computations with helium. Left: Hartree-Fock radial wavefunction; dashed: first iteration, dotted: Hartree-Fock result. Right: V_{HF} by iteration: solid: first, dashed: second, short dashed: third, dotted: final result.*

Table 23.2 *Ground state energies for helium by various methods.*

method	value (hartrees)
zeroth order	−4
pert. thy.	−2.75
variational	−2.8477
Hartree-Fock	−2.86163(4)
exact	−2.903724375
experimental	−2.9033

generated, which permits monitoring dependence on the selected algorithmic (basis and box size in this case) parameters. For reference, ground state energies for helium are given in Table 23.2.

The Hartree-Fock formalism can become complex when orbitals other than s-waves are considered. In this case it is useful to write the generic orbital in terms of spin, radial, and spherical harmonic wavefunctions:

$$\varphi_j(r) \to \chi(s_j) Y_{\ell_j m_j} u_{n_j \ell_j}(r)/r.$$

Similarly, the Coulomb interaction can be expanded according to the well-known formula:

$$\frac{1}{|\vec{x} - \vec{y}|} = 4\pi \sum_{LM} \frac{1}{2L+1} Y^*_{LM}(\hat{x}) Y_{LM}(\hat{y}) \frac{r^L_<}{r^{L+1}_>} \tag{23.39}$$

where $r_<$ =min(x,y) and $r_>$ =max(x,y). In general this leads to coupled equations involving various angular momenta. For example orbitals of angular momentum ℓ couple to those of angular momentum $2\ell, 2\ell-2, \ldots, 0$, which greatly complicates the numerical solution to the Hartree-Fock equations. However, in the case of closed subshells, the sums over spin and angular momentum quantum numbers are unrestricted and can be performed. Thus the spin sum simply yields a factor of two while the m_ℓ sum can be done with the aid of the addition theorem:

$$\sum_{m_\ell} |Y_{\ell m_\ell}(\hat{x})|^2 = \frac{2\ell+1}{4\pi}. \tag{23.40}$$

This relationship in turn implies that the $L=0$ term is selected from the expansion in Eq. 23.39 and that the direct interaction is (Bransden and Joachain, 2003):

$$V_D(r) = \sum_{n_j \ell_j} 2(2\ell_j+1) \int dx \frac{|u_{n_j \ell_j}(x)|^2}{r_>(r,x)} \quad \text{[closed shell average]}. \tag{23.41}$$

Similarly, the exchange potential for the ℓth wave is:

$$V_E^{(\ell)}(r,x) = -\sum_{n_i \ell_i} \sum_L \frac{2\ell_i+1}{2L+1} |\langle \ell 0 \ell_i; 0|L0\rangle|^2 \, u^*_{n_i \ell_i}(r) \, u_{n_i \ell_i}(x) \cdot \frac{r_<^L}{r_>^{L+1}}$$

$$\text{[closed shell average]}. \tag{23.42}$$

Importantly, both the direct and exchange interactions are diagonal in angular momentum, thus the closed shell Hartree-Fock potentials are central, which greatly simplifies the numerical work.

If one peruses the Hartree-Fock literature one will quickly discover an extensive series of acronyms of the sort STO-3G or 6-31G. These refer to different basis expansion methods. For example, the use of Gaussian basis functions is common, and these are often used to form a fixed (rather than variational) approximation to hydrogenic or Slater-type basis functions. At first sight, the decision to use a fixed basis expansion appears to be a curious choice as it explicitly rejects variational optimization of the basis. However, optimizing in the scale factor in the exponential of the Gaussian is nonlinear, and it is sometimes better to put that effort into straight diagonalization. Similarly, experience shows that the effort to optimize the coefficients in a STO-3G basis function is better spent enlarging the basis itself.

23.3 Molecules

An interesting and nontrivial feature of nature is that when two atoms are brought together a merged atom is not necessarily created. Rather, atoms tend to maintain their identities due to a certain "atomic rigidity". This phenomenon underpins the existence of molecules and their wonderful complexity, permitting phenomena such as phosphorescence, DNA replication, and life.

23.3.1 Adiabatic separation of scales

Atomic rigidity exists because nuclear degrees of freedom are nearly hidden by their accompanying electron clouds, which in turn occurs because of the scale separation between electronic and nuclear degrees of freedom. This concept was formalized in 1927 by Max Born and Robert Oppenheimer (Born and Oppenheimer, 1927) and underpins the *Born-Oppenheimer approximation*.

Because electrons and nuclei are bound in a molecular system they have comparable momenta (the total momentum is zero in the center of mass frame), and therefore electrons move much faster than nuclei. Electrons are called *fast degrees of freedom* and essentially "see" nuclei frozen in place. Conversely, the nuclei move in an average electron potential.

Call m the mass of the electron and M the mass of a nucleus in the molecule. The size of a molecule is denoted a and has a scale in terms of angstroms. The mass ratio is:

$$\frac{m}{M} \equiv \kappa^4 \approx 10^{-3} - 10^{-5}. \tag{23.43}$$

Typical energies associated with electronic degrees of freedom are

$$\epsilon_{\text{el}} = \frac{\hbar^2}{ma^2} \tag{23.44}$$

while those assocated with nuclear rotation are

$$\epsilon_{\text{rot}} = \frac{\hbar^2}{Ma^2} \tag{23.45}$$

and for vibrations

$$\epsilon_{\text{vib}} = \frac{\hbar^2}{\sqrt{mM}a^2}. \tag{23.46}$$

Thus one has (Messiah, 2014)

$$\epsilon_{\text{rot}} \approx \kappa^2 \epsilon_{\text{vib}} \approx \kappa^4 \epsilon_{\text{el}}. \tag{23.47}$$

Now consider the molecular Hamiltonian:

$$H = T_N + T_e + V(x, X) \tag{23.48}$$

where

$$T_N = -\sum_j \frac{\nabla_j^2}{2M_j} = -\kappa^4 \sum_j \frac{\lambda_j}{2m} \nabla_j^2 \qquad (23.49)$$

where the λ_j are constants of order unity, and

$$T_e = -\frac{1}{2m} \sum_i \nabla_i^2 \qquad (23.50)$$

Nuclear coordinates are denoted $\vec{X}_j$ and electronic ones by $\vec{x}_i$. We assume that the interaction is only a function of spatial coordinates, which implies that:

$$H_0 = T_e + V \qquad (23.51)$$

commutes with the nuclear coordinates, X_j. Thus H_0 and the nuclear coordinates can be simultaneously diagonalized and definite values can be assigned to the latter. In this way the eigenvalues depend parametrically on the coordinates:

$$H_0 \psi(\vec{x}_i, \vec{X}_j) = U_n(\vec{X}_j) \psi(\vec{x}_i, \vec{X}_j). \qquad (23.52)$$

At this stage, the molecular problem has been reduced to an analogue of the atomic problem but with substantially reduced symmetry. The problem is completed by solving for nuclear motion in what amounts to a variational Ansatz:

$$\Psi_n(x, X) = \psi_n(x; X)\phi(X). \qquad (23.53)$$

The effective nuclear Hamiltonian, H_n, can be obtained by projecting on to ψ_n. Notice that the nuclear Hamiltonian depends on the electronic configuration. Substituting, isolating the nuclear degrees of freedom, and simplifying yields:[3]

$$H_n = T_N + W_n + U_n \qquad (23.54)$$

where

$$W_n(X) = \sum_j \frac{1}{2M_j} \int dx \left(\frac{\partial \psi_n}{\partial X_j}\right)^2. \qquad (23.55)$$

---

[3] The method introduced by Born and Oppenheimer differs from this variational approach; they expanded H in powers of κ and then performed Rayleigh-Schrödinger perturbation theory.

One can estimate (Messiah, 2014) that the error in the wavefunction is order κ^3 and the error in the energy is approximately $\kappa^2 \epsilon_{rot}$. Furthermore, W_n is also small, of order ϵ_{rot}.
An alternative approach is to expand the exact wavefunction in the electronic basis:

$$\Psi(x, X) = \sum_n \psi_n(x; X)\, \phi_n(X) \tag{23.56}$$

obtaining

$$\sum_s \langle \psi_n | T_N | \psi_s \rangle\, \phi_s(X) + U_n \phi_n(X) = E \phi_n(X). \tag{23.57}$$

(the derivatives in T_N act on ϕ in this expression). Expanding the matrix element gives:

$$T_N \phi_n + U_n \phi_n + \sum_s W_{ns} \phi_s = E \phi_n \tag{23.58}$$

where

$$W_{ns} = \sum_j \frac{1}{2M_j} \left(\int \nabla \psi_n^* \cdot \nabla \psi_s + 2 \int \nabla \psi_n^* \psi_s \cdot \nabla \right). \tag{23.59}$$

This represents a set of coupled equations that is equivalent to the Schrödinger equation. In the adiabatic approximation, ψ_n varies slowly with respect to $\vec{X}_j$ and W_{ns} can be neglected, which decouples the nuclear equations. Notice that $W_{nn} = W_n$ since the second term in Eq. 23.59 vanishes in this case.

It is useful to remember that an approximation is being made, and even though it is a good approximation, important physics could be neglected depending on the problem at hand. For example, if one is attempting to understand superconductivity, neglecting nuclear recoil is a disaster since this gives rise to the phonon degrees of freedom that play such an important role in creating Cooper pairs.

23.3.2 The electronic problem

The problem of determining the electronic structure of a molecule can be very difficult. Here we illustrate the issues and methods involved by considering the simple cases of the H_2^+ ion and the H_2 molecule. For more complex molecules it is advisable to use one of the many codes developed by professionals—see (Cramer, 2004) for more information.

H_2^+ Ion

The H_2^+ ion is defined by the Hamiltonian:

$$H = -\frac{1}{2} \nabla_r^2 + \frac{1}{R} - \frac{1}{r_A} - \frac{1}{r_B}. \tag{23.60}$$

We only address the electronic portion of the Born-Oppenheimer method so nuclear kinetic energies are not considered. The electron's coordinate is denoted $\vec{r}$, the nuclei are labelled A and B and the distances between the electron and the nuclei are denoted r_A and r_B. These are written in terms of the electron coordinate and the relative coordinate between the nuclei, $\vec{R}$. Thus $r_A = |\vec{R}/2 + \vec{r}|$ and $r_B = |-\vec{R}/2 + \vec{r}|$.

A typical solution method is to construct a ***molecular orbital*** from a linear combination of atomic orbitals (LCAO). The result is dubbed a MO-LCAO. Thus one forms a LCAO that is centered about each nucleus:

$$\phi_g(r;R) = \frac{1}{2}(\psi(r_A) + \psi(r_B)) \qquad (23.61)$$

with

$$\psi(r) = \frac{1}{\sqrt{\pi}} e^{-r}. \qquad (23.62)$$

Notice that the left-right symmetry is being used to write an Ansatz that is symmetric about reflections in the plane that is perpendicular to the $\vec{R}$ axis and that intersects it at the center of mass. Similarly, an odd parity Ansatz can be constructed[4]

$$\phi_u(r;R) = \frac{1}{2}(\psi(r_A) - \psi(r_B)) \qquad (23.63)$$

It is a relatively simple process to compute the analytical variational estimate, $E_g(R)$ (see Ex. 16). The result is shown as a solid line in Figure 23.2. The exact minimum in the adiabatic potential is shown as a point. Although the MO-LCAO does reasonably well, it predicts a minimal separation that is too large and a binding energy that is too small.

One problem with the method is that as the internuclear distance approaches zero the exact wavefunction should approach

$$\psi = \sqrt{\frac{8}{\pi}} e^{-2r} \qquad (23.64)$$

rather than

$$\phi_g(r;0) = \frac{1}{\sqrt{\pi}} e^{-r}. \qquad (23.65)$$

---

[4] The subscripts refer to the German, *gerade* and *ungerade*, which mean even and odd respectively.

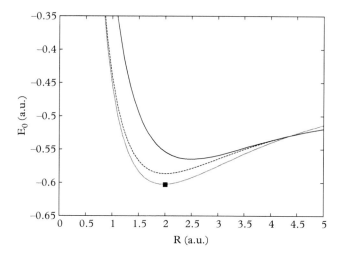

Figure 23.2 H_2^+ *Ion adiabatic potentials. MO-LCAO (solid line), MO-LCAO variational (dashed line), exact (dotted). The point represents the exact minimum at* $R = 2.003$ *a.u. and* $E_0 = -0.603$ *a.u.*

This effect can be simply accommodated by making a variational Ansatz:

$$\phi_g(r; R; \xi) = \frac{1}{2}\left(\sqrt{\frac{\xi^3}{\pi}}\, e^{-\xi r_A} + \sqrt{\frac{\xi^3}{\pi}}\, e^{-\xi r_B}\right). \tag{23.66}$$

Again, the integrals can be evaluated analytically, and the minimum found (it is $\xi_0(R_0) = 1.24$, $R_0 = 2.00$, and $E_0 = -0.5865$). The results are shown as a dashed line in the figure, with a clear improvement in performance.

Let us stop skirting the issue and attempt an exact solution to the electronic problem for the H_2^+ ion. In fact, the ion is simple enough that an exact solution can be obtained with the aid of confocal elliptic coordinates; but because this is presented in many text books, we follow the straightforward, but more numerical approach.

The most direct way in which to obtain an exact solution is to expand in spherical harmonics:

$$\psi(\vec{r}; \vec{R}) = \sum_{\ell m} \frac{1}{r} u_{\ell m}(r; R)\, Y_{\ell m}(\hat{r}). \tag{23.67}$$

Similarly, the interaction can be expanded and one obtains a coupled system of equations:

$$-\frac{1}{2}\left(\frac{\partial^2}{\partial r^2} + \frac{\ell(\ell+1)}{r^2}\right) u_{\ell m}(r; R) + \sum_L V_L^{\ell,m}(r, R)\, u_{Lm}(r; R) = \epsilon\, u_{\ell m}(r; R) \tag{23.68}$$

where

$$V_L^{\ell,m}(r,R) = \sum_a \frac{r_<^a}{r_>^{a+1}} \sqrt{\frac{2L+1}{2\ell+1}} \langle a0L0|\ell 0\rangle \langle a0Lm|\ell m\rangle. \quad (23.69)$$

Notice that only even waves contribute to the ground state, the ground state has $m=0$, and the interaction is diagonal in magnetic quantum numbers (see Ex. 17). The expansion converges relatively quickly for internuclear distances near the preferred bond length of 2.003 a.u. The ground state energy computed in the point basis for the radial wavefunctions, with up to eight partial waves, is shown as a dotted line in Figure 23.2. The black point represents the exact minimum at $R = 2.003$ a.u. with a energy of -0.603 a.u. Evidently, the agreement is quite good, which should please us. More ominously, the exact computation was fairly demanding, even though we have been dealing with the simplest possible molecule.

H_2 Molecule

The Hamiltonian for the H_2 molecule (in atomic units) reads:

$$H = -\frac{1}{2}\nabla_1^2 - \frac{1}{2}\nabla_2^2 + \frac{1}{R} - \frac{1}{r_{1A}} - \frac{1}{r_{1B}} - \frac{1}{r_{2A}} - \frac{1}{r_{2B}} + \frac{1}{r_{12}}. \quad (23.70)$$

The nuclei are denoted by A and B and have a relative coordinate $\vec{R}$. The electrons are denoted 1 and 2 and interact with each nucleus ($1/r_{iN}$ terms) and each other ($1/r_{12}$ term).

As with the H_2^+ ion, the simplest approach that yields a reasonable estimate of the ground state energy is to form MO-LCAO states. Since antisymmetry in the electron coordinates must be implemented, six possible (including the three possible spin-one states) MO-LCAO Ansätze exist:

$$\varphi_A(1,2) = \phi_g(1)\phi_g(2)\chi_{00},$$
$$\varphi_B(1,2) = \phi_u(1)\phi_u(2)\chi_{00},$$
$$\varphi_C(1,2) = \frac{1}{\sqrt{2}}\left(\phi_g(1)\phi_u(2) + \phi_u(1)\phi_g(2)\right)\chi_{00},$$
$$\varphi_D^{(M)}(1,2) = \frac{1}{\sqrt{2}}\left(\phi_g(1)\phi_u(2) - \phi_u(1)\phi_g(2)\right)\chi_{1M}. \quad (23.71)$$

The spin wavefunctions are:

$$\chi_{00} = \frac{1}{\sqrt{2}}\left(\alpha(1)\beta(2) - \beta(1)\alpha(2)\right) \quad (23.72)$$

and

$$\chi_{11} = \alpha(1)\alpha(2), \qquad \chi_{10} = \frac{1}{\sqrt{2}}\left(\alpha(1)\beta(2) + \beta(1)\alpha(2)\right), \qquad \chi_{1-1} = \beta(1)\beta(2).$$
(23.73)

The traditional notation wherein a spin up (down) state is denoted α (β) is being followed here.

Several options are available at this stage. For example, one could form the six-by-six matrix $\langle \varphi_I | H | \varphi_J \rangle$ (which will be block diagonal by total spin) and diagonalize. Here we follow the H_2^+ ion strategy and guess that φ_A minimizes the energy. This yields a variational upper bound:

$$E_A = 2\epsilon_g(R) - \frac{1}{R} + \frac{1}{D}\int d^3 r_1 d^3 r_2 \, \frac{|\phi_g(1)\phi_g(2)|^2}{r_{12}}$$
(23.74)

with

$$D = \int d^3 r_1 d^3 r_2 \, |\phi_g(1)\phi_g(2)|^2.$$
(23.75)

Although analytic methods can be applied to the Coulomb integral, this is a numerical text and we choose to evaluate the five-dimensional integral using the methods of Chapter 5. The result is the solid curve shown in Figure 23.3. See Ex. 19 for details on evaluating the integral.

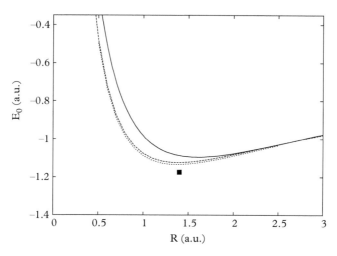

Figure 23.3 H_2 Molecule adiabatic potentials. MO-LCAO (solid line), MO-LCAO variational (dashed line), Hartree-Fock (dotted line). The point represents the exact minimum at $R = 1.4011$ a.u. and $E_0 = -1.1745$ a.u.

Again, it is simple to extend the method with the introduction of a single variational parameter, as in Eq. 23.66. The resulting adiabatic surface is the dashed line of Figure 23.3. In this case the minimum is at $R_0 = 1.3875$ a.u. with a variational parameter of $\xi_0 = 1.180$, yielding $E_0 = -1.1182$ a.u.

The Hartree-Fock computation is more involved than that of helium because the interaction is non-central. The Ansatz wavefunction is φ_A of Eq. 23.71 where ϕ_g is to be determined. In this case the single particle Hartree-Fock equation is:

$$\left[-\frac{1}{2}\nabla_r^2 - \frac{1}{|\vec{R}/2 + \vec{r}|} - \frac{1}{|\vec{R}/2 - \vec{r}|} + V_D(\vec{r})\right]\phi(\vec{r};\vec{R}) = \epsilon\,\phi(\vec{r};\vec{R}). \qquad (23.76)$$

The exchange interaction is zero and the direct interaction is given by:

$$V_D(\vec{r}) = \int d^3x \, \frac{|\phi(\vec{x};\vec{R})|^2}{|\vec{r} - \vec{x}|}. \qquad (23.77)$$

Finally, the total ground state energy is given by:

$$E_0 = 2\epsilon_0 - \int d^3x\,d^3y \, \frac{|\phi(\vec{x};\vec{R})|^2 \, |\phi(\vec{y};\vec{R})|^2}{|\vec{x} - \vec{y}|}. \qquad (23.78)$$

The exact solution is given in terms of a partial wave expansion as in Eq. 23.67; however, in this case the resulting equation is not diagonal in magnetic quantum numbers and all waves must be included. The problem is thus rather daunting. Nevertheless, numerical exploration indicates that nonzero magnetic quantum numbers contribute negligibly. In this case the computation simplifies dramatically and one obtains the dotted line of Figure 23.3. The result is obtained by summing waves up to $\ell = 6$ and yields $R_0 = 1.40$ a.u. and $E_0 = -1.1336$ a.u. This should be compared to the black point, which shows the 'exact' solution obtained with an elaborate large scale computation. Some of these results are given in Table 23.3. The acronyms therein refer to the Hartree-Fock method (HF) and the configuration interaction method (CI). The last column gives the electronic dissociation energy, $D - e$, defined by twice the atomic energy minus the minimal nuclear energy. In this case $D_e = -1 - E_0$.

Table 23.3 H_2 molecule properties. (*) see (Kolos and Wolniewicz, 1968).

method	R_0 (a.u.)	E_0 (a.u.)	D_e (eV)
variational	1.387	−1.118	3.216
HF	1.40	−1.1336	3.636
CI	1.4013	−1.16696	4.54306
precise	1.4011	−1.17447	4.74759 (*)

More complicated molecules

The computation just presented for H_2 is often called working in the ***Hartree-Fock limit*** because it solves (modulo truncations) the Hartree-Fock problem exactly. In general, Hartree-Fock computations are sufficiently complex that additional approximations are often made. For example, a typical basis employs S-wave functions centered on nuclear coordinates. In the case of H_2 these can be:

$$\{u(\vec{r}-\vec{R}/2)\} \cup \{u(\vec{r}+\vec{R}/2)\} \tag{23.79}$$

where the basis functions are, say, various Gaussians. This basis is not complete because it fixes the functional form of the angular dependence to match that of the radial dependence. Nevertheless, reasonably accurate results can be obtained (albeit, still in the single particle approximation).

It is clear that an efficient basis is crucial to making accurate computations for more complex molecules. An obvious choice is a Slater-type basis such as:

$$\{\exp(-\beta(n)r)\}, \tag{23.80}$$

where $\beta(n)$ varies with n in a fixed fashion, or

$$\{r^n \exp(-\beta r)\}. \tag{23.81}$$

The problem with these choices is that correlation integrals, $\langle \phi_i \phi_j | \frac{1}{r_{12}} | \phi_\ell \phi_m \rangle$, are not easy to evaluate. Attempts to evaluate them numerically can lead to trouble for larger bases because errors in normalization appear in the denominator of expressions for eigenvalues.

Alternatively, Gaussian functions are an attractive option because they permit analytic computation of many integrals. This relies on the observation that the product of two Gaussians that are centered at different points is another Gaussian:

$$\exp[-\alpha(\vec{r}-\vec{A})^2] \exp[-\beta(\vec{r}-\vec{B})^2] = \exp[-\mu(\vec{A}-\vec{B})^2] \exp[-(\alpha+\beta)(\vec{r}-\vec{C})^2] \tag{23.82}$$

where

$$\mu = \frac{\alpha\beta}{\alpha+\beta} \tag{23.83}$$

and

$$\vec{C} = \frac{\alpha}{\alpha+\beta}\vec{A} + \frac{\beta}{\alpha+\beta}\vec{B}. \tag{23.84}$$

A natural way to leverage this is to form a basis such as:

$$\{\exp(-\beta(n)r^2)\} \tag{23.85}$$

but this is problematic because elements in the normalization matrix are similar, which means that the determinant of the normalization matrix is very small, which in turn amplifies errors in the computation.

As in atomic physics, a variant that combines the latter approaches is to approximate a Slater basis function with a sum over Gaussians with fixed strengths and scale factors. Modern bases are of this type or a Cartesian variant of the Gaussian basis:

$$\Psi(\vec{r}; \vec{R}_A \ldots \vec{R}_N) = \sum_{\ell m n} a_{\ell m n}(\vec{R}_A \ldots \vec{R}_N)\, x^\ell\, y^m\, z^n\, \exp(-\beta r^2). \tag{23.86}$$

An alternative sticks with S-wave Gaussians and approximates higher waves by considering differences of Gaussians centered at nearby points (Szabo and Ostlund, 1996).

A number of tricks are used to assist the convergence of the Hartree-Fock iteration. These include over-relaxation (see Chapter 19), a variety of improved initial guesses (Hückel theory can be usefully employed here; see Cramer (2004)), or using MOs from neighboring molecular geometries. Finally, Hartree introduced the device of artificially increasing Z to improve convergence and then relaxing it to its nominal value as the computation proceeds.

23.4 Density functional theory

The Hartree-Fock formalism and its extensions are able to reliably compute properties of systems of dozens of particles. Beyond this our aspirations must be ameliorated or the approximations must become more severe. We know that classical physics is the effective theory that describes nature as the number of particles becomes very large—what is required is an approximation that interpolates the quantum and classical regimes. A popular realization of this approximation is called *density functional theory*, which is the topic of this section.

The genesis of density functional theory (DFT) lies in a remarkable theorem due to Pierre Hohenberg and Walter Kohn (1964), namely, many ground state properties of a many-electron system can be uniquely determined from its electron density, $n(\vec{r})$. At first glance this hardly seems credible since the theorem claims that all the information carried in $\psi_0(\vec{r}_1, \ldots, \vec{r}_N)$ is contained in a function of three variables. Plausibility can be recovered once it is realized that the ground state function is not arbitrary but is that function which minimizes $\langle \psi | H | \psi \rangle / \langle \psi | \psi \rangle$. From another perspective, Kato's theorem (1957) states that for Coulombic systems the electronic density has a cusp at the location of nuclear charges:

$$Z_k = -\frac{a_0}{2n(r)} \frac{dn}{dr}\bigg|_{r \to R_k}$$

where a_0 is the Bohr radius. Thus the density permits determining the charges and locations of all the nuclei in the system, while the integral of n gives the number of electrons. We see that the entire Hamiltonian is fixed by the density, and, in principle, all ground state properties are available.

We shall consider the N-electron Born-Oppenheimer Hamiltonian in the following:

$$H = -\frac{1}{2}\sum_{j=1}^{N}\nabla_j^2 + \frac{1}{2}\sum_{i\neq j}\frac{1}{|\vec{r}_i - \vec{r}_j|} + \sum_{i=1}^{N}\sum_{k=1}^{K}\frac{Z_k}{|\vec{r}_i - \vec{R}_k|} \equiv T + U + V_{ext}.$$

As usual we work in atomic units, $\vec{R}_k$ denotes the (fixed) position of a nucleus of charge Z_k, and $\vec{r}_i$ is the position of the ith electron.

Following the lead of Hohenberg and Kohn, we define a functional of the electronic density:

$$E[n] = \frac{\langle\psi|T + U + V_{ext}|\psi\rangle}{\langle\psi|\psi\rangle}\bigg|_{\psi|n} \equiv T[n] + U[n] + V[n] \equiv F[n] + V[n]. \quad (23.87)$$

The subscript $\psi|n$ means that the functional is defined over all wavefunctions subject to the condition

$$n(r) = N\int d^3r_2\ldots d^3r_N|\psi(r, r_2, \ldots, r_N)|^2. \quad (23.88)$$

The external nuclear potential is a one-body operator (see Eq. 23.15) and therefore can be immediately written in terms of the density as

$$V[n] = \int d^3r\, n(r)V_{ext}(r).$$

Thus $V[n]$ contains all of the information that specifies the unique character of the problem at hand (i.e., it differentiates metallic crystals from argon from β-hydroxybutyric acid). Alternatively, the functional $F[n]$ is *universal* in the sense that its only parametric dependence is on the number of electrons. In fact N-dependence is subsumed into the density, and $F[n]$ is equally applicable to atoms, molecules, or solids.

Evidently our goal is to minimize the energy functional subject to the constraint $\int n(r) = N$, determine n_0, and then compute other observables,

$$\mathcal{O}[n_0] = \frac{\langle\psi|\mathcal{O}|\psi\rangle}{\langle\psi|\psi\rangle}\bigg|_{\psi|n_0}.$$

In practice, this avenue to observables is difficult because the functional $\mathcal{O}[n]$ is unknown. However, the ground state energy is readily computable (given $E[n]$), as is the nuclear potential, which gives access to molecular geometry, charge distributions, lattice constants, compressibilities, bulk moduli, etc. Furthermore, comparing systems of different sizes permits computing dissociation energies, electron affinities, and ionization energies.

A serious obstacle is that the route to these riches passes through the functional $F[n]$, which is not known. The standard way to deal with this is to approximate F as the sum of two terms:

$$F[n] \approx F_{KS}[n] + U_{es}[n],$$

where the second term is the electrostatic energy:

$$U_{es}[n] = \frac{1}{2} \int d^3x d^3y \, \frac{n(x)n(y)}{|\vec{x}-\vec{y}|},$$

and F_{KS} is the **Kohn-Sham functional**, which is constructed from a single-particle Hamiltonian in a way that will be specified shortly.

It is traditional to "correct" this approximation by placing all of the missing elements into another unknown functional, dubbed $E_{xc}[n]$, and known as the *exchange-correlation functional*:

$$E[n] = F[n] + V[n] = F_{KS}[n] + U_{es}[n] + V[n] + E_{xc}[n]. \qquad (23.89)$$

At this point the reader may have the feeling that some sort of swindle is happening. We are required to minimize an unknown functional that is being rewritten in terms of known functionals and an unknown functional! Fortunately, the situation is not as bad as it appears: recall that F, and therefore E_{xc}, is universal and can therefore be determined in a well-understood system and applied elsewhere. Furthermore, there is substantial and well-established evidence that the exchange-correlation energy is small.

A technical issue with the method is that it is not known that a potential exists for every possible electron density. This is referred to as *v-representabilty*. Alternatively, any density can always be written as an integral over a wavefunction, as in Eq. 23.88. This is called *N-representability*.

Kohn and Sham (1965) suggested that it is useful to make a *local density approximation* (LDA) for the exchange-correlation functional:

$$E_{xc}[n] \approx \int d^3r \, n(r) \, \varepsilon_{xc}(n)$$

where $\epsilon_{xc}(n)$ is the exchange-correlation energy as obtained from a uniform electron gas of density n. This function has been computed accurately with Monte Carlo techniques and can be parameterized as some convenient function of density. Another common approach is to split the correlation and exchange corrections:

$$E_{xc} = E_x[n] + E_c[n].$$

The local density approximation is rather restrictive as it ignores functional dependence on the derivatives of the density or more complicated expressions such as:

$$E_{xc}[n] = \int d^3x\, d^3y\, f(n(x), n(y)).$$

Nevertheless, the approximation is remarkably accurate for many purposes. Applications and extensions will be discussed in the next few pages.

23.4.1 The Kohn-Sham procedure

The Kohn-Sham functional is an approximation to the (unknown) kinetic energy functional $T[n]$ and is defined as a sum over single-particle wavefunctions:

$$T[n] \approx F_{KS}[n] = -\frac{1}{2}\sum_{i=1}^{N} \langle \varphi_i | \nabla^2 | \varphi_i \rangle.$$

Of course:

$$n(r) = \sum_i |\varphi_i(r)|^2. \tag{23.90}$$

An equation for the Kohn-Sham single particle orbitals can be obtained by differentiating $E[n]$ with respect to $\varphi^*(r)$ subject to $\int n = N$. Using

$$\frac{\delta n(x)}{\delta \varphi_i^*(r)} = \varphi_i(x)\delta(\vec{r} - \vec{x})$$

and recalling that φ^* and φ are treated as independent variables when differentiating gives

$$\left[-\frac{1}{2}\nabla^2 + \phi(r) + V_{ext}(r) + \varepsilon_{xc}(n) + n(r)\frac{\partial \varepsilon_{xc}}{\partial n} \right]\varphi_i(r) = \epsilon_i \varphi_i(r) \tag{23.91}$$

with

$$\phi(r) = \int d^3x\, \frac{n(x)}{|\vec{r} - \vec{x}|}.$$

The similarity to the Hartree-Fock method is evident, and the same solution method can be employed. In particular, one can choose one of the Hartree-Fock bases and iterate the equations until convergence. An algorithm is:

```
select initial n
iterate
   form v_eff(n)
   solve for the Kohn-Sham orbitals
   obtain n
end
compute observables
```

Once the iteration has converged one knows the single particle orbitals and the commensurate ground state electron density. A convenient expression for the ground state energy combines Eqs. 23.89 and 23.91 to obtain

$$E_0 = \sum_i \epsilon_i - U_{es}[n_0] - \int d^3r\, n_0^2(r) \frac{\partial \varepsilon_{xc}}{\partial n}\bigg|_{n=n_0}. \tag{23.92}$$

23.4.2 DFT in practice

It is possible to forego the Kohn-Sham procedure and construct other approximations to the energy functional. For example, the perturbative expression for the one-body kinetic energy operator in the degenerate electron gas (this is a model with N interacting electrons on a positive background of charge $+eN$) leads to the approximation

$$T_D = \frac{3}{10m}(3\pi^2)^{2/3} \int d^3r\, n^{5/3}.$$

A factor of 2 for spin has been included in this expression. Minimizing

$$E = T_D[n] + U_{es}[n]$$

gives rise to the famous **Thomas-Fermi** statistical model of atoms. This is explored in Ex. 11.

A similar perturbative calculation, first made by Bloch in 1929, yields the following expression for the exchange energy:

$$E_x = -\frac{3}{4}\left(\frac{3}{\pi}\right)^{1/3} \int d^3r\, n^{4/3}. \tag{23.93}$$

Slater (1951) has suggested using this expression with a correction factor of 1.05.

Over the years an industry has developed with the goal of fitting functions to numerical computations of properties of the degenerate electron gas. As a result a plethora of functional forms exist for E_{xc}, E_c, and E_x. We seek a correlation functional to add to

E_x; the simplest local form that fits Monte Carlo data we have been able to find is due to Chachiyo (2016):

$$E_c = \int d^3r\, n(r)\varepsilon_c(n)$$

with

$$\varepsilon_c = a \log\left(1 + \frac{b}{r_s} + \frac{b}{r_s^2}\right).$$

The parameters are $a = (\log(2) - 1)/(2\pi^2)$ and $b = 20.4562557$.

An alternative approach is to fit the exchange and correlation functionals together. Of the many models that exist, we present one due to Gunnarsson and Lundqvist (1976) that is relatively compact.

$$E_{xc} = \int d^3r\, n(r)\varepsilon_{xc}(n)$$

with

$$\varepsilon_{xc}(n) = -\frac{0.0402}{x} - 0.0333\, g(x),$$

where

$$g(x) = (1 + x^3)\log\left(1 + \frac{1}{x}\right) - x^2 + \frac{1}{2}x - \frac{1}{3}.$$

Here $x = r_s/11.4$ where r_s is the *Wigner-Seitz radius* given by $r_s = r_0/a_0$ and r_0 is the average interparticle spacing defined by $1 = (4/3)\pi r_0^3 n$. The reader can consult MacLaren et al. (1991) or Parr and Yang (1989) for a list of other local density functionals.

Table 23.4 gives the ground state energy of helium for all of the functionals mentioned above. These results were computed with Hartree-Fock methods using the discretization method of Chapter 21 and extrapolated to the continuum limit. In practice over-relaxation is often implemented in updating the density to avoid oscillations in the solution. Run times were never longer than a few seconds. One sees quite good agreement with experiment, even though helium is about as far removed from a uniform electronic system as possible.

Table 23.4 *Ground state energy for helium (a.u.)*

Gunnarsson-Lundqvist	Slater	Slater + Chachiyo	Hartree-Fock	expt
−2.86024	−2.76635	−2.87457	−2.8616	−2.9033

830 Density functional theory

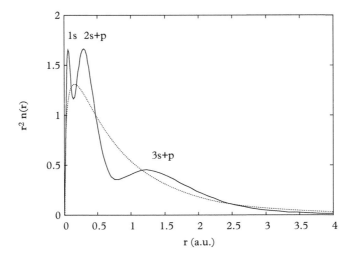

Figure 23.4 *Thomas-Fermi (dotted line) and density functional (solid line) charge densities for argon. After (Argaman and Makov, 2000).*

A more complex computation of the electron density for argon is displayed in Figure 23.4, showing clear shell structure. This level of detail is not reproduced in the Thomas-Fermi model, which is shown as the dotted line. The computed DFT ground-state energy is −525.9 hartrees to be compared to the Hartree-Fock energy of −526.8 hartrees and the experimental value of −527.6 hartrees.

23.4.3 Further developments

For large molecules, density functional theory is essentially the *only* viable tool available. Diagonalization methods grow exponentially in complexity with the number of electrons, whereas the complexity grows as approximately N^3 in the density functional approach (Argaman and Makov, 2000). As a result, properties of molecules with hundreds of atoms can be reasonably computed. The additional local density approximation gives very good results for many quantum problems, including atoms, molecules, and some solid state systems.

In spite of these successes, the LDA does not give the proper dissociation limit of diatomic molecules, and can fail for strongly correlated systems or for systems with van der Waals intermolecular forces. Unfortunately, the latter issue means that biological molecules are difficult to treat with the density functional formalism.

Various strategies for improving the formalism have been implemented over the years. A simple improvement is to allow for spin-dependence in the functionals, thus (Vosko *et al.*, 1980)

$$\int d^3 r f(n) \to F[n_\uparrow, n_\downarrow].$$

This mirrors the additional degrees of freedom that are available to the Hartree-Fock method when open shells are considered.

An obvious additional extension is to permit the functionals to depend on derivatives of the densities

$$\int d^3 r f(n) \to F[n, \nabla n, \nabla^2 n, \ldots].$$

It transpires that too naive an approach to this generalization worsens the agreement with experiment (this happenstance is common in our experience!). However, a more careful development that respects known properties of the exchange-correlation functional is able to improve the accuracy of the density functional approach. This class of functionals is called the ***generalized gradient approximation*** (Perdew *et al.*, 1992).

23.5 Conclusions

The inhomogenous and anisotropic environment created by a typical molecule presents daunting challenges to attempts to describe them quantum mechanically. We are therefore fortunate that nature has simplified the problem somewhat, chiefly through the large scale difference, $m_e/m_p \ll 1$ and because inner electrons are screened and become "inert". These facts engender an extensive and very successful phenomenology that chemistry students spend much time mastering. They also make it possible to accurately (in most cases) represent the quantum mechanical properties of a system with an effective single-particle approximation. The classic exemplars of this approach are the Hartree-Fock and density functional methods, and we are happy to recommend them as starting points.

It is possible, and instructive, to write your own code to obtain properties of simpler atomic and molecular systems. For more complicated ones it is advisable to employ one of dozens of available professional software packages. See Young (2001) or the wikipedia page

`List_of_quantum_chemistry_and_solid-state_physics_software`

for much more information.

Finally, if these methods do not yield sufficiently accurate results, employing coupled cluster, configuration interaction, or perturbative techniques are your best bet for progress. These are mature technologies and one is well-advised to consult the literature before expending serious effort. Try Fetter and Walecka (2003), Cramer (2004), or Szabo and Ostlund (1996).

23.6 Appendix: Beyond Hartree-Fock

Several methods for improving the accuracy of Hartree-Fock computations are used in the theoretical chemistry community. We will briefly discuss three of these in this appendix.

The most direct method to move beyond the Hartree-Fock approximation is to combine the Hartree-Fock and LCAO methods by solving for single particle wavefunctions for a set of configurations and then diagonalizing the Hamiltonian in this basis. As before, the formalism can become tedious because it must be tailored to the atom or molecule under consideration.

An alternative is to leverage the relative accuracy of the Hartree-Fock method by employing perturbation theory around the Hartree-Fock Hamiltonian. This is known as **Møller-Plesset perturbation theory**. Thus one performs Raleigh-Schrödinger perturbation theory in:

$$\mathcal{V} = V - V_{HF} \tag{23.94}$$

taking

$$\mathcal{H}_0 = H_0 + V_{HF}. \tag{23.95}$$

A more elegant approach is to employ the power of the **second quantization formalism**. A brief review of this formalism is presented here; the reader is referred to Fetter and Walecka (2003) for an excellent introduction to the subject and for more information and applications.

Amongst the advantages of second quantization are that statistics (antisymmetrization) is automatically incorporated and an arbitrary number of bodies can be considered (including $N \to \infty$). The latter feature is gained by relabeling states in terms of **occupation numbers** rather than quantum numbers. Thus a state labelled, say,

$$\mathcal{A}\varphi_1(q_1)\varphi_2(q_2)\varphi_4(q_3) \tag{23.96}$$

is written as

$$\langle q_1 q_2 q_3 | 1, 2, 4 \rangle. \tag{23.97}$$

This is interpreted as meaning there is one fermion in state "1", one in state "2", and one in state "4". All other states have no fermions in them. Creation and annihilation operators are introduced to construct the new states from the vacuum. For example,

$$|1, 2, 4\rangle = a_1^\dagger a_2^\dagger a_4^\dagger |0\rangle. \tag{23.98}$$

Quantization for Fermi-Dirac statistics is imposed with the anticommutation relations:

$$\{a_i, a_j^\dagger\} = \delta_{ij} \quad \text{and} \quad \{a_i, a_j\} = 0. \tag{23.99}$$

(as usual, we have set $\hbar = 1$.)

With this formalism in place, the second quantized Hamiltonian is

$$H = \sum_{ij} \langle i|H_0|j\rangle \, a_i^\dagger a_j + \frac{1}{2} \sum_{ijk\ell} \langle ij|V|k\ell\rangle \, a_i^\dagger a_j^\dagger a_\ell a_k. \qquad (23.100)$$

where one- and two-body matrix elements follow the same form as throughout this chapter:

$$\langle i|H_0|j\rangle = \int d^3x \, \varphi_i^\dagger(x) H_0(x) \varphi_j(x) \qquad (23.101)$$

and

$$\langle ij|V|k\ell\rangle = \int d^3x \, d^3y \, \varphi_i^\dagger(x) \varphi_j^\dagger(y) \, V(\vec{x}-\vec{y}) \, \varphi_k(x) \varphi_\ell(y). \qquad (23.102)$$

The indices i and j refer to single particle quantum numbers, as in the Hartree-Fock formalism.

A number of beautiful results lend the second quantized formalism its power. An early theorem due to Brueckner and Goldstone (1959) is a proof that perturbation theory is *extensive* (or exhibits *size-consistency*), meaning that terms in the perturbative expansion of any observable are proportional to the number of particles. This property is important because one is often interested in the *difference* between matrix elements of different sized systems. Size-consistency is expressed by saying that only *connected* expressions contribute to, say, eigenenergies.

Computations in the operator formalism are made dramatically simpler with the use of *Wick's theorem*, (which simplifies the process of computing with anticommuting operators) and made simpler again with diagrammatic methods. The interested reader is directed to Fetter and Walecka (2003) for more information.

Expressions for Green functions and other properties of many-body systems have a graphical representation known as *Goldstone diagrams*. Two such diagrams, representing direct and exchange contributions to the ground state energy (Eq. 23.17), are shown in Figure 23.5. Two diagrams contribute to second order in perturbation theory; these are shown in Figure 23.6. These diagrams evaluate to the second order corrections:

$$E_D^{(2)} = \frac{1}{2} \sum_{ij} \sum_{ab} \frac{\langle ij|V|ab\rangle \langle ab|V|ij\rangle}{\epsilon_i + \epsilon_j - \epsilon_a - \epsilon_b} \qquad (23.103)$$

and

$$E_E^{(2)} = \frac{1}{2} \sum_{ij} \sum_{ab} \frac{\langle ij|V|ab\rangle \langle ab|V|ji\rangle}{\epsilon_i + \epsilon_j - \epsilon_a - \epsilon_b}. \qquad (23.104)$$

834   *Appendix: Beyond Hartree-Fock*

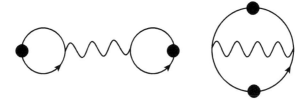

Figure 23.5 *Hartree and Fock diagrams contributing to the vacuum energy.*

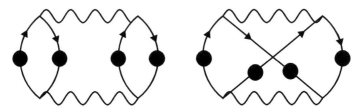

Figure 23.6 *Second order contributions to the vacuum energy.*

The indices i and j refer to occupied orbitals (holes), while a and b refer to virtual orbitals (particles).

It is possible to go to higher order in perturbation theory using the diagrammatic formalism, however the number of diagrams grows exponentially with the order of the expansion, and the method quickly becomes unwieldy. One of the great advantages of the diagrammatic formalism, however, is that it is possible to sum classes of diagrams to all orders in perturbation theory, thereby obtaining nonperturbative, albeit approximate, information.

A convenient way to achieve this is in terms of the **Green function**, which represents the amplitude for a particle to be created at one spacetime point and annihilated at another. Equivalently, a Green function is the inverse of a Hamiltonian:

$$G(\vec{x}, \vec{x}'; \omega) = \int d(t-t')\, e^{i\omega(t-t')} \langle x, t | \frac{1}{i\hbar \partial/\partial t - H} | y, t' \rangle. \qquad (23.105)$$

Green function can be used to obtain the ground state energy and thus diagrammatic methods can also be applied to this quantity. In the diagrammatic formalism a Green function is denoted with a line whose ends correspond to the spacetime points $(\vec{x}, t)$ and $(\vec{x}', t')$. Figure 23.7 displays an equation for the (inverse of) the Green function (denoted by a line with a dot) in terms of the perturbative Green function (a line without a dot) and two other diagrams. The equation involves the full Green function on the right hand side, and therefore must be solved to obtain a self-consistent solution. This solution corresponds precisely to the Hartree-Fock equations considered in this chapter.

Self-consistent equations of this sort sum infinitely many diagrams of perturbation theory and are therefore very powerful. Choosing which diagrams to sum is an art that is strongly dependent on the system under study. For example, Coulomb screening

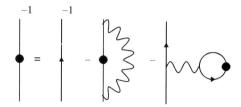

Figure 23.7 *Hartree-Fock equations in terms of the Green function.*

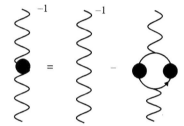

Figure 23.8 *Vacuum polarization corrections to the interaction.*

Table 23.5 *Hartree-Fock energies with second and third order corrections for simple systems. (Wilson, 1983). Atomic units.*

system	E_{HF}	E_2	E_3
Ne	−128.54045	−0.210784	−0.000753
Ar	−526.80194	−0.149200	−0.011657
N$_2$	−108.97684	−0.326887	−0.009201
CO	−112.77551	−0.300509	−0.005083

effects can be incorporated by replacing the interaction in Figure 23.7 with a ***dressed*** interaction obtained by solving an equation such as in Figure 23.8; this is called the ***ring approximation***. While resummation can be powerful, the student is warned that elaborations such as this introduce frequency dependence into the Hartree-Fock equations, which greatly complicates their solution.

Corrections to Hartree-Fock energies tend to be small. This is illustrated in Table 23.5, which shows second and third order corrections to Hartree-Fock ground state energies for a variety of small systems.

23.7 Exercises

1. [T] Electron Density.
   Show that the two expressions given for the electron density (Eqs. 23.88 and 23.90) agree for a product wavefunction.

2. [T] Weizsäcker Functional.
   Argue that the kinetic energy functional for a single electron can be written as:
   $$T = \frac{1}{8} \int d^3r \, \frac{|\nabla n|^2}{n(r)}.$$

3. [T] Roothaan's Relationship.
   Prove that
   $$\frac{1}{R} \frac{dR}{dr}\Big|_{r=0} = -Z$$
   where R is the single particle Hartree-Fock radial wavefunction for an atomic system.

4. [T] Yellowcake.
   Construct the analogue of the Born-Oppenheimer Hamiltonian for purified yellowcake (UO_2) using a two-step adiabatic approximation.

5. [T] LCAO Matrix Elements.
   The LCAO method requires evaluating matrix elements of the type $\langle \beta|H|\alpha \rangle$ where α and β refer to atomic configurations. Show that this matrix element is zero if the configurations differ by promoting a single electron into an excited orbital. This is known as **Brillouin's theorem**.

6. [T] Analytical Helium.

   a) Show that the perturbative ground state energy of helium is given by:
   $$E_0 \approx -Z^2 + \frac{5}{8}Z.$$

   b) By varying the range parameter in the ground state hydrogenic wavefunction obtain the variational upper limit:
   $$E_0 < -\left(Z - \frac{5}{16}\right)^2.$$

7. Hartree-Fock Helium.

   a) Show that the Hartree-Fock equation for helium reduces to:

$$\left[-\frac{1}{2}\nabla^2 - \frac{Z}{r} + V_{HF}(r)\right]\varphi(r) = \epsilon\varphi(r)$$

where

$$V_{HF}(r) = \int d^3x \frac{|\varphi(x)|^2}{|\vec{x}-\vec{r}|}.$$

Simplify this equation for S-waves.

b) Obtain the analytic form of the Hartree-Fock potential after one iteration (assuming one starts with $V_{HF} = 0$).

c) Obtain the Hartree-Fock ground state energy of helium using the point basis method and compare to the exact result.

8. Hartree-Fock Helium in a Gaussian Basis.
   Solve the Hartree-Fock equations for helium in a simple Gaussian basis:

$$\langle r|n\rangle \propto \exp(-\alpha_n r^2).$$

a) Expand the Hartree-Fock orbital as $\phi = \sum_{n=1}^{4} c_n \exp(-\alpha_n r^2)$. You should get something like:

$$\sum_{nm} T_{nm}c_m + \sum_{pqnm} V_{npmq}c_p c_q c_m = \epsilon \sum_{nm} N_{nm}c_m.$$

Provide expressions for T, V and N.

b) Find the expression for the ground state energy in terms of the c_n.

c) You have obtained the Roothaan equations. As usual, they can be solved by iteration. You will want to find a package that can deal with the generalized eigenvalue problem $H\vec{v} = \epsilon N\vec{v}$. In Eigen try

```
#include <Eigen/Eigenvalues>
...
GeneralizedEigenSolver<MatrixXf> ges;
MatrixXf H = ...
MatrixXf N = ...
ges.compute(H, N);
cout << "Eigenvalues: " << ges.eigenvalues().
    transpose();
```

Code up and solve to obtain the ground state energy. Try using $\alpha_n = 0.1 \cdot \exp(-n/0.7)$. Compare to the Hartree-Fock value of -2.8616 a.u.

9. **Bases.**
   Examine alternative bases by solving for the hydrogenic and Hartree-Fock ground state energies for helium. Use:

   a) $\langle r|n\rangle = \delta(r - r_n)$
   b) $\langle r|n\rangle = r^n \exp(\alpha r^2)$
   c) $\langle r|n\rangle = \exp(\alpha n r^2)$

10. **LDA Helium.**
    Write code to evaluate the ground state energy and single particle eigenvalues for helium in the density functional formalism in the following models.

    a) $E[n] = F_{KS} + U_{es} + V$.
    b) $E[n] = F_{KS} + U_{es} + V + E_x$ where E_x is the Bloch exchange functional of Eq. 23.93.
    c) $E[n] = F_{KS} + U_{es} + V + E_x^\alpha$ where E_x^α is the Slater variant of the Bloch exchange functional.
    d) $E[n] = F_{KS} + U_{es} + V + E_x^\alpha + E_c$ where E_c is the Chachiyo correlation functional.

    Compare all your results to the Hartree-Fock and exact values.

11. **[P] The Thomas-Fermi Model.**
    We derive the Thomas-Fermi model from the simplest energy functional mentioned in Section 23.4.2:

    $$E[n] = \frac{3}{10m}(3\pi^2)^{2/3} \int d^3r\, n^{5/3} + \frac{1}{2}\int \frac{n(x)n(y)}{|\vec{x}-\vec{y}|} + \int d^3r\, n(r) V_{ext}.$$

    a) Take the functional derivative with respect to $n(r)$. Account for the constraint $\int n = N$ with a Lagrange multiplier μ.
    b) Let $\phi(r) = \int n(y)/|\vec{r}-\vec{y}|$ and obtain an equation for $\nabla^2\phi$ in terms of ϕ and V_{ext}.
    c) Take $V_{ext} = -Z/r$, let $u = \phi + V_{ext}$, and obtain a differential equation for u. You will need $\nabla^2 V_{ext} = 4\pi Z\delta(r)$.
    d) Determine μ by considering the limit $r \to \infty$.
    e) Let $u = -Z/r \cdot \chi$ ($x = \alpha r$) and simplify your partial differential equation with radial coordinates. Choose α to eliminate annoying constants. You should end up with a second order nonlinear differential equation for χ.
    f) Assume a power law solution and obtain the asymptotic behavior for the density for large and small r. How do these compare to the desired asymptotic behaviors?
    g) Solve the differential equation for $\chi(x)$ with appropriate boundary conditions for a neutral atom. As a check you should obtain $\chi(1.0) = 0.425$.
    h) Use χ to obtain the electron density for argon and compare to Figure 23.4.

12. Lithium.

    a) Obtain the Hartree-Fock equations for lithium using the open shell configuration $1s^2 2s^1$.

    b) Code up and compare your results for the energy levels to those obtained with a central (Hartree) approximation.

13. Beryllium.
    Obtain Hartree-Fock estimates of the ground state and ionization energies for beryllium. Use the $1s^2 2s^2$ configuration.

14. Ionization Energies for Neon.

    a) Obtain the ionization energy of neon ($1s^2 2s^2 2p^6$) for a 1s electron using Koopmans' theorem. This is the statement that the ionization energy of the jth electron is approximately given by ϵ_j.

    b) Obtain the ionization energy of neon by direct computation (i.e., evaluating the difference in energies $E(N) - E(N-1)$).

    c) Compare your results to the experimental value of 870.3 eV.

    d) Repeat for 2s and 2p electrons.

15. H⁻ Ion.

    a) Show that the product Ansatz

    $$\psi(r_1, r_2) = \frac{\alpha^3}{\pi} \exp[-\alpha(r_1 + r_2)]$$

    is incapable of establishing binding for the H⁻ ion.

    b) Following Chandrasekhar, introduce correlation in the variational Ansatz as follows:

    $$\psi(r_1, r_2) \propto \exp(-ar_1 - br_2) + \exp(-br_1 - ar_2).$$

    Obtain $E(a, b) = \langle \psi | H | \psi \rangle / \langle \psi | \psi \rangle$.

    c) Minimize $E(a, b)$ and show that binding occurs for $a \neq b$.

    d) Repeat this procedure with the Ansatz

    $$\psi(r_1, r_2) \propto \exp[-a \min(r_1, r_2) - b \max(r_1, r_2)]$$

    and compare the ground state energy to that from part (c).

16. [T] H_2^+ Ion (i).
    Consider the simplest LCAO trial functions for the H_2^+ ion:

    $$\phi_{g/u}(r; R) = \frac{1}{\sqrt{2}} (\psi_{1s}(r_A) \pm \psi_{1s}(r_B))$$

    where $\psi_{1s}(r) = \exp(-r)/\sqrt{\pi}$.

    a) Show that the corresponding adiabatic surfaces are given by:

    $$E_{g/u}(R) = E_{1s} + \frac{1}{R} \frac{(1+R)\exp(-2R) \pm (1 - 2R^2/3)\exp(-R)}{1 \pm (1 + R + R^2/3)\exp(-R)}$$

    b) Obtain the minimum distance and energy for the gerade configuration. Compare to the exact results of $R_0 = 1.06$Å and $D_e = 0.103$ au $= 2.79$ eV.

    c) The principle defect with the LCAO method is that the trial wavefunction does not approach that of an atom with charge $Z = 2$ as R becomes small. Rectify this problem by constructing a variational LCAO with atomic wavefunctions:

    $$\psi_t(r) = \sqrt{a^3/\pi} \exp(-ar)$$

    and compare your results to those above.

17. H_2^+ (ii).
    Solve the Hartree-Fock equation for the H_2^+ ion by expanding the single particle wavefunction in spherical harmonics.

    a) Show that only even waves contribute to the interaction, and hence the eigenvectors, of the H_2^+ wavefunction.

    b) Show that the interaction is diagonal in the magnetic quantum number. Is this physically sensible?

    c) Show that the ground state belongs in the $m = 0$ sector.

    d) Examine the convergence of the ground state energy in the number of partial waves. You should obtain something like Figure 23.9.

18. [P] Monte Carlo H_2^+ Ion.
    Obtain the ground and first excited state energy surfaces of the H_2^+ ion as a function of the interproton radius using the quantum Monte Carlo guided random walk algorithm. Compare this to the exact result.

19. H_2 Integrals.
    Evaluate the integral of Eq. 23.74 numerically using spherical coordinates.

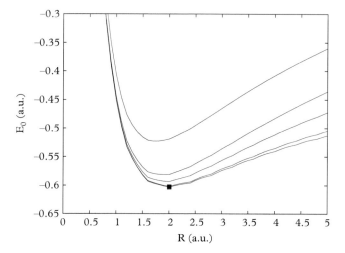

Figure 23.9 H_2^+ Ion. Convergence in the partial wave expansion.

a) Let $z_1 = \hat{r}_1 \cdot \hat{R}$ and $z_2 = \hat{r}_2 \cdot \hat{R}$. Obtain an explicit expression for $\hat{r}_1 \cdot \hat{r}_2$ in terms of z_1, z_2, and another angle.

b) There are two azimuthal angles. Show that the integrand is not a function of a linear combination of these angles and that performing the relevant integration gives rise to a factor of $4\pi - 2\phi$, where ϕ is the orthogonal azimuthal combination.

20. H_2 Molecule.
    Consider the Hartree-Fock equations for H_2.

    a) Argue that the m-dependence in the expansion of the single particle Hartree-Fock wavefunction:

    $$\Psi(\vec{r}; \vec{R}) = \sum_{\ell m} \frac{1}{r} \varphi_{\ell m}(r; R) \, Y_{\ell m}(\hat{r})$$

    is natural in this problem.

    b) Numerically confirm that the waves $\ell 0$ dominate for even ℓ.

    c) Confirm the convergence in the partial wave expansion and determine the minimum ground state energy and interatomic distance.

21. [T] Perturbative Nuclear Matter.
    We seek to describe a uniform gas of nucleons. In this case the Hartree-Fock and perturbation theory merge because the Hartree-Fock orbitals are plane waves.

a) Establish the last statement.

b) Assume a local central two-body interaction describes NN interactions and simplify the Hartree-Fock expression for the system energy.

c) Show that the system collapses at sufficiently high density for an attractive potential.

BIBLIOGRAPHY

Argaman, N. and G. Makov. (2000). *Density Functional Theory—an Introduction*. American Journal of Physics **68**, 69.

Born, M. and J. R. Oppenheimer. (1927). *Zur Quantentheorie der Molekeln*. Annalen der Physik **389** 457.

Bransden, B. H. and C. J. Joachain. (2003). *Physics of Atoms and Molecules*. Prentice-Hall.

Chachiyo, T. (2016). *Simple and accurate uniform electron gas correlation energy for the full range of densities*. J. Chem. Phys. **145**, 021101.

Cramer, C. (2004). *Computational Chemistry: Theories and Models*. John Wiley and Sons.

Fetter A. L. and J. D. Walecka. (2003). *Quantum Theory of Many-Particle Systems*. Dover Publications.

Gunnarsson, O. and B. I. Lundqvist. (1976). *Exchange and correlation in atoms, molecules, and solids by the spin-density formalism*. Phys. Rev. B **13**, 4274.

Hartree, D. (1928). *The Wave Mechanics of an Atom with a Non-Coulomb Central Field. Part II. Some Results and Discussion*. Math. Proc. Cambridge Phil. Soc., **24**, 111.

Hohenberg, P. and W. Kohn. (1964). *Inhomogeneous Electron Gas*. Phys. Rev. **136**, B864.

Hylleraas, E. A. (1929). *Neue Berechnung der Energie des Heliums im Grundzustande, sowie des tiefsten Terms von Ortho-Helium*. Z. Phys. **54**, 347.

Janoschek, M. *et al.* (2015). Science Advances, **1**, e1500188.

Kato, T. (1957). *On the eigenfunctions of many-particle systems in quantum mechanics*. Communications on Pure and Applied Mathematics, **10**, 151.

Kohn, W. and L. J. Sham. (1965). *Self-consistent equations including exchange and correlation effects*. Phys. Rev. **140**, A1133.

Kolos W. and L. Wolniewicz. (1968). *Improved Theoretical Ground-State Energy of the Hydrogen Molecule*. J. Chem. Phys. **49**, 404.

Perdew, J. P. *et al.* (1992). *Atoms, molecules, solids, and surfaces: applications of the generalized gradient approximation for exchange and correlation*. Phys. Rev. B **46**, 6671.

MacLaren, J. M, D. P. Clougherty, M. E. McHenry, M. M. Donovan. (1991). *Parameterised Local Spin Density Exchange-Correlation Energies and Potentials for Electronic Structure Calculations I: Zero Temperature Formalism*. Comp. Phys. Com. **66**, 383.

Messiah, A. (2014). *Quantum Mechanics* vol 2. Dover Publications.

Parr, R. G. and W. Yang. (1989). *Density Functional Theory of Atoms and Molecules*. Oxford University Press.

Pekeris, C. L. (1958). *Ground State of Two-Electron Atoms*. Phys. Rev. **112**, 1649.

Scerri, E. R. (1998). *How Good Is the Quantum Mechanical Explanation of the Periodic System?*. J. Chem. Ed. **75**, 1384.

Slater, J. C. (1951). *A Simplification of the Hartree-Fock Method*. Phys. Rev. **81**, 385.

Szabo, A. and N. S. Ostlund. (1996). *Modern Quantum Chemistry*. Dover Publications.
Vosko, S. H., L. Wilk, and M. Nusair. (1980). *Accurate spin-dependent electron liquid correlation energies for local spin density calculations: a critical analysis*. Can. J. Phys. 58, 1200.
Wilson, S. (1983). *Diagrammatic Many Body Perturbation Theory*. pgs 273–298, "Methods in Computational Molecular Physics", NATO ASI vol. 113.
Young, David (2001). *Computational Chemistry: A Practical Guide for Applying Techniques to Real World Problems*. John Wiley and Sons.

24
Quantum field theory

24.1 Introduction   844
24.2 φ^4 theory   845
    24.2.1 Evaluating the path integral   849
    24.2.2 Particle spectrum   853
    24.2.3 Parity symmetry breaking   856
24.3 Z_2 Gauge theory   857
    24.3.1 Heat bath updates   861
    24.3.2 Average Plaquette and Polyakov loop   863
24.4 Abelian gauge theory: compact photons   865
    24.4.1 Gauge invariance and quenched QED   867
    24.4.2 Computational details   869
    24.4.3 Observables   872
    24.4.4 The continuum limit   875
24.5 $SU(2)$ Gauge theory   877
    24.5.1 Implementation   879
    24.5.2 Observables   881
24.6 Fermions   886
    24.6.1 Fermionic updating   890
24.7 Exercises   893
Bibliography   903

24.1 Introduction

The basis of our current understanding of all physical phenomena is the quantum field. In fact, the successes of the quantum field theoretical description of nature are so numerous and precise that many physicists believe that a better description of nature will never be found. At the heart of the method is the belief that the description of nature must be relativistic, local, and causal. These qualities place significant constraints on acceptable theories and confer tremendous predictive powers.

The Standard Model is the field theory that is currently thought to describe most of the universe (it does not include gravity). It postulates that (almost) everything is described in terms of one scalar field, $\varphi(x, t)$, 24 spinor fields, $\psi^\alpha(x, t)$, and 12 vector fields, $A^\mu(x,t)$. Dynamics is specified by a Lagrangian and a procedure to compute with the Lagrangian, namely the path integral and the renormalization prescription.

Unfortunately computing with quantum field theory can be daunting, and one is often forced to rely on numerical methods to make progress. The chief technique employed in the community is ***lattice field theory***, wherein the path integral is represented as a sum over discrete variables and the resulting multidimensional integral is evaluated with stochastic methods. This idea was introduced in Section 7.6.2 and was applied to continuum quantum mechanical problems in Chapter 21 and to discrete problems in Chapter 22.

We shall review the lattice field method for a series of representative theories in this chapter. The basic technique will be sketched for a self-interacting scalar field, then we will apply it to a simple gauge theory, quantum electrodynamics, and non-Abelian gauge theory. The realistic cases of SU(3) gauge theory and the Higgs sector are presented in the exercises. In preparation for reading this chapter we encourage the assiduous student to review material on classical fields (Chapter 19), stochastic methods (Chapter 7), the renormalization group and universality (Chapter 8), and the sections mentioned in the previous paragraph.

24.2 φ^4 theory

The basic goal of any field theoretic computation is to evaluate matrix elements of an observable, generically defined as:

$$O(x,t) = \frac{1}{Z} \int D\varphi \, \mathcal{O}[\varphi; x, t] \, e^{iS[\varphi]}. \tag{24.1}$$

Here the notation $f[\varphi; x, t]$ means that f is a functional of φ and a function of x and t. The partition function is defined by:

$$Z = \int D\varphi(x,t) \, e^{iS[\varphi]} \tag{24.2}$$

The measure $D\varphi$ denotes a sum over all possible fields (with appropriate boundary conditions). A concrete definition will be given in Eq. 24.6. The action given as:

$$S[\varphi] = \int d^d x \, \mathcal{L}(\varphi) \tag{24.3}$$

(we work in d spacetime dimensions for the moment) and the Lagrangian we consider in this section is:

$$\mathcal{L} = \frac{1}{2} \partial_\mu \varphi \partial^\mu \varphi - \frac{1}{2} m_0^2 \varphi^2 - \frac{1}{4!} \lambda_0 \varphi^4. \tag{24.4}$$

Although the parameters m_0 and λ_0 are commonly called the mass and coupling, at this stage the meaning of these parameters is unknown and must be determined in an operational way by comparison with measurable quantities. For this reason, these parameters are called **bare** and a subscript "0" is appended.

The path integral is a rather dubious mathematical entity and must be defined with care. One such definition is in fact the mapping to the numerical procedure we shall employ, namely, the field is defined in Euclidean space. In practice this means that $t \to -i\tau$, precisely as done when mapping the Schrödinger equation to the diffusion equation. This has the benefit of making the path integral convergent for real-valued fields. The next step is to replace the integral over fields with a multidimensional Riemann integral by discretizing spacetime:

$$\varphi(x,t) \to \varphi(n_1 a, \ldots, n_d a) \to \varphi_n \tag{24.5}$$

where n represents the collections of integers $\{n_i\}$. The discretization scale a, is called the *lattice spacing* and serves the dual purpose of defining the path integration and an *ultraviolet cutoff*. At this stage the partition function is *regulated*: it is well-defined and ready for numerical study.

The new form of the partition function is:

$$Z = \int \prod_n d\varphi_n \exp\left(-a^d \sum_m L\right) \tag{24.6}$$

with:

$$L = \frac{1}{2} \sum_\mu \Delta_\mu \varphi_m \Delta_\mu \varphi_m + \frac{1}{2} m_0^2 \varphi_m^2 + \frac{1}{4!} \lambda_0 \varphi_m^4 \tag{24.7}$$

Here μ is a unit vector in the μth direction and is summed over. Notice that L is positive definite and the integral for an observable is absolutely convergent if the operator is less than exponentially divergent. The operator Δ_μ is a differencing representation of the derivative (see Chapter 19 for much more on this). Here we use the forward difference operator

$$\Delta_\mu \varphi_m = \frac{1}{a}(\varphi_{m+a\mu} - \varphi_m). \tag{24.8}$$

At this stage it is convenient to change variables to account for the dimensionality of the terms in the action. Thus we let $(\varphi, m_0, \lambda_0) \to (\phi, g, \kappa)$ as follows:

$$\phi_n = \frac{a^{d/2-1}}{\sqrt{\kappa}} \varphi_n \tag{24.9}$$

$$g = \frac{\kappa^2}{24 a^{d-2}} \lambda_0 \tag{24.10}$$

and

$$m_0^2 = \frac{2-4g}{a^2\kappa} - \frac{2d}{a^2}. \tag{24.11}$$

Finally the action becomes:

$$S = \sum_m \left(-\kappa \sum_\mu \phi_m \phi_{m+a\mu} + (1-2g)\phi_m^2 + g\phi_m^4 \right). \tag{24.12}$$

Confirm that ϕ_n, κ, and g are dimensionless.

Notice that the last two terms of the action can be written as $g(\phi^2-1)^2 + \phi^2 - g$, thus ϕ is driven to ± 1 as the coupling g becomes large, and it appears that the theory approaches the Ising model with coupling $\tilde{J} = \kappa$. This can be checked by expanding

$$\phi_n = s_n + \delta\phi_n$$

where $s_n = \pm 1$. One obtains an Ising model that is coupled linearly and cubicly to the field $\delta\phi(x)$. However $\delta\phi$ has a mass that is proportional to g and hence decouples from the low energy dynamics as g becomes large (this is called the **Appelquist-Carazzone theorem**). We conclude that the Ising limit is indeed recovered and that φ^4 theory is in the same universality class as the Ising model (for large coupling, and for all coupling if no phase transition intervenes).

All computed properties depend on κ and g. Comparison to two measurable quantities, say O_1 and O_2, permits the identification:

$$\begin{aligned} O_1 &= f_1(\kappa, g) = \frac{1}{Z} \int \prod d\phi\, \mathcal{O}_1 e^{-S} \\ O_2 &= f_2(\kappa, g) = \frac{1}{Z} \int \prod d\phi\, \mathcal{O}_2 e^{-S} \end{aligned} \tag{24.13}$$

which can be solved to obtain $\kappa(O_1, O_2)$ and $g(O_1, O_2)$, thereby defining the parameters of the theory.

Notice that all scales have been removed from the theory, thus O_1 and O_2 are dimensionless. If one wants to make a prediction of a dimensionful quantity then the lattice cutoff must be reinstated. For example

$$m_{\text{phys}} = m_{\text{lattice}}/a. \tag{24.14}$$

848   φ^4 theory

An experimental measurement of $m_{\rm phys}$ and a theoretical measurement of $m_{\rm lattice}$ determines the lattice spacing. This then allows the determination of all dimensionful quantities. This procedure is called **setting the lattice scale**.

Spacetime symmetry

At this point the perspicacious (or suspicious!) reader will ponder about the damage we have done to the continuum theory by discretizing it. In particular, Euclidean $O(D)$ symmetry (in D dimensions) has been replaced by **hypercubic symmetry**, which is described by a group that is of dimension $2^D D!$ and is a subgroup of $O(D)$. The most obvious implications are that (i) rotational symmetry is lost (ii) it can be difficult to construct operators of given continuum quantum numbers. The hope, of course, is that the continuum symmetry is recovered as the lattice spacing is taken to zero. In classical field theories one can guarantee that this happens by confirming that the continuum theory is indeed recovered as one removes the lattice spacing. However, the situation is more complicated in quantum field theories because of well-known issues with *divergences* (ultraviolet infinities) that can ruin naive dimensional scaling of things like field operators.

In the end our labor has not been in vain because the discretized theory does successfully reproduce the continuum theory as the lattice spacing is removed. However, this happenstance is nontrivial! Namely, it transpires that the set of all renormalizable and hypercubic symmetric operators *happens* to also be $O(D)$ invariant; thus $O(D)$ symmetry is recovered in the infrared, continuum, limit as an accidental symmetry[1]. This means that differences between lattice and continuum values of correlation functions will vanish as the lattice spacing squared, and the two implementations of the theories agree as the lattice spacing is removed.

From now on we shall consider three dimensional ϕ^4 theory. This case is especially interesting because the theory exhibits parity symmetry breaking, as we will shortly establish.

Analytical results

Debugging Monte Carlo code requires care, a good grasp of statistics, and results with which to compare. Here we choose to compare to exact results computed in the noninteracting case $g = 0$. In this case the action is quadratic in the fields and it is possible to perform the path integral (or, in our case, the many-dimensional Riemann integral) exactly. This is achieved with the aid of the well-known formula for multidimensional Gaussian integrals.

$$\int \prod d\phi_n \, \exp\left(-\frac{1}{2}\phi_i M_{ij}\phi_j + \phi_i \mathcal{J}_i\right) = \frac{1}{\sqrt{\det(M)}} \exp\left(\frac{1}{2}\mathcal{J}_i (M^{-1})_{ij} \mathcal{J}_j\right). \quad (24.15)$$

---

[1] In fact, the full Poincaré symmetry is recovered.

Thus, for example, the correlation function $\langle T\phi(x)\phi(y)\rangle$ is given formally by:

$$\langle T\phi(x)\phi(y)\rangle \equiv \frac{1}{Z}\int D\phi\, \phi(x)\phi(y)\, e^{-S}. \qquad (24.16)$$

With our conventions this is (again, for $g=0$):

$$\langle T\phi_x\phi_y\rangle = \frac{1}{\kappa L^3}\sum_k \frac{\cos(k\cdot(x-y))}{a^2 m_0^2 + 4\sum_\mu \sin^2(k_\mu/2)} \qquad (24.17)$$

where $k_\mu = 2\pi n_\mu/L$ and n_μ runs from 0 to $L-1$.

There is a pole in this expression at:

$$k_t = aE = 2\sinh^{-1}\left(\frac{1}{2}\sqrt{a^2 m_0^2 + 4\sin^2(ap_x/2) + 4\sin^2(ap_y/2)}\right). \qquad (24.18)$$

This defines the dispersion relation for free particles on the lattice and will prove useful later for code validation. Notice that as the lattice spacing becomes small this approaches the familiar dispersion relation $E^2 = m_0^2 + \vec{p}^2$.

24.2.1 Evaluating the path integral

We seek to evaluate matrix elements such as $\langle T\phi_x\phi_y\rangle$ with quantum Monte Carlo methods. These have been described and employed in several places throughout the text so we shall dive right in. In this and all subsequent sections we will impose periodic boundary conditions on the fields.

Metropolis-Hastings method

The Metropolis-Hastings method is the simplest to code and is often the fastest to execute. As with many other applications in this text, we choose to implement it via a local sampling technique. Our approach will be to propose a new field $\phi_n \to \phi'_n$ and accept it with probability:

$$p(\phi \to \phi') = \min(1, \exp(-\delta S)), \qquad (24.19)$$

where:

$$\delta S = S(\phi' \equiv \phi + \Delta) - S(\phi)|_{\text{site } n}$$
$$= \Delta\Big[-\kappa\sum_\mu(\phi_{n+\mu}+\phi_{n-\mu}) + 2(1-2g)\phi_n + 4g\phi_n^3 + (1-2g+6g\phi_n^2)\Delta$$
$$+ 4g\phi_n\Delta^2 + g\Delta^3\Big] \qquad (24.20)$$

It is possible to get fancy with the field update proposal distribution, but simple is often best, and we choose the uniform density:

$$p_\phi(\phi \to \phi') = \begin{cases} \frac{1}{\Delta_0} & -\frac{1}{2}\Delta_0 \leq \phi' - \phi \leq \frac{1}{2}\Delta_0 \\ 0 & \text{elsewhere} \end{cases}. \quad (24.21)$$

The parameter Δ_0 should be chosen to maximize a figure of merit that minimizes errors, taking into account the time to achieve those errors. In practice this means choosing Δ_0 so that about one half of proposed updates are accepted.

Heat bath updates

Although the action looks fairly complicated, it is still possible to implement the heat bath algorithm with a few tricks (Bunk, 1995). The idea is to re-write the action in a form that yields an update density that is mostly tractable. The remainder of the density is accounted for with the rejection method (Section 7.2.6). Some manipulation reveals:

$$S|_{\text{site } n} = \frac{\alpha}{2}\left(\phi_n - \frac{N}{\alpha}\right)^2 + g(\phi_n^2 - A)^2 \quad (24.22)$$

with

$$N = \kappa \sum_\mu (\phi_{n+\mu} + \phi_{n-\mu}) \quad (24.23)$$

and

$$A = 1 + \frac{\alpha - 2}{4g}. \quad (24.24)$$

The new parameter α can be choosen to optimize the efficiency of the updates, as with Δ_0.

The heat bath algorithm may now be implemented by generating a new field variable ϕ'_n with probability:

$$p(\phi') = \sqrt{\frac{\alpha}{2\pi}} \exp\left(-\frac{\alpha}{2}\left(\phi' - \frac{N}{\alpha}\right)^2\right). \quad (24.25)$$

The proposed change is then accepted with probability:

$$p_a = \min\left(1, \exp[-g(\phi'^2 - A)^2]\right). \quad (24.26)$$

Microcanonical updates

The Metropolis and heat bath methods are good at updating the short wavelength modes of the theory, but are less efficient at updating long wavelength modes. This often leads to strong autocorrelation effects which must be addressed. A useful technique is to implement *microcanonical updates* (also called *over-relaxation* in analogue with the technique employed in continuum systems).

The method proceeds by replacing a local variable $\phi_n \to \phi'_n$ where the new variable is chosen to leave the action invariant (Creutz, 1987):

$$S[\phi_n] = S[\phi'_n].$$

We see that no damage is done to the distribution of field configurations that the Markov chain is generating, even though possibly large changes in the configuration can be created. However, microcanonical updating is not ergodic, and thus must be combined with Metropolis or heat bath updates.

In our case one could choose ϕ' by solving the cubic equation $\delta S[\phi] = 0$ for one of the roots. A root is then chosen at random and accepted with probability (Creutz, 1987):

$$p(\phi \to \phi') = \min\left(1, \left|\frac{S'(\phi)}{S'(\phi')}\right|\right). \qquad (24.27)$$

While feasible, solving the cubic equation is rather computationally intensive (it can be done with the aid of the *companion matrix*, see Chapter 3) for something that is repeated as often as a Monte Carlo update. We therefore implement a different strategy that involves making a reflection:

$$\phi \to \phi' = \kappa \sum_\mu (\phi_{m+\mu} + \phi_{m-\mu}) - \phi \qquad (24.28)$$

Notice that the inverse relationship is the same, thus the factor

$$\frac{p_c(\phi' \to \phi)}{p_c(\phi \to \phi')}$$

cancels in the expression for the Metropolis-Hasting update density (see Section 7.5.1).

The action is only partially invariant under this reflection, so the proposed update cannot be accepted unconditionally. The remainder must be implemented as an update probability:

$$p(\phi \to \phi') = \min\left(1, \exp\left(-g[-2\phi'^2 + \phi'^4 + 2\phi^2 - \phi^4)]\right)\right). \qquad (24.29)$$

Autocorrelation

As we have stressed many times, autocorrelation is one of many effects that must be monitored in quality Monte Carlo work. In general, autocorrelation depends on the observable under consideration, model parameters, proximity to phase boundaries, and even the order in which updates are executed. Figure 24.1 (top curve) displays the autocorrelation function:

$$C(\tau) = \frac{\langle \mathcal{O}(t)\,\mathcal{O}(t+\tau) \rangle - \langle \mathcal{O} \rangle^2}{\langle \mathcal{O}^2 \rangle - \langle \mathcal{O} \rangle^2} \qquad (24.30)$$

for the operator

$$\mathcal{O} = \frac{1}{L^3} \sum_{xyt} \phi_{xyt}^2.$$

It is evident that one must generate one or two hundred Markov chain steps to reduce autocorrelation to an acceptable level, which is, of course, very expensive. However, our effort at developing an over-relaxation update pays off handsomely: the lower curve of Figure 24.1 shows the substantial improvement in the autocorrelation when a Metropolis update is followed by an over-relaxation update.

In general one must experiment with different update protocols in the tuning phase of code verification. A typical protocol will be something like:

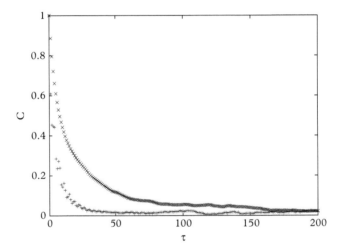

Figure 24.1 *Autocorrelation for $L = 8, \kappa = 0.32, g = 0$. Top curve: Metropolis updates. Bottom curve: Metropolis plus over-relaxation updates.*

```
thermalize for T_therm
loop for T_meas
   perform T_corr Metropolis updates
   perform T_over over-relaxation updates
   measure observables
end loop
```

All the algorithmic parameters in this scheme should be optimized as a function of other algorithmic parameters such as lattice size and the physical parameters (at least to the extent that is practical).

24.2.2 Particle spectrum

With all onerous code validation and tuning accomplished, we are ready to measure observables! Typical first tasks are to measure simple local matrix elements such as $\langle\phi(x)\rangle$. We will instead examine the **mass gap** of the theory, which is the lowest energy difference $E_1 - E_0$. We will focus on the single particle excitations on the following.

The single particle mass gap can be extracted from the two-point function by expanding in a complete set of states. First the field is written in Euclidean time as:

$$\phi(x,t) = e^{Ht}\phi(x,0)e^{-Ht}. \tag{24.31}$$

Thus (ignoring temporal boundary conditions for the moment):

$$C(t) \equiv \langle\phi(t)\phi(0)\rangle = \sum_n |\langle n|\phi(0)|0\rangle|^2 e^{-(E_n-E_0)t} \tag{24.32}$$

where a complete set of (discrete) eigenstates of H has been inserted between the operators and Eq. 24.31 has been used. Observe that it is possible to extract the leading eigenvalue E_1 of the system (defined as the lowest state with nonzero matrix element $\langle n|\phi(0)|0\rangle$) by fitting the behavior of the correlation function in time. Eq. 24.32 makes it clear that the correlation function must be evaluated for times sufficiently large that contamination from excited states has died away. However, it cannot be evaluated for times that are too long because the signal will devolve into noise.

An alternative is to plot the two-point correlation function as an **effective mass** defined by:

$$m_{\text{eff}} = \log\left(\frac{C(t)}{C(t+a)}\right). \tag{24.33}$$

(A slightly more general version of this was used in Section 22.4.)

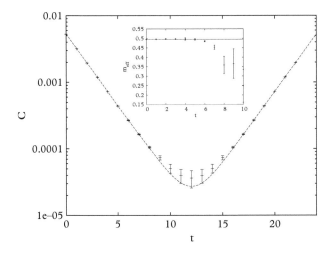

Figure 24.2 *Correlation function* $C(t) = \langle \Phi(t)\Phi(0)\rangle$ *with fit. Inset:* $m_{\it{eff}}(t)$ *with exact result.*

Any operator that one prefers (and that can create the state of interest from the vacuum) can be used with this method. Here we choose to use a zero momentum single-particle operator

$$\Phi(t) = \frac{1}{L^2}\sum_{xy}\phi_{xyt}.$$

The extracted correlation function is displayed on a log scale in Figure 24.2. These results were obtained on a 24^3 lattice with $\kappa = 0.32$ and $g = 0$ with 200000 measurements separated by a single Metropolis sweep followed by a single over-relaxation sweep. The periodic boundary conditions in t are clearly evident in the figure.

The numerical results were fit with a function of the form:

$$c_0\left(\exp(-mt) + \exp(-m(L-t))\right)$$

and yielded $m = 0.4965(5)$. This can be compared with the analytic result of Eq. 24.18, which gives $E_1 - E_0 = 0.4949$. The inset of Figure 24.2 shows the effective mass along with the exact result. As can be seen, the effective mass cannot be trusted as t nears $L/2$: errors grow substantially and the periodic boundary conditions begin to be felt (in fact, the effective mass will turn over past $L/2$ and approach $-m$ for larger t). However, the agreement for $t < L/2$ is very good, lending confidence in the method and in our implementation.

Correlation matrix method

The method just presented is very good for extracting mass gaps. However it does poorly if one is interested in excited states. Attempts to fit the subleading exponential typically fail because the increased importance of statistical noise makes it more difficult to isolate the excited state energy. The accepted method for extracting excited state energies in the field is to employ a *correlation matrix*.

The idea is to define a set of *interpolating operators* $\{\mathcal{O}_j\}$, each of which has non-zero overlap with the states of interest (in fact this overlap can be optimized by building parametric dependence into the operators). Then one computes the correlation matrix:

$$C_{ij} = \langle \mathcal{O}_i^\dagger \mathcal{O}_j \rangle.$$

The idea is that the increased information contained in this matrix permits the extraction of several excited states. Depending on the perspicacity with which the researcher has chosen her operators, this is usually successful and yields about N (typically a fraction like $N/3$) levels for an $N \times N$ correlation matrix.

Although we will not pursue this method any further here, it is useful to see an interpolating operator set. A useful way to construct these in our case is by considering all possible one- and two-particle states of zero total momentum. Thus we define momentum space operators via:

$$\Phi(t, n_1, n_2) = \sum_{x,y} \phi_{x,y,t} \exp(2\pi i n_1/L) \exp(2\pi i n_2/L). \tag{24.34}$$

The sole single particle state is:

$$\mathcal{O}_0(t) = \Phi(t, 0, 0),$$

which we already used to extract the mass gap.

The next lowest five operators can be built from two-particle states as follows:

$$\begin{aligned}
\mathcal{O}_1(t) &= \Phi(t, 0, 0)\, \Phi(t, 0, 0) \\
\mathcal{O}_2(t) &= \Phi(t, 1, 0)\, \Phi(t, -1, 0) \\
\mathcal{O}_3(t) &= \Phi(t, 0, 1)\, \Phi(t, 0, -1) \\
\mathcal{O}_4(t) &= \Phi(t, 1, 1)\, \Phi(t, -1, -1) \\
\mathcal{O}_5(t) &= \Phi(t, 1, -1)\, \Phi(t, -1, 1).
\end{aligned} \tag{24.35}$$

Evidently, the operator $\mathcal{O}_1$ produces a two-particle state in which both particles have zero momentum, and $\mathcal{O}_2$ produces a two-particle state in which one particle has one unit $2\pi/(aL)$ of momentum in the $+\hat{x}$-direction and the other has one unit of momentum in the $-\hat{x}$-direction, etc.

24.2.3 Parity symmetry breaking

We have established that ϕ^4 field theory is equivalent to the Ising model in the large coupling limit. Since the 3d Ising model exhibits a phase transition at $k_B T_* = 4.5115\mathcal{J}$ we expect a similar phase transition at $\kappa_* = 0.22166$. Because the phase transition picks out a "magnetization" direction, we say that parity symmetry has been **spontaneously broken** and one speaks of a **symmetric phase** where $\langle \phi \rangle = 0$ and a **broken phase**, where $\langle \phi \rangle \neq 0$. Lastly, because this phase transition is defined in terms of a coupling rather than a temperature, it is called a **quantum phase transition** (see Chapter 22 for another).

Of course, all of our previous discussion about phase transitions remain relevant. Thus the average field is strictly never nonzero. What happens in practice (say, in a Markov chain, or in an experimental realization) is that the field spends much of its time fluctuating around $\phi = +1$, say, and then makes a transition to the "other vacuum", where it fluctuates about $\phi = -1$. The average over very long times is thus zero. However, as the system gets larger the probability of switching vacua approaches zero and the system is effectively trapped in one vacuum. This is the practical way in which spontaneous symmetry breaking occurs in Monte Carlo computations and in nature.

One way to avoid these subtleties is to employ a robust **order parameter** to distinguish phases. This can be as simple as $\langle |\phi| \rangle$. Here we choose to examine the putative parity breaking phase transition in ϕ^4 theory by measuring the correlation of fields a great distance apart. Since the greatest distance on a periodic lattice is $L/2$ we consider:

$$M \equiv \frac{1}{L^3} \sum_{xyt} \phi_{xyt} \, \phi_{x+L/2, y+L/2, t+L/2} \tag{24.36}$$

and evaluate $\langle M \rangle$. Intuitively, we are asking for the probability that a "spin" a great distance from an "up spin" at the origin is also up. This will be zero if there is no long range order, and nonzero if the vacuum has magnetized.

Running the code for a small lattice and $g = 0.4$ while looping over κ reveals something dramatic happens near $\kappa = 0.4$. Running again near $\kappa = 0.39$ over a variety of lattice sizes and with more statistics yields the curves shown in Figure 24.3. These have been obtained on lattices of size $L = 12, 16, 20, 24$, and 28 (top to bottom). It is clear that a transition is occurring just below $\kappa = 0.4$; furthermore, the transitions become sharper and remain continuous as the lattice volume increases, lending strong evidence that we have discovered a second order phase transition. The inset shows similar curves for larger lattices, indicating that the phase transition occurs at $\kappa_*(g = 0.4) \approx 0.397$.

More accurate methods to obtain critical points have been discussed in Chapter 8 (scaling and likelihood) and Chapter 20 (susceptibility). The applications of these methods in lattice field theory are exactly the same, hence we will not go into them here. Rather we simply present the results of a rough computation of the phase diagram for parity symmetry breaking in the three dimensional ϕ^4 model shown in Figure 24.4. Notice that the large coupling limit does indeed agree with the Ising result.

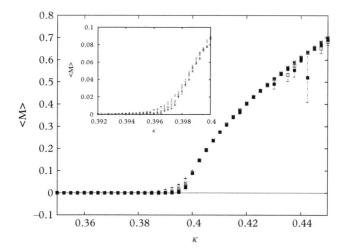

Figure 24.3 *Parity symmetry breaking. $\langle M \rangle$ vs. κ for $g = 0.4$. From top to bottom the curves are for $L = 12, 16, 20, 24,$ and 28. Inset: curves are for $L = 32, 40,$ and 48 from top to bottom.*

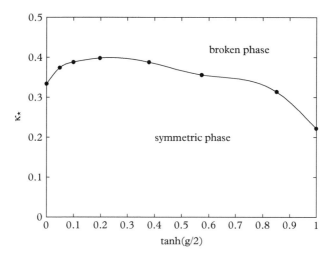

Figure 24.4 *Phase diagram for ϕ^4 in three dimensions (Morningstar, 2009).*

24.3 Z_2 Gauge theory

The Ising model Hamiltonian, $H = -\mathcal{J} \sum_{ij} S_i S_j$, makes it clear that the system is invariant under a global reflection, $S_i \to -S_i$, for all lattice sites i. This symmetry implies that it is impossible for a finite Ising system to magnetize, since that implies that a direction in spin space is preferred. However, as we have discussed, an infinite system can magnetize, essentially because it takes an infinite amount of time for a system that is magnetized in one direction to "realize" that another possibility exists. In this case, we speak of spontaneously breaking the global reflection symmetry.

Symmetries are often described in terms of mathematical constructs called **groups**. A group is a set of objects $\{a, b, \ldots\}$ and an operation $\circ$ that obeys four axioms:

closure if a and b are elements of a group G then $a \circ b \in G$.

associativity for all a, b, c in G, $(a \circ b) \circ c = a \circ (b \circ c)$.

identity there exists an element of G, called e, such that $a \circ e = e \circ a = a$, for all a.

inverse for all a in G there exists an element b of G such that $a \circ b = b \circ a = e$.

A simple example of a group is the numbers $\{1, -1\}$ under the operation of multiplication. This group is called Z_2 and is the group that describes the global reflection symmetry of the Ising model.

In 1971 Fritz Wegner constructed a novel quantum spin model in an attempt to find a theory that did not magnetize but had a phase transition (Wegner, 1971). His idea was to impose a *local symmetry* on the system. Whereas a global symmetry transforms each element of a system in the same way, a local symmetry permits different transformations for each degree of freedom. This is called "gauge" symmetry

A crucial part of Wegner's construction was moving the spin degrees of freedom from sites to *links* between sites. A link variable is thus labelled by its starting point on the lattice n and the direction in which it points, $\hat{\mu}$. Thus Wegner replaced site variables S_n with link variables $L(n, \mu)$ (we shall often omit the "hat" when denoting unit vectors in the following).

With this change, a *local* Z_2 symmetry transformation is implemented by reflecting each spin variable that is attached to a given site, as indicated in Figure 24.5. A given link will be affected by the local gauge transformation made at each end, thus we write:

$$L(x, \mu) \to g(x) L(x, \mu) g(x + \mu). \qquad (24.37)$$

For the Z_2 symmetry being considered here, $g(x)$ is ± 1 at all sites (but not the same at all sites!).

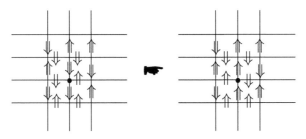

Figure 24.5 *Local gauge transformation. A local gauge transformation is made at the indicated point.*

Requiring that the Hamiltonian is invariant under arbitrary local transformations imposes a severe constraint on the structure of the Hamiltonian. Some thought reveals that the only way to achieve this invariance is if the Hamiltonian is described in terms of *loops of spin*. For example, two links that share a site:

$$L(x,\mu)\, L(x+\mu,\nu)$$

transform to:

$$g(x)\, L(x,\mu)\, g(x+\mu)\, g(x+\mu)\, L(x+\mu,\nu)\, g(x+\mu+\nu).$$

Of course the two intermediate factors of g cancel; and if this is carried around any closed loop of link variables, *all* factors of g cancel. We say that a closed loop is a *gauge invariant* operator.

The solution to Wegner's problem is now apparent: to construct a Hamiltonian that exhibits local gauge invariance work in terms of loops of link variables. The simplest (and most local) loop is the product of four links around a square on a lattice. Because this object is fundamental to all lattice gauge theories it has earned its own name: *plaquette* (Fr. small plaque).

Finally, Wegner's gauged Z_2 spin model can be written as:

$$H = -\frac{\beta}{2} \sum_{\square} L_\square \qquad (24.38)$$

with

$$L_\square = L(n,\mu) L(n+\mu,\nu) L(n+\mu+\nu,-\mu) L(n+\nu,-\nu). \qquad (24.39)$$

The choice of prefactor will be explained in the next section. The notation $\sum_\square$ means sum over all plaquettes, which means $\sum_{n,\mu<\nu}$. The whole thing is called the *gauged Z_2 model*, or Z_2 gauge theory.

Elitzur's theorem

Recall that Wegner's motivation in developing Z_2 gauge theory was an exploration of phase transitions in systems that do not magnetize. This point is underscored by *Elitzur's theorem*, which states that it is impossible for a local gauge symmetry to break spontaneously (Elitzur, 1975). It is clear that the expectation value of a gauge-variant quantity like a link variable must be zero in the same way that an integral such as

$$\int d^3x\, \vec{x}\, f(|\vec{x}|)$$

must be zero because of the rotational symmetry of the function f and the measure. However, this statement applies to *finite* systems. What Elitzur succeeded in doing was showing that the matrix element remains zero even in the thermodynamic limit.

Wegner avoided this issue by considering *gauge invariant* operators such as loops of link variables. He then proved a remarkable result: at large temperatures (small β in Eq. 24.38) the expectation value of a planar rectangular loop behaves as:

$$\langle L_\Box \rangle \sim \exp(-A) \qquad \text{[high } T\text{]},$$

where A is the area enclosed by the loop. Alternatively, at low temperatures one has:

$$\langle L_\Box \rangle \sim \exp(-P) \qquad \text{[low } T\text{]},$$

where P refers to the perimeter of the loop (Kogut, 1979). These results are referred to as **area laws** and **perimeter laws**, respectively.

Much more is known about Z_2 gauge theory, for example the two-dimensional model can be mapped to the one-dimensional Ising model. Exact results are also known for three and four-dimensional systems. The interested reader is directed to the masterful review by Kogut (1979) for more details.

In the following we will study the three-dimensional gauged Z_2 model with Monte Carlo lattice field theory (often called *lattice gauge theory* in this case). We will find that there is a phase transition at the critical point

$$\beta_* = 1.5226$$

(Balian, 1975). Furthermore, perturbative expressions for observables such as the average plaquette can be obtained. We report:

$$\langle \Box \rangle = \frac{1}{2}\beta + O(\beta^2) \qquad \text{[small } \beta\text{]} \qquad (24.40)$$

and

$$\langle \Box \rangle = 1 - 8\exp(-4\beta) \qquad \text{[large } \beta\text{]}. \qquad (24.41)$$

Although these expressions are not difficult to derive, we will forego this exercise and get straight to numerics.

A final note on nomenclature: actions for other gauge theories often yield values of β that are inversely proportional to couplings, thus *large coupling* refers to *small β*. Also, the similarity of Euclidean time field theories with finite temperature classical systems leads to the identification of β with $1/k_B T$. Thus the following are all equivalent:

small β	↔	large coupling	↔	hot
large β	↔	small coupling	↔	cold.

24.3.1 Heat bath updates

As with ϕ^4 theory, we seek to evaluate matrix elements of operators:

$$\langle \mathcal{O} \rangle = \sum_{\{L\}} \mathcal{O}[L] \exp\left(\frac{\beta}{2} \sum_{\Box} L_{\Box}\right), \tag{24.42}$$

where the notation is made less obscure in the description following Eq. 24.38.

Because the action is relatively simple and the variables are discrete numbers, we choose to the employ the Gibbs sampling variant of heat bath Monte Carlo to evaluate these matrix elements. Thus we require the conditional probability for a single link variable. In two dimensions this link will form part of two plaquettes, and in three it will be part of four plaquettes, as illustrated in Figure 24.6. The arrows in this figure imply that the link variables are *oriented* – this will be become relevant in subsequent sections.

Factoring the link variable $L(R, \mu)$ out of the action leaves (in three dimensions) four three-sided objects called **staples**. Thus:

$$S|_{\text{local}} = \frac{\beta}{2} L(R, \mu) \cdot \sum_{\text{staples}} S(R, \mu, \nu). \tag{24.43}$$

More explicitly, the sum over staples is given by the following code:

```
int staple(int x, int y, int t, int mu) {
  // compute the sum of staples wrt to link L(R,mu)
  int st = 0;
  int R[3] = {x,y,t};
  int Rmu[3] = {x,y,t};
  Rmu[mu] = pl[R[mu]];
  for (int nu=0;nu<3;nu++) {
    if (nu != mu) {
      vector<int> Rnu = {x,y,t};
```

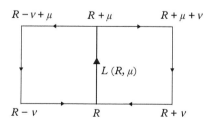

Figure 24.6 *Shared plaquettes in two dimensions.*

```
            Rnu[nu] = pl[R[nu]];
            vector<int> Rnum = {x,y,t};
            Rnum[nu] = mn[R[nu]];
            vector<int> Rnumu(Rnum.begin(),Rnum.end());
            Rnumu[mu] = pl[Rnum[mu]];
            st += latt[Rmu[0]][Rmu[1]][Rmu[2]][nu]    *
                  latt[Rnu[0]][Rnu[1]][Rnu[2]][mu]    *           // ^*
                  latt[x][y][t][nu];                              // ^*
            st += latt[Rnumu[0]][Rnumu[1]][Rnumu[2]][nu] *        // ^*
                  latt[Rnum[0]][Rnum[1]][Rnum[2]][mu]    *        // ^*
                  latt[Rnum[0]][Rnum[1]][Rnum[2]][nu];
        }
    }
    return st;
}
```

We encourage the student to study this code snippet carefully to understand (i) the sum over nu and (ii) the arrangement of indices in the matrices. Notice that the link variables are stored in a four-dimensional matrix `latt[x][y][t][mu]`. Moving around the staple is accomplished by adding unit vectors to $R = (x, y, t)$ in the μ or ν directions. As always, the arrays `pl` and `mn` implement periodic boundary conditions. Finally, we define all unit vectors to be in the positive directions, thus, for example, the lower right link in Figure 24.6 is given by

$$L(R + \nu, -\nu) = L(R, \nu)^\dagger.$$

Again, the complex conjugation indicated here will become relevant in subsequent sections – it can be ignored for Z_2 models.

Because all link variables are ± 1, all staples evaluate to ± 1, and the sum over staples evaluates to one of $S \equiv \{-4, -2, 0, +2, +4\}$. Thus the conditional probability becomes:

$$p(L \to L') = \frac{\exp(L' \frac{\beta}{2} S)}{\sum_{\ell = \pm 1} \exp(\ell \frac{\beta}{2} S)}. \tag{24.44}$$

The code snippet implementing this is quite simple:

```
double r0 = wt[4+staple(x,y,t,mu)];
r0 = r0/(r0+1.0/r0);
if (uDist(mt) < r0) {
    latt[x][y][t][mu] = 1;
} else {
    latt[x][y][t][mu] = -1;
}
```

Random numbers are generated with uniform_real_distribution<double> via a Mersenne twister engine. Probabilities have been pre-computed and stored:

```
for (int i=-4;i<=4;i=i+2) {
  wt[i+4] = exp(beta*i/2.0);
}
```

24.3.2 Average Plaquette and Polyakov loop

As we have stressed, code validation, tuning, and checking autocorrelation and convergence must all be carried out. The techniques are the same as with ϕ^4 theory, so we will not pursue these points further.

Moving on to observables, the simplest gauge invariant quantity is the plaquette and thus it is natural to measure its expectation value as an initial computation. This can be achieved with a small function that computes plaquettes:

```
int plaquette(int x, int y, int t, int mu, int nu) {
  // compute the plaquette at (xyt) in the mu-nu direction
  // mu != nu  !!
  int R[3] = {x,y,t};
  int Rmu[3] = {x,y,t};
  Rmu[mu] = pl[R[mu]];
  int Rnu[3] = {x,y,t};
  Rnu[nu] = pl[R[nu]];
  int plaq = latt[x][y][t][mu] *
             latt[Rmu[0]][Rmu[1]][Rmu[2]][nu] *
             latt[Rnu[0]][Rnu[1]][Rnu[2]][mu] *           // ^*
             latt[x][y][t][nu];                           // ^*
  return plaq;
}
```

Again, the student is encouraged to carefully examine the indexing as this is a typical source of errors in lattice gauge code. An additional routine is used to sum all plaquettes in all orientations over the lattice. These are then averaged over the Markov chain to arrive at a final estimate (with error) of the average plaquette. Consistency between hot (random initial spins) and cold (all spins up) starts was used to check thermalization. Typical results on small lattices and with modest statistics (runs lasted for less than a minute) are shown in Figure 24.7. The figure also shows the small and large coupling perturbative results, Eqs. 24.40 and 24.40, as dashed and dotted lines. Checks like this are essential in all Monte Carlo work.

Despite the appearance that something might be going on near $\beta = 1.5$, the figure makes it clear that the average plaquette is not a good order parameter as it is not zero in any phase. The discussion of Elitzur's theorem implies that "global" properties of the

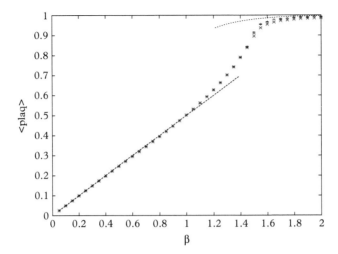

Figure 24.7 *Average plaquette in 3d Z_2 gauge theory. The points are for $L = 10$ and $L = 20$ lattices. Leading results for strong and weak coupling are shown as dashed and dotted lines respectively.*

field theory must be probed to test for possible phase transitions. Of course the operators must also be gauge invariant. A viable choice is a large loop of link variables. Rectangular planar loops of size $R \times T$ are called **Wilson loops** and will be considered later. Here we pursue a simpler option, called a **Polyakov loop**, which consists of a line of links that extend up the temporal axis and traverse the entire lattice. Periodic boundary conditions imply that the line is actually a loop because the link at $t = L - 1$ connects to the link at $t = 0$. Thus a Polyakov loop is gauge invariant and highly nonlocal.

As with the average plaquette, computing the Polyakov loop matrix element is best achieved with a utility function:

```
int polyLoop(int x, int y) {
  // compute the Polyakov loop at location x,y
  int ell = latt[x][y][0][2];
  for (int t=1;t<L;t++) {           // march up the t-axis
    ell *= latt[x][y][t][2];
  }
  return ell;
}
```

Results for $L = 10, 20$, and 40 are shown in Figure 24.8. Unlike the plaquette, the Polyakov loop does indeed serve as an order parameter. The figure indicates that a second order phase transition is occurring near $\beta = 1.50$, in good agreement with the exact result, which is shown by the arrow. Of course, more detailed analysis of the type mentioned earlier would be required to pin the critical coupling down. We do not pursue this here since the techniques have already been presented elsewhere.

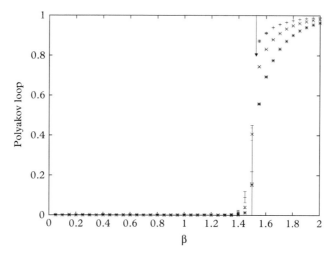

Figure 24.8 *Polyakov loop in 3d Z_2 gauge theory. $L = 10$: plusses; $L = 20$: crosses; $L = 40$: bursts. The arrow shows the analytic critical coupling.*

24.4 Abelian gauge theory: compact photons

The concept of local gauge invariance is likely familiar to the reader in a different context, namely the transformation:

$$A^\mu(x) \to A^\mu(x) - \frac{1}{e}\partial^\mu \Lambda(x) \tag{24.45}$$

that leaves Maxwell's equations invariant. In this case locality is made evident by the space-time dependence of the gauge function, $\Lambda(x)$. In fact, Maxwellian gauge invariance is related to Wegnerian gauge invariance, as we will demonstrate shortly. This has the important implication that it is possible to construct a lattice field theory of electromagnetism that is locally gauge invariant. Before launching into the formalism, let us reminds ourselves about classical relativistic electromagnetism.

The action for electromagnetism is written:

$$S = -\frac{1}{4} \int d^4 x \, F^{\mu\nu} F_{\mu\nu}, \tag{24.46}$$

where the field strength (***Faraday***) tensor is given by:

$$F_{\mu\nu} = \partial_\mu A_\nu - \partial_\nu A_\mu. \tag{24.47}$$

Notice that this tensor is trivially invariant under the gauge transformation of Eq. 24.45 with some modest assumptions about the smoothness of the gauge function. The electric and magnetic fields are given in terms of the vector potential by:

$$\vec{E} = \partial_t \vec{A} - \nabla A^0$$
$$\vec{B} = \nabla \times \vec{A}. \quad (24.48)$$

With these fields the action becomes:

$$\frac{1}{2} \int d^4x \, (E^2 - B^2).$$

We now make the rotation to Euclidean space in the usual way: $t \to -i\tau$, or, more relativistically, $x^0 \to -ix^4$. We also set

$$A^0 \to +iA^4.$$

This is motivated by considering the **pure gauge field**, $A^\mu = \partial^\mu \Lambda$, thus:

$$A_\mu = (A^0, -\vec{A}) = \frac{\partial}{\partial x^\mu} \Lambda \to (\frac{\partial}{-i\partial x^4}, -\nabla)\Lambda = (iA^4, -\vec{A}) \equiv A_E^\mu.$$

Finally, the generating functional for quantum electrodynamics is:

$$Z[\mathcal{J}] = \int DA^\mu \exp\left(-\frac{1}{4}i \int d^4x \, F_{\mu\nu} F^{\mu\nu} + i \int d^4x \, A^\mu \mathcal{J}_\mu\right)$$
$$\to \int DA_E^\mu \exp\left(-\frac{1}{4} \int d^4x_E \, F_E^{\mu\nu} F_E^{\mu\nu} - \int d^4x_E \, A_E^\mu \mathcal{J}_E^\mu\right). \quad (24.49)$$

Here $\mathcal{J}_E^\mu$ is defined in the same way as A_E^μ, thus:

$$A_E^\mu \mathcal{J}_E^\mu = A_E^1 \mathcal{J}_E^1 + A_E^2 \mathcal{J}_E^2 + A_E^3 \mathcal{J}_E^3 + A_E^4 \mathcal{J}_E^4$$

and

$$F_E^{\mu\nu} = \partial_E^\mu A_E^\nu - \partial_E^\nu A_E^\mu.$$

Notice that the path integral is now a sensible convergent integral. Please be aware that all reminders of the Euclidean nature of the formalism will be dropped henceforth!

24.4.1 Gauge invariance and quenched QED

The construction of a lattice version of quantum electrodynamics parallels that of Wegner's gauged Z_2 model exactly. Thus gauge fields are assigned to links while matter fields (electrons) are defined on lattice nodes (electrons will be discussed later). We establish a connection with the lattice gauge transformation of Eq. 24.37 by generalizing Eq. 24.45 slightly:

$$A^\mu(x) \to g(x)A^\mu(x)g^{-1}(x) - \frac{i}{e}g(x)\partial_\mu g^{-1}(x) \tag{24.50}$$

Note that the factors of g cancel in the first term and that the second term implies that:

$$g = \exp(i\Lambda).$$

Now consider the gauge transformation of a Schwinger integral:

$$U(x,y) \equiv \exp\left(ie\int_x^y dz_\mu A^\mu(z)\right) \to g(x)\,U(x,y)\,g^{-1}(y). \tag{24.51}$$

This is precisely the form of the Z_2 link transformation of Eq. 24.37! One is led to identify link variables with Schwinger integrals. If the points x and y are separated by one lattice unit, a, then this relationship establishes a physical meaning for link variables:

$$U(x, x + a\hat{\mu}) \approx \exp\left(ieaA^\mu(x + \mu/2)\right) \equiv L(x, \mu). \tag{24.52}$$

This remarkable connection is strengthened even more when the expression for the plaquette is expanded in powers of a. One obtains:

$$L_\square = L(n,\mu)L(n+\mu,\nu)L(n+\mu+\nu,-\mu)L(n+\nu,-\nu)$$
$$\approx \exp(iea^2 F_{\mu\nu}F_{\mu\nu}) \tag{24.53}$$

This implies that (Rothe, 1992):

$$\frac{1}{e^2}\sum_\square (1 - \operatorname{Re} L_\square) \to \frac{1}{4}a^4 \sum_{n\mu\nu} F_{\mu\nu}(n)F_{\mu\nu}(n). \tag{24.54}$$

Finally, a discretized version of matterless Euclidean quantum electrodynamics can be expressed in terms of a (very) large-dimensional Riemann integral:

$$Z = \int \prod_{n\mu} dL(n,\mu)\,\exp\left(\frac{\beta}{2}\sum_{\square=n,\mu<\nu} \operatorname{Re} L_\square\right). \tag{24.55}$$

with

$$\beta = \frac{2}{e_0^2}. \tag{24.56}$$

In writing this equation we have dropped the uninteresting constant in Eq. 24.54 and have reminded ourselves that the Lagrangian parameters are bare by appending the subscript "0".

The last detail in the model definition is specifying the measure. In the case of $U(1)$ variables this is easy:

$$\int dL(n, \mu) = \int_{-\pi}^{\pi} \frac{d\theta(n, \mu)}{2\pi}. \tag{24.57}$$

Alternatively, we will find it useful to represent an element of $U(1)$ with two numbers, (a_0, a_1), subject to the constraint $a^2 = 1$. In this case one can write:

$$\int dL(n, \mu) = \frac{1}{\pi} \int_{-\infty}^{\infty} da_0 \, da_1 \, \delta(a^2 - 1). \tag{24.58}$$

Why one would take this seemingly retrogressive step is discussed below.

Noncompact QED

It is worthwhile to pause here and reflect on the form of the theory we have constructed. If one were not familiar with Wegner's gauge theory, one might be tempted to discretize spacetime and consider the variables $A(n, \mu)$. Thus the path integral would be defined via the measure

$$\int DA^\mu(x) \to \prod_{n,\mu} \int_{-\infty}^{\infty} dA(n, \mu).$$

Notice, in particular, that the integrals are defined over infinite domains rather than the finite ones of Eq. 24.57. For this reason, our version of quantum electrodynamics is called *compact QED*.

A discretization in terms of $A^\mu(x)$ *can* be used in numerical computations, however it explicitly breaks $U(1)$ gauge invariance. Unlike the case of spacetime symmetry discussed on page 848, gauge invariance is *not* recovered in the continuum limit. In particular, gauge non-invariant operators are created by quantum effects. These include such operators as

$$A^\mu A^\mu \text{ or } A^\mu A^\mu A^\mu A^\mu,$$

which are disasters for the theory: the first is dimension 2 and the second is not $O(D)$ invariant. In principle these operators can be cancelled by the judicious addition of explicitly noninvariant operators, but this requires tuning the couplings to ensure the cancellation, which is extremely tedious. In short, one is well-advised to listen to Wegner and Wilson!

24.4.2 Computational details

With the model in place, the next step is to implement a computational strategy. As usual, the Metropolis-Hastings algorithm provides an efficient and simple method for evaluating many-dimensional integrals. However, compact $U(1)$ gauge theories are nearly as simple as Z_2 gauge theories and it is possible, and instructive, to implement the heat bath algorithm.

Heat bath updates

As with Z_2 gauge theory, the Gibbs conditional probability is given in terms of staples:

$$dp(L) \propto \exp\left(\frac{\beta}{2} \operatorname{Re} L \cdot S\right),$$

where S is the sum of staples at the link in question. This sum is proportional to an element of $U(1)$ (i.e., it is a complex number) and thus we can write $S = kU$ where $k = |S|$ and U is an element of $U(1)$. Call the phase of L, θ and the phase of U, ϕ and change variables from θ to $\theta + \phi$. One obtains:

$$dp(L(\theta)) \propto \exp\left(\frac{\beta}{2} \operatorname{Re} kL\right)$$
$$= \exp\left(k\frac{\beta}{2} a_0\right), \tag{24.59}$$

where $a_0 = \operatorname{Re} L$.

One might be tempted to evaluate this by using the measure shown in Eq. 24.58 but this is something of a disaster, because performing the integral over a_1 yields a probability

$$dp(a_0) \propto \frac{1}{\sqrt{1-a_0^2}} \exp\left(\frac{\beta}{2} k a_0\right),$$

which is awkward at the limits $a_0 \to \pm 1$. It is better to change variables from (a_0, a_1) to θ and generate probabilities according to

$$dp(\theta) \propto \exp\left(\frac{\beta}{2} k(\cos(\theta) - 1)\right).$$

Our final algorithm is:

```
void update(int N) {
  uniform_real_distribution<double> uDist(0.0,2.0*M_PI);
  uniform_real_distribution<double> unDist(0.0,1.0);
  for (int sw=0;sw<N;sw++) {
    for (int x=0;x<L;x++) {
    for (int t=0;t<L;t++) {
    for (int mu=0;mu<2;mu++) {
      U1 S = staple(x,t,mu);
      double bk = beta*S.mag()/2.0;
      repeat: double theta = uDist(mt);
      double wt = exp(bk*(cos(theta)-1.0));
      if (unDist(mt) > wt) goto repeat;
      Latt[I(x,t,mu)] = U1(theta)*S.pinv();
    }}}
  }
  return;
} // end of update
```

There are several things to notice in this code snippet.

1. we work in two dimensions.
2. staples are of datatype U1.
3. there is a `goto` statement. Boudreau rails against these as some sort of coding monstrosity; but Swanson finds it more intuitive than other constructs. We can't settle this so please just develop your own style.
4. lattice indexing appears odd.

We expand on all these points in the next subsection.

Coding considerations

It is of course simple to work in an arbitrary number of dimensions. Here we choose two dimensions because it is possible to solve the theory exactly in this case which makes it a useful pedagogical example.

The `goto` statement forces the algorithm to continue to throw new link configurations until one is accepted. Thus this method is equivalent to the heat bath algorithm. This approach is computationally efficient in this case because the calculation of the sum over staples is the most expensive part of the update routine, and it is best not to waste this effort.

An indexing function $I(x,t,\mu)$ has been introduced. This is implemented with the following method:

```
int I(int i1, int i2, int mu) {
   // indexing for 2d lattice x,y => 0:L-1 and mu => 0:1
   return (i1*L+i2)*2 + mu;
}
```

Of course, this can be easily generalized to any number of dimensions with any number of points along each axis.

Lattices can become quite large (an L^D array of $SU(N)$ objects requires $8 \cdot N^2 \cdot D \cdot L^D$ bytes of memory. Thus a lattice of $SU(3)$ matrices on a 48^4 lattice requires approximately 1.5 GB of memory). Variables that are declared in threaded code are created in a thread stack that can be very limited in size; this limitation can be cirumvented by allocating memory on the heap, e.g.:

```
Latt = new U1[L*L*2];
```

Alternately, use of collection classes like `std::vector` can solve the problem, because even if the `std::vector` is allocated on the *stack*, its internal storage is allocated on the *heap*. Finally, the indexing function is introduced to compress D-dimensional indices to a single super-index. It is possible to invert this process but this is never required in lattice code.

Notice that lattice variables and staples are declared as type U1. The idea here is to leverage the power of (computational) objects in C++ to simplify manipulations with the (mathematical) objects that live on links. So far these have been elements of the group Z_2 or $U(1)$. In general they can be elements of any group. This is precisely why C++ classes exist.

The first task in designing the class for the link variables is choosing the representation. The most memory efficient approach would be to store the phase at each link. However, this requires evaluating trigonometric functions to perform group multiplies, which is very slow. It is better to trade memory for speed and to employ the two-component representation

$$(a_0, a_1) = (\cos\theta, \sin\theta)$$

mentioned above. In this case multiplying group elements becomes linear:

$$a * b = (a_0 b_0 - a_1 b_1, a_0 b_1 + a_1 b_0),$$

greatly increasing computational speed.

Typical tasks that we require of the class are creating elements, adding elements, multiplying elements, and inverting elements. The first is achieved with constructors, while the second two are achieved by overloading the addition and multiplication operators. In the case of $U(1)$ all of this functionality is contained in `std::complex<T>` and this class can be used to implement the necessary functionality.

We have also employed classes to simplify multi-threading. The idea is that all variables should be isolated by thread, and this is simply achieved with the encapsulation property of classes. Thus the `main` code declares a vector of classes called `Markov`:

```
vector<Markov> vecM;
```

These are instantiated and pushed onto the vector

```
for (int th_id=0;th_id<nthreads;th_id++) {
   vecM.push_back(Markov(seed+th_id,L,beta));
}
```

so that they can be run in separate threads:

```
for (int th_id=0;th_id<nthreads;th_id++) {
   th.push_back(thread(&Markov::run,&vecM[th_id],start_op,
      Ntherm,Nmeas,Ncorr));
}
```

Notice that the classes are initialized with a random engine seed, the lattice size (`L`), and the coupling (`beta`). The `run` methods are initialized with a start option (hot or cold starts in this case), and integers that specify the update protocol. The class `Markov` creates the `Latt` array, initializes it, implements boundary conditions, updates link variables, and computes a variety of lattice observables – all of which occur in a safely encapsulated environment.

The remainder of the code in `main` waits for the threads to finish

```
for (auto &t : th) {    // wait for threads to exit
   t.join();
}
```

and then combines the results from each thread to obtain final estimates for observables.

24.4.3 Observables

Let us blunder ahead and compute some observables. The "blundering" is happening because one must carefully consider the lattice spacing dependence of the coupling β before making definitive claims. We will take care of that in the next section.

Computation of the average plaquette is always a good first task because it is relatively simple to code and easy to evaluate. In this case we were able to obtain accurate results with only a few hundred configurations. Recall that we chose to work in two dimensions because analytic results are available thanks to a trick due to Gross and Witten (1980). The result for the average plaquette is:

$$\langle \Box \rangle = \frac{I_1(\beta/2)}{I_0(\beta/2)}, \tag{24.60}$$

where I_1 and I_0 are modified Bessel functions. Since the techniques for evaluating the plaquette have been discussed in the case of Z_2 gauge theory, confirming this result is left as an exercise for the student.

Computing correlators is a more challenging task. One could, for example, compute $\langle \Box(t)\Box(0) \rangle$ where $\Box(t) = \sum_x \Box(x,t)/L$. Using Eq. 24.31 and inserting a complete set of states gives

$$\langle \Box(t)\Box(0) \rangle = \sum_n |\langle \Box(0)|0\rangle|^2 \exp(-(E_n - E_0)t). \tag{24.61}$$

Here the energy levels refer to any state that can be created from the vacuum by the operator $\Box(t)$. Because the plaquette carries no indices, these must be states with vacuum quantum numbers, $J^{PC} = 0^{++}$. Because our theory is *pure gauge* (in other words, there are no electrons), the states $\{|n\rangle\}$ must be interpreted in terms of photons: say weakly interacting two-photon states, or possibly as bound states of photons. The latter interpretation may seem strange but is possible in compact electrodynamics. In non-Abelian theories (to be discussed shortly), such states are called **glueballs**.

Here we shall focus on the correlator of Polyakov loops, which carries interesting information on the long distance dynamics of the theory. Thus we seek to measure:

$$\langle PL(R) \cdot PL^\dagger(0) \rangle$$

as a function of $R = x$ (or $|\vec{R}|$ in more than two dimensions). This correlator is equivalent to a Wegner loop (also referred to as a ***Wilson loop***) that is of extent $R \times T$ where $T = L$. This follows because the Wilson loop can be written as

$$WL(R,T) = PL(R) \cdot L(R \to 0, T) \cdot PL^\dagger(0) \cdot L(0 \to R, 0)$$

where $L(R \to 0, T)$ means the product of link variables from $x = R$ to $x = 0$ at time T. But

$$L(0 \to R, 0) = L(0 \to R, T)$$

because of the periodic boundary conditions, and

$$L(0 \to R, T) = L^\dagger(R \to 0, T).$$

Finally the theory is Abelian (which means that group elements commute) so the temporal portions of the Wilson loop cancel, leaving the correlator of Polyakov loops.

This is an interesting observation because the Wilson loop can be related to the interaction energy of static electrons, $V(R)$:

$$\langle WL(R,T) \rangle = \exp(-V(R)T) \tag{24.62}$$

as T becomes large (Kogut, 1979). Notice that Wegner's area and perimeter laws apply to the same matrix element and therefore carry strong implications on the nature of the static interaction. These will be discussed in more detail shortly.

Maxwell's equation in two dimensions imply that the interaction of a very heavy electron and a very heavy positron is:

$$V(R) = \frac{e^2}{2}R.$$

The unusual R-dependence arises because electromagnetic flux cannot spread out in a single spatial dimension. Alternatively, dimensional analysis shows that the charge e has units of energy and thus a linear potential between static particles must arise. Notice that we have not defined the coupling e! This happens in a few paragraphs.

The Polyakov loop correlator for $\beta = 6.0$ is shown in Figure 24.9. Because this is a log plot, the straight line for small r reveals that the interaction is indeed linear in r. The line is a fit of the form:

$$\frac{1}{2}[\exp(-\sigma\, r\, T) + \exp(-\sigma\, (L-r)\, T)]$$

where the second term is required by the periodic boundary conditions and T is the temporal extent of the lattice. The coefficient of the linear potential has been dubbed σ; this is referred to as the **string tension**, and is a quantity of high interest in the Standard Model.

The results of Figure 24.9 were obtained with $T = L = 10$, which is a fairly small lattice. This choice is forced on us because the exponential decay of the correlator implies

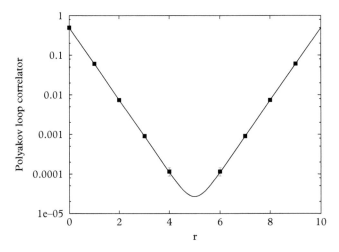

Figure 24.9 *Static potential for 2d U(1) gauge theory at $\beta = 6.0$. The fit is shown as a line.*

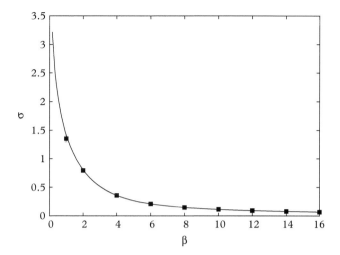

Figure 24.10 *String tension for 2d U(1) gauge theory. The exact result is shown as a line.*

that it is extremely difficult to obtain for large L. In fact, even for L as small as 10, we evaluated the correlation function with 32 million configurations (recall that only a few hundred were required for the average plaquette!). It turns out that smaller values of β yield larger slopes so that even these statistics were incapable of obtaining reliable correlators. In the case of $\beta = 1.0$ we were forced to run on a tiny lattice of size $L = 6$.

The fit yields the result $\sigma(6) = 0.21076(7)$. Repeating this exercise for a collection of couplings gives the string tension shown in Figure 24.10. These results can be compared to the analytic result

$$\sigma = \log \frac{I_0(\beta/2)}{I_1(\beta/2)}$$

which is shown as a line in the figure.

24.4.4 The continuum limit

Interpreting the results just obtained requires that we address the lattice spacing dependence of the coupling, $\beta = \beta(e_0(a))$.

If the universe were discrete at a given scale, say 1 fm, then the numerical procedure is straightforward: one would set $a = 1$ fm in the code, compute some quantity, compare the result to an experimental determination of the same quantity, and thereby determine $\beta(a)$. All other computations then become predictions of the theory. In fact, this is precisely the procedure followed for statistical mechanical or condensed matter systems.

This simple program is disrupted in the case of quantum electrodynamics (and in the rest of the Standard Model) because there is no evidence of granularity in the universe down to scales of 10^{-18} m or lower. There are two possible ways in which the physics can play out:

1. theoretical predictions are *insensitive* to the scale a as this scale becomes very small;
2. theoretical predictions retain a dependence on a, however the effects of this dependence are suppressed by powers of a. The theory must be modified once it probes physical scales of order a.

Theories that follow the first option are called **renormalizable**, whereas those that follow the second are called **effective**.

Our $U(1)$ lattice gauge theory is renormalizable, and hence it is possible to sensibly consider the continuum limit (at this point the reader may want to re-read Section 8.4 on universality and the renormalization group). Of course, in practice this must be carefully obtained by considering a series of measurements at various lattice spacings a. Recall that this dependence is implemented indirectly via the value of the coupling β.

Although we did not state it, the results from the Polyakov loop correlator have been presented in lattice units. Thus R, T, and σ are all dimensionless. Let us call them $\hat{R}$, $\hat{T}$, and $\hat{\sigma}$ to emphasize this fact and retain R etc for quantities with units.

The lattice interaction we obtained can be written as

$$\hat{V}(\hat{R}, \beta, a) = \hat{\sigma}(\beta)\hat{R}.$$

Thus

$$V(R, \beta, a) = \frac{\hat{V}}{a} = \frac{\hat{\sigma}(\beta)}{a^2} R.$$

We appear to have run into a problem: the interaction diverges as the lattice spacing is taken to zero! This nonsensical result is avoided by recalling that:

$$\beta = \frac{2}{e_0^2(a)}.$$

Thus it is possible to tune the behavior of the bare coupling, e_0, so that a sensible interaction arises. In fact this can be achieved simply by setting

$$e = \frac{e_0}{a}.$$

This can be seen with dimensional analysis: β, and therefore e_0, has no units, but the coupling in two-dimensional QED has units of energy.

It is now possible to define a sensible continuum limit of the lattice model. Namely one defines:

$$V(R,e) = \lim_{a \to 0} V(R, \beta(a), a) = \frac{1}{2} e^2 R \qquad (24.63)$$

with

$$\beta(a) = \frac{2}{e^2 a^2}.$$

Notice that the divergence that was in the interaction has been transferred to the coupling. This is fine: bare couplings cannot be measured, only renormalized ones such as in Eq. 24.63.

Scaling windows

In general determining the renormalized theory is not as simple as presented here (if you want to be fancy, you can say that the simplicity is because two-dimensional quantum electrodynamics is *superrenormalizable*). The normal procedure is set β to some value, measure a lattice observable and adjust a so that theory and experiment agree. Repeating this for various β then traces a trajectory $\beta = \beta(a)$ such that physics remains the same.

In practice this procedure cannot be implemented over arbitrary values of β because the lattice becomes incapable of accurately representing physics. Thus, too large a coupling implies a small lattice spacing and a lattice volume that is too small to contain the observable of interest (say, a proton). Alternatively, too small a coupling forces a large lattice spacing which is too coarse to accurately reproduce the observable of interest. The sweet spot in the middle in which physics is accurately reproduced is called the *scaling window*. Employing larger lattices increases the size of the scaling window at the expense of more difficult Monte Carlo computations.

24.5 *SU*(2) Gauge theory

It is instructive to compare the generating functionals for Z_2 and $U(1)$ gauge theories (see Eqs. 24.42 and 24.55). You will see that the forms are identical, the only differences are in the interpretation of the coupling and in the link variables. In the first case they are elements of the group Z_2 and in the second they are elements of the group $U(1)$. This powerful observation was exploited by Yang and Mills (1954) to obtain new field theories that were soon to find application in the description of the weak and strong forces. In particular, Yang and Mills considered *non-Abelian* gauge theories where link variables (or gauge fields) transform as elements of a non-Abelian group (or algebra), such as $SU(N)$. This is the group of $N \times N$ unitary matrices with a determinant of unity. The nomenclature "non-Abelian" refers to the fact that group elements do not commute.

The generalization of gauge theory to non-Abelian groups presents no special difficulties (there were substantial difficulties to overcome in the analytic treatment of these theories, but that is not our concern). To illustrate, we consider $SU(2)$, the ***special unitary group*** of 2×2 unitary matrices. The generators of the group will be familiar to you as the Pauli matrices of spin physics that we labored over so much in Chapter 22.

A general element of $SU(2)$ can be written in terms of three real numbers $\vec{\omega}$ as:

$$a(\omega) = \exp\left(i\vec{\omega} \cdot \frac{\vec{\sigma}}{2}\right) = \mathbb{1} \cos\frac{\omega}{2} + i\hat{\omega} \cdot \vec{\sigma} \sin\frac{\omega}{2}. \tag{24.64}$$

As with $U(1)$, employing this representation requires the computation of many trigonometric functions and it is faster to use a linear representation:

$$a = \mathbb{1}\, a_0 + i\vec{a} \cdot \vec{\sigma}, \tag{24.65}$$

where the four numbers a_i are subject to the constraint:

$$a^2 \equiv a_0^2 + \vec{a} \cdot \vec{a} = 1. \tag{24.66}$$

The generating functional is a slight generalization of the expression given in Eq. 24.55:

$$Z = \int \prod_{n\mu} dL(n,\mu) \, \exp\left(\frac{\beta}{N} \sum_{\square = n,\mu < \nu} \mathrm{Re}\,\mathrm{Tr}\, L_\square\right). \tag{24.67}$$

$U(1)$ link variables are replaced with $SU(N)$ link variables; a trace over plaquettes must be taken, and the coupling is defined as:

$$\beta = \frac{2N}{g_0^2}. \tag{24.68}$$

With these definitions the continuum limit of the Euclidean lattice action is:

$$S = -\frac{1}{2} \int d^D x \, \mathrm{Tr}\, F_{\mu\nu} F_{\mu\nu}.$$

with

$$F_{\mu\nu} = \frac{\sigma^a}{2}(\partial_\mu A_\mu^a - \partial_\nu A_\nu^a - g_0 \epsilon^{abc} A_\mu^b A_\nu^c)$$

for $SU(2)$. Notice that there are now three gauge fields. After some manipulation, these correspond to the $W^\pm$ and Z^0 bosons of the weak force.

The Haar measure

The remaining piece to be defined in the generating functional is the measure. In general one seeks to integrate over the **group manifold**. If one defines a group element in terms of a set of parameters $a(\vec{\omega})$ then this is defined by the group metric

$$M_{ij} = \mathrm{Tr}\left(a^{-1}\frac{\partial a}{\partial \omega_i} a^{-1}\frac{\partial a}{\partial \omega_j}\right)$$

as

$$\int dL \propto \int d\vec{\omega}\, |\det M|^{1/2}.$$

This is referred to as the **Haar measure** (Creutz, 1983).

The formalism is rather simple when applied to our four-component representation of $SU(2)$ elements:

$$\int dL = \int \frac{d^4 a}{2\pi^2}\, \delta(a^2 - 1). \tag{24.69}$$

24.5.1 Implementation

Heat bath updates

Surprisingly, the heat bath algorithm can be applied to $SU(2)$ gauge theory following the same tricks as used in $U(1)$ gauge theory (Creutz, 1983). We seek to draw new elements of $SU(2)$ according to the local conditional probability:

$$dp(L) \propto \exp\left(\frac{\beta}{2}\mathrm{Tr}(L \cdot S)\right) \tag{24.70}$$

(traces of $SU(2)$ matrices are real) where S is the traditional sum over staples associated with the link L. As with $U(1)$, a sum of $SU(2)$ matrices is proportional to an $SU(2)$ matrix:

$$S = k \cdot U \tag{24.71}$$

with $k^2 = \det S$. Now one changes variables to $W = L \cdot U$:

$$dp(W) \propto \exp\left(\frac{\beta}{2}kW\right)$$
$$= \exp(\beta k a_0)$$

where we have used Eq. 24.65 to evaluate the trace of the matrix W. Thus:

$$dp(W) \propto d^4 a\, \delta(a^2 - 1)\, \exp(\beta k a_0)$$
$$= da_0\, d\hat{a}\, (1 - a_0^2)^{1/2}\, \exp(\beta k a_0).$$

Since most of the variation in the probability density is carried by the exponential it is convenient to set $z = \exp(\beta k a_0)$. We obtain:

$$dp(z) \propto dz \left(1 - \frac{\log^2(z)}{\beta^2 k^2}\right)^{1/2}.$$

The strategy is then to draw z uniformly in the domain $[\exp(-\beta k), \exp(\beta k)]$ and use rejection to realize $dp(z)$. Finally, the rest of the group element is determined by choosing a random direction for $\hat{a}$ and the new value for the link is set to be:

$$L \to L' = W(a) \cdot U^{-1}.$$

Microcanonical updates

As we have discussed in Section 24.2.1, autocorrelation can be significantly reduced with microcanonical updates. In the case of $SU(2)$ a particularly simple algorithm due to Creutz (1987) exists: one simply replaces link variables

$$L \to L' = \hat{S}^\dagger L^\dagger \hat{S}^\dagger \tag{24.72}$$

where

$$\hat{S} = \frac{S}{k}$$

is the projection of the staples with respect to the link L onto $SU(2)$. It is easy to verify that this transformation preserves the action (because $\text{Tr}(a^\dagger) = \text{Tr}(a)$ for $SU(2)$). Because this transformation is symmetric and perfectly action preserving, it is always accepted, making it especially efficient. Subsequent numerical results will be obtained with an update protocol that alternates between heat bath and microcanonical updates.

Code

We choose to work in four dimensions in the following. Thus indexing for the lattice array is written as:

```
int I(int i1, int i2, int i3, int i4, int mu) {
  // indexing for 4d lattice 0:L-1 and mu 0:3
  return  (((i1*L+i2)*L + i3)*L + i4)*4 + mu;
}
```

The class that defines the requisite $SU(2)$ algebra is identical to the $U(1)$ class in structure—the essential differences are in the implementation of group multiplication and inversion:

```
SU2 inv() {
   SU2 b(a0,-a1,-a2,-a3);
   return b;
}
```

and

```
SU2 SU2::operator*(const SU2 & other) {
   double c0 = a0*other.a0 - a1*other.a1 - a2*other.a2 - a3*
      other.a3;
   double c1 = a0*other.a1 + other.a0*a1 - a2*other.a3 + a3*
      other.a2;
   double c2 = a0*other.a2 + other.a0*a2 + a1*other.a3 - a3*
      other.a1;
   double c3 = a0*other.a3 + other.a0*a3 - a1*other.a2 + a2*
      other.a1;
   return SU2(c0,c1,c2,c3);
}
```

24.5.2 Observables

As usual, we test the code by computing the average plaquette. Results are shown in Figure 24.11. These were obtained on a lattice of size 8^4 with very few configurations. A small lattice such as this suffices to obtain the average plaquette quite accurately because the plaquette is an "ultraviolet" quantity, meaning that its size approaches zero in the continuum limit. Furthermore, few lattice configurations are required because each lattice contains $6L^4$ plaquettes that are summed to obtain the estimated average. The figure also shows the small and large coupling perturbative results:

$$\langle \Box \rangle = \frac{\beta}{2} \quad \text{small } \beta$$

and:

$$\langle \Box \rangle = 2 - \frac{3}{2\beta} \quad \text{large } \beta.$$

SU(2) string tension and confinement

We now turn to consideration of the string tension for $SU(2)$ pure gauge theory. Recall that this has the interpretation of a linear potential between static charges in $U(1)$ gauge

882 SU(2) Gauge theory

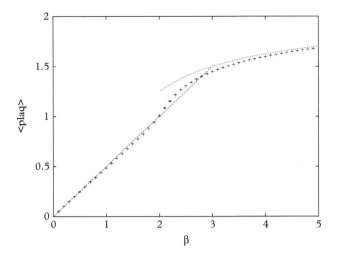

Figure 24.11 *Average plaquette for 4d SU(2) gauge theory.*

theory in two dimensions and is the coefficient of the area law for $Z(2)$ gauge theory (of course, the linear static potentials are also area laws). An interaction energy that rises linearly with distance is termed a **confinement** potential because it implies that opposite charges require infinite energy to separate them asymptotically far. This behavior is not unexpected in two-dimensional electrodynamics since the Coulombic field is necessarily confined to a one-dimensional tube. Somewhat confusingly, confinement also occurs in four-dimensional lattice electrodynamics, which definitely does not agree with experience. The resolution to this conundrum is that a phase transition occurs and the continuum limit of the theory lies on the side of the transition where the string tension is zero. In effect, confining compact electrodynamics evolves into nonconfining noncompact electrodynamics as the continuum is approached.

We seek to explore this same behavior for the case of non-Abelian $SU(2)$ gauge theory, and, in particular, discover if $SU(2)$ gauge theory confines in the continuum limit. We choose to find the string tension by evaluating the Wilson loop, which is expected to behave as in Eq. 24.62 for large T. The interaction is best parametrized as a combination of a weak coupling Coulombic term and a strong coupling linear interaction. Thus we write:

$$V(R) = \sigma R - \alpha/R + C. \tag{24.73}$$

Notice that this is a simple sum of Wegner's area and perimeter laws.

The strong coupling (small β) expression for the Wilson loop is given by:

$$\langle WL(R,T) \rangle = 2\left(\frac{\beta}{4}\right)^{RT}. \tag{24.74}$$

Thus in this limit the string tension is given by:

$$\sigma = -\log\left(\frac{\beta}{4}\right), \qquad (24.75)$$

which is useful for code verification.

Obtaining decent estimates of Wilson loops is not difficult if the loops are not too large and the lattices are small. For small β the linear dependence of the static interaction is nearly exact and thus only a few values of R and T are necessary to extract the string tension. We are thus able to obtain the string tension on lattices of size 10^4 or smaller with only a few hundred configurations. For larger couplings extracting the string tension becomes tougher because the Coulombic interaction begins to become important and larger loops are required to accurately extract the linear term. Useful strategies are to perform a two-dimensional fit of the Wilson loop to $\exp(-\sigma RT - \alpha(T/R) + CT)$, make a one-dimensional fit in the variable RT, or to make a one-dimensional fit of $\langle WL(R,R)\rangle$ to $\exp(-\sigma R^2 - \alpha + CR)$.

Results from the second and third methods are presented in Figure 24.12. These were generated on a 10^4 lattice and were confirmed on a 14^4 lattice. The string tension obtained from the second method is plotted as crosses with error bars. Notice that they follow the perturbative behavior (dotted line) very closely for $\beta \lesssim 2.0$. After this point the string tension obtained with this method disagrees with the third method (which is disconcerting, but is an indication that sometimes numerical work devolves into something of an art). Results for the third method are shown as solid squares for

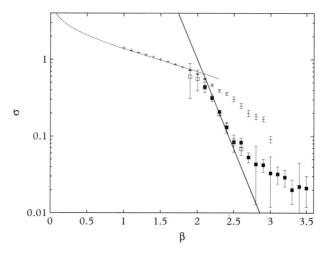

Figure 24.12 *String tension for 4d SU(2) gauge theory. Solid line: renormalization group scaling. Dotted line: perturbative expression. Solid points: fit to $WL(R,R)$ for $L = 10$. Open points: fit to $WL(R,R)$ for $L = 14$. Crosses: fit to $WL(R,T)$ for $L = 10$.*

$L = 10$ and open squares for $L = 14$. As can be seen, these agree quite well, indicating that finite size effects are under control.

> Argue that the second fitting method should work well for small β and the third should work better for β in the scaling region.

Extracting the continuum limit from these results follows a similar path as that used above for $U(1)$ gauge theory except that the formalism must be generalized. As before, we have computed

$$\hat{V}(\hat{R}, g_0, a) \to \hat{\sigma}(g_0)\hat{R}, \quad \text{large } R$$

and we identify

$$a^2 \sigma = \hat{\sigma}.$$

The dependence of the bare coupling on the ultraviolet cutoff is chosen to keep σ constant thus one must have the **renormalization group equation**:

$$a\frac{d}{da}\sigma = a\frac{d}{da}a^{-2}\hat{\sigma} = 0.$$

Or

$$\left(a\frac{\partial}{\partial a} + a\frac{dg_0}{da}\frac{\partial}{\partial g_0} - 2\right)\hat{\sigma} = 0. \tag{24.76}$$

The derivative appearing in the second term can be evaluated in the weak coupling limit (which corresponds to the continuum limit). This is traditionally defined in terms of the renormalized coupling g rather than the bare coupling g_0 and is called the **beta function** (*not* our β!). For pure $SU(N)$ gauge theory it is given by:

$$\beta(g) = a\frac{\partial g}{\partial a} = -\frac{1}{(4\pi)^2}\frac{11}{3}Ng^3 - \frac{1}{(4\pi)^4}\frac{34}{3}N^2 g^5 + O(g^7) \equiv \beta_0 g^3 + \beta_1 g^5 + O(g^7). \tag{24.77}$$

Applying the renormalization group idea to the renormalized coupling gives the relationship (Creutz, 1980)

$$a\frac{dg_0}{da} = -\beta_0 g_0^3 + \ldots.$$

We can now solve Eq. 24.76 to obtain:

$$\hat{\sigma} = a^2 F\left(a \exp\left(\int \frac{dg_0}{\beta_0 g_0^3}\right)\right) = a^2 F\left(a \exp\left(\frac{3\pi^2}{11}\beta\right)\right)$$

where F is an unknown function. But $\hat{\sigma}$ is $a^2\sigma$ so the argument of F must be constant and the dimensionless string tension must be given by:

$$\hat{\sigma} = \exp\left(-\frac{6\pi^2}{11}\beta\right) \cdot const.$$

This is plotted as a solid line in Figure 24.12; the normalization constant was fit to the data and is approximately

$$\exp\left(\frac{12\pi^2}{11}\right).$$

The agreement of the renormalization group result with the lattice result indicates that the scaling window is in the range of approximately $\beta \in (2.0, 2.5)$ for lattices of size 10^4 or so. Most importantly, this agreement implies that the continuum limit can be attained on the lattice with *no intervening phase transition*. Thus the confinement that exists in the strong coupling limit also exists in the continuum limit, and hence is a property of nature. Creutz's 1980 lattice study of $SU(2)$ pure gauge theory was the first to establish this central feature of the Standard Model.

The phases of Gauge theory

Thus far we have seen two ways in which lattice gauge theories can manifest in the continuum limit: as a confining theory, or as theory without confinement. In general there are three possibilities, listed here

confinement phase a mass gap exists and the Wilson loop obeys the area law.

Higgs phase a mass gap exists and the Wilson loop obeys the perimeter law.

Coulomb phase there is no mass gap and the Wilson loop obeys the perimeter law.

These mass gaps are defined in the pure gauge theory and therefore can be interpreted as the mass of the gauge boson. The three phases map directly to Standard Model phenomenology: the strong interaction manifests in the confinement phase, while electromagnetism does so in the Coulomb phase. The "Higgs phase" refers to the famous construction of gauge field theories with massive gauge bosons which are required to explain the weak force.

These phases collapse to two possibilities when the Higgs field transforms in the fundamental representation of the gauge group because the Higgs and confinement phases are smoothly connected (Fradkin and Shenker, 1979). In fact, Elitzur's theorem implies that the Higgs field *cannot* take on a nontrivial vacuum expectation value. Since the Standard Model does indeed place the Higgs field in the fundamental representation, the usual textbook development of the Higgs mechanism is flawed, and the phenomenon must be investigated in far more detail.

Practical observables

The computations presented in this chapter make it appear that obtaining statistically meaningful results is a simple matter of generating sufficiently many field configurations. We feel compelled to inform the reader that this convenient situation is rarely seen in practical lattice gauge computations. In particular, measuring physical observables such as glueball masses, can be extremely difficult. Thus, for example, the plaquette-plaquette correlation function can access the $\mathcal{J}^{PC} = 0^{++}$ glueball mass. But attempting to do this will simply introduce the numericist to persistent and frustrating noise. Some reflection reveals the problem: a glueball is a physical entity with extent of order one fermi, whereas a plaquette is an ultraviolet operator that disappears as the continuum is approached. Thus the plaquette has negligible overlap with the physical glueball (look at the coefficient in Eq. 24.32 or Eq. 24.61). Bashing at the problem with configurations will not be sufficient to overcome a poor overlap in all but the simplest of cases.

Because of this, lattice physicists expend great care in choosing operators that have good overlap with the physical objects that they wish to measure. Thus, in the case of glueballs, it is important to construct gluonic operators with a size of roughly 1 fermi. An elegant way to achieve this is with **link smearing** in which "skinny links", $L(x, \mu)$, are replaced with "fat links", $L_{eff}(x, \mu)$. An automatic way of achieving this is explored in Exercise 14.

24.6 Fermions

The presentation so far has ignored a rather important portion of the universe, namely the fermions that comprise matter. This is not an accident, as incorporating fermions in a stochastic formalism is a notoriously difficult task.

Grassmann fields

As with bosons, fermions are represented by classical fields that serve as variables of integration in the path integral. However, since fermions anticommute, these fields are written in terms of *Grassmann* numbers that behave peculiarly (Berezin, 1966):

$$\eta \xi = -\xi \eta.$$

Not surprisingly, this is a disaster for evaluating the probabilities that underpin stochastic methods.

The resolution to this problem is both straightforward and unfortunate (for reasons we will expand on below). Namely, all fermions in the Standard Model appear quadratically in the action, and hence can be explicitly integrated using the Grassmann Gaussian integral:

$$\int D\bar{\psi} D\psi \, \exp(-\bar{\psi} M \psi) = \det(M). \tag{24.78}$$

We follow this approach as well.

Euclidean fermions

Our immediate goal is to obtain the matrix M of Eq. 24.78 in Euclidean space. The starting point is the Minkowski action:

$$\int d^D x \, \bar{\psi} (i\slashed{D} - m_0) \psi$$

with $D_\mu = \partial_\mu + ie_0 A_\mu$.

Following the same rules as before for the temporal derivative and for the vector potential, the covariant derivative becomes:

$$D_\mu = (iD_E^4, -\vec{D}_E).$$

The Euclidean version of the Dirac matrices is defined via

$$\gamma^\mu = (\gamma^0, \vec{\gamma}) = (\gamma_E^0, i\vec{\gamma}_E).$$

Notice that this implies that

$$\{\gamma_E^\mu, \gamma_E^\nu\} = 2\delta^{\mu\nu}.$$

Putting it together gives

$$i\slashed{D} = -(\gamma_E^4 D^4 + \vec{\gamma}_E \cdot \vec{D}_E)$$

and the fermionic part of the Euclidean path integral becomes

$$\int D\bar{\psi} \, D\psi \, \exp\left(-\int d^4 x_E \bar{\psi} (\gamma_E^\mu D_E^\mu + m_0) \psi \right). \tag{24.79}$$

We drop the subscripts henceforth.

Discretization

If we ignore the gauge fields for the moment, placing the action on a lattice simply involves discretizing the derivative:

$$\partial_\mu \psi(x) \to \frac{1}{2a}(\psi_{n+a\hat{\mu}} - \psi_{n-a\hat{\mu}}).$$

As usual, we will refer to $a\hat{\mu}$ as μ in the following. At this point one might think that using the central difference is as good as the forward or backward difference; however, the latter two are not hypercubic symmetric and hence would not yield the correct continuum limit without tedious tuning.

The full expression for the free fermionic action reads:

$$a^D \sum_{n\mu} \frac{1}{2a}(\bar{\psi}_n \gamma_\mu \psi_{n+\mu} - \bar{\psi}_{n+\mu} \gamma_\mu \psi_n) + m_0 a^D \sum_n \bar{\psi}_n \psi_n. \tag{24.80}$$

The matrix M is then (rescaling a bit):

$$M_{nm} = am_0 \delta_{mn} + \sum_\mu \frac{1}{2} \gamma_\mu (\delta_{m+\mu,n} - \delta_{m-\mu,n}).$$

If we generalize for the moment to $SU(N)$ theory on an L^D lattice this is an enormous, but sparse, matrix of size $4NL^D \times 4NL^D$!

Fermion doublers

It is useful to evaluate the free fermion propagator as we did for the scalar field in Section 24.2. The result is

$$\langle T\bar{\psi}(x)\psi(y)\rangle = \sum_k \frac{\cos(k \cdot (x-y))}{am_0 + i \sum_\mu \gamma_\mu \sin(ak_\mu)} \tag{24.81}$$

which has a pole at

$$aE = \sinh^{-1}\sqrt{a^2 m_0^2 + \sum_i \sin^2(ap_i)}. \tag{24.82}$$

This result is very similar to that for the scalar field (Eq. 24.18) and obeys the continuum dispersion relation $E^2 = \vec{p}^2 + m_0^2$ as desired. Nevertheless, the minor differences in form have a major impact: because of the missing factors of two, there is an additional pole in the propagator at $p_i = \pi/a$. This means that there is effectively an additional identical fermion in the spectrum along each direction, and therefore a total of 16 identical fermions in four dimensions. The 15 extra fermions are called *doublers* and are a generic and serious problem with the lattice formulation of the Standard Model.

When Wilson first formulated quantum chromodynamics on the lattice he solved the doubler problem by adding a term to the dispersion relation that moves the doubler poles to $O(1/a)$ while at same time not affecting the continuum limit. Wilson's term was:

$$r \int \bar{\psi} a \partial^2 \psi$$

where r is an arbitrary dimensionless constant. The result is called a *Wilson fermion*.

Chiral symmetry

The Wilson prescription does great damage to the theory when fermions are massless. In this case the continuum theory is symmetric under a *chiral transformation*:

$$\psi \to \exp(-i\gamma_5 \theta) \psi$$

where

$$\gamma_5 = \gamma_1 \gamma_2 \gamma_3 \gamma_4.$$

The discrete action is also symmetric under this transformation; however the Wilson term explicitly breaks chiral symmetry. This is problematic because the electroweak sector of the Standard Model is a *chiral gauge theory* in which left-handed and right-handed fermions behave differently. The Wilson prescription couples these two types of fermions and therefore mixes up the chirality properties of the weak interaction.

The practical implication of this is that it was possible to place quantum chromodynamics on the lattice, but not the electroweak theory. Fortunately, in the past few decades this problem has been solved with the development of *domain wall* and *overlap* fermions which implement chiral fermions on the lattice, although at great computational expense (Jansen, 1996; Giusti, 2003). In the following we focus on non-chiral gauge theories, which allows us to press ahead with Wilson fermions.

Interacting fermions

The free action of Eq. 24.80 is clearly not gauge invariant. A natural way to make it so is to connect the fermion fields that appear at different sites with a link variable. Thus Eq. 24.80 is replaced with:

$$S = a^D \sum_{n\mu} \frac{1}{2a} (\bar{\psi}_n \gamma_\mu L(n,\mu) \psi_{n+\mu} - \bar{\psi}_{n+\mu} \gamma_\mu L^\dagger(n,\mu) \psi_n) + m_0 a^D \sum_n \bar{\psi}_n \psi_n +$$

$$+ r a^D \sum_{n\mu} \frac{1}{2a} (2\bar{\psi}_n \psi_n - \bar{\psi}_n L(n,\mu) \psi_{n+\mu} - \bar{\psi}_{n+\mu} L^\dagger(n,\mu) \psi_n). \qquad (24.83)$$

In fact this simple procedure reproduces the desired continuum limit, as substituting $L(n\mu) = \exp(ieaA_\mu(x+\mu/2))$ and expanding reveals. The end result is a fermionic matrix $M(L)$ that appears in the Wilson action and is ultimately integrated out using Eq. 24.79.

24.6.1 Fermionic updating

Integrating the fermionic portion of the generating functional yields:

$$Z = \int DA \det M(A) \exp(-S_A) = \int DA \exp(-S_A + \text{Tr} \log M(A))$$

where we have reverted to continuum notation. A Metropolis update thus relies on an update probability:

$$p(A \to A') = \exp\left(\Delta S_A(A \to A')\right) \cdot \frac{\det M(A')}{\det M(A)}.$$

The determinants in this expression make the probability extraordinarily difficult to compute. In fact, for several decades this difficulty was sufficient to induce the entire field to avoid the determinant by setting it equal to unity. This drastic step was called the *quenched approximation* but is actually not an approximation because it ruins the causal structure of the field theory. More recently a number of strategies have been successfully employed to account for the fermionic part of the action.

The simplest approach is to rely on the fact that gauge variables are typically updated locally (i.e., a link at a time) and therefore only a small portion of $M(A)$ changes. This can be leveraged to directly evaluate the ratio of determinants.

Alternatively, the ratio of determinants can be re-written in terms of a complex scalar field as

$$\frac{\det M(A')}{\det M(A)} = \frac{\int D\phi^* D\phi \, \exp(-\phi^* M(A) \phi)}{\int D\phi^* D\phi \, \exp(-\phi^* M(A') \phi)}.$$

One can store $\langle M(A) \rangle = \det M(A)$ and compute the denominator in terms of the change

$$\det M(A') = \langle \Delta M \rangle + \langle M(A) \rangle.$$

This yields the ratio of determinants and the value of $\langle M(A') \rangle$ for the next update step. Notice that ΔM is local and therefore only a few bosonic sweeps should suffice to determine $\langle M(A') \rangle$ accurately.

Hybrid Monte Carlo

The nonlocal nature of the effective action that fermions introduce represents a serious problem for Monte Carlo techniques. What is needed is a method to update the entire lattice with reasonable acceptance rates. A way forward (as discussed in Section 18.3.3) is provided by the observation that the expectation value of operators in canonical ensembles are equivalent to expectation values in microcanonical ensembles if the energy is fixed by the temperature, and these are given by time averages over trajectories (Callaway and Rahman, 1982):

$$\langle \mathcal{O} \rangle (\beta)_{can} \underset{thermo}{\rightarrow} \langle \mathcal{O} \rangle (E = E(\beta))_{micro} \underset{ergodic}{\rightarrow} \lim_{\tau_f \to \infty} \frac{1}{\tau_f} \int_0^{\tau_f} d\tau\, \mathcal{O}(\varphi(\tau)). \qquad (24.84)$$

The degrees of freedom in the action are represented generically as φ here and in the following discussion.

This relationship has the important implication that methods we have studied in Chapter 18 can be applied to quantum field theory. Thus, for example, the Langevin method can be applied to an action $S[\varphi]$ if a fictitious time τ is introduced (Parisi and Wu, 1981). In this case new configurations are generated with the aid of random variables $\hat{\eta}$ as (discrete indices on φ are suppressed):

$$\varphi(\tau + h) = \varphi(\tau) - 2h \frac{\partial S}{2\, \partial \varphi} + \sqrt{2h}\, \hat{\eta}, \qquad (24.85)$$

with $\hat{\eta}$ chosen from a Gaussian distribution:

$$p(\hat{\eta}) = \frac{1}{\sqrt{2\pi}} \exp(-\hat{\eta}^2/2). \qquad (24.86)$$

Alternatively, one can employ molecular dynamics methods directly if a fictitious time and fictitious momenta are introduced (Callaway and Rahman, 1982). In this case the expectation value of an operator is given by:

$$\langle \mathcal{O} \rangle = \frac{1}{\bar{Z}} \int D\pi\, D\varphi\, \mathcal{O}[\varphi]\, \exp(-\beta H[\pi, \varphi]) \qquad (24.87)$$

where $\bar{Z}$ is the usual normalization and the fictitious Hamiltonian is given by:

$$H = \frac{1}{2} \sum_i \pi_i^2 + S[\varphi]. \qquad (24.88)$$

Eq. 24.87 holds, of course, because the path integral over momentum cancels between numerator and denominator.

As with the Langevin approach, Eq. 24.84 permits computing expectation values by following the relevant operator over its trajectory, in this case, as computed with Hamilton's equations of motion.

It is instructive to compare the molecular dynamics and Langevin approaches. In particular, the analogue of Eq. 24.85 is (we employ the symmetric discretization of $\ddot{\varphi}$ and use ϵ as the temporal step size):

$$\varphi(\tau + \epsilon) = \varphi(\tau) - \epsilon^2 \frac{\partial S}{2\, \partial \varphi} + \epsilon \pi(\tau) \qquad (24.89)$$

with:

$$\pi(\tau) = \frac{1}{2\epsilon}(\varphi(\tau+h) - \varphi(\tau-h)).$$

Thus the methods are related by the mapping $2h \leftrightarrow \epsilon^2$ and $\hat{\eta} \leftrightarrow \pi$. It is evident that molecular dynamics method explores phase space much more rapidly than the Langevin method when similar temporal spacings are used. On the other hand, the Langevin method will probe the local environment more closely.

In 1985 Duane (1985) suggested combining the merits of the two approaches by running a molecular dynamics simulation while occasionally choosing the momenta randomly from a Gaussian distribution. While successful, Duane's method suffers from a problem shared by the Langevin and molecular dynamics approach, namely a temporal step size dependence exists. This is a serious issue because lattice gauge computations tend be exceptionally costly, hence performing many of them to permit extrapolation in the temporal spacing is to be avoided if at all possible.

Remarkably, this problem can be overcome by combining everything with occasional Metropolis updates (Scalettar, 1986). A generic algorithm achieving this follows.

```
initialize φᵢ

loop

  generate π according to its Gaussian pdf

  compute π'ᵢ = πᵢ - (ε/2) ∂S[φ]/∂φᵢ

  call the initial configuration (π,φ)

  iterate
      φᵢ(τ+ε) = φᵢ(τ) + επ'ᵢ(τ)
      π'(τ+ε) = π'(τ) - ε ∂S[φ]/∂φᵢ(τ+h)
  end iterate

  call the final configuration (π̃,φ̃)
  accept the final configuration with probability

          min(1, exp(-βH[π̃,φ̃]) + βH[π,φ]))

  continue with either φ or φ̃

end loop
```

Notice that leap-frog updating is used for the molecular dynamics portion of the algorithm.

If the equations of motion were integrated exactly then H would be constant along a molecular dynamics trajectory and the configuration would always be accepted by the Metropolis test. This is not possible because of the finite temporal step size; however, the Metropolis test eliminates the error induced by the step size dependence (Duane, 1987)! Thus a combination of Langevin, molecular dynamics, and Metropolis updating succeeds in combining the superior elements of all methods. The result is called **hybrid Monte Carlo**. This forms the state of the art when combined with methods for updating fermionic determinants discussed above.

24.7 Exercises

1. [T] Wilson Fermions.
   Confirm that the dispersion relation for a free Wilson fermion behaves as expected. What is the $O(a)$ correction to the fermion energy? What is the coefficient of the $O(1/a)$ term in the doubler mass?
2. $U(1)$ String Tension.
   Obtain the string tension for compact $U(1)$ gauge theory in four dimensions as a function of β. You should find a phase transition near $\beta = 1$. Compare your results to the high temperature expansion $\sigma = -\log(\beta/2) + \beta^2/8$.
3. Z_2 String Tension.
   Obtain the string tension for two-dimensional Z_2 theory. Is there a phase transition? The high temperature expression for the string tension is $\sigma = \log \tanh(\beta)$.
4. [T] The Hopping Parameter.
   Rescale free ϕ^4 theory so that:

   $$S = \sum_m \phi_m^2 + \kappa \sum_{mn} \phi_m \phi_n.$$

   Show that in the continuum limit the hopping parameter κ goes to unity at a rate that depends on the bare mass.
5. Mass Gap and Correlation Length.
   Study the mass gap for three-dimensional ϕ^4 theory as a function of κ for $g = 0.4$. How does the mass gap behave near the critical coupling? Discuss this behavior in light of the correlation length.
6. Z_2 Polyakov Loops.

   a) Obtain the average Polyakov loops in two-dimensional Z_2 theory as a function of β using cold starts.
   b) Repeat with hot starts. Explain the observed behavior.
   c) Use hysteresis to evaluate the Polyakov loop. Check Section 20.3 if this is unfamiliar.

7. **Random Gauge Transformations.**
Local gauge transformations leave physics and observables invariant. This provides a powerful test for code.

a) Write code to evaluate the plaquette in $SU(2)$ gauge theory in four dimensions.

b) Implement a random gauge transformation over the entire lattice. The relevant formula is Eq. 24.51, where $g(x)$ is a random element of $SU(2)$. You can generate this by selecting $a_\mu \in (-1, 1)$, rejecting sets with $a^2 > 1$, otherwise normalize a^2 to unity and accept. Making two transformations at a given link is awkward so in your routine you should replace Eq. 24.51 with the following operation:

$$L(x, \mu) \to g(x)L(x, \mu)$$
$$L(x - \mu, \mu) \to L(x - \mu, \mu)g^\dagger(x).$$

Confirm that this is equivalent to the original formulation of the gauge transformation.

c) Compute the value of a particular plaquette in a particular gauge configuration, make the random gauge transformation, recompute the plaquette and confirm that the new value agrees exactly with the old one.

8. **[T] Temporal Gauge Transformations.**
Construct the gauge transformation that makes $U(x, \hat{t}) = 1$. Apply the transformation to a Polyakov loop and show that the value of the loop does not change.

9. **Two-dimensional $U(1)$ Average Plaquette.**
Compute the average plaquette in 2d $U(1)$ gauge theory. Confirm that your results agree with the exact result of Eq. 24.60.

10. **[T] Microcanonical $SU(2)$ Updates.**
Confirm that the microcanonical transformation of Eq. 24.72 preserves the action.

11. **$SU(2)$ Kennedy-Pendleton Heat Bath.**
For values of β that are physically interesting the accept/reject step in the Creutz heat bath algorithm can experience high rejection rates. Kennedy and Pendleton (1985) have addressed this issue with the following update algorithm:

```
compute S = staple and k = √det S
repeat:
D = -(log(η₁) + log η₂ · cos²(2π η₃))/(βk)
if (η₄² > 1 - D/2) goto repeat
a₀ = 1 - D
generate â
new link = a * (S/k)⁻¹
```

All η_i are uniform variates in $(0, 1)$.

Implement this algorithm and compare its run times to Creutz updating.

12. [T] Langevin Algorithm.
Prove that the Langevin algorithm obeys detailed balance in the limit $h \to 0$ by considering the probability for making a transition from φ to φ'. Hint: what must $\hat{\eta}$ be for this transition to occur?

13. [P] Hybrid Adiabatic Energy Surface.
The Wilson loop can be used to obtain the ground state interaction between static fermions:

$$\langle WL(R, T) \rangle \propto \exp(-T\,V(R))$$

for large T. This potential expresses the lowest energy response of the gauge degrees of freedom to the source and sink. It is commonly called the Σ_g^+ adiabatic energy surface. The first excited state, called the Π_u **hybrid adiabatic surface**, represents the interaction between the source and sink when the gauge degrees of freedom are in an excited mode. (The reader may recognize the notation as that for diatomic molecules.) We explore this structure in pure gauge four-dimensional $SU(2)$ theory.

a) Use the Wilson loop to obtain the ground potential. Be sure to work in the scaling window.

b) Obtain the Π_u surface. To do this we need to form a gluonic state that is orthogonal to the ground state (and with Π_u quantum numbers). This can be achieved by replacing the spatial legs of the Wilson loop with large, asymmetric staples, as indicated in Figure 24.13. Sum the staples over all the directions orthogonal to the spatial leg with weights of ± 1 depending on whether the staple moves in the $\pm \hat{v}$ direction. We recommend going two or three lattice spacings in the $\pm \hat{v}$ directions to improve the overlap of the configuration with the physical state.

Your answer will be similar to that for $SU(3)$ gauge theory. For help consult Juge (1998).

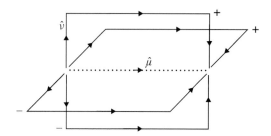

Figure 24.13 Π_u *link configuration.*

14. [P] $SU(2)$ 0^+ and 2^+ Glueball Masses.

   a) The particle excitations in pure gauge theories are called "glueballs". Compute the mass of the $J^P = 0^+$ and 2^+ glueballs in four-dimensional $SU(2)$ gauge theory. You can do this by fitting the correlation function

   $$C_J(t) = \langle \phi_J^\dagger(t)\phi_J(0) \rangle$$

   or plotting the effective mass. Use the following for interpolating fields:

   $$\phi_0(t) = \frac{1}{L^3} \sum_{xyz} \left[\Box_{xy}(R;t) + \Box_{xz}(R;t) + \Box_{yz}(R;t) \right]$$

   and:

   $$\phi_2(t) = \frac{1}{L^3} \sum_{xyz} \left[\Box_{xz}(R;t) + \Box_{yz}(R;t) - 2\Box_{xy}(R;t) \right].$$

   Here $\Box_{xy}(R;t)$ denotes a spatial loop of extent $R \times R$ in the xy plane at time t. The scale R should maximize overlap with the physical state; choosing it to be 3 or 4 is not a bad idea.

   The scalar glueball requires some care since it has the quantum numbers of the vacuum. Thus you should employ the connected correlator:

   $$C_0(t) = \langle \phi_0^\dagger \phi_0(0) \rangle - \langle \phi_0(0) \rangle^2.$$

   Since the two terms in C_0 are strongly correlated it is advisable to use jackknife or similar techniques to estimate the error in the 0^+ glueball mass.

   b) These operators do not have good overlap with physical glueballs, even for optimal R. As discussed in the text, the problem can be ameliorated by using "smeared links" which have transverse size. A simple iterative scheme for smearing links is called **APE smearing** (Albanese, 1987). The algorithm is as follows:

   ```
   iterate n_ζ times
      L(x, μ) → L(x, μ) + ζ S(x, μ)
      project L to SU(2)
   end iterate
   ```

   Here S is the sum of staples with respect to L. The parameters n_ζ and ζ should be tuned to optimize the desired signal. We have found that $n_\zeta = 8$ and $\zeta = 0.3$ do well.

   Repeat your calculation of the 0^+ and 2^+ glueball masses with smeared operators. Fix the scale by setting the string tension equal to 0.18 GeV^2 and determine the physical glueball masses. Compare your results to (Teper, 1998).

15. [P] $SU(3)$ Gauge Theory.
    We seek to generalize the $SU(2)$ heat bath algorithm to $SU(3)$ gauge theory. The action is given by Eq. 24.67.

    The first task is to find a convenient representation of an element of $SU(3)$. It is tempting to write a group element in terms of the group generators as we did with $SU(2)$,

    $$g = \exp(i\vec{\omega} \cdot \vec{T})$$

    where $T_i = \lambda_i/2$ and λ_i is a **Gell-Mann matrix**. However, there is no simple constraint on the determinant of g and this method bogs down. An alternative is to toss in the towel and simply keep track of 18 real numbers (i.e., 3×3 complex numbers). A middle ground would be to store 12 real numbers that correspond to the first two rows of g and then reconstruct the third row as:

    $$g = \begin{pmatrix} \vec{a} \\ \vec{b} \\ \vec{a}^* \times \vec{b}^* \end{pmatrix}.$$

    The vectors $\vec{a}$ and $\vec{b}$ must be normalized and mutually perpendicular. As usual the best choice is dictated by the tradeoff of memory and speed. We note that round off error accumulates in the group elements and it is useful to periodically project link variables to unitarity.

    The $SU(2)$ heat bath algorithm can be extended to $SU(N)$ by updating in $SU(2)$ subgroups (Cabibbo and Marinari, 1982). For $SU(3)$ one uses the Creutz or Kennedy-Pendleton algorithms to update each of:

    $$R = \begin{pmatrix} r_{11} & r_{12} & 0 \\ r_{21} & r_{22} & 0 \\ 0 & 0 & 1 \end{pmatrix}, \quad S = \begin{pmatrix} s_{11} & 0 & s_{12} \\ 0 & 1 & 0 \\ s_{21} & 0 & s_{22} \end{pmatrix}, \quad T = \begin{pmatrix} 1 & 0 & 0 \\ 0 & t_{11} & t_{12} \\ 0 & t_{21} & t_{22} \end{pmatrix}$$

    and forms $g' = RST$.

    a) Compute the average plaquette for four-dimensional $SU(3)$ gauge theory. Compare your results to the large and small coupling expressions:

    $$\langle \Box \rangle = \frac{\beta}{6}, \quad \langle \Box \rangle = 3 - \frac{6}{\beta}.$$

    b) At what coupling does the Polyakov loop take on a non-zero value?

    c) Repeat the finite size scaling analysis of the string tension of Section 24.5.2. Consult (Cabibbo and Marinari, 1982) for help.

16. [P] Center Vortex Confinement Hypothesis.
The *center* of $SU(N)$ is comprised of the elements of that group that commute with all other elements. This subgroup is isomorphic to the cyclic group Z_N that we have studied earlier. It is easy to convince yourself that multiplying all the links on a particular time slice by $z \in Z_N$ is a global symmetry of $SU(N)$ (this has no effect on spatial plaquettes and yields a factor of $zz^\dagger$ in temporal plaquettes. Thus Yang-Mills theory has a global Z_N symmetry.

Notice, however, that the Polyakov loop is *not* symmetric under this operation; in particular it transforms to $z \cdot PL$ since only one link in the loop is modified by the transformation. For this reason the average Polyakov loop serves as an order parameter of the global Z_N symmetry:

$$\langle PL \rangle = \begin{cases} \text{zero, unbroken } Z_N, \\ \text{not zero, broken } Z_N. \end{cases}$$

Interestingly, this observation can be related to *confinement* because the expectation value of the Polyakov loop is related to the free energy of a quark:

$$\langle PL \rangle = \exp(-F_q L)$$

and this is zero when quarks are confined (i.e., their free energy is infinite) (Greensite, 2011).

Z_N transformations are an example of "singular gauge transformations" (which are *not* gauge transformations!); namely they can be generated by performing a gauge transformation that is periodic modulo a Z_N phase factor:

$$L(t,\vec{x},\hat{t}) \to g(t,\vec{x})U(t,\vec{x},\hat{t})g^\dagger(t+1,\vec{x}) \qquad (24.90)$$

with $g(L,\vec{x}) = z^* g(0,\vec{x})$. For example the symmetry transformation $L(t_0,\vec{x},\hat{t}) \to zL(t_0,\vec{x},\hat{t})$ can be achieved by setting $g(t,\vec{x}) = 1$ for $t \leq t_0$ and $g(t,\vec{x}) = z^*$ for $t > t_0$. Singular gauge transformations are of central interest to the study of confinement because (in this case) they are associated with $D-2$ dimensional gauge field structures called *vortices*.

It has been postulated that vortices drive confinement ('t Hooft, 1979) because they are localized field configurations which percolate (Chapter 8!) the lattice. The argument is quite general and relies on the fact that localized field distributions contribute independently to the expectation value of the Wilson loop operator, and hence yield an area law interaction.

Our goal is to test this hypothesis by computing the effect of removing vortices from lattice field configurations on the string tension in four-dimensional $SU(2)$ gauge theory. Vortices can be identified by making a gauge transformation that brings a link variable as close as possible to the center group elements, ± 1. This can be done by maximizing the quantity

$$D = \sum_{x,\mu} \left[\frac{1}{2}\mathrm{tr}L(x,\mu)\right]^2$$

over the space of gauge transformations, $L(x\mu) \to L^{(g)}(x,\mu)$. This is called **direct maximal center gauge** (Del Debbio, 1998).

The projection onto vortex configurations is then made with the replacement:

$$L^{(g)}(x,\mu) \to Z(x,\mu) \equiv \mathrm{sgn}[\mathrm{tr}L^{(g)}(x,\mu)].$$

Alternatively, removing vortices from the field configuration can be achieved by setting:

$$L^{(g)}(x,\mu) \to V(x,\mu) \equiv Z^\dagger(x,\mu) \cdot L^{(g)}(x,\mu).$$

The task of minimizing the quantity D is difficult because a large space must be searched and $D[\{L(x,\mu)\}]$ has many local minima. We therefore recommend minimizing D with the **simulated annealing algorithm**. The idea is to slowly "cool" (i.e., lower D) by making Metropolis updates with random gauge transformations. The (inverse) annealing temperature will be called β_A in the following, and we wish to increase its value as the link variables are updated. An implementation is

```
loop in β_A from a minimum to a maximum value
    loop to N_A
        anneal(β_A, N_touch)
    end loop
end loop
// obtain projected and vortex configurations
loop in (x,μ)
    Z(x,μ) = 1
    if (tr(L(x,μ)) < 0)  Z(x,μ) = -1
    V(x,μ) = Z(x,μ) · L(x,μ)
end loop
```

The annealing algorithm is

```
loop in (x,μ)
    loop until N_touch updates are made
        select g, a random element of SU(2)
        compute δD(L, L^(g))
        if η̂ < min(1, exp(β_A δD))
            transform L(x,μ) and L(x-μ,μ)
        end if
    end loop
end loop
```

That there is art in the simulated annealing algorithm is made evident by the number of parameters that define it. In particular one needs to define the values of β_A to be probed, the number of times to sweep through the lattice at each annealing temperature (N_A), and the number of times a Metropolis update is made at each site. It is very difficult to run the algorithm and declare that an absolute minimum has been obtained. We therefore recommend studying the behavior of observables as a function of N_A. Useful values for the parameters are $N_{touch} = 4$; $N_A = 10, 20$, or 40; and $\beta_A = (1, 2, \ldots, 20)$.

a) Obtain the average plaquette with L, Z, and V configurations.
b) Obtain the Polyakov loop in L, Z, and V configurations.
c) Use the Wilson loop to obtain the static quark interaction in L, Z, and V configurations. Confirm that the string tension goes to zero when vortices are removed and that it remains approximately constant when vortex configurations are used. Compare this behavior to that for the plaquette and Polyakov loop.

17. [P] Z_N Gauge Theory with Z_N Matter Fields.
The analogue of gauge theory coupled to fermions is pursued in the Z_2 gauge model in this project. The "matter" field will be represented as elements of Z_2 that reside on lattice sites. We call these S_x. The action is defined to be:

$$S = \beta_M \sum_{x,\mu}(1 - S_x L(x, \mu) S_{x+\mu}) + \beta_G \sum_{\square}(1 - L_{\square}).$$

This theory is invariant under the gauge transformation:

$$L(x, \mu) \to g_x L(x, \mu) g_{x+\mu}, \quad S_x \to S_x g_x.$$

We seek to explore the phase diagram for this theory in four dimensions using the average plaquette and gauge invariant link defined by:

$$\frac{1}{6}\langle 1 - L_{\square}\rangle$$

and

$$\frac{1}{4}\langle 1 - S \cdot L \cdot S\rangle$$

respectively.
The theory has a first order phase transition at $\beta_G = \log(1+\sqrt{2})/2$ when $\beta_M = 0$. The exact results for the average plaquette and gauge invariant link for $\beta_G = 0$ are

$$\frac{2}{1+\exp(2\beta_M)}$$

and

$$1 - \tanh^4(\beta_M)$$

respectively. These results will be useful for code verification.

Do your best to determine the phase structure of the theory. You will want to restrict your investigation to $\beta_G \in (0, 1)$ and $\beta_M \in (0, 0.6)$. Consult Creutz (1980b) for help.

18. [P] The Higgs Boson Mass.

    The Higgs mechanism is an ideal application of lattice field theory because it couples a scalar field to an $SU(2)$ gauge boson. We shall consider a Higgs doublet in the fundamental representation with continuum Minkowski action:

    $$\int d^4x \left[(D^\mu \Phi)^\dagger (D_\mu \Phi) - m_0^2 \Phi^\dagger \Phi + \lambda_0 (\Phi^\dagger \Phi)^2 - \frac{1}{4} F_a^{\mu\nu} F_{\mu\nu a} \right]$$

    where

    $$D^\mu \Phi = [\partial^\mu + i g_0 \frac{\sigma_a}{2} A_a^\mu] \Phi$$

    and

    $$F_a^{\mu\nu} = \partial^\mu A_a^\nu - \partial^\nu A_a^\mu - g_0 \epsilon_{abc} A_b^\mu A_c^\nu.$$

    The lattice version of this is:

    $$S = -\frac{\beta}{2} \sum_\Box L_\Box + \sum_x \left[\frac{1}{2} \text{tr}\, \phi_x^\dagger \phi_x + \lambda \left(\frac{1}{2} \text{tr}[\phi_x^\dagger \phi_x] - 1 \right)^2 \right] - \kappa \sum_{x\mu} \text{tr}\, \phi_x^\dagger L(x, \mu) \phi_{x+\mu}$$

    where $\lambda = \kappa^2 \lambda_0$, $\beta = 4/g_0^2$, and

    $$\kappa = \frac{1 - 2\lambda}{8 + a^2 m_0^2}.$$

    We use the "SU(2)" notation of (Bunk, 1995) to represent the scalar field: if the original field is defined as

$$\Phi = \begin{pmatrix} \varphi_1 + i\varphi_2 \\ \varphi_3 + i\varphi_4 \end{pmatrix}$$

then

$$\phi = \varphi_3 + i\vec{\sigma} \cdot (\varphi_2, \varphi_1, -\varphi_4).$$

Notice that $\Phi^\dagger \Phi = \text{tr}\phi^\dagger \phi / 2 = \det\phi$.

Fix parameters to $\beta = 8$, $\kappa = 0.131$, and $\lambda = 0.0033$ in the following. Update as follows

```
one gauge heat bath sweep
two scalar heat bath sweeps
one gauge microcanonical update
four scalar microcanonical updates
```

Allow ten of these between measurements to reduce autocorrelation.

Your task is to measure the Higgs and W gauge boson masses. Use the following for the Higgs interpolating operator:

$$H(t) = \frac{1}{L^3} \sum_{\vec{x}} \frac{1}{2} \text{tr}\, \phi_x^\dagger \phi_x.$$

For the W use:

$$W_\mu^a(t) = \frac{1}{L^3} \sum_{\vec{x}} \frac{1}{2} \text{tr}[-i\sigma^a \phi_x^\dagger L(x,\mu) \phi_{x+\mu}].$$

Use your results and the mass of the W to predict the mass of the Higgs boson. For more details consult (Bunk, 1995) or (Wurtz and Lewis, 2013).

19. [P] Symanzik Improvement.

The Wilson action is the simplest action that reproduces the required continuum field theory. It is possible to construct more complicated actions that have better convergence or improve rotational invariance. This idea was first promoted by Symanzik (1983) who used it to construct lattice actions that remove subleading terms in the lattice spacing. In particular, the following action has a lattice spacing error of order a^4, rather than the a^2 of the standard Wilson action[2]:

$$\sum_{\mu > \nu} \text{Tr}\, L_\square \rightarrow \frac{5}{3} \sum_{\mu > \nu} \text{Tr}\, L_\square - \frac{1}{12} \sum_{\mu > \nu} (\text{Tr}\, \square_{\mu\nu} + \text{Tr}\, \square_{\nu\mu}).$$

---

[2] More precisely, if the correlation length is ξ the error is reduced from $a^2/\xi^2 \log(a/\xi)$ to $a^4/\xi^4 \log(a/\xi)$.

The objects in the second term are 1 × 2 plaquettes. Code up and compare the recovery of rotational invariance in the Wilson loop to that obtained with the Wilson action. You might want to use Metropolis updating for this project. A detailed study of the problem can be found in Berg (1985).

BIBLIOGRAPHY

Albanese, M., F. Costantini, G. Fiorentini, *et al.* (1987). "Glueball masses and string tension in lattice QCD". Phys. Lett. B **192**, 163–169.

Balian, R., J. H. Droufe, and C. Itzykson (1975). *Gauge fields on a lattice. II. Gauge-invariant Ising model*. Phys. Rev. **D11**, 2098.

Berezin, F. A. (1966). *The Method of Second Quantization*. Academic Press.

Berg, B., A. Billoire, S. Meyer, and C. Panagiotakopoulos (1985). *SU(2) Lattice Gauge Theory: Standard Action Versus Symanzik's Tree-Improved Action*. Commun. Math. Phys. **97**, 31.

Bunk, B. (1995). *Monte Carlo Methods and Results for the Electroweak Phase Transition*. Nucl. Phys. Proc. Suppl. **42**, 566.

Cabibbo, N. and E. Marinari (1982). *A new method for updating SU(N) matrices in computer simulations of gauge theories*. Phys. Lett. **B 119**, 387.

Callaway, D. J. E. and A. Rahman (1982). *Microcanonical Ensemble Formulation of Lattice Gauge Theory*. Phys. Rev. Lett. **49**, 613.

Creutz, M. (1980). *Monte Carlo study of quantized SU(2) gauge theory*. Phys. Rev. **D21**, 2308.

Creutz, M. (1980b). *Phase diagrams for coupled spin-gauge systems*. Phys. Rev. **D21**, 1006.

Creutz, M. and K. J. M. Moriarty (1982). *Numerical studies of Wilson loops in SU(3) gauge theory in four dimensions*. Phys. Rev. **D26**, 2166.

Creutz, M. (1983). *Lattice Gauge Theory*. Addison-Wesley Publishing Company, New York.

Creutz, M. (1987). *Overrelaxation and Monte Carlo simulation*. Phys. Rev. **D36**, 515.

Del Debbio, L., M. Faber, J. Giedt, J. Greensite, and S. Olejník (1998). *Detection of Center Vortices in the Lattice Yang-Mills Vacuum*. Phys. Rev. **D58**, 094501.

Duane, S. (1985). *Stochastic quantization versus the microcanonical ensemble: Getting the best of both worlds*. Nucl. Phys. **B257**, 652.

Duane, S. and J. B. Kogut (1986). *The Theory of Hybrid Stochastic Algorithms*. Nucl. Phys. **B275**, 398.

Duane, S., A. D. Kennedy, B. J. Pendleton, and D. Roweth (1987). *Hybrid Monte Carlo*. Phys. Lett. **195B**, 216.

Elitzur, S. (1975). *Impossibility of spontaneously breaking local symmetries*. Phys. Rev. **D12**, 3978.

Fradkin, E. and S. H. Shenker (1979). *Phase diagrams of lattice gauge theories with Higgs fields*. Phys. Rev. **D19**, 3682.

Giusti, L. (2003). *Exact chiral symmetry on the lattice: QCD applications*. Nucl. Phys. (Proc. Suppl.) **119**, 149.

Greensite, J. (2011). *An Introduction to the Confinement Problem*. Springer.

Gross, D. J. and E. Witten (1980). *Possible third-order phase transition in the large-N lattice gauge theory*. Phys. Rev. **D21**, 446.

Jansen, K. (1996). *Domain Wall Fermions and Chiral Gauge Theories*. Phys. Rep. **273**, 1.

Juge, K. J., J. Kuti, and C. J. Morningstar (1998). *Gluon Excitations of the Static Quark Potential and the Hybrid Quarkonium Spectrum.* Nucl. Phys. Proc. Suppl. **63**, 326.

Kogut, J. B. (1979). *An Introduction to Lattice Gauge Theory and Spin Systems.* Rev. Mod. Phys. **51**, 659.

Montvay, I. and G. Münster (1994). *Quantum Fields on a Lattice.* Cambridge University Press.

Morningstar, C. J. (2009). *Explorations of Real Scalar Field Theory in 2+1 Space-Time Dimensions using the Monte Carlo Method.* Unpublished notes.

Parisi, G. and Y.-S. Wu (1981). *Perturbation Theory without Gauge Fixing.* Scientia Sinica **24**, 483.

Rothe, H. J. (1992). *Lattice Gauge Theories: an Introduction.* World Scientific.

Scalettar, R. T., D. J. Scalapino, and R. L. Sugar (1986). *New Algorithm for the Numerical Simulation of Fermions.* Phys. Rev. **B34**, 7911.

Symanzik, K. (1983). *Continuum limit and improved action in lattice theories. I. Principles and ϕ^4 theory.* Nucl. Phys. **B226**, 187.

Teper, M. (1998). *Glueball masses and other physical properties of SU(N) gauge theories in D=3+1.* arXiv:hep-th/9812187.

't Hooft, G. (1979). *A property of electric and magnetic flux in non-Abelian gauge theories.* Nucl. Phys. **B153**, 141.

Wegner, F. (1971). *Duality in Generalized Ising Models and Phase Transitions without Local Order Parameters.* J. Math. Phys. **12**, 2259.

Wilson, K. (1974). *Confinement of Quarks.* Phys. Rev. D. **10**, 2445.

Wurtz, M. and R. Lewis (2013). *Higgs and W boson spectrum from lattice simulations.* Phys. Rev. **D88**, 054510.

Yang, C. N. and R. Mills (1954). *Conservation of Isotopic Spin and Isotopic Gauge Invariance.* Phys. Rev. **96**, 191.

Index

abstract base class 415
accessor (class) 44
accessor function 184
action 110
adaptive step size 374
addition theorem 814
ADI method 655
advection equation 645
advection-diffusion
    equation 231
affine map 266
agglomeration 43
aggregate class 587
algorithm
    alternating direction
        implicit 655
    Barnes-Hut 621
    Chan 698
    cluster 243
    Cooley-Tukey 670
    Crank-Nicolson 648
    Graham scan 698
    guided random walk 740
    heat bath 229
    heuristic 238
    Jarvis march 698
    Lanczos 778
    leap frog 598
    Metropolis 223
    Metropolis-Hastings 223,
        237, 849
    Neville's 87
    particle mesh 627
    particle-particle
        particle-mesh 627
    Persson-Strang 683
    quotient-difference 99
    simulated annealing 238,
        899
    Swendsen-Wang 713, 727
    Thomas 649
    Verlet 598
    Visscher 693
    Wolff 714, 727
algorithms (Standard
    Library) 565

alternating chain model 797
alternating direction implicit
    method 655
ambiguous declaration 563
anisotropic Heisenberg
    antiferromagnet 774
anomalous dimension
    255, 706
anonymous function 584
anticommutation relations 832
antiferromagnet 703
antiferromagnetism 774
antisymmetrization
    operator 805
APE smearing 896
Appelquist-Carazzone
    theorem 847
application programming
    interface (API) 17
apt-get 52
ar command 17
area law 860, 882, 885
Argand plot 769
argon 830
argument deduction 559
arithmetic logic unit 36, 275
arrhythmia 691
assignment operator
    (class) 162
assignment to self 181
associative containers 566, 570
asymptotic normality 512
atomic class template 294
atomic configuration 806
atomic units 804
attenuation coefficient 489
attractor 431
    strange 436
auto (keyword) 580
autocorrelation 227, 608, 710,
    852, 902
autocorrelation dynamical
    exponent 712
autocorrelation function 227,
    601
autocorrelation time 227

autonomous differential
    equation 344
avalanche 720, 730
axon 726

B^0 meson 478
backward difference 644
Baker-Campbell-Hausdorff
    formula 654
bandwidth 278
bare couplings 846, 877
Barkhausen noise 720
Barnes-Hut algorithm 621
basin of attraction 262, 428
Baxter-Wu model 728
Bayes' theorem 539
Bayesian statistics 512
beam torsion 697
Bernoulli distribution 202
Bernoulli shift map 451
beta function 884
Bethe Ansatz 775, 796
Bethe-Bloch equation 504
Bhabha scattering 509
bias (double) 34
bifurcation plot 433
bilinear interpolation 89
billiard model 447
binary expression tree
    369, 635
binomial distribution
    202, 246
bit 32
bit reversal 670
Black-Scholes equation 688
blocking 284
Bogacki-Shampine
    method 380
Bohr radius 804
Bohr-Sommerfeld
    quantization 110
boids 636
boilerplate 418
boost library 78
Born-Oppenheimer
    approximation 815

Bose-Einstein
    condensation 116
boson 116
    W 488
    Z 485, 488
bound function 39, 43
boundary condition
    octahedral 628
boundary value problem 353
Box-Müller method 212
Bragg peak 509
Breit-Wigner distribution
    208, 768
bremsstrahlung 491, 492
Brillouin's theorem 836
broken ergodicity 724
broken phase 704
built-in datatypes 32
Burger's equation 689
Butcher tableau 362
butterfly effect 449
byte 31, 32

C++ 3
C++ classes 43
C++ concurrency library 289
C++11 199, 415, 577,
    578, 580
C++14 577, 580
C++17 577
c++filt 7
callback function 335, 338
calorimeter 37
camera coordinates 314
canonical transformation
    387, 444
capture clause 585
cardiac dynamics 691
Cash-Karp method 381
cast operator 174
catch 25
Cauchy distribution 202, 208
caustic 448
Cayley tree 267, 331, 341, 728
center vortex 898
central difference 644
central limit theorem 211, 227,
    272, 743
central processing unit 33,
    37, 275
CFL condition 647, 685
Chachiyo correlation
    functional 829
Chan's algorithm 698
charge conjugation 764

charmonia 147, 764
Chebyshef acceleration 663
Chebyshev polynomials 105
checkerboard updating 663,
    707
chi squared distribution 202
chiral gauge theory 889
class 30
    abstract base 415
    pure interface 416
    template 40
class SU2 881
class accessor 44
class constructor 39
class declaration 39
class instance 39
class interface 39
class method 43
class relationships 406
class scope 160
class template 557
class template
    specialization 561, 562
class tree diagram 137
Clebsch-Gordan
    coefficients 193, 194
client-server architecture 302
closed channel 758
cloud computing 21
cluster algorithm 243, 713
co-moving coordinates 634
coarray fortran 279
coarse graining 260, 270, 271
Coin nodes 315
Coin3d 315
collisionality 618
color perception 103
command line arguments 9
commutation relations 458
compact QED 868
companion matrix 71, 851
compilation error 5
compilation unit 4
compiler 4
completeness relationship 735
Compton scattering 488, 498
Computational Geometry
    Algorithms Library 682,
    698
conditional stability 394
configuration interaction 809
confinement 898
confinement interaction 765,
    882
confinement phase 885

conjugate gradient method 680
const (keyword) 167, 183
constrained dynamics 611
constructor 39
constructor (class) 160
container (Standard
    Library) 565
container adaptors 571
container class 566
continued fraction 98
continuity equation 645
convex hull 698
convolution 210
convolution theorem 695
Cooley-Tukey algorithm 670
Cooper pair 116
Cooper pair 817
copy constructor (class) 162
core 275
core-cusp problem 633
correlation coefficient 214, 515
correlation function 254, 607,
    727, 853, 896
correlation length 253, 705,
    893
correlation matrix 855
cosmological concordance
    model 619
Coulomb phase 885
coupled channel scattering 757
covariance matrix 214, 515
CPU 33, 37
Cramér-Rao bound 512
Crank-Nicolson algorithm 648,
    651, 654
Creutz heat bath 879
critical exponent 255, 705
critical slowing down 710
critical temperature 704
cross section 488
cubic spline 90
cumulative chi-squared
    distribution 215
CVS 21

damped harmonic
    oscillator 344, 398, 438
dark matter 619
data container (Standard
    Library) 565
data parallel model 279
data stack 289
debugger 26
default template argument 563
Delauney triangulation 682

Index 907

delta ray 485
dendrite 726
density effect 504
density functional theory 824
density waves 619
deque 568
destructor (class) 165
detailed balance 222, 715, 895
deuteron 147
diagonalization-variational method 735
differencing 644, 846
diffusion 259
diffusion equation 846
Dirac matrix 887
Dirac matrix 472
Dirac spinor 470–3
Dirac spinor 467
direct interaction 810
direct maximal center gauge 898
Direct3d 315
Dirichlet boundary conditions 644, 658
disease dynamics 400, 688
disk galaxy 633
displacement operator 654
dissociation energy 822
DNA 629
 stacking potential 629
domain wall fermion 889
double Compton scattering 500
double pendulum 399
Duffing oscillator 450
dynamic type 413

eclipse 4
editor 4
Edwards-Anderson model 724, 729
effective field theory 876
EGS 494
Eigen 54, 680
electromagnetic shower 493
electroweak theory 889
Elitzur's theorem 859
elliptic differential equation 644
encapsulation 30, 38, 43
energy drift 383
engine (`coin`) 335
EPR Paradox 503
equipartition theorem 596

ergodicity 222, 426, 851
 broken 724
Euclidean time 231, 846, 866
Euler angles 456, 479
Euler method 359
Euler parameters 456
Euler-Cromer method 388, 393, 439
Euler-Lagrange equations 348
exception 25
 catch 36
 throw 36
exception handling 187
exchange interaction 810
exchange-correlation functional 826
excitable medium 691
explicit differencing 647
exponential distribution 202
expression templates 165
extrapolation 95

Faraday tensor 473, 865
fast Fourier transform 669
Feigenbaum constant 435, 452
Fermi statistics 773, 832
fermion 116
 chiral 889
 domain wall 889
 overlap 889
 Wilson 889, 893
fermion doublers 888
fermion propagator 888
fermion sign problem 787
ferromagnet 703
ferromagnetism 774
Feynman diagram 495
Feynman path integral 231
Feynman-Hellmann theorem 747
FFT 671
fidelity 393, 650
field (`coin`) 335
field tensor 473
FIFO list 571
file (command) 7
final (keyword) 415
fine structure constant 803
fine structure interaction 765
finite element conforming 678, 682, 695
finite element method 673
finite size scaling 727
first order phase transition 717
FitzHugh-Nagumo model 691

five-point stencil 653
fixed point 222, 262, 427, 430, 691
 stable 431
 unstable 431
fixed-point notation 32
flop 275
FLRW cosmology 633
fluctuation-dissipation theorem 604
FLUKA 494
Flynn's taxonomy 278
forward difference 644
forward time, centered space differencing 645
four-tensor 473, 474
four-vector 468, 473
four-vectors 467
 contravariant 467
 covariant 467
 polarization 468
fpclassify 35
FPU model 426, 453
fractal 249, 326, 327, 443
Fredholm equation 753
frequentist statistics 512
Fresnel's equations 310
Friedmann equation 633
friend class 171
friend function 171
frustration (spin) 724
function
 bound 39
function parameter pack 578
function template 557
functor 67, 354
future object 296

gamma distribution 202
gamma matrix 472
gamma ray 488
gauge invariance 859, 865, 868
gauge transformation 894
Gauss-Hermite quadrature 147
Gauss-Legendre integration 132
Gauss-Seidel iteration 662
Gaussian quadrature 132
GEANT4 494, 505
gedit 4
Gell-Mann matrix 897
generalized gradient approximation 831
generic programming 556

generic type parameter 557
generic types 558
geometry transformation 329
ghost bit 34
Gibbs sampling 229, 707, 849, 861
Glauber updates 729
global error 359
glueball 873, 886, 896
GNU scientific library 78
GNU scientific library 18, 422
golden ratio 98
Goldstone boson 721, 796
Goldstone diagrams 833
Goldstone's theorem 775, 796
GPU 314
Graham scan 698
granularity 278
graphical processing unit 276
graphics engine 313, 314
graphics pipeline 314
Grassmann numbers 886
Green function 693, 752, 760, 834
group 858
group center 898
group manifold 879
guard macro 182
guided random walk algorithm 740
Gunnarsson-Lundqvist exchange-correlation functional 829

H_2 molecule 820
H_2^+ ion 818
Hückel theory 824
Hénon map 436, 451
Hénon-Heiles model 446
Haar measure 868, 879
hadrons 488, 494
Haftel-Tabakin regularization 754, 770
Haldane's conjecture 795
Hamilton's equations of motion 350
Hamiltonian 350
Hanna, R. C. 500
hartree 804
Hartree-Fock equation 809
Hartree-Fock method 809
Hausdorff dimension 250
header files 15, 17
heap (memory) 413, 581, 871
heat bath algorithm 229, 706

heat bath method 850, 879, 894
heavy weight code 279
Hebbs' rule 726
Heisenberg ladder model 798
helicity 475
helicity formalism 475
helium 813
Hermite polynomials 82
Hessian matrix 515
heuristic algorithm 238
Higgs boson 485, 901, 902
Higgs mechanism 775, 886, 901
Higgs phase 885
histogram 203
Hohenberg-Kohn theorem 825
Hopfield model 725, 729
hopping parameter 893
Hubbard model 799
Hubble constant 634
Hund's rules 807
hybrid meson 895
hybrid Monte Carlo 890
hyperbolic differential equation 644
hypercubic symmetry 848
hyperfine interaction 765
Hyperion 632
hysteresis 719

image processing 695
implicit differencing 648
in-class initialization 589
include guard 15
inelasticity 758
inertia tensor 114
infinite loop, see loop, infinite
inheritance 340
inhomogeneous differential equation 352
initial value problem 353
initialization (class) 162
initializer lists 586
instantiation 39
integrable Hamiltonian 444
integro-differential equation 688, 812
Interactive Development Environment 4
interpolating operator 855, 896, 902
irreducible representation 457
Ising model 229, 264, 270, 271, 774, 847, 856, 858, 860

ISO C++ standard 9
istringstream 11
iteration, unbounded, see limitless regress
iterative map 429
iterator 569
  Forward 573
  Output 573
  RandomAccess 573
iterator (Standard Library) 565

J/ψ meson 478
Jacobi iterator 659
Jacobi method 659, 685
Jarvis march 698
Java3d 315
Jones matrix 154
Jones vector 154
Julia set 437

Kalman filter 543
Kalman gain matrix 548
Kalman gain matrix formalism 548
KAM theorem 445
KAM tori 444
Kato's theorem 825
Kennedy-Pendleton heat bath 894
Klein-Nishina formula 495
Koch snowflake 266
Kohn-Sham functional 826
Koopman's theorem 839
Kosterlitz-Thouless phase transition 728
Kosterlitz-Thouless topological order 629
Kosterlitz-Thouless topological phase 721
krill 409
Kroupa mass distribution 633
kurtosis 114

l-value 573
lagged logistics map 451
Lagrange interpolating polynomial 85
Lagrange multiplier 611
Laguerre polynomial 761, 808
lambda (anonymous function) 584
lambda declarator 585

lambda expression (anonymous function) 584
lambda transition 630
lambda-introducer 585
Lanczos algorithm 778
Lanczos miracles 779
Langevin
  velocity autocorrelation 629
Langevin method 891, 895
latency 278
latent heat 719
lattice spacing 846
Lax method 647
Lax-Wendroff scheme 690
LD_LIBRARY_PATH 12
ldd 7
leap frog method 598, 648, 893
Legendre polynomial 130
Lennard-Jones potential 597
LHC 37
libraries 17
LIFO 571
light weight code 279
likelihood function 246
limit cycle 428
limitless regress, see infinite loop
Lindemann's law 628
linear combination of atomic orbitals 818
linear congruential generator 323
linear perspective 313
link 5, 18
link error 5
link smearing 886
link variable 858
linker 4
Linux 3
Liouville's theorem 444
Lippmann-Schwinger equation 752
Lipshitz continuity 359
list initialization 567
LM741 37
local density approximation 826, 830
local error 359
local force 676
local stiffness matrix 676
local symmetry 858
logistic map 430
  invariant density 434, 450
long range order 775

isotropic Heisenberg antiferromagnet 790
loop, infinite, see iteration, unbounded
Lorentz force 346
Lorentz group 456, 461
Lorentz transformation 456, 460, 461, 468–74
  orthochronous 460
  proper 460
Lorentz transformation 467
Lorentzian distribution 208
Lorenz attractor 452
Lorenz model 348, 426, 452
Lyapunov exponent 440, 632

macOS 3
machine precision 33, 37
Madelung's rule 806
magnon 721
Mahalanobis distance 215
make 5
make system 6, 19
makefile 6, 19
Mandelbrot set 438, 451
Mandelstam variables 495
Manhatten project 426
MANIAC-I 426
mantissa 34
Markov chain Monte Carlo method 221
Markov process 222
mass attenuation coefficient 489
mass gap 796, 853, 885, 893
matrix exponentiation 458
maximum likelihood estimator 512
Maxwell construction 149
Maxwell-Boltzmann distribution 595, 600, 636
mean 114
mean field theory 706, 724, 809
mean free path 601
member data 157
member functions 157
memory 726
memory effects 608
memory leak 176, 414
memory overwrite 176
Mermin-Wagner theorem 721, 775
Mersenne twister 199, 322, 863

mesh generation 677
message passing model 279
metastability 719
method
  Bogacki-Shampine 380
  Box-Müller 212
  Cash-Karp 381
  conjugate gradient 680
  Crank-Nicolson 648, 651
  diagonalization-variational 735
  Euler 359
  Euler-Cromer 388, 393, 439
  finite element 673
  Hartree-Fock 809
  heat bath 850
  Jacobi 659, 685
  Langevin 891, 895
  Lax 647
  Lax-Wendroff 690
  leap frog 648
  Markov chain Monte Carlo 221
  Metropolis-Hastings 849
  multiscale 664
  Neri 393
  Newton-Raphson 70, 82, 135, 689
  Numerov 734
  rejection 850
  rejection (random variates) 204
  Ritz 675
  Romberg's 127
  Runge-Kutta 362
  Ruth 393
  transformation (random variates) 207
  Verlet 393
  Visscher 658
method (class) 43
metric tensor problem 614
Metropolis algorithm 223, 706
Metropolis-Hastings algorithm 223, 237, 849
microcanonical updates 851, 880, 894
midpoint rule (quadrature) 118
minimum image convention 600
minimum ionizing particle 504
MO-LCAO 818

model
  alternating chain 797
  anisotropic Heisenberg
    antiferromagnet 774
  Baxter-Wu 728
  billiard 447
  cosmological
    concordance 619
  Edwards-Anderson 724, 729
  FitzHugh-Nagumo 691
  FPU 426, 453
  Hénon-Heiles 446
  Heisenberg ladder 798
  Hopfield 725, 729
  Hubbard 799
  Ising 229, 774, 847, 856, 858, 860
  Lorenz 348, 426, 452
  PBD 629
  Potts 718, 728
  random field Ising 729
  SEIR 400
  Sinai billiard 448
  SIR 688
  Thomas-Fermi 828, 830, 839
  Vicsek 638
  XY 721, 728, 774
molecular dynamics 594, 891
molecular orbital 818
molecule
  conformation 616
moment of inertia 115
Monte Carlo renormalization group 263
Monte Carlo reweighting 709
Monte Carlo, prime directives 722
Moore's law 276
Morse potential 148, 609, 629
MPI 279
MPI_Datatype 283
MPI_Reduce 283
Mueller matrices 192
multimodality 228
multiple compilation units 12
multipole expansion 115, 148
multipole moments 467
multiscale method 664
Murphy's eyeballs 450
mutable (keyword) 88
mutation (function) 167
mutex 293
mutual exclusion 293

mysterious problems 6, 7
Møller-Plesset theory 832

N-representability 826
Néel state 774, 782
name mangling 422
namespace 39
NaN 35
  quiet 35
  signalling 35
Navier-Stokes equations 643, 689
Neri method 393
Neumann boundary conditions 644, 658
neural net 729
neuron 725
Neville's algorithm 87
Newton-Raphson method 70, 82, 135, 612, 689
nine-point stencil 653, 685, 693
nm 7
node 275
non-Abelian gauge theory 877
non-aggregate class 587
non-type template parameters 564
nonautonomous differential equation 344
nondimensionalization 142, 252
normal distribution 202
normalized numbers 34
Nosé-Hoover thermometer 604
nullptr 583
Numerov's method 734
Nyquist frequency 669

object code 4, 5
object-oriented analysis and design 421
objective function 530
occupation number 832
on shell 753
one-body operator 805
OOAD 152, 354, 404, 421
open channel 758
Open Inventor 315
Open Inventor Toolkit 421
open multi-processing 285
OpenGL 315
openMP 279, 285, 602
openMP join 285

OpenSceneGraph 315
operator delete 174
operator new 174, 413
operator overloading 167
operator splitting 655
optical filters 153
optimization 517
order parameter 704, 796, 856
orthogonal series density estimation 534
orthographic camera 313
otool 7
over-relaxation 662, 824, 829, 851
overlap fermion 889
overload resolution rules 560
overloading 157
overloading on const 184
override (keyword) 415

Padé approximant 98
pair correlation function 596
pair production 488, 490, 491
parabolic differential equation 644
parity 764, 856
partial specialization 562
partial wave expansion 750, 753, 813
partial wave projection 738
Particle Data Group 764
particle mesh method 627
particle-particle particle-mesh method 627
partition function 703
parton distribution function 485
pass by reference 164
pass by value 163
PATH environment variable 10
Pauli exclusion principle 116, 806
Pauli matrix 193, 458, 470, 776, 878
Pauli principle 773
PBD model 629
percolation 240
perimeter law 860
period doubling 433
periodic boundary conditions 707
perspective camera 313
Persson-Strang triangulation 683
phase diagram 439

phase portrait 427
phase shift 751
phase transition
    first order 717
    infinite order 721
photoelectric effect 488
pipelining 275
pituitary gland 509
plaquette 859, 863, 872, 881, 894
plugins 422
Poincaré map 440, 446
Poincaré section 429
Poisson bracket 391
Poisson distribution 202, 524
Poisson equation 627, 658, 675
polarization vectors 468
Polyakov loop 864, 873, 893, 894, 898
polymorphism 137, 340, 402
positronium 500
posix thread 279
posterior density 539
potential
    Lennard-Jones 597
    Morse 148, 609, 629
    stacking 629
    van der Waals 830
    Woods-Saxon 399
Potts model 718, 728
Prandtl function 697
principal value 754
Principia 425
principle of least privilege 575
principle value 760
prior density 539
private (keyword) 44, 157
process 275
processor 275
product Ansatz 805
prolongation operator 665
promise object 296
proposal distribution 223
protein folding 616
proton therapy 509
pseudorandom number generator 199
public (keyword) 44, 157
public inheritance 406
pull
    statistical 527
pure interface class 416
pure virtual function 415
purify (tool) 176

Qat 76
Qhull 698
Qt 76
quadratic map 437
quantum chaos 449
quantum chromodynamics 889
quantum phase transition 856
quark 898
quasi-ergodic problem 601
quaternions 456
quenched (material) 725
quenched approximation 890
quotient-difference algorithm 99

R matrix 769
r-value 573
radiation length 491
rainbow 310
random engine 199
random field Ising model 724, 729
random walk 688, 740
RANDU 322
range-based for loop 582
rapid mixing 224
rapidity 461
rasterization 314
reaction matrix 755
receive buffer 281
recursion 327, 579, 621, 670, 684
recursive function 327
reference counting 318
rejection method 850
rejection method (random variates) 204
relativistic kinematics 739
rendering
    deferred 313
    real time 313
renormalizable field theory 876
renormalization group 258, 262, 270, 271, 727
renormalization group equation 884
renormalization group flow 262
renormalized couplings 877
replica method 729
repository 21
representation
    irreducible 457
representation theory 456
resonance 208, 767, 769

resonant tori 444, 445
restricted three body problem 397
restriction operator 665
reweighting 709, 727
Reynolds number 689
Richardson extrapolation 96
Richardson, Lewis Fry 662
ring approximation 835
Ritz method 675
Rodrigues formula 762
Romberg's method 127
root system 78
Roothaan equations 812, 837
Roothaan's relationship 836
rotation 458, 459, 463–7
    active 456
    passive 456
rotation curve 620, 631
rotations 456
round-off error 33, 36
Runge-Kutta method 362
Ruth method 393
Ryckaert iteration 612

S-matrix 770
saddle point 427
sample covariance matrix 537
sampling frequency 669
scalability 279
scaling laws 252
scaling relations 257, 705
scaling window 877
scattering 750
    coupled channel 757
scattering matrix 751, 758
scene (graphics) 314
scene graph 316
scene object 316
scene object 318
scheduling 288
Schrödinger equation 351, 655
Schwinger integral 867
scope operator 187
scoping directives 287
second order phase transition 856
second quantization 832
seed (random number) 199
SEIR model 400
self-adjoint matrix 56
self-similarity 255, 327
send buffer 281
separatrix 427
sequence containers 566

sequential decay 478
setting the lattice scale 848
shadow (data) 187
Shank's transformation 95
shared libraries 12
shared library 17, 18
shared node 331
Sherrington-Kirkpatrick
    equation 725, 729
shooting method 734
Sierpinski carpet 250
sigmoid function 247
signed distance function 683
simple harmonic oscillator 237, 761
simplex 674
Simpson's rule
    (quadrature) 121
simulated annealing 238, 899
simulation 595
Sinai billard 448
single particle
    wavefunction 804
sinoatrial node 691
SIR model 688
site percolation 241
site variable 858
size-consistency 833
skewness 114
Slater determinant 805
Slater exchange functional 828
smart pointer 569, 581
smeared links 896
SoCalculator (coin) 337
SoQt 318
SoTimeCounter (coin)
    336
spacetime 455
Spacetime library 459, 470
special unitary group 878
spectral radius 659
spin glass 724
spin wave theory 795
spin-spin correlation
    function 727
spinor 457, 465
    Dirac 470
    higher dimensional 465
    Weyl 468
spontaneous symmetry
    breaking 704, 856
St. Venant theory 697
stability analysis 394, 646
stack 571
stack (memory) 413, 871

staggered magnetization 784, 796
standard C++ library 9, 565
Standard Model 845, 874, 885, 886
Standard Template
    Library 565
staple 861
StarUML 405
state (class) 160
static library 17
static type 413
static typing 580
statistical inefficiency 227
std::accumulate 577
std::array 568
std::cerr 7
std::cin 7
std::complex 39, 403, 871
std::cout 7
std::find 576
std::forward_list 568
std::generate 585
std::iostream 286
std::isfinite 35
std::isnan 36
std::isnormal 36
std::knuth_b 199
std::list 567
std::map 459, 570
std::max_element 576
std::min_element 576
std::mt19937 199
std::numeric_limits 32, 35
std::ofstream 165
std::pair 88
std::priority_queue 571
std::queue 571
std::random_device 200
std::ranlux24 199
std::set 570
std::shared_ptr 582
std::sizeof 32
std::sort 576
std::stack 571
std::string 66
std::uniform_real_distribution 200
std::unique_ptr 582
std::vector 65, 158, 403, 567
stencil
    alternate nine-point 685
    five-point 653
    nine-point 653, 685, 693
    prolongation 666
stiff differential equations 396

STL 565
Stoke's law 616
Stoke's vector 192
strange attractor 436
string tension 874, 882, 893
structure 43
structure factor 799
$SU(2)$ gauge theory 105, 877
$SU(2)$ generators 458, 465
$SU(3)$ group 590, 897
$SU(N)$ 590
subclass 137, 403
subnormal numbers 35
subversion 21–4
successive over-relaxation 662, 687
superclass 137, 403
superposition principle 353
superrenormalizability 877
SVN 21–4
Swendsen-Wang
    algorithm 713, 727
Symanzik improved action 902
symmetric phase 704
symplectic group 387
symplectic transformation 387
synaptic (installer) 52
synchronization 278
system fork 300
system on a chip 276

template 402, 557
template class 40
template parameter pack 578
tensor operator 765
tensors
    spherical 466
textures 321
theorem
    addition 814
    Appelquist-Carazzone 847
    central limit 211
    Elitzur's 859
    equipartition 596
    Feynman-Hellmann 747
    fluctuation-dissipation 604
    Goldstone's 775, 796
    Hohenberg-Kohn 825
    Kato's 825
    Koopman's 839
    Mermin-Wagner 721, 775
    virial 596
    Wick's 833
thermalization 601, 708
this (pointer) 180

Thomas algorithm 649
Thomas-Fermi model 828, 830, 839
thread 275, 871
thread safe 284
three-body problem 425
time development operator 387
transfer matrix 270, 726
transferability 609
transformation method (random variates) 207
transition matrix 752
transition matrix (Markov) 222
trapezoid rule (quadrature) 120
traveling salesman problem 238
tree
  binary expression 635
tree (data structure) 590
tridiagonal matrix 94, 101, 648, 778
  inversion 649, 686
try 25
two-body operator 805
`<typename T>` 558

Ubuntu 3
ulimit 285
ultraviolet divergence 848
ultraviolet regulator 846

unbinned maximum likelihood 529
uniform initialization 587
unit conversion 47
unitarity 656
unitary transformation 793
universal modeling language 405
universality 258, 435, 701
universality class 262, 706, 847
unix pipe 301
upwind differencing 650

v-representability 826
valgrind (tool) 176
van der Pol oscillator 428
van der Waals force 609
van der Waals gas 149
van der Waals interaction 830
Vandermonde matrix 129
variance 114
variate 198
vector 464
vegas (Monte Carlo integration) 142, 234
velocity autocorrelation 629
Verlet method 393, 598
Vicsek model 638
virial theorem 596
virial time 618
virtual function 409
  pure 415
Visscher method 658, 693

Visual Molecular Dynamics 619
Voigt distribution 210, 235
vortex 721

W boson 477, 878, 902
Weinberg angle 508
Weizsäcker functional 836
Weyl spinor 467–70
Wick rotation 231
Wick's theorem 833
Wigner D-matrix 479
Wilson fermion 889, 893
Wilson loop 864, 873, 882, 895
Wilson renormalization 262
WKB approximation 111
Wolff algorithm 714, 727
Woods-Saxon potential 399
world coordinates 314
Wu and Shaknov 500

x-ray scattering 630
XY model 721, 728, 774

Yang-Mills gauge theory 877
Young tableaux 590

Z boson 508, 878
Z_2 (group) 858
Z_2 gauge theory 859, 900
Z_n (group) 719
zombie process 300